Ilyoung Chong (Ed.)

Information Networking

Wired Communications and Management

International Conference, ICOIN 2002
Cheju Island, Korea, January 30 – February 1, 2002
Revised Papers, Part I

Volume Editor

Ilyoung Chong
Hankuk University of Foreign Studies
86, Wangsan, Mohyun, Yongin, Kyonggi-Do, Korea 449-791
E-mail: iychong@hufs.ac.kr

Cataloging-in-Publication Data applied for

Die Deutsche Bibliothek - CIP-Einheitsaufnahme

Information networking : international conference ; revised papers / ICOIN 2002, Cheju Island, Korea, January 30 - February 1, 2002. Ilyoung Chong (ed.). - Berlin ; Heidelberg ; New York ; Hong Kong ; London ; Milan ; Paris ; Tokyo : Springer
Pt. 1. Wired communications and management. - 2002
(Lecture notes in computer science ; Vol. 2343)
ISBN 3-540-44256-1

CR Subject Classification (1998): C.2, D.2.12, D.4, H.3, H.4, H.5

ISSN 0302-9743
ISBN 3-540-44256-1 Springer-Verlag Berlin Heidelberg New York

Springer-Verlag Berlin Heidelberg New York
a member of BertelsmannSpringer Science+Business Media GmbH

http://www.springer.de

Printed in Germany

Typesetting: Camera-ready by author, data conversion by PTP-Berlin, Stefan Sossna e.K.
Printed on acid-free paper SPIN 10869951 06/3142 5 4 3 2 1 0

Preface

The papers comprising Vol. I and Vol. II were prepared for and presented at the International Conference on Information Networking 2002 (ICOIN 2002), which was held from January 30 to February 1, 2002 at Cheju Island, Korea. It was organized by the KISS (Korean Information Science Society) SIGIN in Korea, IPSJ SIG-DPE (Distributed Processing Systems) in Japan, the ITRI (Industrial Technology Research Institute), and National Taiwan University in Taiwan. The papers were selected through two steps, refereeing and presentation review.

We selected for the theme of the conference the motto "One World of Information Networking". We did this because we believe that networking will transform the world into one zone, in spite of different ages, countries and societies. Networking is in the main stream of everyday life and affects directly millions of people around the world. We are in an era of tremendous excitement for professionals working in many aspects of the converging networking, information retailing, entertainment, and publishing companies. Ubiquitous communication and computing technologies are changing the world. Online communities, e-commerce, e-service, and distance learning are a few of the consequences of these technologies, and advanced networking will develop new applications and technologies with global impact. The goal is the creation of a world-wide distributed computing system that connects people and appliances through wireless and high-bandwidth wired channels with a backbone of computers that serve as databases and object servers.

Thus, Vol. I includes the following subjects related to information networking based on wired communications and management:

- New Networking and Applications
- Switching and Routing
- Optical Networks
- Network Performance Issues
- Quality of Service
- Home Networking and Local Access Protocols
- Network Management

And Vol. II includes the following subjects related to wireless communications technologies and network applications:

- Wireless and Mobile Internet
- 4G Mobile Systems
- Satellite Communications Systems
- Network Security
- Multimedia Applications
- Distributed Systems

With great pleasure we take this opportunity to express our gratitude and appreciation to Prof. Chung-Ming Huang, who acted as PC vice-chair of ICOIN 2002, for reviewing and encouraging paper submission in Taiwan and other locations, to Prof.

Leonard Barolli, who acted as PC vice-chair of ICOIN 2002, for organizing paper reviewing in Japan, and to Prof. Changjin Suh, Prof. Sungchang Lee and Prof. Kyungsik Lim, who did excellent reviewing, editing and assembling of contributions for this book.

We are confident that this book series will prove rewarding for all computer scientists working in the area of information networking.

June 2002 Ilyoung Chong

Table of Contents, Part I

I. New Networking and Applications

II. Switching and Routing

III. Optical Networks

IV. Network Performance Issues

V. Quality of Services

VI. Home Networking and Local Access Protocols

VII. Network Management

Table of Contents, Part II

I. Wireless and Mobile Internet

II. 4G Mobile Systems

III. Satellite Communications

IV. Network Security

V. Multimedia Applications

VI. Distributed Systems

I. New Networking and Applications

An Agent-Based Personalized Distance Learning System for Delivering Appropriate Studying Materials to Learners

Akio Koyama[1], Leonard Barolli[2], Zixue Cheng[3], and Norio Shiratori[4]

[1] Department of Informatics, Yamagata University
4-3-16 Jonan, Yonezawa 992-8510, Yamagata, Japan
[2] Department of Computer Sciences, Saitama Institute of Technology (SIT)
Fusaiji 1690, Okabe, Saitama 369-0293, Japan
[3] Department of Computer Software, The University of Aizu
Tsuruga, Ikki-machi, Aizu-Wakamatsu 965-8580, Fukushima, Japan
[4] Research Institute of Electrical Communication (RIEC), Tohoku University
2-1-1 Katahira, Aoba-ku, Sendai 980-8577, Japan

Abstract. In this paper, we propose an agent-based personalized distance learning system for delivering appropriate studying materials to learners by judging learners degree of understanding. The main elements of our proposed system are the agents, which play the role of teacher and based on the learning history they analyze the understanding degree of learners. To evaluate the proposed distance learning system, we carried out three experiments and a questionnaire investigation. The evaluation results show that by making grouping of learners the agents can decide what kind of materials should be given to learners. We show that by adding new features such as mental action of color and the competition with other learners, the learners will increase furthermore the learning efficiency.

1 Introduction

During last years a lot of research is going on for distance learning systems [1, 2] and many large projects such as CALAT [3], CALsurf [4], WebCAI [5], The University of The Air [6], and WIDE University [7,8] have been established.

Recently, some distance learning systems which consider learner's capability and understanding have been proposed [9,10]. In Ref.[9], an evaluation system of historical data based on learning environment supported by educational software record is proposed, and reappearance and analysis are carried out for historical learning data. The analysis from the history can be performed, but in order to get the learner's learning condition, the historical data are needed. Thus, the analysis can not be done in real time. In Ref.[10], a multimedia assisted education system is proposed. The system is able to make the teacher operating cost small and offers fine education by the cooperation of CAI and teacher. The system is able to recognize the learner who needs the assistance, but its main purpose is to support the teacher and not the learner.

I. Chong (Ed.): ICOIN 2002, LNCS 2343, pp. 3–16, 2002.

In order to offer a suitable and efficient study for learners, in this paper, we propose an adaptive personalized distance learning system using cooperative agents. The main purpose of our system is to deliver appropriate studying materials by judging the learner's degree of understanding. The main elements of the proposed system are the agents, which play the teacher's role and based on the learning history they analyze the understanding degree of learners. But, it should be noted that only understanding degree is not enough to get the learner's studying conditions. Therefore, the agent makes a dialog with the learner and more accurate learner's condition can be grasped. This information is used to adapt the individual learning. To evaluate the proposed system, we carried out three experiments and a questionnaire investigation. The evaluation show that our system can achieve a good delivery of studying materials for different learners.

The paper is organized as follows. In Section 2 is introduced the system design. In Section 3, we deal with data handling. In Section 4, we discuss the experimental results. In Section 5, we treat system improvements and future work. Finally, the conclusions are given in Section 6.

2 System Design

The learners can access the system in any place where they can be connected to a network. To build the proposed system, we use World Wide Web (WWW) technology which is very suitable for building distance learning systems. However, the present web browsers have different functions and implementation extensions, so the system is subject to restriction on using different functions. But, if we use only text and image, almost all web browsers can meet the requirements of the proposed distance learning system. In order to have a wide range of applications, we use only standard functions. So, the system can be used easily without depending on computer environment.

2.1 System Structure

The system is built on WWW and the agent is established on the web server. The learners can access the server to refer to the studying materials from a client as shown in Fig.1.

The agent can grasp the learner's information and the relevance of the materials to each learner by checking the learner's network access. The agent also manages the studying materials. The studying materials are prepared on the same server where the agent is established, but they can be distributed in different servers and can be accessed when are needed. After the learning session, a confirmation test is performed to check the learner's degree of understanding. The confirmation test is carried out by using choice-type forms and description-type forms.

The collection of learner's information is necessary to provide appropriate studying materials to each learner. In order to make a right judgment about the appropriate degree of delivered materials, we try to collect a large amount of

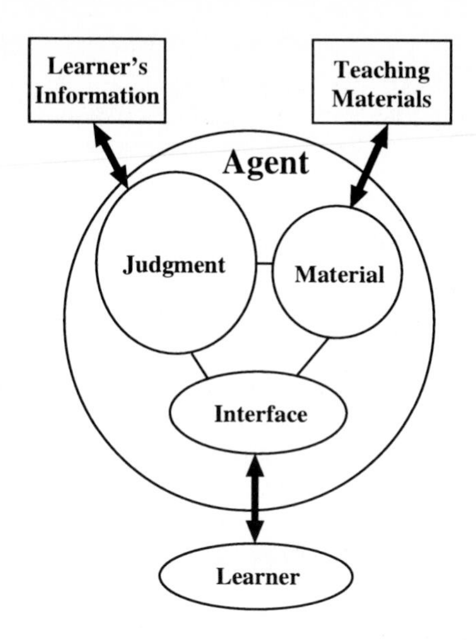

Fig. 1. System structure

information about the learner and then analyze the collected information. From this point of view, a direct test on learner's information and how to combine the test result with indirect information such as learning personal history are important elements on the system design.

2.2 Studying Materials

In this research, we use "network programming" studying materials. They consist of HTML text, GIF and JPEG images. The system treats one page of studying materials as an item and manages the access information item by item. In order to avoid the dependency on computer environment, we do not use materials such as moving picture, voice, etc.

In the page of studying materials, the buttons such as "NEXT", "DETAIL", "SEARCH", and "EXIT" are prepared as shown in Fig.2. The "NEXT" button sends a request to proceed the next studying material. The "DETAIL" button requires more details about studying materials. The "SEARCH" button searches the words and phrases in the studying materials. The "EXIT" button stops learning. When learning is stopped, total learning time is calculated.

The studying materials are prepared to cope with various learners. The system chooses a studying material which is appropriate for each learner and provides the studying material to them.

2.3 Agents

The agents are the main part of the proposed distance learning system. The agents deal with following procedures:

Fig. 2. Screen shot

- collection of learner's information;
- management;
- information analysis;
- learners' understanding judgment;
- studying materials handling;
- communication with learners.

The agents are shown in Fig.3 and they are explained in following.

- Register Agent (RA)
  This agent carries out the learner's authentication and gets the record of the learner's first time access.
- Learner Information Agent (LIA)
  This agent gets the physical information of the reference time of studying materials, the number of reference, and test results from the learner, and changes this information into parameters in order to make the analysis.
- Learner Communication Agent (LCA)
  This agent makes a dialog with the learner. First, the agent asks the learner and after getting the answer uses this information as historical data and makes the analysis.

- Judgment Agent (JA)
  Based on the data from LIA and LCA, this agent makes various judgments and gives orders to other agents.
- Teaching Material Agent (TMA)
  This agent carries out the management of studying materials. The studying materials suitable for the learner are offered by JA order.
- Test Agent (TA)
  This agent manages the test. It offers the test to the learner and marks the test then evaluates the learner by grades. The grades are used to decide whether the learner passed or not the test.
- Question Agent (QA)
  This agent receives the questions from the learner. It records the question history and gives the reply to the learner. When a question does not exist in the question history, an order is given to the following TIA to ask the teacher.
- Teacher Interface Agent (TIA)
  This agent manages the teacher's interface. When the teacher accesses the learner's information and history, this agent supports the teacher.

The agent is implemented with Perl language and the Common Gateway Interface (CGI) technology is used for agent organization.

The Perl language is adopted for the following reasons.

- Perl language is very good for text processing, therefore will have good processing results for studying materials in HTML format.
- Perl is a script language, so the compilation is easy.

When CGI technology is used to build a system on WWW, the agent can not keep conditions in the program. To solve this problem, the agent must put all necessary information in a file and then read the learner's information when it is necessary.

2.4 Processing Flow

The processing flow is showing in following.

- First, a learner accesses the system and tries the authentication.
- Next, the learner requires a studying material.
- The agent receives the learner's requirement.
- The agent reads the information of the learner who has a request.
- A studying material provided for the learner is judged from the learner's information.
- A decision for the next studying material is made and the learner can access the studying material.
- Get the studying material and send to learner.
- Renewal of information.
- The learner starts studying.

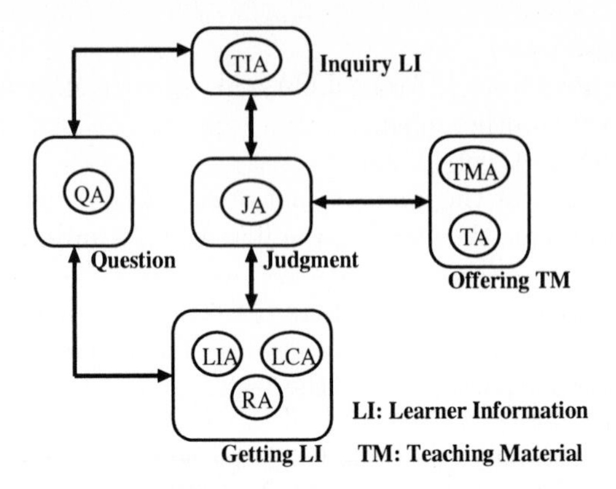

Fig. 3. Organization of agents

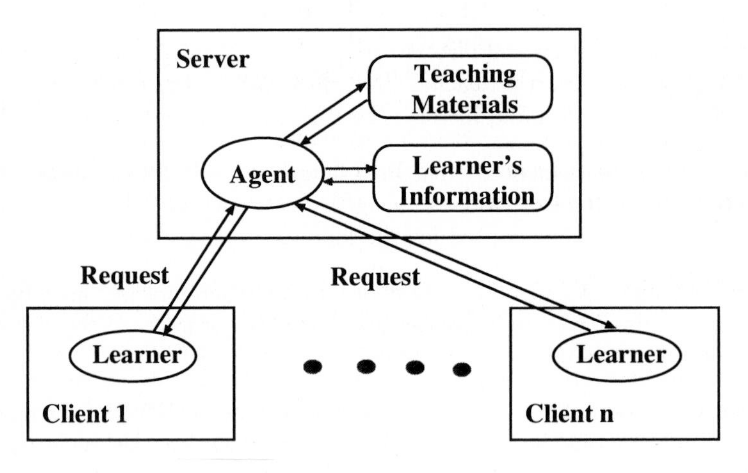

Fig. 4. Processing flow

The processing procedure is shown in Fig.4. The proposed system has five modes which are given in following.

- Registration Mode
- Study Mode
- Test Mode
- Question Mode
- Inquiry Mode (for the teacher)

Registration Mode. First, the learner accesses the system and makes the registration by inputing his name and e-mail address. After that, the learner replies to a questionnaire. RA collects the learner's information based on the knowledge of the studying material contents. Next, the learner goes to the Study Mode.

Study Mode. The learner starts to study appropriate studying materials. JA judges the information of RA and determines the studying materials for first delivery. The learner receives and learns the studying material. In the meantime, LIA collects and analyzes the learning history. LCA starts a dialog with the learner and grasps the learning conditions. JA processes the information from LIA and LCA, and judges the item for the next studying material and sends the studying material to the learner. This procedure is repeated for all items. When the learner has a question, he can shift to the Question Mode.

Test Mode. After learning the studying materials, the learner takes the test. TA sets the problems for each item and the learner gives the answers. TA marks the test and gives the grades then decides whether the learner passed the test or not. If the learner passes the test, the procedure is finished. When the learner did not pass the test, the questions which were not understood are given once again. If the learner passes, then the test is finished. Otherwise, if the learner did not pass again, TA gives an order to return the studying materials to the learner and the system goes to the Study Mode and the learner starts to study the studying materials again. After learning the studying materials, the learner takes the test again. The repetition procedure is carried out the same as in the Study Mode, when the learner has a question, he can shift to the Question Mode.

Question Mode. When there are contents which the learner cannot understand, the learner can shift to the Question Mode by a self-declaration method. QA manages the contents of questions and answers of each learner by using a Question Database (QD). Some of the answers can be sent to learner by using the past examples in QD. When the answer is not in QD, then QA asks the teacher. When the teacher has accessed the system, the teacher and learner can contact each other by using the Question Chat (QC). In the case when teacher has not accessed the system, the learner sends an e-mail and asks him for help.

Inquiry Mode. By using this mode, the teacher can access the system in order to know the learner's information. Thus, the teacher can grasp the learner's study situation. The teacher can also process the learner's questions.

2.5 Judgment Algorithm

Judgment algorithm is as follows.

- Check the progress and learning time and compare with the standard time.
- If the learning time exceeds the standard time, a confirmation test is performed.
- Otherwise, check re-learning item whether exists or not.
- If re-learning item exists, then this item is decided, if not, go to the next item.

- Based on the decision, a studying material is decided and the agent can access the studying material.
- Read the object file from server and provide to the learner.
- Get the reference time of the item, the number of reference, reference time of previous item, learning processing time.
- Calculate the average of reference time and total learning time.
- Examine the learner's information. Make question list of a confirmation test.
- Renew the information and finish the procedure.

The test for a learner is carried out as following.

- A question decided from the question list is sent to the learner.
- The learner answers the question.
- Return the result to the system.
- Marking and recording of results.
- Examination of the learner's information.
- Send the next question.
- If the question list is empty, test is finished. Return to provide studying materials.

3 Data Handling

As learner's information, the agent collects the system access information and after that the information analysis is carried out. By using CGI technology, the referred time of studying material, the host name which a learner uses, the browser, and the last page referred information can be obtained. The reference time of a studying material, the number of reference, and the referred sequence information is known from the previous access information. The agent analyzes the above information to know learner's knowledge and judges the studying material which is appropriate for each learner.

We use the following information as learner's information.

- Learning Progress — The progress of learning.
- Total Learning Time — The total learning time on the system.
- Average of Reference Time — The average of the reference time for each item.
- Tested Time — The time until the test finished. After the test is finished this value is registered.

For each item we use the following information.

- Referred time — The time when the agent delivered a studying material.
- Reference time — The time until a request for the next studying material is sent.
- Number of Reference — The frequency that a studying material was referred to.
- Testing Time — The time when a learner answers the test.
- Test Result — The result of the test.
- Re-learning Item — The item which learner did not understand.

Table 1. RT and NR

Item	RT(min)	NR
1	2	1
2	3	2
3	1	1
4	4	1
5	1	1
6	1	1
7	2	1
8	1	2
9	6	4
10	2	1
11	1	2
12	1	1
13	1	1
14	1	1
15	1	1
16	1	1
17	2	1
18	3	3
19	3	1
20	1	2
21	1	2
22	2	1
23	1	1

4 Experimental Results

4.1 Experiment 1

We examined the user's learning behavior by experiment 1. In this experiment, the learner is a 1-st year undergraduate student. Therefore, he has a few preliminary knowledge. In Table 1 are shown the data collected by agents while the student was using the system. Based on the data of Reference Time (RT) and Number of Reference (NR), the learner's understanding is judged.

When an item has a long RT and a big NR, it is presumed that the studying material has been difficult for understanding. For example, in this experiment, the item 9 was referred for 6 minutes and the NR is 4. This means, this item was more difficult to understand comparing with the other items.

4.2 Experiment 2

In this experiment 5 learners used the system. The system recorded the RT of learners for the studying material and then tested the learner understanding. The Average Reference Time (ART) and Average Score (AS) are shown in Table 2.

Table 2. ART and AS

Item	ART(s)	AS
1	10	100
2	12	80
3	36	100
4	24	100
5	60	100
6	48	50
7	30	60
8	36	50
9	36	80
10	96	40
11	24	40
12	12	100
13	24	100
14	12	100
15	84	20
16	12	100
17	24	75
18	24	100
19	12	100
20	72	75
21	36	50
22	15	100
23	84	66

From Table 2, we see that learners which refer the studying material for a long time have a low score. The learner access time for studying material is different, this is because of reading speed, interest on the material, preliminary knowledge and study desire. Therefore, it is important to consider for material delivery not only the learner's degree of understanding but also the personal differences between the students.

4.3 Experiment 3

In experiment 3, 15 learners used the system and we collected the following data: Test Result (TR), RT of studing materials, and NR of each item. The Average Value (AV) and Standard Deviation (SD) for each parameter are calculated by using formulas (1), (2), (3), (4), (5), and (6), respectively. The deviation values of TR (TRD) are decided based on formula (7), the deviation values of RT (RTD) are decided by formula (8), the deviation values of NR (NRD) are calculated using formula (9), and the Reference Efficiency (RE) values are calculated by formula (10). In these formulas, x_i, y_i, and z_i are the data for each user in TR, RT, and NR columns, as shown in Table 3.

The learners grouping is carried out based on TRD and RE. In order to verify the experimental results, after the experiment, we carried out and investigation for each learner using a questionnaire and compared the experimental results with questionnaire results.

$$AV_{TR} = \frac{1}{N}\sum_{i=1}^{N} x_i \tag{1}$$

$$AV_{RT} = \frac{1}{N}\sum_{i=1}^{N} y_i \tag{2}$$

$$AV_{NR} = \frac{1}{N}\sum_{i=1}^{N} z_i \tag{3}$$

$$SD_{TR} = \sqrt{\frac{1}{N}\sum_{i=1}^{N} x_i^2 - AV_{TR}^2} \tag{4}$$

$$SD_{RT} = \sqrt{\frac{1}{N}\sum_{i=1}^{N} y_i^2 - AV_{RT}^2} \tag{5}$$

$$SD_{NR} = \sqrt{\frac{1}{N}\sum_{i=1}^{N} z_i^2 - AV_{NR}^2} \tag{6}$$

$$TRD_i = \frac{x_i - AV_{TR}}{SD_{TR}} \times 10 \tag{7}$$

$$RTD_i = \frac{AV_{RT} - y_i}{SD_{RT}} \times 10 \tag{8}$$

$$NRD_i = \frac{AV_{NR} - z_i}{SD_{NR}} \times 10 \tag{9}$$

$$RE_i = \frac{RTD_i + NRD_i}{2} \tag{10}$$

The experiment 3 results are shown in Table 3 and the grouping of learners is shown in Fig.5.

The group I has a good RE value, but TRD is not so good. The learners in this group have a low degree of understanding. According to the questionnaire investigation after the experiment, it was proved that degree of understanding was low, because there were many careless mistakes. Therefore, the agent should inform these learners to be more careful during the study.

The group II has good RE and TRD. The learners belonging to this group have high degree of understanding. In the questionnaire, C and J said that content of studying materials was very easy. Therefore, the agents should give more difficult exercises in following learning steps.

The group III has bad RE and TRD. According to questionnaire investigation, the learners in this group wanted more easy materials. Therefore, it is

Table 3. Experiment 3 results

Learner	TR[%]	TRD	RT[s]	RTD	NR	NRD	RE
A	65	-20.1	285	6.9	25	4.9	5.9
B	70	-15.1	364	4.4	24	6.9	5.7
C	85	4.7	542	-1.5	24	6.9	5.4
D	70	-10.2	160	10.9	24	6.9	8.9
E	75	-5.2	321	5.7	24	6.9	6.3
F	75	-5.2	680	-6.0	44	-32.7	-19.4
G	90	9.6	700	-6.7	27	0.9	-2.9
H	75	-5.2	223	8.9	24	6.9	7.9
I	70	-10.2	636	-4.4	29	-3.1	-3.8
J	100	19.5	373	4.1	25	4.9	4.5
K	95	14.6	698	-6.3	28	-1.1	-3.7
L	90	9.6	672	-5.7	27	0.9	-2.4
M	80	-0.3	922	-13.8	31	-7.0	-10.4
N	100	19.5	862	-11.9	25	4.9	-3.5
O	75	-5.2	658	-5.0	31	-7.0	-6.0

necessary to give more illustrated examples such as animations or images in order to get a better understanding.

The group IV has a good TRD, but a bad RE. The learners of this group are considered to be careful learners. From the questionnaire investigation resulted that three persons among 4 learners said that they looked very carefully to the studying materials. Therefore, the agent should inform the learners to make more questions about the items they did not understand. Thus, they can increase the study efficiency.

5 Improvements and Future Work

In this paper, we proposed an agent based distance learning system which can provide appropriate studying materials for learners by collecting information of learners and checking their degree of understanding. However, the proposed system has a passive learning style because does not stimulate the learners volition. It should be noted that keeping and stimulating learners volition is an effective method to increase the learners understanding degree for a studying material.

We are improving now the system by stimulating the learners volition and by making them to be more interested for studying materials. We are adding new features such as mental action of color and the competition with other learners in order to increase the learning efficiency. The performance evaluation of the improved prototype system is for future work.

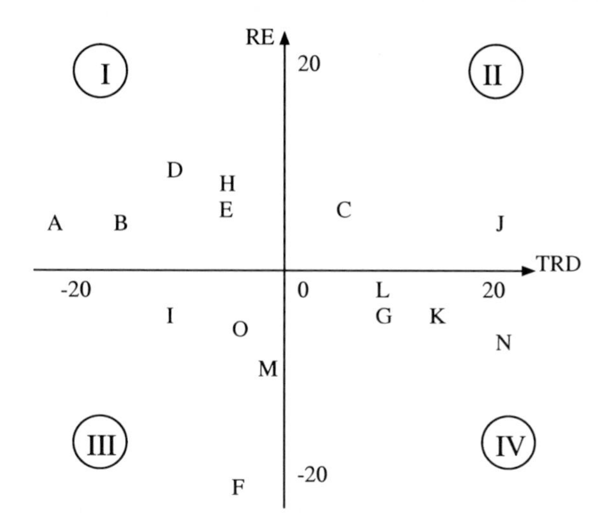

Fig. 5. Grouping of learners

6 Conclusions

In this paper, we proposed an adaptive personalized distance learning system using cooperative agents. The system is able to deliver appropriate materials to learners by collecting and evaluating the learner's information.

In order to evaluate the system performance, we carried out three experiments. From learning behavior results, we conclude: the learners which refer the studying material for a long time have a low score; the learners which progressed to the next item understood all item contents or a part of item contents; the learner's access time for studying material is different, this is because of reading speed, interest on the material, preliminary knowledge and study desire.

The test results show: the learner degree of understanding can be judged; for study materials delivery not only the learner degree of understanding but also the personal differences between the students should be considered; when the learner's study desire is low, the agent should provide an interesting material such as animations or images.

We verified the experimental results by investigating the learners using a questionnaire and compared the experimental results with questionnaire results. The evaluation results show that by making the grouping of learners the agents can decide what kind of materials should be given to learners.

We are improving now the system by adding new features such as mental action of color and the competition with other learners in order to increase the learning efficiency. The performance evaluation of the improved prototype system is for future work.

References

1. Nakabayashi, K., Koike, Y., Maruyama, M., Touhei, H., Fukuhara, Y., Nakamura, Y.: CALAT: An intelligent CAI system using the WWW. Trans. of IEICE, D-II, **80** (4) (1997) 906–914
2. Katayama, K., Kambayashi, Y.: Support of distance lectures using active database. Trans. of IEICE, D-I, **82** (1) (1999) 247–255
3. Welcome to CALAT Project, Nippon Telegraph and Telephone Corporation. http://www.calat.com/ (1998)
4. CALsurf, NTT Software Corporation. http://calsurf.ntts.co.jp/
5. WebCAI, CAI Courseware System, Nihon University. http://iclab.ce.nihon-u.ac.jp/~webcai/index.html
6. The University of The Air. http://www.u-air.ac.jp/
7. WIDE University, School of Internet. http://www.sfc.wide.ad.jp/soi/contents.html
8. Ogawa, K., Ijuin, Y., Murai, J.: School of Internet. IPSJ Journal **40** (10) (1999) 3801–3810
9. Matsumoto, T., Nakayasu, H., Morita, E., Kamejima, K.: An evaluation system of historical data based on learning environment supported by educational software. IPSJ Journal **40** (9) (1999) 3596–3607
10. Tamaki, M., Kuwabara, T., Yamada, K., Nakamura, Y., Mitsunaga, Y., Konishi, N., Amano, K.: Multimedia assisted education system with individual student advance. IPSJ Journal **41** (8) (2000) 2351–2362

A Simple Rate-Based Congestion Control Scheme Using an Enhanced BECN in ATM Networks

Ik-Kyun Kim[1], Koohong Kang[2], Zyn-Oh Kim[1], Jong-Soo Jang[1], and Youngsik Baek[3]

[1] Network Security Department, Information Security Division, ETRI
161 Gajenong-Dong, Yuseong-Gu, Daejeon, 305-350, Korea
{ikkim21, zyno21, jsjang}@etri.re.kr

[2] Department of Information and Communications Engineering, Seowon University
231, Mochung-Dong, Chongju, 361-742, South Korea
khkang@seowon.ac.kr

[3] CEO, PaxComm, Inc.
220 Kung-Dong, Yoseong-Gu, Daejon, 305-764, Korea
ysbaek@paxcomm.com

Abstract. In the rate-based congestion control, three distinct schemes - Explicit Forward Congestion Indication (EFCI), Backward Explicit Congestion Notification (BECN), and Explicit rate based scheme - are mixed within the ATM Forum. With respect to implementation complexity, the EFCI or BECN are better than the explicit rate based scheme. In this paper, we propose an Enhanced BECN scheme to support max-min fairness problem and a method to determine the optimal source recovery time and filter period in order to minimize the maximum queue length and improve the link utilization. Especially the minimization of queue length fluctuation gives better performance over cell delay variation, end-to-end delay, and so on. Moreover the max-min fairness can be guaranteed with simple use of Resource Management (RM) cells containing Current Cell Rate generated by Source-End-Systems without increasing implementation complexity of the switch.

1. Introduction

Asynchronous transfer mode (ATM) is originally a networking protocol with the potential to support Wide Area Networks (WANs) for the public carriers. Unlike the Constant Bit Rate (CBR) and the Variable Bit Rate (VBR) services which are common in the WANs for circuit emulation and video service, respectively, data traffic in the LANs often requires no firm guarantee of bandwidth, but instead can be sent at whatever rate if convenient for the network. So ATM Forum had defined new service classes for such data applications known as the Available Bit Rate Service(ABR) and Unspecified Bit Rate(UBR). To support ABR traffic, the network requires a feedback

I. Chong (Ed.): ICOIN 2002, LNCS 2343, pp. 17-27, 2002.

mechanism in order to tell each source how much data to send because no pre-allocated bandwidth can be reserved to circumvent the congestion in the intermediate network nodes. In the rate-based congestion control, three distinct schemes – Explicit Forward Congestion Indication(EFCI), Backward Explicit Congestion Notification(BECN), and Explicit rate based scheme - are mixed within the ATM Forum.

The choice among various congestion management mechanisms influences an implementation in a significant way. Because ATM is meant to be scalable to much higher speeds, a significant part of the congestion management algorithm needs to be implemented in hardware, both in switches and in the end systems. As a matter of fact, minimum implementation complexity added to switch design is desired to support ABR service because ATM adopts originally the end-to-end transport protocol taking advantage of the high-speed and reliable networking. One-bit feedback rate-based congestion control schemes like BECN or EFCI schemes may be very promising candidates because of their moderate implementation complexity regardless of different perspectives of the switch designers and of those implementing the end system. Hence the EFCI or BECN are better than the explicit rate based scheme in which the switch adjusts the explicit rate field in the Resource Management(RM) cells and allocates bandwidth fairly to all the VCs.

In this paper, we proposed an Enhanced version(EBECN) of BECN scheme, which is originally proposed in the ATM Forum by Newman [1, 2]. Our scheme supports the max-min fairness, and provides a method to find optimal source recovery time and filter period to minimize the queue length fluctuation and high link utilization in the steady state. Since the maximum queue length and the variation of the queue length fluctuation affect the memory size in the switches, cell delay variation(CDV), and maximum end-to-end delay, it is important to find the optimal source recovery time and the filter period. Even if it is little practical to assume the transmission delay and current cell rate of each virtual channel(VC), EBECN scheme provides a simple traffic management mechanism for the ABR traffic with low cost. Note that most existing schemes make use of the same assumption as above.

2. EBECN Scheme

2.1 Basic Operation

The model of the EBECN scheme is illustrated in Fig. 1. For ABR data calls, the Source End System(SES) generates packets that are queued for segmentation process corresponding to a particular virtual connection in the ATM Adaptation Layer(AAL) within the ATM adapter. The transmission delay between each ATM adapter transmitting data or RM cells and the congested node represents the combination of the propagation delay and the cell switching time within the intermediate switches. For simplicity all SESs are assumed to experience the same transmission delay although this assumption might be inadequate to the WANs.

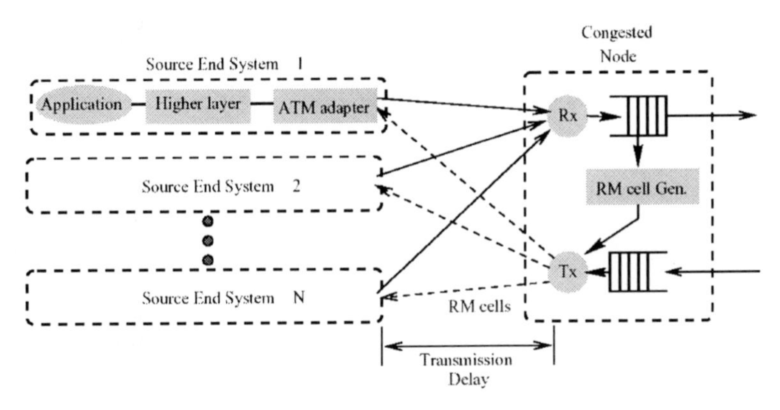

Fig. 1. The EBECN Scheme

If a switch detects the congestion state, the congested switch node will generate a BECN cell, which is a sort of RM cell but we refer to as BECN cell to differentiate it from the RM cell generated by the SES, and send it to the corresponding SES immediately without waiting for the delivery of the received cells in the queue through the up-stream path. When the switch generates the BECN cell, it just tags a single Congestion Indication(CI) bit so that it requires the least complexity for the switch behavior. Hence we can classify EBECN scheme into single bit feedback rate-based scheme. The selection of the measure of resource congestion has not yet been sufficiently investigated in the literature. The simple and most common approach is based on observing the instantaneous length of the queue of cells waiting to be transmitted out of a switch port. Hence in this paper we also use the buffer threshold for the detection of the congestion.

From [1], we know that the employing of BECN cell filter at the RM cell generator is necessary to reduce the amount of feedback information getting back to the SES. The proposed EBECN scheme adopts ‘A Per Virtual Channel Filter’ which transmits only a single BECN cell to each active source during each filter period if the queue is past threshold and ‘Max-min Flow Control’ which only permits VC to generate a BECN cell whose current cell rate (CCR) is greater than or equal to the other VCs’ CCRs. This ‘Max-min Flow Control’ reduces the number of BECN cells in the case of asymmetric peak cell rates (PCRs) of sources and guarantees the max-min fairness. If the filter permits the generation of a BECN cell, after which no further BECN cell be generated until filter period T_{FIL} cell times have passed.

When a SES receives a BECN cell it will reduce its cell transmission rate to one half of the current rate. If further BECN cells are received it will ignore them until it has transmitted at least one cell in the forward direction. A transmission rate recovery mechanism is built into each SES. If no BECN cells are received within the source recovery time period T_{REV} , the current transmission rate for that SES will be restored to the previous level, once each recovery time period, until the transmission restored to its original peak value. Upon receiving the BECN cell, the SES would reflect the RM cell containing its updated CCR back to the every switch along with the corresponding VC down stream path to inform the changed CCR. And then the switches update their CCR table for 'Max-min Flow Control'. Whenever a SES restores its current rate, it also generates a RM cell for the same purpose.

2.2 Optimal Source Recovery Time and Filter Period in the Steady-State

In the steady-state EBECN scheme causes each SES to center at some transmission rates that are the rates above and the rates below the output link transmission rate of the congested node during successive filter periods as the queue length oscillates during a congestion event [8, 9]. But the number of states of each SES's transmission rate in the steady-state depends on the source recovery time and filter period. In other words, as we choose time value for them less than the optimal one, the number of states must be increased. Accordingly the maximum queue size and queue empty period might be increased. For example, each SES attempts to transmit at a peak rate of 100%, where the peak transmission rate is expressed as a percentage and normalized to the output cell rate of the output link of the congested node. When 20 SESs are connected to the congested node, the overall arriving cell rate changes its state as shown in Fig. 2. We expect the overall arriving cell rate alternates between only two states R_{OL} (over-load arrival cell rate) and R_{UL} (under-load arrival cell rate) in the steady-state, and we call this situation an optimal condition with respect to the queue length distribution. Because the upper states more than R_{OL} increase the maximum queue size and the lower states less than R_{UL} decrease the link utilization due to increasing the empty period of the queue. R_{OL} and R_{UL} should be determined as follows;

$$R_{OL} = \max\left(\sum_{i=1}^{N} CCR_i \right) such\ that \left(\sum CCR_i > 1 \right) \tag{1}$$

$$R_{UL} = \min\left(\sum_{i=1}^{N} CCR_i \right) such\ that \left(\sum CCR_i > 1 \right) \tag{2}$$

where N denotes the number of connections and CCR_i denotes the current cell rate of SESi.

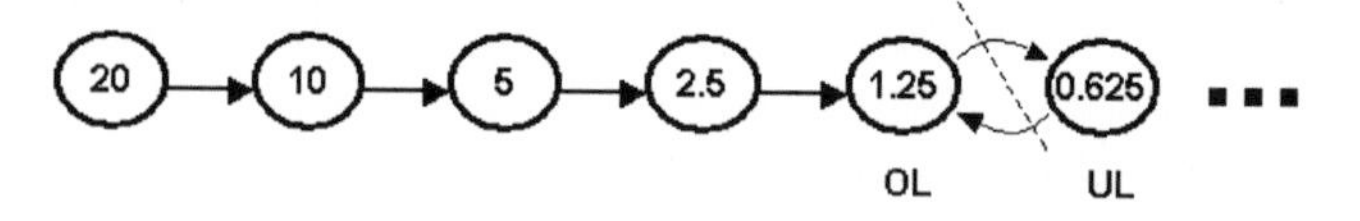

Fig. 2. Rate transit diagram under the congestion control

To bound the states of the arriving cell rate at R_{OL} and R_{UL} in the steady-state, the filter period and recovery time should be determined in the proper manner. A simple "fluid flow" approximation is useful to determine them [7, 8]. Accordingly, the cell flow is considered as a continuous variable as shown in Fig. 3.

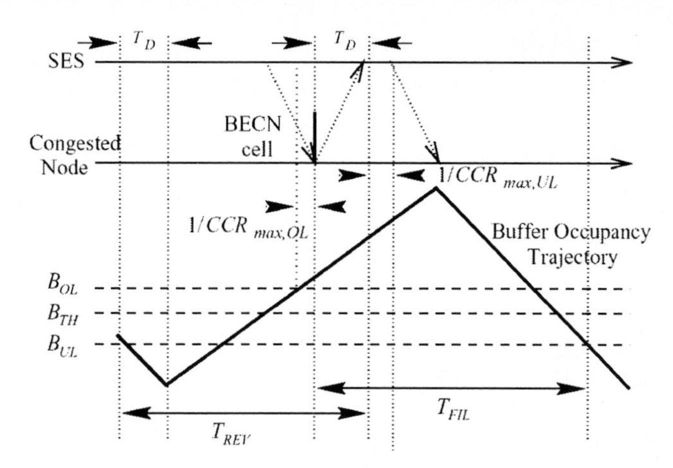

Fig. 3. Optimal filter and recovery period

In order to prevent the overall arrival rate from transition upper state over R_{OL} the recovery period T_{REV} should be determined such that another source recovery must be prohibited after transition to the rate state R_{OL} until all of SESs receive at least one BECN cell from the congested node. If we choose shorter T_{REV} than this, the maximum queue length must be increased exponentially. Unfortunately since the fluid-flow approximation can not reveal the cell level fluctuation, we have to consider the worst case cell level fluctuation as illustrated in Fig. 4. In the worst case, transmission cell rates of all SESs are same and each cell arrives at the congested queue simultaneously. Hence the queue threshold B_{TH} can not guarantee to generate the BECN cells of the incoming data cells in some cases.

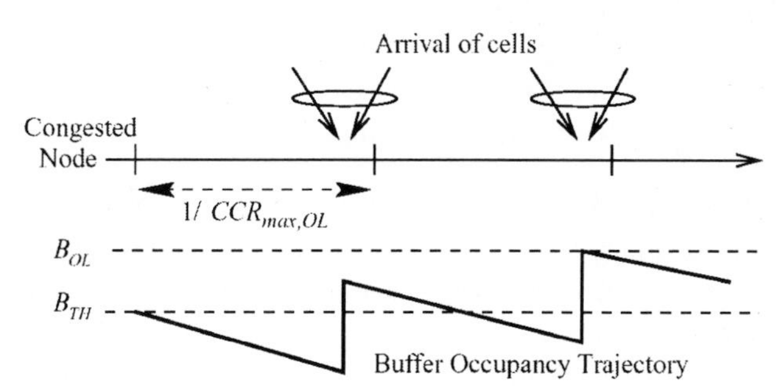

Fig. 4. Worst case condition of recovery period

Finally T_{REV} should be determined as follows,

$$T_{REV} > \left(\frac{B_{OL}}{R_{OL} - 1} + \frac{1}{CCR_{\max,OL}} \right) + 2T_D \tag{3}$$

where, $CCR_{\max,\,OL} = \max\left(CCR_i\right), i = 1, \ldots N$ when overall arrival cell rate is R_{OL}

$$B_{OL} = B_{TH} + 1/CCR_{\max}$$

and T_D denotes the transmission delay between SES and congested node and note that we use B_{OL} instead of B_{TH}.

Therefore we can easily calculate the maximum queue length Q_{max} as follows;

$$Q_{\max} = \max\left(B_{OL} + \left(2T_D + \frac{1}{CCR_{\max,OL}} + \frac{1}{CCR_{\max,UL}} \right)(R_{OL} - 1), Q_{cap} \right) \quad (4)$$

where, $CCR_{\max,\,UL} = \max\,(CCR_i), i = 1,\ldots N$ when overall arrival cell rate is R_{UL}, and Q_{CAP} denotes the given queue capacity.

On the contrary to the source recovery period, we have to prevent the overall arrival rate from transition lower state under R_{UL} in order to reduce the empty period of queue. For this purpose the filter period T_{FIL} has to last until the buffer occupancy level drops under B_{UL} instead of B_{TH} as shown in Fig. 3. If we choose shorter T_{FIL} than this, the state of the overall arrival rate drops down under the R_{UL}. Accordingly the empty queue period increases and then the link utilization drops significantly. As noticed in the case of the source recovery we have to consider cell level fluctuation for the decision of filter period as shown in Fig. 5.

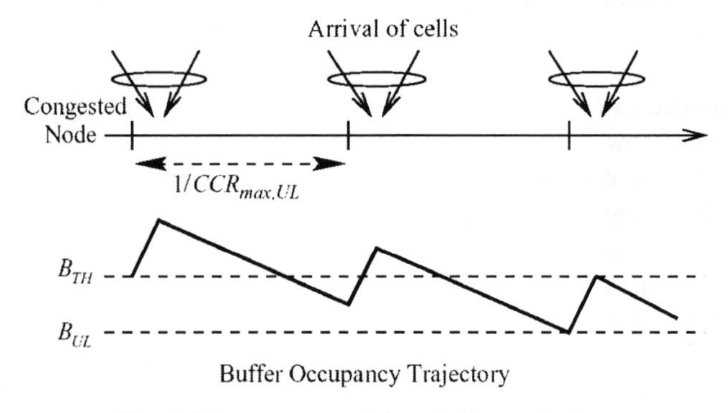

Fig. 5. Worst case condition of filter period

In the worst case all the incoming cells arrive simultaneously, B_{TH} can not guarantee to prevent the congested node from generating a BECN cell. Hence filter period should be determined as follows;

$$T_{FIL} > \left(\frac{Q_{\max} - B_{UL}}{1 - B_{UL}} + 2T_D + \frac{1}{CCR_{\max,UL}} \right) \quad (5)$$

where, $B_{UL} = B_{TH} - N$.

Finally if the T_{FIL} is greater than the T_{REV}, T_{REV} has to be modified based on the T_{FIL} as follows,

$$T_{REV} = \max(T_{REV}, T_{FIL}) \quad (6)$$

Using these expressions, we obtain the BECN cell frequency as follows,

$$\Omega = \frac{1}{T_{FIL} + T_{REV} - T_D} \tag{7}$$

2.3 Max-Min Fairness Problem

One of the most difficult aspects of congestion control is that of treating all users fairly when it is necessary to turn traffic ways from the network. While fairness can be defined in a number of different ways, the ATM Forum has recently converged toward a particular definition called max-min fairness of which any user is entitled to as much as network use as any other user [3, 4]. In [3] max-min fairness leads to the idea of maximizing the allocation of each user i subject to the constraint that an incremental increase in i's allocation does not cause a decrease of some other user's allocation that is already as small as i's or smaller.

To achieve the max-min fairness, the EBECN scheme enables the BECN cell to be generated only if the CCR of the arriving data cell is greater than or equal to the any other CCRs of VCs. Although this simple idea guarantees the max-min fairness easily but every switch keep on tracing the CCRs of VC connections on them. While there are also many methods to let the switches know the CCRs of VC connections, the approach adopted in this paper emphasizes the role of SES because it is difficult for the switches to know them by themselves. So each SES sends RM cell containing its updated CCR to the every switches along with the VC down stream path to inform the changed CCR whenever it changes its transmission cell rate according to the source recovery procedure as well as the reception of BECN cells. Especially upon receipt a BECN cell from the congested node, SES sends RM cell according the above rule and this RM cell also can be used for the acknowledged information for the BECN cell of the congested node. Even if this approach increases the total amount of RM cell traffic, but it is about 0.16% per active SES and 'Max-min Flow Control' reduces the number of BECN cell traffic in case of asymmetric PCRs of sources inhibiting of BECN cell generation of the lower CCR of VC.

2.4 SES and Switch Behavior

We summarize the SES behavior as follows;

1. The SES shall send an RM cell for every Allowed Cell Rate(ACR) transitions. Hence upon receiving an BECN cell from the congested node, it encodes its ACR to the CCR in a RM cell and send it along with the VC down stream path.
2. When any BECN cell is received, the SES decreases its ACR to one half of the current ACR.
3. If no BECN cells are received within the SES recovery time period, the ACR for that SES will be restored to the previous level.

4. The value of ACR is allowed between Peak Cell Rate(PCR) and Minimum Cell Rate(MCR).

The switch behavior is as follows;

1. The switch manages the table containing CCRs for all of the VCs. Hence at the call set up stage the switch keeps the PCR of each VC. And it also handles the RM cells to update the CCR of each VC from the SESs.
2. If the length of the switch queue exceeds a threshold, the switch may generate BECN cells corresponding incoming data cells. But if $CCR_i < CCR_j, \forall j \neq i$ or within the filter period of the VC, the switch would not generate a BECN cell for the VC although the switch queue exceeds a threshold.

3. Simulation Result

The simulation model is illustrated in Fig. 1. We assume that each SES generates traffic at saturation such that cell stream of each VC is stream-like traffic not bursty. This assumption is too tight to apply in some special cases but in the most cases under the congestion state this assumption might be reasonable. In other words, in the congestion state in the ATM-LAN, the traffic of each VC may be regulated or shaped during Segmentation and Reassembly(SAR) process in AAL within ATM adapter. So the traffic characteristic of the congested VC might be stream-like bursty having very high burstiness and very long mean burst duration even if its original is bursty.

From BECN scheme [1], the optimal design of filter is one that transmits only a single BECN cell to each active SES during each filter time period if the queue congested, and the optimal filter period is of the same order of magnitude as the maximum propagation delay for which the system is designed. But as a matter of fact, in order to reduce maximum queue length and increase the link utilization we have to concentrate on the decision of source recovery time and filter period as we mentioned earlier.

In the following simulations the performance of a EBECN scheme is investigated and all the simulation are performed using SIMSCRIPT II.5. And in order to compare with the performance result of the BECN scheme, refer to the study [1].

The BECN traffic and maximum and average queue length against transmission delay are given in Fig. 6 and Fig. 7 respectively. For the per virtual channel filter with up 80 sources each attempting to transmit at a peak rate of 100%. The filter and recovery time periods and maximum queue sizes in terms of the number of sources and propagation delay are shown in Table 1. And the queue threshold is 250 cells.

As we can notice that in case of N=50 in Table 1, T_{REV} is equal to T_{FIL} independent upon the transmission delay because the value of T_{FIL} is greater than that of the calculated T_{REV}.

Curves of BECN cell traffic is given in Fig. 6. But note that the total amount of excess traffic including BECN cell and RM cell is about 2.5 times than that of Fig. 6.

But if the PCRs of each of VCs are different to each other, the BECN cell traffic is reduced compared with that of BECN scheme with guaranteeing the max-min fairness as we will show late. The amount of BECN traffic is about 0.07% per active SES but the total amount of BECN traffic is increased proportional to the number of active SESs. As we can notice that Eq. 7, the frequency of BECN traffic depends on the T_{REV} and T_{FIL}.

The mean and maximum queue length against transmission delay are shown in Fig. 7 up to 80 SESs. As we can expect from Eq. 4 maximum queue length is bounded and it should not be fluctuated like to the previous study in [1]. Especially it dose not depend on the number of SESs but the states of steady-state transmission rate. The bounded maximum queue length minimize the CDV and average queueing delay which are affect seriously the Quality of Service(QoS).

Table 1. T_{REV}, Q_{max}, and T_{FIL} against transmission delay for number of sources with a per VC BECN cell filter and divided-by-two rate throttling

N	Delay	T_{REV}	Q_{max}	T_{FIL}
20	50	1180	295	306
	100	1280	320	472
	150	1380	345	639
	200	1480	370	806
50	50	1046	393	1046
	100	1398	449	1398
	150	1754	504	1754
	200	2114	561	2114
80	50	1420	387	897
	100	1520	412	974
	150	1620	437	1140
	200	1720	462	1307

Table 2. BECN versus EBECN in the case of asymmetric PCR

	Q_{max}	Q_{avg}	Utilization	%BECN cell traffic
BECN	443	106	0.66	0.09
EBECN	305	120	0.88	0.07

Now we consider two different SES groups attempting to transmit at a peak rate of 100% for one SESs group and at a peak rate of 50% for the other SESs group. The performance results are shown in Table 2. The reason of improvement of utilization and decreasing BECN cell traffic is that the EBECN scheme guarantees the max-min fairness such that the states of cell transmission rate of the SESs are bounded optimally.

4. Conclusions

The choice among different congestion control mechanisms for ABR service in ATM-LAN influences the implementations of the switch and source-end-system in a significant way. As a matter of fact, minimum implementation complexity added to design switch is desired to support ABR service because ATM adopts originally the end-to-end transport protocol taking advantage of the high-speed and reliable networking. According to the above reasoning, in this paper we proposed the Enhanced BECN scheme to support ABR service guaranteeing max-min fairness. Moreover we formulate a method to determine the optimal source recovery time and filter period.

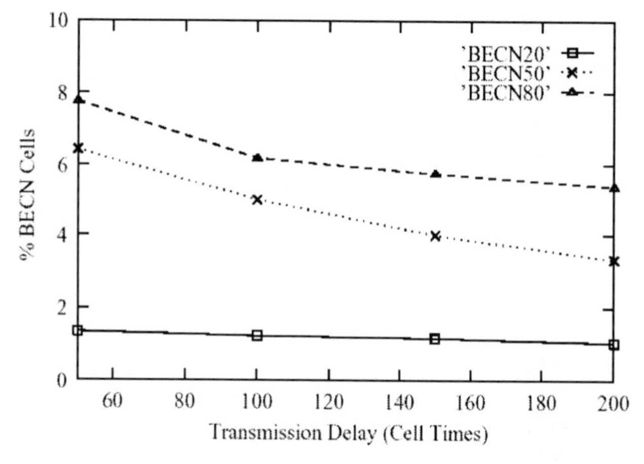

Fig. 6. BECN traffic against transmission delay for number of sources

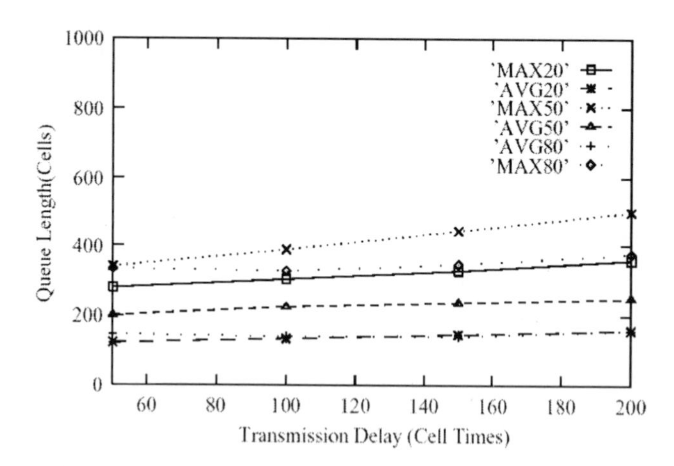

Fig. 7. mean and maximum queue length against transmission delay for number of sources

Even if the assumptions to know the transmission delay and current cell rate of each VC are a little hard constraints, EBECN scheme provides a simple traffic management mechanism for ABR traffic with cost effective. Especially the reduction of queue length fluctuation minimizes the cell delay variation and end-to-end delay.

Acknowledgment. Research of the second author was partially supported by PaxComm. Inc. in Korea

References

[1] Peter Newman, "Backward Explicit Congestion Notification for ATM Local Area Networks," Proc. GLOBECOM'93, pp. 719-723, 1993.

[2] Peter Newman, "Traffic Management for ATM Local Area Networks," IEEE Communications Mag., pp. 44-50, 1994.

[3] D. Bertsekas and R. Gallager, Data Networks, Prentice-Hall, 1987.

[4] Flavio Bonomi and Kerry W. Fendick, "The Rate-Based Flow Control Framework for the Available Bir Rate ATM Service," IEEE Network Mag., pp. 25-39, 1995.

[5] H.T. Kung and R. Morris, "Credit-Based Flow Control for ATM Networks," IEEE Network Mag., pp. 25-39, 1995.

[6] K.K. Ramakrishnan and P. Newman, "Integration of Rate and Credit Schemes for ATM Flow Control," IEEE Network Mag., pp. 25-39, 1995.

[7] M. Butto et al. "Effectiveness of the "Leaky Bucket" Policing Mechanism in ATM Networks," IEEE JSAC, Vol.9, No.3, pp. 335-342, April 1991.

[8] A. Gersht and K. J. Lee, "A congestion Control Framework for ATM Networks," IEEE JSAC, Vol.9, No.7, pp.1119-1130, Sep. 1991

[9] Y. T. Wang and B. Sengupta, "Performance Analysis of a Feedback Congestion Policy," Computer Communication Review, Vol.21(4), Sep. 1991.

Performance Analysis of Media Gateway for Voice Service

Suk-ki Jang[1], Hoon Kim[1], Jong-Dae Park[2], Sang-Sig Nam[2], and Kwang-Che Park[3]

[1] School of Electronics, Information & Communications Eng. Chosun Univ.
#375 Seosuk-Dong, Dong-Gu, Kwangju 501-759, Korea
{chjjan, hongsikh}@hanmail.net
[2] 161 Gajeong-dong, Yuseong-gu, Daejeon, 305350, Korea
Electronics and Telecommumcations Research Institute
Team of system integration
{jdpark, ssnam}@etri.re.kr
[3] School of Electronics, Information & Communications Eng. Chosun Univ.
#375 Seosuk-Dong, Dong-Gu, Kwangju 501-759, Korea
kcpark@mail.chosun.ac.kr

Abstract. Owing to the change of various services requirement and the change of communication paradime as the development of new transit network technology: ATM, MPLS and DWDM, the latest PSTN is evolving into the next generation network, which have the very high speed integrated packet network of open-type structure. But in the next generation network, the existing voice service is being researched for the continuous technology development as the important service from the viewpoint of communication business. This paper suggests the method of efficient bandwidth management in the media gateway that can accommodate the various new services as well as the existing voice services by appling to the next generation network.

1 Introduction

The communication service that can at a time offer the wireless mobile communication service of internet increasing recently as well as the voice service and intelligente network service with the existing circuits network or N-ISDN is required. Up to the present, because these function are separated and offered into each different transport networks: circuit network, N-ISDN, wireless network and IP network, the user must access the different networks to use these services. And from the viewpoint of communication business, in order to accommodate the new service requirements: the next high-speed and large capacity multimedia internet service and the IMT-2000 wireless service economically and rapidly, the necessity to integrate the each transit networks into one very high speed communication network was beginning to make its appearance. But in the present switching system structure, the control system, switch system and integration module like the existing circuit switching system are tightly coupled,

I. Chong (Ed.): ICOIN 2002, LNCS 2343, pp. 28–39, 2002.

and has been developing into the system for only target service network. Because this structure can support only the specific service, in order to accommodate the new service requirement, it has the problem that the development period of high expense and long term must be spent. But at present, the open-type switching system, which the development is being progressed, separates the control system and switch system like the softswitch structure, and the integration module is replaced with media gateway by the service that the system must accommodate, thus the interface among each modules can use the commercial products of third party application by using the standardization protocol. In case that there is the service required in the switching system newly, because such open-type system module can easily add the related media gateway and control module, it has the advantage that can rapidly cope with the various service requirements.

Between MG(Media Gateway) replacing the integration module and the general idea model of MGC(Media gateway Controller) taking charge of its control, the main purpose of media gateway is to transmit to the destination via high-speed packet core network by changing the voice and data information inputted from the inter-connection network in media port into the packet of IP or ATM. For the sake of that, MG must perform the integration function to accommodate the various types of input traffic such as circuit network, N-ISDN, FR network, wireless network etc and the transformation function that can transform these into the packets.

This paper suggests the method of the bandwidth management in the media gateway that can accommodate the various new services as well as the existing voice service by appling to the next generation communication network. Section 2 explains the concept and the sort of the media gateway, and section 3 describes VoIP, VToA and VoDSL researched to accommodate the voice service at present, and section 4 suggests the method of bandwidth management with the media gateway, and section 5 describes the conclusion.

2 Media Gateway

As the key function device of next generation open-type communication network, the main purpose of MG(Media Gateway) is to transmit to the destination via the high-speed packet core network by transforming the voice and data information inputted from the interconnection network via media port into the packet of IP or ATM. For the sake of that, MG must perform the integration function to accommodate the various types of input traffic such as circuit network, N-ISDN, FR network, wireless network etc and the transformation function that can transform these into the packets.

The general idea model of MG and MGC is shown in Fig. 1. MG is composed of the media port module to perform the integration with interconnection network and the transformation function of data format, the bearer control module to manage the bearer connection of packet network, the packet port module to perform the integration function with packet network, the virtual switch function module to perform the switch function between the media port module and the packet port module. And MGC needs the signaling gateway control module to use the existing SS7 signal network or to perform the interconnection function, the media port control and service call control module to perform the management of interconnection network and packet network resources and the process of signal protocol. Meanwhile, the interface protocol standardized between MG and MGC is used the H.248/MEGACO(Media Gateway Control) protocol.

The media gateway function provides a phase to evolve into the next generation communication network naturally by integrating the transit networks such as the existing circuit network, N-ISDN and FR network into the high-speed packet network. The gateway function of next generation switch network system can separate into the residential gateway, access gateway and trunking gateway by the service required in the interconnection network.

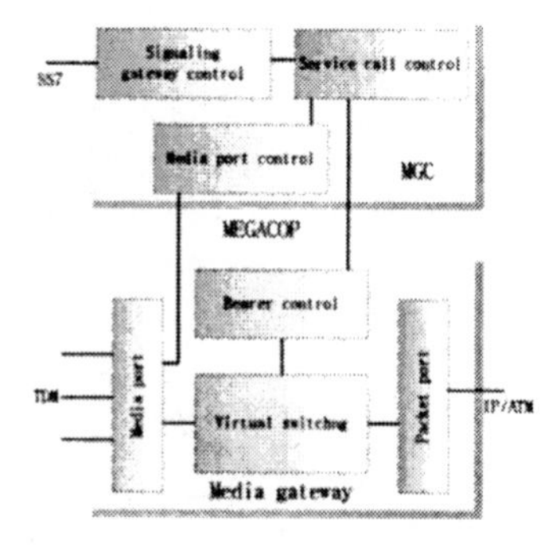

Fig. 1. The general idea model of MG/MGC

As the gateway that can accommodate the PSTN/ N-ISDN and xDSL subscriber of telephone or PC directly by connecting to the subscriber network, the residential gateway performs the function to interconnect the voice service of subscriber or VoIP service into the packet network directly by dealing with the various sort of subscriber interface. The access gateway that can accommodate the integrated subscriber access devices such as the existing access node: PBX or FLC, performs the PRI interface process function. The trunking gateway that can interconnect with the junction line of PSTN or N-ISDN connected to the local switch to replace the existing toll/tandem switch, performs the process of junction line signal protocol and the AALl/AAL2 trunking function for VToA service. Because this was equipped first in the evolutional stage of the existing network, it provides the function that can accommodate the existing PSTN or N-ISDN into the next generation switch network.

3 Support Technology of Voice Service

The packet voice technology is the technology that delivers the existing circuit switch-type voice and voice-band fax and modern data traffic via the

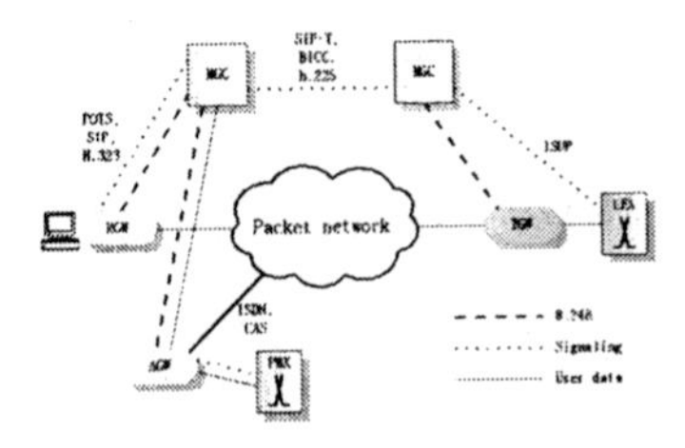

Fig. 2. The sort of gateway

packet switching network. The technology, which supports the fregmentation/incapsulation and real-time voice service quality to transport into the voice signal coding/compression and packet forms, and the integration between directory service and number system to grasp the situation of user and the additional service provided in the existing PSTN, is required. By improving the transmission efficiency about ten times than the offering method of existing circuit switch-type service, the packet voice technology with the voice compression of high efficiency and the silence process can reduce the transmission expense, but has the problem that weak than the existing circuit switching method from the viewpoint of the real-time traffic transport quality' support. Because of integrating and delivering the voice traffic in the data service-centered packet network, from the viewpoint of subscriber, the packet voice technology is achieved the voice service cheaply, and from the view-point of network business, guarantees the security of competitiveness facilities for the reduction of network construction/application expense, the improvement of network resources use efficiency, the intensification of service offer ability.

As the alternative of packet voice technology, VoIP(Voice over Internet protocol) technology via internet, VToA(Voice and Telephony over ATM) technology via ATM network, and VoDSL(Voice over Digital subscriber Line): the service type that the existing xDSL equipment manufacturing industries suggest the next generation communication network are a great interest.

3.1 VoIP

The VoIP technology that transports the voice traffic via IP network began at the gateway/gatekeeper with H.323 in the computer network, and at the early days, didn't come into notice because of the problem of service quality, but recently, VoIP market is being developed rapidly by the diffusion of internet user, the reduction of communication expense, the development of related technology, the additional service development of UMS.

At present, the VoIP has developed up to the type of pc-to-pc or pc-to-phone, and in case of developing into the next generation IP-based communication network, will be served as the type of phone-to-phone. Meanwhile, the protocol for VoIP: the SIP/SAP/SDP for the service control and signaling,

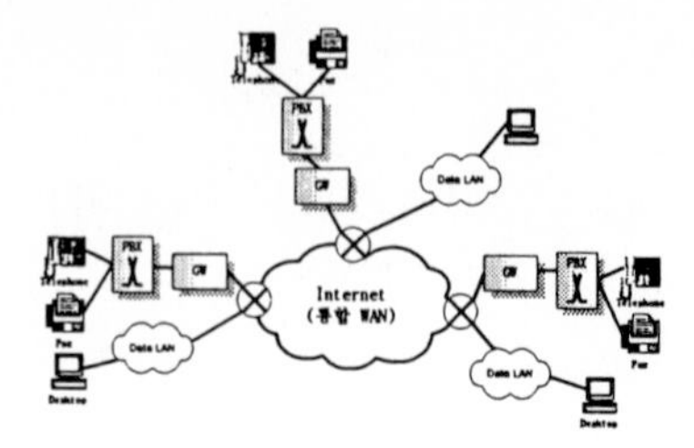

Fig. 3. The data/voice integrated network with VoIP technology

RTP/RTCP/RTSP for the media delivery, MEGACO for the gateway control, SCTP protocol is being standardized.

By appling the residential gateway connected in the subscriber network directly as Fig. 3, the VoIP service must perform the function that changes voice into IP packet, the subscriber signal terminal and call control function, the subscriber call and IP address interconnection function.

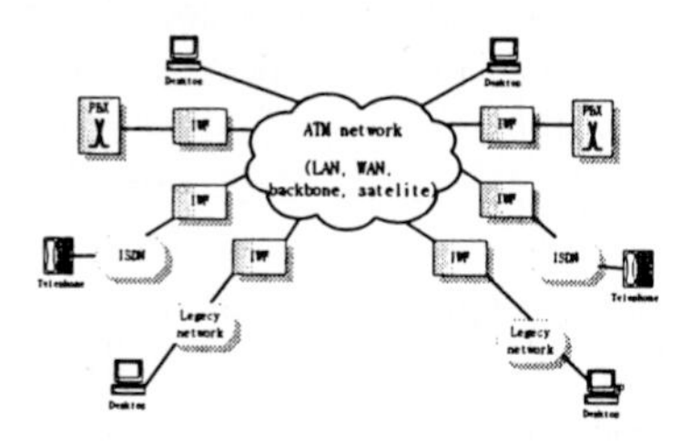

Fig. 4. The application of VToA technology

3.2 VToA

As the technology that the ATM Forum-centered standardization is being progressed, VToA is the method that delivers the voice traffic via ATM network. The trunking method using AALl, the non-switched/switched trunking method using AAL2, the loop emulation service method are suggested. Meanwhile, the media gateway for VToA use the access gateway interconnected to the enterprise network or access network, the trunk gateway interconnected to the PSTN network and junction line as Fig. 4, and the function that changes the TDM voice data into AAL1 or AAL2, the ATM trunking signaling function, the PSTN/N-ISDN signaling message terminal and call control function must be progressed.

3.3 VoDSL

The VoDSL is the method that transports the voice traffic inputted via the DSL circuit of subscriber network to the PSTN via the very high-speed packet network

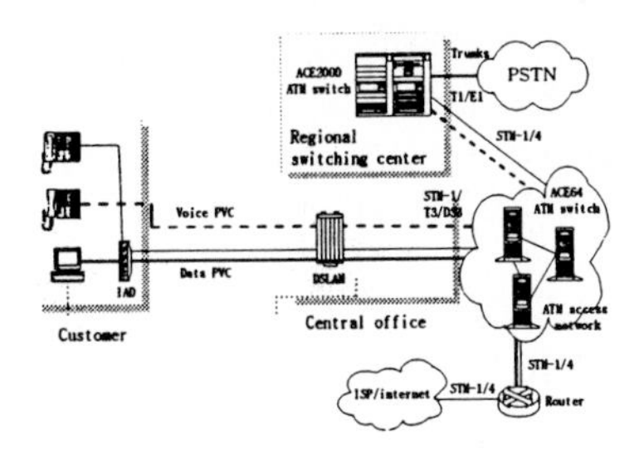

Fig. 5. The architecture of VoDSL service network

with VToA or VoIP. The VoDSL is composed of IAD(Integrated Access Device), DSLAM and MG interconnected to the PSTN as Fig. 5. This VoDSL is the service type suggested into the next generation communication network by the existing xDSL equipment manufacturing industries, and at present, the VoDSL service plan that apply the ATM-based packet network to the SOHO subscribers, is being published. From the viewpoint of media gateway, this VoDSL has no the function to add newly, and must provide the trunking or access gateway function according to the interconnection type with PSTN.

4 Performance Analysis

4.1 The Presentation of Multiplex Traffic Model

To evaluate the three alternatives of CBR, rt-VBR and UBR, the following model is used. In this model, we assume that there are N_s sources for each VC or cell-assembly device. A talkspurt from each source is transformed if need, additional information is added, and the talkspurts are assembled into CPS packets in the base station. When n sources are active in a talkspurt,n/R_p CPS packets are assumed to arrive at the cell-assembly advice in the corresponding time slot. Here, R_p is the time interval for generating a CPS packet at each source. Silence periods are assumed not to be transmitted. Talkspurts and silence periods are assumed to be geometrically distributed with means of $1/\alpha$ and l/β.

The CPS packets from the individual sources arrive at the cell-assembly device, which assembles them into a cell on an FCFS basis. A cell is assumed to be completed immediately after enough packets to make a cell have arrived at assembly device. Here, the number of packets needed to fill a cell is assumed to be fixed at L to simplify the notation.

The cell-assembly device uses a timer to ensure that the cell-assembly delay is less than a certain level. If the timer expires, a partially filled cell is generated. That is, the CPS packets in the assembly device are placed in the cell payload and the remaining payload is filled with padding when the timer expires.

The assembled cells are immediately placed in the transmission queue to await transmission. When CBR or rt-VBR VCs are used, each VC is assumed to have a dedicated transmission queue. When UBR VCs in a CBR VP are used, one transmission queue is shared by the VCs in each VP. The transmission queue has a finite-sized buffer. Let K be the buffer size, including the cell that is currently being transmitted. Cells in the transmission queue are transmitted at the fixed PCR of each VC(CBR VC and rt-VBR VC) or each VP(UBR VC in a CBR VP).

We assumed the time axis to be discrete with units equal to the transmission time of a cell via a VC(CBR VC and rt-VBR VC) or via a VP(UBR VC in a CBR VP) like figure 6. At the beginning of a time slot, CPS packets arrive at the cell-assembly device, and a cell is assembled on an FCFS basis as soon as enough CPS packets have arrived. Partially filled cells may also be generated due to a timeout at the beginning of a time slot. The assembled cells are immediately placed in the transmission queue. A single cell is transmitted on an FCFS basis at the end of a time slot to the VP multiplexing buffer when CBR or rt-VBR is used, or to the next network entity when UBR is used. (Let K' be the size of the VP multiplexing buffer. Cells in the VP multiplexing buffer are scheduled to be sent to the next network entity at the PCR of the VP.) The timer for cell assembly starts at zero, then increments by one just after the beginning of each time slot. If a cell is assembled in the time slot, the timer is updated afterwards. A partially filled cell is generated if the timer is T-1 at the boundary between the time slot and the previous time slot and not enough CPS packets have arrived at the time slot. Here, T is the timeout of cell assembly.

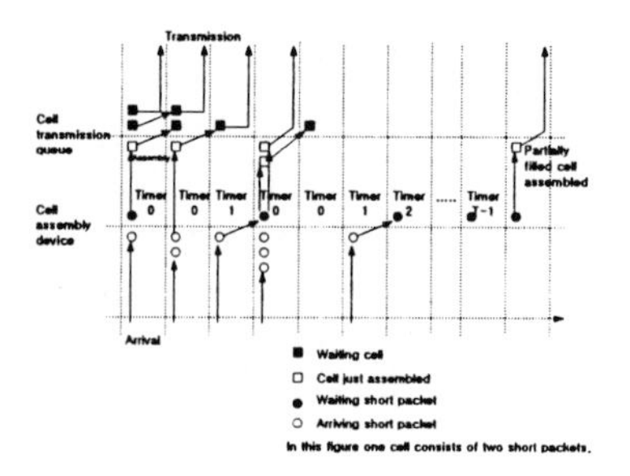

Fig. 6. Discrete time model

4.2 The Analysis of Bandwidth Management

In the following numerical examples, L is 2 and T is 4 ms for all of the three alternatives: CBR, rt-VBR, and UBR. For CBR and rt-VBR, the buffer size K of the transmission queue is 3 ms, the buffer size K' of the VP multiplexing queue is 2ms, and the CLR objectives are $0.5\times 10^{-}4$ in both the transmission and VP multiplexing queues. Thus, the total buffer size is 5 m s and the total CLR objective for these buffers is almost $10^{-}4$. For UBR, K is 5 ms and the CLR objective in the transmission queue is $10^{-}4$. Based on Yatsuzuka's analysis, it is assumed that the mean talkspurt is 169.7 ms and the mean silence period is 123.9 ms. Considering recent low-bit-rate coding methods, such as CS-ACELP, the voice coding rate in a talkspurt is assumed to be 8 kb/s.

Fig. 7 plots the bandwidth needed to accommodate various numbers of VCs under a fixed total amount of offered voice traffic and a fixed CLR objective. Thus, as the number of VCs increases, the traffic in each VC becomes small. As shown in the figure, the CBR VCs request much more bandwidth than the UBR ones do. Bandwidth usage is thus inefficient under CBR VCs, particularly when the VC bandwidth is narrow. Note that even under CBR VCs, a statistical multiplexing gain is achieved among AAL2 connections or the traffic increases, the necessary bandwidth decreases.

The rt-VBR is more effective than CBR, but the effectiveness is limited when the number of VCs is larger than ten. Even when the rt-VBR becomes effective, UBR greatly outperforms rt-VBR. For any of the three alternatives, if there are many VCs, then more VP bandwidth may be required. The increase in VP bandwidth under UBR is, however, much smaller than that under CBR or rt-VBR.

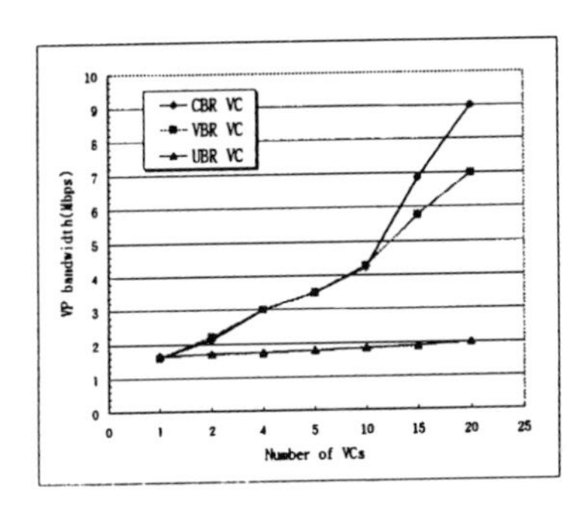

Fig. 7. VP bandwidth for fixed total offered load

The equivalent bandwidth for the fixed offered load in each VC is plotted in Fig. 8. The equivalent bandwidth in this paper is defined by dividing the VP bandwidth by the maximum number of AAL2 connections that can be accommodated in the VP while still satisfying the objectives. For CBR, the equivalent bandwidth is kept constant. This is because the traffic for each VC is fixed and, as a result, the bandwidth for each VC is fixed. For rt-VBR, the equivalent bandwidth decreases when the number of VCs is larger than about 10. This is because the difference in the statistical multiplexing gain between rt-VBR and CBR is the gain among VCs, and also because the gain among VCs is achieved when the number of VCs is larger than 10. For UBR, the equivalent bandwidth decreases sharply even when the number of VCs is less than 10. This is because a statistical multiplexing gain among AAL2 connections in different VCs is achieved under the UBR alternative.

The number of AAL2 connections in a VC is more than 50 for a 250-kb/s voice signal and more than 20 for a l00-kb/s voice signal. Therefore, the statistical multiplexing gain in the case of two VCs is equivalent to the gain of more than 100 or 40 AAL2 connections, respectively. On the other hand, the statistical multiplexing gain under the rt-VBR alternative is separated into two Parts: the gain among AAL2 connections in a single VC and the gain among VCs. Thus, the gain using the rt-VBR is less than that using UBR. The difference in equivalent bandwidth between 250 kb/s and 100 kb/s under CBR is the difference in the statistical multiplexing gain among AAL2 connections within a VC, and this gain is achieved for each of the three alternatives, including the CBR one. When the number of VCs is 1, the equivalent bandwidth for each AAL2 connection is less with 250 kb/s in a VC than it is with 100 kb/s in a VC under every alternative. This is because the statistical multiplexing gain for 250-kb/s traffic in a VC is larger than that for l00-kb/s traffic in a VC.

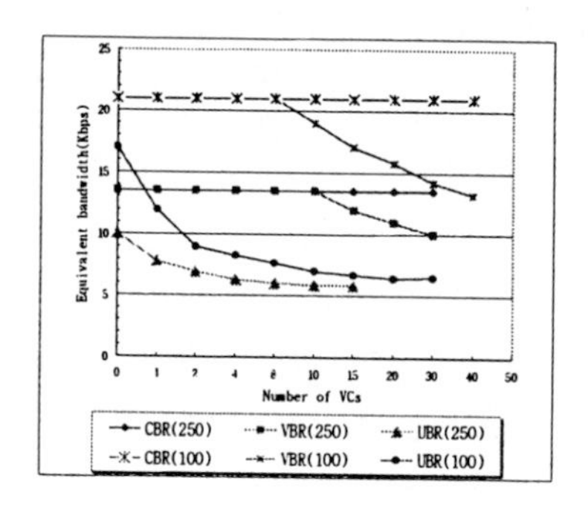

Fig. 8. Bandwidth for each source with a fixed offered load in a VC

The reason that the curve plotting the equivalent bandwidth under rt-VBR has a steeper decrease with 100 kb/s than with 250 kb/s that the PCR of t he VC with 100 kb/s is lower than that with 250 kb/s. Due to the property that low PCR VCs achieve high statistical multiplexing gain, the statistical multiplexing among VCs under rt-VBR gains more for 100 kb/s than for 250 kb/s. (However, the statistical multiplexing within a VC is larger for 250 kb/s than for 100 kb/s. The total statistical multiplexing effect is larger for 250 kb/s than for 100 kb/s under rt-VBR, so the equivalent bandwidth under rt-VBR is smaller for 250 kb/s than for 100 kb/s.)

There are two reasons why the equivalent bandwidth can be larger than (the voice coding rate (8 kb/s) + AAL2 overhead): the existence of partially-filled cells and the unused VC/VP bandwidth. Unused VC bandwidth can arise as follows: Many CPS packets can arrive in a short time from different sources. During this short time, the cell rate can be larger than the voice coding rate plus the overhead. For example, consider the situation that there are ten voice sources and each cell includes two CPS packets. If these ten sources generates

CPS packets in a short time, five cells are assembled in this short time. The cell rate during this short time is higher than the voice coding rate plus the overhead. The buffer in the transmission queue may lose some of these assembled cells in this short time due to buffer overflow because the buffer size is very limited in order to reduce the delay due to the buffer waiting time. To limit the CLR below a certain level under this situation, the VC bandwidth must increase and can result in bandwidth larger than the voice coding rate (plus overhead). In particular, a small-bandwidth VC is likely to provide a small buffer, and the effect of this cannot be neglected. Traffic aggregation and large VC/VP bandwidth are important for efficient AAL2 transport.

One method of aggregating traffic is to use an AAL2 switching node. Its introduction is especially efficient in a mobile network for the following reason: In a mobile network on an ATM backbone, each pair of base stations is connected via a VC connection. Different traffic streams whose originating and destination AAL2 end Points are in the same pair of base stations can be multiplexed in the VC. Unfortunately, however, the load of such traffic is normally too small to use a single VC efficiently. Since AAL2 switching nodes terminate ATM connections, however, they tan aggregate traffic streams that, for example, have originating AAL2 end points in the same base station, but destination AAL2 end points in different base stations.

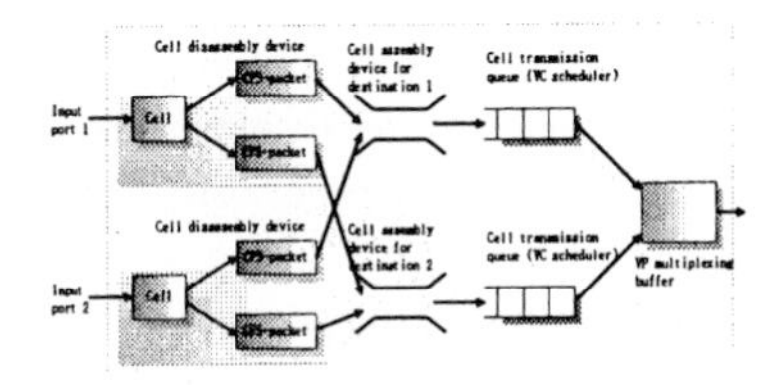

Fig. 9. AAL2 switching node.

This paper focuses on a single VP and compares its bandwidth when the termination point oft he VP connection is located in an AAL2 switching node or in a VCH. An AAL2 switching node is shown in Fig. 9.

The three alternatives tan be applied to a network that introduces the AAL2 switching node, and they affect the VP bandwidth. Fig. 10 plots the VP bandwidth with and without an AAL2 switching node for each of the three alternatives. Since an AAL2 switching node causes cell assembly delay, the allocated timeout and buffer size are reduced when an AAL2 switching node is used. Under the UBR alternative, the impact of introducing an AAL2 switching node is very small for any number of sources in a VC and for any total number of VCs.

This is because a large statistical multiplexing gain has already been achieved under the UBR alternative even without AAL2 switching nodes. Under the CBR alternative, the effectiveness of introducing an AAL2 switching node is significant. The reduction in VP bandwidth is very large, particularly when the number

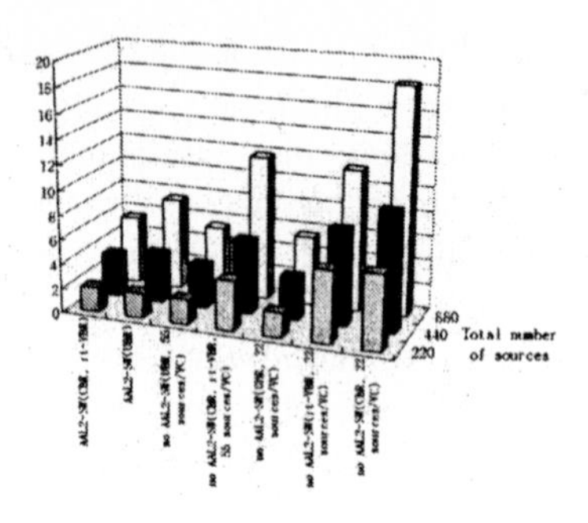

Fig. 10. Effectiveness of AAL2 switching nodes.

of sources in a VC is small. It is almost the same as but slightly larger than that under the rt-VBR alternative. Under the rt-VBR alternative, introducing an AAL2 switching node is also effective, but not as effective as under the CBR alternative, because the statistical multiplexing effect among VCs can reduce the bandwidth when the number of sources in a VC is small.

5 Conclusion

In the next generation telecommunications network, we described the voice over packet related technologies to support voice service required the continued technology development as the important service, and described the function of media gateway. and we considered the three alternatives: CBR, rt-VBR, UBR in the bandwidth management of media gateway.

From the viewpoint of bandwidth efficiency, the alternative using UBR VCs is the best. In this alternative, admission of AAL2 connections is judged based on whether AAL2 connections can be accommodated in a VP. This requires VCH's to have an admission control function for AAL2 connections, whereas the other alternatives do not need VCH's to have that control function. The alternative using rt-VBR is better than CBR, but the advantage is limited when the number of VCs is larger than 10. The admission control for UBR VCscan be implemented easily because it is unnecessary to take account of the differences among VCs. Introducing AAL2 switching nodes is effective for the CBR and rt-VBR-alternatives. In particular, the effectiveness is significant when the rt-VBR alternative with a small number of VCs or the CBR alternative is used.

References

1. C. Waller, Draft Inplementation Agreement for MEGACO/H.248 Profile for Media Gateway Controller/Access Gateway using ATM Trunks, *MC WG IA., MSF-2000.147.0*,2000.9.
2. CSELT, "phone Service Architecture Options," *ETSIVoIP Workshop*, Mar. 1997.

3. M. Taylor, "Convergence in Local Telephone Networks," Coppercom while paper, June 2000.
4. ATM Forum af-vtoa-0113.000, ATM Trunking using AAL2 for Narrowband services Jan. 1999.
5. ATM Forum. AF-VMOA-0145.000, "Voice and Multimedia Over ATM-Loop Emulation Services Using AAL2" 2000. 7.
6. J. G. Gruber, "A comparison of measured and calculated speech temporal parameters relevant to speech activity detection," IEEE Trans. Commun, vol. COM-30, no. 4, pp. 728–738, 1982.
7. K. Sriram and W. Whitt, "Characterizing superposition arrival processes in packet multiplexers for voice and data," IEEE J. Select. Areas Commun., vol. JSAC-1, no. 6, pp. 1124–1132, 1983.
8. S. Li, "Study of packet loss in a packet switched voice system,", 1988, pp. 1519–1526.
9. K. Sriram, R. S. McKinney, and M. H. Sherif, "Voice packetization and compression in broadband ATM networks", IEEE J. Select. Areas Commun. Networks. Vol. 9, no. 3, pp. 294–304, 1991.
10. U. Brown, T. H. Thinner, and H. Kroner, "A general discrete-time queueing model: Analysis and applications," in ITC13 Copenhagen, Denmark, 1991, pp. 13–19.
11. A. G. Valko, A. Racz, and G. Fodor, "Voice Qos in third-generation mobile systems," IEEE J. Select. Areas Commun, vol. 17, no. 1, pp. 109–123, 1999.
12. D. W. Petr, R. R. Vatte, P. Sampath, and Y.-Q. Lu, "Efficiency of AAL2 for voice transport: Simulation comparison with AAL1 and AAL5," Vancouver, Canada, 1999.

E-ONIGOCO: Electronic Playing Tag Support System for Cooperative Game

Jun Munemori[1], Tomohiro Muta[2], and Takashi Yoshino[2]

[1] Wakayama University,
Center for Information Science,
930 Sakaedani, Wakayama 640-8510, Japan
munemori@sys.wakayama-u.ac.jp
[2] Wakayama University,
Department of Design and Information Sciences,
Faculty of Systems Engineering,
930 Sakaedani, Wakayama 640-8510, Japan
yoshino@sys.wakayama-u.ac.jp

Abstract. There has been few location-aware system which have two-way system. The playing tag is a very popular game in Japan. This game has a two-way communication in a sense. We have developed an electronic playing tag support system for cooperative game, named E-ONIGOCO. The system consists of a PDA, a GPS and a mobile phone. We performed five types of experiments at two universities 24 times. Four types are the electronic playing tags and the remainder is a conventional playing tag. The results of experiments were showed below. (1) The devised services based on the electronic playing tag support groupware were better estimation than that of the conventional playing tag. The mutual positioning information using a GPS was effective for the playing tags. (2) If the accuracy of a GPS decrease, we spend a little more time to catch. But the time did not exceed the playing time of the conventional play tag. (3) We can realize the electronic playing tag between remote universities. But we should add the reality using the remainder of five senses for lack of seriousness.

1 Introduction

PDA (Personal Digital Assistant) is highly portable, and can collect data in anytime and anywhere [1]. The positioning data would become important for a PDA. Location awareness from positioning data may improve the richness of the communication.

GPS is useful equipment to get positioning data in the open air. The accuracy of the GPS was improved last year. We could treat the positioning data flexibly. The modification of positioning data would produce a new service of PDA.

Most of location-aware systems using PDAs have one-way communication [2]. The two-way communication would be important in location-aware system. The word of two-way communication means that clients of the system should move freely and get and present positioning data of other clients easily.

I. Chong (Ed.): ICOIN 2002, LNCS 2343, pp. 40–49, 2002.

The playing tag is a very popular game in Japan. This game has a two-way communication in a sense. The game consists of runaways and chasers. If chasers catch all runaways, then the game is over. The playing tag is known as "onigokko" in Japan. Fox hunting (or radio direction finding) is a similar game, but the game has a one-way communication basically.

We have developed an electronic playing tag support system. The system is named E-ONIGOCO (the ElectrONIc playing taG suppOrt system for COoperative game). E-ONIGOCO is a two-way location aware system. The system consists of a PDA, a GPS and a mobile phone. In Japan, we can use an electronic mail and a web browser by a mobile phone easily. We call the function, for example, as "i-mode." This system has a two-way communication using an electronic mail and a Web browser. We applied it to the four types of playing tag at two universities (Kagoshima Univ. and Wakayama Univ.).

First of all, runaways are one group and chasers are one group. We call it one-to-one experiment. Secondly, runaways are one group and chasers are two groups. We call it one-to-two experiment. Thirdly, screens of map do not represent accurate place of participants. We call it a blurred experiment. Last of all, experiments are performed between Kagoshima Univ. and Wakayama Univ. Runaways and chasers cannot see each other, because they are playing tag in distant places. But runaways and chasers seem to be existed at the same university on the screen of a mobile phone. We call it virtual playing tag experiment.

This paper mainly describes the effectiveness of the mutual positioning data for the playing tag, the effectiveness of its accuracy for it, the effectiveness of virtual playing tag, and the new service which was introduced from the positioning data.

2 Methodology

We propose three hypotheses about the electronic playing tag, which is based for our experiments. We perform five types of the playing tag under a rule.

2.1 Hypothesis

Hypothesis 1 : If chasers will be two groups, runaways might be caught quickly comparing with one group.

Hypothesis 2 : If the presentation of the position will be blurred, runaways might not be caught easily.

Hypothesis 3 : If runaways and chaser exist between distant places and runaways and chasers cannot see each other, they might feel different impression from other experiments.

2.2 Playing Tags

We performed five types of experiments. Four types are the electronic playing tags and the remainder is a conventional playing tag. The contents of experiments are shown below.

1. One-to-one experiment
   Runaways are one group and chasers are one group.
2. One-to-two experiment
   Runaways are one group and chasers are two groups.
3. Blurred experiment
   Screens of maps represent places of participants roughly.
4. Virtual playing tag experiment
   Experiments are performed between Kagoshima Univ. and Wakayama Univ. Runaways and chasers cannot see each other, because they are playing tag in different places. But runaways and chasers seem to be existed at the same university on the screen of a mobile phone. They do not know the information of map on the other university. Participants know who are runaways or chasers except for the virtual playing tag experiment.
5. Conventional playing tag without equipment
   Runaways are one group and chasers are one or two groups. They do not use electronic equipment.

2.3 Rules

We show the rules of electronic playing tag.

1. Runaways and chasers send information of their position by an electronic mail every ten minutes. But runaways run away ten minutes before when chasers start.
2. Runaways keep their position 2 minutes when they send the information of their position by an electronic mail.
3. Participants are prohibited from running. They must move within the area of a university.
4. If chasers catch runaways, or chasers and runaways seem to be the same position, the game is over.

We performed five types of experiments. Four types are electronic playing tags and the remainder is a conventional playing tag. The rule of the conventional playing tag is basically the same as the electronic playing tag.

3 Electronic Playing Tag Support System (E-oNIGOCO)

E-ONIGOCO consists of the mobile system and the information processing system for map.

3.1 Mobile System

The mobile system consists of a PDA (3Com, Palm III), a GPS (GARMIN, etrex), a modem (I·O Data, SnapConnect), and a mobile phone. The accuracy of the GPS is 15 meters. The program of the mobile system was developed on

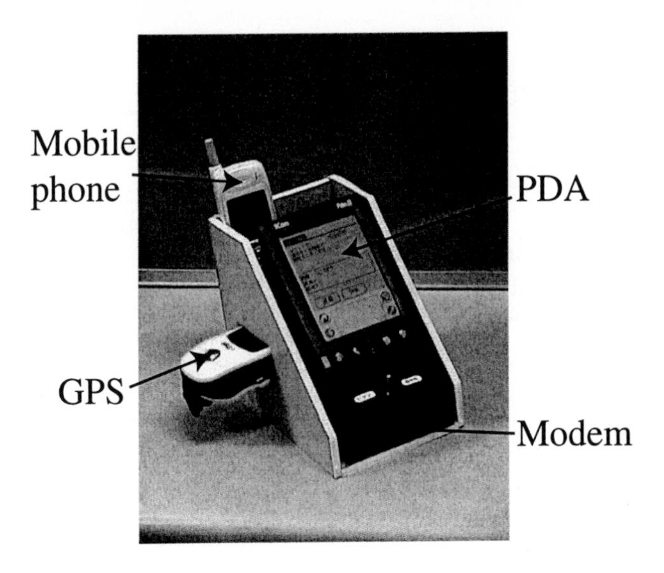

Fig. 1. The mobile system.

a Macintosh using CodeWarrior for the Palm OS Release 5 (Metrowerks). Now, the number of program lines is about 1700 lines.

Figure 1 shows the mobile system. If a participant wants to see the screen of a mobile phone, he or she can lift up the phone from the box. The total weight of the system is about 700g. Figure 2 shows a screen of a PDA. A participant chooses his or her role (a runaway or a chaser) from a menu of a screen on a PDA. If a runaway or a chaser pushes the start button, passing time is shown on the button from the beginning. The latest longitude and latitude data of GPS is presented.

The longitude and latitude data of GPS is also recorded every one minute in the PDA. The data recorded every one minutes and the data of present time are composed for the sending data. If we cannot collect a longitude and latitude data from GPS, the data will be skipped. But the recent data was recorded within every 5 seconds. So, we will be able to send a GPS data to the information processing system for map surely.

Participants send information of their positions by an electronic mail of a mobile phone. They can view maps by a Web browser of a mobile phone. URL of the map is indicated by an electronic mail when the map is updating. The mail also indicates who is renewed.

3.2 Information Processing System for Map

The information processing system for map processes the data from the mobile system. Now, the number of program lines is about 17000 lines by HyperTalk (Apple Computer). The system decides and makes a corresponding map from longitude and latitude data.

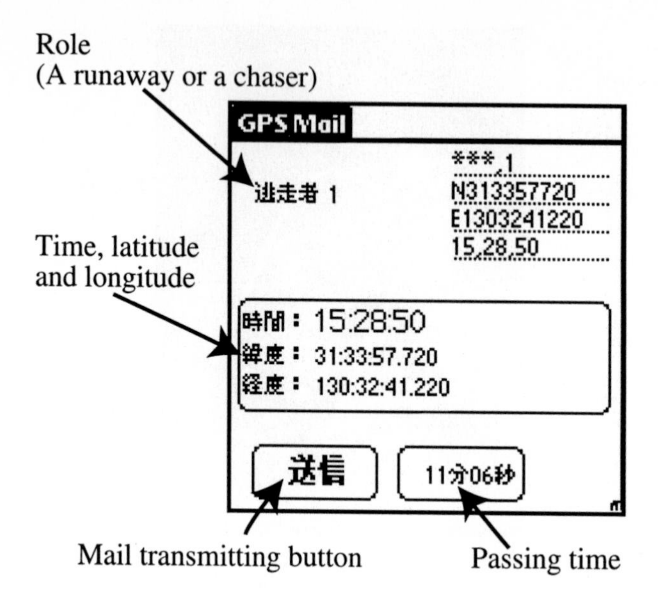

Fig. 2. An example of a screen on a PDA.

Figure 3 shows an example of a map on a mobile phone. A runaway or a chaser is positioned the center of the map and marked as a kind of symbol marks. The system also shows a locus, building name near their position, and a scale of a map. If a runaway or a chaser is not presented within one map, they are displayed on different maps.

The system has a blurred function. This function hides accurate place of participants. That is, only adequate area (without symbol mark) around them is displayed on a map. The size of an area is 87.5 meters on all sides.

This system also has a virtual playing tag function. Figure 4 shows an image of the virtual playing tags. In figure 4, the runaway is existed in Wakayama Univ. and the chaser is existed in Kagoshima Univ. But a runaway and a chaser seem to be existing in the same university on the screen of a mobile phone. This function transforms the positioning data of a GPS and corresponds to the longitude and latitude data of both two universities. The longitude of a place (E 135°9'1") at Wakayama Univ. is corresponding to that of a place (E 130°32'47") at Kagoshima Univ. The latitude of a place (N 34°15'46") at Wakayama Univ. is corresponding to that of a place (N 31°34'1") at Kagoshima Univ. If the difference between the position of a runaway and that of a chaser will become below a certain fixed value at the same time, we decide that a runaway seems to be arrested. Figure 5 shows a screen of the map information processing system for the virtual playing tag. Figure 5 shows loci of a runaway and a chaser in Wakayama Univ. The runaway walks the road around the building of Wakayama Univ. The chaser, he walks the road of Kagoshima Univ. in reality, chases the

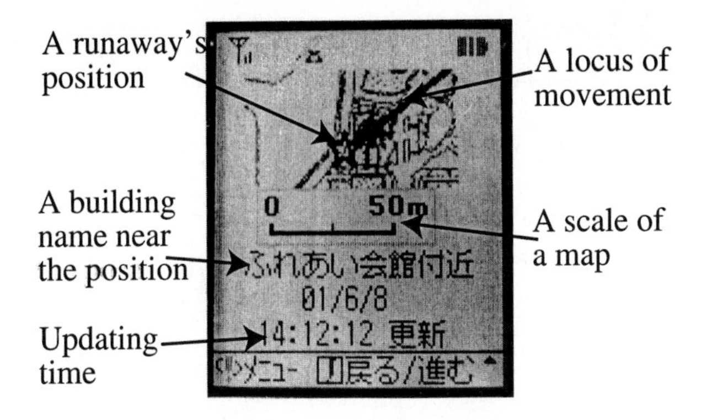

Fig. 3. An example of a map on a mobile phone.

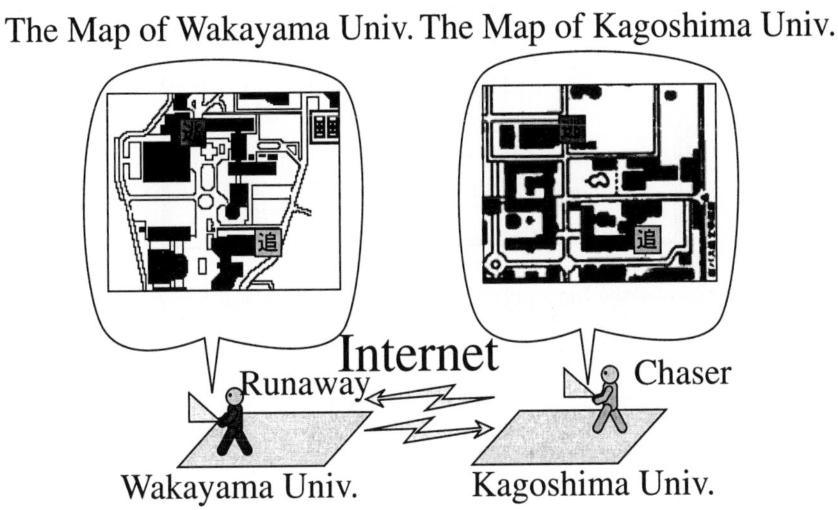

Fig. 4. An image of the virtual playing tag.

runaway on the screen. The chaser seems to be broken through the building on the upper part of the screen.

4 Experiments and Results

4.1 Experiments

We performed experiments 24 times from December 2000 to June 2001 at two universities. Experiments were carried out in Kagoshima Univ. and Wakayama Univ. The area of Kagoshima Univ. is about 480,000 square meters, which is including 85 building and the area of Wakayama Univ. is about 100,000 square

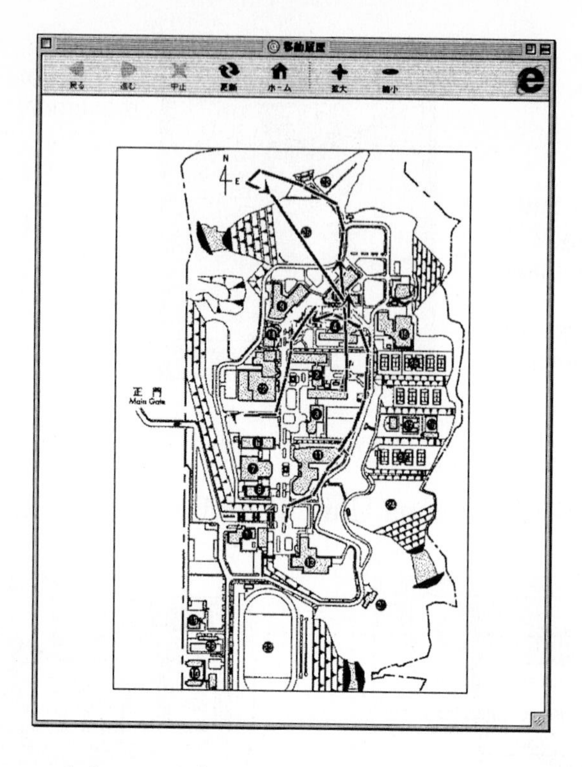

Fig. 5. The screen of the map information processing system for the virtual playing tags.

meters, which is including 26 building. Runaways or chasers are consisted of one to three persons (except for the virtual playing tag experiment).

The number of each experiments was indicated below:

1. One-to-one experiment: Three times in Kagoshima Univ. and four times in Wakayama Univ.
2. One-to-two experiment: Three times in Kagoshima Univ.
3. Blurred experiment: Four times in Wakayama Univ.
4. Virtual playing tag experiment: Five times in both universities.
5. Conventional playing tag without equipment: Three times in Kagoshima Univ. and two times in Wakayama Univ.

4.2 Results of Experiments

Table 1 shows the results of experiments. The value of Table 1 (from 1. to 8.) is a 5-point scale (with answers "1: very bad", "2: bad", "3: neutral", "4: good" and "5: very good") for participants to evaluate each item on the questionnaire.

Table 1. Results of experiments.

Kind of experiments		Electronic Playing Tag				
		Normal	Normal	Normal	Blurred	Virtual
	Runaways and chasers	One-to-one	One-to-one	One-to-two	One-to-one	One-to-one
	Place	Kagoshima Univ.	Wakayama Univ.	Kagoshima Univ.	Wakayama Univ.	Kagoshima Univ. and Wakayama Univ.
1.Was it interesting?		3.6	2.9	4.1	3.9	2.9
2.Was the location information useful?		-	3.7	-	4.2	4.3
3.Was the map easy to see?		2.4	3.2	3.0	3.1	2.7
4.Was the locus easy to see?		2.8	3.0	3.1	-	2.6
5.Was the position display by the character useful?		4.8	3.5	4.9	4.0	4.4
6.Was it easy to use screen operation of a mobile phone?		3.4	3.4	3.4	3.5	3.2
7.Was the screen of a mobile phone small?		1.9	2.2	2.1	2.1	2.2
Required time (minutes)		80	24	23	20	66
		110	21	29	36	39
		90	14	48	14	107
			30		64	34
						28
Average (minutes)		93	22	33	34	55

Kind of experiments		Conventional playing tag experiments		
	Runaways-to-chasers	One-to-one	One-to-two	One-to-one
	Place	Kagoshima	Kagoshima	Wakayama
1.Was it interesting?		3.8	3.5	3.5
Required time (minutes)		61	22	46
			33	62
Average (minutes)		61	28	54

4.3 Discussion

1. One-to-one experiment:
   It takes over 80 minutes to catch in Kagoshima Univ. The area of Kagoshima Univ. is five times as broad as that of Wakayama Univ. It seems too long to catch. Participants were somewhat tired, but they felt interesting (The estimation value of "interesting" was 3.6/5.0 in Table 1). But it seems to be too short to catch about 20-30 minutes in Wakayama Univ. Participants felt that the experiment was not interesting (The estimation value of "interesting" was 2.9/5.0 in Table 1).
2. One-to-two experiment:
   This experiment was estimated very high (4.1/5.0 in Table 1), because chasers estimated the experiment very high. Participants were satisfied with shorter the playing. The devised service based on the electronic playing tag is estimated highly. They felt that they need strategy to catch in the experiment. For example, each chaser communicates each other using mobile phones and makes a plan to catch runaways.

3. Blurred experiment:
   This experiment was estimated considerably high (3.9/5.0 in Table 1). Participants were satisfied with longer the playing. The devised service based on the electronic playing tag is also estimated highly.
4. Virtual playing tag experiment:
   This experiment was not good estimation (2.9/5.0 in Table 1). Most of all participants could not feel the reality of the other party without real world. Because their existence were only showed in the screen of the mobile phone. We must add the rest of five senses, for example, sound or vibration, if the other party will approach.
5. Conventional playing tag without equipment:
   In conventional playing tag without equipment, runaways run away longer time than in the electronic playing tag in Wakayama Univ. They took a plan to run away from the other party, because they wanted to run away for a long time. But runaways were caught relatively short time in Kagoshima Univ., because they did not feel good to carry out the playing tag without support system. They did not satisfy the playing tag without equipment. So, these results caused to suppose that location-awareness improved the richness of the communication.
6. The overall electronic play tag and the conventional play tag:
   Participants of playing tags were fond of mutual positioning information using a GPS. The estimation value of the "interesting" about the one-to-two experiments is better than that of the conventional playing tag in Kagoshima Univ. The estimation value of the "interesting" about the blurred experiment is also better than the conventional playing tag in Wakayama Univ. So, the electronic playing tag, which includes the devised services, was better estimation than the conventional playing tag. The mutual positioning information using a GPS was effective for the playing tags.
   In Wakayama Univ., the mean value of the time to be arrested is 22 minutes in the one-to-one experiment, 34 minutes in the blurred experiment, and 54 minutes in the conventional playing tag experiment, respectively. Judging from these experiments, the time to be arrested is dependent on the accuracy of the GPS. But the time using the inaccurate GPS (i.e. the blurred experiment: mean value is 34 minutes) did not exceed the playing time (mean value is 54 minutes) of the conventional playing tag.

4.4 Related Work

The typical system of using a PDA and positioning information of a GPS is a location-aware tourist guide system. The early work on developing a location-aware tourist guide was Cyberguide [3]. The outdoor Cyberguide used a PDA and a GPS unit. The GPS unit sends a position in latitude and longitude which was then translated into a pixel coordinate representing the user's current position on the map [3]. In Japan, LOCATIONWARE was proposed and applied to a city guide [4]. Both systems are using position information, but they are one-way communication system.

Tarumi et. al. proposed the SpaceTag and active SpaceTag [5]. SpaceTag is also a location-aware system. SpaceTag is a virtual object that can be accessed only from limited location and period. SpaceTag objects will be able to communicate with other objects or users. The system is said to be a two-way communication system. But the system has not been applied and estimated yet.

5 Conclusion

We have developed the electronic playing tag support system. We performed five types of experiments at two universities 24 times. Four types are the electronic playing tags and the remainder is a conventional playing tag. The results of experiments were showed below. (1) The devised services based on the electronic playing tag support groupware were better estimation than that of the conventional playing tag. The mutual positioning information using a GPS was effective for the playing tags. (2) If the accuracy of a GPS decrease, we spend a little more time to catch. But the time did not exceed the playing time of the conventional play tag. (3) We can realize the electronic playing tag between remote universities. But we should add the realty using the remainder of five senses for lack of seriousness.

In the future, we will use the system for various experiments, and thus repeated evaluations will be required. Furthermore, the system will be improved on the based on the results obtained in the present study.

References

1. R. Davis, J. Landay, V. Chen, J. Huang, R. Lee, F. Li, J. Lin, C. Morrey, B. Schleimer, M. Price, B. Schilit, "NotePals: Lightweight Note Sharing by the Group, for the Group",in Proceedings of ACM CHI'99, ACM Press, 338-345,1999.
2. Y. Ito, K. Morishita, H. Tarumi, Y. Kambayashi, "Design and Application of Active Objects in an Overlaid Virtual System: SpaceTag", DICOMO2000, pp.595-600, 2000.
3. G. D. Abowd, C.G. Atkeson, J. Hong, S. Long, R. Kooper and M. Pinkerton, "Cyberguide: A mobile contex-aware tour guide", Wireless Networks, 3, pp.421-433, 1997.
4. A. Kurashima, S.Ichimura,and K. Sakata, "Development and Applying of Location-aware Mobile Com-munication Service Middleware -experiment of applying to a GIS service system to support sightseeing at Matsue-city-",The 63th IPSJ Conference, Special track A-1, 2001.
5. H. Tarumi, K. Morisita, Y. Ito, and Y. Kambayashi, "Communication through Virtual Active Objects Over-laid onto the Real World", CVE 2000, pp.155-164, 2000.

Approximated Overpassing Period to Virtual Capacity of LSP in MPLS

Ilyoung Chong[1] and Sungchang Lee[2]

[1]Dept. of Information and Communications Eng., Hankuk Univ. of FS
Seoul , Korea
iychong@hufs.ac.kr
[2]Hankuk Aviation University, Goyang, Korea
sclee@mail.hau.ac.kr

Abstract. In MPLS network multiple LSPs shares bandwidth of a physical link. In order to perform a traffic engineering function it is necessary to estimate allowed and available bandwidth within assigned virtual bandwidth for a LSP. The paper suggests a novel algorithm to compute the expected period exceeding assigned virtual bandwidth, which computes packet loss probability in transport system shared with multiple LSPs in MPLS. The algorithm is based on the discrete-time Markov BD process at burst level. And it uses the devised concepts from first passage time, and it is efficiently applied to the computation of packet loss statistics and call admission control in burst scale traffic model of MPLS network.

1 Introduction

In MPLS network, a LSP(Label Switched Path) based on RSVP-TE (Resource Reservation Protocol–Traffic Engineering) or CR-LDP (Constraints Routed Label Distribution Protocol) may contain multiple traffic sources with the same traffic characteristics, and reserves virtual bandwidth within physical transport link. Multiple traffic sources with the same traffic characteristics are multiplexed into a virtual bandwidth. In this case it is necessary to perform traffic engineering and to estimate allowed and available bandwidth within assigned virtual bandwidth for a LSP since multiple LSPs with FEC shares bandwidth of a physical link, particularly in ATM based MPLS. The paper suggests a novel algorithm to compute the expected period exceeding assigned virtual bandwidth, which the algorithm will be greatly useful to compute packet loss probability in transport system shared with multiple LSPs in MPLS.

In LSP with virtual capacity, traffic sources are modeled as periodic bursty *on-off* sources, and a traffic has been characterized into packet level and burst level. In bursty traffic, and packets are generated at a peak rate during burst (*on*) period, while no cells are generated during silent (*off*) period. Define *a frame* as the inter-arrival time of cells within a burst, namely D. If we consider the number of active sources during frame period (D) at burst level, the state of arriving traffic sources is characterized by discrete-time Markov birth-and-death(BD) process. A bursty traffic

I. Chong (Ed.): ICOIN 2002, LNCS 2343, pp. 50-63, 2002.

source is also described by the average burst length ($1/\mu$) and average silence length ($1/\lambda$). An additional parameter to characterize a bursty traffic source is the burstiness ($\beta = 1/\mu/(1/\lambda+1/\mu)$), which is the ratio between a burst a burst length and a burst cycle. If the number of sources that are on at any instant is larger than the frame size(D), a system goes into an overflow period and the incoming cells are discarded. The interval of time during which incoming cells are discarded is termed the loss period. Let the state of the system be defined as the number of on sources at a frame boundary. When k bursts from N traffic sources are in progress at frame, the system changes from k to (k+1) in the next frame with probability of $(N-k)\lambda$. Similarly, the transition rate from state k to state (k-1) is $k\mu$. Define $\lambda k = (N-k)\lambda$ and $\mu k = k\mu$ for number of on sources.

The paper uses a discrete-time Markov BD process model and proposes a novel concept derived from a conventional first passage time [3], which is a time duration that state i reaches to state ($i \neq j$) in Markov BD process. The proposed passage time in the paper is conditioned by initial state, and its transition path is diversified rather than a conventional first passage time. The novel concept, which is a special case of the first passage time, the first up-passage time (FUT) and the first down-passage time(FDT), is introduced in this paper.

2 Revised Concept of Passage Time

The conventional first passage time is not sufficient to approximate a period lasting the continuous upper state i or lower state i. The section introduces the novel concept based on passage time lasting upper or lower state i. The algorithm introduced at this section shows two novel concepts based on the discrete-time Markov BD process, the first up-passage time (FUT) and the first down-passage time (FDT). They are conditioned by restricted transitions from state i to j, while the conventional first passage time takes a journey from one state to another state without restriction. Before turning to the details of computations of FUT and FDT, the fundamental properties of Markov BD process are reviewed.

[Definition 1] First Up-passage Time (FUT):

The time duration until the starting state reaches a given state in BD process is called the first-passage-time [3]. The first-up-passage time $\left(FUT_{ij}^{(i)}\right)$ is the time period to reach a given state (j) from the initial state (i), such that any intermediate state (k) is not smaller than the initial state i. FUT from state i to state j, $\left(FUT_{ij}^{(i)}\right)$ (see Fig. 1) for $(i < j)$, can be expressed as

$$\left(FUT_{ij}^{(l)}\right) = \inf\{t : (X(t) = j \mid X(0) = l) and (X(t') = k, t' < t, l \leq k < j)\} \quad (1)$$

Where, the superscript (i) means that the initial state is i and X(t) indicates the state at time t (t^{th} frame period in discrete-time Markov BD process).

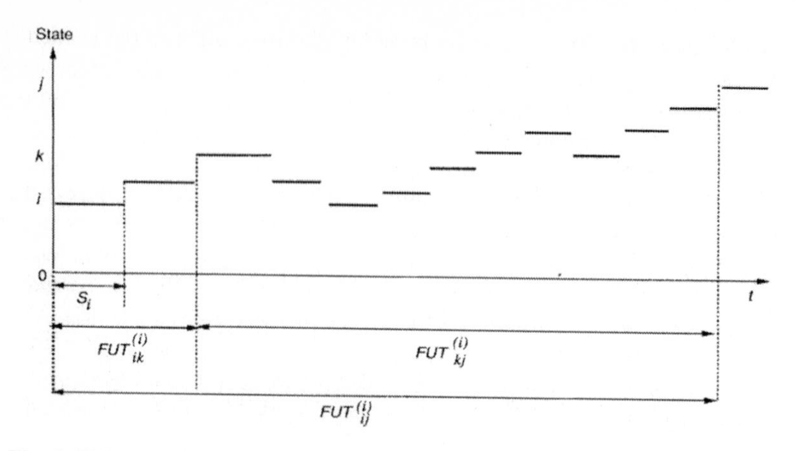

Fig. 1. FUT: At an intermediate state, it must not visit a lower state than initial state.

[Definition 2] First Down-passage Time (FDT): The FDT is the time period to reach a given state (j) from initial state (i) for i > j, such that any intermediate state (k) cannot be larger that the initial state. The FDT from i to j, FUT $_{ij}^{(i)}$ for (i > j), by

$$\left(FUT_{ij}^{(i)}\right) = \inf\{t : (X(t) = j \mid X(0) = i) and (X(t') = k, t' < t, i \le k < j)\} \quad (2)$$

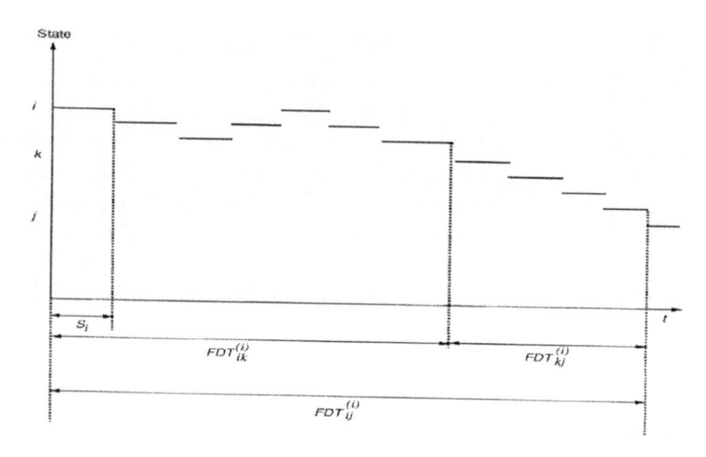

Fig. 2. FDT: At an intermediate state, it must not visit a higher state than initial state.

3 Computation Algorithm of FUT and FDT

The number of active sources of FEC in LSP with virtual capacity will fluctuate by going up and down in the intermediate state as shown in Fig. 2, but by definition 1, no intermediate state is below the starting state. It characterizes the fluctuations of number of active sources in increasing phase. The fluctuations in the number of active sources during frame intervals are assumed to have only one step transition from i to $(i + 1)$ state, or from state i to $(i - 1)$ in discrete-time Markov birth-and-death process.

In particular, we introduce the discrete-time Markov BD process with finite population (N). The number of active sources in discrete-time Markov BD process, undergoes on increasing phase and decreasing phase, and state **0** is recurrent.

If we draw the transitions of the number of active sources at each frame interval by a step function in a discrete-time BD process, it will go up down and finally return to state **0**. And after some period it will start an another fluctuation from state **0** to state k, and finally go down to state **0**. We call the fluctuations starting at state **0** and returning to state **0** to *Heap*. We define a heap h_k as the one which the number of active sources increases from **0** to k and then decreases to **0**. A heap h_k is defined by a maximum value k. One thing to remind is that the heap h_k cannot be expressed by the conventional first passage time but by FUT and FDT. Fig. 4 shows and example of heaps generated by active traffic sources in the system. Since the change of state have unit magnitude in a Markov BD process, FUT is divided into two parts as shown in Fig. 1. (2) defines the FUT property to be used for the computation of FUT.

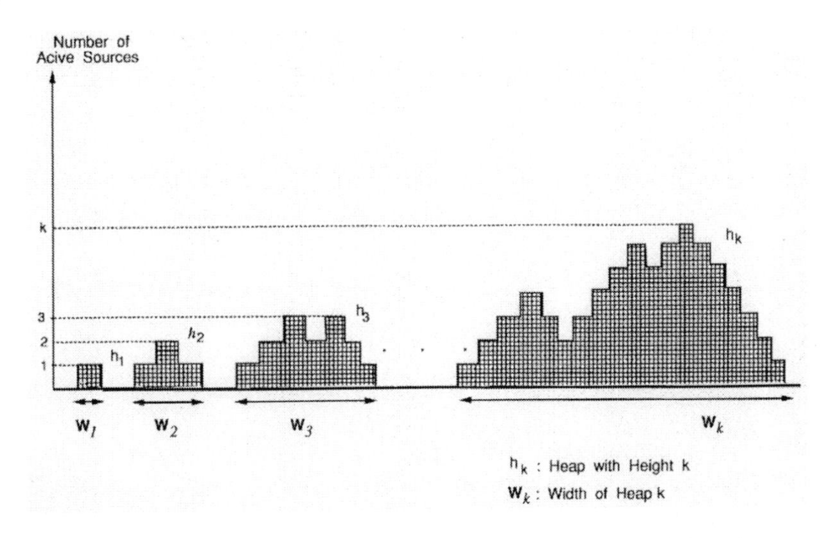

Fig. 3. An Example of Heap Generation in the LSP with Virtual Capacity

3.1 FUT Computation

FUT Property: Consider the initial state l in the first up-passage time. Given a heap of maximum height h, suppose $i < j$, by the Definition 1 $FUT_{i,j}^{(l)}$ can be expressed by

$$FUT_{i,j}^{(l)} = FUT_{i,k}^{(l)} + FUT_{k,j}^{(l)}, \quad l \leq i \leq k < j \leq h \tag{2}$$

Iterating for $k = i+1, i+2, i+3, \cdots, j-1$ yields

$$FUT_{i,j}^{(l)} = \sum_{k=i}^{j-1} FUT_{k,k+1}^{(l)}, \tag{3}$$

where, $FUT_{i,j}^{(l)}$ passes through all possible paths between state l and state $k+1$ after starting state k as defined at Definition 1. One important thing to be reminded is that a *heap* can be represented rather than a conventional first passage time. A first up-passage from i to j can be divided into two parts: the first transition out of state i (S_i) followed by a first passage from $(i + 1)$ to j. That is,

$$FUT_{i,j}^{(l)} = S_i + FUT_{i+1,j}^{(l)}, \quad l \le i < j \tag{4}$$

Using (4) let $\Gamma_k^{(l)}$ be the expected value of $FUT_{k,k+1}^{(l)}$ $(l \le k < j)$

$$\Gamma_k^{(l)} = \mathrm{E}\left[FUT_{i,j}^{(l)}\right] \tag{5}$$

The superscript value l of (5) denotes the initial state. By the Definition 1 and the homogeneity of Markovian birth-and-death process, we consider that $FUT_{k,k+1}^{(l)}$ will have the following possible transition paths:
state k to state $(k + 1)$ directly, or
state $k \to$ state $(k - 1) \to \cdots \to$ state $l \to \cdots$ state $k \to$ state $(k + 1)$ with transition probability q_k. The state below an initial state l $(l < k)$ cannot be reached by the definition 1.
Consider the above case 1 and 2 in the transition paths described in Fig. 3. By Markov property, we consider the transitions from state k to adjacent $k - 1$ and state $k + 1$. In the transition from state k to state $k - 1$, let q_k denote the transition probability of making a transition from state k to state $(k - 1)$. They are given as

$$q_k = \frac{\mu_k}{\lambda_k + \mu_k}$$

By using .(5), we can obtain the following recursive relationship for initial state l directly.

$$\Gamma_k^{(l)} = \frac{1}{\lambda_k + \mu_k} + \frac{\mu_k}{\lambda_k + \mu_k}\left(\Gamma_{k-1}^{(l)} + \Gamma_k^{(l)}\right) = \overline{S}_k + q_k\left(\Gamma_{k-1}^{(l)} + \Gamma_k^{(l)}\right) \tag{6}$$

where,

$$\overline{S}_k = \frac{1}{\lambda_k + \mu_k}$$

Similarly with a conventional first passage time, by using a next condition of transition path, we can obtain a recursive equation of (4) from (5) and (6). But, a transition of (6) has a restriction at the boundary barrier of starting state rather than a conventional first passage time. That is, the next transition is not permitted for the states below initial state. The boundary condition restriction will be used for defining a heap in fluctuations of system state.
Especially, we can find out that the conventional first passage time from state i to state j (j > i) is a special case of $FUT_{i,j}^{(l)}$ (l = 0, 1, 2, ···, i). That is, the conventional first passage time from state i to state j is $FUT_{i,j}^{(0)}$. The expected first up-passage time (FUT) from i to j will be computed by using (6) as:

$$\mathrm{E}\left[FUT_{i,j}^{l}\right] = \sum_{k=i}^{j-1} \Gamma_k^{(l)}$$

$$= \sum_{k=i}^{j-1} \frac{1}{1 - q_k}\left(\overline{S}_k + \Gamma_{k-1}^{(l)}\right) \quad l \le I < j \tag{7}$$

When the initial state is 1, the general closed formula to compute $\Gamma_i^{(1)}$ will be derived as follows:

$$\Gamma_i^{(1)} = \frac{1}{1-q_i}\overline{S}_i + \frac{q_i}{(1-q_{i-1})(1-q_i)}\overline{S}_{i-1} + \frac{q_{i-1}q_i}{(1-q_{i-2})(1-q_{i-1})(1-q_i)}\overline{S}_{i-2} + \cdots$$

$$+ \frac{q_2 q_3 \cdots q_i}{(1-q_1)(1-q_2)\cdots(1-q_i)}\overline{S}_1$$

$$= \sum_{j=1}^{i}\left(\prod_{k=j}^{i}\frac{q_k}{(1-q_k)}\right)q_j\overline{S}_j \tag{8}$$

And if we would like to include initial state l (including l =0 as initial state), the equation of (8) will be as follows:

$$\Gamma_i^{(l)} = \sum_{j=i}^{i}\left(\prod_{k=j}^{i}\frac{q_k}{(1-q_k)}\right)q_j\overline{S}_j \tag{9}$$

From (3), (8) and (9), the expected value of $FUT_{0,m}^{(0)}$ of a heap with height m (h_m) is computed as:

$$\mathrm{E}\left[FUT_{0,m}^{(0)}\right] = \sum_{i=0}^{m-1}\Gamma_i^{(0)}$$

$$= \sum_{i=0}^{m-1}\sum_{j=0}^{i}\left(\prod_{k=j}^{i}\frac{q_k}{(1-q_k)}\right)q_j\overline{S}_j \tag{10}$$

We note that the (10) is a function of the first transition duration $(\overline{S}_j)$ and , q_k. Let Xi,(0,m) be the duration occupied by state i within the ascending phase of a heap hm (with height m). The expression of $\mathrm{E}[FUT_{0,m}^{(0)}]$ in (10) can be represented by Xi,(0,m) since (10) is a function of $\overline{S}_i$.

$$\mathrm{E}[FUT_{0,m}^{(0)}] = \sum_{i=0}^{m-1} X_{i,(0m)} \tag{11}$$

From (8) and (10), Xi(0,m) can be expressed in terms of $\overline{S}_i$, the general closed expression of Xi,(0,m) is derived as:

$$X_{i,(0,m)} = \sum_{j=i}^{m-1}\left(\prod_{k=i}^{j}\frac{q_k}{(1-q_k)}\right)q_i\overline{S}_i \tag{12}$$

And from (10), we can derive general formula of $\mathrm{E}[FUT_{i,m}^{(l)}]$ for $1 \le i < m$ as:

$$\mathrm{E}\left[FUT_{i,m}^{(l)}\right] = \sum_{n=1}^{m-1} X_{n,(l,m)} = \sum_{n=l}^{m-1}\sum_{j=n}^{m-1}\left(\prod_{k=n}^{j}\frac{q_k}{(1-q_k)}\right)q_n\overline{S}_n \tag{13}$$

As seen (12), the expected length of state k in FUT can be represented by two important components, the average length of the duration of at state i $(\overline{S}_i)$ and the probabilistic occurrence frequencies of state k. This concept will be significantly used for the computation of loss period and loss rate during the load period.

3.2 FDT Computation

As mentioned in Definition 2, the FDT is the path length descending from the top state k to state 0. The similar method used for FUT can be also applied to the computation of the path length of FDT. A first down-passage from i to j (i > j) can be also divided into two parts, the first transition of state i and a first passage from (i – 1) to j, and its initial state is l. So, we can have a similar form of (4) for *FUT* as:

$$FDT_{i,j}^{(l)} = S_i + FDT_{i-1,j}^{(l)} \quad l \geq i > j \tag{14}$$

Let Yi, (N,0) be the period occupied by state i within the period of $\mathrm{E}\left[FDT_{N,0}^{(N)}\right]$ from state N to 0. From The general form of Yi,(N,0) will be derived.

$$Y_{i,(N,0)} = \sum_{j=1}^{i}\left(\prod_{k=j}^{i}\frac{p_k}{q_k}\right)q_i\overline{S}_i \tag{15}$$

And it can be rewritten in terms of Yi,(N,0).

$$\mathrm{E}\left[FDT_{N,0}^{(N)}\right] = \sum_{i=1}^{N} Y_{i,(N,0)} \tag{16}$$

And from (15) and (16), we can obtain the general form of $\mathrm{E}\left[FDT_{l,m}^{(l)}\right]$ for l > i ≤ m as

$$\mathrm{E}\left[FDT_{l,m}^{(l)}\right] = \sum_{i=m+1}^{l} Y_{i,(l,m)} \tag{17}$$

where,

$$Y_{i,(l,m)} = \sum_{j=m+1}^{i}\left(\prod_{k=j}^{i}\frac{p_k}{q_k}\right)q_i\overline{S}_i \quad m < i \leq l$$

Again, we remind the difference between a conventional first-passage time and FDT by using the results of (17) and the conventional approach in [3]. In the computation of a first passage time in Markov BD process with N finite population model, the one-step transition period of a conventional first-passage time from state i to state i – 1 is expressed in terms of $\overline{S}_i, \overline{S}_{i+1}, \cdots, \overline{S}_N$. On the other hand, (15) and (16) are conditioned by the maximum state M as shown in (17), and is expressed in terms of $\overline{S}_{i+1}, \cdots, \overline{S}_M$, where M ≤ N. FDT is also an useful concept to denote descending phase in a heap.

3.3 Computation of Occurrence Probability of Heap Height *i* (U_i)

We define the observation period, that is *Load Period* (T_R), to compute the expected load in the system. The expected *load period* is defined as:

$$\mathrm{E}[T_R] = \sum_{i=1}^{N}(W_i + I_i)\cdot U_i$$

where, N is the total number of sources in the system and I_i is the idle period followed by heap h_i. We will compute the expected overpassing period to virtual capacity of LSP from each heap occurrence probability. In order to capture the expected path lengths for all heaps in the system, it is necessary to estimate the probability that heap h_i occurs in the system as shown in Fig 3.

Let U_i be the probability that heap h_i occurs in the system. Consider a discrete-time Markov BD process having Markov chain with finite population (N) in the computation of the heap occurrence probability U_i. In Markov BD process if the process is currently in state i, then the next transition moves the process to either state $(i + 1)$ or $(i - 1)$.

Suppose the initial state, $X(0) = i$. Then the time to the first transition is obtained by taking a random sample of a population with an exponential distribution having a mean of $1/(\lambda_i + \mu_i)$ (see Theorem 3.1). The effect of transition is obtained by flipping a coin which lands heads with probability $\lambda_i /(\lambda_i + \mu_i)$. If the coin lands heads, then the transitions is to state $(i + 1)$, otherwise it is to state $(i - 1)$. When the new state is entered, these experiments are repeated with the appropriate parameters. This procedure generates most BD process of interest [3]. Suppose that the system enters state k and returns to state 0. This constructs various *connected digraphs* with intermediate nodes. At each node, a coin is tossed to determine the path to go, and if it goes back to initial state 0 it makes a connected digraph. In order to find the heap occurrence probability U_i that a heap with maximum height k occurs in the system, we consider the probability to obtain a connected digraph with maximum state k. Fig. 4 shows a connected digraph constructed by the available paths in Markov BD chain with maximum N states and an example for the maximum height of 3.

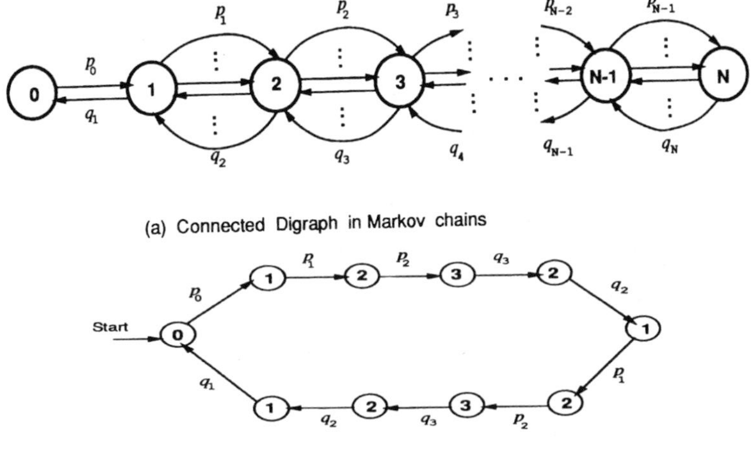

Fig. 4. Struction of Digraph for h3 heap in Markov Chain

Let Ψ_i denote the probability that a graph with maximum state i is created in the system during state transition. The graph doesn't contain state 0 at intermediate transition nodes. In the computation of Ψ_i, it will be necessary to check whether the transition time at each state affects the transition probability or not. This property is based on the property of Markov BD process. A time period is required to reach the next state. The required time period will be computed by using the algorithms of FUT and FDT. However, in the discrete-time Markov BD process, given the current state, the identity of the next state visited is independent of the time required to get there. Here, the procedure to compute probabilities for all possible paths in Markov BD chains in shown. And let $\alpha_i = p_i q_{i+1}$.

1) At first, consider a digraph with maximum state 1. Its transition path is $0 \to 1 \to 0$, and the probability to obtain this graph is,

$$\Psi_1 = p_0 q_1 = \alpha_0. \tag{18}$$

2) For maximum state = 3, the possible paths in digraph chain $0 \to 1 \to 2 \to 1 \to 0$, $0 \to 1 \to 2 \to 3 \to 2 \to 1 \to 0$, $0 \to 1 \to 2 \to 1 \to 2 \to 1 \to 2 \to 3 \to 2 \to 1 \to 0$, $0 \to 1 \to 2 \to 3 \to 2 \to 3 \to 2 \to 3 \to 2 \to 3 \to 2 \to 1 \to 0, \cdots$. And in each path, the term of $\alpha_0\alpha_1\alpha_2$ is common, and we can compute the sum of probabilities of possible paths:

$$\Psi_3 = \alpha_0\alpha_1\alpha_2\left[1 + (\alpha_1 + \alpha_2) + (\alpha_1^2 + \alpha_2^2 + \alpha_1\alpha_2) + (\alpha_1^3 + \alpha_2^3 + \alpha_1^2\alpha_2 + \alpha_1\alpha_2^2) + \cdots\right]$$

$$= \alpha_0\alpha_1\alpha_2\left[1 + \frac{\alpha_1^2 - \alpha_2^2}{\alpha_1 - \alpha_2} + \frac{\alpha_1^3 - \alpha_2^3}{\alpha_1 - \alpha_2} + \frac{\alpha_1^4 - \alpha_2^4}{\alpha_1 - \alpha_2} + \cdots\right]$$

$$= \alpha_0\alpha_1\alpha_2\left[1 + \frac{1}{\alpha_1 - \alpha_2}\left(\frac{\alpha_1^2}{1-\alpha_1} - \frac{\alpha_2^2}{1-\alpha_2}\right)\right]$$

3) For maximum state = 4, the sum of probabilities of possible paths in digraph chains is computed similarly to the above. In order to express Ψ_i simply, let define the *Combination Function*, $H_i^{(m)}(\alpha_1, \alpha_2, \cdots, \alpha_{m-1})$,, which denotes the combinations of α_i's, $1 < i < m-1$ with i different α variables. The superscript m indicates the maximum state and subscript i does the number of a variables of *Combination Function*. That is,

$$H_i^{(m)}(\alpha_1, \alpha_2, \cdots, \alpha_{m-1}) = \alpha_1\alpha_2\alpha_3\cdots\alpha_{i-1}\alpha_i + \alpha_1\alpha_2\alpha_3\cdots\alpha_{i-1}\alpha_{1+1} + \cdots$$
$$+ \alpha_{m-i-1}\alpha_{m-i}\cdots\alpha_{m-2} + \alpha_{m-i}\alpha_{m-i+1}\cdots\alpha_{m-1} \tag{19}$$

For example, $H_3^{(m)}(\alpha_1, \alpha_2, \cdots, \alpha_{m-1})$ is represented as:

$$H_3^{(m)}(\alpha_1, \alpha_2, \cdots, \alpha_{m-1}) = \sum_{i=1}^{m-3}\sum_{j=i+1}^{m-2}\sum_{k=j+1}^{m-1}\alpha_i\alpha_j\alpha_k \tag{20}$$

For the maximum state = 6 , Ψ_6 are shown, and its general formula is derived

$$\Psi_6 = \frac{\left(\prod_{i=0}^{5}\alpha_i\right)\left[1 + \sum_{i=1}^{4}\sum_{j=i+1}^{5}\frac{1}{\alpha_i - \alpha_j}\left(\frac{\alpha_i^2}{1-\alpha_i} - \frac{\alpha_j^2}{1-\alpha_j}\right) - \sum_{i=1}^{5}\frac{\alpha_i}{1-\alpha_i}\right]}{1 - H_3^{(6)}(\alpha_1,\cdots,\alpha_5) + 3\cdot H_4^{(6)}(\alpha_1,\cdots,\alpha_5) - 11\cdot H_5^{(6)}(\alpha_1,\cdots\alpha_5)} \tag{21}$$

Finally, from (19) through (21), we obtain the general formula of Ψ_M ($M \geq 4$), and it is as follows :

$$\Psi_M = \frac{\left(\prod_{i=0}^{M-1}\alpha_i\right)\left[1+\sum_{i=1}^{M-2}\sum_{j=i+1}^{M-1}\frac{1}{\alpha_i-\alpha_j}\left(\frac{\alpha_i^{\ 2}}{1-\alpha_i}-\frac{\alpha_j^{\ 2}}{1-\alpha_j}\right)-\sum_{i=1}^{M-1}\frac{\alpha_i}{1-\alpha_i}\right]}{1-H_3^{(M)}(\alpha_1,\cdots,\alpha_{M-1})+\sum_{i=2}^{M-3}\left[(-1)^i\left(1+\sum_{j=1}^{i-1}2^{2j-1}\right)H_{i+2}^{(M)}(\alpha_1,\cdots,\alpha_{M-1})\right])} \tag{22}$$

and for M = 1, 2 and 3, Ψ_M is as follows:

$$\Psi_M = \alpha_0\alpha_1\alpha_2\left[1+\frac{1}{\alpha_1-\alpha_2}\left(\frac{\alpha_1^{\ 2}}{1-\alpha_1}-\frac{\alpha_2^{\ 2}}{1-\alpha_2}\right)\right] \qquad \text{for M = 3}$$

$$\Psi_M = \frac{\alpha_0\alpha_1}{1-\alpha_1} \qquad \text{for M = 2}$$

$$\Psi_M = \alpha_0 \qquad \text{for M = 1}$$

In Ψ_M for large M, if we are going to compute the exact values of all *combination functions*, $H_k^{(M)}(\alpha_1,\alpha_1,\cdots,\alpha_1)$, the computational complexity is not neglectable. One important feature of the *combination function* $H_k^{(M)}(\alpha_1,\alpha_1,\cdots,\alpha_1)$ is a function of only user defined parameters (λ, μ) and number of sources. And Ψ_M is also a function of only user define parameters and number of sources. Therefore, $H_k^{(M)}(\alpha_1,\alpha_1,\cdots,\alpha_1)$ and Ψ_M can be computed in advance and stored in table. The pre-computation and tabulation of $H_k^{(M)}$ and Ψ_M should be eminent discovery in this paper. The mechanism will be effectively used to approximate available resource of system with real time. In order to minimize its computation cost for large M, [6] indicates an algorithm to compute $H_k^{(M)}(\alpha_1,\alpha_1,\cdots,\alpha_1)$. Its computation cost shows linear with *k*.

From, these results, we approximate the *heap occurrence probability*, U_i. The state fluctuation in the heap with maximum height *i* is described as same with the digraph with maximum state *i*, Therefore, equation (22) provides the formula to U_i for $i \geq 4$:

$$U_i \cong \Psi_i$$

$$\cong \frac{\left(\prod_{k=0}^{i-1}\right)\left[1+\sum_{k=1}^{i-2}\sum_{j=k+1}^{i-1}\frac{1}{\alpha_k-\alpha_j}\left(\frac{\alpha_k^{\ 2}}{1-\alpha_k}-\frac{\alpha_j^{\ 2}}{1-\alpha_j}\right)-\sum_{k=1}^{i-1}\frac{\alpha_k}{1-\alpha_k}\right]}{1-H_3^{(i)}(\alpha_1,\cdots,\alpha_{i-1})+\sum_{k=2}^{i-2}\left[(-1)^k\left(1+\sum_{j=1}^{k-1}2^{2j-1}\right)H_{k+2}^{(i)}(\alpha_1,\cdots,\alpha_{i-1})\right]} \tag{23}$$

The equations, (22) and (23), show a close result with simulation as depicted at Fig. 5. Fig. 5 shows the comparison between analytical and simulation results for the model with N = 5, and indicates the comparison for the model with N = 8. In this simulation, the steady state result of U_i requires very long simulation time, and the comparison range in the number of sources is shown for N = 5 and N = 8.

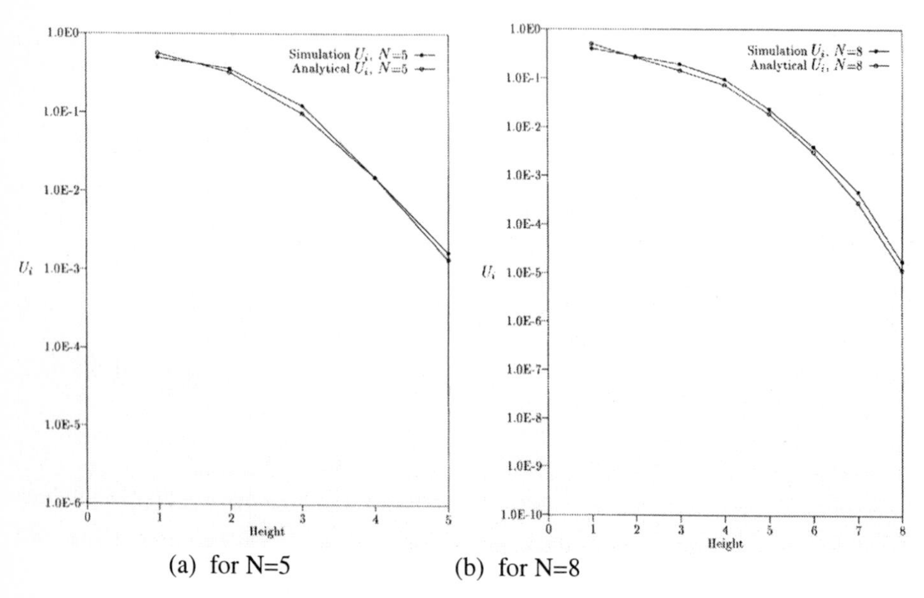

(a) for N=5 (b) for N=8

Fig. 5. Comparison between Analytical and Simulation for Ui (1/μ = 66 cells, 1/λ=267 cells, burstiness (β) = 5)

4 Expected Overpassing Period Computation

The overpassing period to the virtual capacity of LSP in MPLS (particularly in ATM based MPLS) is computed from the results, FUT and FDT and U_i, obtained at previous sections. The expected offered load period ($\overline{W}$) is computed by using the heap occurrence probability (U_i) and the property of probabilistic similarity in expected path length computation.

$$E[W] = \sum_{i=1}^{N} \overline{W}_i \cdot U_i = \sum_{i=1}^{N} \left(\overline{W}_{i,up} + \overline{W}_{i,down}\right) \cdot U_i$$

$$= \sum_{i=1}^{N} E[FUT_{1,i}^{(1)}] \cdot U_i + \sum_{i=1}^{N} E[FUT_{i,0}^{(N)}] \cdot U_i \tag{24}$$

Define $E[W_{up}]$ and $E[W_{down}]$ an expected upward-path length and an expected downward-path length for all heaps, respectively. The upward-path length indicates the total path length of when the number of active sources takes journey from state **0** to the top state *k*. There may be fluctuations going down and up, and the total length of upward-path is a summation of those paths. The downward-path length is also the total path length of while the number of active sources reach to top state from state 0. First $E[W_{up}]$ is rewritten by using (13), (15) and (23) as:

$$E[W_{up}] = \sum_{i=1}^{N-1}\left(\sum_{k=1}^{i}\Gamma_k^{(1)}\right)\cdot U_{i+1} = \sum_{i=1}^{N-1}\left(\sum_{k=1}^{i} X_{k,(1,i)}\right)\cdot U_{i+1}$$

$$= \sum_{i=1}^{N-1}\left[\sum_{k=1}^{i}\sum_{j=k}^{i}\left(\prod_{l=k}^{j}\frac{q_l}{p_l}\right)\cdot q_k\cdot \overline{S}_k\right]\cdot U_{i+1} \tag{25}$$

From (25), we can see that $\mathrm{E}[\mathrm{W}_{up}]$ is a function of $\overline{S}_k\,(k=1,2,\cdots,N-1)$. So, we can rewrite (25) by $A_k(\overline{S}_k)$ as:

$$E[W_{up}] = A_1(\overline{S}_1) + A_2(\overline{S}_2) + A_3(\overline{S}_3) + \cdots + A_{N-1}(\overline{S}_{N-1}) \tag{26}$$

where,

$$A_1(\overline{S}_1) = \sum_{i=1}^{N-1} X_{1,(1,i)}\cdot U_{i+1},\; A_2(\overline{S}_2) = \sum_{i=2}^{N-1} X_{2,(1,i)}\cdot U_{i+1},$$

$$A_k(\overline{S}_k) = \sum_{i=1}^{N-1} X_{k,(1,i)}\cdot U_{k+1}$$

$$A_{N-1}(\overline{S}_{N-1}) = X_{N-1,(1,N-1)}\cdot U_N \tag{27}$$

and,

$$X_{k,(1,i)} = \sum_{j=k}^{i}\left(\prod_{l=k}^{j}\frac{q_l}{p_l}\right)\cdot\frac{1}{q_k}\cdot\overline{S}_k \tag{28}$$

In the computation of $\mathrm{E}[\mathrm{W}_{down}]$,the similar procedure with $\mathrm{E}[\mathrm{W}_{up}]$ will be applied. The calculation of total downward-path length is done through the computation of FDT.

$$E[W_{down}] = \sum_{i=1}^{N}\overline{W}_{i,down}\cdot U_i = \sum_{i=1}^{N} E[FDT_{i,0}^{(N)}]\cdot U_i$$

$$= \sum_{i=1}^{N}\left(\sum_{k=1}^{i}\Theta_k^{(N)}\right)\cdot U_i$$

$$= \sum_{i=1}^{N}\left(\sum_{k=1}^{i} Y_{k,(i,1)}\right)\cdot U_i$$

$$= \sum_{i=1}^{N}\left[\sum_{k=1}^{i}\sum_{j=k}^{i}\left(\prod_{l=k}^{j}\frac{p_l}{q_l}\right)\cdot p_j\cdot\overline{S}_j\right]\cdot U_i \tag{29}$$

As we see(29), $\mathrm{E}[\mathrm{W}_{down}]$ is also a function of $\overline{S}_k$ (k=1,2, · · · ,N-1), (29) is rewritten by $\mathrm{A}_k(\overline{S}_k)$, which is the summation of $\mathrm{Y}_{k,(i,1)}$.

$$E[W_{down}] = B_1(\overline{S}_1) + B_2(\overline{S}_2) + B_3(\overline{S}_3) + \cdots + B_N(\overline{S}_N) \tag{30}$$

where,

$$B_1(\overline{S}_1) = \sum_{i=1}^{N} Y_{1,(i,1)}\cdot U_i,,\; B_2(\overline{S}_2) = \sum_{i=2}^{N} Y_{2,(i,1)}\cdot U_i,\; B_k(\overline{S}_k) = \sum_{i=k}^{N} Y_{k,(i,1)}\cdot U_k$$

$$B_N(\overline{S}_N) = Y_{N,(N,1)}\cdot U_N \tag{31}$$

and

$$Y_{k,(i,1)} = \sum_{j=1}^{k} \left(\prod_{l=j}^{k} \frac{p_l}{q_l} \right) \cdot \frac{1}{p_k} \cdot \overline{S}_k \tag{32}$$

As shown so far, the expected offered load period ($\overline{W}$) is expressed by the expected path length ($\overline{S}_i$) of each state. The expected offered cells during load period can be computed by product of arriving cells at each state during one frame period (D). And It will be noted that the expected *Loss Period* (*E[T$_{overpass}$]*) at this model is simply calculated without any further computation labor. That is, from (33), the overflow period is counted. The expected *Loss Period* (*E[T$_{overpass}$]*) is as follows:

$$\begin{aligned} E[Toverpass] &= \sum_{i=D+1}^{N-1} A_i(\overline{S}_i) + \sum_{i=D+1}^{N} B_i(\overline{S}_i) \\ &= \sum_{i=D+1}^{N-1} \left(\sum_{k=1}^{i} X_{k,(1,i)} \right) \cdot U_{i+1} + \sum_{i=1}^{N} \left(\sum_{k=1}^{i} X_{k,(1,i)} \right) \cdot U_i \\ &= \sum_{i=D+1}^{N-1} \left[\sum_{k=1}^{i} \sum_{j=k}^{i} \left(\prod_{l=k}^{j} \frac{q_l}{p_l} \right) \cdot q_k \cdot \overline{S}_k \right] \cdot U_{i+1} + \sum_{i=D+1}^{N} \left[\sum_{k=1}^{i} \sum_{j=k}^{i} \left(\prod_{l=k}^{j} \frac{q_l}{p_l} \right) \cdot q_j \cdot \overline{S}_j \right] \cdot U_i \end{aligned} \tag{33}$$

5 Conclusion

We have proposed the approximation algorithm to compute *overpassing period* of LSP with virtual capacity in MPLS. In traffic engineering of MPLS, a estimation of virtual capacity is very necessary to accommodate required quality of service. For example, MPLS L2 or L3 VPN strongly requires to sustain a predefined bandwidth during communication. When multiple VPN LSPs share a limited bandwidth a stable bandwidth provision in the system is very important requirement in VPN QoS. In order to sustain a minimum required bandwidth, a computation of overpassing period to assigned virtual capacity of a VPN LSP is useful to define a regulation to keep the user required or network bandwidth requirement.

Many approximation algorithms for bursty traffic sources have been proposed, but they take limited applications in real world due to complexity in computation or due to too much approximation. As shown in paper, the proposed approximation algorithm provides a less computational cost for real-time application than the fluid flow approximation approach.

In our approximation algorithm, $A_{N-1}(\overline{S}_{N-1})$ and $B_k(\overline{S}_k)$ are used as function of λ, μ and N. The values of $A_{N-1}(\overline{S}_{N-1})$ and $B_k(\overline{S}_k)$ can be computed and tabulated according to input traffic parameters in advance. The tabulation is constructed as a function of a number of sources. The approximation algorithm can make a table considering few adjacent possible situations in advance, and it reduces the computation cost (e.g., *O(n)*) to estimate and reserve system resource for newly incoming traffic. It should be useful approximate for LSP virtual capacity with fixed-sized packets, particularly ATM based MPLS implementation.

References

1. D. Anick , D. Mitra and M. M. Sondhi, " Stochastic Theory of a Data-Handling Systems with Multiple Sources," Bell Systems Technical Journal vol 61. No. 8 pp 1971-1894, Oct, 1982
2. S P Miller and C B Cotner, "The Quality Challenege ... Now and For the Future ...", ICDSC, May 1995
3. Danel P. Heyman and Matthew J. Sobel, "Stochastic Models in Operations Research - Volume I," McGraw Hill Book Company pp 38-104, 1982
4. Fumio Ishizaki and Testuya Takine, "Cell Loss probability Approximation and Their Application to call Admission Control," Advances in Performamce Analysis Vol. 2 No. 3, pp225-258, 1999
5. Ilyoung Chong, "Cost Minimization Allocation (CMA) Algorithm", Technical Report, Univ. of Massachusetts, 1992.
6. Ilyoung Chong, "Traffic Control at Burst Level", Ph.D. Thesis, Univ. of Massachusetts, 1992
7. T. Kodama and T Fukuda, "Customer Premises Networks of the Future," IEEE Communications Magazine, Feb. 1994
8. Roger C. F. Tucker, "Accurate Method for Analysis of a Packet Speech Multiplexer with Limited Delay," IEEE Trans. On Communications, vol 36, No. 4, April 1988
9. Danel P. Heyman and Matthew J. Sobel, "Stochastic Models in Operations Research - Volume I," McGraw Hill Book Company pp 38-104, 1982
10. Daniel P. Heyman and Mathew J. Sobel, "Stochastic Models in Operations Research – Volume I," Mcgraw-Hill Book Company, pp 38-104, 1988
11. Hans Kroner, "Staictical Multiplexing of Sporadic Sources – Exact and Approximate Performance Aalysis, " Proceedings of 13th ITC '91, pp723-729, 1991
12. Roch Guerin, Hamid Ahmadi and Mahmoud Naghshineh, "Equivalent Capacity and iys Aplication to Bandwidth Allocation in High Speed Networks," IBM Research Report RC 16317, 1990

Enhanced Communications Transport Protocol for Multicast Transport

Seok Joo Koh, Juyoung Park, Eunsook Kim, and Shin-Gak Kang

Electronics and Telecommunications Research Institute,
161 Kajeong-Dong, Yusung-Gu, Daejeon, KOREA
{sjkoh, jypark, eunah, sgkang}@etri.re.kr

Abstract. This paper proposes a new multicast transport protocol, called the Enhanced Communications Transport Protocol (ECTP). The proposed protocol is currently being standardized in the ITU-T SG7 and ISO/IEC JTC 1/SC 6. The ECTP is designed to support tightly controlled multicast connections. The sender is at the heart of one-to-many multicast group communications. The sender is responsible for overall connection management such as connection creation, termination, pause, resumption, and the join and leave operations. For tree-based reliability control, ECTP configures a hierarchical tree during connection creation. Error control is performed within each local group defined by a control tree. Each parent retransmits lost data in response to retransmission requests from its children. ECTP has been implemented and tested on Linux machine, along with Application Programming Interfaces based on Berkeley sockets.

1. Introduction

This paper proposes a new protocol for tight control of multicast transport connections, named the Enhanced Communications Transport Protocol (ECTP). ECTP operates over IP networks that have IP multicast forwarding capability [1].

ECTP is targeted for tightly controlled multicast services. The sender is at the heart of multicast group communications. The sender is responsible for overall connection management such as connection creation/termination, connection pause/resumption, and user join/leave operations.

The sender triggers the connection creation process. Some or all of the enrolled receivers will participate in the connection, becoming designated "active receivers". Any enrolled receiver that is not active may participate in the connection as a late-joiner. An active receiver can leave the connection. After the connection is created, the sender begins to transmit multicast data. If network problems (such as severe congestion) are indicated, the sender suspends multicast data transmission temporarily, invoking the connection pause operation. After a pre-specified time, the sender resumes data transmission. If all of the multicast data have been transmitted, the sender terminates the connection.

ECTP provides reliability control for multicast data transport, which has been designed to keep congruency with those being proposed in the IETF [2]. To address

I. Chong (Ed.): ICOIN 2002, LNCS 2343, pp. 64-74, 2002.

reliability control with scalability, the IETF has proposed three approaches: Tree based ACK (TRACK), Forward Error Correction (FEC), and Negative ACK Oriented Reliable Multicast (NORM). ECTP adopts the TRACK approach, because it is more similar to the existing TCP mechanisms and more adaptive to the ECTP framework.

The ECTP has been designed based on the preliminary works defined in [3], [4], and so far standardized in the ITU-T SG7 and ISO/IEC JTC 1/SC 6, as a joint work item [5], [6]. The ECTP has been implemented over Linux machine and tested on the Asia-Pacific Advanced Networks (APAN) testbed.

This paper is organized as follows. Section 2 provides overall operations of the ECTP protocol. In Section 3, we discuss implementation details together with packet format and the associated Application Programming Interfaces. Section 4 presents some preliminary experimental results for ECTP and multiple TCP connections. Section 5 concludes this paper.

2. Protocol Overview

The ECTP is a transport protocol designed to support Internet multicast applications. ECTP operates over IPv4/IPv6 networks that have IP multicast forwarding capability.

ECTP supports the connection management functions, which include connection creation and termination, connection pause and resumption, and late join and leave. For reliable delivery of multicast data, ECTP also provides the protocol mechanisms for error, flow and congestion controls. To allow scalability to large-scale multicast groups, tree-based reliability control mechanisms are employed which are congruent with those being proposed in the IETF RMT WG.

Figure 1 shows an overview of the ECTP operations. Before an ECTP transport connection is created, the prospective receivers are enrolled into the multicast group. Such a group is called an enrolled group. The IP multicast addresses and port numbers must be announced to the receivers. These enrollment operations may rely on the SAP/SDP, Web Page announcement and E-mail. An ECTP transport connection is created for the enrolled receivers.

ECTP is targeted for tightly controlled multicast connections. The ECTP sender is at the heart of the multicast group communication. The sender, designated as connection owner, is responsible for the overall management of the connection such as connection creation and termination, connection pause and resumption, and the late join and leave operations.

The ECTP sender triggers the connection creation process by sending a connection creation message. Each enrolled receiver responds with a confirmation message to the sender. The connection creation is completed when the sender receives the confirmation messages from the all of the active receivers, or when a pre-specified timer expires. QoS negotiation may be performed in the connection creation.

Throughout the connection creation, some or all of the enrolled group receivers will join the connection. The receivers that have joined the connection are called active receivers. An enrolled receiver that is not active can participate in the connection as a late-joiner. The late-joiner sends a join request to the sender. In response to the join request, the sender transmits a join confirm message, which indicates whether the join request is accepted or not. An active receiver can leave the connection by sending a leaving request to the sender. A trouble-making receiver, who cannot keep pace with the current data transmission rate, may be ejected.

After a connection is created, the sender begins to transmit multicast data. For data transmission, an application data stream is sequentially segmented and transmitted by means of data packets to the receivers. The receivers will deliver the received data packets to the applications in the order transmitted by the sender.

To make the protocol scalable to large multicast groups, ECTP employs the tree-based reliability control mechanisms. A hierarchical tree is configured during connection creation. A control tree defines a parent-child relationship between any pair of tree nodes. The sender is the root of the control tree. In the tree hierarchy, local groups are defined. A local group consists of a parent and zero or more children. The error, flow and congestion controls are performed over the control tree.

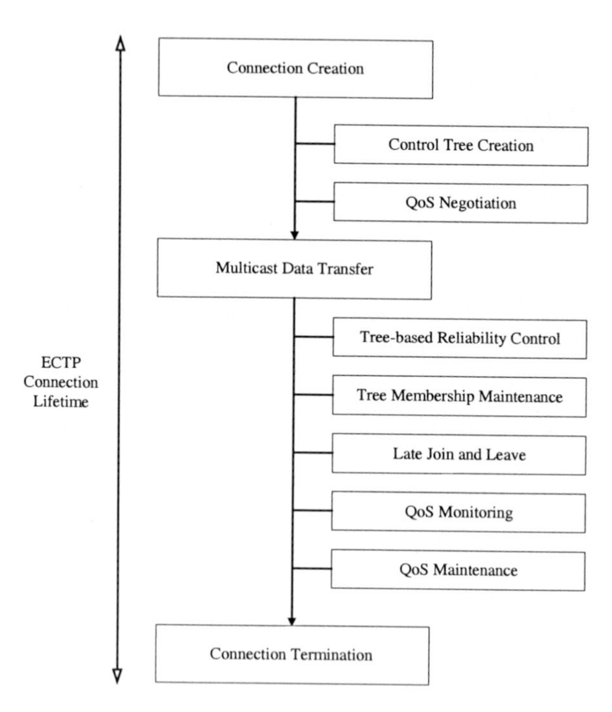

Fig. 1. ECTP Protocol Operations

Figure 2 illustrates a control tree hierarchy for reliability control, in which a parent-child relationship is configured between a sender (S) and a receiver (R), or between a parent receiver (R) and its child receiver (R).

In the tree creation, a control tree is gradually expanded from the sender to the receivers. This is called a top-down configuration [7]. On the other hand, the IETF RMT WG has proposed a bottom-up approach, where the receivers initiate a tree configuration. Those schemes may be incorporated into the ECTP as candidate tree creation options in the future.

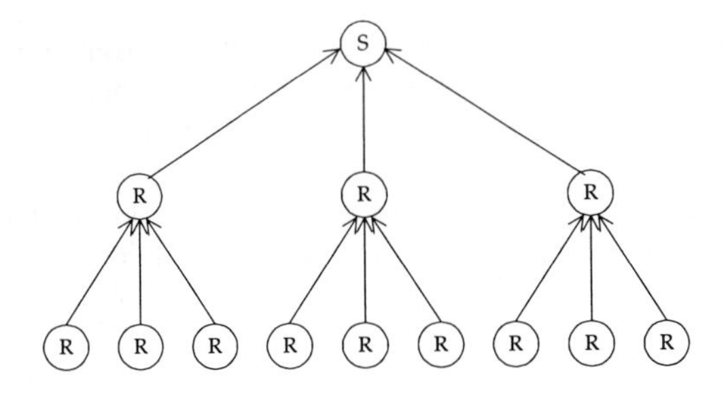

Fig. 2. Control Tree for Reliability Control

Tree-membership is maintained during the connection. A late-joiner is allowed to join the control tree. The late-joiner listens to the heartbeat messages from one or more on-tree parents, and then joins the best parent. When a child leaves the connection, the parent removes the departing child from the children-list. Node failures are detected by using periodic control messages such as null data, heartbeat and acknowledgement. The sender transmits periodic null data messages to indicate that it is alive, even if it has no data to transmit. Each parent periodically sends heartbeat messages to its children. On the other hand, each child transmits periodic acknowledgement messages to its parent.

In ECTP, error control is performed for each local group defined by a control tree. If a child detects a data loss, it sends a retransmission request to its parent via ACK.

An ACK message contains the information that identifies the data packets, which have been successfully received. Each child can send an ACK message to its parent using one of two ACK generation rules: ACK number and ACK timer. If data traffic is high, an ACK is generated for the ACK number of data packets. If the traffic is low, an ACK message will be transmitted after the ACK timer expires.

After retransmission of data, the parent activates a retransmission back-off timer. During the time interval, the retransmission request(s) for the same data will be ignored. Each parent can remove the data out of its buffer memory, if those have been acknowledged by all of its children.

During data transmission, if network problems (for example, severe congestion), the sender suspends the multicast data transmission temporarily. In this period, no new data is delivered, while the sender transmits periodic null data messages to indicate that the sender is alive. After a pre-specified time has elapsed, the sender resumes the multicast data transmission.

After an ECTP connection is created, QoS monitoring and maintenance operations are performed for the multicast data transmission. For QoS monitoring, each receiver is required to measure the parameter values experienced. Based on the measured values, a receiver determines a parameter status value for each parameter. These status values will be delivered to the sender via ACK packets. Sender aggregates the parameter status values reported from receivers. If a control tree is employed, each parent LO nodes aggregates the measured values reported from its children, and forwards the aggregated value to its parent via its ACK packets.

Sender takes QoS maintenance actions necessary to maintain the connection status at a desired QoS level, based on the monitored status values. Specific rules are pre-configured to trigger QoS maintenance actions such as data rate adjustment, connection pause and resume, and connection termination. Those rules are based on observation that how many receivers are in the abnormal or possibly abnormal status.

The sender terminates the connection by sending a termination message to all the receivers, after all the multicast data are transmitted. The connection may also terminate due to a fatal protocol error such as a connection failure.

3. Implementations

3.1 Packet Structure

ECTP packets are classified into data and control packets. Data and Retransmission Data are the data packets. All the other packets are used for control purposes.

Table 1 summarizes the packets used in ECTP. In the table, the transport type 'multicast' represents global multicast using a multicast data address, while the 'local multicast' does local multicast using a multicast control address. The Retransmission Data and Heartbeat packets are delivered from a parent to its children by local multicast.

Table 1. ECTP Packets

Packet	Transport Type	From	To
Creation Request	Multicast	Sender	Receivers
Creation Confirm	Unicast	Child	Parent
Tree Join Request	Unicast	Child	Parent
Tree Join Confirm	Unicast	Parent	Child
Data	Multicast	Sender	Receivers
Null Data	Multicast	Sender	Receivers
Retransmission Data	Multicast	Parent	Children
Acknowledgement	Unicast	Child	Parent
Heartbeat	Multicast	Parent	Children
Late Join Request	Unicast	Receiver	Sender
Late Join Confirm	Unicast	Sender	Receiver
Leave Request	Unicast	Parent/Child	Child/Parent
Connection Termination	Multicast	Sender	Receivers

Each control or data packet consists of a header part and a data part, and the header part can contain zero or more extension elements as illustrated in Figure 3.

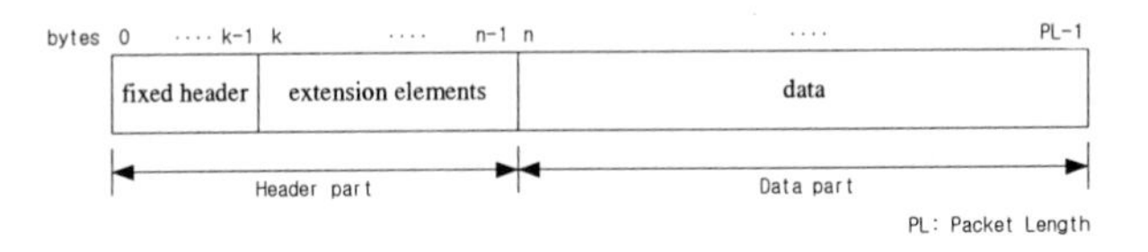

Fig. 3. Packet Structure

In the figure, '*k*', '*n*' and '*PL*' represent the length of the fixed header, the header part and the total packet.

3.1.1 Fixed Header

The fixed header contains the fields of the parameters frequently used in the protocol. An example of the fixed header with 16 bytes is depicted below:

0	8		16	24	31
next element	version	packet type	checksum		
destination port			source port		
sequence number					
payload length			reserved		

Fig. 4. Fixed Header

- *Next element* indicates the type of the next component immediately following the fixed header. The extension element also has the next element field, and thus the header part can chain multiple extension elements.
- *Version* defines the current version of the fixed header usage. Since each extension element has its own version field.
- *Packet type* indicates the type of the current packet. The maximum number of packet types to be defined is $2^8 = 256$.
- *Checksum* is used to check segment validity of a packet.
- *Destination* and *source ports* are used to identify the sending and receiving applications. These two values, together with the source and destination IP addresses in the IP header, uniquely identify each transport connection.
- *Sequence number* is the sequence number of a packet in a series of segments. This sequence number is a 32-bit unsigned number that wraps back around to '0' after reaching '$2^{32} - 1$'.
- *Payload length* indicates the length of the data part in bytes following the protocol header part. It can be used to indicate presence or absence of the data part.

3.1.2 Extension Elements

The extension elements can follow the fixed header, and thus the header part of a packet is composed of a fixed header and zero or more extension elements. Each extension element has a next element field, as shown in Fig. 5, which indicates the type of the next extension element. The header part can thus chain multiple extension elements.

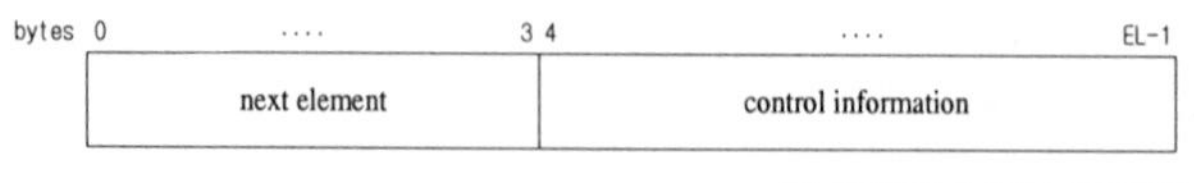

Fig. 5. Structure of Extension Element

The next element field can be encoded to indicate which type of the extensions element follows immediately. The next element field of the last extension element must be encoded as '0000', indicating "no further element".

According to the extension element type, its next element field is encoded as shown in Table 2. The next element field of the last extension element MUST be '0000'.

Table 2. Encoding table of the extension elements

Element	Encoding
Connection Information	0001
Acknowledgment	0010
Tree Membership	0011
Timestamp	0100
QoS	0101
No element	0000

Each element specifies the following:

- *Connection information element*: This element contains information on generic characteristics of the connection including the connection type, tree configuration option, connection creation timer, and ACK bitmap size, etc.
- *Acknowledgment element*: This element can be used for acknowledgment of the data packets and for report of the perceived connection status at the receiver side. A bitmap is used to indicate the selective acknowledgements of the received data.
- *Tree information element*: This element describes information on the local group defined by the control tree, etc.
- *Timestamp element*: This contains the timestamp information.
- *QoS element*: This extension element specifies QoS-related parameters: *throughput, transit delay, transit delay jitter, and packet loss rate.*

3.2 Kernel Structure for Implementation

The ECTP is currently being implemented on Linux RedHat 7.0 platform, with the C language. Some libraries are used such as LinuxThreads for ECTP Timer and Gtk+ for the ECTP applications with enhanced Graphic User Interface.

The current ECTP implementation is targeted to operate on top of UDP (UDP port: 9090 temporarily), with ECTP daemon process. Figure 6 shows the structure of ECTP kernel. Each application is assumed to use IPC (Inter Process Communication) for communications to ECTP.

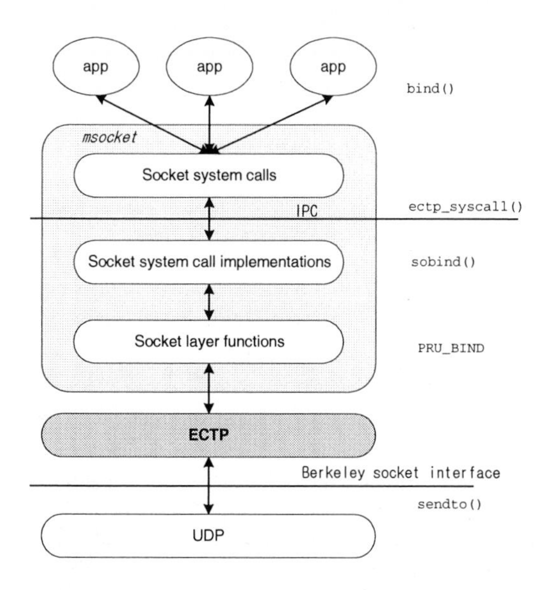

Fig. 6. ECTP Kernel Structure

3.3 Application Programming Interfaces

The ECTP API functions are designed based on the well-known Berkeley socket API in the fashion that the Berkeley socket API functions are used as wrapping functions in ECTP API. For indication of difference from the Berkeley socket functions, ECTP API functions are named with a prefix '*m*'. The following API functions are invoked by applications to communicate to ECTP:

- msocket(): This is used to create a socket for ECTP communications.
- mbind(): This function is to used to bind a pair of an IP address and a port to the socket.
- mconnect(): This is used by sender to initiate the connection creation, or by late-joiner to connect to the sender.
- maccept(): Each receiver waits for the connection creation indication signal from the sender by invoking this function.
- msend(): This is used by sender to transmit the multicast data.
- mrecv(): This is used by receivers to receive the multicast data.
- mclose(): This is used by sender to terminate the connection, or by a receiver to leave the connection.
- msetsockopt(): This is used to configure a set of socket options necessary for ECTP communications.
- mgetsockopt(): This is used to obtain the currently configured socket options.

Figure 7 illustrates an example use of ECTP API functions in terms of the sender, early and late joining receivers. Sender invokes *mconnect()* after *mbind()* and *msetsockopt()*. A receiver waits for the connection establishment message from the sender. In case of late-joiner, it tries to connect to the sender by invoking *mconnect()* function.

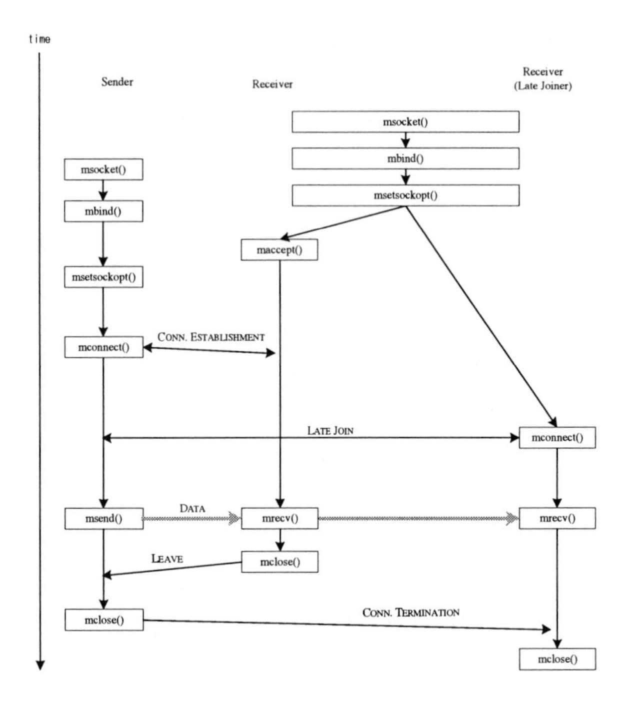

Fig. 7. Use of ECTP API

4. Experimental Results

This section shows some preliminary experimental results for the ECTP protocol performance test. This test is mainly targeted to evaluate how much bandwidth gains the ECTP provides over multiple TCP connections.

To test ECTP and TCP connections, we configure a test network consisting of one sender and 10 receivers. The sender generates the data stream until totally 100, 200, and 300 data packets have been generated. For each test instance, the total number of data and control packets flowing in the network is calculated for the ECTP and TCP connections.

Figure 8 shows the test results for ECTP and TCP connections, in terms of the total number of data and control packets generated in the network. From the figures, we see that the ECTP connection generates almost an equal number of data and control packets, independently of the number of receivers. On the other hand, in multiple TCP connection, the number of data and control packets generated gets larger, as the number of receivers increases.

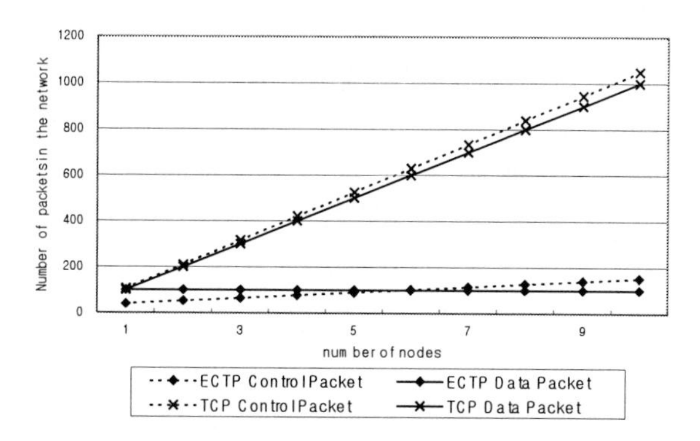

(a) Test result for transmission of data 100 packets

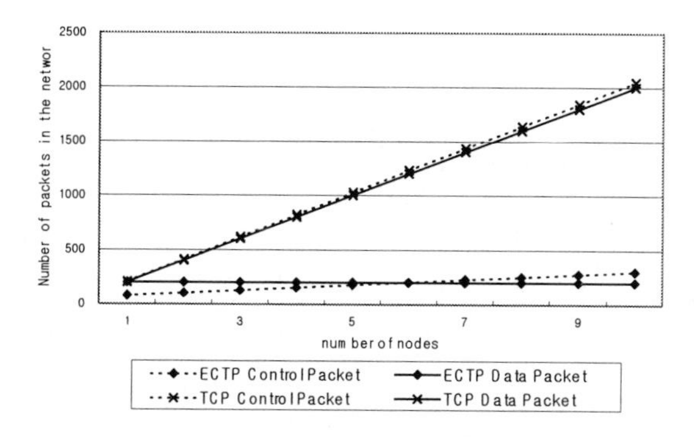

(b) Test result for transmission of data 200 packets

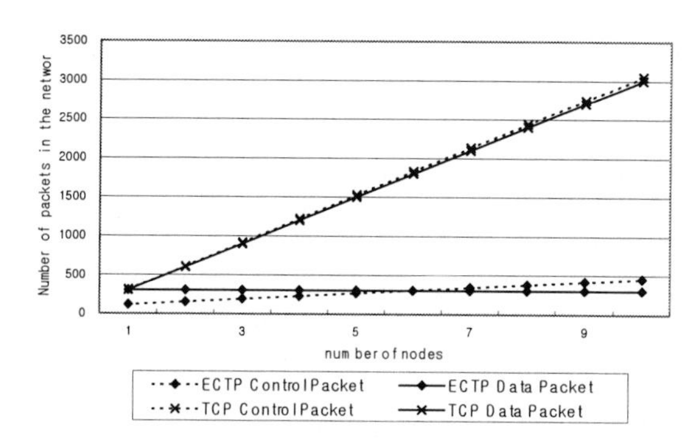

(c) Test result for transmission of data 300 packets

Fig. 8. Performance Comparison with TCP Connections

5. Conclusions

This paper has discussed a new multicast transport protocol, ECTP. We presented the protocol operations, implementation issues, and some preliminary results. The ECTP is designed for tightly controlled one-to-many multicast transport connection. The proposed protocol has been standardized in ITU-T SG7 and ISO/IEC JTC 1/SC 6.

Differently from the IETF RMT WG approaches, the ECTP is designed to support tightly controlled multicast connections. The sender is at the heart of multicast group communication. The sender is responsible for overall connection management such as the connection creation and termination, the connection pause and resumption, and the join and leave operations. For tree-based reliability control, a hierarchical tree is configured during connection creation. Error control is performed for each local group defined by a control tree. Each parent retransmits lost data, in response to retransmission requests from its children.

References

1. Enhanced Communication Transport Protocol, Available from http://ectp.etri.re.kr, 2001
2. IETF Reliable Multicast Transport (RMT) Working Group, 2001
3. ITU-T and ISO/IEC JTC1, "Enhanced Communication Transport Services," ITU-T Recommendation X.605 and ISO/IEC International Standard 13252, 1999.
4. ITU-T, "Multi-peer Communications Framework", ITU-T Recommendation X.601, 2000
5. ITU-T SG7 and ISO/IEC JTC 1/SC 6, "ECTP: Specification of Simplex Multicast Transport," ITU-T X.606 | ISO/IEC FDIS 14476-1, 2001
6. ITU-T SG7 and ISO/IEC JTC 1/SC 6, "ECTP: Specification of QoS Management for Simplex Multicast Transport," ITU-T X.606.1 | ISO/IEC CD 14476-2, Working in progress, 2001
7. S. Koh, E. Kim, J. Park, S. G. Kang, et al., "Configuration of ACK Trees for Multicast Transport Protocols," ETRI Journal, Vol. 23, No. 3, pp. 111 – 120, September 2001
8. S. Koh, et al., "Minimizing Cost and Delay in Shared Multicast Trees," ETRI Journal, Vol. 22, No. 1, pp. 30 – 37, March 2000

A Look-Ahead Algorithm for Fast Fault Recovery in MPLS Network Using GSMP Open Signaling Architecture

Hyun Joo Kang[1], Seong Gon Choi[2], Jun Kyun Choi[2], and Kyou Ho Lee[1]

[1] Electronics and Telecommunications Research Institute (ETRI)
161 Gajong-Dong, Yusong-Gu, Daejon, 305-350, Korea
{hjkang, kyou}@etri.re.kr
http://www.etri.re.kr
[2] Information and Communications University (ICU)
58-4 Hwa-am-Dong, Yusong-Gu, Daejon, 305-732, Korea
{sgchoi, jkchoi}@icu.ac.kr

Abstract. In this paper, we propose a fast and reliable fault recovery algorithm, named a resource look-ahead algorithm, for Multiprotocol Label Switch (MPLS) network. The proposed algorithm can alleviate the notification time according to the number of hops, and improve the mean fault recovery time. The algorithm uses a resource look-ahead table, which contains pre-reserved resource information. With using this table, a General Switch Management Protocol (GSMP) controller manages switches included in a given domain of open network architecture. Thus, the propagation delay of reservation message delivery can be alleviated. The restoration time is also improved comparing to the MPLS-based fault recovery.

1 Introduction

The MPLS network can support various protocols and integrated services with guaranteeing the Quality of Service (QoS) differing from existing best-effort services of Internet. The IETF is establishing common agreements on a framework for MPLS to integrate various implementations of IP switching that use an ATM-like "label swapping" technique [1].
The programmable network requires an open signaling architecture, which separate the control plane from the forwarding plane [2]. With the programmable network services and products can be developed independently with Internet Service Providers (ISPs) and device vendors [3]. Therefore, new contents can be not only immediately standardized but also integrated easily with various lower layer devices.
In the programmable and MPLS network environment, a fast fault recovery mechanism is needed in order to stabilize the turbulent network, especially for the guarantee of QoS.

I. Chong (Ed.): ICOIN 2002, LNCS 2343, pp. 75-83, 2002.

This paper proposes a resource look-ahead [4]algorithm which can provide fast and reliable fault recovery for the open Multiprotocol Label Switch (MPLS) network. We believe the reliability and recovery time can be improved by using the proposed algorithm for both data plane and control plane. It is assumed first of a hierarchical architecture in the open network to improve scalability and fast recovery time of centralized management. Traditional fault restoration mechanisms mainly depend on a centralized network manager to detect faults and reroute manually traffic to effected areas. The proposed recovery mechanism, however, is operated in hierarchical and centralized network. A GSMP controller manages such resources as LSRs with a centralized manner, and the controllers exist hierarchically for the scalable network.
Even if a controller in the assumed architecture is centralized, the proposed algorithm can reflect network states for fault recovery in a distributed method [5]. In a switch part, a GSMP [6] slave detects the fault alarm, and reports it to the controller with both event message and state and statistics messages. We introduce a resource look-ahead recovery algorithm for searching a backup path.
The algorithm uses a resource look-ahead table, which contains pre-reserved resource information. With using this table, a General Switch Management Protocol (GSMP) controller manages switches included in a given domain of open network architecture. Thus, the propagation delay of reservation message delivery can be alleviated. The restoration time is also improved comparing to the MPLS-based fault recovery.
Chapter 2 shows the proposed network architecture using GSMP open programmable interface protocol. In Chapter 3, we introduce the resource look-ahead recovery algorithm. Chapter4 analyzes the fault recovery for the open network architecture numerically and then presents the results. This paper is concluded in Chapter5.

2 GSMP Based Open Architecture in MPLS Network

2.1 Proposed Network and Architecture

An initial deployment scenario for the fault recovery is envisioned to the open signaling architecture. We set up a hierarchical and centralized GSMP domain architecture, which consists of more than one LSRs. A GSMP controller in this architecture manages LSRs using a manager-slave based management model. The controllers communicate each other using an interface protocol, named cc-gsmp, of GSMP extension. Fig. 1 shows the proposed open network architecture for the fault recovery.
The aspect that one GSMP controller manages mulitple switches in a domain is similar to the centralized control management network model. The controller is interfaced with switches through virtual switch control (vsc). Although cc-gsmp has never been established yet, communication among controllers in the proposed hierarchical environment is needed. This is the same as the distributed control management network model. All of the LSRs are controlled through the functions of connection management, configuration management, resource reservation, and etc. The GSMP controller commands LSRs directly in the MPLS network. Existing functions of the GSMP protocol has an ability to network management as well as connection management.

We consider an extension of management function to fault management. The GSMPv3 adjacency protocol is used for synchronization between a controller and a LSR as well as controllers.

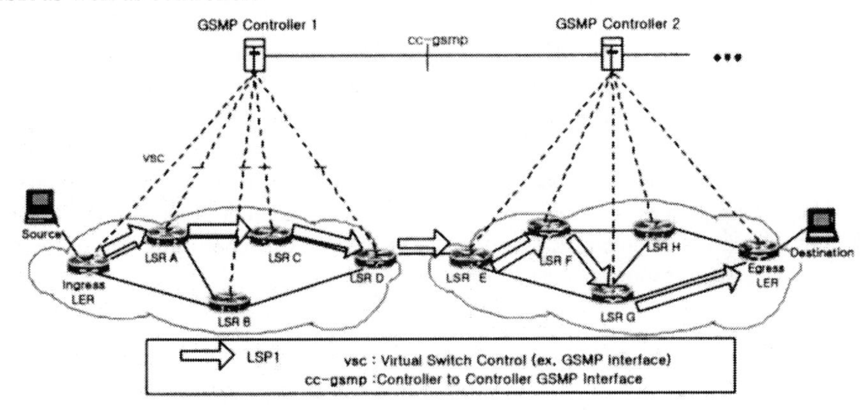

Fig. 1. An Architecture using GSMP on MPLS Network

2.2 Functional Architecture

In this clause, we introduce the functional architecture in GSMP protocol and apply the proposed fault recovery algorithm to that. When a single link fault occurs, corporation of each functional block is described through a block diagram. There are connection management, state and statistics, configuration, and event management blocks, defined in GSMPv3, in the GSMP controller. We add a fault recovery block and a cc-gsmp block in order to process fast fault recovery.

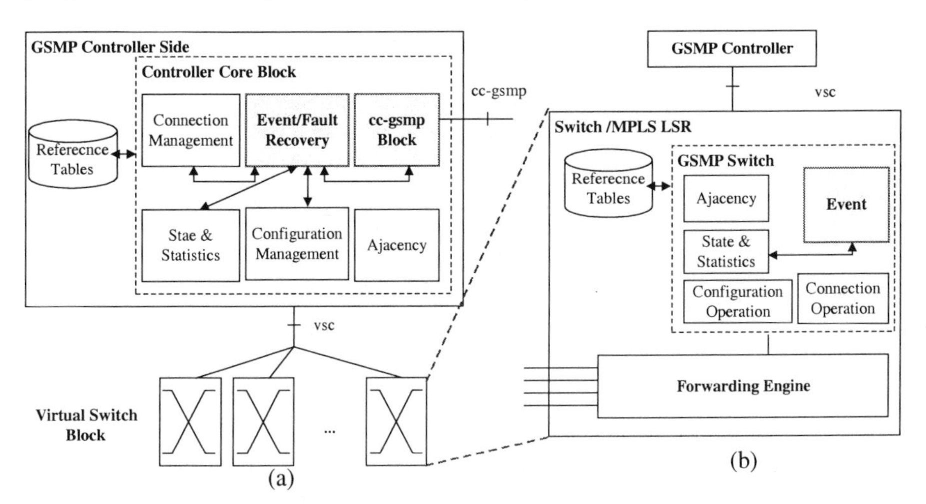

Fig. 2. Block Diagram of Controller Architecture: (a) Controller Side and (b) Switch Side

The fault recovery block performs backup path selection supported by other management blocks. The backup path is stored in a reference table database, which is created before occurring fault. (a) of Fig. 2 shows the relationship of management blocks, database, virtual switches, and a neighbor controller for fault recovery.
In the switch side, there are mainly four blocks and one resource reference table. An operational block has a role of processing the command from the controller side. The state and statistics block is to store information of related LSR's state and statistics into a table. The adjacency protocol block is important for synchronizing with the controller. If synchronization is failed, any message is not transmitted without an adjacency message. For fault recovery, the switch has an event block, which receives fault alarm, reports fault notification, and acknowledges to controller. The reference table includes such resource information as bandwidth etc.

3 Resource Look-Ahead Recovery Algorithm

Fig. 3 shows a controller processor model diagram for look-ahead fault recovery. The GSMP switch in Fig.3 detects the link fault through hardware fault alarm, and then sends a GSMP event message to a controller.

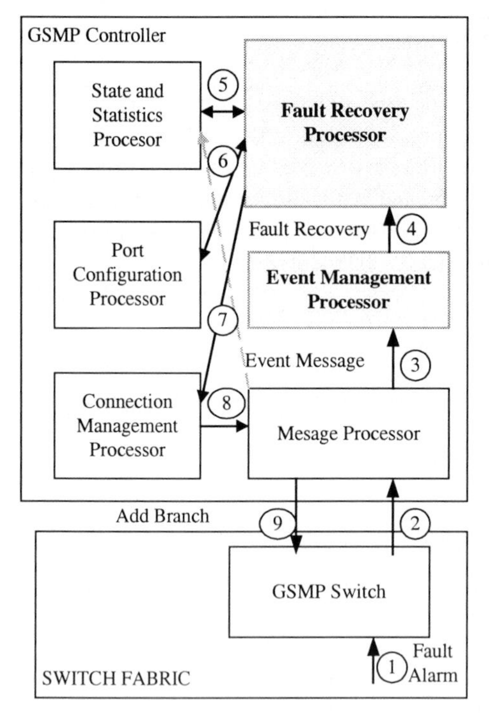

Fig. 3. Look Ahead Recovery Model through Processor Diagram.

In the model, the message processor classifies sort of input messages. If a message is an event message, the message processor sends it to the event message processor. Otherwise the message processor sends to the related processor. The event processor decides the necessary of fault recovery. If necessary, the fault recovery processor gets the right to control. The fault recovery processor prosecutes a backup path by referring several resource tables, state and statistics processor, and configuration processor. It is important in the proposed resource look-ahead model that the fault recovery processor looks for the availability of resource from referring resource table and reservation table.
After selecting the backup path, the fault recovery processor sends the right to the connection management processor, which sends Add Branch messages to all LSRs related to the backup path in order to command to reroute.
Fig. 4 shows a reference table for the resource look-ahead algorithm.

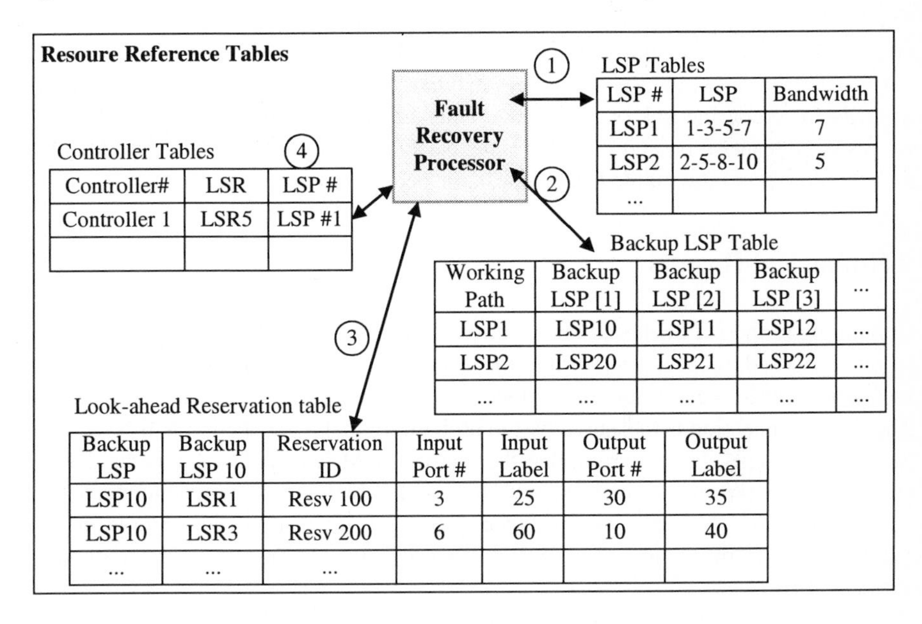

Fig. 4. Resource Look-ahead Reference Table

The fault recovery processor knows the affected LSP(Lable Switched Path)s by looking up LSP tables. A LSP table has the LSP number, a series of LSR numbers, and required bandwidth. The LSR table and the Backup LSP table help the fault recovery processor to select a backup path, and the controller table is for the cc-gsmp interface related to controller information. Controller information has LSR and LSP information for its own domain. Finally, the reservation table is the key table for the resource look-ahead recovery model. It has reservation ID, input port, input label, output port, and output label. The backup path can be reserved with a reservation ID. When the Add Branch message is delivered form controller to switch, the reservation ID is specified inside of the message. Therefore, a reservation request message and a label mapping message are not used in this recovery model.

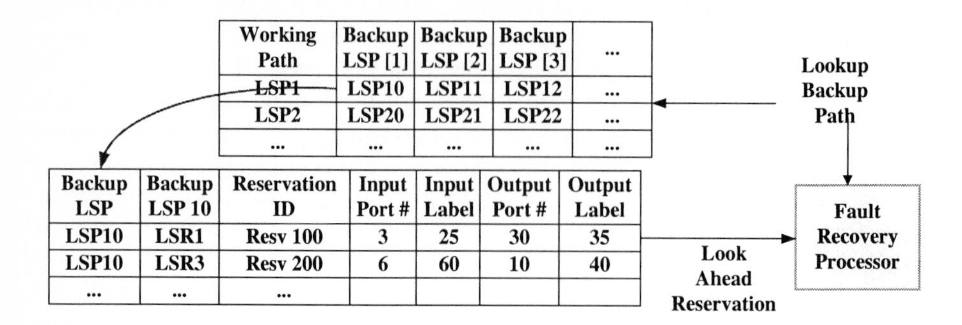

Fig. 5. Resource Look Ahead Fault Recovery Mechanism.

The fault recovery processor searches the backup table and looks ahead the reservation table. We called the backup path searching mechanism as a resource look-ahead recovery algorithm. During the idle time, the controller requests for resource reservation to LSRs for preparing link or node fault. Thus we can alleviate the propagation delay for resource reservation and mapping time. Such an algorithm, however, requires that the resource reservation table has to be established before occurring fault.

4 Numerical Results

Comparing to the MPLS-based fault recovery mechanism, we adapted equation of [7] to our proposed recovery algorithm for obtaining numerical result. For this entire recovery process, we are interested in the degree of restoration of a failed connection. The failed connection is measured by the probability that a connection will be restored within M attempts, P_M, assuming that there are only M feasible backup paths. Restorability may be expected to increase with larger M, which is a design parameter for the network provider. Another performance metric of interest is the restoration time measured in the mean time to complete the procedure, E(Tr), and mean number of attempts, E(m).

We assume that a random probability of rejection q by an LSR cause of resource insufficiency. The probability will vary between the first attempt and consecutive attempts; the probability of success should presumably be higher on the first attempt if the path selection algorithm is correct, but the analysis here will make the simplifying assumption that each the probability q is the same on each attempt. The Probability of rejection q will be complicated function of many factors such as active traffic load, available bandwidth and buffers, requested QoS and traffic characteristics. In this paper, we only determine the probability of q by bandwidth.

Additionally, we assume that each of M feasible backup paths will be h hops. The length of backup routes may range between 1 and H with uniform probability. For analysis, we assume that a typical connection (called label switched path or LSP) consists of H hops and a fault is equally likely with probability 1/H to the located at any hop.

An important metric of performance is the probability of successful restoration on the first attempt:

$$P_1 = (1-q)^H \tag{1}$$

Under the assumptions, each attempt has the same probability of success, and the probability of eventual restoration within M attempts is clearly

$$P_M = 1-(1-P_1)^M \tag{2}$$

The mean number of unsuccessful attempts is therefore

$$E(m) = \sum_{m=1}^{M-1} mP_1(1-P_1)^m = [1(1-P_1)^{M-1}(1+(M-1)P_1)]/P_1 \tag{3}$$

The mean delay experienced by fault notification message will be 2*Tp. In this case Tp is message processing time. This delay must be constant of the number of hops multiplied by packet processing time. The restoration time Tr for a successful attempt is the sum of store and forwarding delay and processing delay at each LSR.
The mean total delay involved in each unsuccessful attempt is therefore

$$E(T_{delay}) = \sum_{i=1}^{H} i(T_r + T_p)q(1-q)^{i-1} \tag{4}$$

The probability of exactly m unsuccessful attempts before a successful restoration is $P_1(1-P_1)m$, $0 \le m < M$, and the mean number of successful attempts is given by (3) where P1 is now (1). Combining (3) and (4), the total mean recovery time including fault notification message and failed attempts is

$$E(T_{recovery}) = 2*T_p + H*T_p + E(m)E(T_{delay}) \tag{5}$$

The restoration blocking probability caused by insufficient resource in each LSR is reduced by the reattempt counts, M, increase. The following graph is drawn when the reject probability of each LSR, q, is assumed 0.02. And fig. 6 shows the restoration blocking probability to the reject rate caused by resource insufficiency.
The mean fault recovery time rises when the reject probability increases. We can get less mean recovery time comparing to MPLS based fault recovery in [7]. The reason why we enhance the performance of mean recovery time is proposed model can reduces the propagation delay of resource reservation and resource availability.

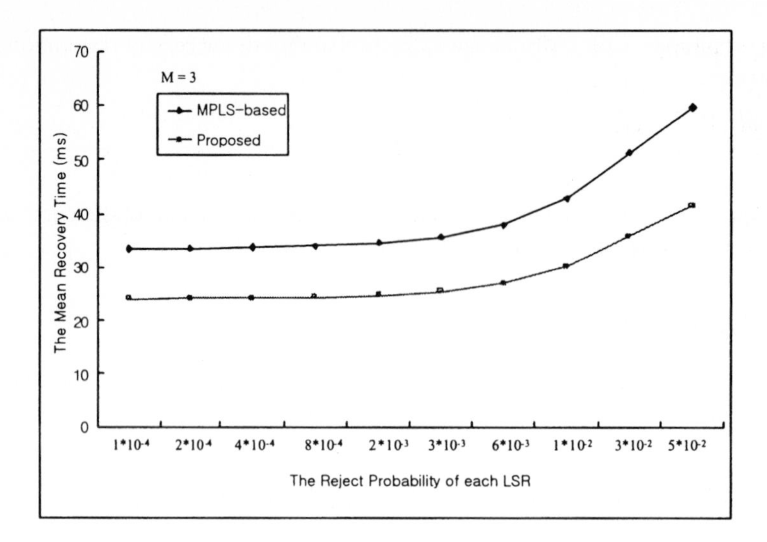

Fig. 6. The Behaviors of Mean Recovery Time

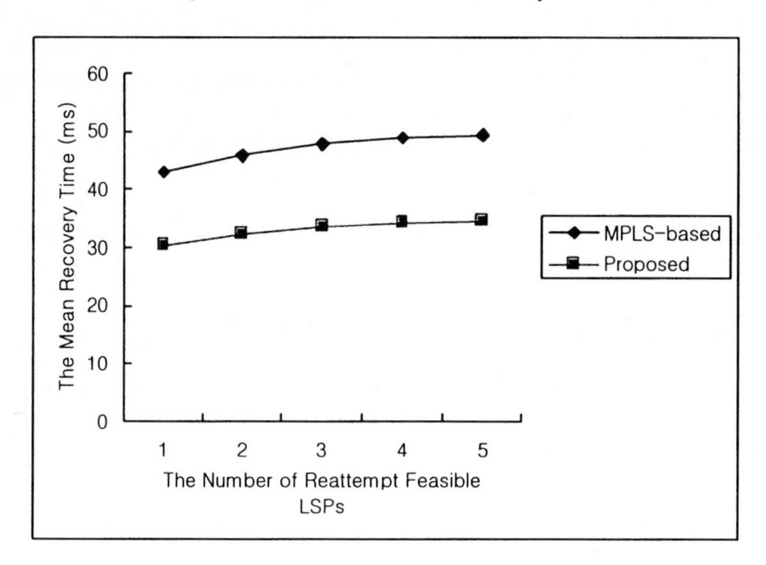

Fig. 7. The Behaviors of Mean Recovery Time

The mean fault recovery time rises according to the number of feasible backup path, as shown in Fig.7. The difference between two graphs results from the resource look-ahead mechanism. We obtain nearly two times of improvement of mean recovery time. In addition, the recovery time is very acceptable comparing to requirement of 50 ms to recover.

5 Conclusion

In the results, we verified that the proposed resource look-ahead recovery algorithm can alleviate the time of selecting a backup path with a small propagation delay. It can also make less of the control message volume and take less notification time according to increment of the number of hops. Finally, we can obtain faster mean recovery time through the proposed fault recovery algorithm.

Acknowledgment. This research was supported in part by KOSEF (Korea Science and Engineering Foundation) and OMIC (Ministry of Information and Communication) of Korean government through OIRC (Optical Internet Research Center) project.

References

1. Viswanathan, N. Feldman, Z. Wang and R. Callon, "Evolution of multi-protocol label switching," IEEE Commun. Mag., vol. 36, May 1998.
2. Biswas, J., Lazar, A.A., et. al., "The IEEE P1520 standards initiative for programmable network interfaces", IEEE Communications Magazine, vol. 36 Issue 10 , Oct. 1998.
3. Bjorkman, N., et. al. MSF System Architecture Agreement (http://www.msforum.org/). April.2000.
4. van der Zee, D.-J., "Look-ahead strategies for controlling batch operations in industry-overview, comparison and exploration", in Proceedings of Winter Simulation Conference 2000, vol. 2 , Dec. 2000.
5. Vishal, et al., "Framework for MPLS-based Recovery", (draft-ietf-mpls-recovery-frmwrk-03.txt), July 2001.
6. General Switch Management Protocol "draft-ietf-gsmp-08.txt" (http://www.ietf.org/html.charters/gsmp-charter.html), May 2001
7. Oh, T. H., Chen, T. M. and Kennington, J. L., "Fault restoration and spare capacity allocation with QoS constraints for MPLS networks," in Proceeding of IEEE Global Communications Conference, vol. III , Nov. 2000.

A Router Assisting Control Tree Configuration Mechanism for Reliable Multicast

Eunsook Kim[1], Seok Joo Koh[1], Juyoung Park[1], Shin-Gak Kang[1], and Jongwon Choe[2]

[1] Electronics and Telecommunications Research Institute,
161 Kajeong-Dong, Yuseong-Gu, Deajeon, KOREA
{eunah, sjkoh, jypark, sgkang}@etri.re.kr
[2] Sookmyung Women's University,
53-12 Cheongpa-Dong 2Ka, Yongsan-Gu, Seoul, KOREA
choejn@sookmyung.ac.kr

Abstract. For reliable multicast service, the mechanism based on hierarchical control tree can be a promising solution to avoid well-known feedback implosion. However, configuration of an efficient control tree is very difficult for IP Multicast because it does not provide explicit membership and routing topology information to upper layer protocol. If the transport layer tree and the network layer tree are very different, it may take large cost to handle control messages. Especially, when a node at a downstream link of the network routing tree becomes a parent node of an upstream link at the control tree of transport layer, the discrepancy between routing tree and control tree causes to waste network resources by redundant messages. This problem can be solved if router that knows the information on routing topology can support configuration of a control tree and reliable delivery. However, the change of router function embraces deployment problem. Thus, this paper proposed a very simple method of router assist to minimize change of router functions. With this method, routers are only required to recognize message types of control messages in order to forward the messages to correct direction: upstream or downstream.

1. Introduction and Problem Statement

As current IP multicast[2] provides only best-effort service, reliable multicast transport service should be implemented on it. Over the past several years, many studies on the reliable multicast transport have been made[4], [7], [16], and of those research works, the Tree-based ACK(TRACK) protocol is one of the promising RMT protocols of which the error recovery, congestion control and other functionalities are provided over a pre-configured logical tree[9], [15]. This mechanism is known to provide high scalability as well as reliability[10] since it reforms a multicast group into a logical tree rooted by a sender. It classifies its receivers into Service Nodes(SN) and receivers. SNs are intermediate nodes which take charge of local retransmission, while receivers only denotes leaf nodes of the tree.

As the performance of this mechanism may highly depend on the efficiency of the control tree, many researchers have been making their efforts to build an efficient control tree[9]. However, to configure a logical tree is difficult because IP Multicast does not provide the information of explicit membership and routing topology to upper layer protocols.

I. Chong (Ed.): ICOIN 2002, LNCS 2343, pp. 84-93, 2002.

Because of this problem, a transport layer tree may not be congruent with a underlying routing tree, and in this case, reliable multicast services may take large cost. The Fig. 1 shows an original multicast tree, and Fig. 2 draws a possible control tree derived from it. We should notice that due to the lack of information on the underlying network topology, node *a* would bind to *f* lying in its downstream link rather than *b* placed on its upstream link, as *f* and *b* have the same hop distance from *a*. It is also noticeable that *f* can choose *c* as its SN, while *b* and *f* have the same hop distance. Let's suppose that link *e3* is failed for a moment. Children nodes of *c* including *d*, *e* and *f* will request retransmission to *c*, while *a*, *g*, and *h* request the error repair to *f*. The node *c* also sends repair request to *b*. In this situation, some links should carry the same data on several times, because of the discrepancy between the routing tree and the control tree.

For example, *e8* should deliver the data three times for one repair, when *b* multicasts the repair data, *c* repairs the loss for *f*, and *f* transmits the retransmission data to *a*. The node *a* is put on the most severe situation among the receivers. It undergoes redundant retransmission as well as long delay of repair. When we deliberate the case of *a*, it could have been considerably reduced the overhead if it bound to its upstream SN, *b*. It can be possible when it knows the underlying network topology. Eventually, we can say that the best control tree is to match the underlying multicast routing tree topology[14].

There have been some researches to support underlying routers to assist reliable service like GRA [1] or PGM [4]. However, router functions should be modified in order to apply the router assist mechanism, in which the process takes times to deploy. It can be used as an assisted method when it is supported. So, an easy and simple solution which helps to prompt fast deployment should be studied.

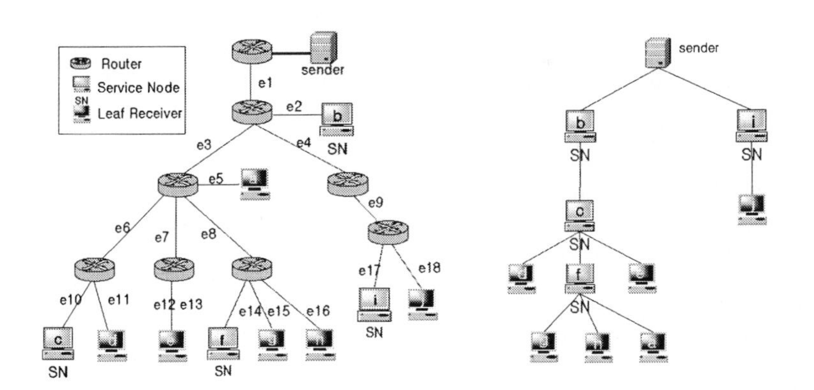

Fig. 1. Multicast Routing Tree

Fig. 2. A Possible Control Tree

Thus, this paper proposes a simple router assist mechanism aimed at configuration of a scalable, efficient, and robust control tree that keeps a close shape with multicast tree. The proposed mechanism does not require the underlying routers to keep the state information for reliable service unlike the previous works[4]. In addition, it does not require routers to be aware of error recovery[14]. Routers simply provide the path for receivers to reach other receivers. We believe this simple way may facilitate router to assist reliable service.

This paper is organized as follows: Section 2 provides a brief summary of related works and Section 3 describes the proposed scheme. The evaluation and analysis of the proposed scheme are shown in Section 4. In Section 5, we conclude this paper.

2. Related Works

Addressable Internet Multicast(AIM) [11] is a scheme that uses forwarding services that require routers to assign per-multicast group labels to all routers participating in that group. AIM uses these labels to send a request towards the sender which get redirected to the nearest upstream member. If data is available, the NACK receiver responds with a retransmission which is also forwarded according to the router labels. However, to keep the label information may cause scalability problem when the group size grows.

PGM [4] peeks into transport headers to filter messages. NACKs create state at the routers that are used to suppress duplicate NACKs and guide retransmissions to receivers that requested them. PGM creates a hierarchy rooted at the source, but provision is made for suitable receivers to act as Designated Local Retransmitters(DLSs) if desired. However, this work requires that routers keep and check the packet sequence number from transport layer. For this operation, routers should maintain buffer and it may add overhead of routers.

LMS [14] proposes to use minimal router support not only to make informed parent/child allocation, but also to adapt the hierarchy under dynamic conditions. With LMS, each router marks a downstream link as belonging to a path leading to a *replier*. A replier is simply a group member willing to assist with error recovery by acting as a parent for that router's immediate downstream nodes. Because repliers are selected by routers, parents are always upstream and close to their children. The forwarding services introduced by LMS allow routers to steer control messages to their replier, and allow repliers to request limited scope multicast from routers.

Active Error Recovery(AER) [7] is very similar to LMS. In AER, each router that has a repair server attached periodically announces its existence to the downstream routers and receivers, and serves as a retransmitter of the lost data on the subtree below it, or collects and sends NACKs upstream.

However, in these works, the function which routers select and maintain repliers should be added in routers. It may not be easy to routers to be largely changed and give some overhead to routers to function the operation.

Generic Router Assist(GRA) [1] gives guidelines to design router assist mechanism for reliable multicast. It proposes to use minimal router support for loss recovery. In these router assisted schemes, hierarchy construction is achieved by routers keeping minimal information about parents for downstream receivers, then carefully forwarding loss recovery control and data messages to minimize implosion problem. With this approach, the constructed control tree can keeps congruency with the underlying multicast routing tree topology. To do this, each GRA router collects the routing tree information and delivers it to the downstream receivers for the concerned multicast group. In the GRA, hierarchy construction requires little explicit mechanism at the expense of adding router functionality. The proposed work would be designed to keep the principles of the guidelines.

3. The Proposed Router Assist Mechanism

The control trees are used for forwarding control information towards the root or data towards the leaf nodes. A generic tree configuration of Reliable Multicast Transport(RMT) proceeds in the order of session advertisement, SN discovery, and binding to the best SN.

In the session advertisement, each receiver realizes the existence of a session and the sender by using an out-of-band mechanism such as SAP[6], or Web announcement. In this process, the receiver will obtain the multicast group address, the sender's address, and the other information needed for the construction of a control tree. When the sender indicates creation of a control tree, each receiver begins SN discovery process to find one or more candidate SNs that are active in the session. Among the candidate SNs discovered, a receiver selects and binds to the best SN by using a pre-configured rule such as TTL distance or IP address.

The message types used for control tree creation are *BEACON*, *ADVERTISE* (or *QUERY* and *QUERY-RESPONSE)*, *BIND*, and *ACCEPT* or *REJECT*. *BEACON* message is multicast to the all group members to indicate a control tree creation. When a *BEACON* message is transmitted to the group, SN discovery process will be started. There is two ways to seek an SN for a receiver. One is that each SN seeks its child receivers by sending *ADVERTISE* messages within a scoped area[12]. In the other way, each receiver solicits its SN with *QUERY* messages as it widens the message scope until it receives a response message(*QUERY-RESPONSE*)[16]. For both of methods, each receiver who receives the message tries to bind to a SN with *BIND* message, and each SN should respond with *ACCETP* or *REJECT* message about the bind request.

This paper chooses the first method because it has shorter procedures of message exchanging, and may guarantee less message overhead, because the message overhead is not inclined to grow depending on the number of receivers unlike the latter method which *QUERY* messages are used.

With the message exchanging scheme, the most important fact of control tree to be efficient is that it should be congruent with underlying routing tree as possible as it can, in order to reduce redundant control messages. As Fig. 2 shows, a control tree would suffer from redundant messages if a downstream node becomes a parent SN of an upstream node. To prevent from making such relationship, the proposed mechanism configures a control tree as follows:

Step 1. The sender multicasts *BEACON* messages for indication of the tree creation into the multicast session. Each router recognizes the input link of the message as upstream link.

Step 2. Each SN sends *ADVERTISE* messages to invite receivers after a *BEACON* message is indicated. It forwards *ADVERSITE* messages only to its downstream links. For example, in the Fig. 1, we assume that the node *b*, *c*, *f*, and *i* are designated as SNs. In this case, *ADVERTISE* messages from *b* are forwarded to all downstream receivers, and *ADVERTISE* messages from *c* are forwarded to *d*, the messages from *f* passes to *g* and *h*, and the messages from *i* are delivered to *j*.

Step 3. Receivers now can explicitly send a *BIND* message to their closest parent. Closeness is estimated by hop distance in this stage. RTT or other metrics can be considered in later. In Fig. 1, node *c*, *e*, *f*, and *i* only receive *ADVERTISE* messages from *b*, whereas node *d* receives *ADVERTISE*s >from *b* and c, *g* and *h* get from *f*, and *j* receives them from *b* and *i*, respectively. Now, it is clear that *c*, *e*, *f* and *i* try to bind to node *b*, and *d* will try to bind to *c*. The nodes *g* and *h* will try to bind to *f*, and *j* will do to *i* as estimating hop distance.

Step 4. Each SN determines whether the BIND request must be accepted or rejected. In the *REJECT* case, it should notify the reason.

Fig. 3 illustrates the forwarding process of *ADVERTISE* messages. We see that the operation of tree configuration can be simplified with this router assist, as there is no modification of router functions except choosing downstream or upstream by message types. In this mechanism, we do not need to calculate *Sender Distance*[1] and *Neighbor Distance*[2], or tree level of each node which is needed to select an SN and to avoid a loop under no router assist mechanisms, which may add complexity in the process of a control tree creation. Moreover, it can avert that a SN placed on a downstream link from a receiver becomes the parent SN, which causes redundant data delivery and inefficient link usages.

Fig. 4 shows the control tree established from Fig. 1 with the proposed router assist mechanism. As we see, it maintains congruency from the underlying routing tree as every parent is an upstream node of its children, where the link overhead for exchanging control messages can be minimized.

When a receiver reports losses to its SN via *ACK* packet, the SN sends *RDATA*, retransmitted data, corresponding the *DATA* packet. The only change for this router assist mechanism is that each router forwards *RDATA* only to the downstream links.

With this mechanism, the TTL-scope, which is controversial of the stability on the current Internet, is not needed. Furthermore, error request and repair processes are efficiently performed because SN of each receiver is located in the close area on real network topology.

For instance, if we suppose again that link *e3* is failed for a moment. In Fig. 2, children nodes of *c* including *d*, *e* and *f* will request retransmission to *c*, while *a*, *g*, and *h* request the error repair to *f*. The node *c* also sends repair request to *b*. In this case, each link should carry the same data several times, such as *e8* should deliver the data three times for one repair, when *b* multicasts the repair data, *c* repairs the loss for *f*, and *f* transmits the retransmission data to *a*. The node *a* is put on the most severe situation among the receivers. It undergoes redundant retransmission as well as long delay of repair.

However, in Fig. 4, *e8* only gets request of retransmission from *f*, and delivers retransmitted data one time towards *f*, *g*, and *h*. We can obviously see how the case of node *a* can be improved in the proposed scheme. It is simply served by *b* without unnecessary transit path.

---

[1] Distance >from an SN to the sender

[2] Distance from a receiver to a neighboring SN

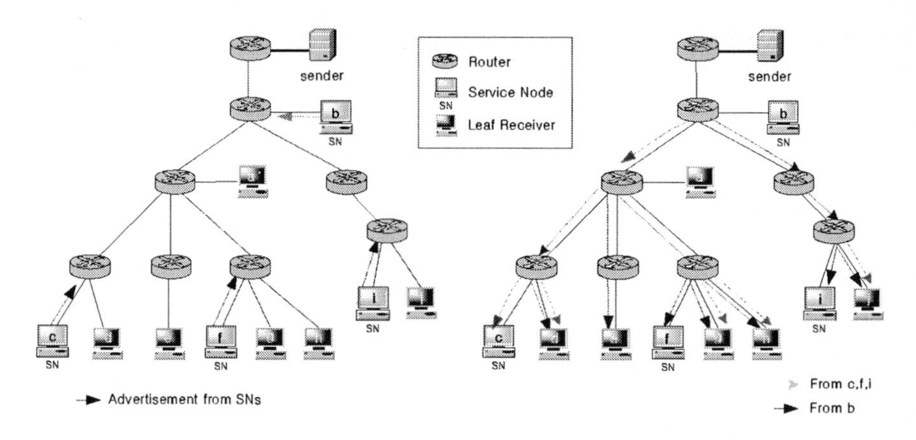

Fig. 3. Forwarding path of the ADVERTISE message

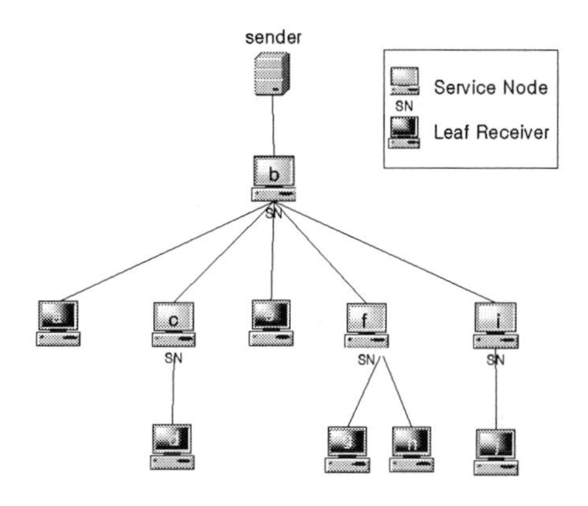

Fig. 4. A configured control tree from Fig. 1

4. Evaluation and Analysis

In section 3, we insist that a parent SN always lies on upstream link or at least in the same level of its children in the proposed mechanism. With the feature, we see the configured control tree keeps congruent with routing tree and reduce inefficient usage of network links We will evaluate the proposed mechanism with both of mathematical evaluation and simulation.

4.1 Mathematical Evaluation

We will show that the proposed mechanism keeps congruency with the given routing tree as we verify the following proposition:

Proposition. A parent SN is always in the upstream link or at least in the same network of its children.

To prove this, the notations below are defined first.

$N = (V, E)$, where V = a set of node, and E = a set of link: N indicates network which is illustrated by a directed graph.

$e_{ij} = (v_i, v_j) \in V$, where $i \neq j$, $v_i \geq v_j$ which means v_i is an upstream node of v_j: link e implies a path between v_i and v_j;

$Cost(e_{ij}) : E \rightarrow I+$, where $i \neq j$: link cost of e from v_i to v_j. The cost value is assigned as hop counts between the two nodes;

$G = \{v_1, \ldots\ldots, v_k\} \subseteq V$: a set of multicast group members;

$R = \{r_1, \ldots\ldots, r_n\} \subseteq G$, where $n \leq k$: a set of multicast group receivers;

$SN = \{sn_1, \ldots\ldots, sn_m\} \subseteq G$: a set of SNs;

$CSN(r) = \{sn_1, \ldots\ldots, sn_p\} \subseteq SN$: a set of candidate SNs for a receiver, r;

$Child(sn) = \{c_1, \ldots\ldots, c_q\} \subseteq R$: a set of children for sn;

$Deliver(e_{ij}) \in \{0, 1\}$: Delivery vector of *ADVERTISE* messages, indicating if a message is forwarded via the link. *0* implies no message delivery.

Now, we can define the following rules of SN measurement and selection.

$$r \in Child(sn_k),$$

$$\text{iff } Cost(sn_k, r) = min\,(Cost(sn_m, r)),$$

$$\text{where } \forall\, sn_k \in CSN,\ sn_m \in CSN,\ m = 1 \text{ to } p, \text{ and } 1 \leq k \leq m \quad ⓐ$$

$$Cost(sn_i, r) < Cost(sn_j, r),\ \forall\, sn_i < sn_j,\ \textit{iff}\ r \leq sn_i \quad ⓑ$$

Together with the above formulas we can prove the Assertion 1 as follow:

Proof.

We suppose that there is a receiver, $\{r_i, r_j\} \subseteq R$ and $CSN(r) = \{sn_k, sn_l\}$, where $r_i < sn_k < r_j < sn_l$, and we know that an *ADVERTISE* message is always forwarded by downstream of a router. Then,

$$Deliver(sn_k, r_i) = 1,\ Deliver(sn_l, r_i) = 1,\ Deliver(sn_k, r_j) = 0,\ Deliver(sn_l, r_j) = 1 \quad ⓒ$$

$$\text{From ⓒ, } CSN(r_i) = \{\ sn_k, sn_l\}, \text{ and } CSN(r_j) = \{\ sn_l\}$$

$$\text{Then, } Cost(sn_k, r_i) < Cost(sn_j, r_i),$$

Thus, $r_i \in Child(sn_k)$ from ⓐ and ⓑ, and $r_j \in Child(sn_j)$ because the sn_j is the only SN available to r_j.

Now, we see that both of r_i and r_j have their SNs with upstream nodes from themselves.

Thus, we can say that the proposed mechanism guarantee the receivers choose their parents among the upstream nodes.

4.2 Simulation Results

In order to estimate the performance of the proposed mechanism, we perform experimental simulations by network simulator *ns*[9]. Control traffic overhead and data recovery time are evaluated under conditions that 1024byte fixed sized packets are generated with 10ms CBR(Constant Bit Rate) stream, a packet loss is forced to occur in 3 consecutive packets, and it is supposed that every loss happens in only data packets not in control packets.

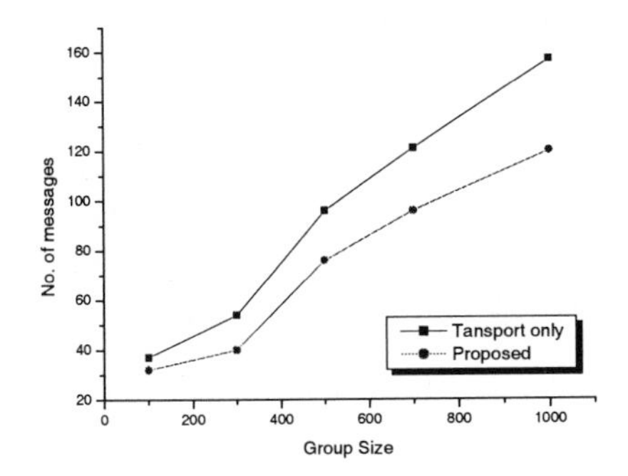

Fig. 5. Control traffic overhead

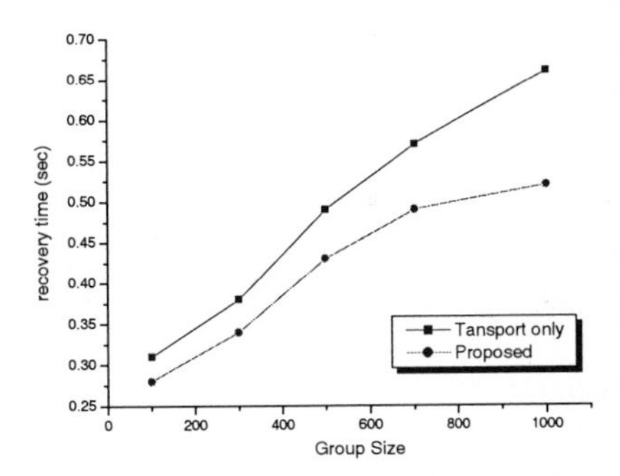

Fig. 6. Data recovery time

Fig. 5 and Fig. 6 shows the evaluation results of control traffic overhead and data recovery time respectively. 'Transport only' means a control tree built in transport layer. We model HiRM[8] for our simulation. In both metrics, proposed mechanism shows better performance. Especially, the gab between two mechanisms has become bigger when the group size gets growing.

In short, these simulation results stand for stability and scalability of the proposed mechanism, because the performance that this mechanism shows does not much depend on the group size.

4.3 Relationship with Network Layer

The proposed mechanism is independent from a special network layer, but it will be well-matched with SSM[5] or IPv6[3]. Current reliable multicast protocols targets one-to-many multicast service, and SSM provides simple and efficient routing mechanism for one sender based multicast service. It is well matched with the proposed mechanism.

However, the router assist mechanism has a deployment problem. It is not easier to be changed for already-existing routers. So, if we put this functionality into IPv6 routers which is not much deployed yet, it can be smoothly deployed as the network grows. In addition, its address scheme is hierarchically designed, so the each address reflects in which network it belongs to. It is very useful to choose an SN when there exists more than one SN in the same region.

5. Conclusions

In tree-based reliable multicast protocols, it is very important to design an efficient logical tree. However, to configure a logical tree is difficult because IP Multicast does not provide explicit membership and routing topology information to upper layer protocols.

Thus, there have been some researches to use routers to build a control tree which is to match the underlying multicast routing tree topology. However, it has deployment problem, so an easy and simple solution to router to assist reliable delivery.

This paper proposes a simple router assist mechanism aimed at configuration of a scalable, efficient, and robust control tree that keeps a close shape with multicast tree. The proposed mechanism does not require the underlying routers to keep the state information for reliable service and to be aware of error recovery. Routers simply provide the path for receivers to reach other receivers.

With this simple way, we see that the proposed mechanism guarantee that the receivers choose their parents among the upstream nodes. In addition, the simulation results show that this mechanism stably operates in different group sizes. We believe this simple way may facilitate router to assist reliable service.

To evaluate the proposed mechanism more specifically, now we are simulating the proposed scheme with various topology and metrics. In addition, we put our efforts to enhance the mechanism.

References

1. B. Cain, T. Speakman and D. Towsley, Generic Router Assist(GRA) Building Block – Motivation and Architecture, IETF Internet Draft, March 2000
2. S. Deering, Host Extensions for IP Multicasting, RFC1112, August 1989
3. S. Deering and R. Hinden, “Internet Protocol, Version 6(IPv6), Specification,” RFC 2460, December 1998
4. D. Farinacci, A. Lin, T. Speakman and A. Tweedly, PGM reliable transport protocol specification, IETF Internet Draft, Auguest 1998
5. H. Holbrook and B. Cain, Source-Specific Multicast for IP, IETF Internet Draft, November 2000
6. M. Handley, C. Perkins and E. Whelan, Session Announcement Protocol, IETF Internet Draft, March 2000
7. S. K. Kasera, S. Bhattacharyya, M. Keaton, et al., Scalabel Fair Reliable Multicast Using Active Services, IEEE Network Magazine (Special Issue on Multicast), January/February 2000
8. E. Kim, Design and Analysis of a Hierarchical Reliable Multicast, Ph. D Dissertation, June 2001

9. M. Kadansky, B. Levine, D. Chiu, et al., Reliable Multicast Transport Building Block: Tree Auto-Configuration, IETF Internet Draft, draft-ietf-rmt-bb-tree-config-01.txt, November 2000
10. B. Neil Levin and J.J. Garcia-Luna-Aceves, A Comparison of Known Classes of Reliable Multicast Protocols, Proceedings of International Conference on Network Protocol (ICNP-96), 1996
11. B. N. Levine and J. J. Garcia-Luna-Aceves, Improving Internet Multicast Routing with Routing Labels, IEEE International Conference on Network Protocols (ICNP '97), October 1997
12. S. McCanne, S.Folyd, NS (Network simulator), http:://www-nrg.ee.lbl.gov/ns, 1995
13. S. J. Koh, E. Kim, J. park, S. G. Kang, et al., Cofiguration of ACK Trees for Multicast Transport Protocols, ETRI Journal, Vol.23, September 2001
14. C. Papadopoulos, G. Parulkar and G. Varghese, An Error Control Scheme for Large Scale Multicast Applications, In proceedings of IEEE INFOCOMM '98, March 1998
15. B. Whetton, D. Chiu, M. Kadansky, and G. Taskale, Reliable Multicast Transport Building Block for TRACK, IETF Internet Draft, draft-ietf-rmt-bb-track-00.txt, November 2000
16. R. Yavatkar, J. Griffioen and M. Sudan, A Reliable Dissemination Protocol for Interactive collaborative Applications, University of Kentucky, 1995

A Proposal of Multicast for Personalized Media Stream Delivery

Katsuhiko Sato[1,2] and Michiaki Katsumoto[3]

[1]Japan Radio Co., Ltd, Network laboratory,
5-1-1 Shimorenjaku Mitaka Tokyo 181-8510, Japan
sato@lab.jrc.co.jp
[2]The University of Electro Communication, Department of Information Network Science,
1-5-1 Chofugaoka Chofu Tokyo 182-8585, Japan
ksato@hn.is.uec.ac.jp
[3]Communications Research Laboratory, High-Speed Network Section,
4-2-1 Nukui-Kitamachi Koganei Tokyo 184-8795, Japan
Katumoto@crl.go.jp

Abstract. This paper describes a scheme of network distribution that enables innovative broadcast content delivery service over the next-generation broadband Internet. The scheme enables producing diverse video/audio contents customized according to the user's specifications and delivering such contents via streaming with guaranteed quality. The broadcasting of diverse media contents results in a massive consumption of network resources. We developed two multicast techniques, based on asynchronous and layered multicasting. One is an algorithm for statistical traffic control based on asynchronous multicasting that enables flexible bandwidth control to meet the network bandwidth design requirement for QOS. The other is a technique for the delivery of diverse media segments/objects based on a layered multicast technique, improved for the delivery of media with different levels of quality. We prove the implementability of the two techniques by designing network models and protocols using the latest Internet technologies and show their effectiveness by using numerical analysis.

1 Introduction

Recently a number of Internet broadcast services called Web casting [1] have been developed, which may become the basis for the new application of the Internet. These Web casting services provide live broadcasting and on-demand content delivery, and they are usually used with real-time streaming transfer, and are resorted to in some multicast or decentralized cache techniques to reduce the load on the network. Meanwhile, a broadband infrastructure for content distribution based on Internet technologies has been growing rapidly so that the quality of video/audio delivery may soon become equal to or exceed that of current TV broadcasting. Some experiments have shown that this may soon become a reality [2].

We have been trying to create new applications and propose new techniques on the premise that the broadband infrastructure for content distribution will become a reality. For innovative broadcast content delivery over the next-generation broadband

I. Chong (Ed.): ICOIN 2002, LNCS 2343, pp. 94-108, 2002.

Internet, we have proposed a scheme that enables producing diverse video/audio contents customized to the user's needs and delivering such contents via streaming with guaranteed quality [3].

In this paper, we focus on a scheme for delivering content over the Internet. When video/audio contents customized for each user are delivered by streaming with guaranteed quality, a number of network resources are consumed. In this case, multicast techniques can be used, however, such techniques cannot be easily applied to the delivery of diverse video/audio content. Here, we take two multicast techniques, an asynchronous multicast and a layered multicast technique, and try adapting them to our proposed scheme. Then, we propose two new techniques. One is an algorithm for statistical traffic control based on asynchronous multicasting that enables flexible bandwidth control to meet the network bandwidth design requirements for QOS. The other is a technique for the delivery of diverse media segments/objects based on a layered multicast technique improved for the delivery of media with different levels of quality. We prove the implementability of the two techniques by designing network models and protocols using the latest Internet technologies, and show the effectiveness of traffic retrenching in our techniques as compared with the currently used unicast method by using numerical analysis.

2 Personalized Media Stream Delivery

In the future, both communication service and broadcast service will be unified on the same infrastructure of the Internet. Based on the bilateral feature of the Internet, a personalized broadcasting service will come to be offered providing the delivery of individualized content at the time of the user's own choosing. Personalized Media Stream Delivery (PMSD) is a system endowed with the following features.

- A wide range of video/audio contents based on the user's interests or lifestyle can be produced;
- These contents can be delivered individually by streaming with an assured level of quality.

Producing and delivering a diverse selection of video/audio content means, for example, that several kinds of video segments (scenes) and objects prepared in the server system are compiled based on the specific information about each user, and are delivered with an assured level of quality based on the computational ability of the receiver system to play back the content. PMSD will make broadcasting efficiency and will enable the transmission of important information to the target user.

3 General Problems with Multicasting

The easiest way to enable 'individualized-contents distribution' and 'on-demand contents distribution' is by using a unicast method in witch contents are carried to each user in their own flows. But this method consumes a lot of network resources. Therefore, other solutions are needed to save network resources even though networks are becoming broadbanded.

With a multicast method, the contents are carried to each user in a single flow that is copied at appropriate network nodes depending on the location of the receiving user. Therefore, a good deal of network resources can be saved, but it will not be easy to apply real-time multicasting to PMSD because the same content must be delivered at the same time to the same multicast-group members.

Some works have discussed a multicast technique for on-demand content delivery, which is called asynchronous multicasting. Other works have discussed a multicast technique for delivering content with a level of quality determined according to the user's processing ability. This technique is called layered multicasting. Here, we will see if the two can be applied to PMSD, and if the two can function co-operatetively (previously they have been studied independently). In addition, we will try to create a way to implement them on the real Internet.

4 Related Works

4.1 Asynchronous Multicasting

The principle of asynchronous multicasting is that a portion of overlapping data in the delivery of a certain content is aggregated into a multicast flow among users whose requests are made at around the same time, and the data not included in the multicast flow (i.e. the portion between the time when the multicast flow starts and the time when a request occurs) are individually delivered by unicast flows. The data arriving at the receiver system, which are aggregated into a multicast flow, are not played back on the spot, but are buffered until the data delivered by a unicast flow is completely played back.

Several methods have been studied for asynchronous multicasting. A fast-transfer method is described in [4], [5], where the content are divided into small units of data and the data are transmitted a few times as fast as they are played back. A stream-transfer method has been proposed in [6]. In [7], the portion of data delivered by a unicast flow is transmitted in a burst manner.

4.2 Layered Multicasting

Layered multicasting has been studied for the distribution of multi-layered coding video. In layered multicasting, each piece of layered encoding data is mapped onto a different multicast group, and the receiver system joins multiple multicast groups according to its computational ability. There have been a number of works published about layered multicasting. An experimental implementation to develop a standard is reported in [9]. S.McCanne et al and others have proposed that receiver systems dynamically change to obtain the number of multicast groups depending on the network congestion.

5 Multicast Techniques

5.1 Asynchronous Multicasting

Asynchronous multicast techniques have not been taken into consideration about bandwidth control. Some traffic control is required for bandwidth allocation in the network to guarantee QOS.

In [4], [5], [7], the premise is that every receiver system has a broadband interface. Therefore, stream-based asynchronous multicasting reported in [6] can be used for PMSD.

Here, we approach the relationship between the cycle of aggregating flows into a multicast flow and the arrival rate of requests for delivery (and the length of the content). We propose an algorithm for statistical traffic control for multiple flows, which is based on the principle in [6], to derive the maximum effect of stream-based asynchronous multicasting and perform flexible bandwidth control.

5.2 Layered Multicasting

Layered multicast techniques have been used to deliver content with different levels of quality in multi-layered coding video. We extend this idea to the delivery of diverse video segments and objects. A segment here means a unit that constitutes a video program. An object here means a unit that constitutes a video picture (e.g. they are called "background object" and "foreground moving objects").

For the delivery of diverse segments, the segments being common among users are aggregated, and each is mapped onto a different multicast group. The receiver systems dynamically change to join the multicast groups and play back the segments of their own choosing.

For the delivery of diverse objects, all objects that every user possibly needs are contained in one segment mapped onto one multicast group. Objects to be played back are chosen from all objects in the segment based on the user's needs. This method does not require complicated implementation such as grafting and pruning multicast trees per object.

5.3 Algorithm for Statistical Traffic Control

Let's consider the traffic intensity in stream-based asynchronous multicasting. We assume that the provided content is transmitted at a constant bit rate, that its length is h, and that delivery requests occur randomly (i.e. the average arrival rate of requests is λ, the average arrival interval is $1/\lambda$). Figure 1 shows that one delivery to a receiver is made by multicasting as a shared flow, and subsequent deliveries to other receivers are made by unicasting as an individual flow.

There are two ways to determine the length of individual flows: one is that we make the length of every individual flow the same, which equals to the interval of generation of shared flows; the other is that we make the length of each individual flow correspond to a period of time between starting a shared flow and starting an individual flow. Figure 1 shows the latter. The latter can acquire more effect for the traffic reduction instead of more complicated implementation.

The length of an individual flow can be expressed as the number of segments (or subdivided segments) that is the same unit used for the delivery of diverse segments. Therefore, the server and receiver systems do not require highly accurate time processing to determine the length of individual flows.

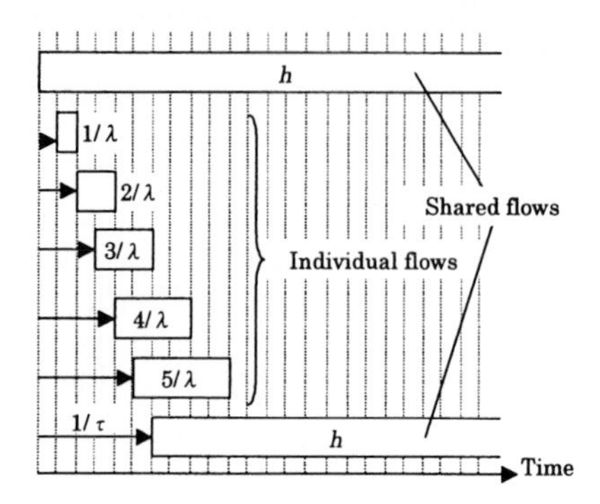

Fig. 1. Asynchronous multicasting

Here, the generation rate of shared flows is expressed as τ ($1/h < \tau \leqq \lambda$). In case that the length of every individual flow is the same, it is equal to the interval of generation of shared flows (i.e. $1/\tau$), so that the traffic intensity, ρ, is

$$\rho = \tau h + (\lambda - \tau)\frac{1}{\tau} = \tau h + \frac{\lambda}{\tau} - 1 \quad \text{[erl]} \tag{1}$$

In case that the length of each individual flow is different, the number of individual flows between two shared flows is $\lambda/\tau - 1$, the length of each individual flow is $1/\lambda, 2/\lambda, 3/\lambda, \cdots$, the average length is then $1/2\tau$. Hence, the traffic intensity is

$$\rho = \tau h + (\lambda - \tau)\frac{1}{2\tau} = \tau h + \frac{\lambda}{2\tau} - \frac{1}{2} \quad \text{[erl]} \tag{2}$$

In both equations above, $\rho = f(\tau)$ is a downward convex curve, ρ takes the minimum value when $\tau = \sqrt{\lambda/h}$ in eq.(1), or $\tau = \sqrt{\lambda/2h}$ in eq.(2). Hence, it is possible to maintain the traffic intensity at a minimum all the time by updating τ to the most appropriate value according to the observed λ and h.

The average use of buffer memory at the receiver (the buffer is used for the data of the shared flow and its size corresponds to the length of the individual flow) becomes small as τ increases. Therefore, in determining τ, there is a trade-off between minimizing the traffic intensity and reducing the use of buffer memory.

Provided that the upper bound of the traffic intensity (i.e. the available bandwidth) is given as A, τ is determined as follows. Hereafter, we describe the case of the length of each individual flow is different.

In case of the observed $\lambda \times h$ is below A, the traffic intensity does not exceed the upper bound even if all contents are delivered by unicasting. Therefore, τ is set to λ

(i.e. All contents are delivered as a shared flow), so that the use of buffer memory at the receiver is zero.

In case of the observed $\lambda \times h$ is greater than A and the minimum value of ρ (i.e. $f\left(\sqrt{\lambda/2h}\right)=\sqrt{2\lambda h}-1/2$) is smaller than A, τ is set to the largest value satisfying $f(\tau) \leqq A$. Here, the equation is

$$A = \tau h + \frac{\lambda}{2\tau} - \frac{1}{2} \quad \text{[erl]} \tag{3}$$

and the solutions for τ are

$$\tau = \frac{1/2 + A \pm \sqrt{(1/2 + A)^2 - 2\lambda h}}{2h} \tag{4}$$

The larger one is selected as τ.

In case of A is smaller than the minimum value of ρ (i.e. $f\left(\sqrt{\lambda/2h}\right)=\sqrt{2\lambda h}-1/2$), τ is set to $\sqrt{\lambda/2h}$ to minimize the traffic intensity.

5.4 Technique for Delivery of Diverse Segments

To enable the delivery of diverse segments, we must solve one problem. The receiver system dynamically changes to join multicast groups and plays back their segments. Then, the network must dynamically build multicast trees for each segment before the server sends corresponding data. But there is no time to build the multicast trees because each segment is transmitted at the same rate as it is played back and a series of segments are played back continuously.

The nature of asynchronous multicasting can solve this problem. The server system produces several patterns of a series of segments and relates them to user groups that represent typical users. The server determines which user group the user requesting content delivery belongs to. For the user that is determined to receive only the shared flow in asynchronous multicasting (here, referred to as the first user), the server system maps the same multicast-group onto all of the series of segments to be delivered. For the users that are determined to receive both the individual flow and the shared flow (here, referred to as the following users), the server system maps a new multicast group onto only the segments that are different from those that already have been determined to be delivered to previous users.

This means the following. For the first user, the transience of multicast groups is not allowed while the content is being delivered because the user plays back the received segment as soon as it arrives at the receiver system. For the following users, the transience of multicast groups is allowed because there is some extra time between the arrival of the segment at the receiver and its playback (i.e. segments in the shared flow are buffered until all segments in the individual flow is completely played back), which enables the network to build multicast trees.

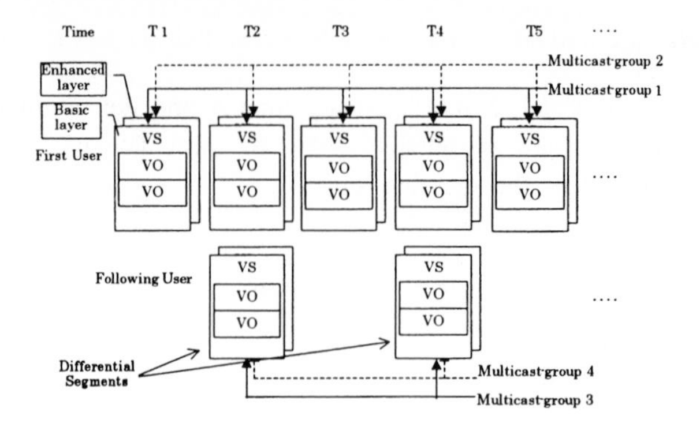

Fig. 2. The delivery of diverse segments and objects using MPEG-4

Figure 2 shows the delivery of diverse segments and objects using MPEG-4, which is based on improved layered multicasting used in conjunction with asynchronous multicasting. MPEG-4 video consists of video object sequences (VSs) that correspond to segments. A VS consists of video objects (VOs). A VO consists of video object layers (VOLs) that are layers of time-spatial resolution. A VOL consists of video object planes (VOPs). Here, we stress that layers of resolution are determined per VS rather than per VO.

6 Implementation

6.1 Requirements

A major premise in the delivery of video/audio content is that the network must provide a consistent QOS while streaming all the data of the content. Policies like those for best-effort forwarding in the network and quality manipulation at the receiver system are not used here. This is because we assume that content delivery is served with any charge or advertisement, like in current broadcasting TV services.

6.2 Basic Concept of Network Control

Generally, there are two network models for network control: 'A concentrated control' and 'A distributed control' model. In the former, one specific control system maintains the only database and executes routing and signaling functions to control the whole network. In the latter, each network node autonomously exchanges information about topology and resources, maintains the same databases and executes routing and signaling functions collaboratively.

To guarantee the QOS of real-time streaming delivery, let's consider two strategies: 'A provisioning based QOS guarantee' and 'A scrambling based QOS guarantee'. The former guarantees QOS by estimating the statistical traffic for multiple flows and overinvesting network resources. The latter guarantees QOS by executing signaling (e.g. admission control and resource reservation) for each delivery before sending data, where users scramble for network resources.

6.3 Network Design

We described some models and design considerations in [3]. We believe that concentrated network control would be easier at an early stage of implementation. In core network where the traffic is concentrated, a multicast tree is statically constructed and the provisioning based QOS guarantee is used. In access network where there are limited resources, multicast trees are dynamically constructed and the scrambling based QOS guarantee is used. In addition, the MPLS (Multi Protocol Label Switch)[11] technology is used. MPLS enables high-speed forwarding by using a cut-through mechanism for various granularities of the packet flow called FEC (Forwarding Equivalence Classes) and it provides traffic engineerings by establishing connections called LSP (Label Switch Path) for each FEC to meet various kind of requirements such as QOS.

Our network model (Figure 3) consists of a single core network and multiple access networks. Both type of networks are physically decoupled into a data-forwarding plane and a control plane. RSs (Relay Server systems) are deployed at the contact points between the networks. The S (Server system) initiates shared flows only, and the RSs initiate individual flows and flows for differential segments (for the delivery of diverse segments) and relay shared flows. Thus, the traffic in the core network can be further reduced.

6.3.1 Core Network

The data-forwarding plane consists of LSs (Label Switch systems) that can execute Diffserv (Differentiated Services)[10]. Diffserv provides a scheme of traffic control within a closed network called a DS domain by setting a DSCP (Diffserv Code Point) on every packet header, which indicates behaviors for forwarding packets on network nodes. Diffserv guarantees QOS by estimating the traffic for multiple flows as a unit of service.

A static multicast tree is made by setting up a permanent LSP, and every shared flow sent from the S is carried onto it. EF-PHB (Expedited Forwarding-PHB) is set as the highest class of forwarding service. The EF-PHB is specified as a class to provide the highest QOS (low-delay, low-jitter, low-loss) in ref. [13]. Every LS carries out queuing control to minimize the queue length for this class, and the incoming rate at ingress LS must be less than the minimum guaranteed outgoing rate configured at LSs along the path.

The statistical traffic control for multiple flows (described in section 5.3) is practical here. The S monitors the arrival rate of requests for delivery and dynamically updates the appropriate generation rate of the shared flows to meet the upper limit of the transmitting rate on the LSP, which is configured for EF-PHB.

In the control plane, an NC (Network Controller system) controls the policy of utilizing the network, in particular, it constructs a static multicast tree and determines the allocation of network resources. The TM (Traffic Manager system) maintains a topology database with QOS parameters and performs admission control and resource reservation for links and nodes. The AM (Address Manager system) manages multicast addresses within local scope. The S obtains multicast addresses from the AM and forwards them to the R (Receiver system) before sending any segments of content. The PC (Path Controller system) distributes label information to the LSs along the multicast tree to establish the LSP.

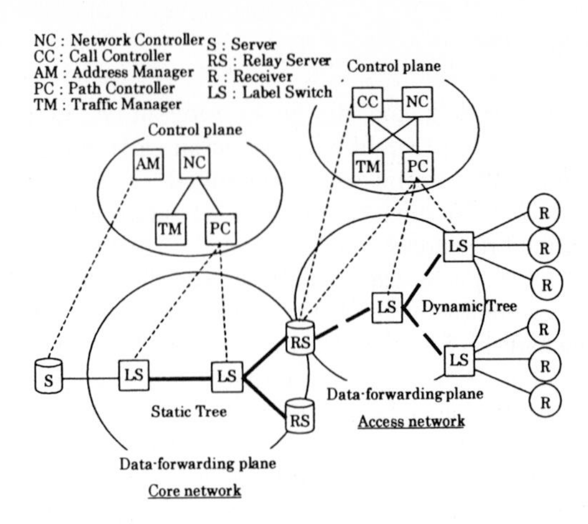

Fig. 3. Network model

6.3.2 Access Network

An FEC is assigned per micro flow, be it a shared flow, an individual flow or a flow for the differential segments (for the delivery of diverse segments), and an LSP is dynamically set up by signals of the control plane whenever there is a micro flow. To ensure the QOS, every LS grasps the utility of its links (available bandwidth) and perform policing and shaping per micro flow. The CC (Call Control system) is deployed in the control plane to process signaling functions that establish and release LSP within the access network.

For the delivery of diverse segments, the RS does not send a differential segment until a corresponding multicast tree is completely built. Therefore, the CC coordinates the RS sending differential segments and the PC setting up LSPs .

To consistently guarantee the QOS in delivering the whole content data, the bandwidth of the links along the tree for all multicast groups to which one R is supposed to join must be secured beforehand. Therefore, the TM maintains a resource-management database on a time axis, and admission control and resource reservation are made in a span of time rather than at a point of time.

6.4 Protocol Design

A basic sequence between end systems is shown in Figure 4. First, the R knows presence of content and obtains the location of the S from a WS (Web Server). The R requests the S to provide some basic information about the content (media type, etc). Next, the R requests the S to set up transport functions, and the S selects sequence patterns of segments (i.e. user group) based on the user's specification and maps a multicast address onto each segment and determines if the content will be sent by a shared flow only or by both a shared and an individual flow. Then, the S sends these sets of information to the RS in a request to setup a relay point. The RS requests the CC to perform admission control and the RS replies to the S if the task has been done. The S return the response for the request of setting up transport functions to the R. Next, the R requests the S to start transmitting data, and the S requests the same to the RS, and the RS requests the CC to reserve resources and set paths, then, the CC replies to the RS and the RS replies to the S. In case that the content is sent by a shared flow only, the S return the response for the request of starting transmitting data to the R and it transmits the streaming data by multicasting. In case that the content is sent by both a shared flow and an individual flow, the RS return the response for the request of starting transmitting data to the R and it transmits the streaming data by unicasting.

A basic sequence among network node systems is shown in Figure 5. Having received a request to set up a relay point, the RS requests the CC to establish a call, and the CC requests the TM to compute the path route and perform admission control. Having received a request to start transmitting data, the RS notifies the CC, and the CC requests the TM to reserve resources and the CC provides the PC with path information computed by the TM. the PC requests LSs along the path to establish an LSP.

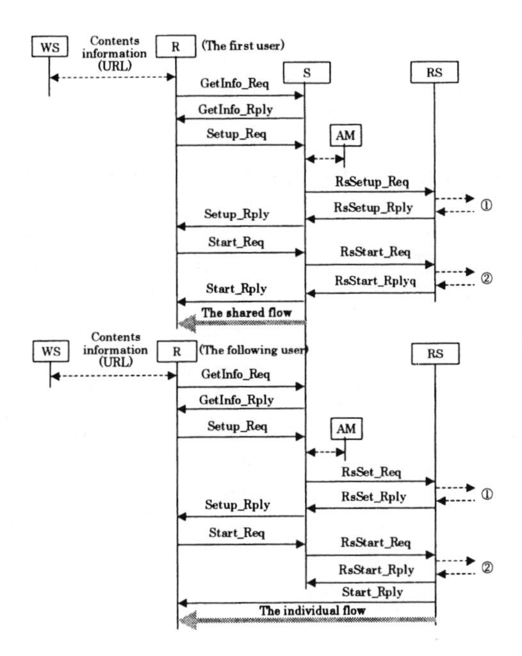

Fig. 4. Sequence between end systems

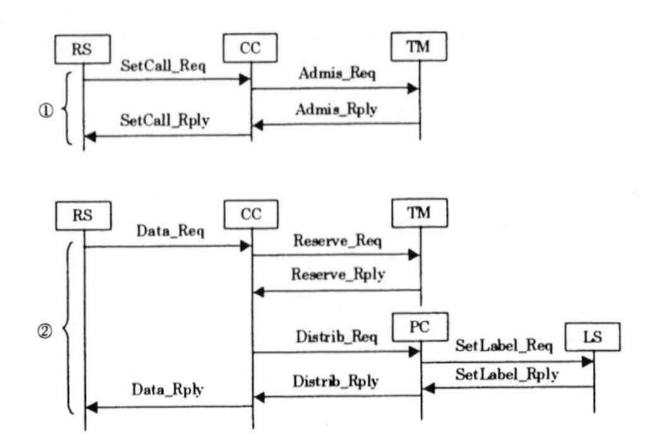

Fig. 5. Sequence among network node systems

Table 1. Messages and parameters

Messages (meaning)	Parameters
GetInfo_Req (Request program information)	URL
GetInfo_Rply (Reply program information)	Program descriptor, Media descriptor, IOD (Initial Object Descriptor)
Setup_Req (Request to setup transport)	Transport protocol, Unicast address, Port number, User descriptor, System descriptor
Setup_Rply (Reply to setup transport)	Scenario descriptor (Multicast address, Port number, User-group, QOS class), Session identifier
RsSetup_Req (Request to setup RS)	Scenario descriptor (Multicast address, Port number, User-group, QOS class), Flow type, Session identifier
RsSetup_Rply (Reply to setup RS)	Session identifier
Start_Req (Request to start)	Session identifier, Period
Start_Rply (Reply to start)	Session identifier, The number of segments between the two flows
RsStart_Req (Request to start RS)	Session identifier, Period
RsStart_Rply (Reply to start RS)	Session identifier
SetCall_Req (Request to establish call)	Session identifier, Time-scaled traffic descriptor
SetCall_Rply (Reply to establish call)	Session identifier, Call identifier
Data_Req (Request to start sending data)	Call identifier
Data_Rply (Reply to start sending data)	Session identifier
Admis_Req (Request admission control)	Call identifier, Time-scaled traffic descriptor
Admis_Rply (Reply admission control)	Call identifier, Reservation Identifier
Reserve_Req (Request to reserve resouces)	Reservation identifier
Reserve_Rply (Reply to reserve resouces)	Call identifier, Path-node list descriptor
Distrib_Req (Request Distribution)	Call identifier, Distribution sequence number, Path node list descriptor
Distrib_Rply (Reply Distribution)	Call identifier, Distribution sequence number
SetLabel_Req (Request to set labels)	Node identifier, Flow identifier, Forwarding cache entry, label
SetLabel_Rply (Reply to set labels)	Node identifier, Flow identifier

6.5 Receiver System Model Design

The receiver system model is based on a standard decoder model specified in MPEG-4. Each ES (Elementary Stream) for media objects and the scene descriptor corresponds to one RTP (Real-time Transport Protocol) session [15]. Functions higher than DMIF (Delivery Multimedia Integration Framework)[16] are the same as in the standard model.

The received ESs in an individual flow are decoded and played back as soon as they arrive. The ESs in a shared flow are buffered until all the ESs in an individual flow are completely played back. A coordinate function called 'asynchronous multicast control' is installed between the RTP and the DMIF. The time stamp used for MPEG-4 is calculated from the time stamp in an RTP packet and the number of segments between the two flows, which is a parameter included in the message 'Start_Rply' from the RS to the R.

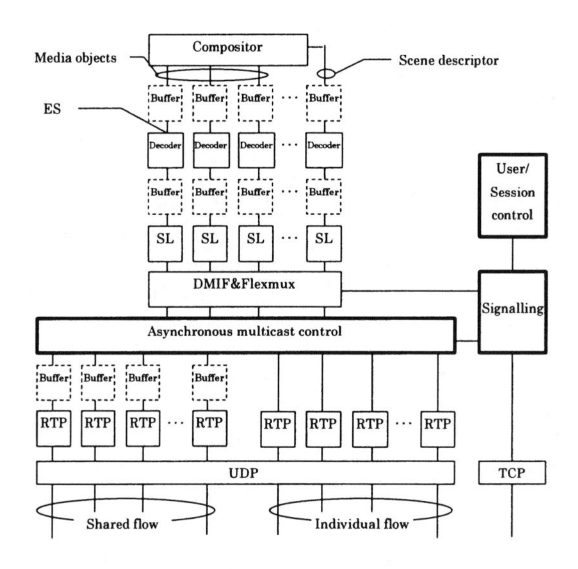

Fig. 6. Receiver system model

6.6 Numerical Analysis

We calculated the traffic intensity in the core network and the access networks implemented with an algorithm for statistical traffic control, and evaluated our network design. We show the effectiveness of retrenching the traffic intensity at a trunk link in the access networks and at trunk and branch links in the core network by comparing the traffic intensity in our model with that in unicast delivery. Although the length of individual flows is expressed as the number of segments (or sub-segments), it is assumed that a segment is very small and the length of individual flows can be flexibly changed. The number of access networks attached to the core network is expressed as m, the average rate of requests for delivery in one access-network is expressed as λ.

In the access network, both shared flows and individual flows are transmitted. When $\tau = \sqrt{\lambda / 2h}$, the traffic intensity is the smallest. Substituting this into eq.(2), we get the traffic intensity as,

$$\rho = \sqrt{2\lambda h} - 1/2 \quad \text{[erl]} \tag{5}$$

In unicast delivery, the traffic intensity at the trunk link in the access network is λh. Therefore, the ratio of traffic intensity in our multicast technique to that in unicasting, R_{at}, is

$$R_{at} = \frac{\sqrt{2\lambda h} - 1/2}{\lambda h} \tag{6}$$

In the core network, only the shared flows are transmitted, and the traffic intensity at trunk and branch links is the same. When $\tau = \sqrt{\lambda / 2h}$, the delivery is the most efficient, so the traffic intensity is

$$\rho = m\tau h = m\sqrt{\frac{\lambda h}{2}} \quad \text{[erl]} \tag{7}$$

In unicast delivery, the traffic intensity at the trunk link is $m\lambda h$, and the traffic intensity at the branch link is λh. Therefore, the ratio of traffic intensity in our multicast technique to that in unicasting at the trunk link, R_{ct}, and at the branch link, R_{cb}, are

$$R_{ct} = \frac{m\sqrt{\lambda h / 2}}{m\lambda h} = \frac{1}{\sqrt{2\lambda h}} \tag{8}$$

$$R_{cb} = \frac{m\sqrt{\lambda h / 2}}{\lambda h} = \frac{m}{\sqrt{2\lambda h}} \tag{9}$$

Figure 7 shows the ratio of traffic intensity in our multicast technique to that in unicasting in the core network and the access network. When m is 4 and the requests are few (λh is below 8), the effect of retrenching the traffic is the opposite at the branch link in the core network. But otherwise, the effect of retrenching the traffic increases as the rate of requests gets higher. When λh is 100, the ratio is only 7% at the trunk link in the core network and 13% in the access network.

7 Conclusions

In this paper, we described PMSD as a new broadcast content delivery service. Focusing on the scheme for delivering the content over the Internet, we developed multicast techniques and proposed an algorithm for statistical traffic control based on stream-based asynchronous multicasting. We also developed a technique for the delivery of diverse media segments/objects based on layered multicasting. Both techniques enable effective distribution of diverse content with guaranteed QOS. To implement both techniques, we designed concentrated control based network models and protocols by using MPLS technology and proved the implementability of the techniques. Finally, we presented the effect of retrenching the traffic intensity on the designed network models by using numerical analysis.

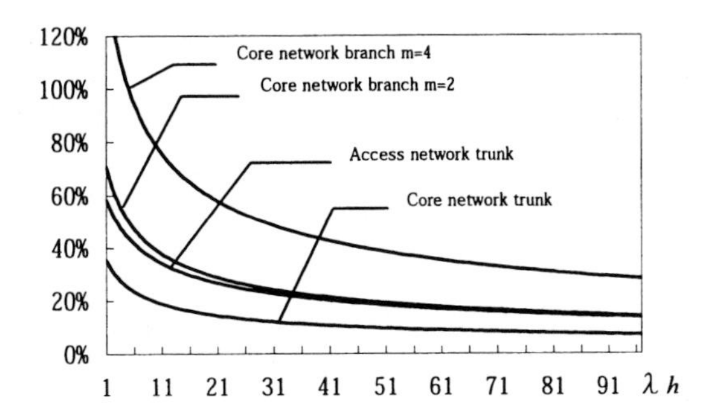

Fig. 7. Ratio of multicast traffic intensity in proposed technique to unicast traffic intensity

In our future studies of schemes to deliver video/audio content over the Internet, we will explore the possibility of developing a scheme of resource management on the time-axis. We will also explore the possibility of developing a method of profiling to reflect users' interests and lifestyles, and a method of gathering individual information without encountering any legal problems such as copyright infringement.

References

1. Peggy Miles: Internet World Guide to Webcasting, John Wiley & Son, 1998
2. S.Nakagwa, M.Katsumoto: The future of digital media by IP communication, IPSJ Magazine VOL.41 No.12, Dec.2000
3. K.Sato, M.Katsumoto: A proposal for personalized media stream delivery, ISPJ Report, DPS103-14, Jun.2001
4. H.Woo, C.K.Kim: Multicast scheduling for VOD services, Multimedia Tools and Applications 2(2) pp157-171, Mar.1996
5. H.Kalva, B.Fuhrt: Techniques for improving the capacity of video-on-demand systems, Proc.29th Annual Hawaii International Conference on System Sciences, pp308-315,Wailea, HI,USA,IEEE Computer Society Press, Jan.1996
6. S.W.Cater, D.E.Long: Improving Video-on-demand Server Efficiency Through Streaming Tapping, Proc.the International conference on Computer Communication and Networks,pp200-207 Las Vegas, Sep.1997
7. S.Uno, H.Tode, K.Murakami: Performance evaluations on video distribution system with multicast and burst transmission, IEICE Technical Report, IN99-82, Nov.1999
8. S.McCanne, V.Jacobson, M.Vetterli: Receiver-driven Layered Multicast, Proceedings of ACM Sigcomm, 1996
9. T.Suzuki, T.mimura, et al: Development of middleware for multi QOS over IP network, Proc.2000, TAO Jun.2000
10. S.Blake, D.Black, et al: An Architecture for Differentiated Services, RFC 2475, 1998

11. E.Rosen, A.Viswanathan, R.Callon: Multiprotocol Label Switching Architecture, RFC3031, 2001
12. Francois Le Faucheur, Liwen Wu,et al: MPLS Support of Differentiated Services, IETF Internet Draft, Feb.2001
13. V.Jacobson, K.Nichols, et al: An Expedited Forwarding PHB, RFC2598, 1999
14. D.Ooms, B.Sales, et al, Framework for IP Multicast in MPLS, IETF Internet Draft, Jan.2001
15. H.Schulzrinne, S.Cater, et al: A Transport Protocol for Real-Time Applications, RFC1889, 1996
16. Delivery Multimedia Integration Framework, ISO/IEC FDIS 14496-6
17. H.Schulzrinne, A.Rao, et al: Real Time Streaming Protocol, RFC2326, 1998
18. D.Singer, Y Lim, et al: A Framework for the delivery of MPEG-4 over IP-based Protocols, IETF Internet Draft, Nov 2000

A New Delivery Scheme for 1-to-N Multicast Applications

Juyoung Park[1], Seok Joo Koh[1], Eunsook Kim[1], Shin-Gak Kang[1], and Dae Young Kim[2]

[1] Electronics and Telecommunications Research Institute,
161 Kajeong-Dong, Yuseong-Gu, Deajeon, KOREA
{eunah, jypark, sjkoh, sgkang}@etri.re.kr

[2] Chungnam National University,
200 Kung-Dong, Yusong-Gu, Daejeon, KOREA
dykim@cnu.ac.kr

Abstract. We propose a new delivery scheme for 1-to-N multicast applications such as webcasting service used for the web-based broadcasting of contents streams in Internet. The proposed scheme is based on the unicast transport from a remote sender to a subnet host and the subsequent multicast transmissions to the other receiving hosts within the subnet. The main design goal of this scheme is to improve transmission efficiency over the existing unicast transports, without help of multicast-capable networks. The proposed scheme has been tested and compared with the unicast-only transports in terms of the amount of traffic generated in the subnetworks. From experimental results, we see that the proposed scheme provides a nearly same performance as IP multicasting does in the subnetwork environments. Recognizing that IP multicast seems to be a little more delayed, it is expected that the proposed scheme can be used as an alternative short-term solution for one-to-many multicast services over the unicast networks.

1. Introduction

Webcasting, as known as Internet broadcasting, has recently been focused as a multicast killer application service. Webcasting literally means the web-based distribution of multimedia contents to Internet users. Webcasting is expected to realize a variety of the commercial multimedia services such as Internet TV and movies, remote education, and stock tickers [1].

Recognizing that the webcasting service can be viewed as one-to-many multicast application service for numerous users, it is reasonable to use IP multicasting, rather than the replicated unicast transport to each of the webcasting users. Nevertheless, IP multicasting has not yet been widely deployed in the public Internet [2]. Actually, there still exist a lot of issues to be addressed for rapid deployment of IP multicasting, as pointed out by Christope Diot, et al. in [3]. In reality, most of the current webcasting services are being provided over the unicast networks. The recently

I. Chong (Ed.): ICOIN 2002, LNCS 2343, pp. 109-118, 2002.

focused Contents Delivery Networks (CDN) service is also dependent on unicast transports in the networks [4].

In this paper, we proposed a new delivery scheme for one-to-many multicast applications. The proposed scheme is based on unicast transport from a remote sender (contents provider) to a subnet host (client) and the subsequent multicast transmissions to the other receiving hosts located in the subnet. The proposed scheme is designed to exploit the trivial multicast capability in a subnet environment which is called subnet multicast [5].

The main design goal of the scheme is to improve efficiency of the replicated unicast transports for webcasting services without modification of the current network infrastructure. The study of the scheme is also motivated from recognition that a large number of Internet users are usually located in the local area network (LAN) environments, as shown in the examples of the private enterprise networks and Digital Subscriber Lines access networks, in which the subnet multicast can easily be achieved.

This paper is organized as follows. In Section 2, the webcasting service is briefly presented as an example of one-to-many multicast application services. Section 3 describes the proposed scheme for the unicast transports with subnet multicast. In Section 4, more detailed operations for subnet multicast transport are presented along with the associated control messages. Section 5 discusses the analytical and experimental results that have been performed for comparison of the proposed scheme with the unicast-only transports. In Section 6, we conclude this paper.

2. Webcasting Services

Webcasting is a typical example of one-to-many multicast services and can be considered as a multicast killer application service. The webcasting represents the web-based distribution of streaming multimedia contents to Internet users. The webcasting system established in contents providers roughly consists of the various servers and storage equipments. Each client requests transmission of streaming contents to the sender via the web server. The requested contents are delivered over the network.

Figure 1 illustrates an example webcasting system that has widely been deployed. The contents provider generates live or video on demand (VoD) data streams by encoding raw audio/video materials. Some of those contents may be recorded into the storage devices. Each client or user contacts a web server located at the contents provider so as to request contents that it wishes to receive. The web server guides the user onto a suitable media server. Then a connection is established between user and media server to deliver the contents stream. The streaming data will be transmitted to the user by unicast or multicast transport in Internet.

Most of the current webcasting systems use the replicated unicast transports for contents delivery to numerous users. This incurs severe traffic overload at the webcasting (processing) system as well as at the network access link. This inefficiency gets severe, as the number of simultaneous access users increases. In this paper, we discuss a simple and realistic delivery scheme so as to improve efficiency of the current unicast-only transports.

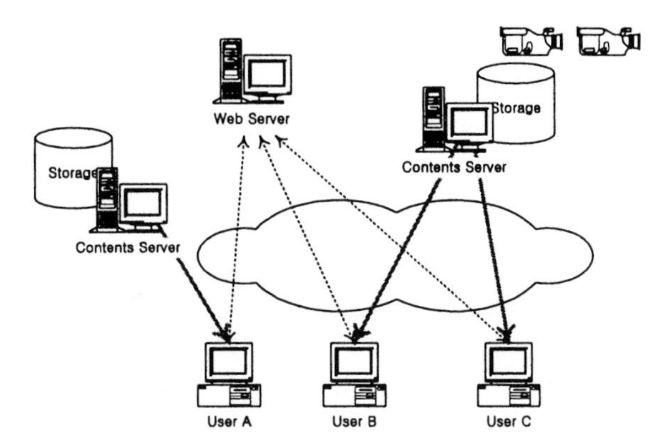

Fig. 1. Webcasting system

3. Unicast Transport with Subnet Multicast

In this section, we describe the scheme for the unicast with subnet multicast transport, and also present the associated operational procedures for the webcasting service.

Differently from the unicast-only transport, the proposed scheme exploits the multicast transmissions in a subnet. To do this, a new entity called 'feeder' is introduced. One of the receivers located in the subnet is dynamically configured as the feeder for the associated webcasting session. The feeder is a receiver and also plays a role of relaying the webcasting contents received from the remote sender to the other receivers in the subnet.

Figure 2 illustrates the unicast-only transport and the proposed delivery scheme. In the unicast-only transport, as shown in Figure 2(a), the sender has to send a data stream to each of the receivers by establishing multiple (replicated) unicast connections. In the proposed scheme, as depicted in Figure 2(b), only a single feeder receives the data stream from the remote sender, while the other receivers in the subnet receive the same data stream from the feeder, not the remote sender.

A new receiver that wishes to join a webcasting session first checks whether or not there is a feeder in the subnet. The first-joining receiver becomes the feeder for the webcasting session. The feeder establishes a unicast connection to the remote sender. The other late-joining receiver, after it realizes that there is a feeder in the subnet, will receive the streaming data from the feeder in the same subnet over the subnet multicast channel. The detailed operations related to the feeder configuration will be described in the next section.

Figure 3 compares the connection setup procedures for webcasting. In the unicast-only transport, as shown in Figure 3(a), each client downloads the requested contents from the media server, just after it gets information on the stream data via web access. In the unicast transport with subnet multicast, as illustrated in Figure 3(b), the first-arriving client becomes a feeder in the subnet by way of the feeder configuration process, and then establishes a unicast connection to the sender. The feeder relays the data stream to the other late-joining clients in the subnet, if any.

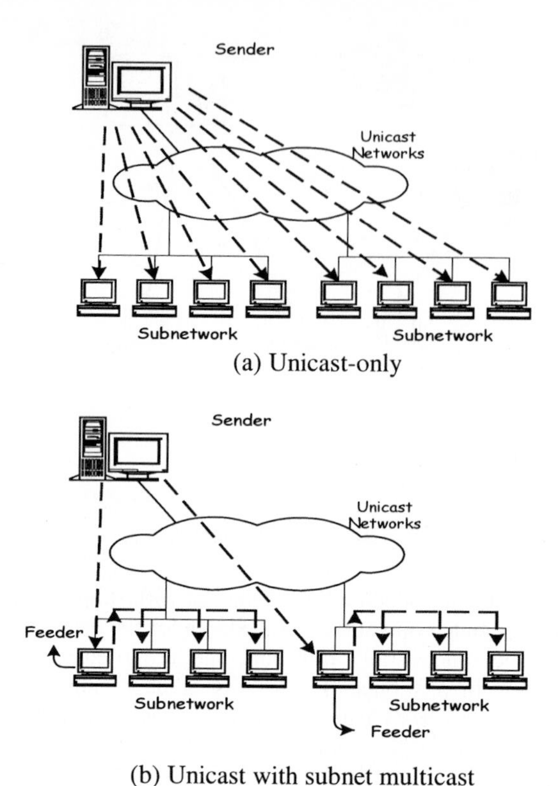

(a) Unicast-only

(b) Unicast with subnet multicast

Fig. 2. Transport schemes based on unicast-only and unicast with subnet multicast

4. Subnet Multicast Transport

In the proposed scheme, each new receiver checks first checks whether or not there is a feeder in the subnet, and the first-arriving receiver becomes a feeder in the subnet for the webcasting session. After that, the feeder forwards the data streams received from the remote sender toward the other receivers in the same subnet, if any. To do this, a feeder needs to be configured and announced to the other receivers in the subnet. The feeder reconfiguration is also required against the feeder release in the event that the existing feeder leaves the webcasting session.

For configuration and maintenance of a feeder, three kinds of control messages are employed: Feeder Solicitation (FS), Feeder Announcement (FA) and Feeder Release (FR). Those messages are summarized in Table 1.

All the control messages are delivered within the subnetwork via a well-known multicast address such as 224.0.0.x/24 or 224.0.1.x/24. Note that those addresses are reserved by IANA [5] for the multicast delivery of control messages in the subnetwork.

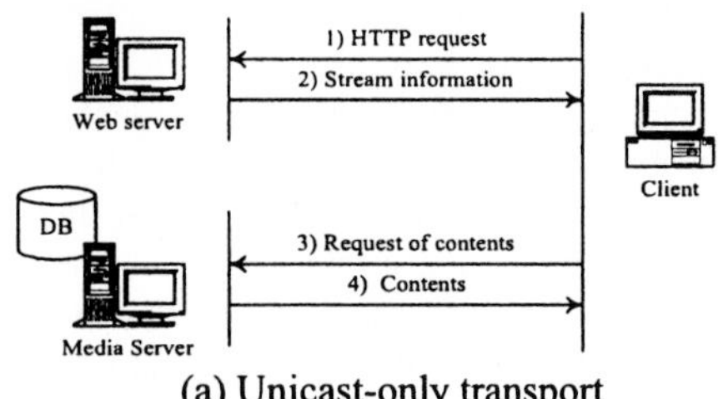

(a) Unicast-only transport

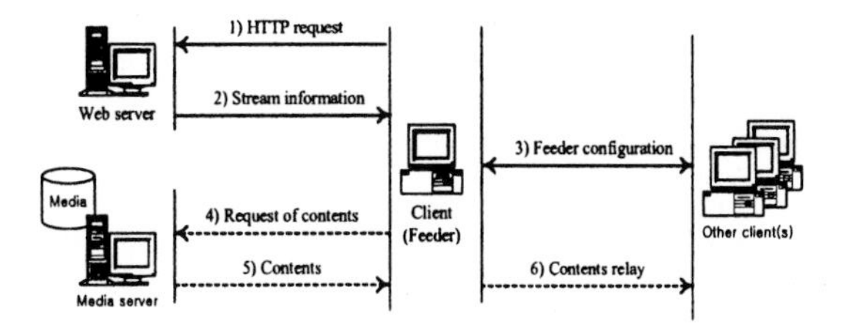

(b) Unicast transport with subnet multicast

Fig. 3. Connection setup procedures for webcasting

The data stream will be forwarded from the feeder to the other receivers by subnet multicast. For this purpose, a multicast (IP class D) address is required. For the scoping of the multicast traffic to the subnet, it is recommended to set the TTL (time to live) of the corresponding IP packets to '1'.

Table 1. Control Messages used for Subnet Multicast Transport

Type	Direction	Contains
FS	Receiver → 224.0.0.x	Sender's address
FA	Feeder → 224.0.0.x	Sender's address
FR	Feeder → 224.0.0.x	Feeder's address Data forwarding address

4.1. Feeder Configuration

Before establishing a unicast connection to the remote sender, each new receiver sends an FS message to the subnet, and checks whether or not there exists a feeder for the webcasting session in the subnet, as shown in Figure 4. The FS message includes information on the IP address of the remote sender so as to indicate a webcasting session (see Table 1). If there is no response for the pre-configured FS time, (i.e., there is no feeder in the subnet), then the receiver becomes a feeder for the webcasting session. Otherwise, the receiver must receive an FA message from the existing feeder already configured in the subnet. In the example of Figure 4, the host F becomes a feeder after the FS timer expires. The feeder then connects to the remote sender and receives the data stream over the unicast connection.

If another receiver (host A in the figure) joins the webcast session by sending an FS message, then the feeder (host F) responds with an FA message to the host A. The FA message includes IP addresses for the sender, feeder and multicast data forwarding, as shown in Table 1. The feeder now begins to forward the data stream to the subnet. The other late-joining receivers such as host C will also receive the data stream from the feeder, after exchanging an FS message with the corresponding FA message. In case that there is no other receiver, the feeder does not need to forward the data stream.

4.2. Multicast Data Forwarding

If one or more other receivers are detected in the subnet, then the feeder begins the multicast forwarding of the data streams to the subnet. The destination address of the forwarded IP data packets will be set to the multicast address, and TTL (time to live) is set to '1' for restricting propagation to the subnet only.

During the data forwarding, the feeder continues to check whether there still exist the receivers in the subnet. If not, the feeder will not forward the data streams any more. For this purpose, each receiver sends periodic FS messages every FS time. In response to FS messages, the feeder sends the corresponding FA messages to the subnet receivers. If no FS message has arrived for some time, the feeder stops multicast forwarding of the data streams. Note that if a receiver has already sent an FS message, the other receivers in the subnet may cancel their FS messages so as to avoid implosion of the FS messages.

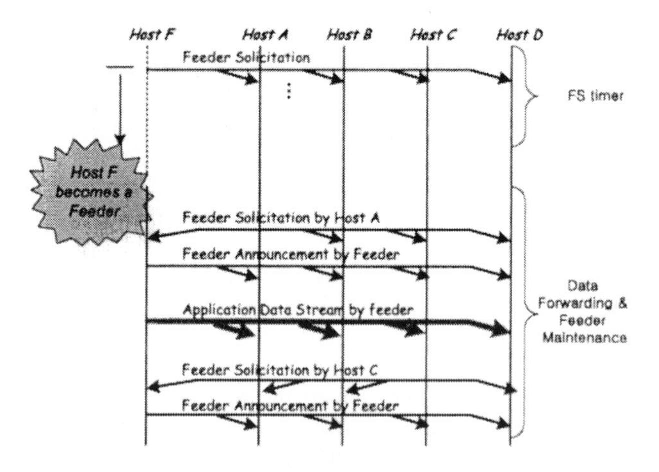

Fig. 4. Feeder Configuration and Data Forwarding

4.3. Feeder Reconfiguration

If the current feeder releases the webcasting connection to the remote sender, then a new feeder must be selected among the receivers in the subnet. To do this, the current feeder first sends an FR message to the receivers in the subnet. When an FR message arrives from the feeder, each receiver activates the 'FR timer'. One of the receivers will send the FA message, after its FR timer expires. The receiver that has first responded with an FA message will be configured as a new feeder.

The new feeder then established a unicast connection to the remote sender. Once a new feeder is configured, the other receivers cancel the FR timers. They now receive the data stream from the new feeder. Figure 5 depicts the feeder reconfiguration. In the figure, the host A is configured as a new feeder.

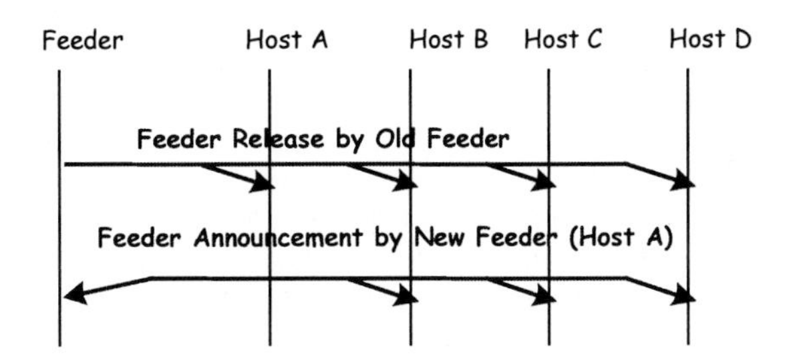

Fig. 5. Reconfiguration of a New Feeder

5. Numerical Results

Table 2 compares the proposed scheme with the unicast and multicast transports in terms of the amount of traffic generated at the sender and receiver sides. In the table, R and S represent the total number of the receivers and the number of the subnets participating in the webcasting session.

Table 2. Comparison of the schemes in terms of the traffic amount generated

	At Sender	In a Subnet
Unicast	$\propto$ R	$\propto$ R
Multicast	= 1	= 1
Proposed	$\propto$ S	$\leq$ 2

In the unicast-only transport, the traffic amount is proportional to the number of receivers at the sender and receiver sides. In IP multicasting, the traffic amount is fixed to one unit, regardless of the number of receivers. In the proposed scheme, the traffic amount at the sender depends on the number of subnets including the receivers.

The traffic amount generated in the subnet is reduced to one (for the subnet with only the feeder) or two units (for the subnet with the other receivers as well as the feeder).

From the table, we see that the proposed scheme provides a nearly same performance as the conventional IP multicasting in terms of traffic amount generated. In particular, the proposed scheme does not require any modification of the network infrastructure, and thus it can easily be deployed in the network. On the other hand, the IP multicasting requires the multicast-capable routers and multicasting routing protocols together with a reasonable control of multicast traffic in the network.

We implemented the proposed scheme on top of FreeBSD 4.1 Unix machines [6] in the form of a user-level library. The corresponding Application Programming Interface (API) was based on the Berkley socket Interfaces in the form of the wrapping function, which was employed by the each host receivers in the subnet. The API functional modules include the processing of the FS, FA, and FR messages, and the forwarding and reception of application stream over multicast data address.

The proposed scheme is compared with the unicast-only scheme. To do this, two subnetworks are configured, as shown in Figure 6. The number of receivers in a subnet is increased from one to nine. In the figure, the sender generates a random traffic stream. In Subnet A, after exchanging an FS with the corresponding FR message over a multicast address (224.0.0.18), a feeder is configured and it then begins to forward the data streams by subnet multicast. In Subnet B, the data stream is transferred over the unicast connections between the sender and all the receivers.

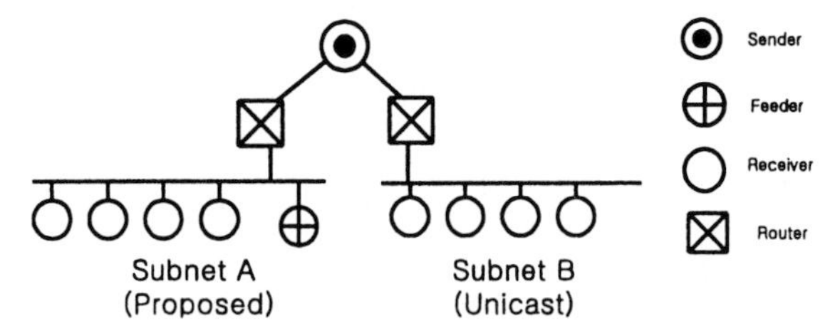

Fig. 6. Test Networks

Figure 7 shows the amount of traffic generated in the subnetworks with four receivers over a given time interval. On the average, the proposed scheme provides three or four times bandwidth utilization gains over the unicast-only scheme. Note that the sender temporarily stops transmitting data at the time of 40 and 50 seconds.

In Figure 8, the required network bandwidths for those two schemes are plotted for the different number of the receivers. For each test subnetwork, the number of receivers is increased by one, every 25 second, to nine. In the figure, we note that the proposed scheme requires a relatively fixed amount of bandwidth independently of the number of the receivers located in the subnetwork. On the other hand, the required network bandwidth in the unicast-only scheme increases linearly to the number of the receivers.

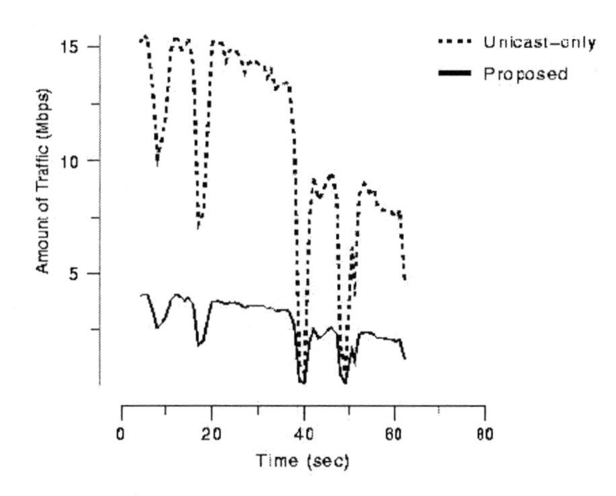

Fig. 7. Comparison in terms of traffic generated in the subnetwork

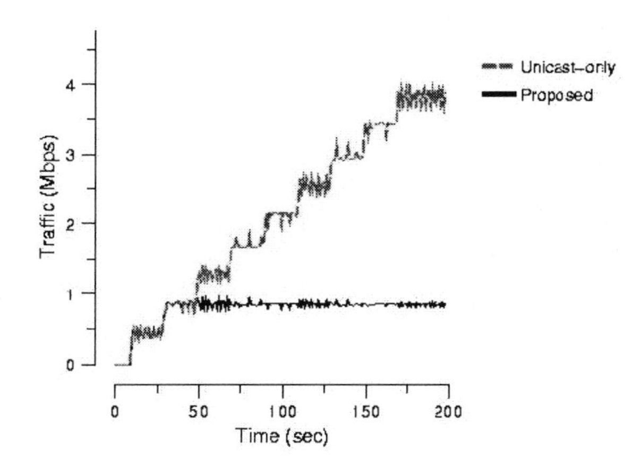

Fig. 8. Comparison of the required bandwidth as the number of receivers increases

Figure 9 plots the traffic generated by feeder(s), when the feeder reconfiguration occurs. In the figure, the feeder reconfigurations occurs at the time 50, 100, and 150 ms. In case of the first reconfiguration at the time 50 ms, the connection setup time between a new feeder and the sender elapses larger than the feeder reconfiguration time interval. At the time 100 ms, the connection setup time interval is smaller than the feeder reconfiguration interval. On the other hand, for the third reconfiguration, the connection setup time occurs exactly after the feeder reconfiguration, and thus neither data losses nor duplications are made.

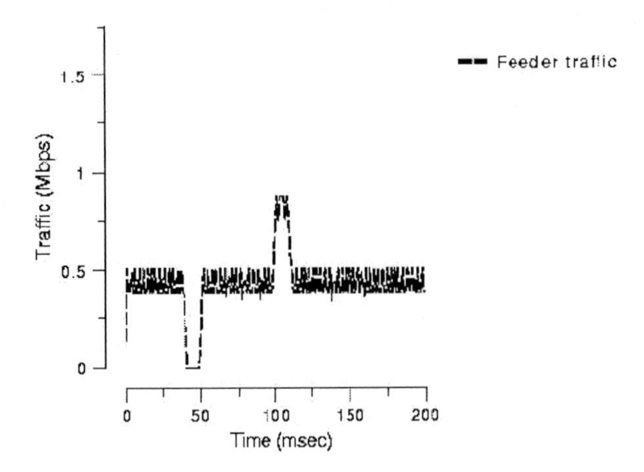

Fig. 9. Traffic discontinuity during feeder reconfigurations

6. Conclusions

Until now, we have discussed a new delivery scheme for webcasting, which is based on the unicast transport with subnet multicast. The proposed scheme has been tested and compared with the existing unicast-only transport in terms of the amount of traffic generated in the subnetworks. From the numerical and experimental results, we have seen that the proposed scheme provides a nearly same performance as the IP multicasting in the subnetwork environments.

It is noted that the proposed scheme does not require any change of the underlying network infrastructure for IP multicasting and thus can easily be deployed in the network. Recognizing that the IP multicasting still have much concern for the wide deployment, it is expected that the proposed scheme is used as an alternative short-term solution for one-to-many multicast application services.

References

1. Content Alliance, Available from http://www.content-peering.org/, 2001.
2. S. Koh, et al., "Minimizing Cost and Delay in Shared Multicast Trees", ETRI Journal, Vol. 22, No. 1, pp.30 - 37, March 2000.
3. C. Diot, et al., “Deployment Issues for the IP Multicast Service and Architecture”, IEEE Networks Magazine’s Special Issue on Multicast, January 2000.
4. J. Park, et al., "Multicast Delivery Based on Unicast and Subnet Multicast", IEEE Communications Letters, Vol. 5, No. 4, pp. 181 - 183, April 2001
5. IETF RFC 1700, Assigned Numbers, 1994.
6. W. R. Stevens, Unix Programming, 2nd Ed., Addison-Wesley, 1998.

Multicast Routing Debugger (MRD) – A System to Monitor the Status of Multicast Network

Jaeyoung Lee[1], Heonkyu Park[2], and Woohyung Choi[2]

[1] SK Teletech, 729-1, Bongchun 1-dong, Kwanak-gu, Seoul, Korea, 151-709
jy@skteletech.co.kr
[2] Dept. of EECS, Korea Advanced Institute of Science and Technology
GuSeong-Dong, YuSeong-Gu, Daejon, Republic of Korea, 305-701
{hkpark, choi}@cosmos.kaist.ac.kr

Abstract. IP Multicast can efficiently provide enormous bandwidth savings by enabling sources to send a single copy of a message to multiple recipients who explicitly want to receive the information. But due to the complexity of IP multicast and its fundamental differences from unicast, there are not very many tools available for monitoring and debugging multicast networks, and only a few experts understand the tools that do exist. This paper proposes a Multicast Routing Debugger (MRD) system that monitor the status of a multicast network. This system is aimed to multicast-related faults detection. In this paper, first, we define the set of information that should be monitored. Second, the method is developed to take out such information from multicast routers. Third, MRD system is prototyped to collect, process information from heterogeneous routers on a multicast network and to display the various status of the network comprehensively. The prototype of MRD system is implemented and deployed. We perform experiments with several scenarios. Experimental results show we can detect various problems as information that we define is monitored. The MRD system is simple to use, web-based and intuitive tool that can monitor the status of a specific multicast network.

1 Introduction

The exponential growth of the Internet combined with multimedia content is increasing the average size of the data transfers and pushing bandwidth constraints to their limits. Unfortunately multicasting has been slow to catch on, except for the research-oriented Multicast Backbone [1][2]. There are a variety of reasons for the difficulties in widespread multicast deployment [12]. One of largest current barriers is the difficulty in managing multicast traffic [13]. Due to the complexity of IP multicast and its fundamental differences from unicast, there are not many tools available to assist in monitoring the various status such as member-subnets, connectivity, statistics and forwarding state in a multicast network in order to diagnose above problems. In this paper, Multicast Routing Debugger (MRD) system is introduced.
The primary goal was to create a simple to use, web-based and intuitive tool that can monitor the status of a specific multicast network. The system would be able to

I. Chong (Ed.): ICOIN 2002, LNCS 2343, pp. 119-126, 2002.

collect information from heterogeneous multicast routers and process it online to give a current snapshot of the multicast network. The MRD system was implemented and deployed in the Asia-Pacific Advanced Network-Korea (APAN-KR)[3] research network.

2 Related Work

2.1 SNMP-Based Tool

SNMP-based tools provide the information to be queried to SNMP-enabled router and visualize it. This information should be defined in MIB of the router. Mstat[9] allows an SNMP-enabled router to be queried for information, including routing tables and packet statistics. Mtree[10] attempts to use cascading SNMP-enabled router queries to determine an entire multicast tree rooted at a given router. Mview[11] is a tool for visualization Mbone topology and monitoring and collecting performance information. SNMP-based multicast tools have limited value outside a particular administrative domain, because most multicast routers are not configured to respond to public SNMP queries. As a result, mstat, mtree, and mview are only useful for debugging and managing the local domain portion of a multicast group.

2.2 Multicast Route Tracing Tool

Mtrace[4] is a tool designed to provide hop-by-hop path information for a specific source and destination. For a specific group, mtrace will tell a user hop-by-hop packet loss, multicast path and round trip information. The utility of mtrace is often limited by the multicast topology. Where multicast and unicast topologies are not aligned (as is the case in many multicast-enabled networks) mtrace may not function. The mtrace provides hop-by-hop path information between a source and a receiver, so several mtraces are issued in order to know the network status.

2.3 Multicast Reachability Monitor (MRM)

The MRM[5] is a network fault detection and isolation mechanism for administering a multicast routing infrastructure. An MRM based fault monitoring system consists of two types of components: an MRM manager that configures tests, collects and presents fault information, and MRM testers that source or sink test traffic. For monitoring the network status, MRM has architecture involved by several routers and hosts. But the packet loss information provided by MRM Test Receivers lacks for more detailed fault detections.

3 Model

Figure 1 illustrates overview of the proposed MRD system. A multicast router obtains the routes information from its neighbor routers using the routing protocol and updates its routing table. And it obtains the group membership information from its local hosts using IGMP or from neighbor routers by *join/prune* messages. It locates in a regional network, local network or subnetwork i.e., in all levels of networks. A collector gathers the multicast status information from one or more routers and locates in one domain for reflecting the status of each domain in time. A storer stores the status information gathering from one or more collector in a database and locates in one or more regional networks. A viewer retrieves the status information from the database system stored by the storer and displays it to users.

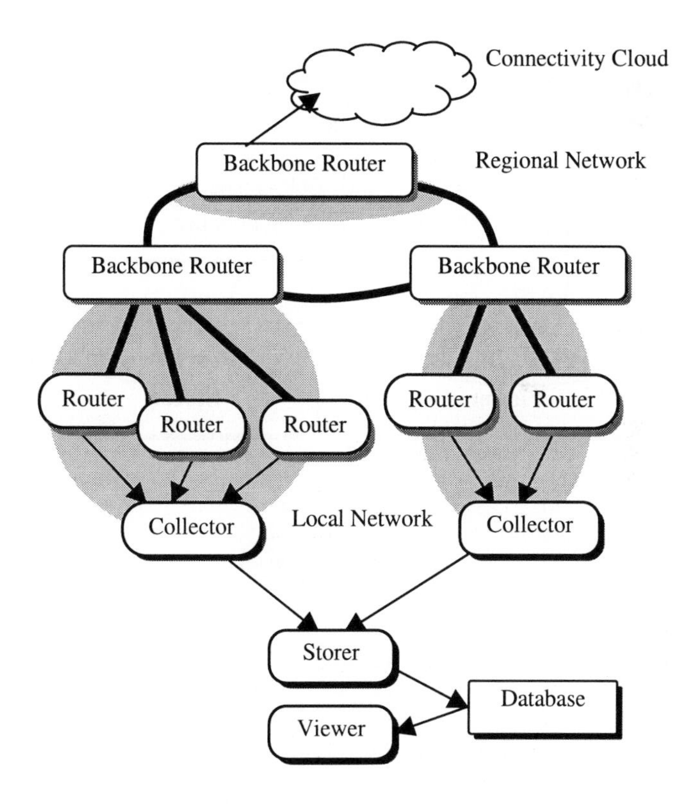

Fig. 1. System Overview

The system, as shown in Figure 2 is divided into four components based on functionality. We describe four parts in more detail:

1. Status report module accepts request and provides the status information obtained from a routing table and a forwarding table.
2. Collect module does a request for the status report to one or more routers and gathers data from them.
3. Store module stores the status information collected by one or more collectors in a database.
4. View module displays the multicast status based on the contents of the database on demand.

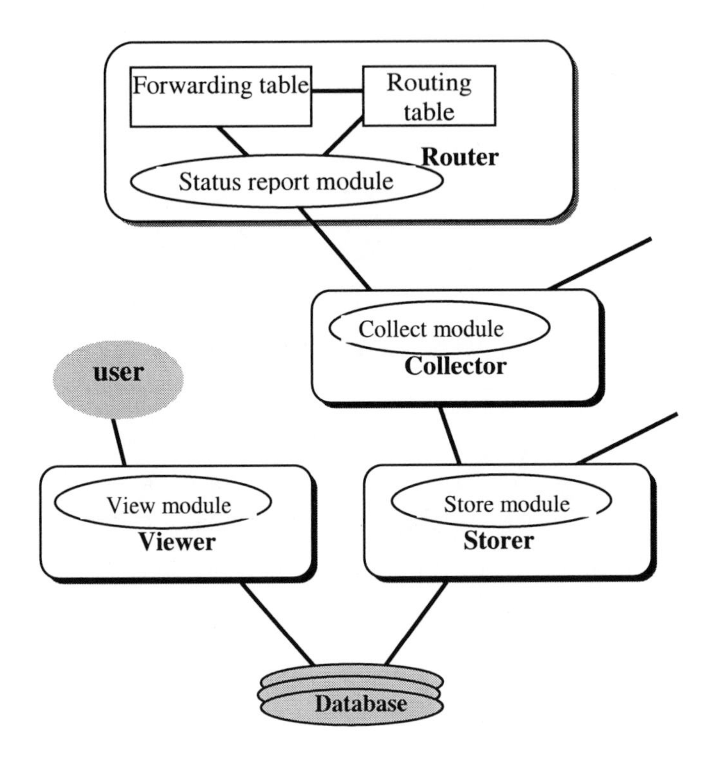

Fig. 2. Basic Components

4 Design and Implementation

MRD system consists of five programs, which are four programs prototyped from four basic modules of model and an additional status request program. Figure 3 illustrates the process architecture of our system.

The MRD-capable mrouted program is responsible for multicast routing and status report method. To support status report method, we modified mrouted source files, whose version is 3.9 beta 3.

The MRD program is responsible for doing request to MRD-capable mrouted program and making it available to users on the local host and remote host. Its functionalities are similar to Cisco's IP Multicast Routing commands.
Now MRD-capable mrouted and MRD support five commands to users. :

1. show ip mroute
2. show ip mroute active
3. show ip mroute count
4. show ip mroute summary
5. show ip igmp groups

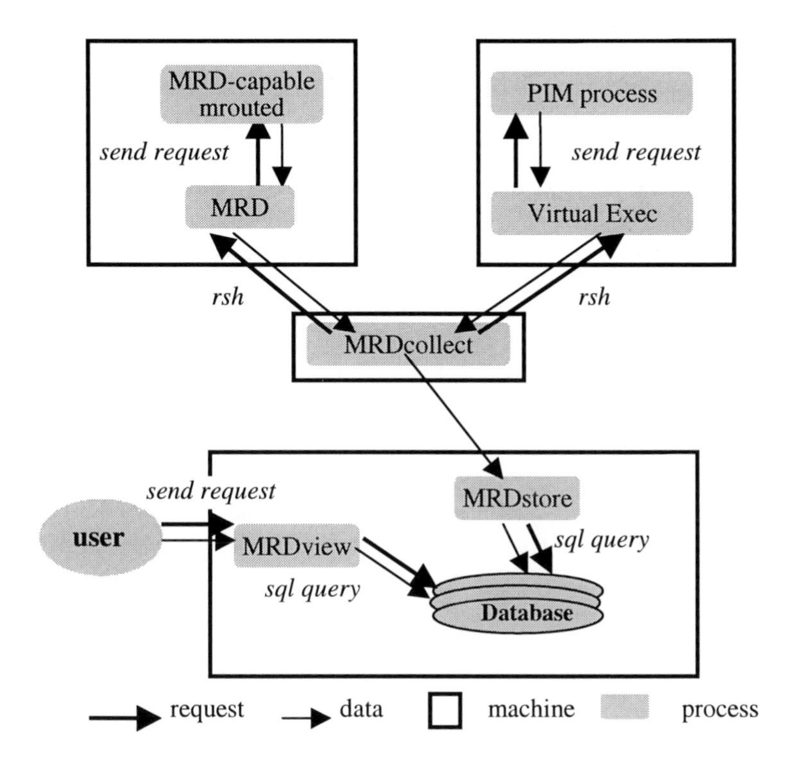

Fig. 3. Process Architecture

The MRDcollect program is responsible for collecting data from MRD-capable mrouteds by executing MRD program and from Cisco's router by executing Cisco's IP Multicast Routing commands. The MRDcollect is executed every 5 minutes. For reflecting the status of multicast network in time, it locates in one domain.
The MRDstore program is responsible for storing data collected by MRDcollect program in the database system. We use mSQL (Mini SQL) whose version is 2.0.11 made by Hughes Technologies[6] to make and query a database.
The MRDview program is responsible for displaying the status of multicast network based on sql query of the database system.

The MRDview is actually the set of Lite[6] scripts on W3-mSQL[6] in order to display the result of mSQL query on Web pages.
The MRDview consists of three Lite scripts, session.msql, reception.msql and forwarding.msql. The status which they display is as follows:

1. Session Announcement Status
2. Multicast Traffic Reception Status
3. Multicast Forwarding Status

5 Experiment and Analysis

5.1 Experiment

We perform an experiment to measure the effect of our system. We show MRD usages based on scenarios and the various statistics. Based on our experimental results, we compare the related works with our system.

Our testbed is an Asia-Pacific Advanced Network-Korea (APAN-KR). We choose Seoul-XP, Seoul-AP, Taejon-AP which are the backbone routers of APAN-KR as monitored routers. And the routers of Seoul National University (SNU[7]) and Korea Advanced Institute of Science and Technology (KAIST[8]) are selected since they have the most traffic on APAN-KR. Figure 4 illustrates our testbed.

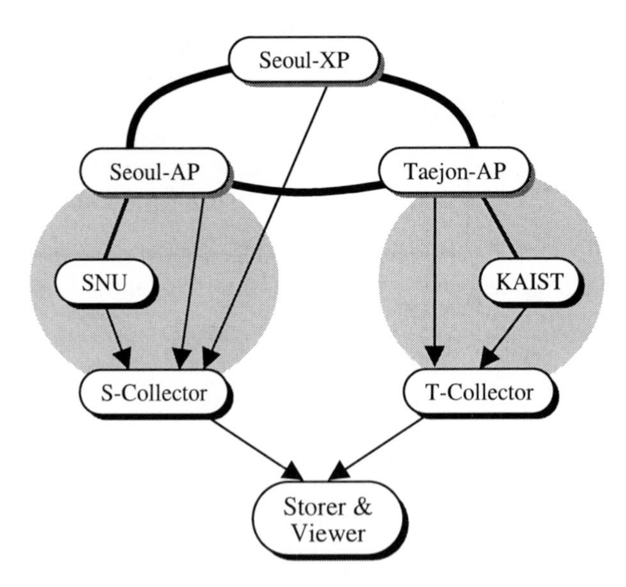

Fig. 4. Testbed for experiments

5.2 Experimental Results

MRD system is used to detect various multicast-related problems based on various scenarios. These scenarios are as follows:

- "I can't see the IMJ-Channel 1."
- "I have some traffic but ..."
- "Where does traffic flow?"

The first scenario is to detect the session announcement problems. We could detect the problem, the session information was not announced.
The second scenario is to detect the multicast traffic reception problems. We use the example, IMJ-Channel 1 session is not announced at Seoul-AP for detecting this problem. We could find the reason of traffic reception problem.
The third scenario is to detect the forwarding problems. When the bursty traffic incomes, we wonder whether this traffic is justifiable or not.

5.3 Analysis

We compared the related works with MRD system based on experimental results. The comparison factors are the required information mentioned earlier, portability and scalability. Table 1 summarizes the result of comparison. SNMP-based tool provides all required information except session information. Since SNMP-enable router is required for SNMP-based tool, portability is middle. Scalability is also middle. Multicast route tracing tools provides the routing and forwarding information from source to receiver, hop-by-hop statistics. Since all routers support multicast route tracing, portability is high. Scalability is middle since where multicast and unicast topologies are not aligned (as is the case in many multicast-enabled networks) multicast route tracing tools may not function. Multicast Reachability Monitor provides packet loss information. Since Cisco's router supports Multicast Reachability Monitor, portability is middle. Scalability is also middle. Multicast Routing Debugger provides all required information. Since Cisco's router and MRD-capable mrouted support Multicast Routing Debugger, portability is middle. Scalability is low, since routers in the static domain are monitored.

6 Conclusion

By reviewing commonly encountered multicast problems, we could define the information which should be monitored in order to detect the problems. Although there exist several types of multicast management tool, any tool isn't satisfied with these requirements. Our proposed system was designed to collect and process these required information from heterogeneous routers.
Our proposed system gave the various status of a multicast network with an easy to use, web-based and intuitive viewer tool. Since the status information was stored in a database system, it was comfortable to maintain and manage it. We could detect various multicast-related problems by using our system.

Table 1. Comparison with related works

Tool \ Factor	Scope of Monitoring					Portability	Scalability
	Session	Forwarding	Routing	Statistics	IGMP Group		
SNMP-based Tool	X	O	O	O	O	△	△
Multicast Route Tracing Tool	X	O	O	O	X	O	△
MRM	X	X	X	X	X	△	O
MRD	O	O	O	O	O	△	X

References

[1] H. Eriksson. The multicast backbone. *Communication of ACM*, 8, 1994

[2] S. Casner. Frequently Asked Questions (FAQ) on the Multicast Backbone (mbone), December 1994. ftp://ftp.isi.edu/mbone/faq.txt.

[3] Asia-Pacific Advanced Network-Korea (APAN-KR) Consortium. http://www.kr.apan.net/

[4] W. Fenner and S. Casner. A traceroute facility for IP Multicast, February 1999. IETF internet-draft.

[5] L. Wei and D. Farinacci. Multicast Reachability Monitor (MRM), March 1999. IETF internet-draft.

[6] Mini SQL (mSQL). http://www.Hughes.com.au

[7] Seoul National University. http://www.snu.ac.kr

[8] Korea Advanced Institute of Science and Technology. http://www.kaist.ac.kr/

[9] D. Thaler. Mstat. http://www.merit.edu/net-research/mbone/mstat.html

[10] D. Thaler and A. Adams. Mtree. http://ww.merit.edu/net-research/mbone/mtree_man.html

[11] D. Thaler. Mview. http://www.merit.edu/mbone/mviewdoc/Welcome.html

[12] C. Diot, et. al, "Deployment Issues for the IP Multicast Services and Architecture," IEEE Network, January 2001.

[13] K. Almeroth, "Managing IP Multicast Traffic: A First Look at the Issues, Tools, and Challenges," IP Multicast Intuitive white paper, February 1999.

Wapplet: A Media Access Framework for Wearable Applications

Takeshi Iwamoto[1], Nobuhiko Nishio[1], and Hideyuki Tokuda[1,2]

[1] Graduate School of Media and Governance, Keio University, Japan
{iwaiwa, vino, hxt}@ht.sfc.keio.ac.jp
[2] Faculty of Environmental Information, Keio University, Japan

Abstract. This paper presents a new framework for constructing applications for wearable computers. We do not consider wearable computing merely as a single self-confined system. Rather, wearable computing can be treated as a cooperative application with information surrounding a person and various devices attached to him/her. We have developed a framework of such wearable computing called "Wapplet." Wapplet provides adaptation to the availability of devices as well as a systematic scheme of constructing applications for wearable computing. In this paper, we describe the design and implementation of Wapplet, and show its prototype system.

1 Introduction

The popularity of personal digital devices has increased rapidly in the last few years. The devices range from Personal Digtial Assistances (PDAs) to cellular phones. Although they differ in sizes and styles, they share common characteristics: wireless connectivity and limited computing capability. To fully employ computing with these devices, two key techniques are required.

First, worn devices themselves need to be configured to be efficiently connected together. These devices can be used to complement traditional computing systems to provide easier access method for resources of information such as WWW, e-mails, and multimedia data. Generally, a mobile user holds several devices simultaneously. Hence, if the worn device is connected together without any configuration, the benefit of the user can increase beyond the limitation of the capability of each device alone. Furthermore, it is important to provide a communication mechanism among the worn devices and other non-worn devices and appliances that are surrounding the user to fulfill a complicated user's demand.

Second, it is also necessary to adapt for changes of device availability because devices that are currently worn by the user may change frequently due to user's activity. Especially, in case of using the appliances equipped with a room, they may become unavailable when the user moves.

In this paper, we focus on such a cooperation of devices to access multimedia data. We have designed and implemented a multimedia application framework,

I. Chong (Ed.): ICOIN 2002, LNCS 2343, pp. 127–137, 2002.

named Wapplet Framework, which aims at cooperation of devices and adaptation for changes of environment. To realize the cooperation, we utilize a wearable computer as a controller of devices, and we have adopted a centralized approach to control devices and to obtain information about changes of environment for adaptation. First, we address issues of this approach and our wearable computing environment which is the basis of our framework. Second, we propose Wapplet Framework as an application framework for a wearable applications. Finally, we describe its implementation and compare our work with related works.

The remainder of this paper is structured as follows; Section 2 shows design issues of our research , and section 3 provides a brief overview about Wearable Network. Requirements for applications are discussed in section 4. Section 5 describes design of Wapplet Framework in detail. The related work is summarized in section 6.

2 Design Issues

Several researches have been done on building a framework to provide cooperation of devices to access multimedia data. From the viewpoint of flows of data, the methodology can be classified into two groups: centralized and distributed approaches.

In centralized approaches, such as Universal Interaction[1] and Document Based Approach[2], all multimedia data control data are maintained by centralized controller such as a PDA. These approaches can allow the system to build a control mechanism easily. However, it does not provide the scalability in accomodating aggregation of multimedia data such as video/audio stream.

In distributed approaches such as Hive[3], each device can behave and cooperate with another device autonomously without having a central controller. This approach provides scalable communications because a bottleneck node does not exist. However, it is more difficult to achieve system-wide adaptation to a user's state and the availability of devices since decision making is done at each node independently.

Considering the advantages and disadvantages of the two approaches, we design a system on the principle of separation of control and data paths. The control paths are centralized at a single node, wearable computer. The wearable computer collects all the information about devices and controls them. In contrast, the data paths are distributed. Therefore there is no single bottleneck node. This is desirable especially for multimedia data. With this approach, we can expect both scalability of data stream and easy adaptation to changes in environment.

3 Wearable Network

A wearable computer is not the only device a user wears. The user will carry other small electronic devices such as a music player, PDA, and a cellular phone. These devices can constitute a network in which they communicate with each other. Since these devices as well as the wearable computer are worn, we call the network "wearable" network. 1 illustrates overview of the wearable network.

The key characteristic of the wearable network is the existence of a single controller of all other nodes. However, as considered in the previous section, data paths are distributed unlike control paths. This design principle is reasonable because the wearable computer is always worn by the user but other devices may change depending on situations. Thus we can design a system of wearable computing by splitting concerned problems into one inside a wearable network and one between the wearable network and the outer environment. Our goal is to establish a suitable software framework for wearable networks.

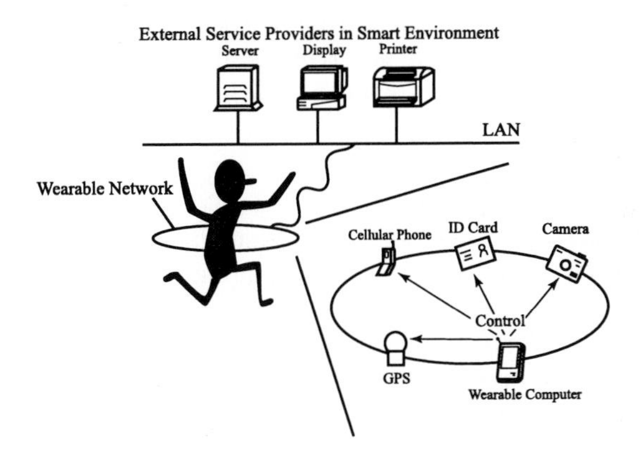

Fig. 1. Wearable Network

Wearable Network architecture consists of several parts, this paper particularly focused on its application framework, named Wapplet Framework. In the next section, we describe the requirements for this application framework in detail.

4 Requirements for Applications

In this section, we consider requirements for applications that run over a wearable network. Based on the consideration, we will determine and design the system for wearable networks. The requirements are summarized as follows: identification, coordination, and adaptation.

4.1 Identification

In a wearable network, a single entity is responsible for controlling all other devices. Therefore an application running over the network needs to identify the status of the devices and the environment surrounding a user of the application.

Although the information of the status may be only availability of devices, it includes in general the sensed values of physical environments. All data should be represented in a format that can be handled in computer software.

4.2 Coordination

Given the correctly identified information about the devices and the environment, the application needs to coordinate all the devices. The coordination spans over a wide spectrum. The simplest side of the spectrum is creating a data path between two different devices. Each device is associated with the capability and kind of input and output; an audio input should be connected to an audio output, and a JPEG output data should be converted when it is connected to a GIF input. The other side of the spectrum is allocating limited resources such as CPU execution time and network bandwidth.

4.3 Adaptation

The status of devices and the environment can change when a user of the wearable network moves. For instance, let us assume that the user is watching a stream video in a small screen in a room where no other display devices are available. When the user moves to a room where a larger screen is equipped, then he can watch the stream video on the larger screen. Thus, the application of the wearable network needs to adapt to the identified status.

5 Design of Wapplet Framework

In this section, we describe a system architecture of Wapplet Framework in detail to satisfy all design goals.

Wapplet Framework consists of three components: Wapplet as application, Mission Mechanism as a control mechanism of Wapplet and Service Providers as the run-time of Wapplet on devices. Figure2 illustrates the outline of these components and the relationship between them.

5.1 Wapplet

In Wapplet Framework, Wapplet uses multiple devices to accomplish a user's task. Hence, these devices should be integrated and their function should be provided to the application. To realize this, we adopt a distributed application method to build Wapplet in which, one application is separated into several modules, named Wapplet module. They are scattered on the devices that are required for the task. Each module is responsible for the task which should be performed using the device where the module is located on.

Furthermore, in order to achieve adaptation, we have made the module as migratable object. When adaptation becomes needed, a module should move from unavailable device to an alternative while preserving the state of the module.

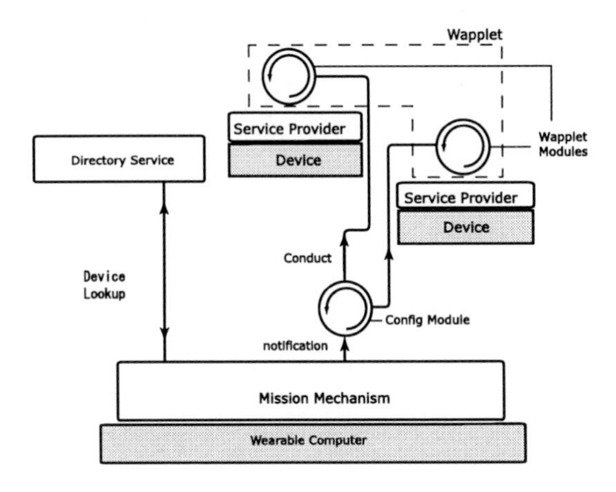

Fig. 2. Software Organization

Wapplet module can access only one data type which is defined in classes of devices(See 1). Hence, one Wapplet module can handle several devices which belong to an identical class.

Adaptation. We can divide changes of service availability into two cases. Wapplet Framework can handle both cases.

- a case in which an alternative service is available
  In this case, another alternative service can be found although a service becomes unavailable. A Wapplet module is able to start using the new service.
- a case without an alternative service
  In this case, no alternative can be found. A Wapplet module attempts media translation to compensate the unavailable service. If the media translation cannot be done properly, the task was using the service is guaranteed to be suspended correctly.

Mission Mechanism decides which way to adopt. The decision and adaptation are hidden from Wapplet programmers. Therefore, it is possible to build Wapplet applications without considerating the changes of service availability in detail.

5.2 Service Provider

Service Providers are executed on each device to provide Wapplet modules with both their execution environment and an interface to access a function of the device. Service Providers are classified from the viewpoint of providing an alternative service to Wapplet modules. Based on the top-level media-type of MIME[10] definition, Service Providers are categorized by media types as shown in 1.

For example, a speaker and a headphone are considered to provide an identical interface, sound output. Therefore a Wapplet module can use these devices alternately through one common interface.

Table 1. Device Classes

class of devices		example
media type	I/O	
video	output	display
video	input	video camera
image	output	display, printer
image	input	digital still camera
sound	output	speaker, headphone
sound	input	microphone
text	output	display, printer
text	input	keyboard, scanner

5.3 Mission Mechanism

Mission Mechanism is a middleware on the wearable computer and it has two purposes: controlling a Wapplet and obtaining information about a device availability. For the first purpose, Mission Mechanism executes Configuration module for each Wapplet to recognize required Service Providers and to notify the changes of Service Provider. A Configuration module has functions to control Wapplet modules and to register a list of Service Providers to Mission Mechanism. To realize this, Mission Mechanism should provide an execution environment for Configuration module. Configuration module can be automatically generated from a description of Wapplet which is written by a Wapplet programmer.

Mission Mechanism looks up Service Providers that are needed to execute the Wapplet. If availability of a Service Provider has changed, Mission Mechanism notifies the Wapplet module which is executed on the Service Provider to migrate the module.

Configuration Module. Configuration module is a part of Wapplet, and it is executed on Mission Mechanism. As described before, it is generated from the description of Wapplet requirements and capability to allow Mission Mechanism to obtain Wapplet information. Wapplet programmers can write a description for Configuration module in XML[6] document which is based on a DTD we have designed.

Furthermore, programmers can write information about adaptation in the document, such as a module structure of the Wapplet and information about Serivce Provider needed for execution.

Configuration module can communicate with Wapplet modules and conduct them based on rules of adaptation written in itself. Mission Mechanism can detect changes of environment, such as service availability and user's state. In this paper, we are not concerned about the user's state but adaptation to service availability.

Regarding adaptation to service availability, Configuration module needs to maintain the following information.

- Location of Wapplet modules
  In order for Wapplet modules to communicate with each other, Configuration module should maintain their location.
- Service Provider availability
  Configuration module should obtain changes of Service Provider availability for adaptation. Accordingly, Configuration module needs to receive the notification about Service Provider availability from Mission mechanism. When availability has been changed, Configuration module conducts migration of Wapplet modules.

Service Provider lookup. Mission Mechanism uses a directory service to recognize Service Providers that can be used in the current environment. The directory service can maintain Service Providers in a particular area , and it replies to a look up request from Mission Mechanism. The Area that is maintained by the Service provider is decided by the administrator of the system or network. However, we assume a directory service exists in each room or floor.

Each Service Provider should register its media type as an attribute to the Directory Service in its proximity. Due to this registration, Mission Mechanism can look up Service Provider by media type. Therefore, the look up does not depend on a device name.

5.4 Wapplet Messages

Wapplet Framework need to handle several types of message to control applications and to achieve adaptation. We classify messages that are used in Wapplet Framework into the following types.

Adaptation Message
This message is used to control Wapplet modules from Configuration module on Mission Mechanism. this message causes migration of Wapplet to adapt to a changes of service availability. It contains new service provider name, which will be used alternatively.

Inter-module Message
This message is used to exchange information among Wapplet modules that consititute one Wapplet to accomplish a task required to each application. This message can be designed by a Wapplet programmer to realize application behavior. Because Wapplet modules can migrate over several Service Providers, the Wapplet module which desire to send a message need to know the location of destination Wapplet modules. This location is maintained by Configuration module of the Wapplet. Therefore, the Wapplet module requests the location of destination before sending a message.

Look up Message
This type of message is exchanged between Wapplet module and Configuration module to solve location of the Wapplet module which is the destination of the inter-module message. This messages contain the name of the Wapplet module which is requested location and its reply contain the location of the Service Provider.

A Wapplet programer must consider only application-specific data that is implemented in the inter-module message. Other types of messages, adaptation and look up messages, are maintain by Wapplet Framework to achieve adaptation and coordination of modules.

6 Implementation

In this section, we describe the implementation of Wapplet, Service Provider and Mission Mechanism. All components are written in Java using JDK 1.3.

We have implemented Java class libraries to provide API for building Wapplet and Service Provider.

6.1 Wapplet

We have built class libraries to provide basic facility to switch devices and to exchange several types of message that are described above. For building multimedia applications, the API for handling media data is also provided.

Device Switching. In order to adapt to changes of device availability, Wapplet modules should be migratable as described above. In our implementation, we use object serialization and Remote Method Invocation(RMI) that are provided by JDK. When adaptation is needed, the Wapplet module requests migration to the Service Provider where it is executed on. Subsequently, the Service Provider serializes the Wapplet module and sends the object to correspondent Service Provider.

Wapplet API. Wapplet API is used for handling multimedia data by a Wapplet programmer. Furthermore, it is also used to build messages for exchanging application-specific data. We have implemented the following two types of API.

Device Control API
We have implemented interfaces written in Java for each device type. The following code represents an example of using an audio interface.

```
setRTPAddress("224.2.253.125","49150","1");
//Configure RTP source adress

AudioOut = createAudioOutPlayer();
//Obtain audio out interface
 --- skip ---

//Handling an Event from such a GUI
// or another Wapplet module
public void MessageHandler(WappletEvent ev){
   if(ev.getContents().equals("PLAY")){
       AudioOut().play();
       //Using the interface to start playing
   }else if(ev.getContents().equals("STOP")){
       AudioOut().stop();
       //Using the interface to stop playing
  --- skip ---
   }
}
```

Message Exchange API
Message Exchange API is used for handling inter-module messages. We provide the following two method to allow a programmer to handle the message.

```
public void sendMessage(WappletMessage msg, String wapplet_name);
// Method for sending the message
public WappletMessage ReceiveMessage();
// Method for reading the message from
// a queue assigned for this Wapplet
```

6.2 Service Provider

The current implementation of Service
Provider can support migration of Wapplet modules (including functions of receiving and executing Wapplet modules), media data handling, and cooperation with directory service.

Migration of Wapplet module. Service
Providers need to provide the run-time for Wapplet modules and interface to be requested for migration of Wapplet module. In our implementation, Service Provider has facility to receive Wapplet module from another Service Provider or Mission Mechanism. In addition it can assign several threads to execute module for each Wapplet module.

Media Data Handling. For handling multimedia data, it is possible to process data which can be obtained through Real-time Transport Protocol(RTP) and from files. Moreover, representation of multimedia data has been built by using JMF2.1(Java Media Framework)[7], which provides `javax.media` package.

6.3 Mission Mechanism

Mission Mechanism, which should play a role of controlling Wapplet applications, is executed on a wearable computer. It should also manage the current state of Service Providers. Mission Mechanism can provide mainly two functions, looking up Service Providers and execution of Configuration module to control Wapplet.

Service Provider Lookup. Mission Mechanism communicates with an LDAP server as a directory service to look up Service Providers to search for required devices to execute Wapplet. Media type, which is shown in 1 , can be used as attribute for look-up. As a result of the look-up, Mission Mechanism can obtain the name of Service Provider and its IP address that are used to migrate Wapplet module. This information is notified to Configuration module, if it is necessary. After this notification, Configuration module may decide where Wapplet module will migrate.

7 Related Work

Hive[3] realizes a cooperation of various devices using an agent framework. The agent can collaborate with each other using message exchange, and a middleware named "Shadow" provides an abstracted interface to access device-specific functions. As described before, this research adopts distributed approach, hence, Hive can provide enough scalability. However, it is more difficult that all agents are conscious of the changes of environment due to their autonomy.

Under Universal Interaction[1], the device can change its source of service when the device moves into a new environment in order to continuously utilize the facility of service. This research is similar to our system in that both system uses various devices which are appropriate for the environment. However, our framework separates control and data paths. Our system uses a wearable computer to control devices in a centralized fashion. We assume that it may also manage a situation infomation of a user. Hence our system geves better support for the adaptive application.

8 Conclusion

In this paper, we presented a Wapplet framework, which enable programmers to build multimedia application in Wearable Network environment.

Wapplet Framework can provide great flexibility and adaptability to changes in environment using the wearable computer as control center. To realize the application in envirnment where various worn device are destributed, we propose the principle of separation of control and data paths. Finally, we described current state of implementation of Wapplet Framework.

References

1. Hodes, T. D., Katz, R. H. et al.: Composable Ad-hoc Mobile Service for Universal Interaction, *proceedings The Third Anuual ACM/IEEE Internatioanl Conference on Mobile Computing and Networking*, pp. 1–12 (1997).
2. Hodes, T. and Katz, R. H.: A Document-based Framework for Internet Application Control, *2nd USENIX Symposium on Internet Technologies and Systems*, Boulder, CO (1999).
3. Minar, N., Gray, M., Roup, O., Krikorian, R., and Maes, P.: Hive: Distributed agents for networking things *Proceedings of ASA/MA'99, the First International Symposium on Agent Systems and Applications and Third International Symposium on Mobile Agents* (1999).
4. Yeong, W. et al.: Lightweight Directory Access Protocol (1995). Internet Request For Comments RFC 1777.
5. The Official Bluetooth Website. http://www.bluetooth.com/
6. W3C: Extensible Markup Language (XML). http://www.w3c.org/XML/.
7. Java Media Framework (JMF). http://java.sun.com/products/java-media/jmf/
8. Bennett, F. et al.: Piconet: Embedded Mobile Networking (1997).
9. Schilit, W, N., Adams, N. and Want, R.: Context-Aware Computing Applications, *Proc. Wksp. Mobile Comp. Sys. and Appls*, pp. 85–90 (1994).
10. Borenstein, N. and Freed, N.: *MIME (Multipurpose Internet Mail Extensions)* (1992). RFC 1341.

Reliable Multicast Delivery Based on Local Retransmission

Jung-Hoon Cheon, Jin-Woo Jung, and Hyun-Kook Kahng

Department of Electronics Information Engineering, Korea University
#208 Suchang-dong Chochiwon Chungnam, Korea 339-700
{grcurly, grjjw, kahng}@tiger.korea.ac.kr

Abstract. Recently, the mechanisms of multicast tree construction for retransmission have been introduced. However, it is not appropriate so that such mechanisms construct a multicast tree using complex metrics or receivers select the designated node. Because all node know complex metrics to construct tree or selection of intermediate node is accomplished through the choice of session creator or receivers. This paper proposes construction mechanism for efficient retransmission using link information and introduces architecture for reliable multicast communication. This mechanism is called Level-limited Tree Management Protocol (TMP). In this mechanism, the join message used to construct a tree is a based on the ICMP extended message.

1 Introduction

In IP Multicast, each source's data flow is delivered efficiently to all interested receivers according to a multicast routing tree. In reliable multicast, the efficiency of data delivery is of paramount importance. Multicast technologies have to provide this efficiency by incurring lower network and end-system costs than typical unicast and broadcast. In most of the current approaches to the multicast issue, the multicast network is tree-structured. For reliable delivery of data, a sender retransmits data for the failed receiver. If the sender retransmits to all receivers by multicast, successful receivers will receive duplicated data. If a sender retransmits to each failed receivers via unicast, a sender has to retransmit large volumes over large-scale network. Thus we need a scalable mechanism for retransmission, i.e. an algorithm designed to reduce the sender's load.

This paper proposes an algorithm of multicast tree construction for retransmission of data. This is a tree construction algorithm through the hop counts from sender to receivers. That is, receivers wanting to join the multicast session send join messages, including ICMP extended messages, to the sender. Senders receiving join message from receivers are informed of hop counts from sender to receivers. The sender builds hierarchical levels using hop counts and constructs a multicast tree for retransmission. The configured tree by the multi-level get to decreases the number of intermediate node (i.e. local server, LS). Also, each intermediate node has less processing overhead for receivers.

I. Chong (Ed.): ICOIN 2002, LNCS 2343, pp. 138-145, 2002.

2 Overview

A sender must retransmit user data to unsuccessful receivers in reliable multicast communication like as in unicast. However, multicast retransmission consumes much bandwidth, as unicast retransmission requires heavy load to the sender. So, the proposed tree algorithms decrease much bandwidth consumption by the multicast retransmission and much load of sender by the unicast retransmission. Following is a new definition for the proposed algorithm.

We redefined terminologies used in conventional reliable multicast.

- Local Retransmission Domain: Scope of partial Retransmission.
- Local Server (LS): When a sender configures multicast tree based on the hop counts, designated router or host have to retransmit to one's retransmission domain.
- Secondary Local Server (SLS): When local server is faulty or leaves from the session, a router or host has to retransmit data packet within retransmission domain instead of the local server.

Figure 1 shows the overview of a multicast retransmission tree. One or more routers may be exist both between sender(S) and local server(LS), and between local server and local server.

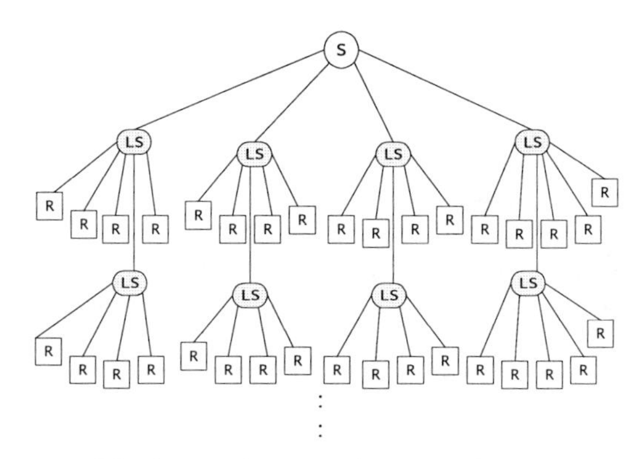

Fig. 1. Overview of multicast topology

3 Proposed Tree Construction Algorithm

This section describes detailed tree construction procedures. The creator must define session characteristics before session creation. That is, the creator defines information concerning each local retransmission domain (i.e. the maximum and minimum number of member within retransmission domain). Because when the number of participants is below the minimum number, the sender can transmit by unicast. Therefore, the creator need not create multicast tree. Conversely, when the number of participants greatly exceeds the maximum number as a result of many late joiners, the creator has to adjust the number of participants within each domain to reduce the load of the local server.

3.1 Initial Construction of Multicast Tree

We assume that receivers know that they have to create an ICMP extended message in order to join the multicast session. Also, we assume that the creator already knows information concerning routers around the creator and can see inside received messages. Initial trees are constructed by following procedure.

1. Receivers have to send a join message to the sender (creator) to join in the multicast session. The extended ICMP message marks both back and forward link information until it arrives at the sender.
2. Sender looks for the join messages received by the receivers and classifies receivers by several levels according to the hop counts. Then, sender divides receivers with the same hop counts.
3. The sender searches receivers with the same link information among receivers of each level and binds them into a group.
4. After tree construction, the sender chooses the local server of each classified group. At this time, secondary local servers are selected to provide reliability when local servers leave the session or otherwise fail to retransmit.
5. The selected local server multicast to the receivers within its retransmission domain to notify that it is the local server. It is instead of ACK message that you can join session. At this time, multicast addresses to be used are determined according to session characteristics. That is, it may be used as a SSM (Source Specific Multicast) address, or as a defined address within each retransmission domain by administratively IP multicast, ZAP (Zone Announcement Protocol).

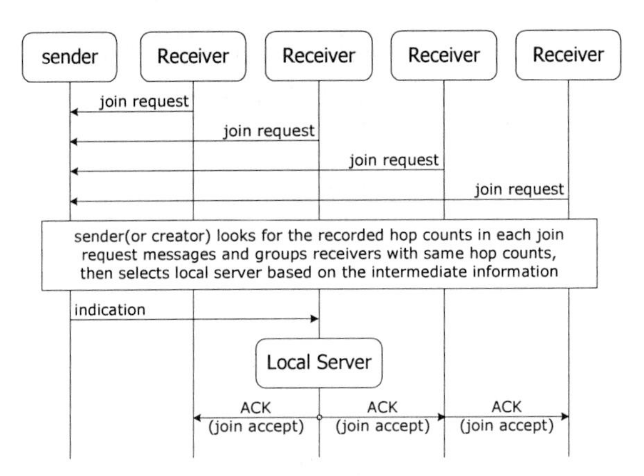

Fig. 2. Initial Construction of Multicast Tree

3.2 Reconstruction of Multicast Tree

The late join of receivers, the leave of session, or leave of the local server reconstructs multicast trees. Following is the procedure of tree reconstruction.

3.2.1 Late Join

Late joiners want to participate in multicast sessions send a join message, including extended ICMP message to the sender. The sender receiving join messages observes the hop counts and IP address of intermediate links, and selects a proper level and an appropriate local server. In the case of an existing local server, the sender notifies the IP address of the local server to the late joiner. In case where no local server exists, the sender selects a new local server, and sends information about the parent (or child) node.

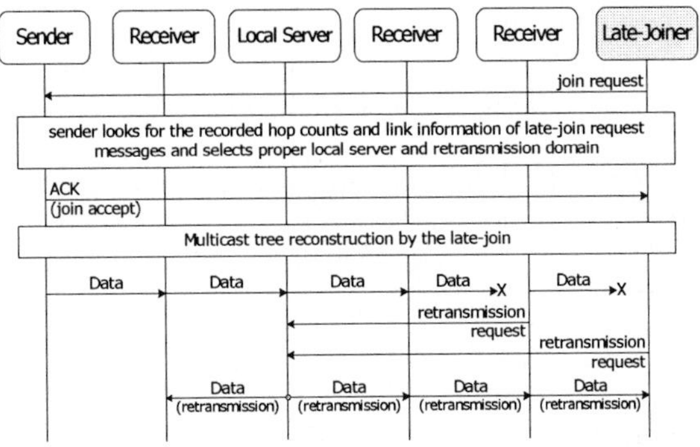

Fig. 3. Tree Reconfiguration by the Late Joiner

3.2.2 Session Leave of Receiver

Receivers want to leave from the multicast session send leave message to the sender. At this time, the local server need not know about the leave of receiver within its retransmission domain, because local servers retransmit using multicast. If the retransmission domain has fewer members less the minimum number of the retransmission domain after the leave of the receiver, the sender must combine with an adjacent retransmission domain and re-select both the retransmission domain and the local server.

3.2.3 Session Leave of Local Server

Local servers want to leave the multicast session send leave message to the sender. The sender receives the leave message of current local server and notifies that the secondary local server is new local server from now. Also the sender selects a new secondary local server within the retransmission domain. The new local server multicast to the receivers within its retransmission domain to notify them that it is now the local server.

3.3 Flow of Tree Construction

When the sender has received join message from receivers, it operates a following procedure for classification of link information (intermediate link address and hop counts). Figure 6 shows a classification of the link information and a decision of retransmission domain for the multicast tree construction. Sender finally compares the receiver number of each final group with N and n, and determines the retransmission domains.

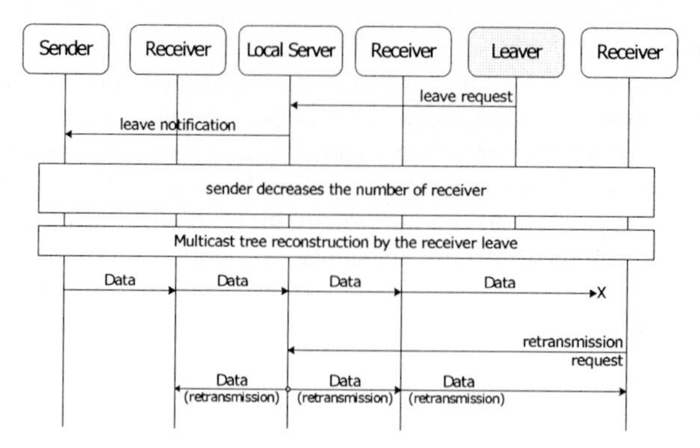

Fig. 4. Tree Reconstruction by the receiver leave

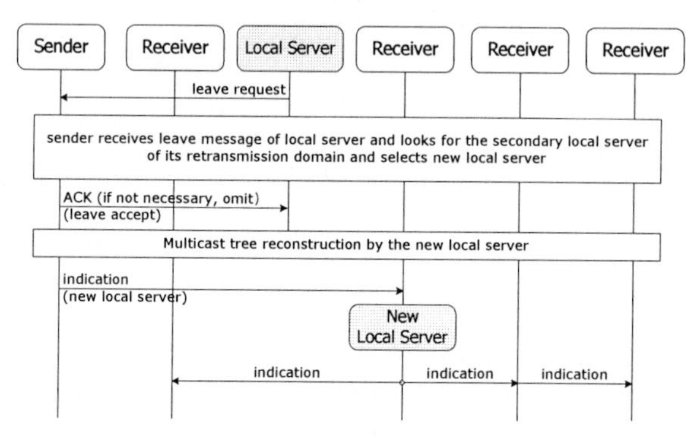

Fig. 5. Tree Reconfiguration of a Session where a Local server is leaving

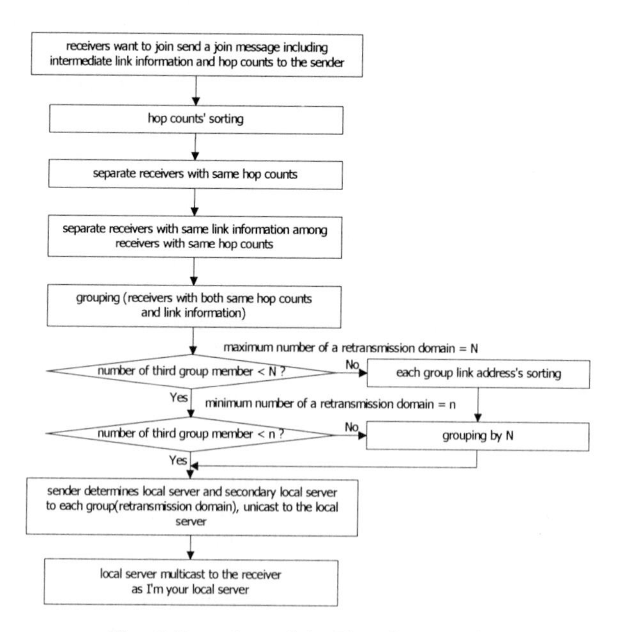

Fig. 6. Procedure of the Tree Construction

3.4 Protocol Stack for Reliable Multicast

Reliable multicast communications require functions concerned with multicast session such as following; an initiation of the multicast session, management of the multicast session, and reliable data retransmission for all receivers. Initiation function of the multicast session defines session's characteristics and allocates multicast address. It also advertises the multicast session. Reliable multicast protocol has to provide a reliable delivery service to all participants. Therefore, session management function creates a multicast tree based on the participants to retransmit user data and maintains participant's status (i.e. late-join of non-participants, session leave of participants, and re-join of leavers). The configured multicast tree could be used in other protocols.

Figure 7 shows a protocol stack included above mentioned. Each protocol requires information stored in the other function or relates with the other function. In this paper, we propose an algorithm for a retransmission tree as part of the session management. Efficient tree facilitates efficient management of resources, the database and membership.

Each protocol is following:

- Initiation Control Protocol (ICP): ICP defines session's characteristics and allocates multicast address. It also advertises the multicast session.
- Membership Management Protocol (MMP): Multicast protocols need to have changeable function according to dynamic members' status so that information about membership could change any time.
- Tree Management Protocol (TMP): Multicast tree for retransmission is created based on the join message of receivers. TMP acts for initial construction and reconstruction by the change of membership information.
- Reliable Multicast Datagram Protocol (RMDP): To provide reliable service, RMDP function retransmits user data to failed receivers using multicast tree, which were configured by the TMP.
- Information Exchange Protocol (IEP): Each protocol has to share information with the other protocol about membership and the session. IEP stores management information used in the session, and delivers the information to the other peer protocol.

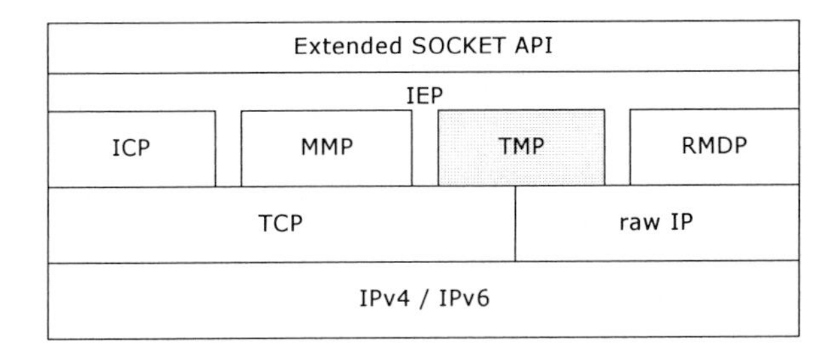

Fig. 7. Protocol Stack for the Reliable Multicast

4 Evaluation of This Mechanism

Reliable multicast protocol based on this tree construction places the responsibility for ensuring reliable packet delivery on the receivers. This mechanism helps sender to decrease sender's processing load.

Table 1. Average message processing cost

Algorithm	Message processing cost
This algorithm	$m \cdot H_{avg} + d \cdot H_{avg}$
Centralized	$2m \cdot H_{avg}$
Jia's	$2m \cdot h_{avg}$
QoSMIC Centr.	$m(w \cdot (w\text{-}1)^{(t\text{-}1)} + c \cdot h_{avg}) \cdot x$
QoSMIC Distr.	$m(w \cdot (w\text{-}1)^{(t\text{-}1)} + \|T\|) \cdot x$

In table 1, h_{avg} is the average hop count of join path from receiver to designated node, while H_{avg} is the average hop count from source to destination. Jia's algorithm only requires that message processing cost is $2m \cdot h_{avg}$. However, each message actually contains the whole tree information, since intermediate nodes have all information in a Jia's algorithm. As a result, the load of intermediate nodes is equal to that of the sender. The result for QoSMIC is from [1](the x factor is added to reflect the fact that messages have to be processed by more than one node). For QoSMIC, w is the average degree of a router; t is the maximum TTL used for search; $|T|$ is tree size; c is the number of candidates for BID-ORDER session(BID-ORDER is message to join by a new member)[1].

5 Conclusion

The efficiency of data delivery is highly important. IP multicast supports efficient multicast data forwarding through the use of multicast trees, but it guarantees only best-effort delivery and can cause a feedback implosion at the multicast sender. So, recently the mechanisms of efficient multicast tree for retransmission were introduced. However such mechanisms construct a multicast tree using complex metrics and the selection of the intermediate node (i.e. designated node) is not appropriate, since selection is accomplished through the choice of session creator or receivers.

In multicast communication, a sender accomplishes data retransmission. But we select a local server and a retransmission domain for retransmission and delivery of control message. That is, this algorithm is designed to reduce sender's load. Also, it defines including maximum number of member and minimum number of number within retransmission domain. Because when number of participants is less than minimum number, sender transmits by unicast, the creator cannot create a multicast tree. And when the number of participants greatly exceeds the maximum number by many late joiners, the creator has to divide the retransmission domain to reduce the load of the local server. We can verify decreased processing overhead through our simulation. Future work will add an algorithm to provide QoS and implement group management protocol based on this algorithm.

References

1. A.Fei, M.Gerla, "Receiver-Initiated Multicasting with multiple QoS Constraints", IEEE INFOCOM 2000, Vol 1, pp 62-70, Mar. 2000.
2. S. K. Kasera, S. Bhattacharyya, M. Keaton. J. Kurose, etc., "Scalable Fair Reliable Multicast Using Active Services", IEEE Network, Vol .14 No. 1, pp48-57, Jan/Feb. 2000.
3. S. Angelopoulos, I. Katzela, "Fast Rearrangement of Multicast Trees for Applications with Bandwidth and Delay QoS Requirements", IEEE CCECE 2000, Vol 2, pp 680-686, May. 2000.
4. Yatin Chawathe, Steve A. Fink, Steven McCanne, Eric A. Brewer, "A Proxy Architecture for Reliable Multicast in Heterogeneous Environments", ACM Multimedia '98, pp151-159, Dec 2000.
5. H-Y Lee, C-H Youn, "Scalable Multicast Routing Algorithm for Delay-Variation Constrained Minimum-Cost Tree", IEEE ICC 2000, pp 1343-1347, Jun. 2000.
6. Bellovin, "ICMP Traceback Messages", Internet-draft, draft-bellovin-itrace-00.txt, Mar. 2000.

NADIA: Network Accessible Device on the Internet Architecture

Shunsuke Shinomiya[1,2], Yutaka Kidawara[1], Takeshi Sakurada[1], Seiji Tsuchiike[1,2], and Shin-ichi Nakagawa[1]

[1] Communications Research Laboratory
4-2-1, Nukui-Kitamachi, Koganei, Tokyo 184-8795, Japan
{shinomiya, kidawara, take-s, tuchiike, snakagaw}@crl.go.jp
[2] yggr-drasill technology Inc.
2-303, Kita-Norimonocho Kanda, Chiyoda-ku, Tokyo 101-0036, Japan
{shinomiya, tsuchiike}@yggr-drasill.com

Abstract. Being accompanied by the recently widespread Internet, computer networks are becoming an indispensable tool for communication. On the other hand, more and more devices other than computers, such as Personal Digital Assistants(PDAs) or cellular phones are connected to the Internet are as well as computers. Moreover, even cameras to take real-time images have being connected to. This tendency seems to grow stronger. Especially on the Next Generation Internet established by IPv6 networks, more and more various devices could be connected to the Internet as network devices with numerous global addresses.
This paper defines devices connected to the network and proposes a system to realize complicated processing by combining these devices.

1 Introduction

With the Internet being developed, computer networks have become diversified and various infrastructures such as wired or wireless networks have been established. Web servers and image distribution servers provide information at clients' request and distribute pre-accumulated images on demand. On the other hand, making positive use of interactive networks enables remote devices to be controlled and information to be obtained. A variety of researches for them have been committed. The remote-controlled robot technology, which has been researching from long ago, is expected to enhance its possibility by using the Internet and its researches have been conducted actively. These remote devices, however, are to control devices with single function by a specific application. On the Next Generation Internet in the future, a number of remote devices could be connected to because of providing numerous IP addresses by introducing IPv6. These devices could realize more complicated processing by unifying each device's process through networks as well as realize given functions in advance.

I. Chong (Ed.): ICOIN 2002, LNCS 2343, pp. 146–156, 2002.

2 Motivation

2.1 Related Works

For the research by Kaplan et al.[1], the controller of an ordinary radio-controlled car is connected to the IP link with 56kbps, the images out of the camera equipped with the radio-controlled car are transferred to an image distribution server by a transmitter, and the video is captured and IP transferred. A similar system with ATM networks[2] has been developed at MIT.

However, in this case, the remote-controlled car merely moves on command value and does not move once it loses control signals from the controller. To cope with these problems, Simmons et al. developed the Xavier[3,4] that has a function to execute the first given command by autonomous operation when the control signals are lost as well as be operated with command values by the interface of Web Screen. Though this function is necessary in the unstable connection environment to the remote devices, to develop autonomous devices adaptive to diverse environments is not easy because various sensors and control functions to determine the movement should be installed. So the environment for it should be specified to design and to develop. In the meantime, Paulous et al. have developed the Web Blimp[5], an airship controlled by the interface of Web screen. The camera equipped with the blimp could be referred on the Web monitor.

Doherty et al. calls these remote-controlled devices Ubiquitous Telerobotic Devices and describe their capability and applications that can be realized by the devices[6]. The device called micronode, which accommodates active IPv6 networks, has already developed[7].

2.2 Proposed Capabilities

The remote device described previous chapter has already been available as a complete function body, but data exchange with other remote devices or coordinated operation has not been in consideration yet. On the Next Generation Internet, however, more complicated processing becomes available by operating these remote devices cooperatively because many of them will be connected to the Internet.

To realize such objectives, data models that can combine network devices would be needed on the condition that what is each network device is defined. And how to let such devices as being very primitive and with no access to the networks work as a network device is also one of important issues.

Considering all pre-mentioned issues, we defined the device connected to the networks as the device with a function and one or more inputs or outputs, or one or more inputs and outputs. In the same way, we defined the remote device composed of multiple network devices as a network device. The network device is named NAD, the Network Accessible Device. The NADIA, the Network Accessible Device on the Internet Architecture, is proposed as a system to manage the NAD on the Internet environment.

3 NAD: Network Accessible Device

3.1 Concept

Once such devices as cameras and microphones are connected to the network, various information in the real world could be digitized and exchanged on line. On the contrary, the digitized information could work with the real one for all geographical distance. The devices, however, vary and the information and its operating instructions are also in wide variety. For instance, the video camera can obtain the image of the place where the camera is set up. The devices of this kind measure the information in the real world and are passive to the real world where the information is converted into digital data. In the meantime, the devices such as motor work according to the indicated value and have active function to approach real objects.

Each device needs to input data and output data in order to make use of the device's function. The camera mentioned above, for example, zoom value and focus adjustment need to be input to the camera in addition to notifying the start and end of filming. The obtained image would be output based on the input data. Thus, the camera makes a device with both input and output functions. As for a motor, it operates with pulse as an input value or a voltage value but does not have an output value. However when frequency measurement is needed, for the actual use, the motor's frequency is measured with rotary encoders. The rotary encoder does not have input data but frequency as an output. If such a motor with a rotary encoder is considered a device, the motor makes a device with both input and output functions. Thus, there are some complex devices that realize one function when combined with other devices.

The NAD that is proposed in this research can be connected to the network and can communicate over the network. On top of that, the NADIA is a system that connects NAD to the Internet and makes it available.

3.2 NAD Element

The minimum units of logical data models that describe the NAD are the following three primitive elements; they are an element that defines input to the device, an element that defines output from the device, and an element described by the element that defines the device's processing function. The following is descriptions of each NAD element.

NAD Input Element. The NAD Input Element indicates the function to receive input data from the network or the device. The NAD Input Element receives data through the network and forwards it to the NAD Function Element. It can also receive input data from the NAD Output Element and transfer data between the NADs. The NAD Input Element has its name, ID, and connected to that should be input as property values.

NAD Output Element. The NAD Output Element indicates the function to output data from the device to the network or other devices. The NAD Output Element receives data from the NAD Function Element. It can also output data from to the NAD Input Element and transfer data between the NAD elements. The NAD Output Element has its name, ID, and connected to that should be output as property values.

NAD Function Element. Each device has functions such as processing depending on input data and providing the result that is obtained by prescript processing. The NAD Function Element indicates the processing function of such devices. The Element describes the processing function, input values, and output values. When the device is used, the function processed by the element's property value, the parameters needed, and the output data can be available as references.

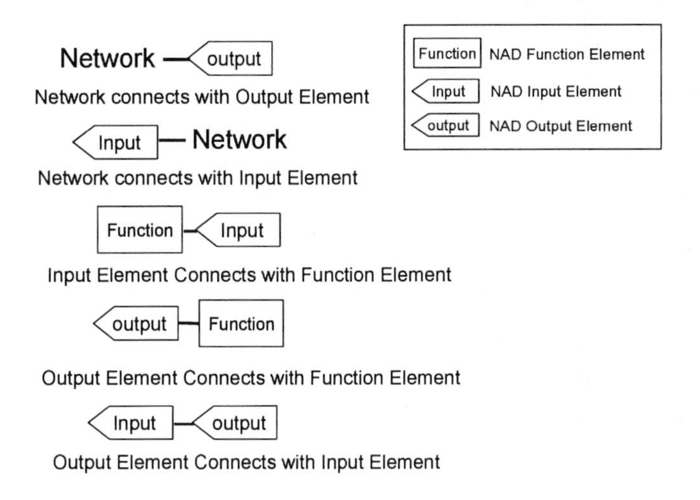

Fig. 1. NAD Elements

3.3 Basic NAD

Basic NAD is a minimum unit what is available as a NAD. In this paper, the device connected to the network is defined as which has one and more Input Element or one and more Output Elemnt or both of them, and has a Function Element. The most fundamental device is called Basic NAD. The Basic NAD includes ones called Input NAD, Output NAD, and Input-Output NAD. Unlike the after-mentioned Composite NAD, the Basic NAD does not contain another NAD.

Input NAD. The Input NAD has one and more NAD Input Element and a NAD Function Element. Users can know functions available and input parameters based on the functions according to the descriptions in the NAD Function Element. Once obtaining the information, users who can access the NAD Input Element become able to operate the targeted Input NAD by specific a function and inputting parameters. The motor that can be connected to the network is one of the Input NAD.

Output NAD. The Output NAD has one and more NAD Output Element and a NAD Function Element. Users can understand functions available and output parameters based on the functions according to the descriptions in the NAD Function Element. Being informed with procedures above, users who can connect to the NAD Output Element would be able to access to the information of the targeted Output NAD by reading functions sought for and parameters out of the Output Element. The rotary encoder that can connect to the network is one of the Output NAD.

Input-Output NAD. The Input-Output NAD has one and more NAD Input Element, one and more NAD Output Element, and a NAD Function Element. It has both characteristics of the Output NAD and the Input NAD. As for this NAD, input value from the NAD Input Element is processed with the function designated by the NAD Function Element and output value is read out of the NAD Output Element. The camera when it specifies start and end of filming according to command value and obtain the image as an output. Therefore, the video camera is one of the Input-Output NAD.

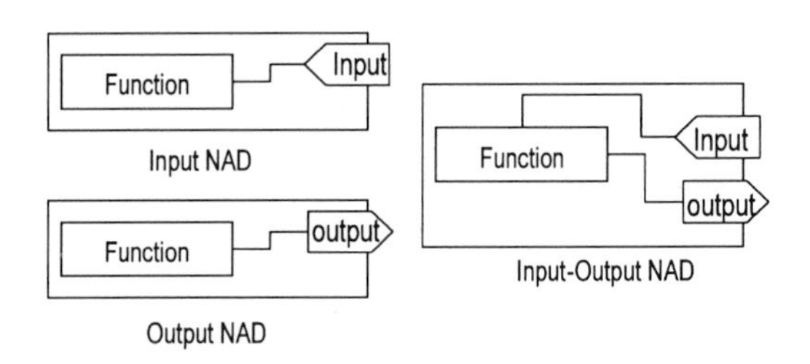

Fig. 2. Basic NAD

3.4 Composite NAD

To implement complicated processes, composite devices combined with more than one device are set up. The composite device is described as a composite NAD, which is a combination of NADs.

There are two methods to configure Composite NAD. One is configuring it with inner element of the NAD such as Input or Output Element which are accessible from outside, and the other is configuring it as Function Elements are exclusively used and access from outside is prohibited. When configuring it with access from outside allowed, NAD, an element of Composite NAD, can be used as NAD. When Configuring it with exclusive use, an element NAD cannot be used as NAD from outside Composite NAD because no access from outside is allowed and it is used through Composite NAD that includes appropriate NAD. When no access to element NAD is allowed, the NAD looks like a Basic NAD from outside.

3.5 Example of Composite NAD

As an example of Composite NAD, a network-controled car controlled by wireless network is taken up. For the network-control car, the device controlled by network consists of a servo controller and a motor amplifier. The amplifier corresponds an accelerator of the network-control car to control output from the motor. The servo controller corresponds a steering to control horizontal moves of the front wheels. The figure 3 shows the NAD configuration of the network-control car.

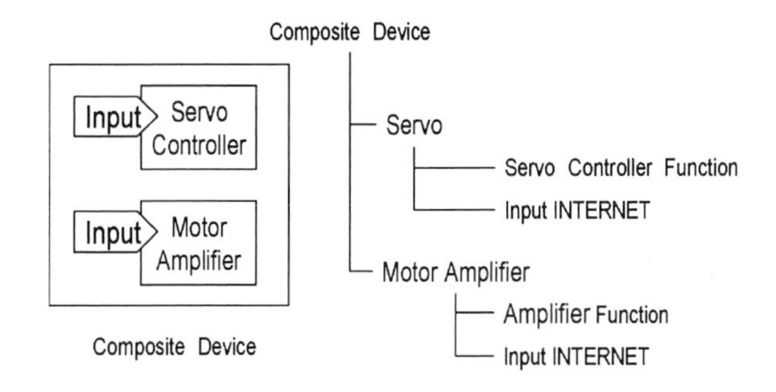

Fig. 3. The NAD which constitutes network-controlled car

This example, however, is nothing more than a combination of just two devices, a Servo Controller and an Motor Amplifier. In order to have elements of Composite NAD, these devices should be added some functions to be operated cooperatively. For instance, operations named Steering Control and Speed Control are provided as functions to operate the NAD shown in the figure 3 cooperatively. To control the NAD named Network Control Car, just using its functions of Steering Angle and Speed Control is enough and there is no need to know how to control Servo Controller and Motor Amplifier. Therefore, the Network Control Car is described as a device of a Function Element that has functions of Steering Angle and Speed Control and an Input NAD that has input data from networks as shown in figure 4.

With a camera installed in the network-controlled car, the car become a Composite NAD, which consists of two devices: Network Control Car NAD and Camera NAD. The Composite NAD, Self-Movable Car NAD can be used as a self-movable camera with which users may control Network Control NAD while monitoring the video sent from Camera NAD. Furthermore, a Chase Car NAD that chases an objective can be configured after what to do as follows. First a Searcher NAD that searches an objective from images filmed by the camera is configured using Camera NAD and Image Analysis NAD. Then the Searcher NAD is combined with Network Control Car NAD.

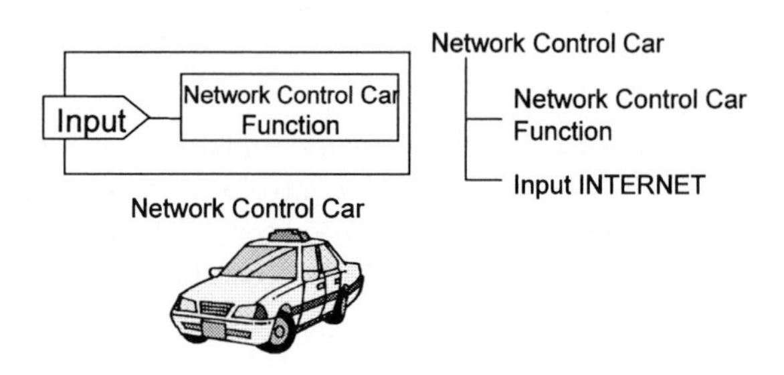

Fig. 4. Network Control Car NAD

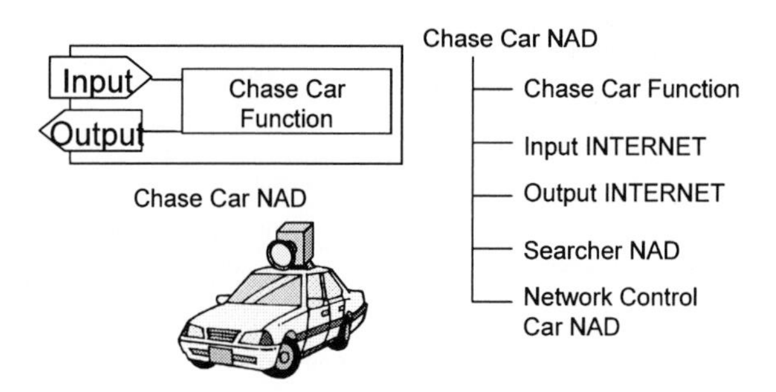

Fig. 5. Chase Car NAD

4 Network Accessible Device on the Internet Architecture(NADIA)

4.1 The Outline of NADIA

The NADs as described have various possible uses when connected to the Internet. Once the Internet being available, numerous computers and devices around the world that are connected to the Net can be dealt with and be available from anywhere in the world. By combining them, the system that has access to the real world with a unified method and how to use the real world device can be provided.

The NADIA is designed to enable direct access to any given NAD by allocating IP address to each NAD on the premise that the communications protocol is IPv6. By allocating an IP address to each NAD, all NADs such as Basic NAD, Composite NAD and element NAD included by Composite NAD have direct unified way of access.

4.2 Access to NAD

As mentioned before, the NAD is described as a logical function device. And some NAD consists of more than one device. In our research, the access to the NAD will be given with URI. The NAD is specified in the host name and its function is specified in the path name. It will be described as follows.

Specification of NAD. In NADIA, the name space the same as the one of existing DNS is used to identify NAD with host names and domain names. Aside from Basic NAD, the NAD includes Composite NAD composed by a combination of more than one NAD. Thus, when the NAD configuration have the hierarchical structure, Element NAD included by Composite NAD is specified with <name of Element NAD><name of Composite NAD><domain name>. A Motor Amplifier NAD, for example, is an Element NAD for Network Control Car of APII and is described as "MotorAmplifier.NetworkControlCar.apii.net.".

Use of the function which NAD has. As mentioned before, We consider the device characteristic after specifying the NAD. For instance, We can specify the Steerfing Control function of Network Control Car as "NetworkControlCar.apii.net/SteeringControl/?angle=30". Element NADs configuring the NAD can be accessed in the same way. If somebody want to watch the video image image of Chase Car NAD mentioned before, they can be specified the functionality "Camera.ChaseCar.apii.net/Moniter". Presently, URI notation is a popular to specify resources on the Internet. The NAD also can be accessed by URIs. Therefore, users can access NAD and obtain the real-world access method through the Internet in the same way of accessing conventional Internet resources such as image, html and XML files.

4.3 Pseudo NAD

Converting various devices to IPv6 NADs, we can access to real-world usefully using NADIA. However some network devices might be difficult to implement IPv6 protocol and might be unsuitable because of the device consists of many devices like a sensor. When these devices connect to the Internet, machines implemented IP protocols mediate conventional devices between the Internet.

The mediator gives pseudo IP addresses to devices and make them behave as NADs. The NAD given pseudo IP address by the mediator calls Pseudo NAD.

Pseudo NAD is not only giving pseudo IP addresses to the conventional primitive devices, but also giving IP addresses to the Internal devices of composite devices because of useful access from the Internet. Attaching pseudo IP address to internal device of composite devices, the device can be accessed independently from Internet and used it as different Basic NADs or NAD elements of the other composite devices.

5 Implementation

We developed a network contol car attached camera called "mini0". mini0 is modified a ordinary radio-controled car. The prototype is consist of single board computer, programmable one-chip micro controller,IEEE802.11b unit and web camera unit . single board computer have control software, receives control command via IEEE802.11b network and servo controller unit and motor amplifier unit.

This prototype is a Composite NAD, consists of one Basic NAD(Network Camera NAD) and Composite NAD(Network Control Car NAD). Network Control Car NAD also consists of two Pseudo NAD(Servo Controller NAD and Motor Amplifier NAD). Servo Controller NAD can change front wheel direction of Network Control Car NAD. Motor Amplifier NAD can control motor output power to drive Network Control Car NAD .

Fig. 6. View of mini0

5.1 Composition

Configuration of mini0 shows figure 7. mini0 is a composite NAD,called Network Camera Car NAD, which consists of Network Control Car NAD and Network

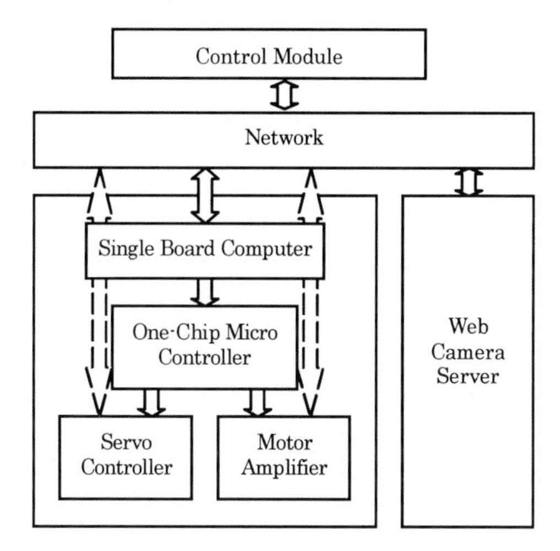

Fig. 7. Consist of mini0

Camera NAD. Each NAD are connected by IEEE802.11b wireless LAN unit and can be accessible from control computers via Network and the Internet.

Network Control Car NAD consists of Single Board Computer installed Linux, one-chip Micro Controller (PIC16F84A),Servo unit and Motor Amplifier unit. The CPU can send commands to the controller, which controls Servo Controller and Motor Amplifier unit. The CPU gives pseudo IP adresses to servo unit and motor amplifier and gives them Pseudo NAD functionality.

Camera NAD has functionality which can show front view and change camera direction. Computers connected with network can access to Camera NAD and browse its video image on web browsers or viewer programs and can change camera direction.

mini0 has a composite functionality. The CPU of Network Control Car NAD gives new function as Network Camera Car NAD. It operates servo unit and camera direction together. Camera of Network Camera Car NAD can view the running direction automatically. When users operate it as a Network Camera Car NAD, camera direction of Camera NAD changes to running direction. However, when they operate it as a Network Control Car NAD, the video image of Camera NAD can be shown front view without changing direction.

To control Network Camera Car NAD and Network Control Car NAD, PC attached Steering wheel and accelerator/brake pedal can communicate with Network Camera Car NAD ,Network Control Car NAD and Network Camera NAD and have interface to watch video image and to show NAD status.

When a user operate Network Camera Car NAD, other people cannot control it as a Network Camera CarNAD. However, they can access to Camera NAD directly and can watch driving video image.

6 Conclusion

Connecting various devices with network which can sense real world status and work on real world. The Next Generation Internet is constructed by IPv6 network, which provide enormous numbers of IP addresses. The specification enables network connection of various devices all around; therefore, it will becomes more and more important to well-manage Internet-connected devices , especially, they should be managed in the unified ways and decided standardization ways.

To use network device usefully, both users and other devices have to know their specifications in some ways. Presently, we are designing and developing NAD describe meta-language to describe specifications and configuration of NADs. The language can describe device functionality, input-output data type, network connection type and connection devices etc. Using our NAD specification describe meta-language, we can understand the device specification and author new network device as a Composite NAD such as XML content.

Furthermore, NADIA enables access to real world resources in the same way of accessing to conventional digital resources stored in hard-disks. As a result, we can develop a seamless information distribution system without distinction between real-world resources and digitized information. We are discussing the real information system used NADIA as a future work.

References

1. A.E. Kaplan, S. Keshav, N.L. Schryer, J.H.Venutolo : An Internet Accessible Telepresence, Multimedia Systems
2. V.Bose:The TNS video Rover.Work by the Telemedia Networks And Systems Group, MIT, http://www.tns.lcs.mit.edu/vs/rover/rover.html
3. R.Simmons, J.Fernandes, R.Goodwin, S.Koening, J.O'sullivan Xavier : An Autonomous Mobile Robot on the Web,Robotics and Automation Magazine, 1999, http://www.cs.cmu.edu/People/Xavier/.
4. L.R.Lab: Where in the world is Xavier, the robot?., http://www.cs.cmu.edu/afs/cs.
5. E.Paulos and J.Canny : Ubiquitous tele-embodiment : Applications and implications, in Proceedings of International of Human-Computer Studies, 1997
6. M. Doherty, M. Greene, D. Keaton, C. Och, M. Seidl, W. Waite, and B. Zorn : Programmable Ubiquitous TeleroboticDevices , in Proceedings of SPIE Telemanipulator and Telepresence Technologies III, Oct. 1997
7. Micronode, Internet Node Inc., http://www.i-node.co.jp/e/index.html

Database Conceptual Modeling for Integrating E-A Model and E-R Model

Jae-Woo Lee, Jong-Ha Hong, and Doo-Kwon Baik

Software System Lab., Dept. of Computer Science & Engineering,
Korea University,
1, 5-ka, Anam-dong, SungBuk-ku, 136-701, Seoul, Korea
Jwlee@Kyungbok.ac.kr, {Sam, Baik}@Swsys2.korea.ac.kr
http://swsys2.korea.ac.kr/index.html

Abstract. Efficient database design and construction is one of the most difficult and hard tasks for constructing information systems. During the database designing process, we have performed data modeling for understanding real world and business process easily. There is E-R model using entity and relationship between entities for supporting data modeling. Unlike E-R model, there is E-A model using specific aspects. Both of these models define entities and are modeling with relationship, so we have understood easily complex real world. E-R model is good for defining entity and relationship, but restrict concepts of abstraction and inheritance. In this paper, we propose a database conceptual modeling procedure, called I-ER modeling, through analyzing similarity and difference between E-A and E-R, applying abstraction concepts of E-A model. Using the I-ER(Improved E-R) modeling we can understand easily real world complexity for modeling entities and relationship.

1. Introduction

We live in information society. There is no place that information systems do not reach on our society whole by the brilliant growth of information technology recently. Software applications are repeating the brilliant development including hardware, network new technologies. However, various business processing and data processing of real world is so complex that the development of information technology does not satisfied various needs because user's information desire also enlarges exponentially. Various information technologies are applied at many corporation's information process, but it may be one of the most difficult and hard works that design efficiently and construct database in information system construction actually. It is necessary that modeling process performed by database designers for various information processing and this result should be represented to data model even if the information technology develops at high grade.

Database design technologies are one of the most essential factors in use of computer and occupy the most important part of data processing in various corporations'

I. Chong (Ed.): ICOIN 2002, LNCS 2343, pp. 157-168, 2002.

businesses, whole industries [1]. In this way, a database should be designed efficiently to support well corporation management and designed rightly according to actual business process. Then, it becomes important information property that can manage easily and supply good information applying the database[2]. Procedures that design database are process that is modeling to represent to understand easily facts of present real worlds or complicated business processing rules etc. conceptually.

There are many data models to support modeling real world, that wish to understand by finding entities as information unit of existence and nonexistence style that exist in real world, and represent complexity according to relation between these entities or particular viewpoint. Among those model for data representation, it is E-R(Entity-Relationship) model that support modeling to describe more easily current real world by extracting entities and defining relationship between these entities. The other, called E-A(Entity-Aspect) model, defines entities in present real world and approaches according to some particular aspects. And there is REA(Restricted EA) model extended by a more applicable data model for improving E-A model[3]. These models are modeling tool that wish to represent easily and understand complicated current real world by defining all entities and defining relationships between these entities.

E-R model is a good data model for defining entities and relationship but has some restriction that it can’t represent generalization, specialization, aggregation, decomposition and concepts of inheritance in many levels of abstraction. While, in E-A model, it has some difficulties to represent relationship because of it's hierarchical structure but it is good for supporting concept of abstraction and inheritance that can not represent variously in E-R model[4].

Ultimately, database design should be integrated data modeling that reflect current real world just as it is and therefore, must be able to represent real world more definitely and need various data model for this. In this paper, we propose I-ER(Improved E-R) model that applies advantage of various aspects of information representation which is offered in E-A model through advantages and disadvantages analysis and comparing representation methods with each component about E-A model and E-R model. We can perform data modeling in advantage point of various abstraction as well as relation definition or inheritance through I-ER model, and I-ER model can support to understand complicated current real world more easily.

This paper is as following. In section 2, we introduce briefly about data modeling and explain related works on data modeling and compare E-R model with E-A model. In section 3, we examine approach for integrating E-A model and E-R model and propose I-ER(Improved E-R) model applying various abstraction level of E-A model. And examine I-ER model's property and strong point through comparison with current E-R model. In section 4, explain I-ER modeling procedures and evaluate I-ER model's practical use. Finally, in conclusion establish modeling assignment and plan future works.

2. Related Works on Data Modeling and Modeling Comparison

2.1 Related Works on Data Modeling

It may not be so easy work to represent definitely through direct observation because real world is very complicated and various. However, we can do various data processing and simulation through using a model if we can represent some knowledge or information using a suitable reasoning procedure or a detailed tool. Current real world can be examined closely by some data processing entities and by relationship between them that we have defined, and we can represent by using entities actions(behavior) and specify the behavior even though not perfectly. It is data model that make specification about phenomenon or real world that do not appear well in eye by some procedures. Data modeling refers to procedure or process that represent current real world to database[2].

Data model is that represent various information by using a modeling tool which can change complicated current real world to entities and relations between the entities, support to understand easily states of present real world. A modeling tool that represent knowledge or information of present real life had been studied variously and can be classified by procedural model and declarative model procedure according to modeling approach method. And declarative model is classified by logic based model, graph based model and rule based model again[5]. In figure 2.1, express various models for knowledge and information representation, we can know all graph based entity-oriented model at E-R and E-A model.

There had been many related works and researches on representing knowledge and information of present real world, these are applied to many database/knowledge base design more efficiently by defining structure of knowledge and information, and consistently representing this with current real world[6, 7, 8, 9, 10, 11, 12, 13].

In E-A model, represent knowledge and information by hierarchical data structure. Therefore there are many complex problems such as probability of conflict created by equal entities in top and bottom side, complexity of constructing support system for the E-A model implementation. So, to resolve those problems there is REA(Restricted EA) model that various aspects of entities is limited by 5 aspects, "IS-A", "A-PART-OF", "Attribute", "Role" and "Operation" etc. for improving extensity of practical use[3].

Also, there is related work on REA, propose READL(Restricted EA Definition Language) that is the knowledge definition language to support REA (Restricted Entity-Aspect) modeling, and construct REAPS(Restricted EA modeling Prolog System) that is knowledge base system that support dynamic alteration of knowledge structure creating Prolog knowledge base using READL[4].

And there are many works, propose framework by analyzing various relationships in E-R model that use for conceptual modeling[14], and work on clustering for large size database design[15], and research for information representation on time change(Temporal ER)[16].

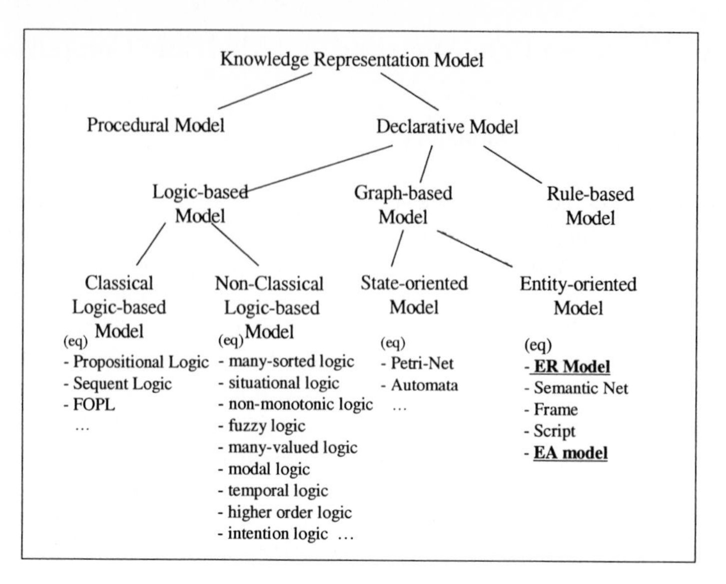

Fig. 2.1 Knowledge information representation model classification-taxonomy

2.2 E-A Model (Entity-Aspect Model)

E-A model applied concept of system entity structure as model that represent hierarchical structure and relationship of knowledge/information systematically and represent to component of entity, aspect, generalization, aggregation and multiplicity etc.[4]. That is, E-A model is easy to understand structure of information systematically by representing hierarchy of knowledge and information as shown in figure 2.2.

There are components of EA model, entity type, aspect type, attribute and relationship(relation link), support to various abstraction concepts, generalization, specialization, aggregation, decomposition and multiplicity.

But in E-A model many conflicts can be occurred in defining schema between entity type and aspect type, and have difficulty of constructing E-A model because of various aspect types. Also, E-A model have some difficulty to consistency because it is very complex model represented in hierarchical structure, so there is related work on REA(Restricted EA) model to overcome these difficulties that limits aspect type to several types, "IS-A", "A-PART-OF", "Attribute", "Role" and "Operation" etc.[4].

2.3 E-R Model (Entity-Relationship Model)

E-R model focuses on representing entities and relationships between entities as graph structure unlike E-A model representing a structure of knowledge/information. A entity can be defined as independent information object of real world and understood as conceptual storing place of data needed in a information system construction.

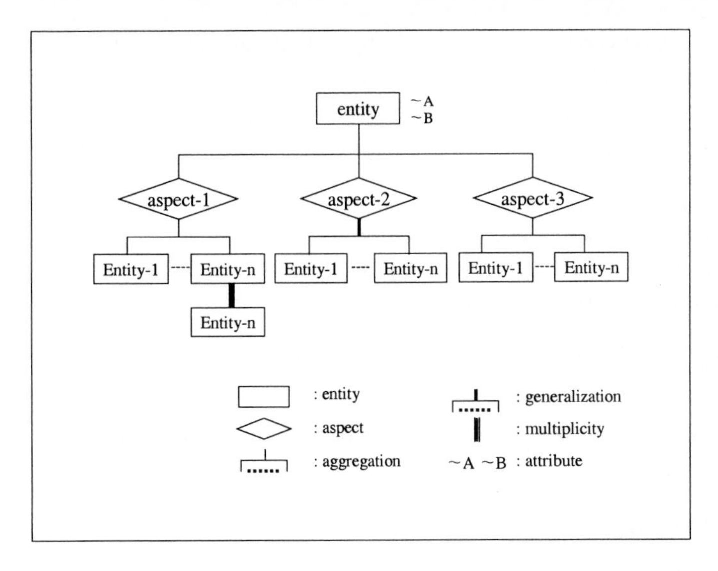

Fig. 2.2 E-A Model's components

To do data modeling using E-R model is process that understand more easily actual problem by defining information composition of present real world by extracting all entities and defining relationships between entities or information as figure 2.3.

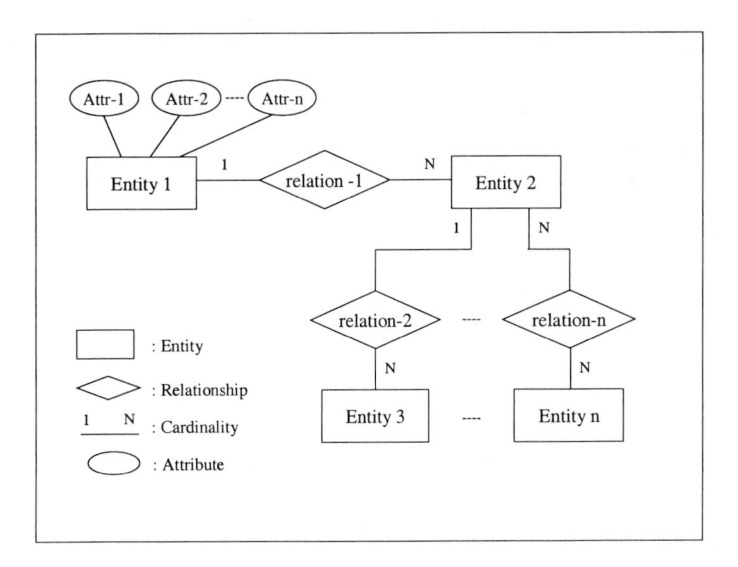

Fig. 2.3 E-R Model's components

As for E-R modeling step, top-down method, classify business areas according to data modeling area, extract entities in the business areas, define relationships between entities, make E-R model, specify attributes in each entities.

E-R model's component consists of entity, relationship and attribute, and only generalization abstraction concept can be provided applied unlike E-A model. Specially, various business processing rule of complicated current real world is applied in definition of relationship by relation link numbers(cardinality) between entities[1].

However, E-R model is easy to representing knowledge and information as graph structure and implementing the model comparatively in E-A model, but there are some difficulty in E-R model because of lacks of abstraction supporting, that is, there is no representation method about various aspect type, specialization, aggregation and decomposition for representing entities differently, and there is difficulty to apply concept of inheritance.

3. Approach for Integrating E-A Model and E-R Model

3.1 I-ER Modeling

When wish to do modeling that specify complicated and various current real world, we usually follow procedure in detail to both way of top-down and bottom-up or generalizing. If a current real world is modeled in those methods, structure of knowledge or information may be hierarchical.

Table 3.1 I-ER Model's notation for integrating E-A and E-R

components(abstraction)	E-A Model	E-R Model	**I-ER Model**
Entity			
Relationship			
Aspect			
Attribute	~		
Aggregation/Decomposition			
Generalization/Specialization		Gs	
Multiplicity			
Cardinality		1 N	1 N
Connector			

But, it is difficult to represent all structure of information hierarchical formally and systematically and even though representing such as hierarchy, those model is high complexity so another new problem will be occurred to manage or implementation. Therefore, in this paper we presented procedure that solve shortcoming of the data abstraction that do not represent variously in E-R model applying E-A model. In extracting entities and defining relationships, we specified two relations. The one is relationship of E-R model, the other is aspect type of applying E-A model. We propose I-ER(Improved E-R) modeling approach that apply both graph based relationship type and frame based hierarchical aspect type.

First in I-ER model symbols, integrated notation is used in case E-R and E-A model using equal notation as shown in table 3.1. General approach for notation of I-ER, it is followed E-R model basically and integrate E-A model's components. In case of collision, relationship and aspect type, we use notation of E-R model's components so aspect type is represented by round quadrilateral. Also, in case of generalization, choose E-A model's notation to represent various abstractions.

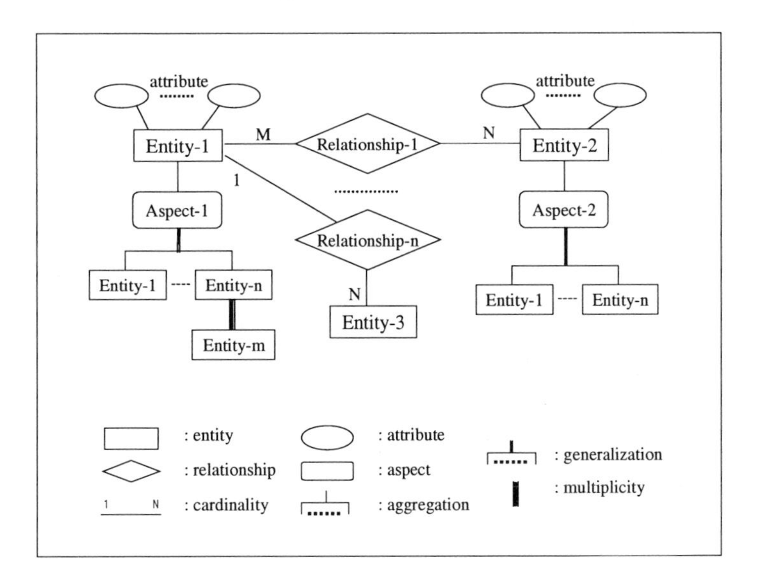

Fig. 3.1 I-ER(Improved E-R) Model's components

Structure of information and knowledge of current real world generally is seems to compounded structure of various abstract concept. The one, relationship that is graph based structure in business rules between entities, the other, hierarchical structure in generalization and specialization about entities. Therefore, in I-ER model, classify relation into 2 divisions. We redefine as follows: the one is relationship between entities, "relationship", the other is hierarchy in entities, "aspect type" as shown in figure3.1.

In I-ER model, there is hierarchical structure of entity type and graph based structure of relationship between entities by abstractions, generalization and specialization, aggregation and decomposition.

3.2 Comparison of E-R Model and I-ER Model

I-ER model applied various abstract concepts that are represented difficult in E-R model using application of E-A model and permitted various representations for complex current real world through those concepts. The major characteristic is that diversified relationship type, as follows 2 summarizing.

First, in case of defining relationship between entities, the relationships explain cardinality between entities and essential/optional participation type of entities.

Second, in case of defining hierarchy in entities, the aspect type is modeled by applying generalization and specialization, aggregation and decomposition and inheritance etc.

Table 3.2 Comparison of I-ER model and E-R model

Content	E-R Model	I-ER Model
Components	Entity Relationship Attribute	Entity Relationship Aspect Attribute
Abstraction Concepts	Generalization	Generalization Aggregation Multiplicity
Structure	Graph Structure	Graph-based Frame Structure
Relationship Representation	Network Representation	Network-based Hierarchical Representation
Inheritance	None	Support

As table 3.2, I-ER model include abstraction concepts that can't represent in E-R model. With components, can classify entity type by particular aspect variously so model is more clear and become formalization in representing knowledge or information by applying various abstraction concepts. Also, in representing structure or relationship, it has graph based tree structure and can support for inheritance.

4. Modeling Procedure for Integration

4.1 Data Modeling Procedure

I-ER modeling process to represent to understand easily complexity of present real world can be approached in the way of two steps. The one is integrated modeling method through defining relationship between entities and hierarchical information structure using aspect type applying I-ER model as appear in figure 3.2, the other is step by step procedure that represent entities and relationship through E-R model and perform I-ER modeling about each particular entities needed more abstraction as appear to figure 4.1.

Explain I-ER modeling procedure in detail. First, extract entities that is information object in real world and define relationship between entities applying business process rules. And extract attributes from relationships and entities so first E-R model is defined. And apply more detailed abstraction about entity type of the E-R model so I-ER model is completed with hierarchical information structure.

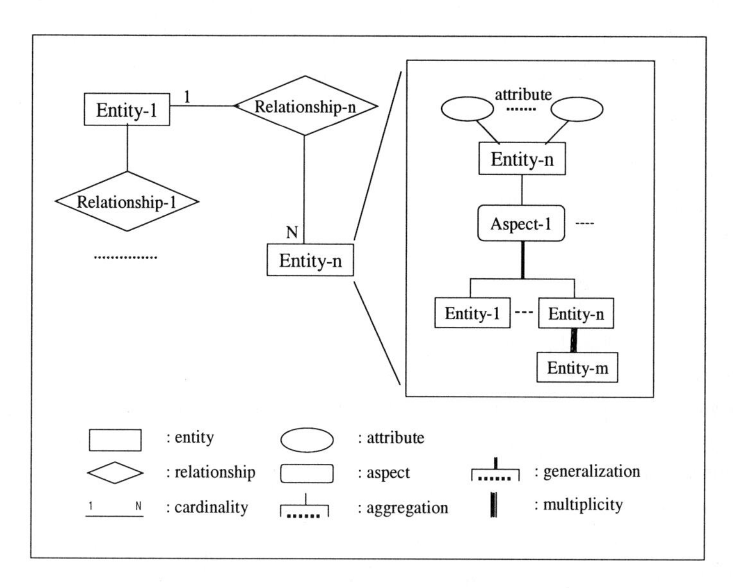

Fig. 4.1 I-ER(Improved E-R) Modeling procedures

I-ER model is used as a tool for conceptual database design. Diagrams applying E-R model includes partly abstraction concept like generalization but other abstraction concept, that is aggregation or inheritance, is not supported. At that point, using I-ER model is proper to representing abstraction and easy to understand real world complexity by appropriate abstraction level.

4.2 An Example of Database Conceptual Modeling Application

Explain example of university management database to examine conceptual modeling procedures using I-ER model. Suppose user's data requirement that have been extracted as following and designed E-R model is drawn in figure 4.2.

1. A student is divided by 3 aspects, that is, first, student being in school and student who stays out of school temporarily according to being in school, second, separated by freshman and other student according to matriculation, third, composed of under-graduate student and graduate student according to school course.

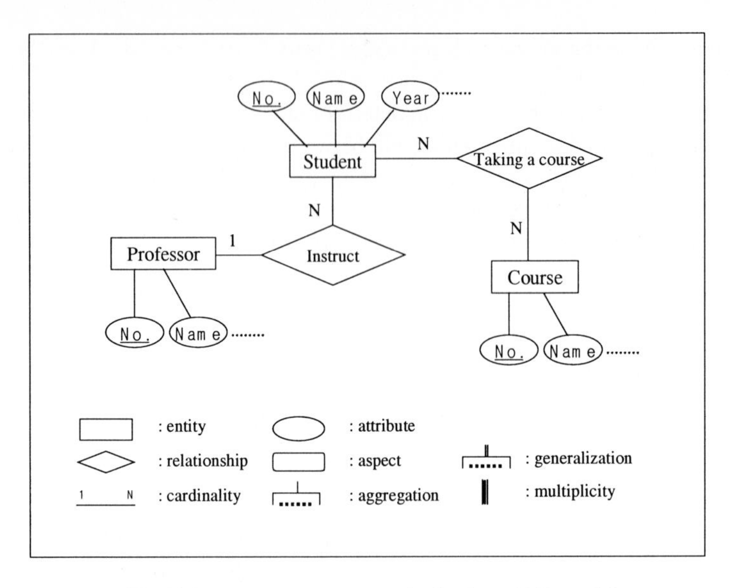

Fig. 4.2 University management database E-R model

2. Courses are separated to indispensable and selection, and divided to major, literacy, optional course according to taking course.
3. A student applies taking a course.
4. A professor instructs several students.

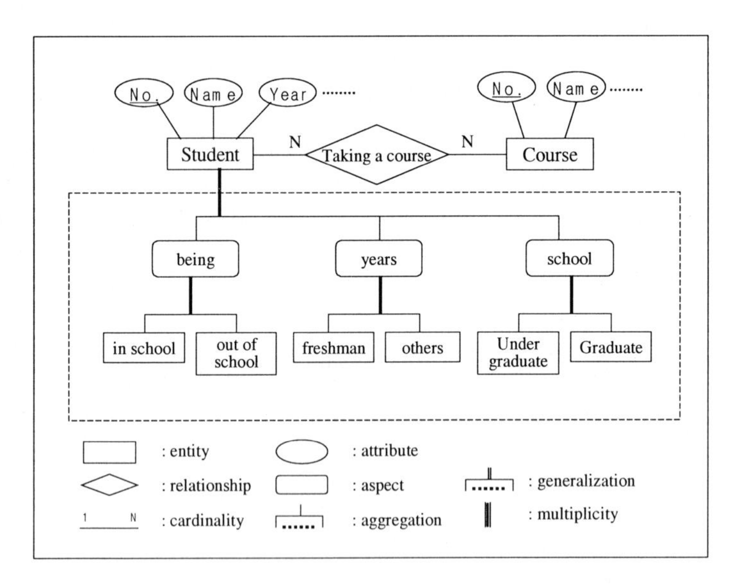

Fig. 4.3 I-ER Modeling with student entity aspect type

In figure 4.3 and 4.4, describe more abstract in detail applying aspect type among entities that appear to E-R model and complete I-ER modeling procedure.

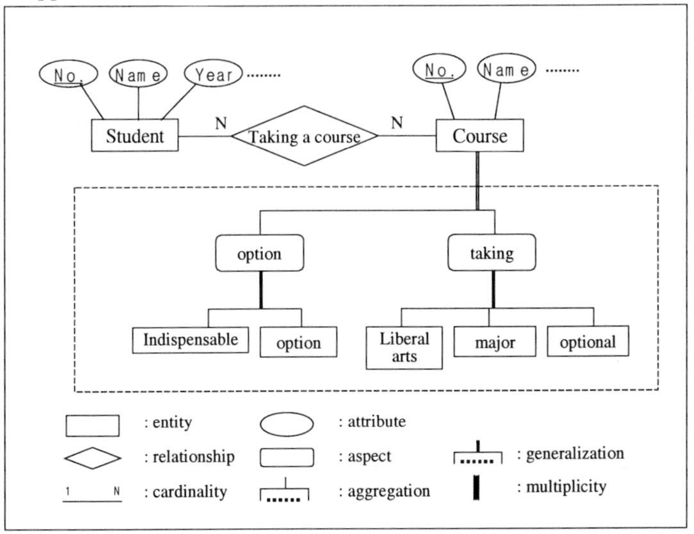

Fig. 4.4 I-ER Modeling with course entity aspect type

5. Conclusion

In this paper, we have shown general concepts of modeling that wish to represent model in order to understand real world's complexity and we propose I-ER(Improved E-R) model in these cases through approach for integrating E-R model and E-A model that is used the most widely. Proposed I-ER model showed that it represent about various abstraction concept of generalization, aggregation and inheritance that can't represent in E-R model. Also, we have shown I-ER modeling procedure that extract entities and define relationship between entities and specify more detailed abstraction about each entity. In this way, a modeling is a need and proper tool to represent current real world clearly and these data models are transformed to physical database at process of database design and construction. By using database designed well we can understand flow and application of information efficiently and get advantage of computer's use.

In the future, we will further research to find integrated model. It is very important that represent structure of information and knowledge more variously, and need research to construct information system easily using various data models.

References

1. Ramez Elmasri and Shamkant B. Navathe, Fundamentals of Database Systems, Addison-Wesley, 1994
2. Peter Rob and Carlos Coronel, Database Systems : Design, Implementation, and Management, International Thomson Publishing, 1993
3. ChangHwa Kim, DooKwon Baik, "REA:Restricted EA model," Proceeding of the KISS Spring Conference, Korea Information Science Society, 16(1):119-122, 1989
4. ChangHwa Kim, DooKwon Baik, "REAPS:Restricted EA modeling Prolog System," Journal of the Korea Information Science Society, 17(5):502-519, September 1990
5. DooKwon Baik, "The EA Model:A Knowledge Representation Model for Expert Systems," Lecture Note(2001.3.12), Dept. of Computer Science & Engineering, Korea University, Seoul, Korea, 2001
6. Jie Cheng, David A. Bell and Weiru Liu, "A Practical Approach to Knowledge Representation and Reasoning in Relational Databases," Proceedings of the 8th International Conference on Tools with Artificial Intelligence(ICTAI '96), 1996
7. Stephan, V., "Extracting symbolic objects from relational databases," Proceedings of the 7th International Workshop on Database and Expert Systems Applications (DEXA), 1996
8. N. Mfourga, "Extracting Entity-Relationship Schemas from Relational Databases:A Form-Driven Approach," Proceedings of the 4th Working Conference on Reverse Engineering(WCRE '97), 1997
9. Morris, R.A. and Khatib, L., "Entities and relations for historical relational databases," Proceedings of the 4th International Workshop on Temporal Representation and Reasoning(TIME '97), 1997
10. Catriel Beeri and Michael Kifer, "An integrated approach to logical design of relational database schemes," ACM Trans. Database Syst., 11(2):134-158, Jun. 1986
11. Carlo Zaniolo and Miachel A. Meklanoff, "On the design of relational database schemata," ACM Trans. Database Syst. 6(1):1-47, Mar. 1981
12. JeongOog Lee, DooKwon Baik, "Automated Integration of Information based on the MultiAspect Semantic Model," In Proceedings of the International Symposium on Future Software Technology(ISFST-98), 259-262, Hangzhou, China, 1998
13. SungKong Park, DooKwon Baik, "An Information Integration Approach Considering Value Heterogeneity," Joint Workshop on Software Engineering, 135-143, Seoul, Korea, April 2001
14. Debabrata Dey, Veda C. Storey and Terence M. Barron, "Improving Database Design through the Analysis of Relationships," ACM Transactions on Database Systems, 24(4):453-486, December 1999
15. Peter Jaeschke, Andreas Oberweis, Wolffried Stucky, "Extending ER Model Clustering by Relationship Clustering," Proceedings of the 12th International Conference on Entity Relationship Approach, Arlington, Texas, USA, December 1993,
16. Heidi Gregersen and Christian S. Jensen, "Temporal Entity-Relationship Models A Survey," IEEE Transactions on Knowledge and Data Engineering, 11(3):464-497, May/June 1999

Design and Implementation of the VoIPv6 Supporting the Differentiated Call Processing*

Chinchol Kim,[1] Byounguk Choi,[1] Yichul Kang,[2] Keecheon Kim,[1] and Sunyoung Han[1]

[1]Department of Computer Science and Engineering , Konkuk University, 1, Hwayangdong, Kwangin-gu, Seoul, 143-701, Korea
{bredkim, buchoi, kckim, syhan}@konkuk.ac.kr
[2]National Computerization Agency, 168, Jukjon-ri, Suji-eub, Yongin-city, Gyonggi-do, 440-717, Korea
kangyc@nca.or.kr

Abstract. We designed and implemented a VoIPv6 system supporting the differentiated call processing using SIP (session initiation protocol) on the IPv6 network. The VoIPv6 system is generally divided into the call control and the voice processing parts. We implemented the SIP6 which is better in simplicity, extensibility, flexibility, scalability, and mobility than H.323 in controlling the calls, as well as IPv6Phone for voice processing. Because our implementation is based on IPv6, it can use QoS (Quality of Service) and security features of IPv6. Particularly, the SIP6 proposed in this paper shows better performance because it uses the differentiated service in the call processing.

1 Introduction

Due to the high speed network and the various Internet services, the number of Internet users are increasing dramatically. But, the current Internet is not sufficient to meet the demand of the users. Especially, there are problems such as duplicate structures of networks (IP network and PSTN), the lack of address space, and QoS. Currently, these problems are considered to be difficult to solve and the several techniques such as VoIP, IPv6, and QoS are considered as solutions to these problems.

In VoIP techniques, H.323 and SIP protocol are used for call control [9][10]. H.323 has been gained a popularity because of its stability and performance through the long time research. However, when you consider it as a global Internet service, it has various problems such as extensibility, mobility, scalability, simplicity, and flexibility as the Internet. However, SIP can overcome these shortcomings [9][10]. Not only it has the function to establish and control the sessions, but also it has a simple structure and the good extensibility as a multicast-based protocol [1][2][9][10]. In addition, It can support the user mobility and interaction with other protocols. As a result, these features have made SIP as a standard in many fields such as 3GPP, VoIP, Messaging systems, and ALL-IP system (next generation mobile system). However, SIP does not

* This work was supported by NCA (National Computerization Agency) of Korea under the contract 2001-hyupyak-wi-02 for Next Generation Internet Infrastructure Development Project.

I. Chong (Ed.): ICOIN 2002, LNCS 2343, pp. 169-179, 2002.

support the differentiated service for the call processing. However it is essential to support differentiated call processing service since it is a default feature of VoIP service [10].

It is likely that the current network system has been unified into a single IP-based system, it leads to the lack of address spaces in IPv4. So it is imperative that IPv6 technologies be developed rapidly and the existing applications must be altered to support the IPv6. IPv6 can retain the innumerable address spaces with 128 bit addressing scheme. And it has an extensive header that supports the QoS and Internet security [4]. Through these features of IPv6, We can expect to gain a performance improvement when we use it in the applications related to the voice and video service.

In this paper, we provide a designed and implemented VoIPv6 system based on IPv6. The SIP proposed by this paper is named as SIP6 as it supports IPv6. The SIP6 shows the better performance in call processing by using a differentiated call processing service. We implemented audio tools including the SIP6 UA (User Agent) for call request and accept.

This paper shows VoIPv6 architecture in section 2 and the result of implementation in section 3. Finally, we present the conclusions and future research in section 4.

2 System Architecture

The VoIPv6 system shown in this paper consists of network server which control session establishment and audio tools with SIP6 UA (SIP6 User Agent Server/Client). For voice service, these two members exchange the information with each other using SIP6 messages. Fig. 1 shows this system components and the following explains each components.

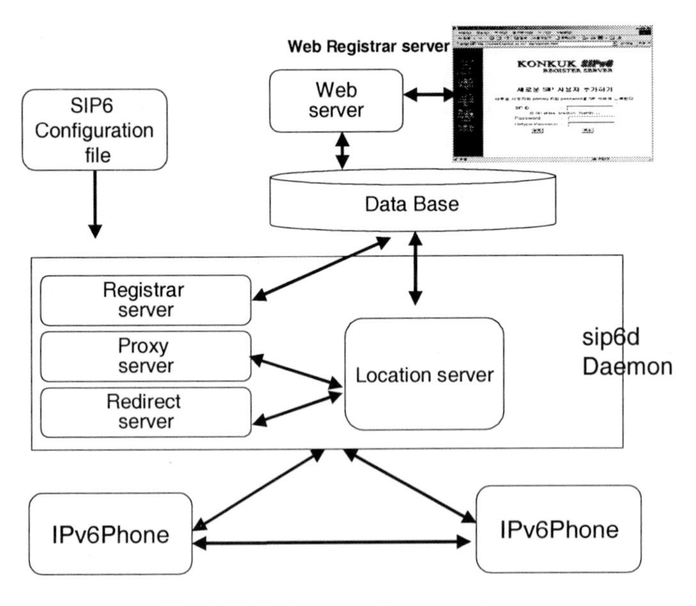

Fig. 1. VoIPv6 System Components

The SIP6 daemon contains Proxy Server, Redirect Server, Location Server, and Registrar Server. It is driven by referencing the configuration file. The Web Registrar Server is responsible for registering the information of user location, It is associated with user database through web interface. The database adds, deletes, or updates the user information.

The IPv6Phone is an audio tool processing the voice service. This is divided into three parts; SIP6 UA for requesting and accepting a session, RTP-UAs (RTP User agent server/client) for audio session control, and CODEC for the voice processing.

2.1 VoIPv6 System Design

2.1.1 SIP6 Server Module Design

SIP6 Server supports the differentiated call processing service for a session establishment through the IPv6 UDP socket interface. It also provides the services for user registration, user management, user location, call-forward, and a call-redirect. Fig. 2 is the SIP6 server structure suggested in this paper.

- Request Receive Module

  This is an UDP receiver which receives SIP6 message producing the IPv6 UDP Socket interface. Those received messages are divided into Request and Response message types and are sent to the Priority Message Parsing Module.
- Priority Message Parsing Module

  It can sort SIP6 messages according to the service level suggested in this paper and applies the differentiated call processing service (see Sect. 2.2.2). The sorted messages are sent to the Method Processing Module for the proper work of the method specified in the message.

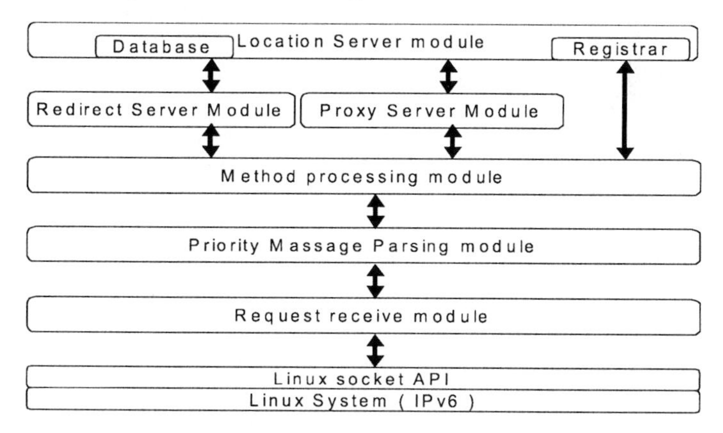

Fig. 2. SIP6 Server Module Structure

- Method Processing Module

  For an INVITE message, the module calls Proxy/Redirect Server Module according to the Server Action information. In the case of REGISTER message, it calls Register Server to register the user information. It also performs a proper work for OPTION, BYE and CANCEL messages.

- Proxy Server Module

  This module manages the forwarding function for the received calls and the other Response messages. Depending on the received methods, it invites a session by forwarding an INVITE method after getting the information of user location from the Location Server Module in the case of INVITE method. The transmitted messages set the service timeout during which the Proxy Server processes the message for a given period. When the timeout is occurred, it will cancel the Request and send 487 Respond message. If it is ACK, 'ACK' method is sent to the destination client.

- Redirect Server Module

  It processes the redirect function for the received Call. When it receives an INVITE method, it initializes the connection header for the Response message. After setting the service timeout, it gets the information of user location. If it receives the address as a domain, it will send URI (user id, host, port) to the client who has requested INVITE after mapping the IP through the address resolving function of the system.

- Location Server Module

  This module is responsible for enrolling the user information and tracking the user location. The following elements are required for managing and enrolling the user information.

  - ***Registrar*** is the registration gateway receiving the user enrollment.
  - ***User Input table*** manages the user ID authorized for accessing the system.
  - ***User Identity table*** is used to give different IDs per user. So there is an unique User ID used in the session establishment in this table.
  - ***User Contact table*** manages the user's current location.

  User information from the Registrar can be transformed to the unique User ID which can be found in the User Identity table and registered into the User Contact table in the database. The registered data are used to serve the user location requests from the Proxy and Redirect Server. The following is an example of user entry information.

  {User ID, SIP6 UA Port, Current URL, Server Action, Service Level}

2.1.2 Differentiated Call Processing

We applied the differentiated call processing in establishing the SIP6 session. We define three different service levels for QoS.

Fig. 3 shows the Priority Call Processing Processor supporting the differentiated call processing.

- Classify Processor

  Those received messages are classified and stored at the High quality buffer, Medium quality buffer and Normal quality buffer according to the service level specified in the Flow Label field of IPv6 packet.

In this paper, the service level is defined as follows;

- ***High quality*** : the message having a superior priority, it supports the fast session establishment and uses the High quality buffer.
- ***Medium quality*** : the message having a middle priority, it uses the Medium quality buffer.
- ***Normal quality*** : the message having a low level priority. It uses the Normal quality buffer.

● Priority Scheduling Processor

This processor continually examines the messages stored in the Multiple buffers and applies the differentiated call processing service according to the service level. It can dynamically process three buffers at a time when a message is arrived. In the highest level priority buffer, processor directly calls the Message Parsing Processor to process the message. After processing the message, it looks for the buffer of next level priority. While processing the message in the lower priority buffer, it can be stopped and handle the message of the higher priority. By this mechanism, It can provide a differentiated call processing service.

● Message Parsing Processor

In order to process the message received by the Priority Scheduling Processor, it can be divided into several parts.

- Request Message Processing

  To avoid the duplicate message, the Request message can be mapped into an unique ID by hash function, and the message with the Request-Uri, TO, FROM, Call-ID and Cseq must be stored in the „execution buffer". After confirming the uniqueness of the message, it can be parsed to detect the errors. At last, Method Processing Module is called for the suitable services for the matching methods of BYE, INVITE, OPTIONS, and REGISTER.

- Response Message Processing

  It performs a response processing for a received Request message and a received Response message. In processing the Response message, it sets up a Response status code for the Request message. And it then generates a socket to send the Response message to the corresponding client.

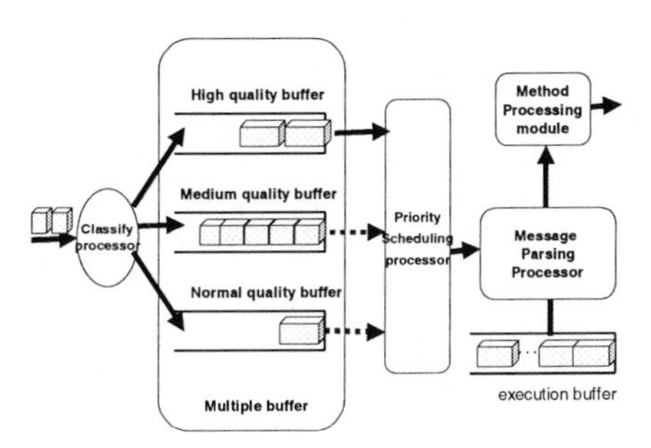

Fig. 3. Priority Call Processing Processor

2.1.3 IPv6Phone Module Design

IPv6Phone uses the IPv6 UDP socket interface. We proposed to a unified user interface that is consisted of the SIP6 UA and RAT (open audio tool). Fig. 4 is the IPv6Phone Module Structure used in our research.

IPv6Phone is has a GUI for user interface and SIP6 UA (SIP6 User Agent Server/Client) for SIP6 session control, and audio tool.

GUI includes QoS marker for setting the service level and the unified GUI with RAT and SIP6 UA. The SIP6 UA has a function to send and receive calls in order to establish the sessions. The open software RAT is used to process the voice data and to establish the RTP session.

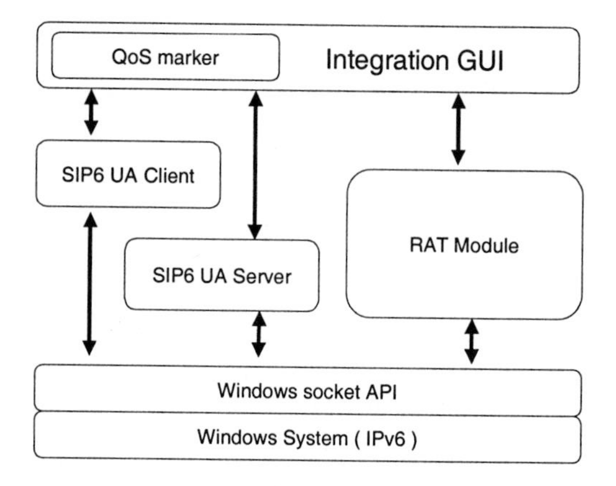

Fig. 4. IPv6Phone Module Structure

2.2 Call Processing Procedure and Message

The VoIPv6 System suggested in this paper follows the call processing procedure and message formats specified in IETF RFC 2543.

3 Implementation and Results

3.1 SIP6 Server Implementation

In this paper, SIP6 Server Module is implemented with C language on Linux Kernel 2.2.x supporting the IPv6. The programs are designed to compose the Proxy Server, Redirect Server, Location Server, and Registrar Server Module. It is executed by the daemon process named SIP6d. Fig. 5 shows the operation of SIP6 Server by using main functions.

When we start a SIP6 Server, it creates a thread that can open an UDP socket and it receives an INVITE Request message. When SIP6 Server receives a message, the message is classified by checking the service level specified in the Flow Label field of the IPv6 header. SIP6 Server also creates another thread to process the classified request message sequentially using the service level proposed in this paper. Finally, this thread selectively calls Proxy, Redirect, Registrar Server Module.

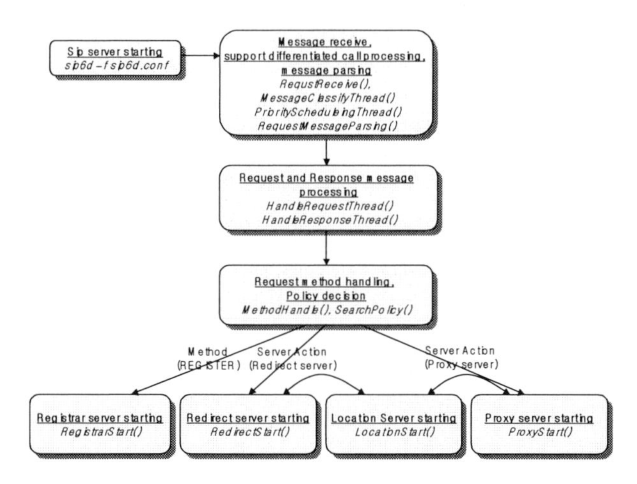

Fig. 5. Procedure of SIP6 Server

Fig. 6 shows Registrar Server window on the web page. It is implemented with the PHP programming language. This Registrar Server has various functions such as user information registration, User Identity table update, and user location registration.

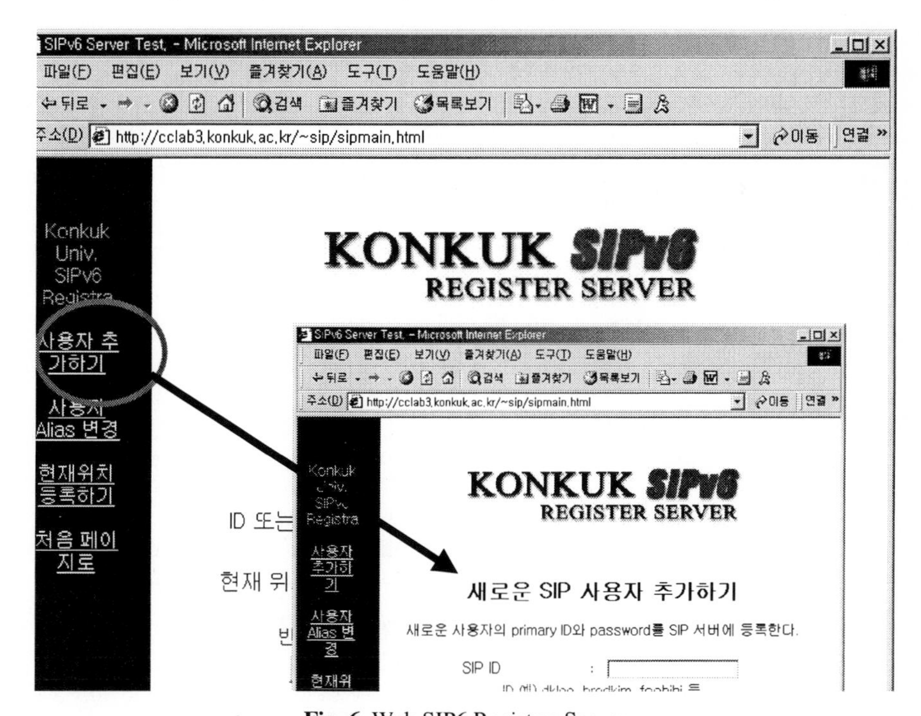

Fig. 6. Web SIP6 Registrar Server

3.2 Differentiated Call Processing

We used the flow label field of the IPv6 header to specify the service levels in applying the differentiated call processing scheme. Currently, the Flow Label field is 24-bit size to support QoS. But it has not been used yet. In this paper, we defined the following code values for service levels.

Table 1. Code value for service levels

Service level	Code value
High quality	1(hexadecimal)
Medium quality	2(hexadecimal)
Normal quality	0(hexadecimal)

In the implementation, the Flow Label field in IPv6 socket structure is used to specify the code value for the service level.

```
struct sockaddr_in6 {
        uint8_t          sin6_len;
        sa_family_t      sin6_family;
        in_port_t        sin6_port;
        uint32_t         sin6_flowinfo;
        struct           in6_addrsin6_addr;
}
```

When the session is established, SIP6 UA specifies a service level in in6_flowinfo field of socket_in6 structure. If SIP6 Server receives an INVITE Request message, it checks a service level in sin6_folwinfo field and dynamically process the INVITE message using the service level specified by the user.

Fig. 7 and Fig 8 describe the functions MessageClassifyThread() and PrioritySchedulingThread() in pseudo codes.

```
/* Classify Processor */
IF ( client_sock.sin6_flowinfo is 1 )
         Save message in High quality buffer
ELSE IF (client_sock.sin6_flowinfo is 2)
         Save message in Medium quality buffer
ELSE
         Save message in Normal quality buffer
```

Fig. 7. The pseudo-code of MessageClassifyThread()

```
/* Priority Scheduling Processor */
...
while(1)
{
/*Message processing in high quality buffer */

Label 1:
WHILE ( High quality buffer have request message )
     Massage parsing and processing
/*Message processing in medium quality buffer */

Label 2:
WHILE (Medium quality buffer have request message)
  {
     IF (High quality buffer have request message)
          Goto Label 1;
     Else
          Massage parsing and processing
  }
```

```
/*Message processing in normal quality buffer */
WHILE (Normal quality buffer have request message)
  {
     IF (High quality buffer have request message)
          Goto Label 1;
     IF ELSE (Medium quality buffer have request message)
          Goto Label 2;
     ELSE
          Massage parsing and processing
  }
}
```

Fig. 8. The pseudo-code of PrioritySchedulingThread()

3.3 IPv6Phone Implementation

IPv6Phone is implemented with TCL/TK language to support the user interface and Winsock API to control the IPv6 messages.

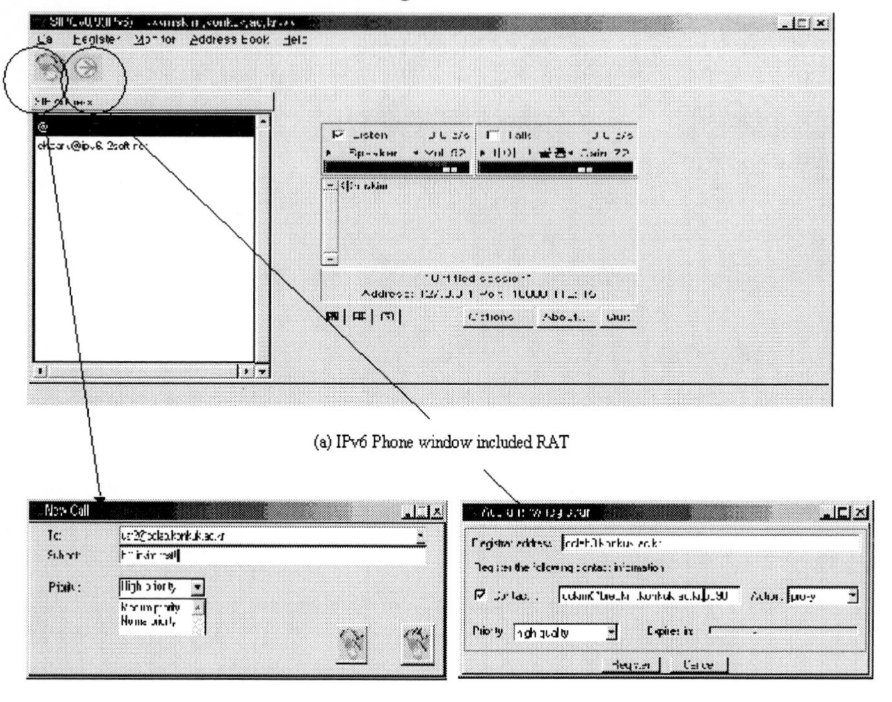

(a) IPv6 Phone window included RAT

(b) New session invite window

(c) New register window

Fig. 9. IPv6Phone Running Figure

IPv6Phone includes RAT (open audio tool) for the voice service and SIP6 UA to establish a session. Also, IPv6Phone implemented in this paper supports a unified user interface integrated in RAT and SIP6 UA.

Fig. 9 shows the running IPv6Phone windows. In the new session invite window, priority field for service level is included.

4 Conclusions and Future Work

In this paper, we designed and implemented the VoIPv6 using SIP6 protocol on the IPv6 network. First, We implemented the SIP6 Server which is more powerful in simplicity, extensibility, flexibility, scalability, and mobility than H.323, as well as a IPv6Phone (Audio Tools) for voice processing. Because this implementation was done on IPv6, it can use the QoS and the security features of IPv6 and resolve the lack of address space on the Internet. We applied the differentiated call processing technique to improve the performance.

In the future, we will analyze the various functions to increase the performance for SIP6 and IPv6Phone.

References

1. Handley, H.Schulzrine, E.Schooler, and J.Rosenberg, "SIP: session initiation protocol", RFC 2543, March1999.
2. A.Johnston, S.Donovan, et al, "SIP Telephony Call Flow Examples", Internet Draft, November 2000.
3. R.Gilligan, S.Thomson, J.Bound, W.Stevens, "Basic Socket Interface Extensions for IPv6", RFC 2553, March 1999.
4. R. Hinden, S. Deering, „Internet Protocol version 6(IPv6) specification", RFC 2460, December 1998.
5. K. Nichols, S. Blake, F. Baker, D. Black, „Definition of the Differentiated Services Field (DS Field) in the IPv4 and IPv6 Headers.", RFC 2474, December 1998.
6. 6. P. Ferquson and G. Huston, „Quality of Service", John Wiley & Sons, 1998.
7. S. Donovan, J. Rosenberg, „The SIP Session Timer", Internet Draft <draft-ietf-sip-session-timer-08.txt>, August 24, 2001.
8. ITU-T Recommendation H.323.
9. I. Dalgic and H. Fang, „Comparison of H.323 and SIP for IP Telephony Signaling", Photonics East, Proceeding of SPIE'99, Boston, Massachusetts, Sep. 1999.
10. Henning Schulzrinne and Jonathan Rosenberg, „Signaling for Internet Telephony", January 11, 1998.

Application-Oriented Flow Control in Dynamic Networking Architecture

Gen Kitagata, Takuo Suganuma, and Norio Shiratori

Research Institute of Electrical Communication / Graduate School of Information Science Tohoku University 2-1-1 Katahira, Aoba-ku, Sendai, 980-8577, Japan

Abstract. In this paper, we propose a new architecture of the global communication networks, the Dynamic Net-working Architecture. In the proposed architecture, a new functional layer called Flexible Network Layer (FNL) is introduced between the application layer and the transport layer to enhance the capabilities of communication networks by dealing with various changes detected in human users, applications, platforms and networked environment. To realize the FNL, we adopt an agent-based computing framework as a software infrastructure to develop and manage various components and related knowledge of the FNL. In this paper we give an internal architecture and agent-based design of the FNL. We also show an experimental application using the FNL, the Dynamic Flow Control Application, which performs the user-oriented flow control, to discuss the characteristics and effectiveness of the proposed architecture.

1 Introduction

In recent years, many useful communication services and applications have been provided for users over the global communication networks such as the Internet. The traditional network applications usually use the standardized transport services such as TCP/UDP services directly, however, several limitations are pointed out against the inflexible architecture.

One is the limitation come from a user-centered viewpoint. By using the transport services, application services can be realized in a restricted way depending on the provided services of the networks. This means that applications and users have to have many additional tasks, such as seeking logical network location of communication peer, connection establishment operations, user level QoS controls, recovery from faults and so on. Because these burdened tasks are common among many applications, heuristic knowledge of users/designers/operators can be utilized effectively in an intensive manner.

The other is the limitation from viewpoints of heterogeneity of the networked environment. Many systems and protocols have been proposed for the dedicated environment ranging from the mobile communication with personal digital devices to the high speed and resource allocable networks. However, the maintenance of consistency of systems, including networks, devices, and applications,

I. Chong (Ed.): ICOIN 2002, LNCS 2343, pp. 180–190, 2002.

that have heterogeneous functions and different performances is not easy to attain.

To overcome these limitations, a new network architecture is required based on the user orientation and the software engineering perspectives. Therefore we propose a new architecture of the global communication networks, the Dynamic Networking Architecture, based on the concept of Flexible Network[1]. In the proposed architecture, a new functional layer called Flexible Network Layer (FNL) is introduced as an adaptive middleware between the application layer and the transport layer[2]. The dynamic functions of the layer enhance the capabilities of communication networks to deal with various changes detected in human users, applications, platforms and networked environment.

We have been working on stepwise development, evaluation and improvement of the proposed architecture by implementing some prototype applications based on this architecture. The Dynamics Flow Control Application (DFCA) is one of these prototypes, that allows us to evaluate the user orientation property of the Dynamic Networking Architecture. The DFCA detects the user level priority among network applications in run-time, and performs application oriented flow control in FNL.

In this paper we introduce the concept, design, and one of the applications of the Dynamic Networking Architecture. Firstly, we give an internal architecture and agent-based design of the FNL. Subsequently, we show the detail of an experimental application of the FNL, the Dynamics Flow Control Application, to discuss the characteristics and effectiveness of the proposed architecture.

In the following sections, firstly, we give the concept and an architecture of Dynamic Networking. In section 3, we describe an agent framework and a design of FNL based on the framework. Then, in section 4, we explain the design and implementation of DFCA. We also show some experimental results and evaluation on the prototype system, and discuss the effectiveness of the proposed architecture in section 4

2 Architecture of Dynamic Networking

An architecture of Dynamic Networking is shown in Fig. 1. A new functional layer called Flexible Network Layer (FNL) is introduced as a middleware layer placed between the application layer and the transport layer. By inserting the middleware layer, network related functions used in common by applications can be intensive in the layer. This means that applications can access to the advanced network functions without understanding the underlying networked environment's complexity. On the other hand, another approach exists that uses the powerful network-level functions provided by IPv6 and so on. These approaches are well studied but issues in scalability are still open. The middleware approach can absorb the heterogeneity of network functions and can give a solution to the scalability problem.

Compare to the traditional middleware for distributed computing[3], the FNL has the following characteristics, i.e.,

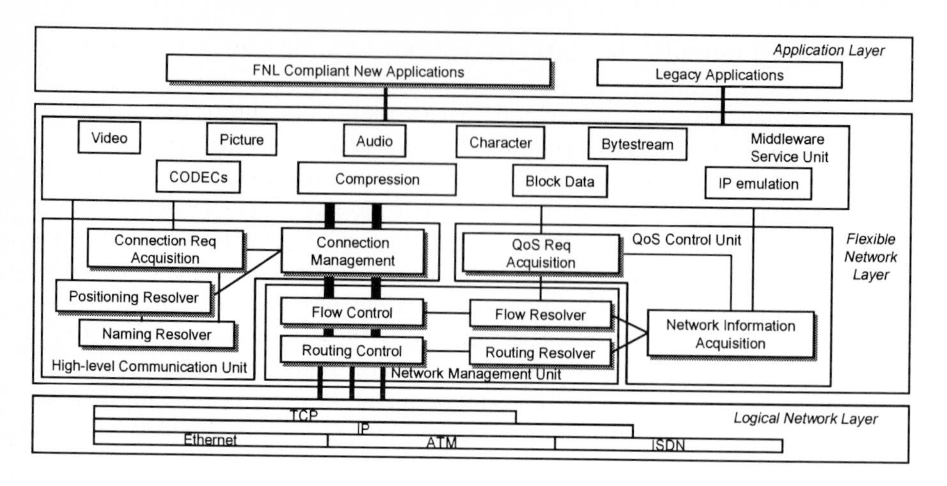

Fig. 1. FNL Architecture

(C1) Collect and accumulate knowledge regarding the global networked environment such as information of users, users' requirements, functional specifications of services, configuration of both platforms and networks and so on,

(C2) Monitor the quality of service (QoS) at the user's requirement level and the application level in order to grasp the operational situations of the networked environment,

(C3) Manage and control the QoS of the network functions,

(C4) Coordinate the interactions between the application functions and the network functions in order to organize and reorganize the services provided for users,

(C5) Provide an integrated way to access to the functions of heterogeneous networks,

(C6) Assimilate and utilize new services/functions of both the applications and the networks to enhance the capability of global networked environment, and,

(C7) Tune the internal configuration of FNL automatically based on the platform's static/dynamic properties.

Based on these properties, the FNL consists of the following function units, i.e., Middleware Service unit, High-level Communication unit, QoS Control unit and Network Management unit, as depicted in Fig. 1.

(U1) Middleware Service Unit (MSU). The MSU maintains the multimedia service components such as video service, picture service and audio service provided by the FNL. The most suitable components of MSU are selected and configured in the runtime automatically. The applications are also constructed dynamically using components of MSU as their communication part.

(U2) High-level Communication Unit (HCU). The HCU deals with the high-level naming and the location maintenance. HCU has to keep tracks of users' logical locations over the networks, and establishes a associations between a user and a service based on the location information. HCU also maintains the associations' intermittent connectivity under unstable network environment such ad wireless communication networks.

(U3) QoS Control Unit (QCU). The QCU is responsible for the QoS controls based on both the user requirements and the network status information. QCU decides the allocations of resources for respective flows of applications by using network information collected from various network nodes.

(U4) Network Management Unit (NMU). The NMU manages the application level connections defined over the transport layer of logical network. NMU finds and selects the appropriate intermediate gateways defined in the FNL and configures the application-oriented connections based on the users' requirements.

3 Agent Framework for Dynamic Networking

To realize the Dynamic Networking Architecture, flexible and robust software infrastructure is required to enable the dynamic property of the architecture. In addition to the basic capability of the infrastructure including facility of the interaction among distributed software modules, the following technical requisites are taken into consideration,

1. Component based computing: Dynamic connection of a set of small-size software components in order to construct large-scaled distributed systems,
2. Legacy software connectivity: Effective reuse of matured and refined software modules,
3. Programming model independency: Flexible selection of programming models compare to the single model architecture such as the Server/Client based synchronous RPC model,
4. Service coordination: Task decomposition to the components, load balancing, conflict detection and resolution, QoS monitoring, component place arrangement and so on,
5. Service continuity: Effective reuse of case knowledge that is used in the past sessions.
6. Design and development support: Cooperation among components that are designed and developed by different people under different situations.

To accomplish these requisites, the whole system should consist of a unified framework and behave in coordinative manner. Moreover each component of the system should have intelligent processing capability such as cooperative behavior and autonomy. From these points of view, the Dynamic Networking Architecture should be constructed as a multiagent system.

Many multiagent platforms have been developed by researchers and software companies. From careful investigation of existing agent platforms based on the requisites above, we decided to apply the ADIPS framework[4]. ADIPS framework mainly consists of the following subsystems; i.e., 1) Workspace which provides the agents' operational environment on the distributed platform, and 2) Repository which manages the reusable agents and send them onto the Workspaces specified by the users' requirements.

Receiving a request from an agent on the Workspace, the Repository retrieves reusable agents, creates an organization of agents which satisfies the request, and sends them to the Workspace by using the ADIPS Organization/Reorganization protocol (AORP). On the other hand, the agents on the Workspaces can communicate with each other by using the ADIPS Communication/Cooperation protocol (ACCP) which holds a set of performatives customized on the basis of KQML.

According to the agent framework given by the ADIPS framework, we have designed an agent-based Dynamic Networking Architecture. We have two types of Repository, i.e., AP-Repository and FN-Repository. AP-Repository contains function components for applications in forms of agent (AP-func agents). FN-Repository contains function components for FNL as FN-func agents as well. AP-func agents and FN-func agents are created and registered by agent developers. In this phase legacy software modules are also agentfied and utilized effectively in the agent organization. On the other hand, the Application Layer (APL) and the FNL are constructed as Workspaces.

When a user requirement is issued to the APL, it is transferred to both of the repositories with additional information such as conditions of network and the target platform on which the layer is working. According to these information, agents in the repositories make agent organization based on the AORP, and necessary and sufficient agents are instantiated to the APL and the FNL. In APL and FNL, agents try to keep contract-based relationships to maintain the functional and QoS requirements from users. The agents negotiate each other to perform the service coordination tasks using the ACCP.

4 Application-Oriented Flow Control Mechanism

4.1 Concept of DFCA

IntServ[5] and DiffServ[6]are well known QoS reservation framework specified by the IETF, which can guarantee bandwidth and delay required by applications. However, implementation of signaling protocol such as RSVP[7] and the ability of handling the DS field of each IP packet are required for all IP routers on particular route. Therefore, a scalability problem is pointed out, i.e., network environment on which these QoS mechanisms can be available is very limited (P1). Moreover application is required to handle the signaling protocol or DSCP setting to the DS field in order to use the framework. Consequently, another problem from an aspect of application reuse is also pointed out, i.e., it is needed to modify an application to make it able to use the framework (P2).

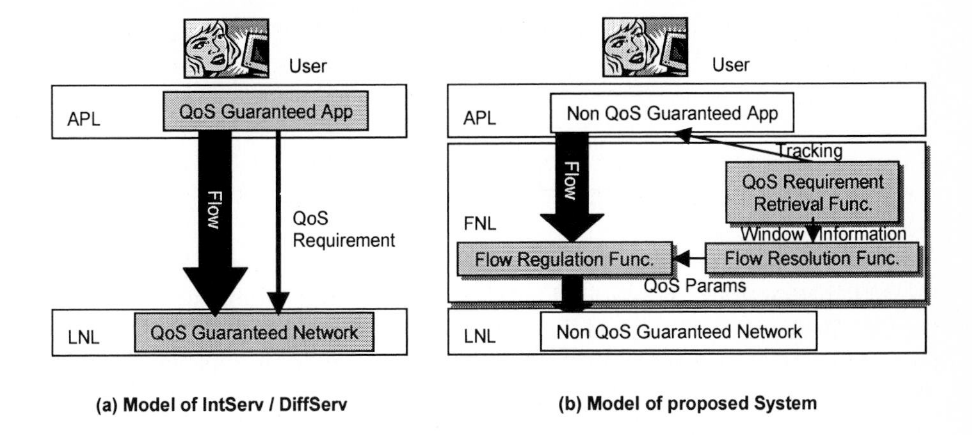

Fig. 2. Comparison of QoS control models

To deal with above problems, we propose a Dynamic Flow Control Application (DFCA), which is one of the services of FNL. DFCA is realized by close cooperation between NMU and QCU of the FNL described in section 2.

IntServ and DiffServ control application flows in the logical network layer, while DFCA controls application flows in the application layer. Since DFCA can be used regardless of the availability of specific QoS guarantee techniques of the network, it would solve the problem (P1). For instance, DFCA stores and forwards application flow between the application layer and the logical network layer using FCF, and controls frequency to take out data stream from each queue according to priorities. As a result, DFCA does not depend on flow control mechanisms of IP routers in the logical network layer and can be used in various network environment.

Additionally, by using QCU of the FNL, DFCA can retrieve the status of U/I window that a user is operating, and decides flow priorities based on the status. As a result, DFCA solves the problem (P2). For instance, QCU watches window status such as order of cascading application windows and geometry of the windows. Based on the status, DFCA calculates priorities of application flows. Thus, a user can control flow of each application without any modifications of existing applications.

Fig. 2 shows a comparison of network QoS control model of proposed system and existing systems such as Intserv/DiffServ. The remarkable advantage of our QoS control model is that we put all functions needed for QoS control inside the FNL, thus no special implementations are required for applications as well as the underlying network.

4.2 Design of DFCA

In this section, we show the design and implementation of following three functions to compose the proposed system DFCA, these are, QoS Requirement Re-

Table 1. A representation of QoS parameters

QoS parameters	
handle	Window ID
classname	Application ID
ratio	Flow Priority ratio

trieval Function, Flow Resolution Function and Flow Regulation Function. Here, the QoS Requirement Retrieval Function is an implementation of 'QoS Req Acquisition' part in QCU. The Flow Resolution Function is an implementation of 'Flow Resolver' part in NMU, and Flow Regulation Function is that of 'Flow Control' part in NMU as well. Actually, these functions are implemented as a single agent or multiple agents.

Design of QoS Requirement Retrieval Function. We assume that the window I/F status on the display represents the user's requirement to the applications precisely. This function obtains status of application windows to recognize the application priority preferred by the users. This function obtains the following information: Window ID Application ID Z order of windows Window size (geometry) Maximize and minimize status This information is forwarded to the Flow Resolution Function.

Design of Flow Resolution Function. This function generates network QoS parameters using window information obtained by QoS Requirement Retrieval Function, and then informs the QoS parameters to Flow Regulation Function described in next section. The window status and the network QoS parameters are represented as shown in Table 1 and 2 respectively. Flow Resolution Function has a flow resolution knowledge in the simple form of IF-THEN rules. There exist window status in the condition part and network QoS parameters in action part of a rule, respectively. The value of QoS parameters is represented by inter-application priority ratio.

Table 2. A representation of window status

Window information	
handle	Window ID
classname	Application ID
zorder	Z order of window
width/height	Window size
max/min	Maximize and minimize status

```
//Network Information//
(network_state :width low :delay normal :reliability high)

//Window Information//
(window_info :classname IEFrame :handle 8D5 :activity no
             :zorder 2 :max no :mini no :width 223 :height 156)
(window_info :classname VTwin32 :handle D77 :activity yes
             :zorder 1 :max no :mini no :width 303 :height 236)

//Rules//
(window_info :classname ?c :handle ?h :activity yes
             :zorder 1 :max ?max :mini ?mini :width > 100 :height > 100)
(network_state :width low :delay normal :reliability high)
-->
(qos_param :classname ?c :handle ?h :ratio 0.8 ) //QoS Parameters//
```

Fig. 3. Example of network information, window information and QoS parameter reasoning rules

Fig. 3 shows network information, window information, and network QoS parameter generating rules. Network information in Fig. 3 shows the network environment information. The network QoS parameters are derived by applying the reasoning, i.e., rule-based knowledge is dynamically activated by obtaining network and window information.

Design of Flow Regulation Function. This function controls each application data flow directly according to the network QoS parameters generated by Flow Resolution Function. The function has queues for each application, and manipulates individual application flow. Additionally, the function dynamically changes the slice, which is a size of streaming data taken out from a queue at one time. If quick response is preferred then it makes the slice smaller, otherwise makes it larger. So that, it achieves flexible application flow control, not only for throughput, but also for response time.

4.3 Implementation

According to the above design, we implemented a prototype of the proposed system. We implemented the system with Visual C++, Java, and TAF[8] on Microsoft Windows Me operating system. To retrieve application window status, we used message hook API on Microsoft Windows. The Flow Regulation Function is implemented as HTTP and Telnet proxy.

4.4 Experiments

To evaluate the proposed functionalities and efficiency, we have two experiments using the prototype system.

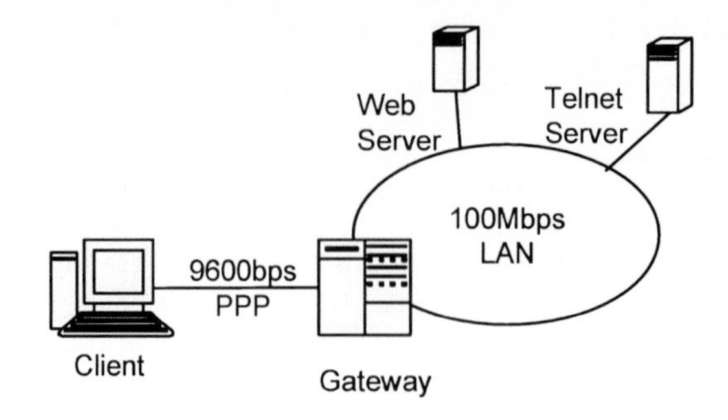

Fig. 4. Experiment environment

Table 3. The results of experiment1: response time

	Response time
QoS control off	9.35[sec]
QoS control on	Approx. 1.00[sec]

Experiment Environment. Fig. 4 shows the experiment environment. The Client and the Gateway are PCs (each CPU is Pentium4 1.4 GHz and OS is Microsoft Windows Me). The Web Server and the Telnet Server are Sun Microsystems Ultra SPARCStations (each OS is Solaris7). The Client and the Gateway is connected by no QoS guaranteed network (Serial line PPP at 9600bps). On the Client, a Web browser (Internet Explorer) and a Telnet client (TeraTerm) are used as applications. Both applications do not support QoS control as they are. We decided connection speed between the Client and the Gateway at 9600bps so as to be much lower than backbone network.

Experiment1: Response Time. In this experiment, we examine response time becomes faster by using the proposed system. We measured the Telnet echo response time under condition that HTTP flow is dominated.
Experiment Conditions:

- A large file is being downloaded by HTTP, and HTTP flow occupies PPP connection line.
- When QoS control by DFCA is activated, priority ratio of Telnet and HTTP flows is 1:1.
- When QoS control by DFCA is activated, slice size to take out from queue is 1 byte.
- Telnet character is sent by 1 character with enough interval.

Table 3 shows the results of this experiment. The results are mean time of 10 times trials. By the result, it is confirmed that when QoS control is turned on, response time becomes faster than it is off. It means that Telnet and HTTP

Table 4. The results of experiment2: throughput

Priority ratio(HTTP:Telnet)	Telnet throughput
1:2	117.80[bps]
1:20	274.89[bps]

flows are transmitted at 1:1 by Flow Regulation Function, and response time of the Telnet flow becomes guaranteed even though HTTP flow is dominated.

Experiment2: Throughput. In this experiment, we examine Telnet throughput becomes higher according to the change of HTTP/Telnet flow priority ratio from 1:2 to 1:20. To confirm the Telnet throughput, we measured time to download a large file by Telnet.
Experiment Condition:

- A large file is being downloaded by HTTP, and HTTP flow occupies PPP connection line.
- QoS control by DFCA is activated.
- Priority ratio of Telnet and HTTP flows is changed from 1:2 to 1:20.
- Slice size to take out from queue is 1 byte.

Table 4 shows the results of this experiment. The results show the average throughput of 10 trials. By the results, when priority ratio is changed from 1:2 to 1:20, it is confirmed that Telnet throughput becomes higher (274.89bps). The result shows that Flow Regulation Function of proposed system can control flow ratio according to the QoS parameters calculated by Flow Resolution Function.

Evaluation. By the results of above experiment 1 and 2, even though network and applications don't have QoS guarantee mechanisms such as IntServ or DiffServ, it is confirmed that our proposed system DFCA can controls each application's flow and the problems (P1) and (P2) are resolved.

5 Conclusion

In this paper, we proposed a new architecture of the global communication networks, the Dynamic Net-working Architecture. In the proposed architecture, a new functional layer called Flexible Network Layer (FNL) is introduced between the application layer and the transport layer to enhance the capabilities of communication networks by dealing with various changes detected in human users, applications, platforms and networked environment. To realize the FNL, we adopted an agent-based computing framework as a software infrastructure to develop and manage various components and related knowledge of the FNL. We gave an internal architecture and agent-based design of the FNL. We also showed an experimental application using the FNL, the Dynamic Flow Control Application, which performs the user-oriented flow control, to discuss the characteristics and effectiveness of the proposed architecture.

In the future work, we will continue to develop rest of the Dynamic Networking Architecture, and to apply to the advanced applications such as virtual reality environment and multimedia communication systems.

References

1. N. Shiratori, K. Sugawara, T. Kinoshita and G. Chakraborty: Flexible networks: Basic concepts and architecture. IEICE Trans. Commun, E77-B(11), (1994) 1287–1294
2. T. Suganuma, T. Kinoshita and N. Shiratori: Flexible Network Layer in Dynamic Networking Architecture. Proc. of International Workshop on Flexible Network and Cooperative Distributed Agents (FNCDA2000), (2000) 473–478
3. K. Geihs: Middleware Challenges Ahead. IEEE Computer, Vol.34, No.6, (2001) 24–31
4. S. Fujita, H. Hara, K. Sugawara, T. Kinoshita and N. Shiratori: Agent-Based Design Model of Adaptive Distributed Systems. Applied Intelligence, Vol.9, No.1, (1998) 57–70
5. J. Wroclawski: The Use of RSVP with IETF Integrated Services. IETF Request for Comments 2210, (1997)
6. S. Blake, D. Black, M. Carlson, E. Davies, Z. Wang, W. Weiss: An Architecture for Differentiated Service. IETF Request for Comments 2475 , (1998)
7. R. Braden, Ed., L. Zhang, S. Berson, S. Herzog, S. Jamin: Resource ReSerVation Protocol (RSVP) – Version 1 Functional. IETF Request for Comments 2205 , (1997)
8. H. Hara, S. Konno, K. Sugawara and T. Kinoshita: Design and Implementation of Training-system for Agent Framework TAF. Proc. of Workshop on Software Agent and its Applications (SAA2000), (2001) 183–190 (In Japanese)

II. Switching and Routing

An Optimal Path Selection Algorithm for Static and Mobile Multicast Routing Problems

Shigeto Tajima[1], Masakazu Fujii[2], Nobuo Funabiki[3], Tokumi Yokohira[3], Kazufumi Tsunemura[1], and Teruo Higashino[1]

[1] Department of Informatics and Mathematical Science Osaka University, 1–3 Machikaneyama, Toyonaka 560–8531, Japan
[2] Kyushyu Matsushita Electric Co. Ltd., 4–1–62 Minoshima, Hakata, Fukuoka 812–8531, Japan
[3] Department of Communication Network Engineering Okayama University, 3–1–1 Tsushimanaka, Okayama 700–8530, Japan

Abstract. In this paper, we present a new approximation algorithm called OPSAM (Optimal Path Selection Algorithm for Multicast) for the static multicast routing problem and the newly defined mobile multicast routing problem. Given a graph with the cost and the delay associated to each edge, a source node, a destination node set, and a delay time limit, the first problem requires finding a multicast tree such that the total cost is not only minimized, but also the delay along any path from the source to a destination does not exceed its limit. OPSAM first extracts plural path candidates for each destination to satisfy the delay constraint simultaneously. Then, it iteratively selects better paths among candidates for finding a low-cost tree within a short computation time. The performance of OPSAM is verified through simulations in Waxman model and Tiers model, where OPSAM is better than the best existing algorithm BSMA, especially for Tiers model instances. The second problem requires finding a sequence of multicast trees to follow changes of mobile user locations, such that the sum of total tree costs and the difference between two consecutive trees is minimized. The simulation results show the performance of OPSAM is better than BSMA.

1 Introduction

Recently, a number of multicast communication applications including Video-on-Demand and videoconference have been studied among academics and industries. The multicast communication transmits the same set of packets from one source node to several destination nodes through the network. In order to realize many multicast communication services simultaneously, the required network cost in each session should be minimized. This significant problem of finding a multicast routing tree to satisfy both conditions is called a multicast routing problem. This problem is known to be NP-complete [1].

In the multicast routing problem, the topology of the network is described by a graph. Each link is usually associated with two weights corresponding to a cost

I. Chong (Ed.): ICOIN 2002, LNCS 2343, pp. 193–202, 2002.

and a delay time respectively. Besides, a source node to initiate packets, a group of destination nodes to receive them, and a delay time upper limit are given as the input to the problem. Then, the static multicast routing problem requires finding a multicast tree to satisfy both conditions above-mentioned for fixed locations of destinations. When destination nodes move around dynamically, a sequence of multicast trees should be composed to follow them. This problem is called the dynamic multicast routing problem. In this paper, we formulate a special case of this problem, where the same set of mobile users continuously changes their locations. In order to minimize the packet loss through modifications of multicast trees, the change of the trees must also be minimized, in addition to the cost minimization of each tree. Actually, for the penalty of a tree change, we consider the sum of delay times on the links, which are newly removed from the tree, because the packets existing on such links are lost in the new tree. We call it the mobile multicast routing problem in this paper.

Several heuristic algorithms have been reported for the static multicast routing problem. In [3], Parsa, Zhu, and Garcia-Luna-Aceves proposed BSMA (Bounded Shortest Multicast Algorithm), which improves a least-delay tree by iteratively replacing superedges in the tree. Among existing algorithms including them, the performance comparisons in [4][9] showed that BSMA provides the best solution quality. However, BSMA usually suffers from the large time complexity and thus, requires a lot of computation time to solve large-scale instances.

This paper presents a new approximation algorithm called OPSAM (Optimal Path Selection Algorithm for Multicast) for the static multicast routing problem. OPSAM is a two-stage algorithm that is composed of the path candidate generation stage and the iterative tree search stage. In the first stage, OPSAM generates plural path candidates from the source node to each destination node at the same time by modifying the k-path generation algorithm in [10]. OPSAM composes an initial routing tree from generated path candidates. In the second stage, OPSAM iteratively improves the multicast tree by selecting better candidates. The performance of OPSAM is evaluated through extensively solving randomly generated instances with two types of topologies, Waxman and Tiers model. As the performance reference, BSMA is also implemented with the modification for the mobile problem, and is executed for the same set of instances.

2 Static and Mobile Multicast Routing Problems

2.1 Static Multicast Routing Problem

The inputs to the static multicast routing problem are a directed connected graph $G = (V, E)$, a source node s, a set of destination nodes $D = \{d_i | 1 \leq i \leq |D|\} \subseteq V - \{s\}$, and an upper limit on the delay time Δ. Each edge or link in $E(G)$ is associated with two weights. One weight to the link l is called a cost $C(l)(l \in E)$, which may represent the consumed network bandwidth on the link, and another one is a delay time $D(l)$, which represents the transmission time through the link. The required output to this problem is a multicast routing tree $T = (V_T, E_T)$, which includes the source s and every destination node in D. Then, the constraint of the problem is that the total delay time along the routing path from the source to any destination does not exceed the given upper limit Δ. It is described by

$$\sum_{l \in P} D(l) \leq \Delta \tag{1}$$

where $P(s, d_i) \in T$ represents a routing path to the destination d_i. The objective function of the problem is to minimize the sum of the costs associated to links in the multicast tree T, which is given by

$$\sum_{l \in E_T} C(l) \to min \tag{2}$$

2.2 Formulation of Mobile Multicast Routing Problem

In the mobile multicast routing problem, mobile users are moving around in the network, and their locations are reported in a regular time interval called a time slot. In each time slot, users' closest nodes become the destination nodes, from where packets are transmitted to users by radio communication. Since a set of destination nodes is obtained at each time slot, a sequence of sets of destination nodes is actually given for some time slots in this problem. Then, the output is a sequence of multicast routing trees where each tree realizes the multicast communication corresponding to each configuration of destination nodes. Here, we note that the time slot interval may be sufficiently large to compute the reconfigurations of multicast routing trees, because as in [13], mobile users are usually migrating locally, and the local networks recently visited by users should be active for a while. The objective function is slightly modified so as to minimize the packet transmission loss due to changes of multicast routing trees, in addition to the cost minimization. The objective function is given by

$$\sum_{t} \{ \sum_{l \in E_{T_t}} C(l) + K \times \sum_{\substack{l \in E_{T_{t-1}} \\ l \notin E_{T_t}}} D(l) \} \to min \tag{3}$$

where t represents a time slot, T_t does a multicast routing tree at time slot t, and K does a constant coefficient to adjust the weight balance between two factors respectively. In our simulations in Section 4, K is fixed to 3.

3 Proposal of OPSAM

3.1 Design Concept of OPSAM

In order to realize efficient computation of providing high quality solutions within short time, OPSAM consists of two stages:

1. path candidate generation stage, and
2. iterative tree search stage.

The existing BSMA applies the Dijkstra shortest path algorithm [11] in each iteration to replace superedges. To avoid the repeated applications of the shortest path algorithm, OPSAM finds plural path candidates for each destination in the same run of this algorithm, and memorizes them in the first stage. Then, the second stage iteratively improves the solution by replacing path candidates in the tree.

3.2 Path Candidate Generation Stage

By modifying the k-path generation algorithm in [10], this stage generates K_{d_i} path candidates for the destination d_i such that any path candidate satisfies the delay constraint of the problem. This number of path candidates K_{d_i} is given by

$$K_{d_i} = A \times \big((\text{number of nodes}) + (\text{number of destination nodes}) + (\text{distance between source and } d_i)\big) \tag{4}$$

where A is constant coefficients.

Usually, a destination can generate more than K_{d_i} different path candidates. Thus, OPSAM memorizes only K_{d_i} path candidates to save computation time and memory in ascending order of the selection function F:

$$F = (\text{sum of costs on links along path}) - \alpha(\text{sum of costs on links shared by path candidates for other destinations}) \tag{5}$$

where α is constant coefficients.

In OPSAM, path candidates are extracted sequentially in descending order of the minimum delay time to the destination node, because distant destinations from the source may not have degree of freedom for path candidates due to the delay constraint, while near destinations can have more freedom so that they can trace existing path candidates. After path candidates are generated, one candidate among them that minimizes the cost increase is selected for the initial solution of the multicast routing tree. The detailed procedure of the path candidate generation stage in OPSAM is as follows:

1. Find the shortest path from the source to the destination in terms of the delay time, and find the minimum delay time by using Dijkstra algorithm [11].
2. Execute the following procedure for each destination $d_i \in D$ in descending of the minimum delay time:
   a) Generate K_{d_i} path candidates in ascending order of F by using the k-path generation algorithm.
   b) Select one path candidate that minimizes the cost increase for the initial solution.

3. Memorize the initial solution T and the link cost $C_T = \sum_{e \in T} C(e)$ as the best-found solution: $T_{best} = T$ and $C_{best} = C_{T_{best}}$, where T_{best} represents the best-found multicast routing tree and C_{best} does the corresponding link cost respectively.
4. Check the possibility of closed path occurrence between any pair of path candidates for different destinations.

3.3 Iterative Tree Search Stage

This stage iteratively improves the solution by replacing the path candidates in the tree T. Actually in each step, after randomly selecting one destination, OPSAM finds another path candidate, which leads to the minimum cost of the tree among candidates different from the currently selected one, and replaces it for a new tree. Then, if this new tree includes closed paths, the path candidates for different destinations from the replaced one that compose closed paths are replaced by other candidates so that the resulting tree has no closed path. The following procedure describes the details of this stage:

1. Randomly select one destination d_i.
2. Find a new path candidate to minimize the tree cost, and replace it to compose a new tree T.
3. Check the existence of closed paths in T, and if there exist, apply the following procedure to remove them:
   a) Remove the path candidates except one in 2 that compose the closed paths.
   b) Select new path candidates to recompose a tree, which do not only make no closed path but also minimize the tree cost among available candidates. If any path candidate makes a closed path for some destination, select path candidates minimizing the cost tree without considering a closed path.
4. Set $T_{best} = T$ and $C_{best} = C_T$, if $C_T < C_{best}$.
5. Repeat 1–4 in L times, and then output T_{best} and C_{best}.

3.4 Time Complexity of OPSAM

Let n be the number of nodes in the network ($n = |V(G)|$), $|D|$ be the number of destination nodes, L be the number of iterations in the second stage, and K be the maximum number of path candidates K_{d_i}. Then, in the first stage, the path candidate generation procedure requires $O(|D|n^2)$ [10] time, and the closed path detection procedure does $O(n|D|^2K^2)$ time. The second stage requires $O(LKn)$ time, because each iteration requires $O(Kn)$ time. As a result, OPSAM requires $O(|D|n^2 + n|D|^2K^2 + LKn)$ time. In our simulations, K and L in OPSAM are proportional to n, and $|D|$ is constant for each instance, while k in BSMA is proportional to n. In this case, the total time complexity of OPSAM is $O(n^3)$, which is better than $O(n^4 \log n)$ for BSMA.

3.5 Extension of OPSAM for Mobile Multicast Routing Problem

OPSAM can be applied to the construction of the initial multicast routing tree in the first time slot without modifications. However, for tree modifications from the second time slot, OPSAM needs to consider the following modifying points because of changes of user locations:

1. new destination nodes may appear,
2. existing destination nodes may disappear, and
3. the objective function is modified.

The modified procedure of OPSAM for the mobile multicast routing problem is as follows:

1. Find destination nodes from the initial user locations.
2. Apply OPSAM to obtain the initial multicast routing tree.
3. Repeat the following procedure in *tslot* times, where *tslot* is the number of time slots:
   a) Find destination nodes from user locations.
   b) If existing destination nodes disappear, remove the corresponding path candidates from the tree.
   c) If new destination nodes appear, apply the following procedure:
      i. Generate path candidates for the new destinations as in the first stage in OPSAM, and check the possibility of closed paths between any pair of path candidates for different destinations.
      ii. Select one candidate among them, which minimizes the increase of the objective function, and insert it into the tree.
   d) Iteratively improve the tree as in the second stage.
4. Output the sequence of trees.

4 Simulation Results

In order to evaluate the performance of the proposed OPSAM, both of OPSAM and BSMA have been implemented on Pentium III 866 MHz with FreeBSD. The solution quality and the computation time are compared between two algorithms.

4.1 Simulated Network Topology

Two types of random graph topologies, namely Waxman model [6] and Tiers model [7][8], are adopted. In Waxman model, n node coordinates are randomly generated on an $n \times n$ Euclid plane, and any pair of nodes, u, v, are connected by a link with the probability:

$$P(u,v) = \beta \exp(\frac{-d(u,v)}{\alpha L}) \tag{6}$$

where $d(u,v)$ represents the Manhattan distance between u and v, and α and β represent real numbers between 0 and 1 which control the link density of the network.

Tiers model has been known as a random graph model to well approximate a real communication network topology. In this model, nodes and links are partitioned into three layers of WAN, MAN, and LAN, and connections between different layers become fewer than connections inside the same layer.

4.2 Simulations for Static Waxman Model

Parameters to generate problem instances for Waxman model are selected by following those in [3]:

1. Average node degree is 5.
2. Each link cost is given by the Manhattan distance between corresponding two nodes.
3. Each delay time is given by (link cost) $\times$ (random real number between 0 and 1).
4. The source node and destination nodes are randomly selected without overlapping.
5. The delay time upper limit is given by d_{max}, $d_{max} \times 1.2$, and $d_{max} \times 1.5$, where d_{max} is the largest one among minimum delay times for destinations.
6. The number of nodes n in the network is given by 25, 50, 75, and 100.
7. The number of destinations is varied by 18 different numbers between 4 and 50.
8. The number of instances generated with different random numbers per each size is 100.

Besides, the number of path candidates k for BSMA, which is important to determine the performance, is given by $n/2$.

The following parameters are empirically selected for OPSAM through simulations: $\alpha = 15$, $A = 0.2$, and $L = 200n$. Table 1 shows the simulation results, where average costs and computation times are summarized for the case of $|D|$ = 4–25 and for the case of $|D|$ = 38–50. This table shows that the computation time for OPSAM is 1.6–8 times faster than that for BSMA. The solution quality (costs) of OPSAM is better or comparable for smaller size instances of $|D|$ = 4–25, whereas that is 0.4–2.1 % worse for larger size instances of $|D|$ = 38–50. The reason is that BSMA can generate new path candidates dynamically by repeatedly applying the shortest path algorithm, whereas OPSAM generates path candidates only in the first stage and they may not be sufficient for large size instances.

4.3 Simulations for Static Tires Model

Network topologies are generated by the existing Tires model generation program in [12]. The instance parameters are selected as follows:

1. Each link cost is given by the inverse number of occupied bandwidth generated in the program.
2. Each delay time is given by the delay time generated in the program

Table 1. Simulation results for static Waxman model.

Δ	$\|D\|$	OPSAM		BSMA	
		Cost	Time(sec)	Cost	Time(sec)
d_{max}	4-25	474.6	0.239	476.4	1.957
d_{max}	38,50	1483.7	3.456	1471.9	9.215
$1.2d_{max}$	4-25	442.7	0.370	444.3	2.032
$1.2d_{max}$	38,50	1429.8	4.715	1412.9	9.621
$1.5d_{max}$	4-25	428.2	0.526	426.4	2.073
$1.5d_{max}$	38,50	1409.7	5.889	1380.7	9.633

Table 2. Simulation results for static Tiers model.

Δ	$\|D\|$	OPSAM		BSMA	
		Cost	Time(sec)	Cost	Time(sec)
d_{max}	4-25	1624.1	0.537	1637.3	6.531
d_{max}	38	2575.4	1.671	2605.9	10.345
$1.2d_{max}$	4-25	1498.8	0.689	1519.8	7.588
$1.2d_{max}$	38	2483.3	2.016	2537.0	11.523
$1.5d_{max}$	4-25	1497.1	0.813	1516.7	7.606
$1.5d_{max}$	38	2484.1	2.313	2534.4	11.565

3. The number of nodes is given by 130, 195, and 260.
4. The number of destinations is varied by 15 different sizes between 4 and 38.
5. The other parameters for instance generations are the same as those for Waxman model.

The following parameters are empirically selected: $\alpha = 75$, $A = 0.04$, and $L = 120n$. Table 2 shows the simulation results. This table shows that the computation time for OPSAM is 5–12 times faster than that for BSMA, and the solution quality is 0.8–2.1 % better. For this model, BSMA may be easily stuck at local minimum because superedges are limited due to the few connections between different layers. On the other hand, OPSAM can find more variety of path candidates for destinations to satisfy the delay constraint in the first stage. Thus, OPSAM is considered the better algorithm for practical network topologies.

4.4 Simulations for Mobile Multicast Routing

The instance parameters for network topologies follow those for Waxman and Tiers models in the static multicast routing problem. User locations are generated by the following method:

1. Initial locations of users (x_i^0, y_i^0) $(1 \leq i \leq |U|)$ are randomly generated on the $n \times n$ Euclid plane with allowing overlapping. Moving velocities of users

Table 3. Simulation results for mobile instances by Waxman.

		OPSAM				BSMA			
Δ	$\|D\|$	Cost	Tree change	Objective function	Time (sec)	Cost	Tree change	Objective function	Time (sec)
d_{max}	4-25	2841.3	158.4	3316.5	0.631	2818.8	178.3	3353.7	6.617
d_{max}	38,50	8182.7	342.6	9210.3	5.585	8215.4	408.8	9441.9	28.886
$1.2d_{max}$	4-25	2673.6	149.5	3122.1	0.927	2658.4	162.4	3145.6	6.764
$1.2d_{max}$	38,50	7900.5	330.6	8892.3	7.471	7855.8	378.6	8991.7	29.209
$1.5d_{max}$	4-25	2606.7	143.6	3037.6	1.265	2565.2	153.2	3024.8	6.800
$1.5d_{max}$	38,50	7797.8	320.2	8758.5	9.218	7663.5	355.6	8730.4	29.531

v_i are given by random integers between 1 and 5 for Waxman, between 50 and 200 for Tiers, and their moving angles to the horizontal axis θ_i are given by random integers between 0 and 360.

2. Locations of users are updated by:

$$x_{i+1}^t = x_i^t + v_i \times \cos(\theta_i \times \pi/180),\ y_{i+1}^t = y_i^t + v_i \times \sin(\theta_i \times \pi/180) \quad (7)$$

   When any user moves out of the $n \times n$ plane, the moving angle is modified by adding a random integer between 90 and 180 so that it can move to a place inside the plane.
3. Moving angles is slightly modified by adding a random integer between -10 and 10.

The same parameter set for the static multicast routing problem is adopted here except for $L = 1/5 \times$ (L for static problem) in tree modifications. Tables 3 and 4 summarize the simulation results, where the average of total tree costs, tree changes, objective function values, and computation times are shown. These tables show that the computation time for OPSAM is always 3–10 times for Waxman model and 12–22 times for Tiers model faster than that for BSMA. The solution quality for OPSAM is 0.7–2.5 % better then that for BSMA except for $\Delta = 1.5d_{max}$ for Waxman. BSMA needs to apply the shortest path algorithm for every destination whenever new destinations are added. On the other hand, OPSAM only needs to apply the shortest path algorithm for newly added destinations. As a result, as the number of time slots increases, the difference in the computation time also increases, and BSMA becomes impractical.

5 Conclusion

This paper presents a new approximation algorithm called OPSAM for the NP-complete static multicast routing problem, and the newly defined mobile multicast routing problem. OPSAM consists of two stages for efficient computations. We verify the performance of OPSAM through simulations for static and mobile multicast routing problem where our algorithm finds better solutions in shorter time and better solution quality than BSMA.

Table 4. Simulation results for mobile instances by Tiers.

		OPSAM				BSMA			
Δ	$\|D\|$	Cost	Tree change	Objective function	Time (sec)	Cost	Tree change	Objective function	Time (sec)
d_{max}	4-25	9242.4	2353.5	16303.0	1.853	9326.0	2356.8	16396.5	41.122
d_{max}	38	14883.2	5380.0	31023.1	4.547	15088.7	5399.0	31285.9	62.088
$1.2d_{max}$	4-25	8639.6	2340.0	15659.5	2.037	8787.0	2334.3	15790.0	43.968
$1.2d_{max}$	38	14353.6	5368.9	30460.2	4.834	14686.5	5367.4	30788.7	65.476
$1.5d_{max}$	4-25	8626.8	2338.6	15642.8	2.231	8770.2	2333.7	15771.2	44.104
$1.5d_{max}$	38	14347.3	5367.5	30450.0	5.343	14677.4	5366.4	30776.7	65.995

References

1. V. P. Kompella, J. C. Pasquale and G. C. Polyzos, "Multicast routing for multimedia communication", IEEE/ACM Trans. Networking, Vol. 1, No. 3, pp.286–292(1993).
2. R. Sriram, G. Manimaran and C. Siva Ram Murthy, "Algorithms for delay-constrained low-cost multicast tree construction", Computer Commun., Vol. 21, pp.1693–1706(1998).
3. M. Parsa, Q. Zhu and J. J. Garcia-Luna-Aceves, "An iterative algorithm for delay-constrained minimum-cost multicasting", IEEE/ACM Trans. Networking, Vol. 6, No. 4, pp.461–474(1998).
4. X. Jia, "A distributed algorithm of delay-bounded multicast routing for multimedia applications in wide area networks", IEEE/ACM Trans. Networking, Vol. 6, No. 6, pp.828–837(1998).
5. Q. Zhang and Yiu-Wing Leung, "An orthogonal genetic algorithm for multimedia multicast routing", IEEE Trans. Evolutionary Computation, Vol. 3, No. 1, pp.53–62(1999).
6. B. M. Waxman, "Routing of multipoint connections", IEEE J. Select. Areas. Commun., Vol. 6, pp.1617–1622(1988).
7. K. L. Calvert, M. B. Doar and E. W. Zegura, "Modeling Internet Topology", IEEE Commun. Magazine, June, pp.160–163(1997).
8. M. B. Doar, "A Better Model for Generating Test Networks", Proc. GLOBECOM '96 (1996).
9. H. F. Salama, D. S. Reeves and Y. Viniotis, "Evaluation of multicast routing algorithms for real-time communication on high-speed networks", IEEE J. Select. Areas. Commun., Vol. 15, No. 3, pp.332–345(1997).
10. T. Baba, N. Funabiki, and S. Nishikawa, "A proposal of a greedy neural network for route assignments in multihop radio networks", IEICE Trans., Vol. J81-D-I, No. 6, pp700–707(1998).
11. T. H. Cormen, C. H. Leiserson and R. L. Rivest, "Introduction to algorithms", MIT press(1990).
12. M. B. Doar, "A Random Network Topology Generator", http://www.pobox.com/~doar/.
13. C.-C. Tseng, K.-H. Chi, and T.-L. Huang, "A new locality-based IP multicasting scheme for mobile hosts", Computer Communications, Vol. 24, pp.486–495(2001).

Performance Evaluation of Combined Input Output Queued Switch with Finite Input and Output Buffers

Tsern-Huei Lee and Ying-Che Kuo

Institute of Communications Engineering
National Chiao-Tung University
Hsinchu, Taiwan 300, Republic of China
{thlee,mike}@atm.cm.nctu.edu.tw

Abstract. It has recently been shown that a combined input output queued (CIOQ) switch with a speedup factor of 2 can exactly emulate an output-queued (OQ) switch [1]-[6]. In particular, the maximal matching algorithm, named Least Cushion First/Most Urgent First (LCF/MUF) algorithm presented in [6], can be executed in parallel to achieve exact emulation. However, the buffer size at every input and output port was assumed to be of infinite size. This assumption is obviously unrealistic in practice. In this paper, we investigate via computer simulation the performance of the LCF/MUF algorithm with finite input and output buffers. We found that, under uniform traffic, a CIOQ switch behaves almost like an OQ switch if the buffer sizes at every input and output ports are 3 and 9 cells respectively. For correlated traffic, to achieve similar performance, the input and output buffer sizes have to be increased to about 7 and 11 times of the mean burst size, respectively.

1 Introduction

Over the years, many service scheduling algorithms have been proposed to provide quality of service (QoS) guarantees in an integrated services network [7]-[10]. Most of these algorithms were designed to be used at the output ports of an output queued (OQ) switch. The main problem of OQ switches is that the switching fabric of an $N \times N$ switch must run N times as fast as its line rate in the worst case. As such, OQ switches have serious scaling problem because the advancement in memory bandwidth is much slower than the advancement in transmission speed. Consequently, input queuing is unavoidable in building a large capacity switch.

On the other hand, input queued switches suffer from head-of-line blocking which limits the maximum throughput to about 0.586 (under uniform traffic assumption) [11]. It can be improved to approach 100% if cells are delivered from input ports to output ports based on maximum matching [12]. However, the high computational complexity of currently known algorithms prohibits maximum matching from being used in a high-speed switch. Although many maximal matching algorithms (e.g., PIM

I. Chong (Ed.): ICOIN 2002, LNCS 2343, pp. 203-214, 2002.

[13], LPF [14], and *i*SLIP [15]) have been proposed, none of these algorithms can provide QoS guarantee even though the complexity is lower than maximum matching. Another approach to improve the performance of an input queued (IQ) switch is to speedup the switching fabric. Because of speedup, an output port may receive cells faster than it can transmit. As a result, buffering at output is necessary and the switch becomes a combined input output queued (CIOQ) switch.

Obviously, to obtain good performance in a CIOQ switch, one has to wisely schedule the usage of switching fabric. Several algorithms had been proposed to achieve the goal. It was shown that a CIOQ switch with a speedup factor of 2 and well-designed matching algorithm can exactly emulate an OQ switch. In particular, a maximal matching algorithm named Least Cushion First/Most Urgent First (LCF/MUF) was proposed in [6]. The LCF/MUF algorithm can be executed with a parallel procedure of complexity $O(N)$, where N denotes the number of input and output ports. It was proved that, using the LCF/MUF algorithm, a CIOQ switch with a speedup factor of 2 can exactly emulate an OQ switch. However, to achieve exact emulation, the buffer size at each input and output port was assumed to be of infinite size. This assumption is obviously unrealistic in practice. In this paper, we investigate via computer simulations the performance of the LCF/MUF algorithm with finite input and output buffers. In our study, the input traffic is characterized by uniform or correlated model. With finite input and output buffers, the property of exact emulation is lost. However, simulation results show that, under uniform traffic, a CIOQ switch behaves almost like an OQ switch if the buffer size at every input port and every output port are 3 and 9 cells respectively. For correlated traffic, to achieve similar results, both input and output buffer sizes have to be increased to about 7 and 11 times of the mean burst size, respectively.

2 The LCF/MUF Matching Algorithm

In this section, we review the LCF/MUF matching algorithm proposed in [6]. Since the switching fabric is speeded up by a factor of 2, there are two scheduling phases in each slot, where a slot is defined as the time duration to transmit a cell. In each scheduling phase, the LCF/MUF scheduling algorithm matches each nonempty input with at most one output, and conversely, each output with at most one input. Cells are then delivered to output ports based on the matching outcome.

Before describing the LCF/MUF algorithm, we introduce some definitions. Let $x_{i,j}$ denote a cell at input port i destined to output port j.

Definition 1 : The cushion of cell $x_{i,j}$ at input port i, denoted by $C(x_{i,j})$, equals the number of cells currently residing in output port j which will depart the emulated *OQ* switch earlier than cell $x_{i,j}$.

Definition 2 : The cushion between input port i and output port j, denoted by $C(i,j)$, is the minimum of $C(x_{i,j})$ of all cells at input port i destined to same output port j. If there is no cell destined to output port j, then $C(i,j)$ is set to ∞.
Definition 3 : The scheduling matrix of an $N\times N$ switch is an $N\times N$ square matrix whose $(i,j)^{th}$ entry equals $C(i,j)$.

The LCF/MUF matching algorithm is described below:
Step 1. Select the $(i,j)^{th}$ entry of the scheduling matrix which satisfies $C(i,j) = min_{k,l}\{C(k,l)\}$ (Least Cushion First). If the selected entry is ∞, then stop. If there are more than one entry with the least cushion residing in different columns, then choose the most urgent cell $x_{i,j}$ among those input ports which correspond to the selected entries (Most Urgent First).
Step 2. Eliminate the ith row and the jth column (i.e., match output port j to input port i) of the scheduling matrix. If the reduced matrix becomes null, then stop. Otherwise, use the reduced matrix and go to *Step 1.*

3 System Model

As a benchmark with which the CIOQ switch we studied is compared, we assume there exists a shadow OQ switch. The traffic arrived at the CIOQ switch is identical to that arrived at the shadow switch. Our goal is to arrange for each cell to depart from the CIOQ switch at exactly the same time it departs from the shadow OQ switch, i.e., to exactly emulate the OQ switch with the CIOQ switch.

3.1 The Shadow OQ Switch Which Adopts Strict Priority Service Scheme

For ease of simulation and feasibility in implementation, we select the strict priority scheme as the service discipline of the shadow OQ switch. We assume that every output port j maintains K FIFO queues named as $Q_{j,k}$ with priority *1* to priority K respectively (priority *1* has the highest priority). At each output port j, the strict priority service algorithm is described below.

```
for k=1 to K {
        if queue Q_{j,k} is non-empty then
                the head-of-line cell of Q_{j,k} is served and break for-loop.
}
```

3.2 The CIOQ Switch We Studied

Fig. 1 illustrates the conceptual model of an $N\times N$ CIOQ switch with finite buffers. Every input port maintains per output port, per priority queues. In other words,

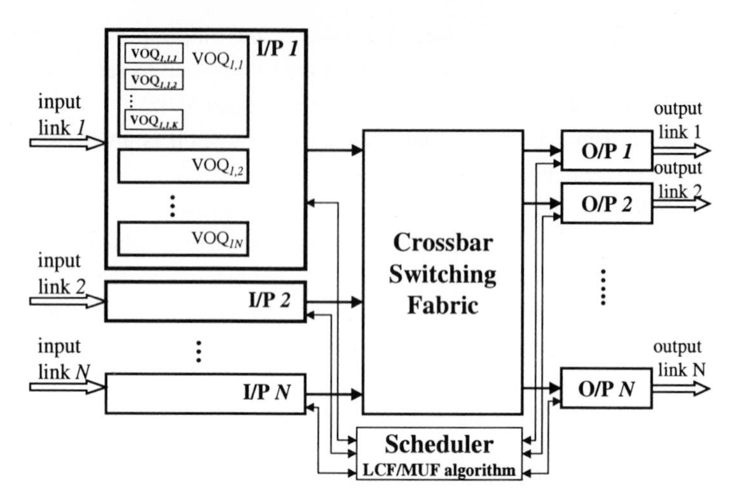

Fig. 1. The CIOQ switch scheduled by LCF/MUF algorithm.

each input port i ($1<=i<=N$) maintains a separate set of FIFO queues for each output port j, named as $VOQ_{i,j}$ for $1<=j<=N$. Therefore, there are N sets of $VOQ_{i,j}$ queues at each input port and each set of $VOQ_{i,j}$ consists of K priority FIFO queues, named as $VOQ_{i,j,k}$ for $1<=k<=K$. As a result, there are $N \times K$ queues to be managed at each input port. A new arrival cell is placed at the tail of its appropriate queue.

Since the service discipline of the shadow OQ switch is assumed to be the strict priority scheme, one can easily determine the service order of cells in the set of $VOQ_{i,j}$ queue. In other words, the entries of the scheduling matrix for the LCF/MUF algorithm can be easily determined.

4 Input Traffic Model

To evaluate switch performance quantitatively, we describe practical input traffic models that will be sent simultaneously to both the shadow OQ switch and the CIOQ switch in each simulation experiment. The following introduces two input traffic models — uniform and correlated models.

4.1 Uniform Input Traffic

The uniform input traffic model is used to characterize the interactive behavior of data arrivals in computer networks [11]; there is no time correlation between arrivals. Cells arrival on the N input links are governed by independent and identical Bernoulli processes. Specially, in any given time slot, the probability that a cell will arrival on a particular input port is r. Each cell has equal probability $1/N$ of being addressed to any given output port, and successive packets are independent.

Consider a particular output port (the "tagged" port) and define random variable A as the number of cell arrival at the tagged port during a given time slot. It follows that A has the binomial distribution as shown in Equation 1 that can be used to estimate the utility of output buffers.

$$\Pr[A=i]=\binom{N}{i}(r/N)^i(1-r/N)^{N-i}, \quad i=0,1,\dots,N \tag{1}$$

4.2 Correlated Input Traffic

The correlated traffic model we studied is characterized by a 2-state Markov process alternating between active and idle states with probabilities p and q, respectively [16][17] (see Fig. 2). This model is often used to describe IP packets fragmented into ATM cells. Based on this model, the traffic source will generate a cell every slot when it is in the active state. Note that there is at least one cell in a burst. Define random variable B as the number of time slots that the active period (burst) lasts. The probability which the active period lasts for a duration of i time slots (consists of i cells) is

$$\Pr[B=i] = p(1-p)^{i-1}, \qquad i \geq 1 \tag{2}$$

The mean burst length is given by

$$E[B] = \sum_{i=1}^{\infty} i \cdot \Pr(B=i) = 1/p \tag{3}$$

Similarly, define random variable I as the number of time slots which the idle period lasts. The probability that an idle period lasts for j time slots is

$$\Pr[I=j] = q(1-q)^{j}, \qquad j \geq 0 \tag{4}$$

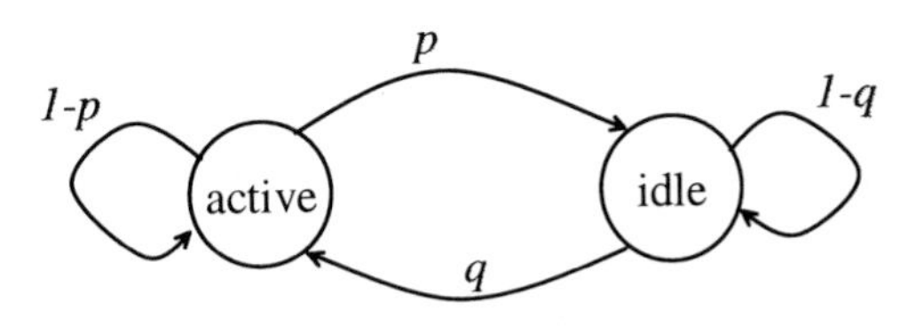

Fig. 2. The 2-state Markov chain process.

And the mean idle period is given by

$$E[I] = \sum_{i=1}^{\infty} j \cdot \Pr(I = j) = (1-q)/q \tag{5}$$

Given p and q, the offered traffic load ρ can be found by

$$\rho = \frac{E[B]}{E[B]+E[I]} = \frac{q}{p+q-pq} \tag{6}$$

We assume there is no correlation between different bursts and the destination of each burst is uniformly distributed among the output ports.

5 Numerical Results

Our simulation experiments are divided into three parts. In the first two parts, we measure the latency of cells which are simultaneously fed to both the shadow OQ switch and the CIOQ switch with infinite input buffer and finite output buffer under uniform and correlated traffic models, respectively. We define $d(x) = | d_{OQ}(x) - d_{CIOQ}(x) |$ as the deviation index of cell x. Here $d_{OQ}(x)$ and $d_{CIOQ}(x)$ denote, respectively, the departure times of cell x for the shadow OQ switch and the CIOQ switch. Given a fixed d, we measure the percentage of cells that has a deviation index smaller than or equal to d. This percentage is denoted by P_d. For example, $P_{d=0}$ equals 100% means that the CIOQ switch exactly emulates the shadow OQ switch. In the third part, we study the effect of finite input buffer. We measure the cell-loss probability (denoted by P_{loss}) of a CIOQ switch with finite input buffer and finite output buffer.

Simulations are performed for N =4, 8, 16, and 32, with 4 priority levels under both uniform and correlated traffic models. For correlated traffic, the mean burst length ℓ = 4, 8, 16, 32, and 64 are considered at the same time. The statistics of about one million cells are collected (over all queues in the switch) and thus loss probabilities smaller than 10^{-6} are not measurable.

5.1 Results of Uniform Input Traffic

For uniform input traffic load, simulation results are shown in Fig. 3 to Fig. 5. Fig. 3 shows that the CIOQ switch almost behaves like an OQ switch under moderate loads if sufficiently many output buffers are installed. For N =16 with infinite input buffer ($B^i = \infty$) and finite output buffer B^o = 9 cells, the value of $P_{d=0}$ is larger than 90% for the 1st priority traffic cells under an offered load up to 0.8. Fig. 4 shows the values of $P_{d=0}$ and $P_{d<=2}$ for all the four priority cells. We observe that the performance behaves similarly for all the four priority cells and the values of $P_{d<=2}$ are close

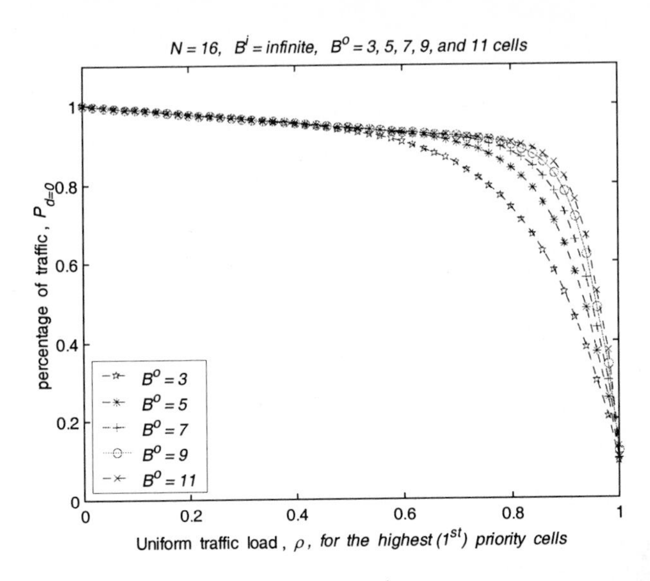

Fig. 3. Performance of the 16×16 CIOQ switch with $B^i = \infty$ and B^o = 3, 5, 7, 9, and 11 cells. The curve is shown for the highest priority cells only.

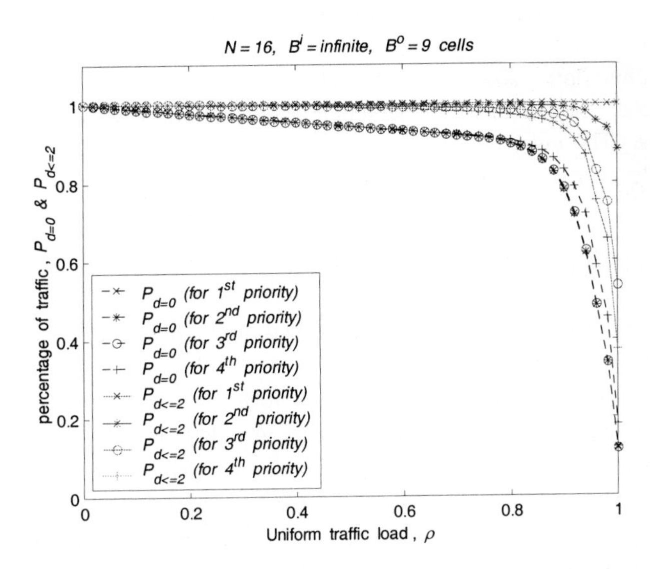

Fig. 4. Performance of the 16×16 CIOQ switch with $B^i = \infty$ and $B^o = 9$ cells. The curves show the performance of values of the $P_{d=0}$ and $P_{d<=2}$ for all the four priority cells.

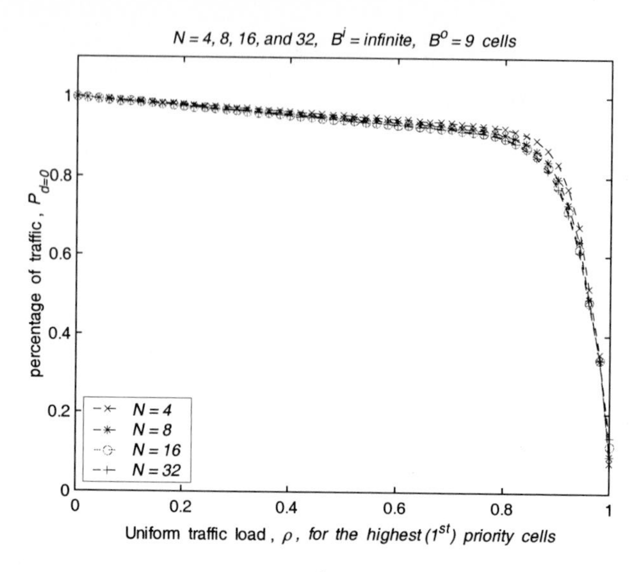

Fig. 5. Performance of the CIOQ switch with $B^i = \infty$ and $B^o = 9$ cells. The curve shows the performance of the highest priority cells as a function of the switch size N=4, 8, 16, and 32.

to 100% even under an offered load up to 0.8. As is shown in Fig. 5, the performance degrades as the number of ports increases. But it degrades slowly when N is larger than 8. Based on these results, we conclude that a CIOQ switch performs like an OQ switch if output buffer size is at least 9 cells.

Note that, the maximum queue length at all input ports is also estimated in these simulations. The input buffer size of 60 cells is sufficient under the uniform traffic model.

5.2 Results of Correlated Input Traffic

For the correlated traffic model, results depend on the mean burst size ℓ. In Fig. 6, we show the results for $\ell = 16$ cells for $N = 16$ with $B^o = 5\ell, 7\ell, 9\ell, 11\ell$ and 13ℓ cells. The value of $P_{d=0}$ is larger than 90% for the 1st priority traffic under an offered load up to 0.8 when $B^o = 11\ell$ cells are installed. From Fig. 7, one can see that the performances are roughly the same for various switch sizes. In Fig. 8 we performed similar simulations for other values of ℓ. Our observation is that the performance degrades as a function of the value of ℓ. But the degradation is became slow once ℓ is larger than 16. Based on these results, we suggest to allocate $B^o = 11\ell$ cells at each output port for correlated traffic.

Note that the maximum queue length at all input ports is smaller than 237 cells (about 15ℓ cells) which is also estimated in simulations.

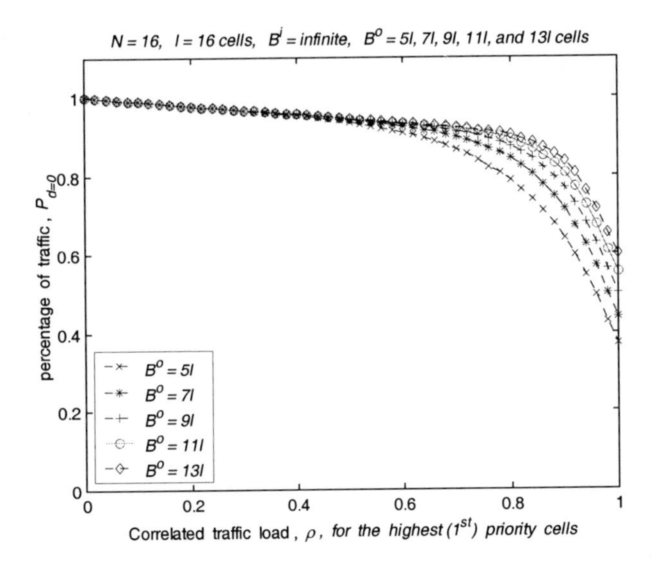

Fig. 6. Performance of the 16×16 CIOQ switch with $B^i = \infty$ and $B^o = 5\ell$, 7ℓ, 9ℓ, 11ℓ and 13ℓ cells (ℓ =16). The curve is shown for the highest priority cells only.

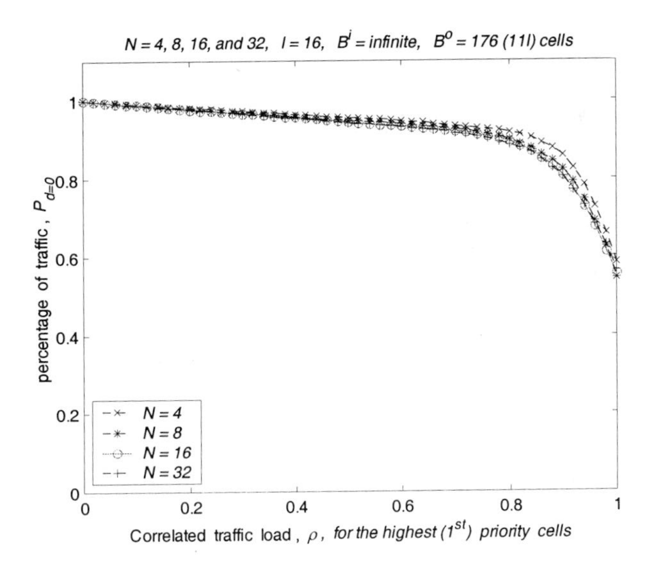

Fig. 7. Performance of the CIOQ switch with $B^i = \infty$ and $B^o = 11\ell$ cells (ℓ =16). The curve shows the performance of the highest priority cells as a function of the switch size N=4, 8, 16, and 32.

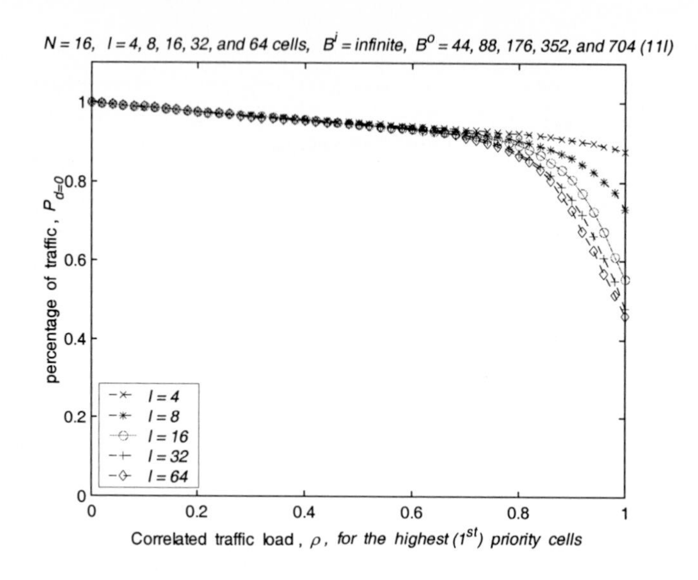

Fig. 8. Performance of the 16×16 CIOQ switch with $B^i = \infty$ and $B^o = 11\,\ell$ cells. The curve shows the performance of the highest priority cells as a function of the mean burst length ℓ = 4, 8, 16, 32 and 64 cells.

5.3 Effect of Finite Input Buffer

In this sub-section, we study the effect of finite input buffer. The output buffer size is chosen to be 9 cells for uniform traffic and 11 ℓ cells for correlated traffic. The switch size and the mean burst length investigated in our simulations are N=16 and ℓ =16. We measure cell loss probability (P_{loss}) due to finite input buffer. Fig. 9 shows the curves of P_{loss} versus ρ for input buffer size B^i=2, 3, 5, 7 and 9 cells under uniform traffic. Fig. 10 shows the curve of P_{loss} versus ρ for input buffer size B^i=3 ℓ , 5 ℓ , 7 ℓ , 9 ℓ and 11 ℓ cells under correlated traffic. It can be seen that P_{loss} of the CIOQ switch with B^i=3 and B^o=9 is smaller than 10^{-5} under the uniform traffic model for an offered load up to 0.8. For correlated input traffic, one can achieve similar performance with B^i=7 ℓ and B^o=11 ℓ .

Note that, based on our simulation results, the CIOQ switch needs to maintain a buffer of 60 cells at each input port to achieve zero cell loss probability under uniform traffic (or 15 ℓ cells under correlated traffic) for an offered load up to 0.9.

6 Conclusions

We have investigated the performance of a CIOQ switch with finite input and output buffers. The LCF/MUF algorithm is selected as the matching algorithm to deliver cells from input ports to output ports. To implement the LCF/MUF algorithm, a switch has to know the cushion of all cells and the relative departure order of cells destined to the same output port. The complexity of cushion calculation strongly depends on service discipline. In this study, we choose strict priority service discipline because it allows cushions to be easily calculated. Based on simulation results, we found that a CIOQ switch can mimic an OQ switch quite well for both uniform and correlated traffic up to ρ=0.8.

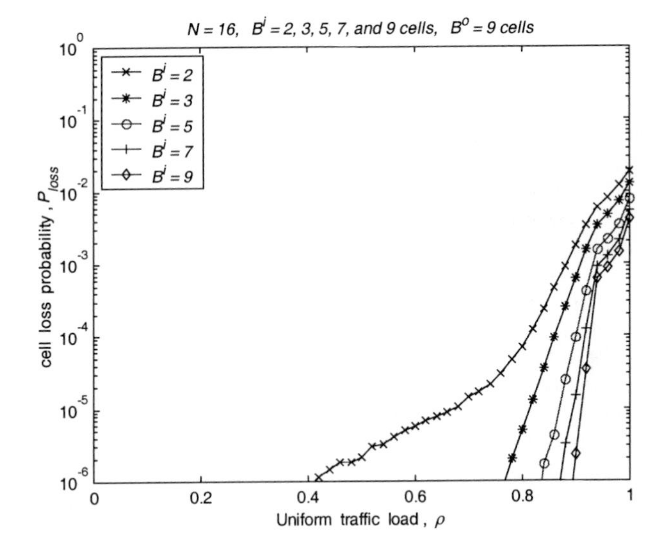

Fig. 9. P_{loss} of a 16×16 CIOQ switch with B^i=2, 3, 5, 7, and 9 cells, and B^o=9 cells under uniform input traffic.

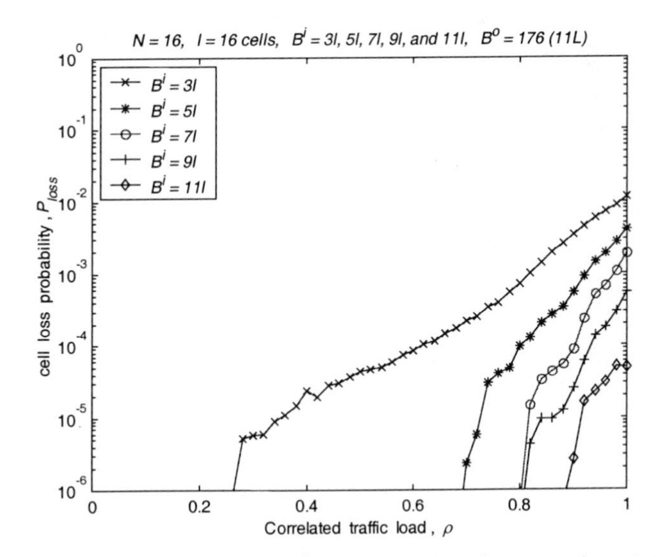

Fig. 10. P_{loss} of a 16×16 CIOQ switch with B^i=3 ℓ, 5 ℓ, 7 ℓ, 9 ℓ, and 11 ℓ, and B^o=11 ℓ cells (ℓ =16 cells) under correlated input traffic.

Further applications may require a more complicated service discipline (such as WFQ) than strict priority. Unfortunately, a complicated service discipline may induce dramatic complexity in cushion calculation. Therefore, how to design a scheme which simplifies cushion calculation and mimics the complicated service discipline is worth to be further studied.

References

1. B. Prabhakar and N. McKeown, "On the speedup required for combined input and output queued switch," Stanford University, Stanford, CA, *Tech. Rep., STAN-CSL-TR-97-738*, 1997.
2. A. Charny, P. Krishan, N. Patel, and R. Simcoe, "Algorithms for providing bandwidth and delay guarantees in input-buffered crossbar with speedup," *Proc. of IWQoS'98*, pp.235-244.
3. I. Stoica and H. Xhang, "Exact emulation of an output queueing switch by a combined input output queueing switch," *Proc. of IWQoS'98*, pp.218-224.
4. P. Krishna, N.S. Patel, A. Charny, and R. Simcoe, "On the speedup required for work-conserving crossbar switches," *Proc. of IWQoS'98*, pp.225-234.
5. S.T. Chuang, A. Goel, N. McKeown, and B. Prabhakar, "Matching output qeueueing with a combined input/output-queued switch," *IEEE J. Selected Area in Commun.*, Vol.17, No.6, pp.1030-1039, June 1999.
6. T. H. Lee, Y. W. Kuo, and J. C. Huang, "Quality of Service Guarantee in a Combined Input Output Queued Switch," *IEICE Trans. on Commun.*, pp.190-195, Feb. 2000.
7. H. Zhang, "Service disciplines for guaranteed performance service in packet-switching networks," *Proc. of the IEEE*, 83(10), pp.1374-1399, October 1995.
8. D. Ferrari and D.C. Verma, "A scheme for real-time channel establishment in wide-area networks," *IEEE J. Selected Area in Commun.*, Vol. 8, No. 3, pp.368-379, April 1990.
9. A.K. Parekh and R. G. Gallager, "A generalized processing sharing approach to flow control in integrated services networks: the single node case," *IEEE Trans. on Networking*, Vol.1, No.3, pp.344-357, June 1993.
10. J.C.R. Bennett and H. Zhang, "WF^2Q: worst-case fair weighted fair queueing," *Proc. of IEEE INFOCOM'96*, pp.120-128.
11. M. Karol, M. Hluchyj, and S. Morgan, "Input verse output queueing on a space division switch," *IEEE Trans. on Commun.*, Vol.35, pp.1347-1763, Dec. 1987.
12. N. McKeown, V. Anantharam, and J. Walrand, "Achieving 100% throughput in an input-queued switch," *Proc. of IEEE INFOCOM'96*, pp.296-302.
13. T.E. Anderson, S.S. Owicki, J.B. Saxe and C.P. Thacker, "High speed switch scheduling for local area networks," *IEEE/ACM Trans. on Computer Systems*, Vol.11, No.4, pp.319-352, Nov. 1993.
14. A. Mekkittikul and N. McKeown, "A practical scheduling algorithm to achieve 100% throughput in input-queued switches," *Proc. of IEEE INFOCOM'98*, pp.792-799.
15. N. McKeoen, "The *i*SLIP scheduling algorithms for input-queued switches," *IEEE/ACM Trans. on Networking*, Vol.7, No.2, pp.188-201, April 1999.
16. S.Q. Li, "Performance of a non-blocking space-division packet switch with correlated input traffic," *Proc. of IEEE Globecom'89*, Vol.3, pp.1754-63.
17. S.C. Liew, "Performance of various input-buffered and output-buffered ATM switch design principles under bursty traffic : simulation study," *IEEE Trans. on Commun.*, Vol.42, pp.1371-79, Feb/Mar/Apr. 1994.

Performance Prediction Methods for Address Lookup Algorithms of IP Routers

Ryo Kawabe[1], Shingo Ata[2], and Masayuki Murata[3]

[1] Graduate School of Engineering Science, Osaka University
1–3 Machikaneyama, Toyonaka, Osaka 560–8531, Japan
r-kawabe@nal.ics.es.osaka-u.ac.jp

[2] Graduate School of Engineering, Osaka City University
3–3–138 Sugimoto, Sumiyoshi-ku, Osaka 558–8585, Japan
ata@info.eng.osaka-cu.ac.jp

[3] Cybermedia Center, Osaka University
1–30 Machikaneyama, Toyonaka, Osaka 567–0043, Japan
murata@cmc.osaka-u.ac.jp

Abstract. Many address lookup methods for an IP router have been recently proposed to improve a packet forwarding capability, but their performance prediction is very limited because of lack of considering actual traffic characteristics. It is necessary to consider actual traffic to predict more realistic performances on routers, specially in case of layers 3 and 4 switches whose performances are more influenced by flow characteristics. In this paper, we propose new methods for predicting the router's performance based on the statistical analysis of the Internet traffic. We also present an example of its application to the existing table lookup algorithms, and show that simulation results based on our method can provide accurate performance prediction.

1 Introduction

A rapid growth of the Internet traffic with the spread of multimedia applications such as streaming media leads to an explosive demand of high-speed packet transmission technologies. For this purpose, it is necessary to improve the packet forwarding capability at IP routers as well as to increase the link capacity. One main task of routers is to determine the output interface for forwarding the arriving packet according to the packet header information, e.g., the destination address. Other information such as the source address, source/destination port numbers, and protocol number may also be used for policy routing and/or the layer 4 switching. IP routers perform the following two steps for every arriving packet.

1. Look up the next-hop of the packet from the routing table.
2. Forward the packet to the outgoing interface determined by step 1.

Step 1 heavily influences the performance of routers because the longest prefix matching [1] in Step 1 is complicated after CIDR (Classless Inter-Domain Routing) [2] is introduced. Therefore, an address lookup method easily becomes a performance bottleneck at high-speed routers.

I. Chong (Ed.): ICOIN 2002, LNCS 2343, pp. 215–225, 2002.

To overcome the performance bottleneck of address lookup, many approaches have been proposed in recent years. See, e.g., [3,4]. However, performance studies on their proposed methods are very limited. Roughly speaking, there are two metrics for the performance evaluation of the address lookup algorithms; the worst-case and the average-case (or actual-case). The worst-case performance is easy to derive from the complexities of lookup algorithms, and by using them, one can easily compare the proposed algorithm with other existing algorithms. The worst-case performance is also a useful index to describe its basic capability. Most of papers on the longest prefix matching thus use the worst-case performance because of above reasons.

However, the actual performance is heavily affected by the time–dependent behavior of destination addresses of arriving packets. Furthermore, the worst-case performance is not always the best metric. It is likely that the design choice according to the worst-case performance leads to very expensive algorithms. A closer look at the performance behavior of the target method may allow a more elegant solution. For example, we may be able to have a much cheaper solution at the sacrifice of limited performance degradation (e.g., introducing a small packet loss probability). For that purpose, we need a realistic address generation method for evaluating the performance of IP routers while the past researches only considered the simulation technique using the random address generation [3]. Or, a small amount of traced data is used [4]. Since the amount of the traced data is limited, the actual performance behavior is likely to be missed. Furthermore, because traced data is very few in the public domain, the simulation result is lack of generality.

In this paper, we propose performance prediction methods of address lookup algorithms with consideration of characteristics of the actual traffic. We first analyze the behavior of the address generation pattern based on the traced data in Section 2. It is then used to generate the pseudo traffic using the mathematical model. Results of simulation experiments are presented in Section 3 to compare our proposed method with the trace–driven simulation method and we show that it is possible to predict more realistic performance using our method. In Section 4, we conclude our paper with future research works.

2 Statistical Address Generation Methods

In this section, we propose two performance prediction methods of address lookup algorithms of IP routers. The one is for generating the destination address of IP packets, i.e., for layer 3 routers. The other is for generating the flow, where the flow duration (the number of packets contained in the flow) is obtained by the statistical analysis of traced-data. Then, the destination address of the flow is generated by the statistical model. See Figure 1 for an outline of our methods.

We will first introduce the model of the destination addresses in Subsection 2.1. The address generation method is then described in Subsection 2.2. It is applied to the traffic generation method, which can be used for simulating the table lookup algorithms.

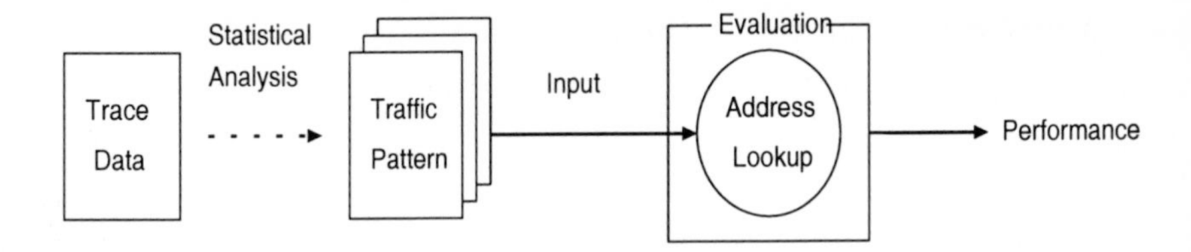

Fig. 1. Outline of Performance Prediction

2.1 Modeling Destination Addresses

We use the approach of the address generation algorithm proposed in [5], where the addresses are generated by a LRU (Least Recently Used) stack and ISGF (Inverse Stack Growth Function).

For briefly summarizing ISGF, we introduce t_i, denoting the arrival time of ith packet (or flow). We define $f(t,T)$ as the expected value of the number of distinct addresses of packets arrived during a period $(t-T,t)$. Assume that $f(t,T)$ is independent from time t, i.e., $f(t_i,T) = f(t_j,T)$ for all i,j. Then, $f(t,T)$ can be denoted by $f(T)$. This assumption is called a time–translation invariance, and makes it for us to obtain the same value $f(T)$ whenever trace data are gathered. ISGF is a power low function given by

$$f(T) \simeq T^\alpha \ (T \gg 1) \tag{1}$$

where α $(0 < \alpha < 1)$ is a constant value.

While ISGF was originally used for predicting the cache hit ratio of computer architectures, the authors in [5] have found that the number of distinct destination addresses also satisfies ISGF. Namely, when we collect T packets from the traced data, the number of distinct destination addresses can be estimated as T^α. By using ISGF, it is possible to derive the probability that the destination address of the arriving packet has already appeared. However, applying only ISGF is insufficient in some situations.

Let us consider the sequence of packet arrivals for the TCP flow. Because the TCP flow is divided into packets, the probability that the next packet of the TCP flow arrives at the router tends to be decreased as the time passes. This tendency cannot be modeled by ISGF solely. Therefore, an LRU (Least Recently Used) stack model is applied to the probability structure as follows.

The probability a_i is defined as

$$\begin{aligned} a_i = &\{f(g(i-1)+1) - (i-1)\} \\ &-\{f(g(i)+1) - i\}, \end{aligned} \tag{2}$$

where $g(T) = f^{-1}(T)$. From Eq. (1), we have

$$\begin{aligned} a_i = &\{((i-1)^{\frac{1}{\alpha}} + 1)^\alpha - (i-1)\} \\ &-\{(i^{\frac{1}{\alpha}} + 1)^\alpha - i\}. \end{aligned} \tag{3}$$

2.2 Address Generation Method

In this subsection, we explain the procedure of destination address generation based on ISGF. We prepare the set of distinct destination addresses of trace data with their numbers of references from the traced data. Then, if N_p packets include N_a distinct destination addresses, α is calculated from Eq.(1). Since Eq.(1) holds when N_p is large, we need to calculate α with the adequately large number of packets. We use a least–square method to calculate α.

Moreover, we define a list of destination addresses as $A(i)$ $(i = 1, 2, \ldots, N_a)$. The following procedure generates time–dependent destination addresses using the above quantities; N_p, N_a, α and $A(i)$.

1. LRU stack size S is set to N_a.
2. Store $A(1), A(2), \ldots, A(N_a)$ sequentially in the LRU stack.
3. Choose a random number p $(0 \leq p < 1)$.
4. Calculate the minimum k if $p \leq \sum_{i=1}^{k} a_i$.
5. Assign the address of a new access as k-th element of the LRU stack.
6. Move k-th element to the top of the LRU stack.
7. Return to Step 3.

The above procedure gives a series of destination addresses, which can be embedded in the simulation program for packet generation. More specifically, either of the following procedures is performed as a packet generator in the simulation.

- Address Generation per Packet (AGP)
  1. A packet is generated according to, e.g., a Poisson process. A heavy-tailed distribution may also be applied to determine the interarrival times of packets.
  2. The destination address of the generated packet is decided by the address generation method described above.
- Address Generation per Flow (AGF)
  1. Generates a flow, and determines the number of packets in the flow.
  2. The destination address of the flow is decided by the address generation method mentioned above.
  3. Determines the interarrival times of packets during which the flow is active. All the packets in the flow use the same destination address.

We note here that in AGF, the number of packets within the flow may be determined from the statistical analysis of the actual data. For example, in [6], the entire flow duration can be well approximated by the log-normal distribution while the tail–part has a heavy-tailed distribution. Thus, the combination of the two distributions can be used for the flow duration and is used in our experiments. To fit the combination of the two distribution function accurately, we estimate parameters using the maximum-likelihood-estimator (MLE) method [7]. Once the flow is accurately characterized, the packet interarrival times may be modeled by a Poisson distribution [6].

On the other hand, AGP cannot model the heavy-tailed distribution for representing the number of flows because the probability that the generated

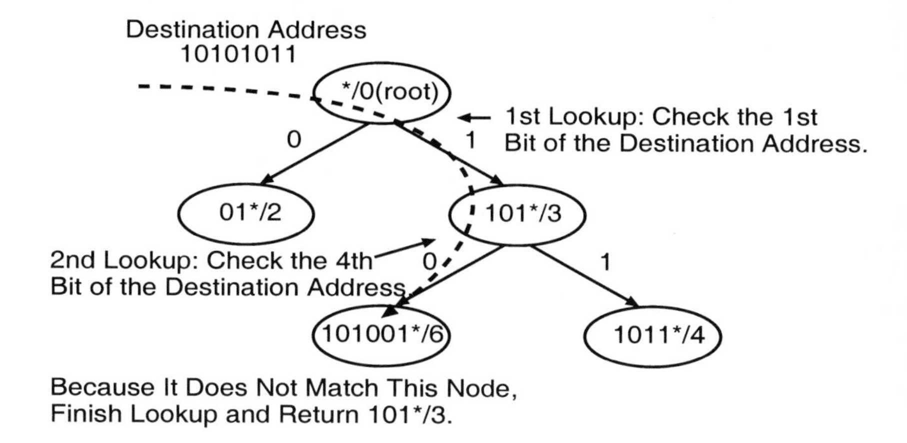

Fig. 2. An Example of the Patricia Tree Search

address would appear again in the future depends only on the position of the LRU stack, not the number of references. However, AGP has some advantages. It requires lower processing overhead and the smaller number of parameters than AGF.

3 Performance Evaluation

In this section, we evaluate the performance of existing address lookup algorithms by using our proposed methods. We use the Patricia Tree search [8] and the pointer cache method proposed in [3].

3.1 Target Algorithms

We briefly summarize the Patricia Tree and the pointer cache method in this subsection. At legacy routers, the longest prefix matching has been performed by the binary tree. The Patricia Tree search [8] is a popular one, which eliminates the node having only one child node. See Figure 2 for an example of the Patricia Tree search. In this case, the nodes for the fourth and fifth bit of the destination address are removed because the routing table has only the entries beginning with 101001 when the fourth bit of the destination address is 0. The Patricia Tree search is simple and hence easy to implement, but relatively slow because the number of removable nodes becomes decreased when the number of entries is increased. In the worst case, the Patricia Tree still requires up to 32 or 128 memory references in IPv4 or IPv6, respectively.

To resolve the problem, several algorithms have been proposed by using the CAM (Content Addressable Memory), which can decrease the number of memory accesses (see [9] and references therein). The pointer cache method [3] is one

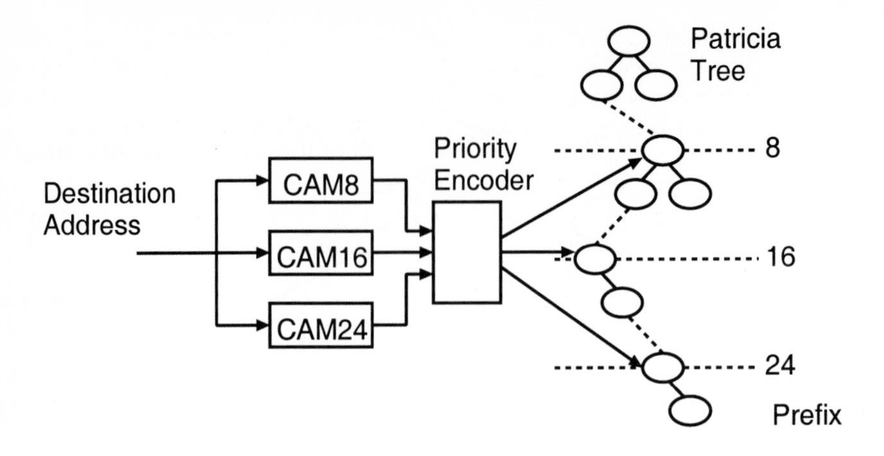

Fig. 3. Composition of Pointer Cache Method

of such CAM-based algorithms, combining CAM and the Patricia Tree. Figure 3 shows its composition. The pointer cache method first divides the Patricia Tree into some parts by borders with predefined values as a prefix length of the Patricia Tree. Then the set of pointers to the root node of each part is stored into CAM. The address lookup for the arriving packet is performed by the following procedure;

1. Search the CAM for the pointer to the objective part which includes the longest prefix matched entry for the destination address.
2. Look up the address from the part of Patricia Tree.

Because CAM is used like a cache memory for the Patricia Tree, the performance of the pointer cache method depends on entries stored in CAM, which must be strongly affected by the traffic characteristics such as the distribution of addresses of arriving packets. This is why we choose the pointer cache method as an example.

3.2 Traffic Models

Before evaluating the performance, we need to determine the parameters N_a, N_d, and $A(i)$ from the traced data. We use one million packet headers gathered by the traffic monitor (OC3MON [10]) at the gateway of Osaka University. Traced data set includes 6,966 distinct destination addresses. Its BGP information (4,669 entries) are used to construct the Patricia Tree and CAM entries in the pointer cache method. The data is collected on January 24th, 2001. The values α of ISGF are 0.64 in AGP and 0.77 in AGF, respectively. For comparison purpose, we also consider the following three kinds of address generation methods.

Actual: A raw sequence of packet headers obtained by OC3MON is used.
Random: A randomly chosen 32-bit value is used as the destination address of each packet.
Trace-Random: Randomly pick up the destination address of the packet from the traced data.

3.3 Performance Metrics

In our experiments, we generate ten million packets according to the Poisson process for the input to address lookup algorithms. Their destination addresses are generated by five traffic patern mentioned above. "Trace-Random" traffic case, AGP and AGF uses one million packet headers of the traced data. The buffer size of the router is set to be 3,000 packets.

We define an address lookup delay S as the time required for a longest prefix matching of an arrival packet. In the Patricia Tree search, S is given by

$$S = t_s \times d_r, \tag{4}$$

where t_s is the number of times of lookup in Patricia Tree and d_r is the delay for read / comparison operations in RAM. Under the present circumstance, since a read requires 5 nsec and a comparison requires 10 nsec in RAM, the value of d_r can be assumed to be 15 nsec. In the case of Figure 2, since two lookups are required to determine the longest prefix entry of the packet, the value t_s is 2 and then the value S is estimated as 30 nsec. An address lookup by using the pointer cache method consists of a search for CAM and a lookup in the Patricia Tree. Since an access to CAM is performed independently of an access to the Patricia Tree, it can be improved by performing in parallel on the pipeline. An arriving packet at the router first searches the longest prefix entry among all entries stored in CAM. After the search has finished, the lookup of the packet begins if no packets are being looked up in the Patricia Tree. Otherwise, it is kept waiting for finishing the lookup in the Patricia Tree and a search of a newly arriving packet for CAM does not start. In the present situation, the search for CAM requires 15 nsec [11,12,13] and we used this value in our experiments.

We use the maximum throughput as the performance metric, which is the reciprocal of the minimum average of packet interarrival time if no packet is lost during the simulation. We also obtain the behavior of the time–dependent queue length (the number of packets queued in the buffer).

3.4 Simulation Results

Patricia Tree. Table 1 compares simulation results of the Patricia Tree search among five traffic patterns (AGP, AGF and three packet generation methods described above). In Table 1, the second column is the maximum throughput and the third column is the relative error ratio to the maximum throughput of the "Actual" case. The table shows that the throughput of "Random" case is 1.8 times larger than the one of the "Actual" case. It is due to the characteristics

Table 1. Maximum Throughput (Patricia Tree Search)

Input Traffic	Maximum Throughput	Error Rate
Actual	4.63 mpps	—
Random	8.33 mpps	79.9%
Trace-Random	4.67 mpps	0.86%
AGP	4.63 mpps	0.00%
AGF	4.52 mpps	2.38%

of the destination address distribution. In the Patricia Tree search, the packet lookup delay is decided by the prefix length of the matched entry in the routing table. If the number of packets whose destination addresses match the longer-prefix entries of the routing table, the average lookup delay becomes larger. Furthermore, the prefix length of the entry shows the size of the organization belonging to it. That is, the address space which includes the longer prefix entries tend to include more organizations. Because the most popular traffic in the current Internet is `http`, packets tend to access the destination address space which includes more `www` servers. From the above reason, the "Actual" traffic case tend to match the longer prefix entries, and the address lookup delay becomes larger. On the other hand, the "Random" pattern generates uniformly–distributed addresses; the number of accesses of the entry is inversely proportional to the prefix length. Namely, in "Random" case, the entry with the short prefix length is preferably accessed. Therefore, "Random" case overestimates the performance of Patricia Tree significantly, when compared with "Actual" case. In contrast to "Random" case, results of "Trace-Random" case, AGF, and AGP provide good estimations with low errors. Among three methods, AGP provides the best prediction with respect to the throughput. Note that since the result of "Random" case is far from other traffic patterns, we will not show the result of "Random" case in the below.

Pointer Cache Method. To see the applicability of our proposed methods to other algorithms, we next examine the pointer cache method [3]. Table 2 compares the maximum throughput. We can observe that the results are almost same as the case of the Patricia Tree search, and our proposed methods can provide good prediction.

We also show the behavior of the time-dependent queue length in Figure 4. From Figure 4(a), it is observed that the behavior of "Actual" case has two properties. Its fluctuation is low (below 50 packets). However, the queue length sometimes increases significantly. It is caused by the characteristic of packet arrivals. Due to the window flow control of TCP, traffic of the TCP connection contains a burst of packets. Of cource, the continuos arrival of packet belonging to the same flow is another reason. Then, a significant increase (called as *spike* below) appears when the packets with the long-prefix-matched address arrive bursty. On the other hand, other three methods show different results. In "Trace-

Table 2. Maximum Throughput (Pointer Cache Method)

Input Traffic	Maximum Throughput	Error Rate
Actual	35.2 mpps	—
Random	58.8 mpps	67.0%
Trace-Random	36.5 mpps	3.69%
AGP	35.0 mpps	0.57%
AGF	35.0 mpps	0.57%

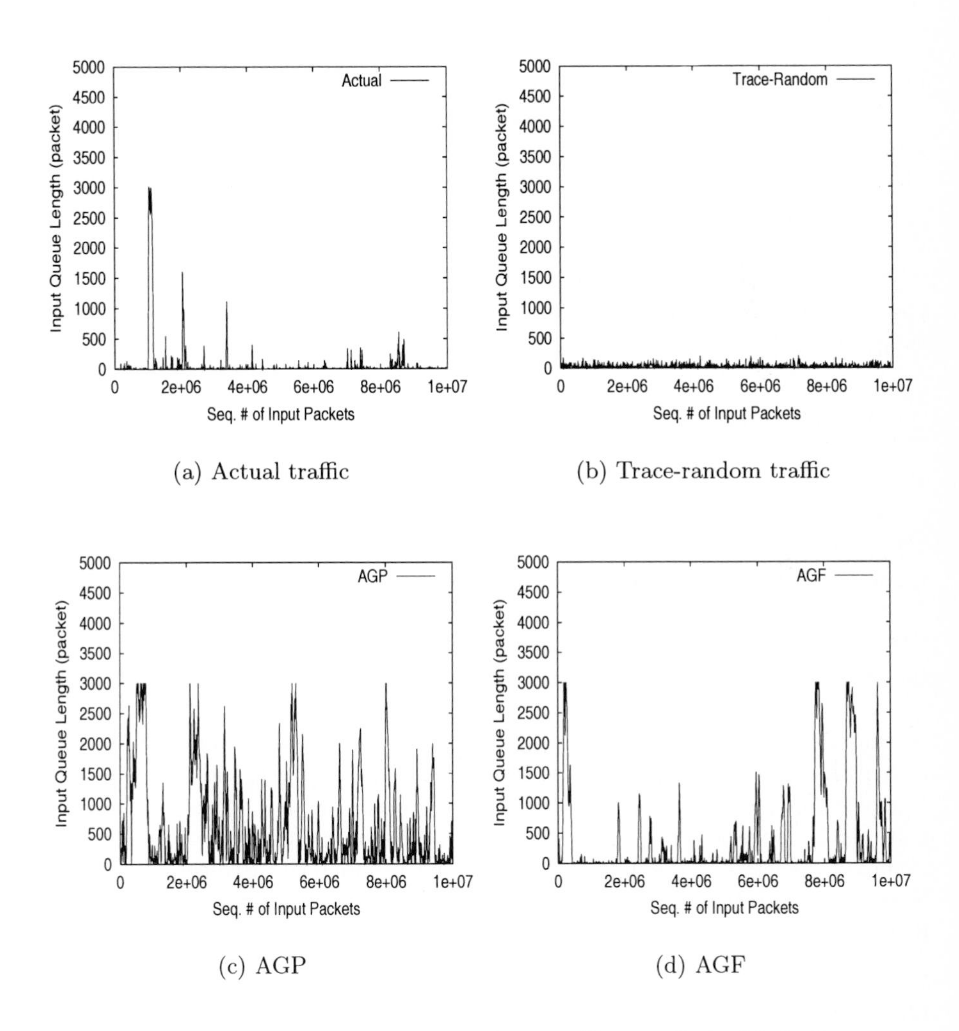

(a) Actual traffic

(b) Trace-random traffic

(c) AGP

(d) AGF

Fig. 4. Transition of Input Queue Length (Pointer Cache Method)

Random" case (Figure 4(b)), any spike does not appear during the simulation. The behavior of AGP (Figure 4(c)) shows many spikes, but its whole fluctuation is not so low, compared with "Actual" case. On the other hand, the result of AGF (Figure 4(d)) shows some spike and its whole fluctuation is low, and its behavior of the time–dependent queue length is similar to the one of the "Actual" case. However, the overall queue length of AGF is longer than the one of "Actual" case. This is because our proposed method cannot generate a new address (i.e., it does not appear in the traced data) that will arrive, for "Trace-Random" traffic, AGP and AGF are based on 6,966 distinct addresses gathered from one million traced packet headers, while "Actual" traffic uses ten million traced packet headers.

This is the open issue of our generation methods that does not consider the new address generation. One solution may be to set the index for each entry of the routing table based on the number of accesses, and to generate the new address according to the index of the entry. A proposal of new address generation algorithm is one of our future research topics.

Recall that AGF explicitly models the flow characteristics (i.e., the number of packets in the flow, interarrival times of flows, and the number of active flows). Thus, AGF can be applied to flow classification algorithms (e.g., [14]) as well as the address lookup algorithms. Its accuracy should be investigated as a future research topic.

4 Conclusion

In this paper, we have proposed performance evaluation methods of the address lookup algorithms based on the statistical model obtained from the real traffic. We have found out that through a packet driven simulation, we can evaluate them accurately with our proposed method, AGF. As a future topic, we are now developing the analysis approach to know the performance of address lookup algorithms more quickly. Also, we need to validate the applicability of AGF to the packet classification algorithm.

Acknowledgements. We would like to thank Dr. Naoaki Yamanaka and Dr. Kohei Shiomoto of NTT Network Innovation Laboratories, and Mr. Masanori Uga of NTT Network Service Systems Laboratories for their comments and suggestions. This work was supported in part by Research for the Future Program of Japan Society for the Promotion of Science under the Project "Integrated Network Architecture for Advanced Multimedia Application Systems" (JSPS-RFTF97R16301).

References

1. M. A. Ruiz-Sánchez, E. W. Biersack, and W. Dabbous, "Survey and taxonomy of IP address lookup algorithms," *IEEE Network*, vol. 15, pp. 8–23, Mar./Apr. 2001.
2. Y. Rekhter and T. Li, "An architecture for IP address with CIDR." RFC1518, Sep. 1993.

3. M. Uga and K. Shiomoto, "A fast and compact longest prefix look-up method using pointer cache for very long network address," in *Proceedings of IEEE ICCCN '99*, pp. 595–602, Oct. 1999.
4. P. Gupta, B. Prabhakar, and S. Boyd, "Near-optimal routing lookups with bounded worst case performance," in *Proceedings of IEEE INFOCOM 2000*, pp. 1184–1192, Mar. 2000.
5. M. Aida and T. Abe, "Pseudo-address generation algorithm of packet destinations for Internet performance simulation," in *Proceedings of IEEE INFOCOM 2001*, pp. 1425–1433, Apr. 2001.
6. S. Ata, M. Murata, and H. Miyahara, "Analysis of network traffic and its application to design of high-speed routers," *IEICE Transactions on Information and Systems*, vol. E83-D, pp. 988–995, May 2000.
7. V. Brazauskas and R. Serfling, "Robust and efficient estimation of the tail index of a one-parameter Pareto distribution," *North American Actuarial Journal*, pp. 12–27, Apr. 2000.
8. D. E. Knuth., *The Art of Computer Programming*, vol. 3. Addison-Wesley, 1973.
9. A. McAuley and P. Francis, "Fast routing table lookup using CAMs," in *Proceedings of IEEE INFOCOM'93*, vol. 3, pp. 1382–1391, Mar. 1993.
10. K. Thompson, G. J. Miller, and R. Wilder, "Wide-area Internet traffic patterns and characteristics," *IEEE Network*, pp. 10–23, Nov. 1997.
11. Lara Network, `http://www.laratech.com/`.
12. NetLogic Microsystems, `http://www.netlogicmicro.com/`.
13. Siber Core, `http://www.sibercore.com/`.
14. P. Warkhede, S. Suri, and G. Varghese, "Fast packet classification for two-dimensional conflict-free filters," in *Proceedings of IEEE INFOCOM 2001*, Apr. 2001.

Shortest and Oldest First Non-interleaving Packet Switching Scheduling Algorithm

Young-Ki Hong[1], Sungchang Lee[1], and Hyeong-Ho Lee[2]

[1] Hankuk Aviation University, Goyang, Korea
{ykhong, sclee}@hau.ac.kr

[2] Electronics and Telecommunications Research Institute, Daejeon, Korea
holee@etri.re.kr

Abstract. A packet-aware non-interleaving scheduling algorithm is proposed and examined in this paper. The non-interleaving scheduling algorithm eliminates the complexity of packet reassembly at the output queue. The characteristics of packet latency and queue occupancies are investigated in comparison with the cell-interleaving scheduling algorithm. The simulated results show that the packet-aware non-interleaving scheduling could be a feasible or better choice for packet switch implementations in the latency, output buffer requirement and implementation complexity viewpoints. The simulated results are obtained using self-similar traffic generated based on the measured Internet backbone traffic that reflects the predominance of short packets.

1 Introduction

The rapid growth of Internet traffic coupled with the availability of fast optical links cause a bottleneck at the switches and routers. Congestion in high speed backbone networks make the real-time applications perform poorly on the Internet, and has created a strong demand for very high speed switches and routers.

A lot of high performance switches or routers are developed to overcome the bottleneck of the network node performance. Output queued switch architecture has been a popular choice for high performance switch implementations, which make use of simultaneous paths from inputs to an output or internal speed-up. But, the recent drastic increase in line speed has made part of those architectures impractical or infeasible [1], [2]. For this reason, input-queued switch using virtual output queue (VOQ) draw a lot of interest these days, because VOQ with appropriate contention resolution scheme can provide 100% throughput with the internal speed that is not higher than the line rate of the connected link [3].

In this paper, we propose a packet-aware non-interleaving scheduling algorithm for the input-queued switch with VOQ, and show that this approach could be acceptable choice for practical implementations in packet latency and other point of view. Most of the existing scheduling algorithms deal with cell level scheduling that is unaware of any information about the packets from which the cells are segmented. The proposed algorithm is distinguished in the fact that the algorithm is packet-aware and the cells

I. Chong (Ed.): ICOIN 2002, LNCS 2343, pp. 226-236, 2002.

from a packet are not interleaved with cells from other packets during the switch. This eliminates the packet reassembly process at the output queue. The packet-aware scheduling simplifies the complexity, and reduces related processing time of the reassembly at the output buffer. Also, the simulation result shows that the amount of buffer that is required for the output queue can be reduced. The performance of the proposed algorithm is compared to that of the exemplary existing cell interleaving scheduling algorithms. The remaining parts of the paper are organized as follows: Section 2 discusses VOQ and a brief summary on scheduling algorithms. The SOFNI scheduling algorithm is explained in section 3. In section 4, simulation results are shown. Finally, we conclude in section 5.

2 Background

In the input-queued switch model, arriving packets or cells are stored in FIFO buffers which reside at the input port until they are signaled to traverse the switching fabric. The scheduling algorithms are supposed to resolve the contention among the inputs that may occur when multiple inputs try to transmit the cells to the same output. FIFO queues of ordinary input-queued switches suffer from head of line (HOL) blocking which is a phenomenon that a blocked cell at the head of a queue prevents other cells behind it in the queue from being transmitted to other unused outputs. Consequently, HOL blocking limits the overall switching throughput to just 58.6% [4], [5]. In Fig. 1, the configuration of the VOQ is shown [2]. In what follows, subqueue(i, j) denotes the j-th subqueue of input-i that is used for packets destined to output j. VOQ can entirely eliminate HOL blocking by maintaining a separate queue for each output at an input port. The performance of a VOQ switch essentially depends on its scheduling algorithm. Many scheduling algorithms were presented in the literature and these can be generally categorized into two classes: maximum size matching (MSM) and maximum weight matching (MWM). MSM finds a match containing the maximum number of input-output pairs, while MWM finds a match so that the sum of weights of the pairs is the maximum.

The most efficient known algorithm for solving the MWM problem takes $O(N^3 \log N)$, and it is too complex to implement in hardware [6]. Even the MSM algorithm has the time complexity of $O(N^{5/2})$. Thus, the practical implementations usually approximate the algorithms.

Several algorithms approximate MSM and MWM to make them practical. PIM[7], 2DRR[8], RRM[9] iSLIP[10][11], FIRM[12] and WFA[13] are the examples of MSM approximations. These reduce the time complexity down to $O(N \log N)$ or $O(i \cdot N \log N)$ serial or $O(\log N)$ parallel time from $O(N^{5/2})$ at the cost of performance degradation [6]. Several weighted algorithms, iLQF[9][14], iLPF[15], iOCF[9], RPA[1] and SIMP[6], approximate MWM. iLQF and iOCF have a complexity of $O(N^2 \log N)$ with the suggested number of iterations and can be run on parallel processors, while RPA, 2DRR and iLPF have a complexity of $O(N^2)$ and cannot be run on parallel processors[1]. SIMP approximates MWM by serializing the search for the maximum of column weights with a complexity of $O(N^2)$ serial or $O(N \log N)$ parallel [6].

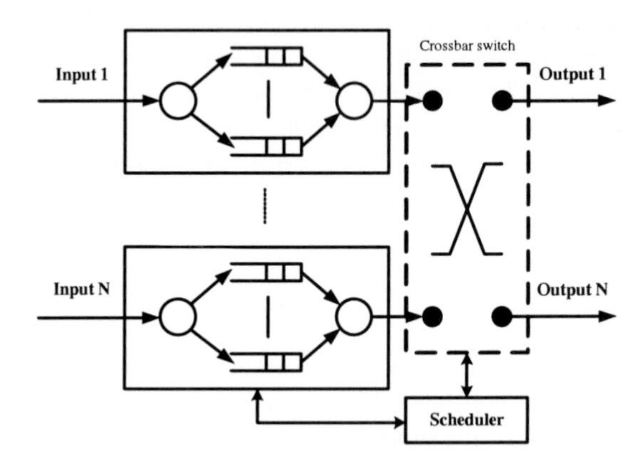

Fig. 1. A VOQ switch, managing N separate FIFO queues for storing arriving cells directed to different output ports in every input port. The switch fabric between the inputs and outputs is assumed to be non-blocking.

These algorithms all deal with the packet-unaware cell scheduling and switching. In the packet-unaware cell scheduling and switching, the cell switching is independent of the packet level. No information about the packets is delivered from packet level to the cell switching level. This is advantageous in the fact that the two levels can be implemented independently. However, the independency also means that the cells of different packets and different inputs can be interleaved during the cell switching through the fabric. Thus, the reassembly of the packets is necessary at the output, and this increases the implementation complexity. Also, output buffer requirement may increase.

On the other hand, in the packet-aware cell scheduling and switching, the information about the packet boundary can be delivered to the cell switching level, so that the cells of a packet may not be interleaved during the switching. This eliminates the complexity of packet reassembly at the output. In this paper, we investigate the characteristics of packet-aware non-interleaving cell scheduling algorithm. The simulation results were obtained using self-similar traffic that reflects the Internet backbone traffic characteristics [16].

3 Packet Switch Model and SOFNI Algorithm

In this chapter, the operation of the packet switch model that is assumed in this paper is described. On the basis of the switch model, the proposed packet-aware scheduling algorithm, SOFNI (Shortest and Oldest First Non-Interleaving) is also described.

3.1 Packet-Aware Switch Model

This paper deals with packet-aware scheduling algorithm and considers packet level latency characteristics. In Fig. 2, the packet level switch model is shown. The packet level handles necessary packet processing, and segments the packets into fixed length cells. It also provides the cell level with the information about packet length and the boundary.

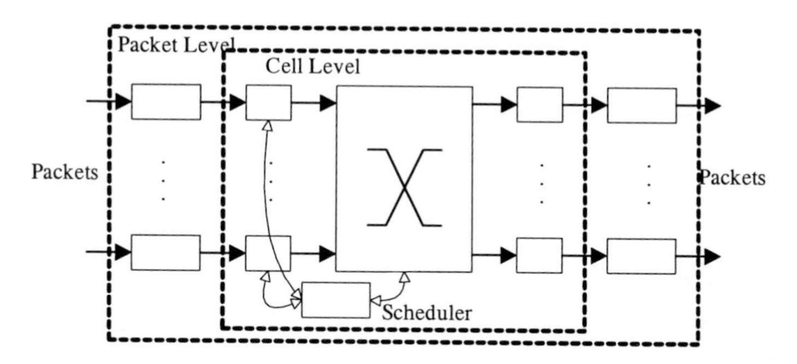

Fig. 2. The packet-aware switch model assumed in this paper

3.2 SOFNI Algorithm

In Fig. 2, the packet switch model that is assumed in this paper is shown. Each input maintains a separate queue for each output, so called VOQ. HOL blocking is eliminated in this VOQ scheme since packets destined to different outputs are queued in different queues. This provides the input-queued switches with the basis to achieve 100% throughput and overcome the 58.6% limitation [3], [4].

SOFNI is a packet-aware algorithm, thus cells of a packet is not interleaved with cells from other packets during the cell switching to the output. This means that an output is held by an input until all the cells of a packet are switched to the output, once so scheduled. This eliminates the reassembly procedure at the output queue, thus simplifies the implementation.

Priority scheme plays an important role in a scheduling algorithm. The priority scheme between traffic classes is out of the scope of this paper, and will be dealt with in our later publication. However, SOFNI uses the priority scheme to prioritize the HOL packets of subqueues in an input. In this priority scheme, SOFNI algorithm adopts the concept of the shortest packet first and the oldest packet first. For this purpose, a priority is assigned to each packet according to its length when the packet moves to HOL of a subqueue (VOQ) in an input.

The priority of a HOL packet of j-th subqueue in input i, $P(i, j)$ is given by

$$P(i, j) = \max\{P_L(i, j) - P_W(i, j), 0\} \quad (1)$$

The priority value, $P(i, j) = 0$ is the highest priority, and $P_L(i, j)$ and $P_W(i, j)$ are given by as follows.

$$P_L(i, j) = f(L_H(i, j), n) \tag{2}$$

$$P_W(i, j) = N_W(i, j) \tag{3}$$

In (2), n is the number of bits used to represent the priority, and $L_H(i, j)$ is the length of the HOL packet of subqueue (i,j). $N_W(i, j)$ is the number cycles(cell time) that the HOL packet has waited at the HOL. Thus, when a new packet moves to HOL position, the value of $P_W(i, j) = N_W(i, j) = 0$. Therefore, the initial priority for a new HOL packet is $P(i, j) = P_L(i, j)$, which is determined by the length of the packet. $P_W(i, j)$ is increased by 1 every cycle when the packet loses the contention in scheduling and wait for next cycle at the HOL. As $P_W(i, j)$ increases, $P(i, j)$ decreases according to equation (1), and as a result, the priority of the packet increases. Varieties of mapping function can be used to determine $P_L(i, j)$. For example, a kind of companding function can be used to map the packet length to n-bit number [6]. In the simulations of this paper, the following mapping function was used for the simplicity's sake.

$$P_L(i, j) = \min\{\text{int}(L_H(i, j)/k), 2^n - 1\} \tag{4}$$

, where k is a constant number.

A cycle of scheduling procedure consists of iterations of 3-step operation like other existing algorithms [2], [7], [11], [12].

In general, if the weights are used in REQUEST, then more abundant information is available for the selection of GRANT and ACCEPT. That could enable the scheduler result in the optimum result or better approximation of the optimum result with respect to its optimization object. The amount of the necessary information in the weight and the algorithm that utilize it affects the complexity of the scheduler. Thus, generally speaking, the tradeoff exists between the amount of the weight information and the simplicity of the scheduler. The weight of SOFNI combines the packet length and the waiting time.

The three step arbitration procedure of SOFNI can be described as follows:

1. ***Request*** : Every unmatched input makes requests to every output for which it has a new packet at HOL of a corresponding VOQ. The inputs that are in the process of sending cells of a packet that was scheduled earlier cycle do not send requests. Every request carries the priority of the HOL packet.
2. ***Grant*** : Each output grants to the highest priority (the smallest one). The ties are broken randomly.

3. ***Accept*** : Each input selects one among the grants that has the highest priority. Every unmatched subqueue in a VOQ updates its priority. Each input has a accept selection pointer, and the first minimum from the pointer is selected when a tie exists. The pointer advances to the one position beyond the accepted one.
4. Repeat step 1 - 3 for ***s*** iterations

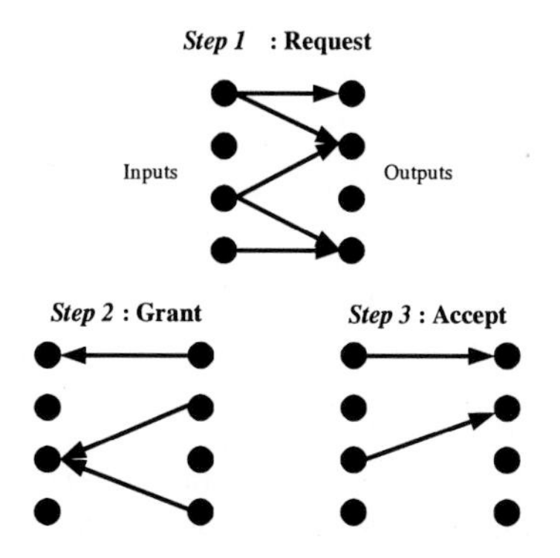

Fig. 3. Three step scheduling consists of request, grant, and accept phases. Many existing parallel MSM, MWM algorithms use the same mechanism.

In the grant select step and accept select step, we can apply different tie breaking rules. First, the ties can be broken randomly. Secondly, the priority can be given in round-robin fashion in which the highest priority advances by one every cycle. Otherwise, pointers with specific update rules may be used as in step 3 and iSLIP, FIRM. The round-robin scheme is simple to implement. Original random selection may not be trivial, but we can make approximation with pseudo-random which can be implemented easily. These two schemes may be simple to implement than the sophisticated schemes; however, we could improve the performance slightly using the pointers. In practical implementation, the random and the round-robin tie-breaking may be used with slight loss of performance.

4 Simulation Results

To evaluate the performance of SOFNI, we compare the performance of SOFNI to those of cell-interleaving (CI) iLQF, and FIRM algorithms. We call the latter algorithms CI-iLQF and CI-FIRM, respectively, as distinguished from non-interleaving (NI) algorithms. In the case of NI-iLQF, the length of each subqueue in cell unit is used as weight. Also, packet-aware non-interleaving versions of iLQF and FIRM were tested to make comparison with SOFNI. FIRM and iLQF are chosen for the comparisons since these two are known to show good performance among the existing algorithms. FIRM was shown to improve the latency performance of iSLIP slightly especially for heavily loaded condition [12].

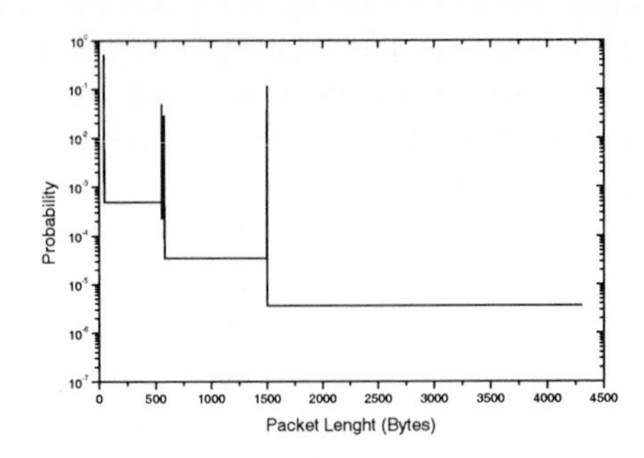

Fig. 4. Probability distribution of packet length

The simulations were performed assuming 16x16 switch. The number of iterations $s=4$ for all algorithm because log2 N iterations is known to be sufficient for NxN switch to achieve the reasonable performance for high speed switch routers. In the case of SOFNI, the number of bits used to represent the priority, is n=8 and k=2.

Since we compare the performance of packet-aware scheduling algorithms, a packet stream is used as input arrivals. The packet stream is modeled by self-similar traffic [16]. The packet inter-arrival time was modeled by Pareto distribution with Hurst parameter, H=0.9.

The packet length distribution used in our simulation is as follows. The average packet length is 389.5 bytes with peaks at the 44, 552, 576 and 1500 bytes. Packet length probability, Pr is $Pr[L_P=44] = 0.5$, $Pr[L_P=552] = 0.05$, $Pr[L_P=576] = 0.03$, $Pr[L_P=1500] = 0.12$, $Pr[45 \leq L_P \leq 551] = 0.25$, $Pr[553 \leq L_P \leq 575] = 0.005$, $Pr[577 \leq L_P \leq 1499] = 0.035$, and $Pr[1501 \leq L_P \leq 4300] = 0.01$. Where L_P is packet length in bytes. This packet length distribution is from [17] which is an approximation of measured Internet backbone traffic [18]. It reflects the fact of predominance of short packets in the Internet backbone. In Fig. 4, the packet length distribution is shown.

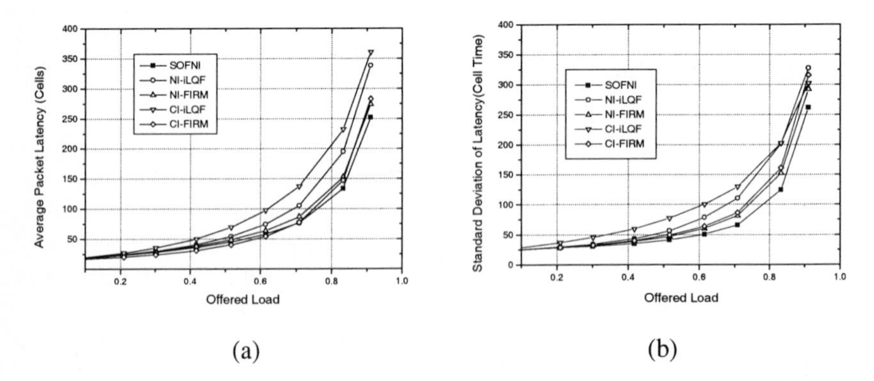

Fig. 5. Average packet latency (a) and standard deviation of packet latency (b)

It is assumed that the internal switching speed is aligned with the line speed. The packets arrive cell by cell in consecutive cell time slots. The packet arrival time is the instance that the last cell of the packet has arrived. Also, packet departure time is the instance that the last cell of the packet has departed an output queue. The packet latency is defined as the time between the arrival of a packet to an input and the departure of the packet from an output queue. A packet is input to an input queue when the last cell of the packet arrives. Also, at the output queue, a packet can be transmitted only when complete packet has been assembled.

Fig. 5 shows the average packet latencies and the standard deviation of packet latencies. The average packet latencies of non-interleaving algorithms are more or less lower than those of cell interleaving algorithms. The average latency of SOFNI is similar to cell-interleaving (CI-) FIRM. It is slightly higher when the offered load is low, but is lower than CI-FIRM as the offered load increases above 0.7. In the region the offered load is over 0.7, the latency of SOFNI is lower than CI-iLQF, CI-FIRM, NI-iLQF, NI-FIRM. Also, SOFNI shows better performance in terms of the standard deviation of packet latency. Note that we deal with the packet latency, not the cell latency, and the packets have variable lengths with the distribution of Fig. 4.

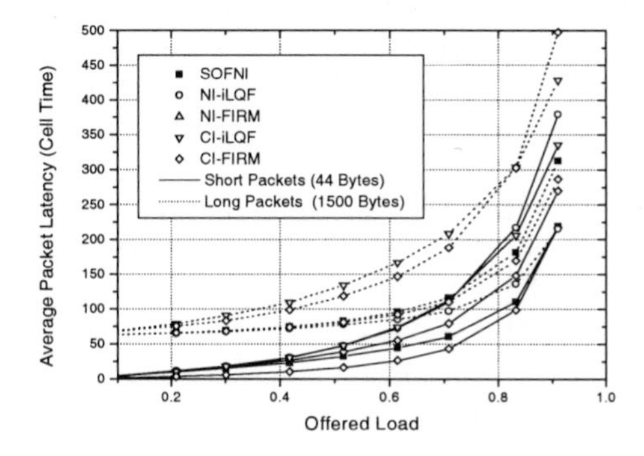

Fig. 6. Average latency of short packets (44 bytes long) and long packets (1500 bytes long)

In Fig. 6, the latencies of the long packets (1500 bytes long) and short packets (44 byte long) are sampled and compared. The latency of short packets is plotted with solid lines, while the latency of long packets is plotted with dot lines. It is shown that the latencies of the long packets of the non-interleaving scheduling are lower than those of the cell interleaving scheduling when the same algorithm is used. However, the latencies of the short packets of the non-interleaving scheduling are greater than those of the cell interleaving scheduling. This means that the cell-interleaving scheduling gives preferential treatment to short packets while penalizing the long packets in terms of latency. We can observe that the performance of SOFNI is good in both cases as compared with other scheduling algorithms.

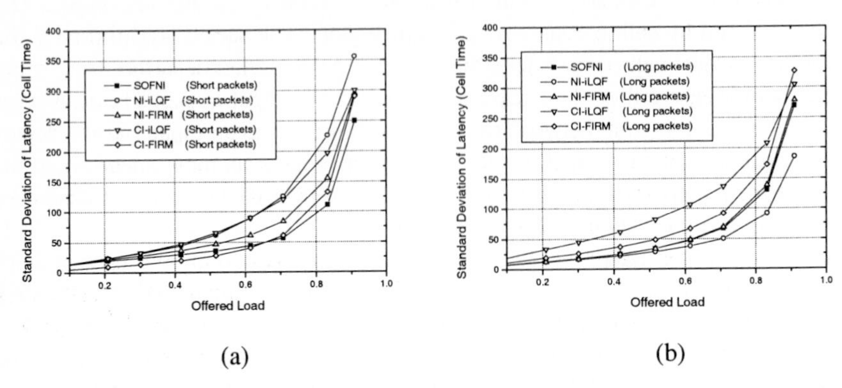

Fig. 7. Standard deviation of packet latencies: Short packets (a) and long packets (b)

The standard deviations of the latencies are compared in Fig. 7. SOFNI also has relatively small standard deviation of long and short packets as compared to other scheduling.

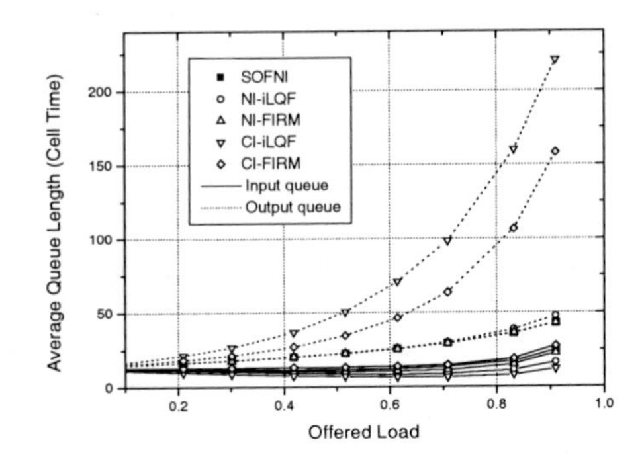

Fig. 8. Average length of input and output queues

Average occupancies of input queues and output queues are illustrated in Fig. 8. As shown in the figure, the output queue occupancies of the cell interleaving algorithm increase rapidly as the load increases. On the other hand, the increase of output queue occupancies of non-interleaving algorithms due to the load increase is small. Thus, we can anticipate that the output buffer requirement of cell-interleaving scheduling algorithms will be much greater than that of non-interleaving scheduling algorithms, assuming the same packet loss probability.

5 Conclusion

We introduced a packet-aware non-interleaving scheduling algorithm, called SOFNI, and investigated the characteristics in comparison with other cell-interleaving scheduling algorithms and non-interleaving algorithms. The packet-aware scheduling algorithm can be implemented using only the packet boundary and length information that are given from the packet level. The non-interleaving operation eliminates the parallel reassembly of multiple packets at the output queue. The simulation result shows that the average packet latency performance of the SOFNI is good as compared to the existing cell interleaving algorithms. SOFNI also showed good performance in terms of latencies of the short packets and the long packets. On the whole, we could conclude that the non-interleaving SOFNI could be feasible choice for a practical implementation. Also, the result shows that the required output buffer size of non-interleaving scheduling algorithms including SOFNI is much less than that of the cell-interleaving scheduling algorithms. The simulation results are obtained using self-similar traffic that reflects the predominance of the short packets in Internet backbone to better grasp the performance of packet scheduling algorithms against packet level traffic.

References

1. M. A. Marsan, A. Bianco, E. Leonardi, and L. Mila, "RPA: A flexible scheduling algorithm for input buffered switches," *IEEE Trans. On Communications*, vol. 47, no. 12, Dec. 1999, pp. 1921-1933.
2. A. Mekkittikul, and N. McKeown "A starvation free algorithm for achieving 100% throughput in an input queued switch," in *Proc. ICCCN'96*, Rockville, MA, Oct. 1996, pp. 226-229
3. N. McKeown, V. Anantharam, and J. Walrand, "Achieving 100% throughput in an input-queued switch," in Proc. IEEE INFOCOM '96, San Francisco, CA. pp.296-302.
4. M. J. Karol, M. G. Hluchyj and S. P. Morgan, "Input versus output queueing switch," *IEEE Journal on Selected Areas in Communications*, Vol. 9, No. 7, Sep. 1991, pp. 1347-1355.
5. M. Hluchyj, and M. Karol, "Queueing in High Performance Packet Switching," *IEEE SAC*-6.9, pp.1587-1597, Dec. 1988.
6. R. Schoen, G. Post, and G. Sander, "Weighted arbitration algorithms with priorities for input-queued switches with 100% throughput," *in Proceedings of IEEE Broadband Switching Systems*, 1999.
7. T. Anderson, S. Owicki, J. Saxe, and C. Thacker, "High speed switch scheduling for local area networks," *ACM Trans. On Computer Systems*, Nov. 1993.
8. R. LaMaire and D. Serpanos, "Two dimensional round-robin schedulers for packet switches with multiple input queues," *IEEE/ACM Trans. Networking*, vol. 2, pp. 471-482, Oct. 1994.
9. N. McKeown, "Scheduling algorithms for input-queued cell switches," Ph.D. Thesis, University of California at Berkeley, 1995.
10. N. Mckeown, P. Varaiya, and J. Walrand, "Scheduling cells in an input-queued switch," *Electron. Lett.*., vol. 29, no. 25, pp. 2174-2175, Dec. 1993.
11. N. McKeown, "The iSLIP Scheduling Algorithm for Input-Queued Switches," *IEEE/ACM Trans. Networking,* vol.7, no.2, pp.188-201, April 1999.

12. D.N. Serpanos and P.I. Antoniadis "FIRM: A Class of Distributed Scheduling Algorithms for high-speed ATM switches with Multiple input queues," *IEEE INFOCOM 2000*, pages 548-555, 2000
13. Y. Tamir, H.-C. Chi, "Symmetric Cross Bar Arbiters for VLSI Communication Switches," *IEEE Transactions on Parallel and Distributed Systems*, vol.4, no.1, Jan. 1993, pp. 13-27.
14. A. Mekkittikul, N. McKeown, "A practical scheduling algorithm to achieve 100% throughput in input-queued switches", *IEEE INFOCOM'98*, vol. 2, 1998, pp. 792-799
15. A. Mekkittikul, "Scheduling Non-uniform Traffic in High Speed Packet Switches and Routers," Ph.D. Thesis, Stanford University, Nov.1998
16. V. Paxson, "Fast approximation of self-similar network traffic," Tech. Rep. LBL-36750/UC-405, April 1995
17. Yijun Xiong, Marc Vandenhoute, and Hakki C. Cankaya, "Control Architecture in Optical Burst-Switched WDM Networks," *IEEE Journal on Selected Areas in Communications*, Vol. 18, No. 10, Oct. 2000.
18. K. Claffy, G. Miller, and K. Thompson, The nature of the beast: Recent Traffic measurements from an Internet backbone.[Online]. Available: http://www.caida.org/outreach/papers/1998/Inet98/)

A Packet-Aware Non-interleaving Scheduling Algorithm with Multiple Classes for Input-Queued Switch

Sungchang Lee[1] and Ilyoung Chong[2]

[1] Hankuk Aviation University, Goyang, Korea
sclee@hau.ac.kr

[2] Hankuk University of Foreign Studies, Seoul, Korea
iychong@hufs.ac.kr

Abstract. A packet-aware non-interleaving scheduling algorithm that has multiple classes according to the packet lengths is proposed, and the latencies and buffer requirement are investigated in comparison with other algorithms. The non-interleaving scheduling algorithm eliminates the complexity of packet reassembly at the output queue. The simulated results show that the packet-aware non-interleaving scheduling inherently improves the latency of the long packets, while the proposed multi-class scheduling complementarily improves the latency of the short packets. The results show that the non-interleaving algorithm is not only feasible for the practical implementation but may have better performance in the latency, output buffer requirement, and implementation complexity viewpoints, if appropriate scheduling algorithm is used. The simulated results are obtained using self-similar traffic generated based on the measured Internet backbone traffic that reflects the predominance of short packets.

1 Introduction

Output-queued switch has been a popular choice for switch architecture for a long time. This is mainly because the output-queued switch can achieve 100% throughput. However, an ideal output-queued switch needs N simultaneous paths from inputs to each output, or the internal speedup rate of N, or the combination of the space and time domain solution. The implementation of N simultaneous paths is impractical when N is large due to the complexity associated with it. Thus, realistic output-queued switches make compromise between the complexity and the throughput and/or cell[1] loss rate. Internal speedup is a preferable solution if possible because of the low complexity, but the line speed of the contemporary network is rapidly increasing, making it unfeasible solution in the future [1], [2].

On the other hand, in this context, there are different approaches that focus on the input-queued switch. The most popular approach is the input-queued switch architec-

[1] Switching by units of fixed-size is advantageous for the high-speed switch implementation, and contemporary high-speed switches and routers use this cell switch technology resulting from ATM [13]. We assume the cell switch fabric throughout the paper.

I. Chong (Ed.): ICOIN 2002, LNCS 2343, pp. 237-247, 2002.

ture with virtual output queue (VOQ) and scheduling algorithms. It is well known that the throughput of original input-queued switch is limited to 58.6% due to the head-of-line (HOL) blocking [3]. The approach of VOQ with scheduling algorithm resolves the throughput limitation problem of input-queued switch. It was shown that input-queued switch using VOQ and an appropriate scheduling algorithm can achieve 100% throughput [4].

Most of existing VOQ scheduling algorithms deals with cell level switching. In this paper, we investigate the packet level characteristics of input-queued switch with VOQ and scheduler. We propose our multi-class packet-aware non-interleaving scheduling algorithm, called multi-class (mc)-SOFNI (Shortest and Oldest packet First Non-Interleaving). Generally speaking, from our earlier works, the non-interleaving scheduling may improve the total latency of the packets and the long packets, but the latency of the short packets tends to increase. The proposed mc-SOFNI improves the latency of the short packets by using segregated queues according to the packet lengths, while the non-interleaving scheduling inherently improves the latency of the long packets. The performance of mc-SOFNI is compared to the well-known cell-interleaving algorithm, iLQF [5] and non-interleaving (NI) version of iLQF that is a modified by us to adapt to non-interleaving scheduling.

The simulated performance is obtained using self-similar traffic [6] reflecting short packet predominance in Internet backbone [7]. The structure of the paper is as follows. Chapter 2 briefly overview related works in this area. We describe our proposed algorithm in chapter 3. The simulation results are presented in chapter 4, and we conclude in chapter 5.

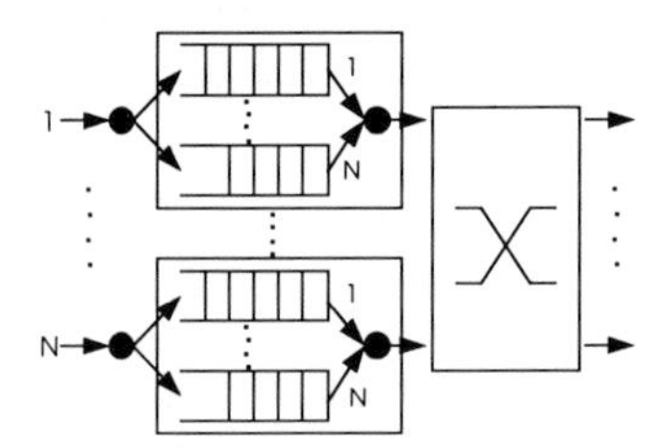

Fig. 1. In VOQ configuration, each input has N subqueues each of which is for one of N outputs. Thus, a VOQ for an output consists of N subqueues each of which is located in each of N inputs. The switch fabric between the inputs and outputs is assumed to be non-blocking.

2 Related Work on Cell Level Scheduling

In Fig. 1 the configuration of the VOQ is shown [2]. The inputs and the outputs are connected by an NxN non-blocking switch fabric. Each input has N subqueues each of which is for each of N outputs. When a cell arrives at an input, the cell is queued into the corresponding subqueue for its destination output. During every cell switching time, an input sends at most one cell that is selected among HOL cells of the N subqueues. During the scheduling, logically, N subqueues of each input make request to corresponding destination outputs, respectively. Consequently, N^2 requests are sent from input side to output side, however, each input can select at most one of its N subqueues for switching. Thus, this is basically bipartite matching problem.

A number of algorithms exist to solve this kind of problems optimally. Those algorithms can be categorized into two kinds, maximum size matching (MSM) and maximum weight matching (MWM). MSM algorithms find the matching so that the number of connections (input-output pairs or edges) is maximized. On the other hand, in MWM, every edge is assigned a weight. MWM finds the matching so that the sum of the all weights of matched connections is maximized. The weight can be chosen among varieties of metrics to achieve the aim, for example, QoS (quality of service, fairness, better throughput. Well known examples of weights are the queue length as in iLQF[5], port occupation as in iLPF[8], waiting time as in iOCF[5], or specially designed metrics on their purpose, and so on. In this context, MSM can be considered as a special case of MWM, where the weight assigned for each edge is simply 1.

The optimal solutions for MSM and MWM have very high complexities. Thus, most of the existing scheduling algorithms realize the approximation of optimal solutions to reduce the time complexity and the hardware complexity at the cost of more or less sacrificed performance. Some examples of MSM approximations are PIM[9], 2DRR[10], MUCS[11], iSLIP[12], [13], FIRM[14]. Although MSM is known to take $O(N^{5/2})$, these approximation algorithms reduce the time complexity to $O(N \log N)$ or $O(i \cdot N \log N)$ serial or $O(\log N)$ parallel steps [15]. MWM takes high complexity of $O(N^3 \log N)$, and the approximation algorithms make compromise between the performance and complexity. The examples are iLQF, iLPF, iOCF, RPA [1], SIMP [15]. SIMP has a complexity of $O(N^2)$, and shows a latency performance close to ideal maximum weight matching.

These algorithms all deal with the cell level switching scheduling. In the cell switching level, the switching is independent of the packet level. No information about the packets is delivered from packet level to the cell switching level. This may be advantageous in the fact that the two levels can be implemented independently. However, the independency also means that the cells of different packets and different inputs can be interleaved during the cell switching through the fabric. Thus, the classification of the cells and the reassembly of the packets are necessary at the output, and this increases the implementation complexity. Also, output buffer requirement may increase due to the simultaneous reassembly of multiple packets.

On the other hand, some information about the packet can be delivered to the cell switching level, so that the cell level may recognize the packet boundaries and the cells of a packet may not be interleaved during the switching. This will eliminate the complexity of packet reassembly at the output. In this paper, we investigate the characteristics of mc-SOFNI, a packet-aware non-interleaving cell scheduling algorithm that complements the weak point of the ordinary non-interleaving scheduling. mc-SOFNI utilizes the segregated queues for this different packet lengths group. This enables mc-SOFNI to improve the latency of the short packets that could otherwise be deteriorated with non-interleaving scheduling. The simulation results were obtained using self-similar traffic that reflects the Internet backbone traffic characteristics [7].

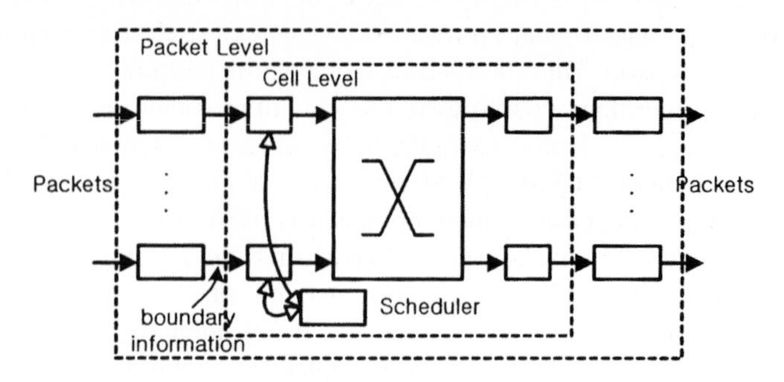

Fig. 2. In the packet-aware switch model assumed in this paper, the packet level provides the cell level with the information of packet boundary and packet length.

3 Packet Switch Model and mc-SOFNI Algorithm

In this chapter, the operation of the packet switch model is described. On the basis of the switch model, the proposed packet-aware scheduling algorithm mc-SOFNI is also described.

In Fig. 2, the packet level switch model is shown. The packet level handles necessary packet processing, and segments the packets into fixed length cells. It also provides the cell level with the information about packet length and the boundary.

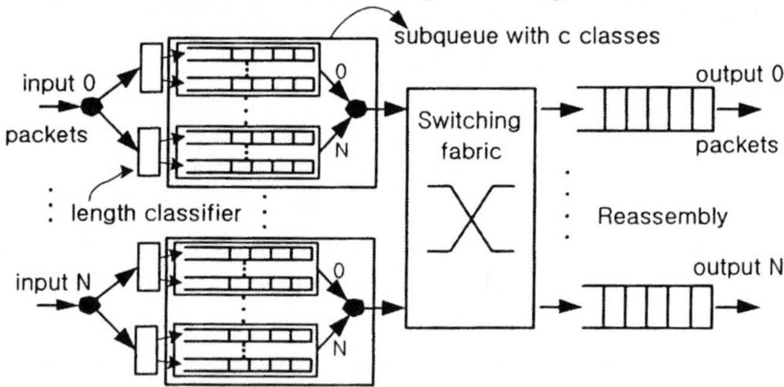

Fig. 3. Each input consists of N subqueues, each of which is for the packets destined to one of the N outputs. Also, each subqueue has c class-subqueues.

In Fig. 3, the packet switch model that is assumed in this paper is shown. A VOQ for an output consists of N subqueues that are located in each of input queues. Each input maintains separate queues for each output. In other words, an input has N subqueues each of which is a component of a VOQ of an output, and buffers the cells that are destined to the corresponding output. HOL blocking is eliminated in this VOQ scheme since packets destined to different outputs are queued in different queues. This provides the input-queued switches with the basis to achieve 100% throughput and overcome the 58.6% limitation [4], [16]. Also, in mc-SOFNI, we use multi-class subqueue for each subqueue as shown in Fig. 3. That is, a VOQ consists of c class-subqueues. The classes can be classified in varieties of criteria like QoS (quality of service). QoS is out of the scope of this paper, and we only deals with the general packet latency characteristics in this paper. The packets are classified into classes by their lengths. The principal concept of this paper's multi-class is shortest job first. The shorter packets are assigned the higher initial priority. But, the long packets may experience starvation if only the predetermined length priority is used. Thus, the oldest first concept is combined in mc-SOFNI.

mc-SOFNI is a packet-aware algorithm, thus cells of a packet is not interleaved with cells from other packets during the cell switching to the output. That means an output is held by an input until all the cells of a packet are switched to the output, once so scheduled. This eliminates the reassembly procedure at the output queue, thus simplifies the implementation.

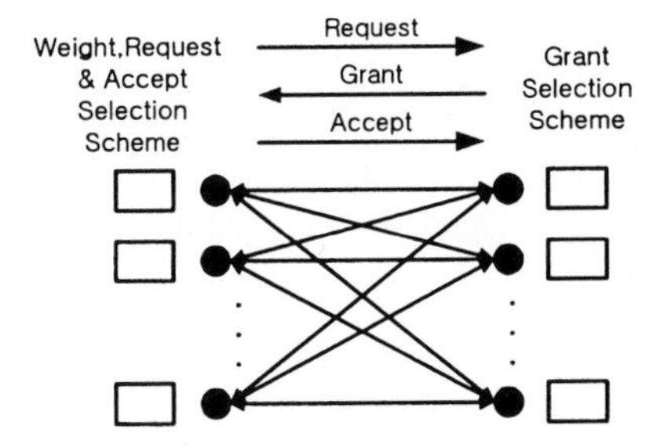

Fig. 4. Three step scheduling consists of request, grant, and accept phases. Many existing parallel MSM, MWM algorithms use the same mechanism. The differences are weights used in communications between input and output side, the grant selection scheme, and the accept scheme

A scheduling cycle consists of iterations of 3-step operation like other existing algorithms as shown in Fig. 4 [2], [4], [8], [12]. Most of existing distributed MSM and MWM algorithms use the 3-step procedures as shown in Fig. 4. The differences are the schemes that the weights are assigned, the grants are selected, and the acceptances are selected.

In general, if the weights are used in REQUEST, then more abundant information is available for the selection of GRANT and ACCEPT. That could enable the scheduler result in the optimum result or better approximation of the optimum result with respect to its optimization object. The amount of the necessary information in the weight and the algorithm that utilize it affects the complexity of the scheduler. Thus, generally speaking, the tradeoff exists between the amount of the weight information and the simplicity of the scheduler.

mc-SOFNI uses the priority of each HOL packet as the weight for the approximated MWM. The priority of a HOL packet of k-th class in subqueue(i, j) (subqueue(i,j,k)), $P(i, j, k)$ is given by

$$P(i, j,k) = \max\{P_L(i, j,k) - N_W(i, j,k), 0\} \quad (1)$$

The priority value, $P(i, j,k) = 0$ is the highest priority. $P_L(i, j,k)$ and $N_W(i, j,k)$ are given by as follows.

$$P_L(i, j,k) = f(L_H(i, j,k),n) \quad (2)$$

In (2), n is the number of bits used to represent the priority, and $L_H(i, j,k)$ is the length of the HOL packet of class k of subqueue (i,j). $N_W(i, j,k)$ is the number cycles(cell time) that the HOL packet has waited at the HOL. Thus, when a new packet moves to HOL position, the value of $N_W(i, j,k) = 0$. Therefore, the initial priority for a new HOL packet is $P(i, j,k) = P_L(i, j,k)$, which is determined by the length of the packet. $N_W(i, j,k)$ is increased by 1 every cycle when the packet loses the contention in scheduling and wait for next cycle at the HOL. As $N_W(i, j,k)$ increases, $P(i, j,k)$ decreases according to equation (1), and as a result, the priority of the packet increases. Varieties of mapping function can be used to determine $P_L(i, j,k)$. For example, a kind of companding function can be used to map the packet length to n-bit number [6]. In the simulations of this paper, the following mapping function was used for the simplicity's sake.

$$P_L(i, j,k) = \min\{\mathrm{int}(L_H(i, j,k)/d), 2^n - 1\} \quad (3)$$

, where d is a constant number.

In the request step, one of the c classes that have new HOL packets is selected from every subqueue. Thus, the weight for a subqueue that has one or more new HOL packets in the class subqueues is,

$$P(i, j) = \min\{P(i, j,k), \ k = 0,...,c-1\} \quad (4)$$

, and the class-k that has a new HOL packet and has minimum $P(i, j,k)$ is selected for the request step. The three step arbitration procedure of mc-SOFNI scheduling can be described as following.

1. *Request*: Each input selects at most one class from each subqueue that has a new HOL packet and has the highest priority as explained above. Each input makes requests with the weight to every output for which it has a new packet at HOL of corresponding subqueue. The inputs that are in the process of sending cells of a packet that was scheduled earlier cycle do not send requests.
2. *Grant*: The outputs that are receiving cells of a packet that was scheduled earlier cycle do not participate in the grant procedure, and send no grant to any input. Other outputs search the request that has the highest priority weight among the requests. Ties are broken randomly.
3. *Accept*: The inputs that are sending cells of a packet that was scheduled earlier cycle do not participate in the accept procedure, and send no accept to any output. Each input selects the grant that has the highest priority weight, and sends

accept to the corresponding output. Each input has an accept selector pointer, and the first one from the pointer is selected when tie occurs..

4. Repeat 1, 2, 3 s iterations.
5. All HOL packets that lose the contention in the scheduling increase $N_W(i, j, k)$, by 1, and consequently, the priority gets higher by 1.

The algorithm is basically different with the cell level scheduling in that the cells of a packet is sent consecutively when the packet is once scheduled. The ties are broken by the accept select pointer in step 3, but other selection scheme like random selection or round-round selection can be used in practical implementation. These schemes are shown to have similar performance in the packet level with the one described above in our simulation.

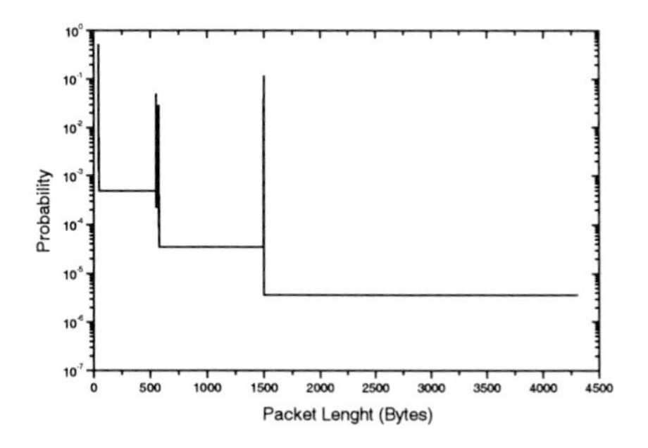

Fig. 5. Probability distribution of packet length

4 Simulated mc-SOFNI Scheduling Performance

Performance of mc-SOFNI scheduling has been simulated for 16x16 switch fabric. Also, simulation results for cell interleaving (CI) iLQF and non-interleaving (NI) iLQF have been also obtained, and compared with mc-SOFNI. The results were compared in terms of the packet level latency. The number of iterations for all algorithms, is s =4. The number of classes in mc-SOFNI is 2. The packets that consist of less than or equal to 3 cells are classified into class-0, and the other packets that consists of 4 or more cells are classified into class-1.

Since we deal with packet-aware scheduling algorithms, a packet stream is used as input arrivals. The packet stream is modeled by self-similar traffic [6]. The packet inter-arrival time was modeled by Pareto distribution with Hurst parameter, H=0.9. The packet length distribution used in our simulation is as follows. The average packet length is 389.5 bytes with peaks at the 44, 552, 576 and 1500 bytes. Let L_P be the packet length in bytes, then the probabilities are $Pr[L_P=44] = 0.5$, $Pr[L_P=552] = 0.05$, $Pr[L_P=576] = 0.03$, $Pr[L_P=1500] = 0.12$, $Pr[45 \le L_P \le 551] = 0.25$, $Pr[553 \le L_P \le 575] = 0.005$, $Pr[577 \le L_P \le 1499] = 0.035$, and $Pr[1501 \le L_P \le 4300] = 0.01$. This packet length distribution is from [17] which is an approximation of measured Internet backbone traffic [7]. It reflects the fact of predominance of short packets in the Internet backbone. In Fig. 5, the distribution of packet length is shown.

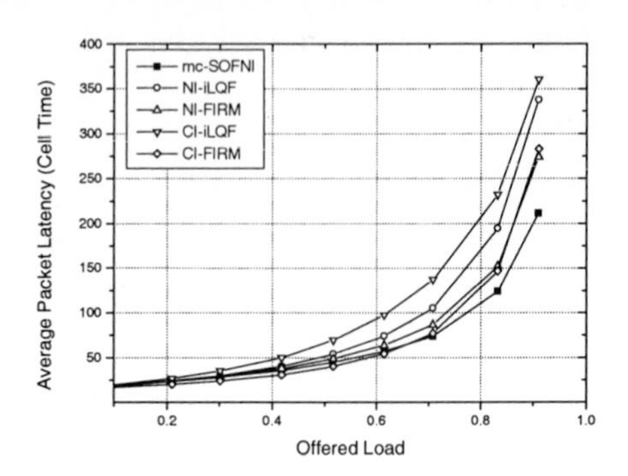

Fig. 6. Average packet latency of scheduling algorithms.

It is assumed that the internal switching speed is aligned with the line speed. The packets arrive cell by cell in consecutive cell time slots. The packet arrival time is defined as the instance that the last cell of the packet has arrived. Also, packet departure time is the instance that the last cell of the packet has departed an output queue. The packet latency is defined as the time between the arrival of a packet to an input and the departure of the packet from an output queue. The cells of a packet are input to a cell level input queue when the last cell of the packet arrives. Also, at the output queue, a packet can be transmitted only when complete packet has been assembled.

In Fig. 6, the average packet latencies of the scheduling algorithms are shown. CI-FIRM and NI-FIRM have similar total average delay. However, it is shown that the average packet latency of CI-iLQF is quite greater than that of NI-iLQF. It is also shown in the figure that mc-SOFNI outperforms other scheduling algorithm in terms of average packet latency. There may be some other overhead in processing time in scheduling like reassembly of cell-interleaving scheduling, but other processing times except queuing and transmission delays are not counted in the results shown in this paper.

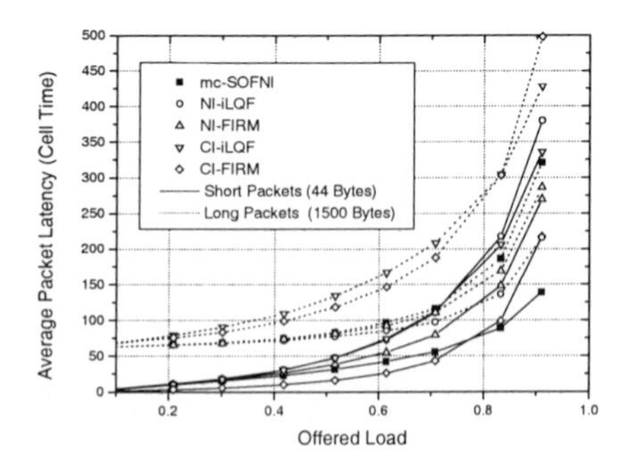

Fig. 7. This figure shows the comparison of the average latencies of long packets (1500 bytes long) and short packets (44 bytes long).

In Fig. 7, the latencies of long packets (length: 1500 bytes) and the short packets (length: 44 bytes) are sampled and compared. We can interpret the results as follows. When the load is light, the queueing latency is very small and the switching time (transmission through the fabric and to the output line) dominates the latency time especially for long packets. However, as the load increases, the queueing latency increases for both the short packets and the long packets. In the region that the load is high, the latencies of the long packets increase drastically for the cell interleaving scheduling. This is because more cells from other packets interleaves as the load increases before the complete transfer of a long packet through the fabric. On the other hand, we can observe the latencies of the long packets of non-interleaving algorithms increase rather slowly.

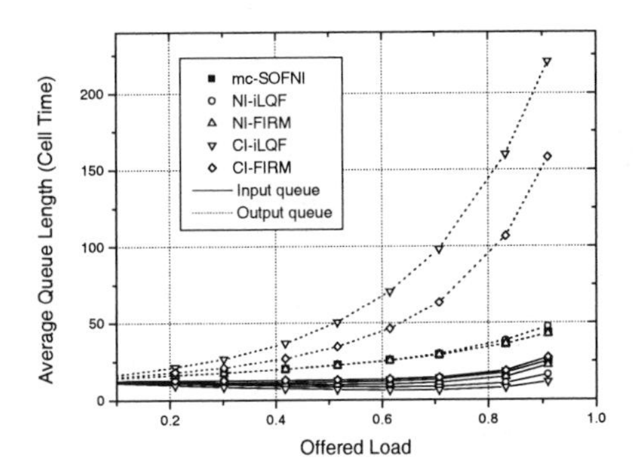

Fig. 8. Average input queue and output queue occupancies of the algorithms are shown in this figure.

Average Occupancies of input queue and output queues are compared in Fig. 8. As shown in the figure, the output queue occupancy of the cell interleaving scheduling algorithms increase rapidly as load increases. In contrast, the output queue occupancies of non-interleaving scheduling algorithms increase smoothly with the load increase. mc-SOFNI is a packet-aware scheduling algorithm, thus the cells of a packet is not interleaved with cells of other packets during the cell switching. This not only eliminates the reassembly of the packet at the output queue but also reduce the output buffer requirement. In cell interleaving scheduling, multiple packets have to be reassembled simultaneously at an output queue. The cells switched to the output have to be identified and distributed to the corresponding temporal packet reassembly queues. The amount of output buffer requirement of the cell interleaving algorithms is shown to be higher than the packet-aware algorithms. The reason for this is that more output buffers are occupied by the waiting packets for the reassembly as load increases.

5 Conclusion

We introduced a packet-aware non-interleaving scheduling algorithm, mc-SOFNI, and investigated the characteristics in comparison with other non-interleaving and cell interleaving scheduling algorithms. The packet-aware scheduling algorithm utilizes the information about the packet boundary that is given from the packet level. The non-interleaving operation eliminates the reassembly of multiple packets in parallel at the output queue. From the result of the simulation, the latency of the short packets in cell interleaving scheduling is smaller than that of non-interleaving scheduling, while the long packets are penalized by the interleaved short packets. mc-SOFNI classifies the incoming packets into the classes according to their length, and give preferential treatment to the shorter packets. This provides mc-SOFNI with the mechanism that can reduce the latency of the short packets, while the non-interleaving scheduling inherently improves the latency of the total packets and the long packets. The average latency of the short packets in mc-SOFNI is comparable to best known cell interleaving scheduling algorithm, and also the latency of the long packets outperform other scheduling algorithms. On the whole, we could conclude that the latency performance of non-interleaving scheduling is feasible for the implementations if an appropriate scheduling algorithm is used. Also, the result shows that the required output buffer size of non-interleaving scheduling is much less than the cell-interleaving scheduling. The simulation results are obtained using self-similar traffic that reflects the predominance of the short packets in Internet backbone to better grasp the performance of packet scheduling algorithms against packet level traffic.

References

1. M. A. Marsan, A. Bianco, E. Leonardi, and L. Mila, "RPA: A flexible scheduling algorithm for input buffered switches," *IEEE Trans. On Communications*, vol. 47, no. 12, Dec. 1999, pp. 1921-1933.
2. A. Mekkittikul, and N. McKeown "A starvation free algorithm for achieving 100% throughput in an input queued switch," in *Proc. ICCCN'96*, Rockville, MA, Oct. 1996, pp. 226-229
3. Karol, M., Hluchyj, M. and Morgan, S. "Input versus output queueing on a space dividion switch," *IEEE Trans. Communications*, Dec., pp. 1347-1356, 1987.
4. N. McKeown, V. Anantharam, and J. Walrand, "Achieving 100% throughput in an input-queued switch," in *Proc. IEEE INFOCOM '96*, San Francisco, CA. pp.296-302.
5. N. Mckeown, Scheduling algorithms for input-queued cell switches, Ph.D. thesis, UC Berkeley, 1995.
6. V. Paxson, "Fast approximation of self-similar network traffic," Tech. Rep. LBL-36750/UC-405, April 1995
7. K. Claffy, G. Miller, and K. Thompson, The nature of the beast: Recent Traffic measurements from an Internet backbone.[Online]. Available : http:// www.caida.org/ outreach/papers/ 1998/Inet98/)
8. N. McKeown and A. Mekkittikul, "A practical scheduling algorithm to achieve 100% throughput in input queued switches," in *Proc. IEEE INFOCOM' 98*, San Francisco, CA, April 1998, vol. 2, pp. 792-799.
9. T. Anderson, S. Owicki, J. Saxe, and C. Thacker, "High speed switch scheduling for local area networks," *ACM Trans. On Computer Systems*, Nov. 1993.

10. R. LaMaire and D. Serpanos, "Two dimensional round-robin schedulers for packet switches with multiple input queues," *IEEE/ACM Trans. Networking*, vol. 2, pp. 471-482, Oct. 1994.
11. H. Duan, J. Lockwood, S. Kand, and J. Will, "A high performance OC12/OC48 queue design prototype for input buffered ATM switches," in *Proc. IEEE INFOCOM '97*, Kobe, Japan, Mar. 1997, vol. 1, pp. 20-28.
12. N. Mckeown, P. Varaiya, and J. Walrand, "Scheduling cells in an input-queued switch," *Electron. Lett.*, vol. 29, no. 25, pp. 2174-2175, Dec. 1993.
13. N. McKeown, et. al, "The tiny Tera: A small high-bandwidth packet switch core," *IEEE Micro*, Jan-Feb. 1997.
14. D.N. Serpanos and P.I. Antoniadis "FIRM: A Class of Ditributed Scheduling Algorithms for high-speed ATM switches with Multiple input queues," IEEE INFOCOM 2000, pages 548-555, 2000
15. R. Schoen, G. Post, and G. Sander, "Weighted arbitration algorithms with priorities for input-queued switches with 100% throughput," *Proceedings of IEEE Broadband Switching Systems*, 1999.
16. M. J. Karol, M. G. Hluchyj and S. P. Morgan, "Input versus output queueing switch," *IEEE Journal on Selected Areas in Communications*, Vol. 9, No. 7, Sep. 1991, pp. 1347-1355.
17. Yijun Xiong, Marc Vandenhoute, and Hakki C. Cankaya, "Control Architecture in Optical Burst-Switched WDM Networks," IEEE Journal on Selected Areas in Communications, Vol. 18, No. 10, Oct. 2000.

A Design of the IS-IS Routing Protocol for the ATM Based MPLS System

Mijeong Yang, Byoungchun Jeon, and Youkyoung Lee

Internet Technology Department, ETRI
Yusong P.O. Box 106, Daejon
305-600, KOREA
mjyang@etri.re.kr

Abstract. This paper proposes an architecture of Intermediate System-to-Intermediate System (IS-IS) for the ATM based Multiprotocol Label Switching (MPLS) system. IS-IS is a link state routing protocol designed to provide routing in a network layer protocols with datagram service. IS-IS has favored scalability in the aspect of minimizing storage and computing in level 1 routers. Therefore, it is important to support IS-IS for the MPLS system used in backbone networks. We propose the architecture of IS-IS routing protocol and extensions for traffic engineering in MPLS system. Also, we describe a packet transmission scheme and a configuration procedure.

1 Introduction

The traffic carried by the Internet is rapidly increasing together with imposing requirements for reliability, quality of service, and manageability. This trend forces the network technology to come up with new approaches and solutions. MPLS has emerged as a technology that can provide many of the functionalities associated with ATM[1]. In MPLS packets are encapsulated, at ingress points, with labels that are then used to forward the packets along Label Switched Paths (LSPs). These LSPs are virtual traffic trunks that carry flow aggregates generated by classifying the packets into Forwarding Equivalence Classes (FECs). These flow aggregates and explicit routing of bandwidth guaranteed LSPs enables service providers to traffic engineer their networks. These LSPs are setup by a signaling protocol such as the Label Distribution Protocol (LDP). The network layer routing protocol is responsible for providing route information to LDP. Currently, IS-IS[2] and Open Shortest Path First(OSPF) are mainly used for this purpose.

IS-IS and OSPF were both designed to support routing in a network layer protocol with datagram service. OSPF was designed for Internet Protocol (IP) which is a network layer protocol in the TCP/IP suite. IS-IS was originally designed for Connectionless Network Layer Protocol (CLNP), the datagram network layer protocol in the International Organization for Standardization (ISO) suite. However, It can support the Internet Protocol (IP) network layer as well as the ISO network layer. IS-IS with the IP specific fields defined in [3]

I. Chong (Ed.): ICOIN 2002, LNCS 2343, pp. 248-256, 2002.

is referred to as Integrated IS-IS. Integrated routing means the use of a single routing protocol to support multiple network layer protocols.

OSPF[4] has favored optimizing routing, whereas IS-IS has favored minimizing storage and computing in level 1 routers. That is, IS-IS is more scalable than OSPF. Also, IS-IS is able to migrate parameters in a running network. These are important factors in the backbone network. In this paper, we propose the architecture of IS-IS routing protocol and extensions for traffic engineering in MPLS system of which the national ATM based Internet backbone consists.

2 Overview of the ATM Based MPLS System

An MPLS domain consists of a kind of routers called Label Switching Routers (LSRs) having label switching capability. Fig. 1 shows the MPLS over ATM based Internet backbone with the edge LSR. The edge LSR consists of MPLS controller block, non-MPLS network interface block, MPLS network interface block, and ATM switch block including FE(Forwarding Engine).

The MPLS controller block performs managing and controlling the ATM switch as well as establishing and releasing LSPs. This block has several modules as follows:

The traditional routing protocol daemon is OSPF/BGP and IS-IS modules. The OSPF/BGP module is using IP encapsulation whereas the IS-IS module is using Layer 2 encapsulation. These routing protocols generate Routing Information Base (RIB). Based on the RIB, the Label Distribution Protocol (LDP) creates Label Information Base (LIB). The LDP is a set of procedures and messages by which LSRs establish LSPs through a network by mapping network layer routing information directly to data link layer switched paths. The LDP associates an FEC with each LSP it creates. Incoming packets are mapped to an FEC.

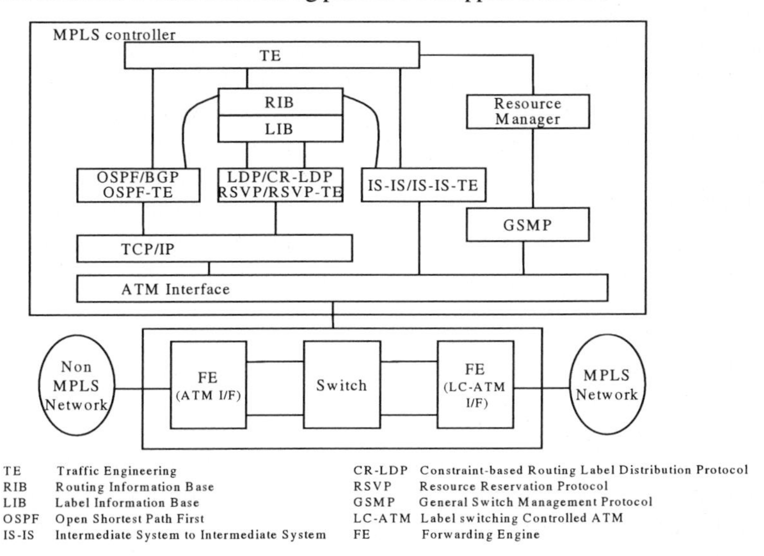

Fig. 1. ATM based MPLS System Architecture

This behavior is performed only once when packets enter an MPLS domain. Subsequently, packets are processed and forwarded strictly according to their labels. The label is removed by the egress LSR. The explicit routes are established by constraint based signaling module (CR-LDP and RSVP-TE). An explicit route has specified all or some of the downstream nodes of that route and provides some of the functionality needed for Traffic Engineering (TE). Also, extensions of routing protocols deliver additional link state information for TE.

The resource manager keeps track of available bandwidth and VPI/VCI values for the switch interfaces. The General Switch Management Protocol (GSMP) module allows the MPLS controller to control the ATM switch. The GSMP is a rather simple master/slave protocol, with the slave running on the ATM switch and the master running on the MPLS controller. Only the slave potion has to know anything that is specific to the ATM switch hardware. Only for edge LSR, the layer 3 forwarding block is needed. This block is responsible for forwarding unlabeled traffic.

3 Comparison between IS-IS and OSPF

IS-IS and OSPF have common characteristics as well as different those[5][9]. There are two types of routing protocols in use in networks. The one is distance vector routing and the other is link state routing. Both IS-IS and OSPF are link state routing protocols. Therefore, two protocols basically have following properties. In a link state routing protocol, each router is responsible for determining the identity of its neighbors and constructing a special packet known as a Link State Packet (LSP). The LSP lists the node's neighbors, broadcasts the LSP to all routers, stores the most recently generated LSP from each router, and computes routes to all destinations based on the stored LSP database.

IS-IS and OSPF support a hierarchical routing architecture. Routers can only support a limited size network. If a network grows beyond the size that can be practically supported by the router, the common technique is to add hierarchy to the network. That means the network is partitioned into pieces known as "areas." However, IS-IS allows multiple levels whereas OSPF allows only two levels. IS-IS terminology is slightly different from OSPF terminology. "Level 2" is called "backbone" in OSPF, "level 1 routing" is "intra-area routing," and LSPs are known as Link State Advertisements (LSAs). In the remainder part of this chapter, we discuss the differences between IS-IS and OSPF.

- Area Configuration

In IS-IS, an area can have multiple area addresses. Routers are part of one area only. This serves to allow migrating an area from one address to another, merging two areas, or splitting an area into pieces. The backbone area can be assigned any address. In OSPF, each area has only a single address. Different interfaces of a router may have different areas. Area boundaries fall within routers, rather than between routers, as in IS-IS. The value 0.0.0.0 means the backbone area. OSPF does not have the ability to merge and split addresses dynamically as IS-IS has. Fig. 2 shows the example of area configuration for IS-IS and OSPF.

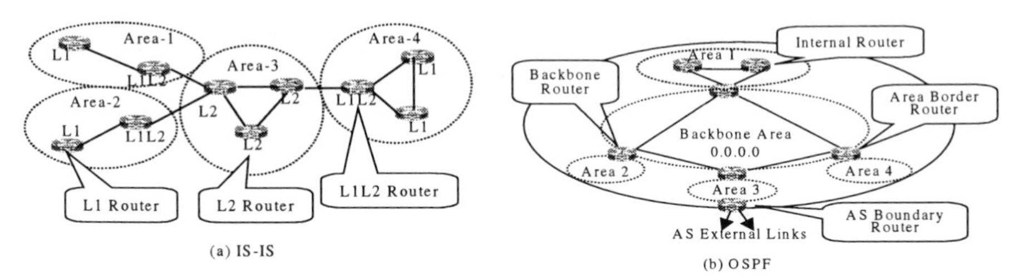

Fig. 2. Area Configuration

- Injection of Level 2 Information

In IS-IS, level 1 routers only know information about their own area. If a level 1 router receives a packet with an address not reachable within the area, this router forwards the packet to the level 2 router nearest to the level 1 router. In OSPF, level 2 information is fed into the area. Thus, OSPF yields more optimal inter-area routers than IS-IS. The cost of providing more optimal routing is increased bandwidth usage, memory, and CPU requirements.

- Neighbor Establishment

Both IS-IS and OSPF routers report holding time in its hello message. In IS-IS, the holding time do not need to be the same on neighbors for them to be adjacent. The receiving router uses the holding time in hello message as the hold time for the neighbor. This enables each neighbor to use a different hello interval. This behavior differs from that of OSPF, where neighbors must agree on the Hello interval and the hold time. So, this scheme makes it difficult to change timers without disruption. In the neighbor adjacecy establishment, OSPF uses complex, multistate process to synchronize databases between neighbors, and IS-IS essentially uses its regular flooding techniques to synchronize neighbors.

- Encapsulation

OSPF runs on top of IP. That means an OSPF packet is transmitted with an IP data packet header. In contrast, an IS-IS packet is transmitted directly on top of the data link layer.

- Packet Encoding

In IS-IS, LSPs lists all the neighbors of a node. OSPF issues a separate LSA for each destination. Because each LSA requires its own sequence number and age, total OSPF database will be bigger than the size of the IS-IS database required for the same information. However, when a single destination changes, IS-IS requires that the entire fragment containing the changed information be transmitted, whereas OSPF can update just the single piece of information that changed. All fields in OSPF are of fixed length and the packet formats specify which fields are present. Most fields in IS-IS, however, are of variable length fields with type-length-value (TLV) format.

4 Design of IS-IS for MPLS System

In this chapter, we propose an IS-IS protocol architecture, a packet transmission scheme, and extensions for TE to support IS-IS routing protocol for the ATM based MPLS system.

4.1 IS-IS Protocol Architecture

Fig. 3 shows the IS-IS protocol architecture. The IS-IS routing protocol consists of a kind of databases, packet processing modules, and SPF algorithm. An administrator specifies the area address and system identifier and registers the circuits on which the IS-IS protocol is worked. The area address and system identifier may be automatically allocated by the IS-IS protocol. The circuit information is stored in the circuit database and managed. The elements in the circuit database are exchanged between routers using hello messages. The router that receives hello messages initializes adjacencies between neighboring routers and updates the adjacency database. This adjacency database can be updated by the administrator as well as by the hello message.

Level 1 LAN IIH(IS to IS Hello) and level 2 LAN IIH are used to acquire the information of a neighboring router on broadcast network. Point-to-point IIH exchanges the level information of a neighboring router on point-to-point network. The router takes the other information through ISH (ISO 9542 Intermediate System Hello PDU). The reachable address database contains the routing information except the neighbor information specified by an administrator.

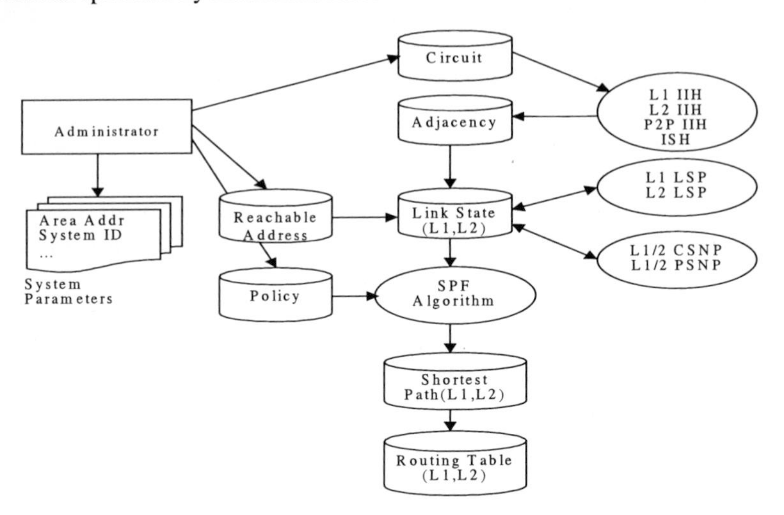

Fig. 3. IS-IS Block Architecture

LSPs deliver the elements in the adjacency database and the reachable address database. The router receiving LSPs updates the link state database. The link state database maintains the newest LSPs that are locally generated or received from other routers. Level 1 LSP is generated by level 1 and level 2 router and propagated throughout an area. The contents of level 1 LSP indicates the state of the adjacencies to neighbor router and end system of the router that originally generated the PDU. Level 2 LSP is generated by level 2 router and propagated throughout the level 2 domain. The contents of level 2 LSP indicates the state of the adjacencies to neighbor level 2 router and to reachable address prefixes of the router that originally generated the PDU.

CSNP (Complete Sequence Number PDU) and PSNP (Partial Sequence Number PDU) are used to synchronize LSPs information with neighboring routers. These PDUs describe the link state database in a compact format so that neighbor routers can ensure that their database stays consistent. They are never forwarded but transmitted only between neighbors. SPF algorithm calculates the shortest path based the link state database. The route results are reflected in the routing table.

4.2 Packet Transmission Scheme

The IS-IS PDUs are transmitted shown in Fig. 4. The PDUs that are received via the ATM interface are decoded in Forwarding Engine[7]. The FEs purely support packet forwarding. A FE receives and transmits labeled packets through the LSP or IP packets through the ATM connection using the IP packet forwarding path. However, there are special cases that an FE could not process packets by itself; packets corresponding to exceptional conditions such as mismatching, routing transients, and so on, and packets being processed by MPLS System Controller(MSC). In that case, the FE must transfer the packets to the MSC.

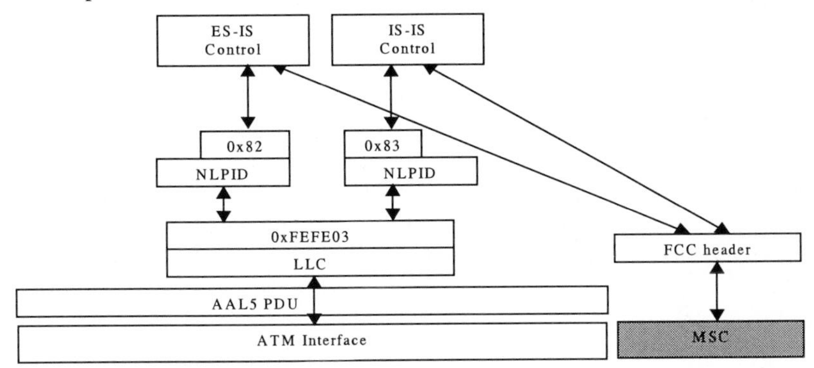

Fig. 4. Encoding and Decoding Procedure of the Forwarding Engine

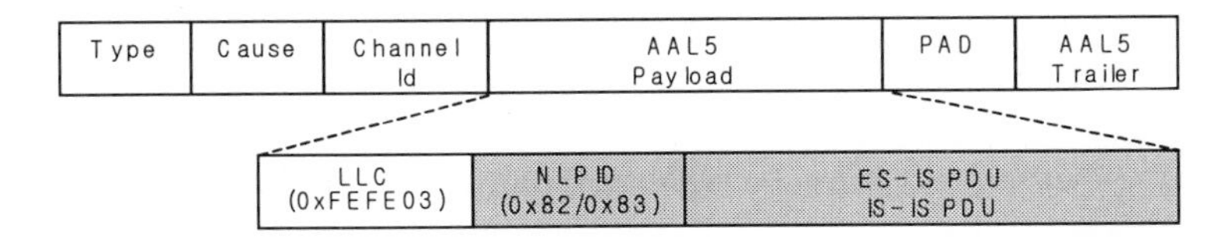

Fig. 5. Encapsulation Packet Format in Forwarding Control Channel

When an FE receives the IS-IS PDU, it checks the AAL 5 PDU format and the link layer headers for LLC/NLPID. If LLC/NLPID is 0xFEFE03/0x82, the FE transmits the PDU to the IS-IS block in the RCP for processing ES-IS (End System to Intermediate System) protocol[8]. If LLC/NLPID is 0xFEFE03/0x83, the FE transmits the PDU to the RCP for processing IS-IS protocol. The PDUs are transmitted through the Forwarding Control Channel (FCC) with forwarding control frame format. The FCC is a control path between the FE and the RCP to process the extension.

Fig. 5 shows the frame format on the FCC. The AAL5 payload and trailer follow the header part with a one octet type field, a one octet cause field, and two octet channel identifier field. ES-IS and IS-IS PDUs are encapsulated into the AAL5 payload with LLC header value 0xFEFE03. This value means that a routed ISO PDU follows.

4.3 Extensions for Traffic Engineering

The MPLS technology has many advantages for traffic engineering. However, MPLS does not address the issue of how to find paths with constraints. So, the existing routing protocol must be extended for TE. To enable the computing of routes with constraints, the new extensions for IS-IS include features carrying additional information about links and removing the 6-bit limit on link metrics. Link information contains maximum link bandwidth, maximum reservable bandwidth, current bandwidth reservation at each of eight priority levels, a default TE metric, and the resource class or color of the link[6]. The default TE metric is provided so that the constraint-based routing process is not restricted to using the metric used by the existing IS-IS routing protocol for shortest path calculation. The color of the link is an administrative attribute assigned to the link to implement policy-based routing.

To serve these purposes, IS-IS extensions define three new TLVs. The TE router id TLV contains the 4-octet router identifier of the router originating the LSP. The extended IP reachability TLV describes reachable IP prefixes and the extended IS reachability TLV describes link information. These new TLVs must replace the existing IP reachability TLV and IS neighbor TLV. Therefore, mechanisms and procedures to migrate to the new TLVs are considered avoiding routing loops.

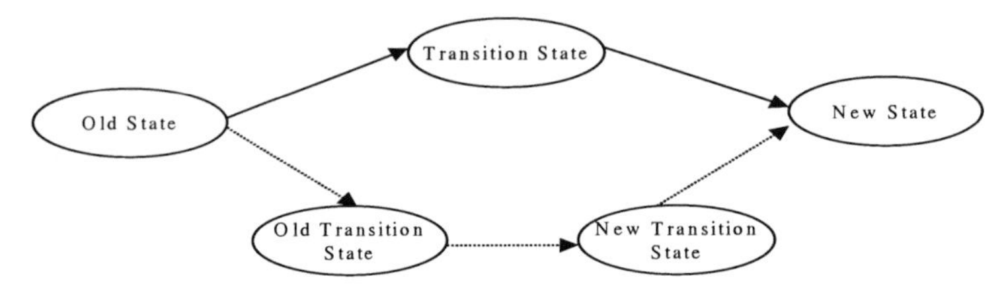

Fig. 6. Transition Scheme to Configure an IS-IS Network using New TLVs

We provide five states and two trasition schemes to migrate from old TLVs toward new TLVs shown in Fig. 6. The old state and the new state generate and accept only old TLVs and only new TLVs repectively. The transition state generates and accepts both old and new TLVs. The old trasition state generates only old TLVs and accepts both types of TLVs, and the new trasition state generates only new TLVs and accepts both types of TLVs. If all routers want to use new TLVs, the administrator can configure each router such as the old state to trasition state to new state scheme or the old state to old trasition state to new transition state to new state scheme.

Fig. 7 shows the block interfaces for TE. The IS-IS extension block stores the local TE link information configured by the administrator and advertises this link state database to neighbors. Also, TE link information in extended link state database must be transferred to the constraint path computation block in the TE block. When the path computation block receives the request of establishing constraint path from the administrator, it computes the path based the TE link information. After the signaling block establishes the computed constraint path by the triggering of the TE block, the signaling block informs the resource manager block of the changed resource information. The resource manager informs the IS-IS extension block of the amount of available bandwidth. Then, the IS-IS extension block advertises the updated link state database.

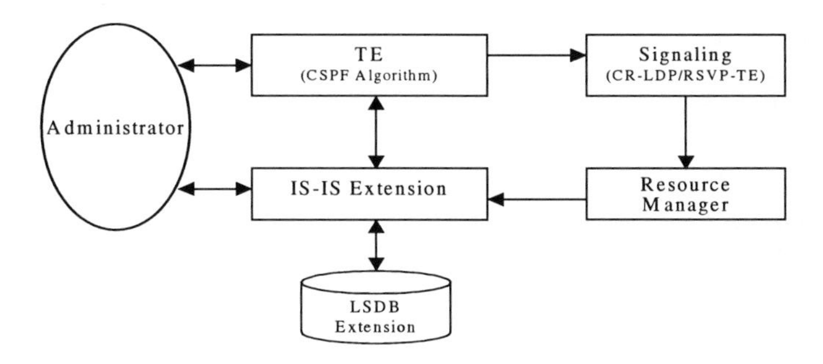

Fig. 7. Interface Architecture for TE Extension

5 Summary

In this paper, we propose the architecture of IS-IS routing protocol for the ATM based MPLS system. We describe the advantages compared with another prevalent link state routing protocol. For developing IS-IS in the ATM based MPLS system, we suggest the IS-IS protocol architecture, the PDU transmission scheme, and the extension architecture for traffic engineering. Now, the ATM based MPLS system with the suggested architecture has been implemented. We will gradually test and upgrade this architecture for the efficiency and the stability. Also, we will provide new features for supporting of generalized MPLS[10][11] and IPv6.

References

1. T. Li, "MPLS and the Evolving Internet Architecture," IEEE Communications Magazine, Vol. 37, December 1999, pp. 38-41.
2. "Information technology – Telecommunications and Information exchange between systems – Intermediate system to Intermediate system Intra-domain routeing information exchange protocol for use in conjunction with the protocol for providing the Connectionless-mode Network Service (ISO 8473)," ISO 10589, 1990.
3. R. Callon, "Use of OSI IS-IS for Routing in TCP/IP and Dual Environments," RFC 1195, December 1990.

4. J. Moy, "OSPF Version 2," RFC 1247, July 1991.
5. O. Sharon, "Dissemination of Routing Information in Broadcast Networks : OSPF versus IS-IS," IEEE Network, January/February 2001, pp. 56-65.
6. T. Li and H. Smit, "IS-IS extensions for Traffic Engineering," Internet Draft, draft-ietf-isis-traffic-04.txt, August 2001.
7. J. You, S. Kang, and W. Chun, "Design of the Packet Forwarding Architecture of the ATM Based MPLS Edge Node," ICOIN 2000, pp. 431-433.
8. "Information processing systems – Telecommunications and Information exchange between systems – End system to Intermediate system routeing exchange protocol for use in conjunction with the protocol for providing the Connectionless-mode Network Service (ISO 8473)," ISO 9542, 1991.
9. R. Perlman, *Interconnection, Second Edition*, Addison-Wesley, 1999.
10. K.Kompella et al., "Routing extensions in Support of Generalized MPLS," Internet Draft, draft-ietf-ccamp-gmpls-routing-02.txt, February 2001.
11. K.Kompella et al., "IS-IS extensions in Support of Generalized MPLS," Internet Draft, draft-ietf-isis-gmpls-extensions-04.txt, September 2001.

A New On-Demand Source Verification Method in Multi-domain Multicast Networks

Chang-jin Suh and Doo-hyun Han

*Soongsil University, Dep. of Computer Science,

cjsuh@computing.ssu.ac.kr, dhhan@kingdom.ssu.ac.kr

Abstract. MSDP is the most practical inter-domain multicasting protocol. It uses periodic broadcast to inform of source location. This periodic broadcast not only consumes considerable network bandwidth but also causes potential delay for new subscribers who request the ongoing multicast service until inter-domain routing path is setup.

This paper proposes an "On-Demand Source Active"(ODSA) method that relies on broadcasting much less than MSDP. ODSA disseminates source location to the distributed database servers while multicast groups refer the source location in response to customers' request with no delay. Our simulation shows ODSA saves more than half bandwidth for all cases we tested.

1 Introduction

A new type of point-to-multipoint transmission, called multicasting, has emerged as multimedia technologies develop in modern networks. Multicast packets are sent once in the source node, and are copied as many as needed in junction routers, and finally arrives at tens or thousands of receivers simultaneously. Multicast is beneficial because it enables a single server to handle even humongous users with enormously reduced bandwidth.

If a node wants to receive multicast packets, it has to join multicast group according to Internet Group Management Protocol (IGMP). Local multicasting routers cast query message at intervals to the receivers in its domain. The nodes that belong to the multicast group respond to the router's message. Routers then exchanges group information to its neighbor routers.

Multicast routing is the technology that provides the multicast tree along which data and control packets routes. At primitive ages a single domain constitutes a whole network, and Distance Vector Multicast Routing Protocol (DVMRP) runs over it. Sooner we are able to multicast in multi-domain networks.

---

1) This work was supported by grant No.(R01-2001-00362) from the Korea Science & Engineering Foundation.

I. Chong (Ed.): ICOIN 2002, LNCS 2343, pp. 257–267, 2002.

There are two ways of intra-domain multicasting. Dense Mode(DM) multicast protocols are designed to use when there are large receivers thickly distributed at each domains. Sparse mode(SM) protocols such as Core Based Tree (CBT) and Protocol Independent Multicasting - Sparse Mode (PIM-SM) [1] are suitable to use when small users scatter in wide area. We focus on sparse mode protocols, because user distribution tends to be sparse and in wider range, and SM protocols have better scalability.

PIM-SM is the most common SM protocol. PIM-SM utilizes a central router called Rendezvous Points (RP's). PIM-SM includes the RP in multicasting trees along which multicast packets route. This constraint causes two problems of triangular routing and the third party dependency. Triangular routing is named after long inefficient routing path established when source and receivers are all closely located but RP are far apart from them. The third party dependency points out the problems caused by less responsible RP's. As RP is determined by a mapping function, sometimes RP is chosen independently of source and receiver nodes. The unluckily chosen router reluctantly saves bandwidth for the third-party multicasting source and receivers, because it should have been used for its customer nodes.

IETF has released various protocols for inter-domain multicasting protocol. Currently Multicast Source Discovery Protocol (MSDP) [2], which we mention in depth in section II.1, is a practical solution. Also as a long-term solution Border Gateway Multicast Protocol (BGMP) [3] were proposed. It distributes multicasting addresses that are assigned hierarchically. We introduce a new inter-domain protocol as a middle-term solution - On Demand Service Active message (ODSA, pronouncing "odyssey"). It modifies MSDP by reducing broadcasting messages and adding on-demand source verification commands.

This paper consists of the following sections. Section two introduces the current standard MSDP. Section three proposes a new inter-domain protocol ODSA. In section two and three we observe how MSDP and ODSA behave differently for a certain given condition and suggest the way to enhance the performance of MSDP. Section four describes assumptions of simulation and analyzes the simulation results. Then conclusions follow.

2 Inter-domain Multicasting

Inter-domain multicast protocols perform multicasting in the wide area network that is decomposed to subnetworks called domains. Currently PIM-SM, MBGP and MSDP are used for this purpose. RP's are selected to control over a domain and multicasting tree is established in accordance with PIM-SM at each domain. MSDP merges separate trees built in each domain into a global inter-domain tree. MBGP is also used to digress from routers that do not support PIM-SM or MSDP. We focus on MSDP and analyze its behavior in this chapter.

2.1 Behavior of MSDP

MSDP uses inter-domain channels between a pair of RP's that have established peering relation through TCP connection. We define two kinds of RP's. RP_S is an RP which is responsible for the source domain to which source node belongs. RP_R is an RP that controls the receiver domain in which there are receiver nodes that receive the multicast packets. Be aware that this classification is per-service based. So an RP can be RP_S for one multicast service as well as RP_R for another at the same time. In MSDP RP_S declares the source domain to all RP_R using Source Active messages, simply SA messages.

SA messages are issued in two cases. When multicast first occurs or the source node changes to another, RP_S issues SA messages to all RP's instantaneously. We represent this type of SA messages as SA_{trg} implying *trigger*. Second, each RP collects all multicast packets whose source nodes are in its domain and puts them altogether into a long SA messages. We call this SA message SA_{prd} meaning *periodic*. SA_{trg} is provided for all current receivers while SA_{prd} is issued for the nodes that try to join the on-going multicast service to notify the source domain.

If an RP_R is given request to begin a certain multicast service, it first checks whether there is a routing tree in its domain that has delivered the requested service. If there is, the RP_R completes its job by adding the requested node to the tree. Otherwise, the following procedure is triggered. The RP_R establishes a new tree to connect from itself to the requested node in the domain. At the same time, the RP_R waits for SA_{prd}. When the RP_R receives SA_{prd} which includes the source domain of the requested multicast, it establishes inter-domain tree to connect RP_S and itself.

MSDP overcomes previously mentioned problems - triangle routing or the third party problem - which were left unsolved when PIM-SM was used alone, because in MSDP RP is chosen in source domain or receiver domain. Thus RP is no more third-party nor located in a long distance from them.

2.2 Analysis of MSDP

This section points out the shortcomings of MSDP as to why it wastes away considerable bandwidth with large service delay. First, it wastes much bandwidth in broadcasting SA messages to announce the source domain to all RP_R, but only a few RP_R's need it. We are going to demonstrate the pitfalls of MSDP by mentioning how SA_{prd} are issued for four transmission periods in Fig 1.

Fig. 1 consists of two parts. At the upper part, the multicasting events are briefly mentioned in both source domain and receiver domain. At below, full entries are shown in the four SA_{prd} messages called SA(1), SA(2), SA(3) and SA(4) issued at time t(1), t(2), t(3) and t(4) respectively. Both parts are drawn with a common horizontal time axis.

Fig. 1 considers two different points (RP_S and RP_R) of view. First look at it from the RP_S's view. At every SA transmission periods RP_S transmits SA messages that include all sources in its domain. At just earlier than t(1), RP_S's domain had eight sources whose multicast addresses are 1, 2, 3, 4, 5, 11, 12 and 20. At between t(3) and t(4) a new source generates. In the first three periods, SA messages have seven entries. The SA message at t(4) includes an additional entry 25. To focus on SA_{prd},

Fig. 1 does not include the SA_{trg} message that is issued when new receiver (entry 25) generates.

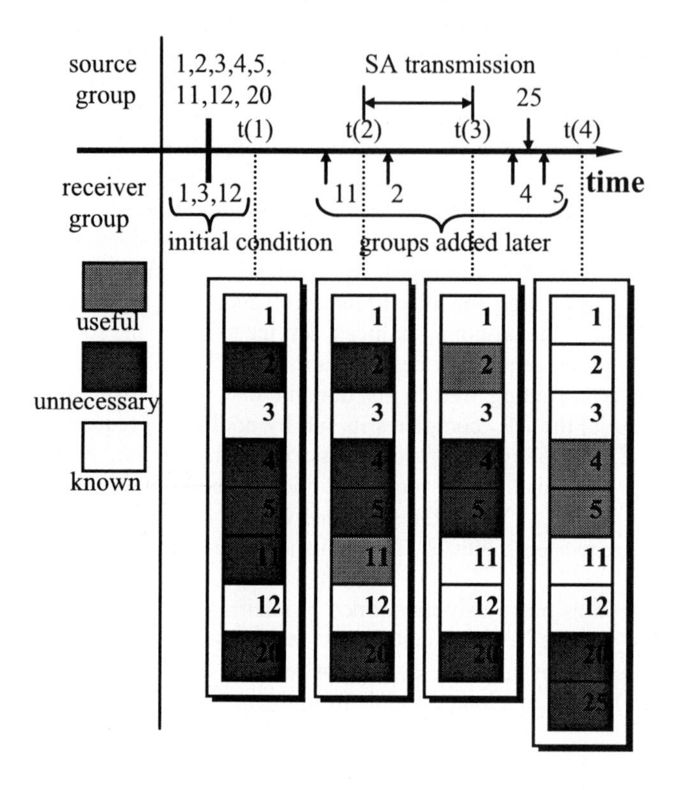

Fig. 1. Operation example of SA_{prd} messages in MSDP

We move to the RP_R's view, and observe how broadcast entries are used in RP_R's domain. Look at the receiver condition just below the time axis in Fig. 1. Initially domain RP_R has received three multicast services (1, 3, 12) among eight or nine services whose source domain is RP_S. And then four services (11, 2, 4, 5) are additionally required to receive. We can classify the received entries into three cases depending on how to use them. First, suppose RP_R received the request to provide a new multicasting whose source domain is not known. The RP_R has to know the source domain and the wanted entry in SA_{prd} messages solves its question, thus SA_{prd} is "useful." The second case is the entry that RP_R does not have to serve multicasting and thus is "unnecessary." The third is the case when the entry contains the multicast service which is being served by RP_R. This information is already "known" to the RP_R. Fig. 1 marks boxes in SA_{prd} that store entries in dark, slashed, or with no markings for the three cases. According to Fig. 1 only four entries are useful out of thirty three issued entries, giving birth to about ten percents of utilization ratio. In this way utilization rate of broadcast messages tends to be very low.

The other problem in MSDP is inappropriately long join delay. If a node joins a multicasting service which is already in service in the domain, the new node can get the service shortly after the request. Otherwise, the node at least has to wait until RP_R receives SA_{prd}. SA_{prd}'s are issued at every period (one minute by default) and join requests occur independently of the period. So the first node to join in the receivers' domain waits for half of the SA transmission period in average (thirty seconds if default value is used) until receiving SA_{prd}. Consumers' magazine reported that most Internet users do not stick to a site if it does not respond for a little more than a few seconds. Commercial service providers strongly eager for short service join delay in the range of a few seconds.

3 On-Demand SA

3.1 Definition of ODSA

We propose a new inter-domain multicast protocol called ODSA, and mention how it solves the problem that MSDP has faced. ODSA disseminates SA messages differently. It follows ODSA's way in using SA_{trg}, but uses new types of SA - SA_{req} and SA_{rep} instead of SA_{prd}.

ODSA adds a new component - RP_D. RP_D is an RP that has database to store source information. RP_S broadcasts SA_{trg} especially to RP_D and RP_R sends SA_{req} to RP_D. The major idea in ODSA is how RP_S and RP_R can choose the same RP_D with no conversation between them. ODSA uses predefined RP_D calculation function to be discussed later.

When a new multicast service begins or a new source occurs, RP_S generates SA_{trg} and broadcasts as in MSDP. Instead, RP_S includes the address of RP_D to SA_{trg} so that the appointed RP_D stores the delivered information to its database. When RP_R needs the location of source domain, RP_R no longer waits for SA_{prd}. It decides RP_D by pre-defined calculation and issues SA_{req} to the calculated RP_D asking what the source domain is. The RP_D that receives SA_{req} searches for its own database and transmits SA_{rep} to the requested RP_R. After all, RP_R reads SA_{rep} and notices the source RP is RP_S. The later procedure is identical to MSDP.

3.2 Enhancements

In ODSA, SA_{prd} that is broadcast in MSDP is no longer used but SA_{req} and SA_{rep} are tranceived in point-to-point manners. Owing to reduced broadcasting, ODSA achieves short response delay and much savings in bandwidth.

Fig. 2 shows how SA_{rep}'s are sent to RP_D's in ODSA. Fig. 2 assumes the identical traffic as in Fig. 1. (Readers easily notice that upper parts of both figures are identical.) Fig. 2 only shows the entries that are classified as "useful" and represented in dark box in Fig. 1. In this way ODSA generates digested information that is sure to use. Also the RP_R's job gets simpler, because it only checks less number of delivered entries. Instead, ODSA requires RP_D calculation once in transmitting SA_{trg} and once in requesting SA_{req} As whole ODSA requires much less work than MSDP due to simplified RP_D calculation.

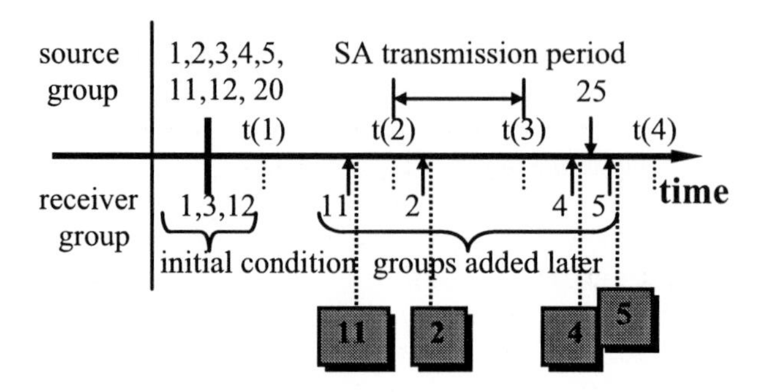

Fig. 2. Operation example of SA_{req} messages in ODSA

The other advantage of ODSA is limited service delay. We measure the delay starting from that RP_R begins to search the source domain until RP_R finds it out. In ODSA the worst-case service delay can be measured from sending SA_{req} until receiving SA_{rep}. This is simply turn-around transmission delay between RP_R and RP_D. Considering that all RP's have peering relation on top of TCP connection, service delay is at most a few seconds. We can explain why ODSA is better from the point of design policies. ODSA is receiver-oriented and event-driven, while MSDP is transmitter-oriented or information-providers-oriented and unidirectional.

Lastly we mention two misunderstandings that readers may easily make. First, does ODSA still have another type of the third party problem because RP_D is decided independently of RP_S and RP_R? No. What RP_D does is to maintain a database and to reply to SA_{req} There is no data path to RP_D, thus this is not third party problem. Second, should RP_S additionally issue SA_{trg} to RP_D for safe transfer when it broadcasts SA_{trg}? Negative. We do not need additional unicast of SA_{trg}, because what we call broadcasting is done on top of TCP connection between a pair of RP in peering, which is safe enough.

3.3 RP_D Selection Function

We represent i_D as the IP address of RP_D that will store the source domain of the multicast service whose address is m. Then the RP_D calculation function $\boldsymbol{F}$ is defined as below,

$$i_D = \boldsymbol{F}(m). \tag{1}$$

Before introducing $\boldsymbol{F}$, we partition all RP's into n groups. n is a predefined number known to all RP's. For simplicity we use the *modular* function. RP_D belongs to the j'th group if the following condition satisfies.

$$j = i_D \ mod \ n \ (j \in \{0,1, \ ..., n\text{-}1\}) \ . \tag{2}$$

F consists of two stages. The first stage decides group number and the second stage specifies an RP in the chosen group. At the first stage the same function in Eq. (2) is done to partition IP address set into n groups. So, the multicast address m belongs to the j'th group if Eq. (3) satisfies.

$$j = m \ mod \ n. \tag{3}$$

RP_D in the group j only takes care of IP's that belong to the same group j. Due to partitioning, RP_D can be in charge of reduced number of IP's.

Now we focus on the j'th group which l RP_D's belong to. Let their IP's be p_k ($k=1, 2, \ldots, l$) respectively. At the second stage, an RP is chosen among l RP's that outputs the largest number in Eq. (4). Highest Random Weight (HRW) algorithm[4] demonstrates an incidence of possible calculation. This method uses a special conversion function C that converts each of all IP addresses – p_k or m – to a floating number f ($0 \le f < 1$). Let us define a temporary variable t_k as

$$\begin{aligned} t_k &= C(p_k) - C(m) && \text{if } C(p_k) > C(m), \ (\text{for } k=1,2, \ldots, l) \\ &= C(p_k) - C(m) + 1 && \text{if } C(p_k) \le C(m) \end{aligned} \tag{4}$$

Then RP whose IP is p_q is selected as RP_D if (5) satisfies.

$$t_q = max_k \ (t_k) \qquad (\text{for } k=1,2, \ldots, l) \tag{5}$$

To run the selection function, all RP addresses in a certain group are required. To achieve this condition, all RP's should maintain the whole RP's address set. As RP is chosen among reliable and stable routers to serves for multicast service in a given area, the RP set does not change often. But once addition or deletion of RP occurs, it should be announced to all current RP's.

4 Performance Evaluation

4.1 Performance Factors

We choose two performance factors – the total consumed bandwidth of SA messages ***C*** and join delay ***D***. MSDP and ODSA differ in the way of disseminating SA messages. To enumerate their relative performance we compare total bandwidths consumed in whole networks, each of which is obtained by applying MSDP or ODSA alone respectively. If we define $P(t)$ as the length of SA messages issued at time t in byte and $L(t)$ as sum of links in which SA message is transmitted, the total throughput ***C*** is defined in Eq. (6). ***C*** includes all kinds of SA messages - SA_{prd}, SA_{trg}, SA_{rep} and SA_{req}.

$$C = \frac{1}{T} \cdot \int_0^T P(t) \cdot L(t) dt \quad (6)$$

The second factor is join delay. The real join delay may be the time lag from user's service request until the service starting time. In our study we define the join delay in narrow sense to focus on the delay caused from inter-domain activity, and declare the join delay as the time interval from the time when RP_R receives join request until it knows the source domain and begins proper action for users' requests.

4.2 Simulation

The simulation assumes that multicast source is always unique. There are two cases. In multiple-sender case the initial sender changes to another during the service. In single-sender case the initial sender continues to send throughout the session. Simulation uses two networks. The ten-node network is derived from the Asia-Pacific Advanced Network (APAN). It is shown in Fig. 3.

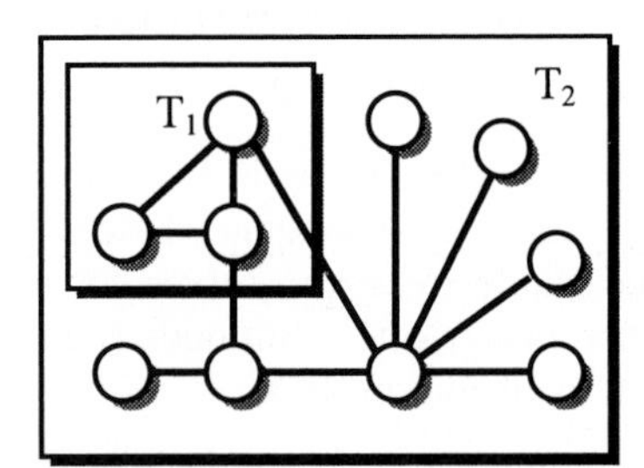

Fig. 3. Two networks (T_1 and T_2) used in simulation.

To simplify the simulation model, we have a few assumptions as followings.

- The SA message transmission period is 60 seconds.
- Each multicast service lasts for 60 minutes.
- There are 50 multicast groups in whole network.
- The size of SA_{trg}, SA_{req}, SA_{rep} SA_{prd} header and an entry in SA_{prd} are 60, 48, 60, 40 and 20 bytes each.
- Each multicast has fixed number of service users throughout the session.

4.3 Simulation Results

Fig. 4 and 5 show the consumed bandwidth in three-node and ten-node networks varying the number of multicast group from single member to 3, 9, 27, 81 and 243. MSDP consumes bandwidth in sending SA_{trg} and SA_{prd}. As we assume the sender is always unique, MSDP transmits fixed rate of SA_{trg} and SA_{prd} independently of receivers' join or leave rate. This makes MSDP curve straight in both figures. The

Single Change Cost (SCC) is the cost caused from a single SA_{trg} which is the theoretically minimum bound. So they locate the lowest position in both figures.

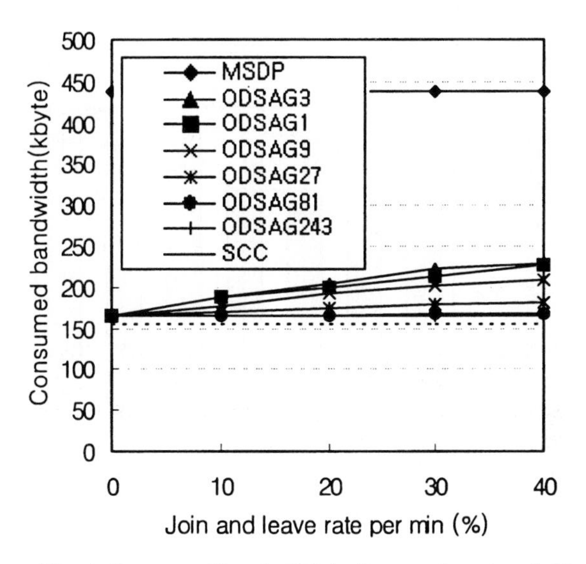

Fig. 4. Consumed bandwidth in three-node network T_1

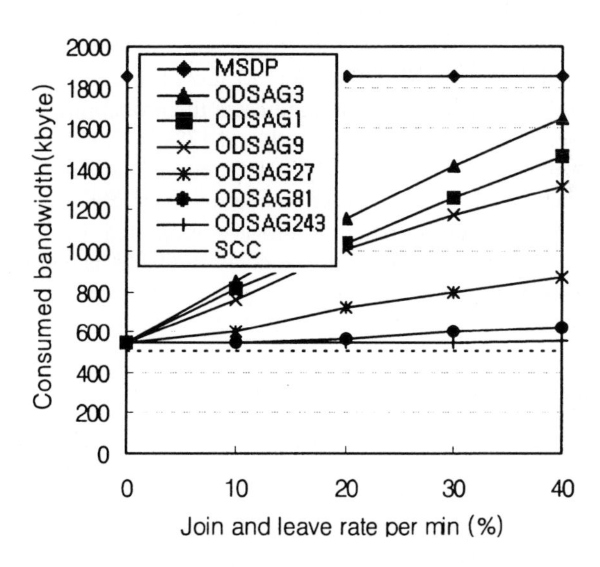

Fig. 5. Consumed bandwith in ten-node network T_2

The results show that ODSA is always better than MSDP and that ODSA and MSDP becomes similar when group members are small and join or leave rate is high. The latter rarely happens because small groups are closely related to each member and have low join or leave rate. Finally we conclude ODSA considerably outperforms MSDP in all cases.

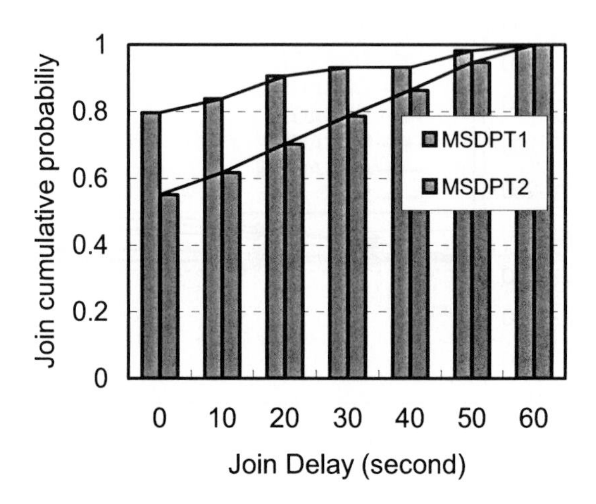

Fig. 6. Cumulative distribution function of Join Delay using MSDP in T_1 and T_2

Fig. 6 draws the cumulative distribution function to show the relation of join delay versus probability when MSDP is used. MSDPT1 and MSDPT2 are obtained in network T1 and T2 respectively. Fig 6 indicates the probability that join delay is longer than ten seconds is 38% in T_2 network and 17% in T_1 network. T_2 has more general network topology than T_1, the MSDPT2 curve may show more practical results. According to this result, users have to wait for a long time with considerable probability even though multicast supported by MSDP becomes practical. The other simulation shows us the delay in ODSA is negligible within 2, 3 seconds in all cases.

5 Conclusions

This paper proposes a new multi-domain multicast protocol ODSA, which does not use the source verification procedure defined in the current standard – MSDP. ODSA reduces broadcasting with distributed database that stores the source domains of the multicast service.

While MSDP broadcasts source verification messages at fixed interval, ODSA provides more receiver-friendly service. If a new join message is issued in ODSA, it triggers source verification procedure on demand without passively waiting for broadcasting period.

Throughout simulation we verified that ODSA is better in consumed bandwidth and join delay. We expect ODSA performs much better in the future Internet because its befit increases when multicasting is more common and multicasting groups are distributed very sparsely.

References

[1] D. Estrin et al. "Protocol Independent Multicast -Sparse Mode (PIM-SM): Protocol Spec.," RFC2117, June 1997

[2] Dino Farinacci et al., "Multicast Source Discovery Protocol (MSDP)," *Internet draft*, draft-ietf-msdp- spec-05.txt. Feb. 2000

[3] D. Thaler, "Border Gateway Multicast Protocol (BGMP): Protocol Spec.", *Internet draft*, draft-ietf- bgmp-spec-01.txt, Mar. 2000

[4] D. Estrin et al., "A Dynamic Bootstrap Mechanism for Rendezvous-based Multicast Routing," *INFOCOM'99*, New York, pp 1090-1098, Mar. 1999

A Balanced Scheduling Algorithm for High-Speed ATM Switch

Shiann-Tsong Sheu , Bih-Hwang Lee*, Hsuen-Wen Tseng*, Chiu-Yun Ko*, and Wei-Zhen Liang**
Department of Electrical Engineering, Tamkang University
*Department of Electrical Engineering
National Taiwan University of Science and Technology
**Department of Civil Engineering
National Taiwan University
E-mail : stsheu@ee.tku.edu.tw

Abstract. As far as asynchronous transfer mode (ATM) is concerned, virtual output queue (VOQ) is an efficient architecture of high throughput performance. ATM is primarily applied in the backbone, and it is necessary to schedule data in an efficient way. In this paper, we will propose balanced scheduling VOQ algorithm (B-VOQ) to resolve the complicated problem of cell scheduling. By utilizing B-VOQ, we can obtain the optimal average delay comparing with other proposed algorithms.

1. Introduction

With the explosive growth of the Internet, it is no doubt that switches and routers play an increasingly important role in data communication. The principal part of ATM is switches, which provide not only high-speed transmission but also complete service. In ATM, each switch possesses several input ports and output ports and transfers a fixed size of cell by using these ports. Nowadays, a crossbar switch is implemented because it allows multiple cells to be transferred simultaneously without cell blocking. Since multiple cells from a number of input lines may be destined for the same output, queuing is required to buffer cells.

Generally speaking, the methods in designing high-speed switch are divided into three parts: 1. input queuing; 2. output queuing; 3. VOQ. In input queuing switch, each input port maintains a first in first out (FIFO) queue to store incoming cells and then designs the sequence of cells. In this method, the switch fabric can run at the same speed as each input link. Due to its scalability, input queuing has always been attractive for high speed switching system. However, it is well known that the maximum throughput of the input-queued switch with a single queue per input port is limited to 58.6% [1] under uniformly distributed traffic condition. This is because of the Head-of-line (HOL) blocking phenomenon. In ATM, it is a cell can be held up by another cell ahead of it in the same queue and is destined to a different output.

I. Chong (Ed.): ICOIN 2002, LNCS 2343, pp. 268–277, 2002.

In output queuing switch, each output port maintains a queue to store incoming cells that request to transmission through the output. In this method, each incoming cell will be transmitted to output port by switch fabric simultaneously, so it can achieve 100% throughput. But the disadvantage of the method is that switch requires N times faster than the speed of each input port. It is too expensive for a high-speed switch.

VOQ is the third architecture and it also achieves high throughput. In an N*N switch, each input maintains N separate FIFO queues as buffer for each output (Figure 1). By utilizing this architecture, not only HOL blocking is eliminated but also no additional speed is required. VOQ has been proved to increase the throughput of an input queuing switch to 100%.

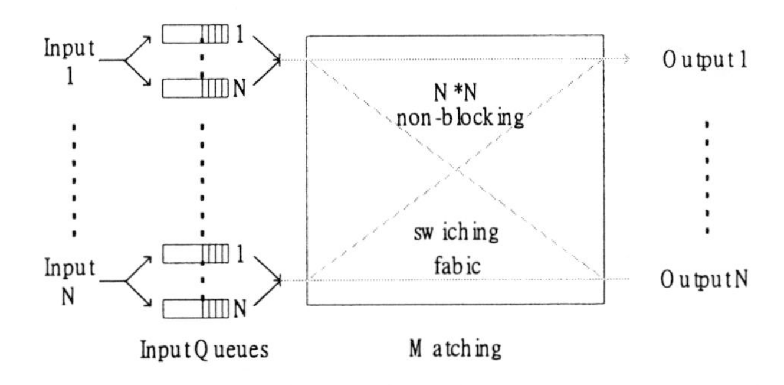

Fig. 1. VOQ switch architecture.

A number of different scheduling algorithms of switch with VOQ have been proposed [2],[4],[5] for finding all optimal matching. These algorithms provide different implementation, and the most important class is distributed scheduling algorithm. In this approach, N*N switch is modeled as N input and output parts, and each part contains N nodes. Using this model, we hope to calculate an optimal matching.

In the literature, four main algorithms, PIM[3], RRM[5], iSlip[6], T-RRM[7] have been proposed, and these algorithms all contain three phases: request, grant and accept (Figure 2).

1. Request phase: each input broadcasts their requests to output.
2. Grant phase: each output picks out one request from the received requests independently, and then transfers a grant to the input that sends request.

3. Accept phase: each input picks out one grant from received grants as accept. When an input receives a grant from output, the input has to send a cell to output.

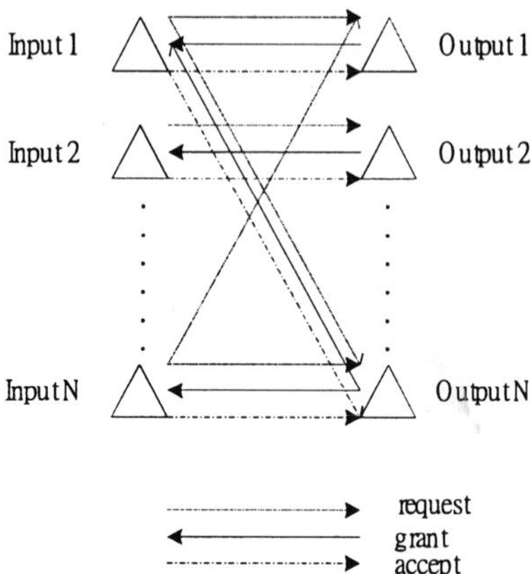

Fig. 2. Distributed Scheduling Algorithm

In various algorithms are only different on how to select a request to grant and how to choose a grant to accept. In PIM algorithm, grant and accept are made randomly. In RRM and iSlip algorithms, grant and accept are based on round-robin order. In RRM, each output informs next request according to a fixed sequence of round-robin way, and moves round-robin pointer to another granted request. In the same way, each output accepts grant. However, in iSlip algorithm, at each output, round-robin pointer is not changed unless the corresponding grant is accepted. Consequently, in RRM, each output which will not wait for the accept acknowledgement from input moves to next location according to round-robin order. On the contrary, in iSlip, each output will wait for an accept acknowledgement from the corresponding input, and then round robin to next point. Above-mentioned algorithms, only iSlip can achieve 100% throughput.

PIM, RRM, iSlip do not concern the state of queues. For instance: one port perhaps queues a number of cells, and the port only turns around the same location delivering one cell during a slot time. This will result in long average delay. Therefore, T-RRM algorithm is proposed for the number of cells sequence in VOQ. In T-RRM, not only round-robin is the principle of cells switching sequence but also request, grant, and accept are the transmitting phases. T-RRM also can achieve 100% throughput. The basic idea of T-RRM is that: if a VOQ has queued more cells than threshold, the corresponding VOQ will send cell to the output first.

In order to avoid the condition that a VOQ has queued more cells, while traffic load is becoming greater, we propose B-VOQ to resolve the balance of each VOQ.

The rest of paper is organized as follows. In section 2, we will introduce the foundation of B-VOQ and describe a brief example to display how it works. In section 3, we will demonstrate simulation results and compare the performance with other algorithms that have been proposed. Finally, some conclusions are given in section 4.

2. B-VOQ Algorithm

Like RRM, iSlip, T-RRM, B-VOQ still utilizes round-robin order and three transmission phases. Before describing B-VOQ, we define marked port first. If an input is selected to send a cell to an output through the crossbar switch, and the number of the corresponding VOQ is greater than a predicted threshold, we mark the corresponding input and output [7]. When the number of cells is served as less or equal to threshold, we take the input and output as unmarked port.

The main idea of B-VOQ is that: when the amount of cells is a lot, it will result that many VOQ queue more cells than threshold. We mark the corresponding inputs and outputs, collect the VOQ over threshold, and calculate the number of cells in corresponding VOQ. The VOQ, which owns the maximum number of cells, delivers cell in this slot time first. In each slot time, we repeatedly calculate the number of cells in VOQ over threshold and select the VOQ, which owns the maximum number of cells to process. If there are more than two VOQ owning the same and maximum number of cells over threshold, we will adopt round-robin order to deliver cells. When traffic load is lower, it results that the cells in VOQ are less than threshold. We send cells by round-robin order. The advantage of B-VOQ is to balance the number of cells in each VOQ. While traffic load is a lot, T-RRM algorithm will result in the unbalance of queue. Some VOQ possess cells over their capacity, and the cells over capacity are lost.

The pseudo code of B-VOQ

```
/* i : number of input ports;   j : number of output ports;   cell_num  : the number of cell
in VOQ ;   Tmport : how many ports own the same cell numbers which is the greatest one
*/

for(all i) request(i ){ /* Request phase */
if(marked_input[i]= = Yes){ /* is it a marked input ? */
        sort;/* calculate the cell numbers of marked input*/
        find max cell number; /* find max cell number in VOQ */
         if(TM port>=2)
         {
          Torr_priority[j]=(i+1); /* select a request over threshold based on the round-robin
of output j */
        }
      accept(i,j);/* use the same matching in this time slot */
        if(cell_num [i,j]<=threshold)
            marked_input[i] =No;/* unmark the input */ }
else for(all j){
if(cell_num [i,j]>0){ /*queued cells for output j ?*/
```

```
        request(i,j);/* request to output j */
}

for(all j) grant(j){ /* Grant phase */
if(num_request[j]>0){ /*received any request?*/
        i=orr_select(j); /* select a request based on the round-robin of output j */
        grant_request(i,j); /*grant request j back to input i */
        }
}
for(all i) accept(i){ /* accept phase */
if(num_grant[i]>0){ /* received any request? */
        j=irr_select(i); /*select a request based on round-robin of input i? */
        accept_grant(i,j); /* accept the grant j */
        irr_priority[i]=(j+1); /* set the highest priority of round-robin of input i higher than
granted output */
        orr_priority[j]=(I+1); /* set the highest priority of round-robin of output j higher
than granted input */
        if(cell_num [i,j]>threshold){ /* queued cells more than threshold ? */
        marked_input[i]=Yes; /* mark input I as marked input */
        }
    }
}
```

Request: Each unmarked input sends a request to every output for which it has a queued cell.

Grant: When an unmarked output receives any requests, it chooses the one to send grant phase by round-robin order.

Accept: If an output receives a grant, it will deliver cell to the corresponding output. Furthermore, when the input queue has queued more cells than threshold, both the input queue and the accepted output queue are set as marked till the number of cells is less or equal to threshold.

Figure 3b shows how B-VOQ works in a 4*4 switch with threshold 3. We use a 2-dimension array L to represent the length of all VOQ of all inputs, and (x, y) represents input and output respectively. In the beginning of the time slot 1, L={{4, 5, 4, 3}, {5, 4, 5, 3}, {6, 6, 7, 3}, {7, 4, 6, 2}}. It means that the number of delivering cells from input 1,input 2, input 3 and input 4 to output 1 are 4, 5, 6, 7(cells) separately. The cells from input 1, input 2, input 3 and input 4 to output 2 are 5, 4, 6, 4(cells), to output 3 are 4,5,7,6(cells), to output 4 are 3, 3, 3, 2(cells). After calculating the number of cells, we find that the number of cells in input 1, input 2, input 3, input 4 are all more than threshold. According B-VOQ, we pick out input 4, which owns the maximum number of cells, to deliver cell to output 1. For output 2 and output 3, we select input 3and input 4 to deliver cell separately. Output 4 will pick out input 1 to send one cell by round-robin order because the number of cells in input 1, input 2, input 3, input 4 are less or equal to threshold.

From the above statement, the number of cells in (1, 4), (3, 2), (3, 3), (4, 1) will decrease 1. In the time slot 2, L={{4, 5, 4, 2}, {5, 4, 5, 3}, {6, 5, 6, 3}, {6, 4, 6, 2}}. It means that the number of delivering cells from input 1, input 2, input 3, input 4 to output 1 are 4, 5, 6, 6(cells). Because input 3 and input 4 possess the same number of cells, we pick out input 3 by round-robin order to deliver cells first. In time slot 2, the number of cells in (1, 2), (2, 4), (3, 1), (3, 3) will decrease 1. The same processing is performed in each time slot. We compare Figure 3a(T-RRM) with Figure 3b(B-VOQ) during 4 time slots, and find that the proposed B-VOQ will perform more balance.

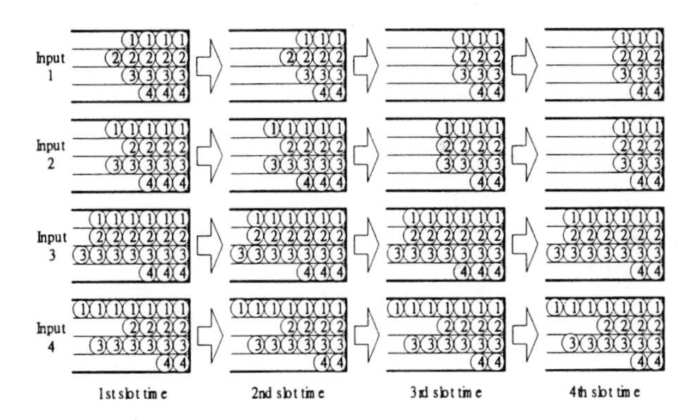

Fig. 3a. T-RRM for 4*4 switch

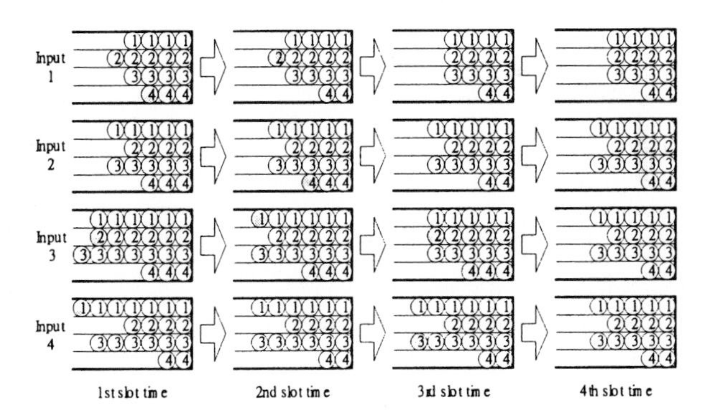

Fig. 3b. B-VOQ for 4*4 switch

3. Simulation Results

In order to prove that B-VOQ has better average delay comparing with other algorithms, we simulate B-VOQ in 16*16 switch. As mentioned above, in all existed scheduling algorithm, only iSlip and T-RRM can achieve 100% throughput, so we only compare delay B-VOQ with iSlip and T-RRM. In all simulations, we assume that the link speed of all inputs and outputs is the same. The independent, identically distributed Bernoulli traffic is used as the input traffic for each input port, and the received cells in a input is uniformly destined for all outputs. According to this typical traffic simulation of network, we will execute 50000 time slots to compare the relationship between the traffic load delay and mentioned algorithm.

By proposed literature, we know that iSlip and T-RRM can reach 100% throughput. In addition, according to the paper [7], comparing with iSlip and T-RRM, T-RRM has lower average delay. Therefore, we will compare delay_improvement of B-VOQ with iSlip and T-RRM.

delay_improvement (iSlip vs B-VOQ) = $(\text{delay}_{\text{iSlip}} - \text{delay}_{\text{B-VOQ}})/\text{delay}_{\text{iSlip}}$.

delay_improvement (T-RRM vs B-VOQ) = $(\text{delay}_{\text{T-RRM}} - \text{delay}_{\text{B-VOQ}})/\text{delay}_{\text{T-RRM}}$.

First, we simulate the average delay of iSlip , T-RRM and B-VOQ at different traffic load and threshold. The results are in figure 4-6.

Figure 4 shows the average delay of B-VOQ has almost no difference with iSlip and T-RRM when the traffic load is lower than 0.5 with threshold 2.

If the traffic load increases, the gap between iSlip and T-RRM separates slowly when the traffic load is 0.6 and it becomes obviously at 0.7. But the delay between the above two methods and B-VOQ will be obvious if traffic load is increasing. It is the most obvious when traffic load is 1.0.

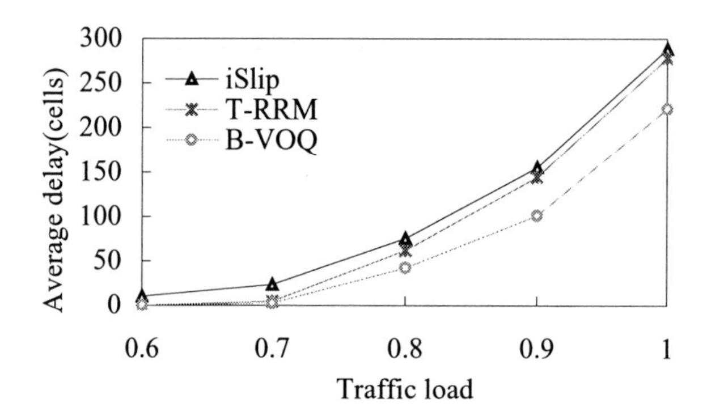

Fig. 4. The value of threshold is 2.

Figure 5 shows that the average delay of B-VOQ has almost no difference with iSlip and T-RRM when the traffic load is lower than 0.5 and the threshold value is 3. If the traffic load increases, the gap between iSlip and T-RRM separates slowly when the traffic load is at 0.7 and it becomes not so obvious at 0.9. However, the delay gap of above two methods and B-VOQ will become more obviously when the traffic load enlarges. The greatest gap is when the traffic load is at 1.0. Therefore, comparing our algorithm to the methods mentioned previously, the average delay of the B-VOQ has better performance.

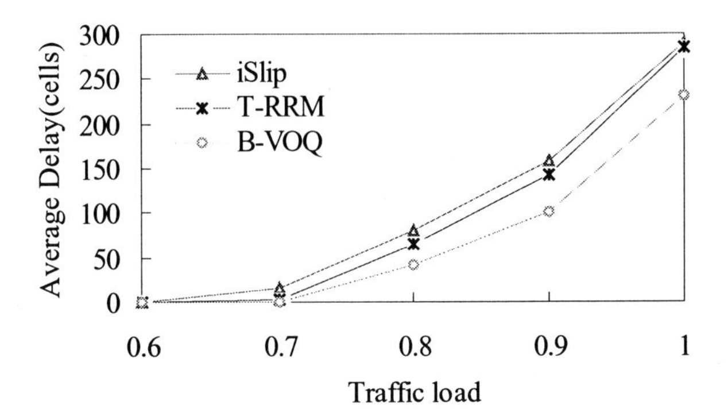

Fig. 5. The value of threshold is 3.

Figure 6 shows that the B-VOQ still has better average delay if the threshold value is 4. Therefore, in Figure 4-6, we can find that whatever the value of T-RRM's threshold is, the average delay comparing to iSlip only depends on the traffic load. When the traffic load is heavier, B-VOQ will have better average delay comparing to the above iSlip and T-RRM.

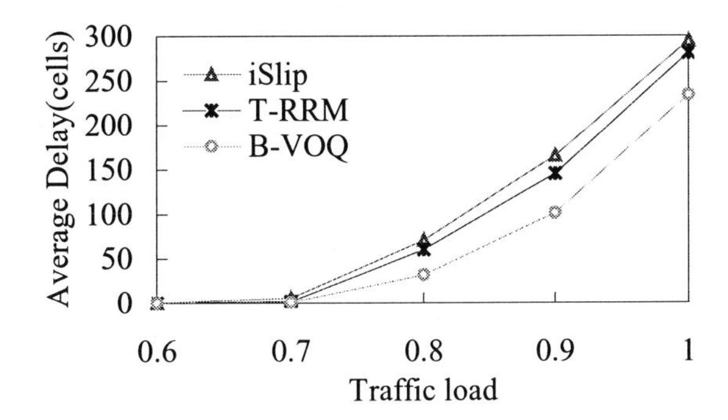

Fig. 6. The value of threshold is 4.

Figure 7 shows the result of B-VOQ's delay_ improvement versus iSlip, and we find that the values of the best improvement rate are 99.89%, 95.21% and 55.05% if we set the threshold values are 2,3,and 4 and the traffic load values are 0.6, 0.7 and 0.8.

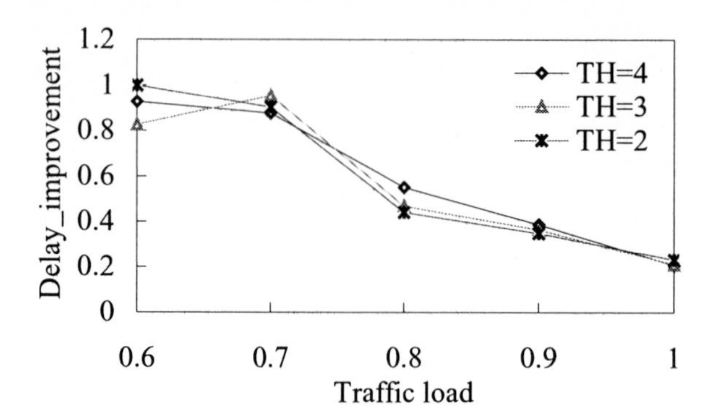

Fig. 7. B-VOQ versus iSlip Delay improvement.

Figure 8 shows the result of B-VOQ's delay_ improvement versus T-RRM, and we also find that it will have a better delay_improvement if traffic load is 0.7. Besides, it also has better improvement rate if traffic load is 0.8 and threshold is 4.

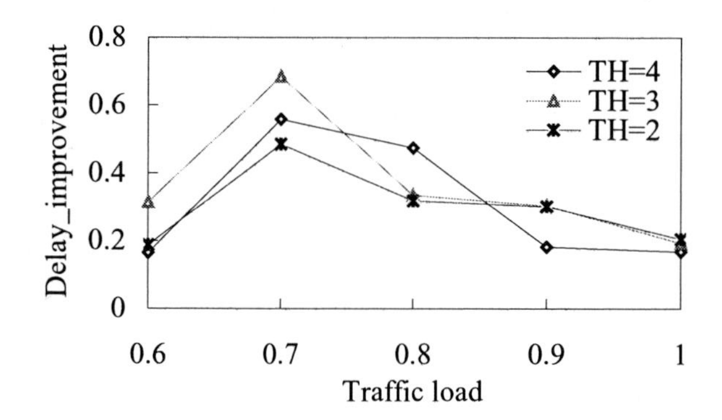

Fig. 8. B-VOQ versus T-RRM Delay improvement.

4. Conclusion

In this paper, we propose a B-VOQ method to reduce the delay time of cells in input queue. By this method, each matching input and output will recalculate the number of cells in VOQ through every slot time. If the number of cells is greater than the threshold, the input and output will be marked. The VOQ owning the maximum cells will be marked and served first, otherwise the round-robin method will be adopted. The result of the B-VOQ simulation shows the average delay comparing to T-RRM will be reduced 16-68%, if the traffic load is larger than 0.6.

5. References

1. Karol, M.; Hluchyj, M.; and Morgan, S.; "Input output queuing on a space division switch," *IEEE Trans. Communications*, vol.35, no.12, pp.1347-1356, 1988.
2. Ajmone Marsan, M.G; Bianco, A.; and Leonardi E.;" RPA: A Simple, Efficient, and Flexible Policy for Input buffered switch," *IEEE Communication Letters*, l93-:83-86, May 1997.
3. Andeson, T.; Owicki, S.; Saxe, J.; and Thacker, C.; "High Speed Switch Scheduling for Local Area Network," *ACM Trans. on Computer System*. pp.319-352. November 1993.
4. McKeown, N; Anantharam, V; and Walrand, J; " Achieving 100% Throughput in an Input-Queued Switch," *Proc. of IEEE '96*, San Francisco, March 1996.
5. McKeown, N;" Switching Algorithm for Input-Queued Cell Switches," PhD Thesis, University of California at Berkeley, 1995.
6. N. McKeown, "The iSlip Scheduling Algorithm for Input-queues Switches," *IEEE Trans. on Networking*, Vol. 7, No.2, pp. 188-201, April 1999.
7. Wenzhe Cui, Hanseok Ko and Sunshin,"A Threshold Based scheduling Algorithm for Input Queue Switch," Dept. of Electronic Engineering, Korea University, Seoul, Korea, *ICOIN 15*, 2000.

Dynamic Constrained Traffic Engineering for Multicast Routing*

Yongho Seok[1], Youngseok Lee[1], Yanghee Choi[1], and Changhoon Kim[2]

[1] School of Computer Science and Engineering, Seoul National University
{yhseok, yslee, yhchoi}@mmlab.snu.ac.kr
[2] Electronic Telecommunication Research Institute
kimch@etri.re.kr

Abstract. This paper presents a new traffic engineering technique for dynamic constrained multicast routing, where routing request of traffic arrives one-by-one. The objective we adopted in this paper is to minimize the maximum of link utilization. Although this traffic engineering is useful to relax the most heavily congested link in Internet backbone, the total network resources, i.e. sum of link bandwidth consumed, could be used when the acquired path is larger(in terms of number of hops) than the conventional shortest path. Accordingly we find a multicast tree for routing request that satisfies the hop-count constraint. We formulate this problem as mixed-integer programming problem and propose a new heuristic algorithm to find a multicast tree for multicast routing request. The presented heuristic algorithm uses link-state information, i.e. link utilization, for multicast tree selection and is amenable to distributed implementation. The extensive simulation results show that the proposed traffic engineering technique and heurisitc algorithm efficiently minimize the maximum of link utilization better than the shortest path.

1 Introduction

The dynamic traffic engineering problem in Internet is how to set up paths between edge routers in a network to meet the traffic demand of a request while achieving low congestion and optimizing the utilization of network resources. In practice, the key objective of traffic engineering is usually to minimize the utilization of the most heavily used link in the network, or the maximum of link utilization. Since the queueing delay increases rapidly as link utilization becomes high, it is important to minimize the link utilization throughout the network so that no bottleneck link exists. It has been known that this problem of minimizing the maximum link utilization could be solved by the multi-commodity network flow formulation[11].

* This work was supported in part by the Brain Korea 21 project of Ministry of Education, in part by the National Research Laboratory project of Ministry of Science and Technology, and in part by Electronic Telecommunication Research Institute, 2001, Korea.

I. Chong (Ed.): ICOIN 2002, LNCS 2343, pp. 278–288, 2002.

However, the present traffic engineering technique assumes that the traffic routing is done in unicast. We extend the scope of traffic engineering to multicast environment. Therefore, the problem is to set up bandwidth guaranteed multicast tree in a network for minimizing the maximum link utilization. Although this traffic engineering scheme is useful to minimize the maximum link utilization, it may require more total network bandwidth resources, i.e. sum of assigned bandwidth at each link of the paths, than the single shortest path. Therefore, the maximum hop-count constraint should be incorporated into multicast routing scheme in order not to waste bandwidth. We formulate this problem to mixed-integer programming(MIP) problem but it is $NP - hard$ problem. So, this paper proposes a practical heuristic algorithm of polynomial running time that finds hop-count constrained multicast tree to minimize the maximum of link utilization for each traffic request, while satisfying the requested traffic demand.

We assume that the traffic request is composed of the source, the destinations set and the traffic demand of multicast session. A traffic demand represents the average traffic volume between edge routers, in bps. For Virtual Private Network (VPN) application, the traffic demand may be the requested amount of bandwidth reservation. Even though the traffic demand varies largely at nodes near end users, it becomes quite stable for the backbone network with aggregated traffic.

This problem is motivated by the need of service providers to quickly setup constrained paths for multicast routing in their networks. An important context in which these problems arise is that of dynamic label switched path(LSP) setup in Multi-Protocol Label Switched(MPLS) networks. In Multi-Protocol Label Switching (MPLS) networks[2] where IP packets are switched through the pre-established Label Switched Path (LSP) by signaling protocols, a multicat tree can be used to forward packets belonging to the same "forwarding equivalent class (FEC)" by explicit routing. We assume that quasi-static information such as priorly known network topology. The only dynamic information available to the routing algorithm is the link utilization which is provided by extension of several routing protocol such as Open Shortest Path First(OSPF)[3] and Intermediate System to Intermediate System(IS-IS)[4].

There are some requirements that a multicast tree setup technique must satisfy in order to be useful in practice. Since the possibility of having new routing request in the future cannot be excluded, the routing algorithm must be an on-line algorithm capable of handling requests in an "optimal" manner when the requests are not all presented at once. Since all traffic requests are not known in advance, the current maximum link utilization is not optimal. To find the optimal maximum link utilization value, it is necessary that all the established paths for the previous requests need to be re-optimized whenever a new request arrives or the traffic characteristics change. However re-routing of existing requests causes a lot of path disruptions and this should not be allowed.

The remainder of this paper is organized as follows. The related works are introduced in section 2. The proposed algorithm is explained in section 3. The results of the performance evaluation by simulation are discussed in section 4, and section 5 concludes this paper.

2 Related Work

The problem of computing the minimum cost tree for a given multicast group is known as a Steiner tree problem. This Steineer tree problem is $NP-complete$. Some recently approximation algorithm of this routing problem in directed networks is proposed in [5].
A minimal hop like algorithm for routing unicast flows which attempts to balance the load of network traffic is proposed in [6]. This widest-shortest algorithm finds a feasible minimal hop path between two node such that the chosen path maximize the residual capacity of the bottleneck link along the path. The enhanced routing scheme for load balancing by separating long-lived and short-lived flows is proposed in [8], and it is shown that congestion can be greatly reduced. In [9], it is shown that the quality of services can be enhanced by dividing the transport-level flows into UDP and TCP flows. However these works did not consider path calculation problem.

For the MPLS network, a traffic engineering method using multiple multipoint-to-point LSPs is proposed in [10], where backup routes are used against failures. Hence, the alternate paths are used only when primary routes do not work. In [11], the traffic bifurcation linear programming (LP) problem is formulated and heuristics for the non-bifurcating problem are proposed. Although [11] minimizes the maximum of link utilization, it does not consider the total network resources and constraints. The authors further showed that the traffic bifurcation LP problem can be transformed to the shortest path problem by adjusting link weights in [16].

The dynamic routing algorithm for MPLS networks is proposed where the path for each request is selected to prevent the interference among paths for the future demands. It considers only a unicast routing and does not include the constraint such as hop-count. [12] proposes a constrained multipath traffic engineering for MPLS networks. Although this assumes unicast and multipath, the traffic engineering with the constraint such as maximum hop count and path count is formulated in mixed-integer programming problem and a heuristic algorithm is proposed. In [15], a minimal interference based algorithm is presented for dynamic routing of multicast request.

[17] proposes an adaptive traffic assignment method to multiple paths with measurement information for load balancing, though this work uses multipath. For differentiated services, finding the traffic split ratios to minimize the end-to-end delay and loss rates is proposed in [13]. However, how to find the appropriate multiple paths is not covered.

3 Hop-Count Constrained Multicast Traffic Engineering

3.1 Problem Definition

In this section, we define the hop-count constrained multicast traffic engineering problem in mixed integer programming (MIP) formulation. For conciseness and ease of terminology, we focus on finding label switched paths(LSPs) for multicast in MPLS networks in the rest of this paper. The network is modeled as a directed graph, $G = (V, E)$, where V is the set of nodes and E is the set of links. The capacity of a directed link (i, j) is c_{ij}.

Each traffic demand $(k \in K)$ is given for a node pair between an ingress router (s_k) and an egress router set(T_k) consisting of multicast group. Ingress router (s_k) is the source of the multicast connection, and egress routers $(t_k \in T_k)$ are destinations. For each traffic demand, there is a maximum number of hop count constraint, H_k.[1] The variable $X_{ij}^k(h)$ represents the traffic demand k that flows through link (i, j), where j is h hops far from s_k. The integer variable Y_{ij}^k tells whether link (i, j) is used or not for the multicast tree rooted at the ingress router s_k and reaching all egress routers t_k. Let d_k be a scaling factor to normalize the total traffic demand from the source to 1. The mixed integer programming (MIP) problem is formulated as follows.

$$Minimize\ \alpha + c \cdot \sum_{(j,i) \in E} \sum_{k \in K} d_k Y_{ij}^k$$

$$Subject\ to$$

$$\sum_{j:(i,j) \in E} X_{ij}^k(h) = \begin{cases} 1, k \in K, i \in s_k, h = 1 \\ 0, k \in K, i \notin s_k, h = 1 \end{cases} \tag{1}$$

$$\sum_{j:(i,j) \in E} X_{ij}^k(h+1) - \sum_{j:(j,i) \in E} X_{ji}^k(h) = 0 \tag{2}$$

$$, k \in K, i \notin s_k, T_k, 1 \le h < H_k$$

$$\sum_{j:(j,i) \in E} Y_{ji}^k = 1, k \in K, i \in T_k \tag{3}$$

$$\sum_{k \in K} d_k Y_{ij}^k \le c_{ij}\alpha, \forall (i,j) \in E \tag{4}$$

$$0 \le Y_{ij}^k - \sum_{h=1}^{H_k} X_{ij}^k(h) < 1 \tag{5}$$

$$where, 0 \le X_{ij}^k(h) \le 1, Y_{ij}^k \in \{0,1\}, 0 \le \alpha, h \in Z$$

The objective is to minimize the maximum of link utilization, α. If there are solutions with same maximum of link utilization, the optimal is to find one with minimum resource utilization among them. Constraint (1) says that the sum of total outgoing traffic over the first hop from the source is 1, and all nodes over the first hop from the source never receive the traffic except the source node. Constraint (2) is the hop-level flow constraint which means that for all

[1] $H_k = H + H_{MH_k}$, H_{MH_k} is the minimum number of hop counts from s_k to T_k for traffic demand k. All destinations $t_k \in T_k$ are reachable from s_k within H_{MH_k} hop. H is additional hop-count that is added to H_{MH_k}.

nodes except source and destination, the amount of total incoming traffic to a node is the same as that of outgoing traffic from the node. Constraint (3) means that all destinations must be connected from the source by using multicast tree. Constraint (4) means that the maximum link utilization for traffic demand k is α. Constraint (5) means that only the link (i, j) being used by multicast tree is computed for the maximum link utilization. This problem is $NP\!-\!hard$ because it includes the constrained integer variable.

3.2 Proposed Heuristic

We propose a heuristic algorithm to find hop-count constrained multicast tree for each traffic demand request between an ingress router and multiple egress routers. The proposed algorithm consists of two parts: 1) modifying the original graph to the hop-count constrained one [12], 2) finding a multicast tree to minimize the maximum link utilization.

Step 1 : Conversion to hop-count constrained graph

The given network, $G = (N, E)$, is converted to H_k hop-count constrained graph, $G' = (N', E')$, where N' and E' are transformed as follows,

$$N' = \cup_{0 \le m \le H_k} N'_m,$$
$$N'_0 = \{s_k\},$$
$$N'_m = \{j_m | (i, j) \in A, i_{m-1} \in N'_{m-1}\}.$$
$$E' = \cup_{1 \le m \le H_k} E'_m,$$
$$E'_1 = \{(s_k, i) | (s_k, i) \in E\},$$
$$E'_m = \{(i_m, j_m) | i_m \in N'_{m-1}, j_m \in N'_m, (i, j) \in E\}.$$

An example of graph conversion is given in Fig. 1. Fig. 1 (a) represents the original network topology. When a traffic demand request from node 1 to node 4 which requires bandwidth of 3 Mbps with the hop-count constraint of one additional hop and the path-count constraint of two arrives, the graph in Fig. 1 (b) is derived after adding redundant nodes and links. It is easily seen that any path traversed from node 1 to node 4 in Fig. 1 (b) does not exceed three hop counts.

Step 2 : Finding multicast tree

On the modified graph G', the link metric(c_{ij}) is given with the current utilization ratio (allocated bandwidth / link capacity). We propose a way of choosing multicast tree on the modified graph, $G' = (N', E')$. For each destination $t_k \in T_k$, we calculate the widest path from s_k to t_k by using Dijkstra's algorithm.

- widest path

  The widest path is selected in order to minimize the usage of the bottleneck link, the link with the maximum utility. In this case, $dist(i)$, the cost of a node i, denotes the maximum link utility from source to the node.

$$dist(i) = min_{j \in S}(dist(i), max(dist(j), c_{ji})).$$

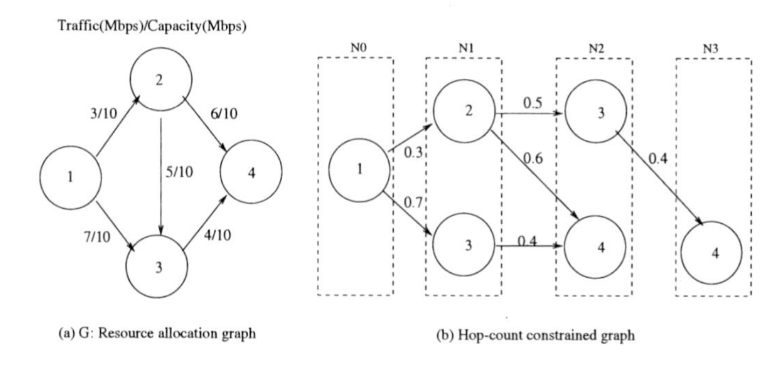

(a) G: Resource allocation graph

(b) Hop-count constrained graph

Fig. 1. Topology conversion example

, where S is the set of nodes whose shortest path from source is already determined.

Next, we determines a destination r_k obeying the following constraint.

– destination r_k

$$(dist(t_k)|t_k \in T_k) \leq dist(r_k)$$

Then, we setup a path from s_k to r_k and reserve the bandwidth at each link along this path. In multicast algorithm, it is important to minimize the cost of the tree, especially network resources(like bandwidth). So we apply a heuristic for the directed Steiner tree problem. After finding this path, we set the cost of all the edges along this path to zero. Setting these edge costs to zero encourages future runs of Dijkstra's algorithm to use them. Until when all destinations are reachable from the source s_k, we repeat this process with the destination set T_k excluding r_k.

The detailed algorithm is explained in Fig. 2. Fig. 3 explains the simple result of the multicast tree selection. It is seen that the proposed heuristic algorithm constructs the multicast tree better than that using Dijkstra's shortest path algorithm. The proposed algorithm minimizes the maximum link utilization, while the resource utilization of each algorithm is the same.

3.3 Complexity Analysis

Proposed algorithm consists of two nested loops, inner loop for computation of widest paths, outer loop for each destination in the set of T_k. For each part, time complexity is bound as follows. First, for the widest path problem, the best known bound is O($nlogn$) in a directed graph, where n is number of vertices in the graph. Algorithm the complexity of inner loop is bound by O($n^2 logn$), because the maximum number of destinations($|T_k|$) is n. The other remained code inside outer loop is bounded by O(n). Second, the outer loop is executed maximum n times. Hence, the worst-case time complexity of the proposed algorithm is bounded by O($n^3 logn + n^2$).

Heuristic : *Find hop-count constrained multicast tree*
- Set d_k to be the traffic demand of traffic request k
- Set s_k to be a source of traffic request k
- Set T_k to be a set of destinations($t_k \in T_k$)
- Modify G to G' satisfying H_k hops;

while(T_k is not empty)

  for each($t_k \in T_k$)
  - Run widest path algorithm from s_k to t_k;
  - Set $\alpha(t_k)$ to be the maximum link utilization when this path is setup;

  endfor
- Set r_k to be the destination of maximum $\alpha(t_k)$;
- Reserve d_k to each link along the path from s_k to r_k;
- Set the link utility of all edges along this path to zero;
- Set T_k to $T_k - r_k$;

endwhile

Fig. 2. The proposed Heuristic.

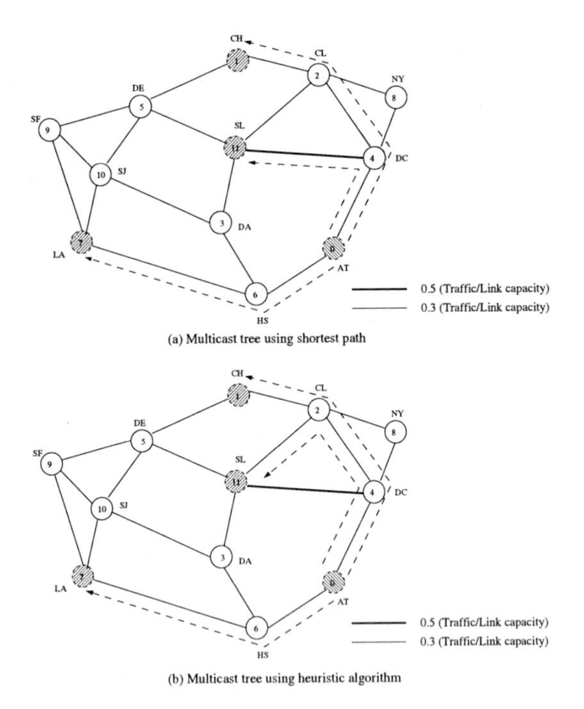

Fig. 3. The result of several multiple path calculation methods

4 Performance Evaluation

4.1 Simulation Environment

The network topology shown in Fig. 3 represents the abstract US backbone topology[18]. In this network condition, we generate ten random requests of traffic demands from one source. For each traffic request, the set of destinations is randomly selected by two cases. In the first case, the multicast session consists of small number of receivers (average 3.3 receivers per source). In the second case, it consists of many receivers (average 7.7 receivers per source). So, we treat both sparse and dense mode. Therefore, 240 requests are tested in total. The duration of each traffic demand is exponentially distributed (ten seconds), and the inter-arrival time is randomly distributed between zero and two hundred seconds. The average traffic demand of each routing request is set to 5 Mbps. Among the simulation, the optimal mixed-integer programing solution is solved with CPLEX tool.

4.2 Simulation Result

The proposed heuristic algorithm is compared with the simple shortest path algorithm(SP) and the optimal MIP solution(OPT). The maximum hop-count constraint(H_k) is given as zero or more additional to that of the minimum hop-count(H_{MH_k}) between an ingress and an egress router set. Table 1 and Table 2 present the average and maximum of α in sparse mode and dense mode respectively. The optimal solution not only almost halves the maximum of α than the shortest path algorithm(SP), but also improves the average α, in the both cases. Also, we can see that the proposed heuristic shows similar results to relatively optimal solution. There is only a small difference below 3% between two solutions.

Table 1. Maximum of link utilization (α) in sparse mode

	SP	OPT	Heuristic
AVG	13.23	8.84	10.24
MAX	30	15.72	18.32

Table 2. Maximum of link utilization (α) in dense mode

	SP	OPT	Heuristic
AVG	18.44	9.78	12.65
MAX	47.85	20.85	23.25

Fig. 4 shows the *normalized* α which was obtained by dividing α of our algorithm in Fig.2 by α of optimal solution of the MIP, simulated in sparse mode. In Fig. 4, we can see that the proposed heuristic constrained on zero, one or three additional hop performs better than the shortest path case. Although the hop-count constraint is increased from one to three additional hops, the difference between (b) and (c) is slight. When the hop-count constraint is changed, the solution of shortest path algorithm is not changed but the *normalized* α is increased largely. It is shown that the α of optimal MIP solution is largely decreased as the hop-count constraint is changed from zero to three.

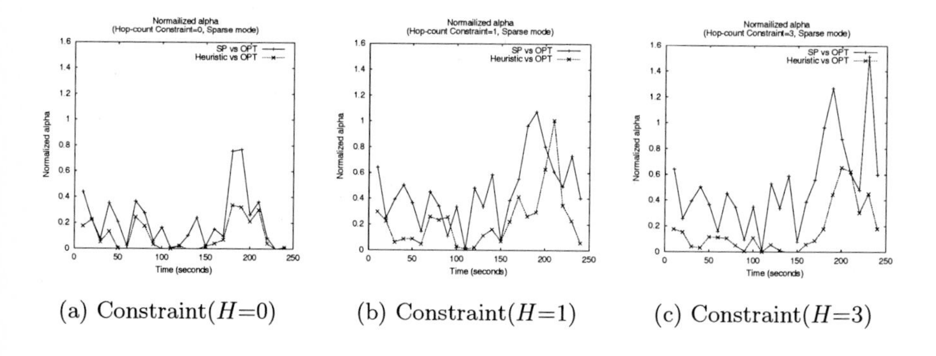

(a) Constraint(H=0) (b) Constraint(H=1) (c) Constraint(H=3)

Fig. 4. Maximum of link utilization (α) with the hop-count constraints in sparse mode

Fig. 5 shows the result of simulation in dense mode. In dense mode, each traffic request has a number of receivers relative to sparse mode. So total resource utility is increased, but we can see that the proposed heuristic also performs better than the shortest path case. Especially, Fig. 5 (b), (c) show the larger performance gap between heuristic and the shortest path algorithm.

Both simulations show similar results that the proposed heuristic algorithm efficiently minimizes the maximum link utilization, α than the shortest path one. Lastly, the performance of both the optimal MIP and proposed heuristic algorithm may not be more enhanced although the number of hop-count constraint increases.

5 Conclusion

In this paper, we propose dynamic traffic engineering schemes for multicast routing that minimize the maximum of link utilization, α by finding multicast tree with the hop-count constraints. We formulate this problem as the mixed-integer programming problem by using network flow model. Because finding solution of this optimal MIP problem is $NP\!-\!hard$, we propose the heuristic algorithm that calculates a constrained muticast tree in polynominal time. The simulation results show that the proposed algorithm solves the problem of multicast routing

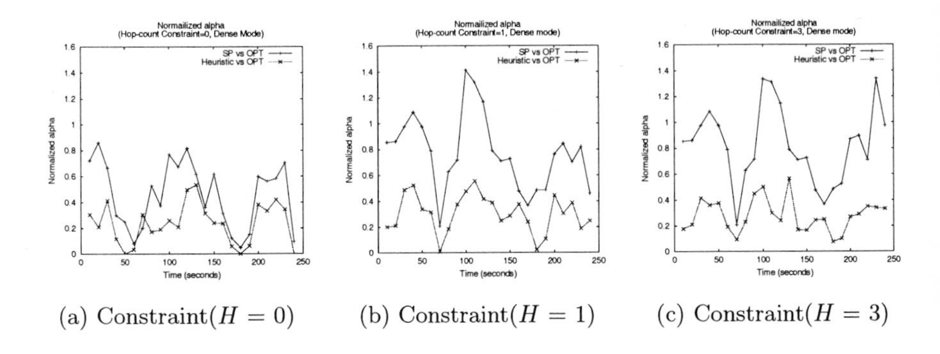

(a) Constraint($H = 0$) (b) Constraint($H = 1$) (c) Constraint($H = 3$)

Fig. 5. Maximum of link utilization (α) with the hop-count constraints in dense mode

with nearly same α as that of the optimal solution. Therefore, the proposed traffic engineering scheme is practical and will be useful for reducing the probability of congestion by minimizing the utilization of the most heavily used link in the network.

References

1. D. Bertsekas, and R. Gallager, Data Networks, Prentice Hall, 1992
2. E. Rosen, A. Viswanathan, and R. Callon, "Multiprotocol Label Switching Architecture," Internet RFC3031, 2001
3. J. Moy, "OSPF Version 2," Internet RFC2328, 1998
4. R. Callon, "Use of OSI IS-IS for Routing in TCP/IP and Dual Environments," Internet RFC1195, 1990
5. M. Charikar, C. Chekuri, T. Cheung, Z.Dai, A. Goel, S.Guha, and M. Li, "Approximation Algorithms for Directed Stiner Problems," SODA, 1988.
6. R. Guerin, D. Williams, A. Orda. "QoS Routing Mechanisms and OSPF Extensions," Globecom 97.
7. N. S. V. Rao, and S. G. Batsell, "QoS Routing Via Multiple Paths Using Bandwidth Reservation," INFOCOM'98
8. A. Shaikh, J. Rexford, and K. G. Shin, "Load-Sensitive Routing of Long-Lived IP Flows," SIGCOMM'99
9. P. Bhaniramka, W. Sun, and R. Jain, "Quality of Service using Traffic Engineering over MPLS: An Analysis," LCN'2000
10. H. Saito, Y. Miyao, and M. Yoshida, "Traffic Engineering using Multiple Multipoint-to-Point LSPs," INFOCOM'2000
11. Y. Wang, and Z. Wang, "Explicit Routing Algorithms for Internet Traffic Engineering," ICCCN'99
12. Y. Seok, Y. Lee, Y. Choi, and C. Kim, "Dynamic Constrained Multipath Routing for MPLS Networks," ICCCN'2001
13. E. Dinan, D. O. Awduche, and B. Jabbari, "Analytical Framework for Dynamic Traffic Partitioning in MPLS Networks," ICC'2000
14. M. Kodialam, and T. V. Lakshman, "Minimum Interference Routing with Applications to MPLS Traffic Engineering," INFOCOM'2000

15. M. Kodialam, T. V. Lakshman, and S. Sengupta, "Online Multicast Routing with Bandwidth Guarantees: A New Approach using Multicast Network Flow,", SIGMETRICS'2000
16. Z. Wang, Y. Wang, and L. Zhang, "Internet Traffic Engineering without Full Mesh Overlaying," INFOCOM'2001
17. A. Elwalid, C. Jin, S. Low, and I. Widjaja, "MATE: MPLS Adaptive Traffic Engineering," INFOCOM'2001
18. Optimized Multipath, http://www.fictitious.org/omp

III. Optical Networks

Efficient Lightpath Routing in Wavelength-Routed Optical Networks

Koji Taira[1], Yongbing Zhang[2], Hideaki Takagi[2], and Sajal K. Das[3]

[1] Doctoral Program in Systems and Information Engineering,
University of Tsukuba, Tsukuba, Ibaraki 305–8573 Japan
ktaira@sk.tsukuba.ac.jp

[2] Institute of Policy and Planning Sciences,
University of Tsukuba, Tsukuba, Ibaraki 305–8573 Japan
{ybzhang,takagi}@sk.tsukuba.ac.jp

[3] Department of Computer Science and Engineering,
University of Texas at Arlington, Texas 76019–0015 USA
das@roop.uta.edu

Abstract. We propose a new lightpath routing algorithm for WDM optical networks that solves the routing and wavelength assignment (commonly known as RWA) problem dynamically and separately. The RWA problem is partitioned into two subproblems, the routing and wavelength assignment problems, and both of them are solved using the shortest path routing technique. For solving the routing subproblem, an auxiliary graph is created whereby the nodes and links in the original network are transformed to the edges and vertices, respectively, and the availability of each wavelength on the input and output links of a node as well as the number of available wavelength converters are taken into account in determining the weights of edges. Furthermore, for solving the wavelength assignment subproblem, an auxiliary graph is also utilized and the cost for wavelength conversion is taken into consideration in the edge weight function. Simulation results show that our algorithm performs much better than previously proposed algorithms in terms of blocking probability, especially if the number of wavelengths is large while the number of converters at each node is limited.

1 Introduction

Wavelength-division multiplexing (WDM) is emerging as the dominant technology for the next generation optical networks [1,2]. Using WDM, multiple signals, distinguished by their wavelengths, can be transmitted on a single fiber and each wavelength operates at its peak speed. All nodes or a limited number of nodes in a WDM-routed network may employ *wavelength converters* that can convert one input wavelength into another different output wavelength in order to increase the wavelength utilization. A route (a set of links) traversed by data between a source-destination (*s*-*d*) pair forms an all-optical path with a wavelength assigned on each link and it is called a *lightpath*. In this paper, we consider that a lightpath is assigned to a connection corresponding to each user connection

I. Chong (Ed.): ICOIN 2002, LNCS 2343, pp. 291–304, 2002.

request for its entire duration. Given a set of connection requests, how to set up lightpaths for them is called the *routing and wavelength assignment* (RWA) problem [3,4,5,6]. The objective of an RWA algorithm is to set up lightpaths and assign wavelengths in a manner which minimizes the amount of the request blocking.

An RWA problem can be partitioned into two subproblems, the routing and the wavelength assignment problems, and are solved separately. For the routing subproblem, the shortest path routing technique can be used to determine the best route for an s-d pair [7,8]. For the wavelength assignment subproblem, heuristic algorithms can be used [9]. A more complicated algorithm, called TRWA in this paper, that considers both of the subproblems at the same time is also possible[8]. This scheme yields good performance but the overhead cost for constructing the auxiliary graph and computation is very high due to the complexity of its auxiliary graph [8,10].

In this paper, we propose a heuristic algorithm that solves the dynamic RWA problem efficiently. The proposed algorithm first solves the routing subproblem and then the wavelength assignment subproblem. Both subproblems are formulated as routing problems and solved using the shortest path routing technique. For creating the auxiliary graph of the first subproblem (routing), the availability of each wavelength on the input and the output links at a node as well as the number of available converters are taken into account. The advantage of this scheme is that the routing decisions can be made based on more accurate information on the possible lightpaths between s-d pairs. For creating the auxiliary graph of the second subproblem (wavelength assignment), the cost of wavelength conversion at each node is taken into account. Under this scheme, if a lightpath uses a converter at a node with less converter availability, it has to pay inevitably higher cost.

The paper is organized as follows. Section 2 introduces the basic system model of our wavelength-routed networks. Section 3 explains our proposed algorithm in details. Section 4 presents the performance evaluation of our proposed algorithm in comparison with other previous algorithms. Section 5 concludes this paper.

2 System Model

A WDM-routed network can be modeled by a directed graph $G(N, L)$, where N and L denote the sets of nodes and communication links, respectively. For simplicity, N and L are also used to denote the numbers of nodes and links, respectively. The bandwidth of each optical fiber link is divided into a set of W wavelengths as communication channels. A connection request of an s-d pair is served by setting up a lightpath that is a series of channels belonging to the immediate nodes along the path from the source s to the destination d. The connection occupies the channels until it terminates. It is assumed that connection requests arrive at each node independently and follow the Poisson process. The occupation time of a lightpath by a connection is assumed to be exponentially distributed.

Besides transmitting and receiving signals, each node provides the optical switching functions such as switching a wavelength of a connection from an input end to an output end and converting an input wavelength to a different output wavelength. Wavelength conversion allows one to resolve wavelength conflicts and to reuse wavelengths, thereby improving network performance. However, the number of converters in a node is usually limited due to the economic, spatial, and efficiency reasons. It is therefore necessary to keep the performance (blocking probability of connection requests) at a reasonable level while to use as less converters as possible. In this paper, it is assumed that each node have a limited number of converters that are shared in the node, as shown in Figure 1, and the wavelength conversion is full range of the waveband; *i.e.*, a converter can convert one input wavelength to any other output wavelength.

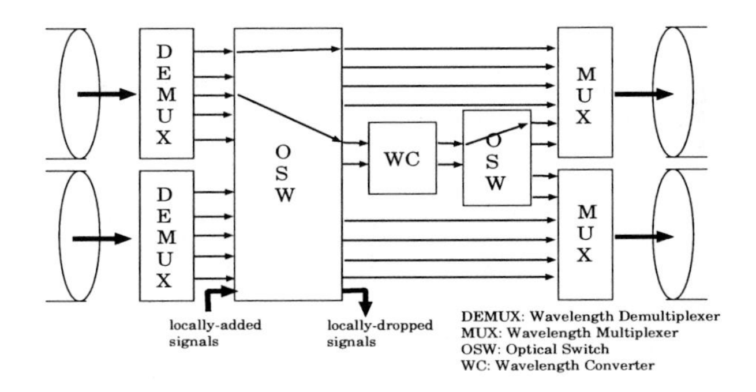

Fig. 1. The wavelength-convertible switch architecture.

It is assumed that the routing and the wavelength assignment (RWA) algorithm is decentralized; *i.e.*, the routing and the wavelength assignment decisions are made at each node autonomously. Each node dynamically broadcasts its state on both the wavelength and the converter availability to all other nodes in the network and receives the state information from other nodes. It then determines independently the best route for each arriving connection request. The routing decision for a new connection request is not allowed to affect the existing connections. Instead, an alternative route that utilizes the currently available channels and converters in the network is selected. Given a network $G(N, L)$ and the wavelength conversion function, the objective of our algorithm under consideration is to find the *best route* from the source to the destination of a connection request, and also to assign the *best wavelength* on each link along the best route so that the *blocking probability* of connection requests is minimized.

3 Proposed Solution

As mentioned earlier, the heuristic algorithm proposed in this paper consists of two components: the routing algorithm and the wavelength assignment algorithm. The routing algorithm approximates a cost function as the sum of individual costs due to using channels and wavelength converters and selects the best route. The wavelength assignment algorithm considers another cost function that sums up the individual costs of using the free wavelengths and wavelength converters along the best route and selects the best wavelength to set up a lightpath. The overall approach includes the following four steps.

1. *Graph transformation for routing*: Transform the original network to the corresponding auxiliary graph.
2. *Determination of the best route*: Solve the routing algorithm using Dijkstra's algorithm to find the best route between the *s*-*d* pair. If no route with finite length is found, reject the connection request; otherwise, go to Step 3.
3. *Graph transformation for wavelength assignment*: Transform the route in the original network determined in Step 2 to its auxiliary graph.
4. *Wavelength Assignment*: Solve the wavelength assignment problem using Dijkstra's algorithm to find the best wavelength on each link in order to set up a lightpath. If no lightpath can be set up, reject the connection request; otherwise, accept the connection request.

In the following subsections, the proposed routing and wavelength assignment algorithms are described.

3.1 Source-Destination Routing Algorithm

The auxiliary graph of the original network is created by considering the specific characteristics of optical networks in order to determine the best route for an *s*-*d* pair. Before constructing the auxiliary graph, two pseudo-nodes are added to the *s*-*d* pair in the original network. The pseudo-nodes s' and d', called the *pseudo traffic input* and *output points*, are connected to the source *s* and the destination *d* with zero cost *pseudo-links*, respectively, as shown in Figure 2(a). The nodes and links in the original network are respectively transformed to the edges and vertices in the auxiliary graph as shown in Figure 2(b). A possible path through an edge with or without wavelength conversion is called an *edge-path* of the edge.

Graph transformation. For a given network $G(N, L)$ and an *s*-*d* pair, an auxiliary graph $G(V, E)$ is created, where V and E denote the sets of vertices and edges, respectively. The procedures of the graph transformation can be listed as follows.

1. For an *s*-*d* pair, add the pseudo-nodes, s' and d', as the pseudo traffic input and output points and connect them respectively to the source and destination nodes with zero cost pseudo-links as shown in Figure 2(a).
2. Create nodes in the auxiliary graph to denote the links in the original network. Note that the pseudo-nodes are treated as links in the original network.
3. Create edges in the auxiliary graph that denote nodes connecting two links in the original network.

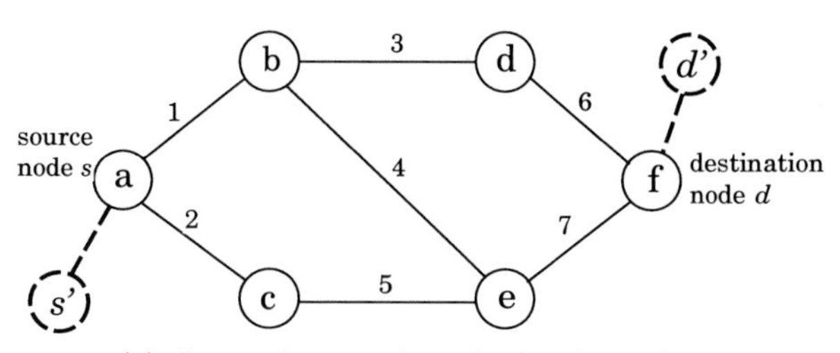

(a) Original network with pseudo-nodes.

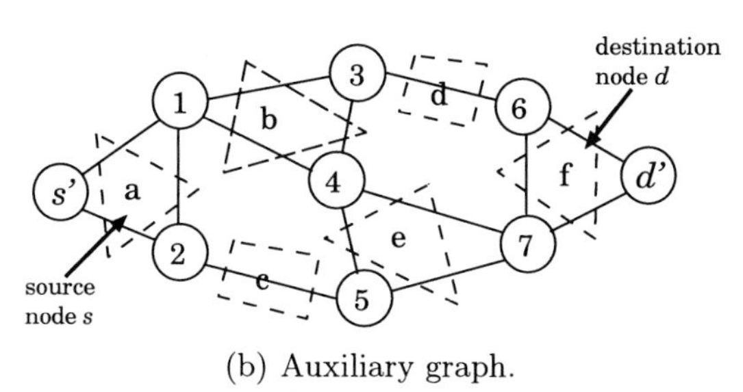

(b) Auxiliary graph.

Fig. 2. Illustration of graph transformation.

Determination of the edge weight. In a wavelength-routed network, a lightpath may not be able to pass through a node for lack of wavelength conversion capacity, even though there are free wavelengths both on the input and output links of the node. It is therefore necessary to consider the states of both the input and output links of a node, *i.e.*, the states of edge-paths of each edge in the auxiliary graph, in routing decisions. In this paper, the states of the available wavelengths on the input and output links at each node, the conversion capacity, and the number of available converters are taken into consideration. In order to introduce the edge weight function, the following notation is used.

C total number of converters at a node
W total number of wavelengths a link or a node handles
c number of converters available at a node
w_{in} number of wavelengths available at the input ends of a node
w_{out} number of wavelengths available at the output ends of a node
w number of available edge-paths of an edge on the route of an s-d pair that a lightpath can pass the edge without wavelength conversion
w' number of available edge-paths of an edge on the route of an s-d pair that a lightpath can pass the edge with wavelength conversion, *i.e.*, $w' = \min\{w_{in} - w, w_{out} - w, c\}$.

The weight of each edge of the auxiliary graph is determined based on the probability of the available edge-paths of an edge. By supposing that the probability of a free edge-path on an edge neither from the source nor to the destination at some time in future is $\frac{w+w'}{W}$, the probability that an edge-path will be occupied is given by $1-\frac{w+w'}{W}$. Therefore, the probability that all the edge-paths will be used is given by $(1-\frac{w+w'}{W})^{w+w'}$. Then, the probability that at least one edge-path will be free on an edge i in future is given by

$$p_i = 1 - (1 - \frac{w+w'}{W})^{w+w'}.$$

The weights of edges originated from the source and targeted to the destination are determined differently from others. By assuming that the probability of a free edge-path on an edge originated from the source in future is w_{out}/W, then the probability that at least one edge-path will be available is given by

$$p_i = 1 - (1 - \frac{w_{out}}{W})^{w_{out}}.$$

On the other hand, since data can always arrive at the destination if its input link is free, the probability that at least one edge-path on an edge to the destination will be available in future is always 1; *i.e.*, $p_i = 1$. For the above three types, the weight of an edge i, ρ_i, in the auxiliary graph is defined by the following function as in [7],

$$\rho_i = -\log p_i.$$

The above equation implies that an edge with a higher probability of free edge-paths has a lower value of the edge weight and that the edge weight becomes infinity if there is no free edge-path. Note that ρ_i is determined by the current status of both a node and a link, and is constantly changing.

Routing algorithm. Given an auxiliary graph $G(V, E)$ and a connection request between the pseudo traffic input and output points, s'-d' pair, Dijkstra's algorithm is used to search for the min-cost path corresponding to the route in the original network. In order for the auxiliary graph to conduct an accurate solution, one additional constraint is needed for the algorithm. In the auxiliary graph, a route may pass through consecutively the same node twice while this situation cannot occur in the original network. It is therefore prohibited in the routing algorithm that the chosen route goes through two edges belonging to the same node successively in the original network. For example, a route is not allowed to pass from vertex 1 to 3 and then from vertex 3 to 4 in Figure 2(b).

3.2 Wavelength Assignment Algorithm

In this paper, three wavelength assignment schemes are considered. They are First Fit Wavelength First (FFW), Least Converter First (LEC), and Least Conversion Cost First (LCC). The first is quite simple but may not be efficient in wavelength utilization. It is usually used for comparison purpose in the literature [9]. On the other hand, the latter two approaches are proposed in this paper.

First Fit Wavelength First (FFW). The source node attempts to find and assign a free wavelength that is found first along the route determined by the routing algorithm. This algorithm searches for a free wavelength on a link in a predefined order and attempts to use a wavelength converter whenever a wavelength conversion is needed. The connection request will be forwarded to the next node along the route when the trial for finding a free wavelength succeeds. If the request fails, the source node gets the feedback and a different wavelength will be chosen. This process is repeated until there is one free wavelength available or the lightpath cannot be set up.

Least Converter First (LEC). Since the number of wavelength converters at a node may be much less than the number of wavelengths on a link, it is natural to take this factor into account in assigning wavelengths. The LEC algorithm treats the wavelength assignment problem as a routing problem and employs Dijkstra's algorithm to find the solution. An auxiliary graph is created by using an approach similar to that in [8], but in LEC only the best route determined by the routing algorithm needs to be considered.

The weight of a channel edge is determined similarly to that in [8]. That is, an idle channel edge has weight f where f is a positive constant while an occupied channel edge has weight of infinity. The weight of a converter edge is g and is larger than the sum of any free path without wavelength conversion, *i.e.*, $g > nf$ where n is the path length from the source to the destinatin. Furthermore the weight of an edge corresponding to a switching operation without wavelength conversion at a node is set to zero. The LEC algorithm therefore attempts to set up a lightpath for a connection request with the lowest cost, *i.e.*, with the least number of converters. If no lightpath can be set up, then the connection request will be rejected.

Least Conversion Cost First (LCC). This algorithm is implemented similarly to LEC. However, the weight function of LCC is different from that of LEC in the sense that LCC employs a nonlinear cost function for using converters. It is assumed that using a free converter at a node where the converter utilization is higher should pay higher cost (higher penalty). The cost function, $c_i(u_i)$, for using a converter at node i is defined as follows.

$$c_i(u_i) = \frac{1}{C - u_i},$$

where u_i denotes the number of wavelength converters in use at node i. The weight, $\rho_i(u_i)$, of an edge i corresponding to a switching operation with wavelength conversion at a node is defined to be the *differential function* of the above cost function; *i.e.*,

$$\rho_i(u_i) = \frac{dc_i(u_i)}{du_i} = \frac{1}{(C - u_i)^2},$$

From the above equation, it can be seen that when u_i is close to the converter capacity the conversion cost becomes very high.

4 Performance Evaluation

Simulation experiments are used to evaluate the proposed heuristic algorithm, denoted by NEW in the figures, and compare it with other algorithms, Hop-based (HW) and Total wavelengths and Available Wavelength (TAW) in [7] and Total Routing and Wavelength Assignment (TRWA) in [8]. The routing algorithm in HW attempts to set up a lightpath from the source s to the destination d using the smallest number of hops. On the other hand, the routing algorithm in TAW attempts to set up a lightpath with the smallest edge weight that are determined by the available wavelengths and the total wavelengths on each link. The network model used in the simulation is the NSFNET model (shown in Figure 3), *i.e.*, there are 14 nodes and 21 duplex links. It is assumed that each link has the same number of wavelengths W and each node has the same number of wavelength converters C.

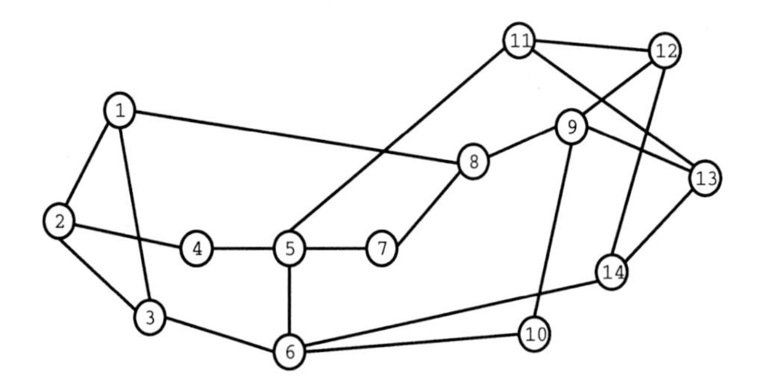

Fig. 3. A wavelength-routed network.

Table 1. Computation times of the algorithms.

Algorithm	Computation time					
	$W = 8, \lambda = 3$		$W = 16, \lambda = 6$		$W = 32, \lambda = 10$	
	FFW	LEC LCC	FFW	LEC LCC	FFW	LEC LCC
HW	30s	2m	50s	5m	1.3m	16m
TAW	40s	2.5m	1m	7m	1.5m	24m
NEW	50s	2.5m	1.2m	6m	1.7m	23m
TRWA	40m		4hrs		17hrs	

Table 2. Auxiliary graph complexity of the algorithms*

Algorithm	Complexity of the algorithms					
	$W = 8$		$W = 16$		$W = 32$	
	# vertices	# edges	# vertices	# edges	# vertices	# edges
HW	(14, 37)	(21, 101)	(14,72)	(21, 435)	(14, 142)	(21,1337)
TAW	(14, 51)	(21, 172)	(14,98)	(21,592)	(14, 188)	(21,2103)
NEW	(21, 49)	(49–53,161)	(21,92)	(49–53,538)	(21,175)	(49–53,1893)
TRWA	364	3720	700	13200	1372	49440

It is assumed that connection requests from a node to each of the others nodes are generated with the equal probability and the request arrivals follow the Poisson process with rate λ. The connection duration time is assumed to be exponentially distributed with mean of 1 time unit. The number of wavelengths W is varied as 8, 16, and 32. However, the results only for the case of 32 wavelengths are shown in the figures because of the space limitation. The wavelength conversion factor, denoted by f, is defined to be the ratio of the number of converters at a node to the number of wavelengths; *i.e.*, $f = C/W$, and is changed as 0, 10%, 20%, 50%, and 100%, where 0 means there are no converters whereas 100% means there are full converters. In each experiment, 200,000 connection requests are generated and the performance metrics are shown as an average in the figures. The performance metrics used are (i) the blocking probability of the connection requests, and (ii) the execution time for running an algorithm.

Table 1 shows the simulation (execution) times of the algorithms under consideration. It can be seen from Table 1 that the computation time of an RWA algorithm using FFW is the faster than that using LEC or LCC because of its simplicity. On the other hand, the computation time of TRWA is the longest among the algorithms and increases sharply if the number of wavelengths increases. The reasons of this result can also be supported by Table 1, which shows the numbers of vertices and edges that the algorithms have to handle.

Figure 4 shows the probability versus traffic load with various wavelengths when the conversion factor, f, is 20% and there are 32 wavelengths. As expected, the TRWA algorithm performs the best among the algorithms, but its computation time is exhaustive as shown in Tables 1. It shows that TRWA may not be practical for a real system, especially in a large network with a large number of wavelengths. On the other hand, our proposed algorithm yields a blocking probability close to that of TRWA while the computation time is close to that of the other algorithms. It can be observed that the new algorithm outperforms significantly both the HW and TAW algorithms over a wide range of traffic load and that the improvement gain becomes obvious as the number of wavelengths increases (*e.g.*, over 40% when λ is below 14). Note that HW may behave bet-

* The two-tuples (x, y) in HW, TAW, and NEW denote the number of vertices or edges in the first and the second subproblems, respectively. The numbers of vertices and edges of the second subproblem in HW, TAW, or NEW are calculated using $\bar{N}$.

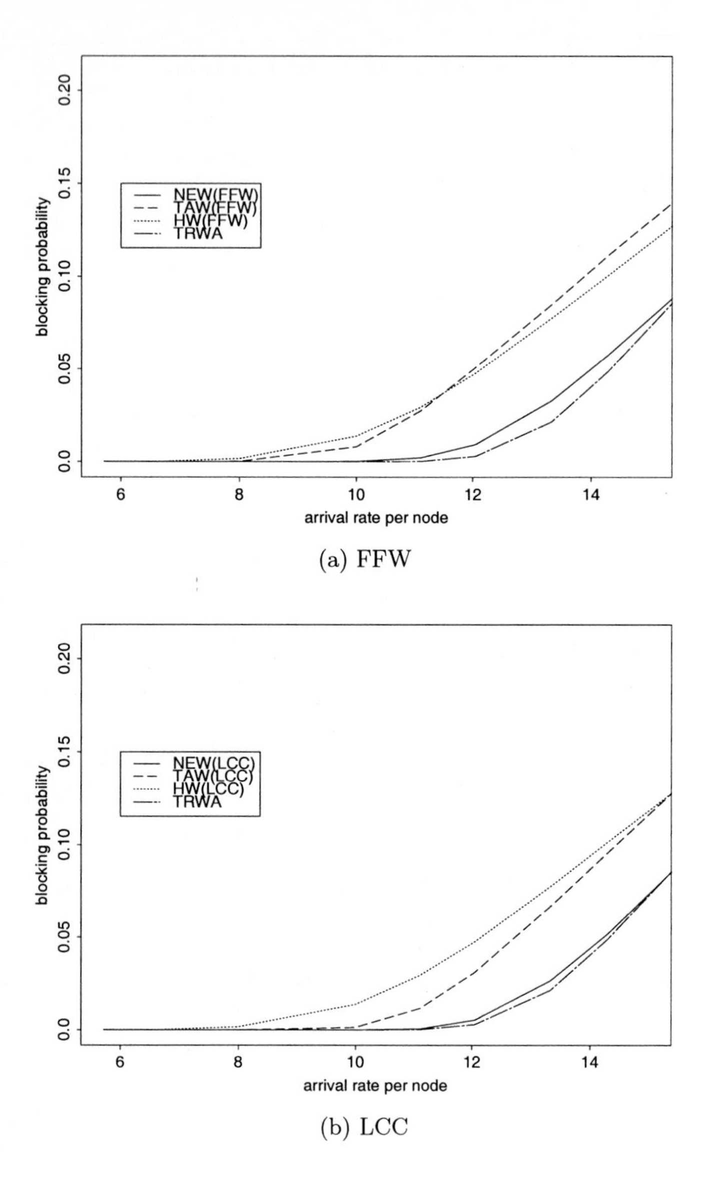

(a) FFW

(b) LCC

Fig. 4. Blocking probability vs. traffic load with 32 wavelengths and 20% converters.

ter than TAW when the traffic load is high. This is because when the traffic load becomes saturated, a route with the least hops yields less waste of system resources and therefore provides better performance.

In addition to Figure 4, it can be observed from Figure 5 that LCC can improve the performance further over FFW even though the performance gain depends on the routing algorithms. Our proposed algorithm provides performance

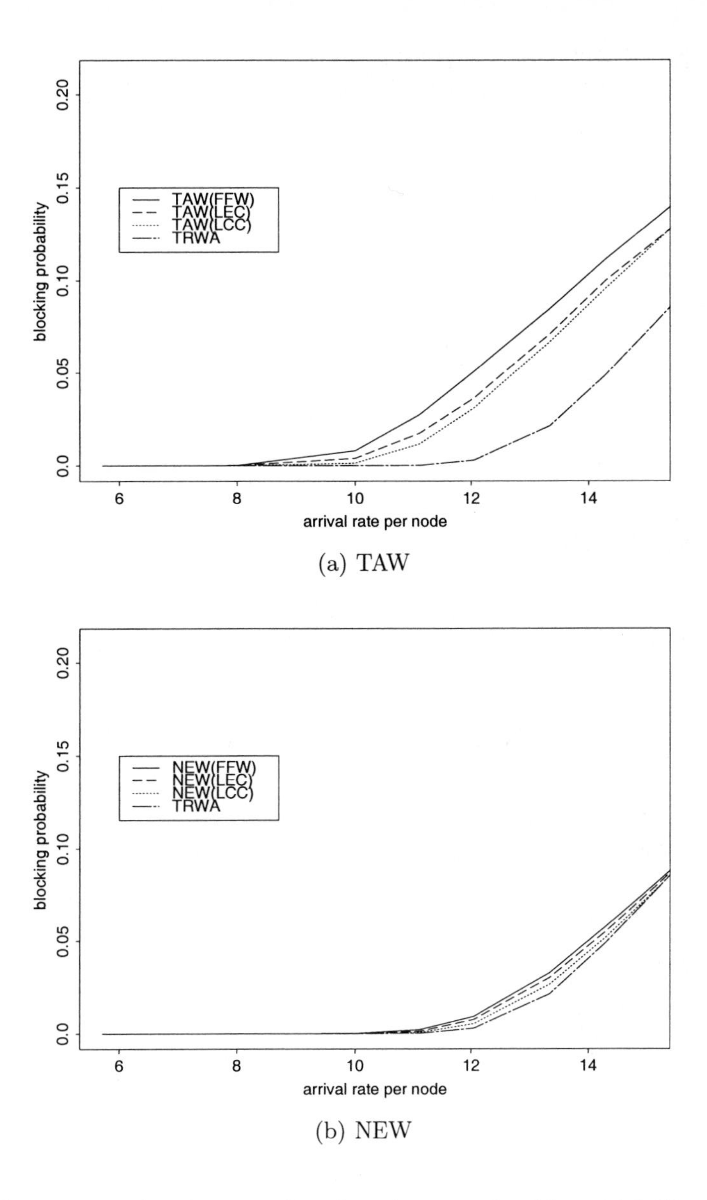

(a) TAW

(b) NEW

Fig. 5. Comparison of various wavelength assignment algorithms with 32 wavelengths and 20% conversion ratio.

improvement large enough and therefore leaves little room to a wavelength assignment algorithm to acquire further performance improvement. Another observation is that a larger performance gain can be obtained for a longer route from the source to the destination. In the NSFNET model under consideration,

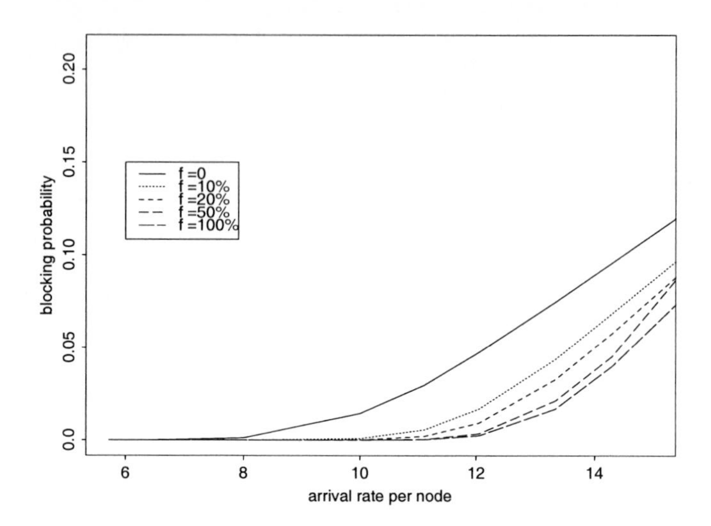

Fig. 6. Blocking probability of the new algorithm vs. traffic load with 32 wavelengths and various conversion factors.

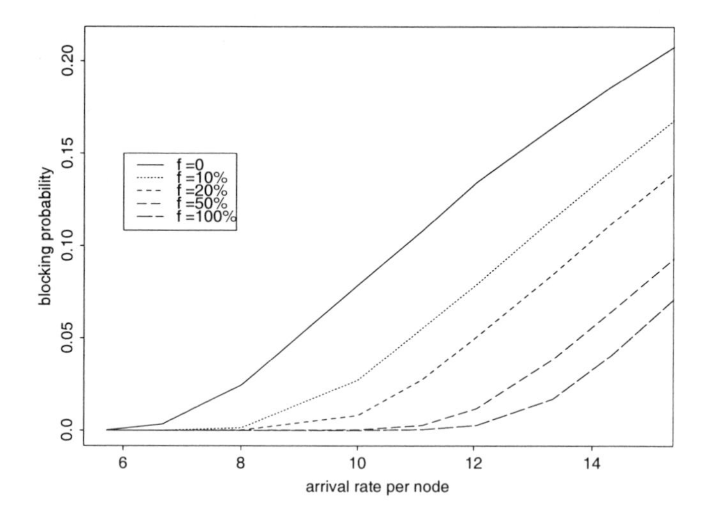

Fig. 7. Blocking probability of TAW vs. traffic load with 32 wavelengths and various conversion factors.

for example, the average lengths of a route of HW, TAW, and NEW are 2.2, 2.9, and 2.7, respectively, and TAW provides the largest performance gain.

Figure 6 shows the blocking probabilities of the new algorithm with various degrees of the conversion factor, f. It can be observed that the performance (blocking probability) is sensitive to f when it is small (*e.g.*, $f = 10$ or 20%); but if f becomes larger (*e.g.*, $f = 50\%$) there is little room for any further improvement over the performance. This trend can be observed clearly when the number of wavelengths becomes large. This coincides with the fact stemmed from the effectiveness of the new algorithm; *i.e.*, only a small number of converters is enough to provide good performance close to that with the full number of converters. On the other hand, the TAW algorithm is much more sensitive to the number of converters as shown in Figure 7 and provides performance comparable to the new algorithm only when the full number of converters is used.

5 Conclusions

In this paper, an efficient lightpath routing algorithm for WDM-routed optical networks is proposed. The algorithm is implemented by partitioning the routing and the wavelength assignment into two subproblems so that the computation time is reduced largely. In determining the best route for an s-d pair, the states of the available wavelengths on the input and output links of a node along with the number of converters are taken into account in the edge weight function. Furthermore, in determining the best wavelengths to set up a lightpath along the best route the cost for using converters is introduced in the weight function of the auxiliary graph so that the selected wavelengths yields the least conversion cost.

The simulation results show that the new algorithm is quite efficient and outperforms significantly other algorithms with comparable computation time in an NSFNET model. Furthermore, the results also show that the new algorithm is specially effective when the number of wavelengths is large while the number of wavelength converters is limited. For example, using the new algorithm the performance can be improved over 40% on TAW or HW when the number of wavelengths is equal to 32 and the conversion factor is 20%. The simplicity and efficiency make our proposed algorithm a practical candidate for real systems.

References

1. R. Ramaswami and K.N. Sivarajan. *Optical Networks–A Practical Perspective.* Morgan Kaufmann Publishers, Inc., San Francisco, 1998.
2. T.E. Stern and K. Bala. *Multiwavelength Optical Networks–A Layered Approach.* Addison-Wesley Longman, Inc., 1999.
3. H.Y. Jeong, S.S. Lee, S.W. Seo, and B.S. Park. An adaptive distributed wavelength routing algorithm in WDM networks. In *Proc. Globecom'00*, pages 1259–1263, 2000.
4. L. Li and A. Somani. Dynamic wavelength routing using congestion and neighborhood information. *IEEE/ACM Trans. Networking*, 7(5):779–786, October 1999.
5. W. Liang, G. Havas, and X. Shen. Improved lightpath (wavelength) routing in large WDM networks. In *Proc. IEEE 16th ICDCS'98*, pages 516–523, 1998.

6. Z. Zhang and A.S. Acampora. A heuristic wavelength assignment algorithm for multihop WDM networks with wavelength routing and wavelength re-use. *IEEE/ACM Trans. Networking*, 3(3):281–288, June 1995.
7. T. Fabry-Asztalos, N.M. Bhide, and K. M. Sivalingam. Adaptive weight functions for shortest path routing algorithms for multi-wavelength optical WDM networks. In *Proc. IEEE ICC'00*, pages 1330–1334, June 2000.
8. K.C. Lee and V.O.K. Li. A wavelength-convertible optical network. *J. Lightwave Tech.*, 11(5/6):962–970, May/June 1993.
9. E. Karasan and E. Ayanoglu. Effects of wavelength routing and selection algorithms on wavelength conversion gain in WDM optical networks. *IEEE/ACM Trans. Networking*, 6(2):186–196, April 1998.
10. A. Mokhtar and M. Azizoglu. Adaptive wavelength routing in all-optical networks. *IEEE/ACM Trans. Networking*, 6(2):197–206, April 1998.

EPSB (Electronic Partially Shared Buffering): A Buffering Scheme for Asynchronous and Variable Length Optical Routing for the Edge Optical Packet Switch

Huhnkuk Lim and Chang-Soo Park

Department of Information & Communications,
Kwang-Ju Institute of Science & Technology,
1, Oryong-dong, Puk-gu, Kwangju, 500-712, Korea
{hklim, csp}@kjist.ac.kr

Abstract. We propose a new buffering scheme for wavelength division multiplexing (WDM) packet switching: the Electronic Partially Shared Buffering (EPSB) scheme. This buffering scheme incorporates separate buffers (i.e., electronic buffers) for all the outputs to share in addition to a prime buffer (i.e., the fiber delay line (FDL) buffer) dedicated to all the outputs. By using this EPSB scheme in cooperation with the proposed outgoing data channel scheduling algorithm, a lower packet loss probability can be achieved with fewer FDLs.

1 Introduction

Based on wavelength division multiplexing (WDM), all-optical networks are increasingly becoming the technology of choice for high-capacity optical networks. By carrying different wavelengths into one optical fiber, WDM has the potential of delivering an aggregate throughput in the order of terabits per second. These networks are likely to use optical packet switching (OPS) to provide flexible bandwidth for internet traffic and to increase the transparency of optical networks. Therefore, WDM optical packet switching has been extensively studied [1] [2]. When optical packets are switched from input fibers to output fibers, however, contention may occur if more than one packet arrives from different input fibers simultaneously destined for the same output fiber. To resolve these contentions, optical buffers realized by FDLs are required. If contention occurs, only one of the contending packets must be allowed to reach the destination and the others must be buffered for later transmission.

FDL buffers have been studied in switching architectures designed for a fixed packet length scenario, in both synchronous and asynchronous cases. To support Internet traffic, however, an optical router should cope with asynchronous variable length packets. The performance of FDL buffers in this case is fairly poor even with intelligent scheduling algorithms [1] [2] to fill the voids due to the discrete time delay of FDLs or inter-arrival time between packets. As a result, a large number of FDLs would be necessary to achieve a low packet loss probability. Because FDL buffers do not provide continuous time delay unlike electronic

I. Chong (Ed.): ICOIN 2002, LNCS 2343, pp. 305–312, 2002.

memory, they are very inefficient to fill the voids if the basic timescale unit to be used for FDL buffers has a large value to increase buffering capacity. On the contrary, if the basic timescale unit of the FDL buffer has a small value to increase time resolution and to fill voids efficiently, buffering capacity will decrease. Obviously, there must be a trade-off between them to provide an acceptable optical packet loss probability.

In this paper, a new buffering scheme named Electronic Partially Shared Buffering (EPSB) is proposed where electronics and optics share the buffering. This buffering scheme is particularly suitable for the edge aggregate optical packet switch (OPS) because of the ability to make use of electronic memory already in place to perform the electronic router functionality [3]. An FDL buffer (named as prime buffer) is used to form the vast majority of storage while electronic buffers are used to provide continuous time delays which the FDL buffer does not support. Hence, the packets that failed to search the available output channel are stored in the electronic buffer and retry to fill idle channels such as voids (the vacant parts between two packets in a time domain) at continuous time delay value.

Simulation results show that the EPSB scheme provides the advantages of low packet loss probability with fewer FDLs (i.e., fewer buffer depth). In addition, using an OPS to interface with electronic routers is expected to give us cost savings in the network because the number of FDLs can be reduced with the help of electronic buffers.

2 The Operation of an OPS for Asynchronous and Variable Length Optical Routing

In this section, we describe the operation of an OPS for asynchronous and variable length optical packets routing. Input packets arrive at many wavelengths at many input fibers asynchronously. Their arrivals are sensed, the headers are separated for processing from the payloads, and the packets are time-stamped. Processing the header involves performing address look-up to determine the output fiber, scheduling to determine the time and wavelength of departure, and switch control for possible reconfiguration of the switch. After these processing operations are completed, a packet is transported into the buffer by means of a strictly non-blocking space switch. The packet is again transported out of the buffer using a strictly non-blocking architecture [2]. The assumption made in our study is that every packet within the buffer should have a distinct wavelength. Thus, the number of wavelengths within the switch is larger than that of the transport wavelengths.

We assume two packets with lengths l_0 and l_1 arriving at times t_0 and t_1 ($t_0 < t_1$), and contending for the same output. l_1 will have to be delayed in the FDL buffer for a duration $i \times D$ (a multiple value of basic timescale D) which can only be $\geq t_0 + l_0 - t_1$. Fig. 1 shows the above situation. If the inequality holds, the output distribution of the packets generates a gap or void between the two packets as shown in Fig. 1 [2]. If nothing is done to fill that void, the process of generation of voids in the output distribution results in an excess load. Thus,

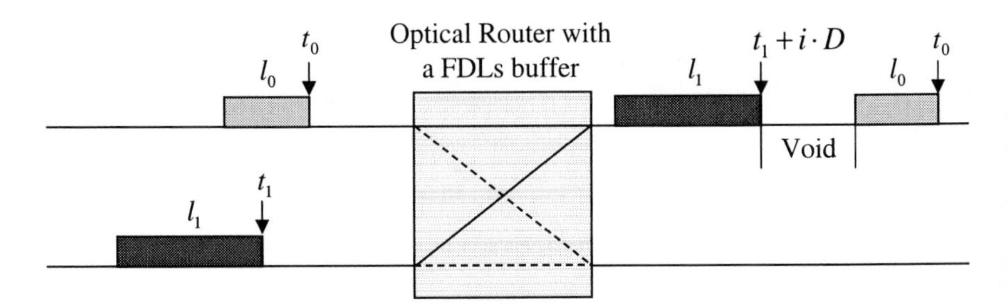

Fig. 1. A void generation process at the output of a 2×2 switch with one wavelength per fiber when the packet processing time is negligible.

for asynchronous and variable length optical packet switching operations, data channel scheduling algorithms to fill voids are required to reduce the excess load.

In fact, the void is necessarily generated regardless of the structure of a FDL buffer (i.e., degenerate or non-degenerate structure of a FDL buffer) because of discrete delay characteristic of FDL. Though, the non-degenerate buffer operates better in tandem with a void filling scheduling algorithm than the degenerate buffer in terms of packet loss probability [4].

Data channel scheduling algorithms can be classified into two categories: with void filling (FF-VF, LAUC-VF) and without void filling (FF, LAUC) [5]. Data channel scheduling algorithms without void filling have the high packet loss probability because they do not try to access voids. However, although an scheduling algorithm with void filling is used, we note that the discrete delay characteristic of FDLs degrades performance of the scheduling algorithm in terms of packet loss probability (i.e., it does not make optical packet to access voids efficiently).

3 Electronic Partially Shared Buffering Scheme

3.1 The Purpose of the EPSB Scheme

For asynchronous variable length optical packet switching operations, data channel scheduling algorithms to fill voids are necessary to reduce excess load, which results in the reduction of packet loss probability. Because FDL buffers do not provide continuous time delays, however, packet loss occurs even when the void size is larger than the packet size. In Fig. 2, although the void size is larger than that of the packet that arrived at the current time τ, the packet that arrived at current time τ will be lost when only the FDL buffer is used.

Table 1 shows the ratio of the number of lost packets corresponding to the case of Fig. 2 to the total number of lost packets. Even when a void filling algorithm is applied, over 70% of the optical packets is lost due to the discrete delay characteristic of the FDL buffer. With the EPSB scheme, the packets that fail to find the available output channel are stored in electronic buffers, while the proposed data channel scheduling algorithm retries to fill idle channels such

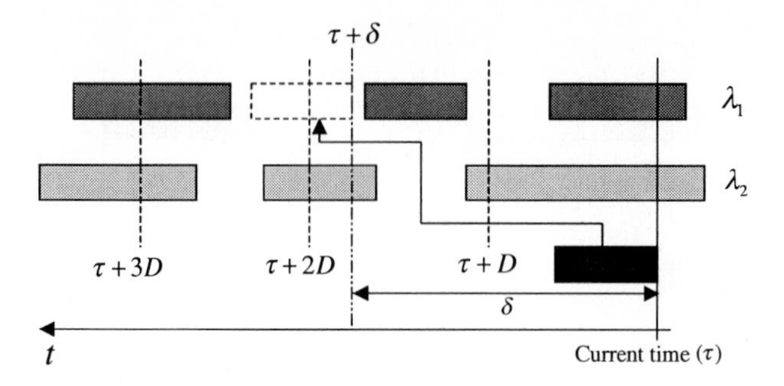

Fig. 2. One of the basic operations of the EPSB scheme.

Table 1. The ratio of the number of lost packets in fig. 2 scenario (void size≥packet size) among the total number of lost packets when FF-VF algorithm is applied. (Number of channels per fiber=8, delay line length=5100, buffer depth=8)

Load	0.5625	0.625	0.6875	0.75	0.8125	0.875	0.9375	1
Ratio (%)	82	75	76	76	77	75	74	75

as voids. Thus, the arrival packet at the current time τ is delayed by as much as δ and served at $\tau + \delta$. The EPSB scheme stores overflow packets that fail to find an available channel in electronic buffers, and retries to fill voids, thus minimizing the total void size as much as possible. Packets stored in electronic buffers can be served at continuous time values when a void is discovered.

3.2 The EPSB Scheme Description

In this subsection we describe the basic concept of EPSB scheme. The scheme uses electronic memory as a minority buffer for storing overflow packets. It still maintains the transparency of optical packets because it can only contribute to serving overflow packets that fail to find available wavelengths. For the EPSB scheme, if two overflow packets that fail to search available wavelengths overlap in the time domain, each packet has to be stored in a different electronic memory to avoid receiver collision. Any electronic buffer structure is possible in the switch if it can prevent receiver collision of the overflow packets to the electronic buffers. In our study, we consider only one electronic buffer for storing overflow packets.

Since the EPSB scheme supports continuous time delays as converting overflow optical packets into electrical packets, it can contribute to the efficient filling of voids in cooperation with the proposed data channel scheduling algorithm. The proposed data channel scheduling algorithm for the EPSB scheme is presented below.

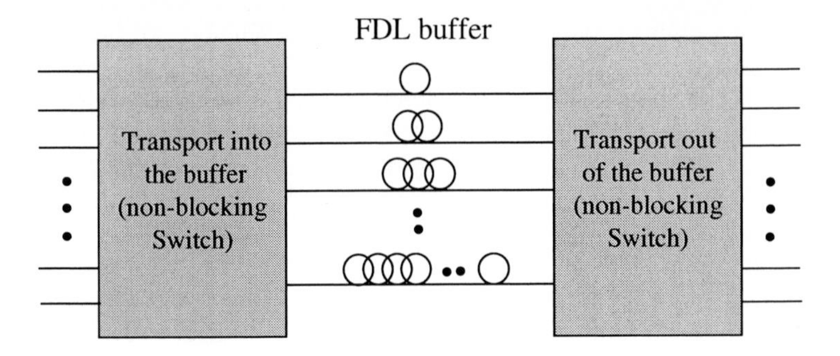

Fig. 3. The assumed switch architecture

Step 1: Create virtual wavelength queues per fiber.

Step 2: Check whether the current time τ is the arrival time of a new packet or the starting time of a void (the serviced time of a packet in the virtual wavelength queues)

Step 3: If the current time τ is the arrival time of a new packet, search an available data channel by FF-VF algorithm [5]. If that packet fails to find an outgoing data channel, it is stored in an electronic memory and time-stamped. If receiver collision is detected (i.e., the packet that failed to find an outgoing data channel overlaps with the previous packet that failed to find an outgoing data channel in the time domain), the packet is dropped without entering to switch.

Step 3': If the current time τ is the starting time of a void (i.e., the serviced time of a packet in the virtual wavelength queues), search a packet capable of filling that void in the electronic memory. If we find an available packet to fill that void in the electronic memory, that packet becomes the head packet of the virtual wavelength queue corresponding to that void. If we do not find an available channel, we add the time value by as much as 1 *unit* to the queuing delay time for all packets in the electronic memory.

Step 4: If the queuing delay time of packets stored in the EPSB is longer than our determined tolerance time value ($\zeta = 100 \cdot \bar{T_{on}}$) , that packet is dropped out of the electronic memory. (This means that the electronic memory has a finite buffer size)

4 Performance Evaluation

In this section, we evaluate the performance of the EPSB scheme in terms of packet loss probability through simulation, and we compare it with the performance of FF and FF-VF algorithms (defined as FF and FF-VF schemes in our study) that are operated only on a conventional FDL buffer.

4.1 Simulation Models

We consider a switch architecture (8×8) in Fig. 3, incorporating an FDL buffer with the degenerate buffer structure and assume a unique wavelength for each packet in the buffer [1] [2] [4] [6]. For the EPSB scheme, we have additional electronic buffers in the assumed switch architecture. The traffic model for the self similar traffic follows the conventional ON/OFF sources model where the traffic is represented by alternating ON (packet presence) and OFF (inter-arrival time) periods. The length T of each period is modeled according to the Pareto heavy-tail distribution and is given by

$$T_{on} = T_{off} = \lfloor bytes/U^{1/\alpha} \rfloor$$

where U is a uniform random variable on [0,1] and $\lfloor\rfloor$ indicates the floor function [1] [2]. Since a Pareto distribution is used, the result is an aggregate traffic exhibiting self similarity with the Hurst parameter defined as follows:

$$H = (3 - \alpha)/2$$

where H has a value on [0.5, 1] for self similar traffic. In our simulation, the $bytes = 400$ parameter is chosen to represent packets with a minimum packet length of 400 bytes. The parameter α expresses the heaviness of the tails of the distribution, and for infinite variance it is necessary that $1 < \alpha < 2$. The traffic is more self similar as H approaches 1 (i.e., α approaches 1) [1] [2].

4.2 Simulation Results

Fig. 4 shows the packet loss probability of three schemes according to the increase of delay line length of the prime buffer when the Hurst parameter (H) and the total offered load (ρ) are 0.9 and 0.75, respectively. In Fig. 2(a), the EPSB scheme gives a lower packet loss probability than the other two schemes. Although the EPSB scheme has a smaller buffer depth ($B = 7$) than those of two legacy algorithms operated on only conventional FDL buffer ($B = 8$), the EPSB scheme provides the lowest packet loss probability. Because the EPSB scheme supports continuous time delay values by converting overflow optical packets into electrical packets, it can contribute to the efficient filling of voids in tandem with the proposed data channel scheduling algorithm. Note in Fig. 2(b) that the EPSB scheme performs better when the number of channels per output fiber (i.e., 4) is reduced. Although the EPSB scheme has a smaller buffer depth ($B = 6$) than two legacy algorithms operated only on an FDL buffer ($B = 8$), the EPSB scheme provides the lowest packet loss probability.

Fig. 5 shows the packet loss probability of the three schemes according to the increase of total offered load. For three schemes, delay line length are determined as values of having the minimum packet loss probability from the results of Fig. 2(a) and Fig. 2(b). As shown in Fig. 3(a), the EPSB scheme also gives the lowest packet loss probability because electronic buffers supports continuous time delay values in searching available channels for overflow optical packets. We also see from Fig. 3(b) that the EPSB scheme performs better when the number of channels per output fiber (i.e., 4) is reduced. Although the EPSB scheme is operated on the smaller buffer depth ($B = 6$) than the two legacy algorithms ($B = 8$), the EPSB scheme gives the lowest packet loss probability.

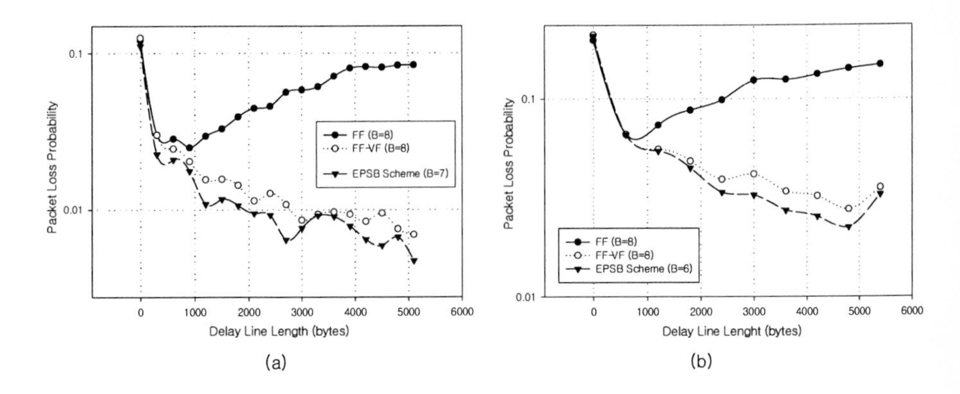

Fig. 4. Packet loss probability as a function of delay line length. (Number of channels per fiber: (a) 8, (b) 4)

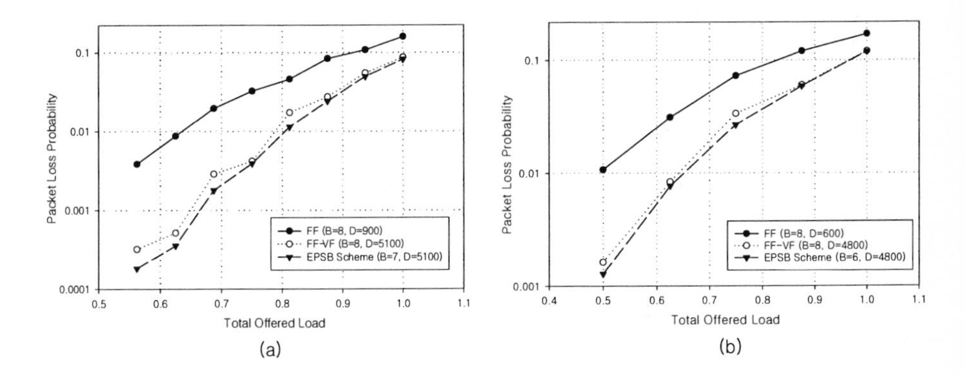

Fig. 5. Packet loss probability as a function of total offered load. (Number of channels per fiber: (a) 8, (b) 4)

Note that the EPSB scheme can efficiently fill idle channels with overflow packets that fail to find available channels, and contributes to the improvement of packet loss probability, as shown in Fig. 4 and Fig. 5.

5 Conclusions

In this paper, we proposed a new buffering scheme (EPSB) suitable for an edge OPS to interface with electronic routers for asynchronous and variable length optical routing. Thanks to the statistical nature of packet arrival and the support of continuous time delay values for overflow packets stored in an electronic memory, more efficient channel utilization could be achieved from the electronic buffer resource. Therefore, a lower packet loss probability was produced without increasing the buffer depth of the FDL buffer (i.e., the number of FDLs).

Acknowledgments. This work was supported in part by the Ministry of Education (MOE) through the Brain Korea 21 (BK21) project.

References

1. L. Tancevski, A. Ge, G. Castanon and L. Tamil, "A New Scheduling Algorithm for Asynchronous, Variable Length IP Traffic Incorporating Void Filling," *Proc. OFC'99*, pp. 112-123, Feb. 1999, Paper ThM7.
2. L. Tancevski, S. Yegnanarayanan, G. Castanon, L. Tamil, F. Masetti and T. McDermott, "Optical Routing of Asynchronous, Variable Length Packets," *IEEE J. Select. Areas Commun.*, Vol. 18, No. 10, pp. 2084-2093, Oct. 2000.
3. Mike J. O'Mahony, D. Simeonidou, David K. Hunter and A. Tzanakaki, "The Application of Optical Packet Switching in Future Communication Networks," *IEEE Commun. Mag.*, Vol. 39, No. 3, pp. 128-135, Mar. 2001.
4. L. Tancevski, L. Tamil and F. Callefati, "Nondegenerate Buffers: An Approach for Building Large Optical Memories,"*IEEE Photon. Technol. Lett.*, Vol. 11, No. 8, pp. 1072-1074, Aug. 1999.
5. Y. Xiong, M. Vandenhoute and Hakki C. Cankaya, "Control Architecture in Optical Burst-switched WDM Networks," *IEEE J. Select. Areas Commun.*, Vol. 18, No. 10, pp. 1838-1851, Oct. 2000.
6. F. Callegati et al., "Architecture and Performance of a Broadcast and Select Photonic Switch," *Optical Fiber Technology*, Vol. 4, pp. 266-284, Aug. 1998

Multicasting Optical Crossconnects with Fault Tolerance Mechanism and Wavelength Routing in All-Optical Networks

Chi-Yuan Chang and Sy-Yen Kuo

Department of Electrical Engineering
National Taiwan University, Taipei, Taiwan
chiyuan@lion.ee.ntu.edu.tw, sykuo@cc.ee.ntu.edu.tw

Abstract. This paper proposes a multicasting and fault-tolerant optical crossconnect (MFOXC) architecture that can support multicasting and fault tolerance. First, a tap-based and two splitter-based MFOXC node architectures are presented for wavelength routed all-optical networks. Compared to the traditional optical crossconnect, the proposed MFOXC node not only has the advantage of multicast capability but also improves the capability of fault tolerance. It could be assigned to critical point in networks to improve the reliability and multicast performance. Furthermore, the communication patterns considered in our algorithms include three general types covering almost all current communication patterns. The MFOXC routing algorithms for point-to-point, multicast and multiple multicasts are presented. We also propose a fault model that is more complete than the existing ones by considering both the active and passive faults. In addition to the fault model, a corresponding restoration mechanism is also proposed.

Keywords: Wavelength Routing, Fault Tolerance, Multicast, All-optical Networks, Wavelength Division Multiplexing.

1. Introduction

As the internet traffic continues to increase exponentially, a wavelength division multiplexing (WDM) network with terabits per second per fiber becomes a natural choice as a backbone in the next generation optical internet. This results in the introduction of wavelength routed all-optical networks (WRAON) [1]. For a WRAON, circuit switching is preferred since the optical technology for implementing the intermediate node buffering, header recognition and processing, which are indispensable for packet switching networks, is not available yet [2-3]. In such a system, network failures could interrupt a large number of communication sessions in progress, such as voice and data. As a result, the design of a WRAON must incorporate mechanisms to protect against potential failures. It is also desirable that these failures should endeavor to be handled within the optical network layer, rather than by higher layers.

I. Chong (Ed.): ICOIN 2002, LNCS 2343, pp. 313-326, 2002.

On the other hand, the multicast (for one-to-many or many-to-many communications) is important and increasingly popular on the Internet (IP over WDM). For a WRAON, the optical crossconnect (OXC) plays an important role to realize switching. The OXC supports point to point connections, as has been intensively investigated [4-6], and point to multipoint (multicasting) connections [7-11] because a lot of future broadband services are multicasting ones.

For WDM multicast, a (optical) switch needs to have the light splitting capability in order to be able to multicast data in the optical domain. To realize optical multicasting, one can utilize optical power splitters [23]. A power splitter is a passive device used to distribute the input signal to all outputs; thus providing multicasting in the optical domain without buffering. In addition, cross-connects which are able to provide all different multicast demands must be equipped with a splitter for every wavelength on every input fiber link. The large number of splitters in a multicast cross-connect has the negative implications of difficulty and high cost of fabrication.

There are many researchers who implemented OXC structures to support multicasting [7-11]. Five classes of multicasting OXCs have been proposed. First, a star coupler that has inherent multicasting capability is used to construct a class of OXC [7]. But wavelength converters and tunable filters are required to work in a large wavelength range. Second, two stage splitters and one stage combiners are used to construct another class of OXCs [7-8]. Third, a splitter-combiner switch that has inherent multicasting capability is employed to construct another class of OXC [9]. Fourth, a splitter-and-delivery switch (SaD) is proposed to build OXCs [10]. Fifth, a Tap-and-Continue switch (TaC) [11] is also built to form OXCs with zero power splitter. All of these OXCs support multicasting. Note that switches with splitting capability are usually more expensive to build than those without. Due to this reason, many researchers consider the constraints on the splitting capability of the switches in a network [12], which means that only a subset of the switches in a WDM network supports light splitting. These researches all focus on designing an OXC with multicasting capability, but without fault tolerance.

The fault-tolerance is one of the most important quantitative measures in optical Quality of Service (QoS). Hence, in order to achieve protection against failures, spares must be provided for the corrupted traffic to be restored. With the advent of WDM techniques, it is possible to provide redundancy by means of spare wavelengths (channels). Several simple failure restoration techniques for WDM mesh networks have been proposed in [13-15]. The required number of wavelengths to guarantee a complete link failure restoration is often larger than that supported by technology. This situation becomes even worse in dynamic wavelength assignment where the connection is set up one by one. Although having multiple bi-directional links between each node pair can overcome the problem of insufficient number of wavelengths, an optical crossconnect will become more complex in order to handle the extra links and the cost of links will also increase. Therefore, in this paper we will propose a series of multicasting and fault-tolerant optical crossconnect (MFOXC) architectures and the corresponding fault tolerant wavelength routing algorithms under the current technology constraint of limited number of wavelengths and the assumption of a single bi-directional link between every node pair.

The rest of this paper is organized as follows. Section II introduces the basic architecture of the MFOXC. Two different types of MFOXC will be presented. Section III describes the routing algorithm for connection establishment in terms of point-to-point requests, multicast, and multiple multicast, respectively. Section IV proposes the fault model and the failure restoration schemes for different fault scenarios. Section V concludes this paper.

2. Node Architecture

2.1 Architecture

Fig. 1 shows a WDM all-optical network employing wavelength routing, which consists of OXCs interconnected by optical links. Each OXC node includes a workstation (*A, B,...,E*) and an optical switch (1, 2, ...,5). Each link is assumed to be bidirectional and consists of a pair of unidirectional physical links.

An optical crossconnect is a device capable of routing a wavelength on an input link to any output link. However, two input links with the same wavelength cannot be routed simultaneously onto an output link, e.g. the link from node 4 to node 3 in Fig. 1. If there are *m* wavelengths on each link, the OXC may be viewed as consisting of *m* independent optical switches, one for each wavelength as shown in Fig. 2. Each optical switch has *N* inputs and *N* outputs where *N* is the number of input/output links.

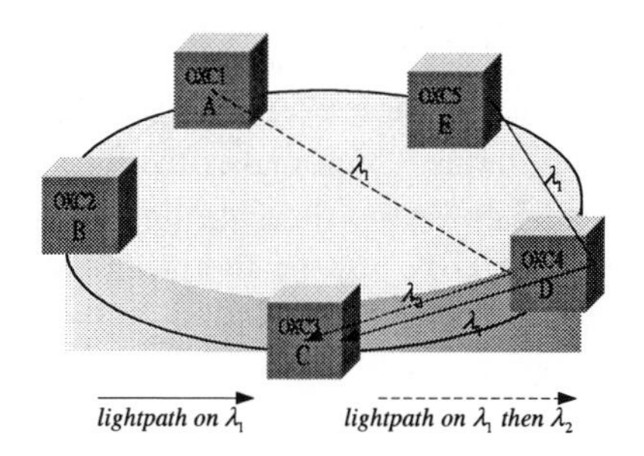

Fig. 1. A WDM network with OXCs interconnected by fiber links.

The effect of wavelength converters on the signal quality bas been investigated by adding converters to the OXC. Wavelength converters are often desired in the OXC to make the network management much easier and reduce the blocking probability because of their signal regeneration and noise reduction capabilities [16-18]. Figure 3 shows an OXC with wavelength converters. The drawbacks of the wavelength converter include not only the higher cost and complexity of the system, but also a big drawback, the single fault problem. If the central optical switch is faulty, the whole OXC will fail if no redundancy is provided at the node level. In addition, the crosstalk problem can be serious when the number of wavelengths is large. In contrast, the OXC without converters in Fig. 2 can tolerate a single fault in an optical switch. This is because any optical switch can replace a faulty optical switch and the resulting OXC can still be functioning with all the wavelengths except the wavelength used by the faulty optical switch. As for fault tolerance capability, an OXC without converters is superior to an OXC with converters. But for network performance (in terms of wavelength reuse), the reverse is true. In order to design a high reliability MFOXC, we adopt the MFOXC architecture with no wavelength converters.

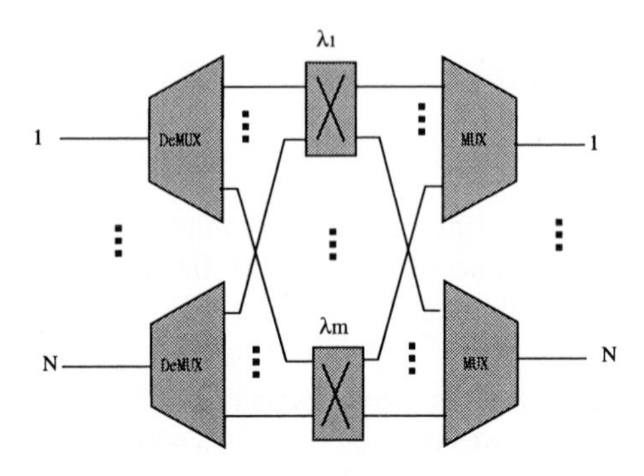

Fig. 2. An OXC without wavelength converters.

On the other hand, mechanical switches usually have very good performance regarding crosstalk and insertion loss, but are sometimes considered less reliable than integrated optical switches, which instead suffer from larger loss and crosstalk. However, owing to their treelike structure, integrated $1 \times N$ switches will only have $\log_2 N$ dominant crosstalk terms which makes the total crosstalk less dependent on the OXC size for interesting crosstalk values [20]. Hence the mechanical switches are good choice for a large size OXC. In the following we will propose two different types of MFOXC, Tap-based and splitter-based.

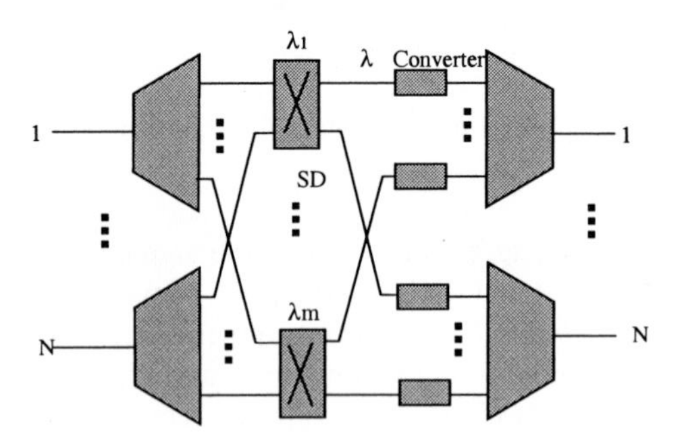

Fig. 3. An OXC with wavelength converters.

2.2 Tap-Based MFOXC

Fig. 4 shows an implementation of an $N \times N$ tap-based MFOXC. The tap-based MFOXC is an OXC with multicasting and fault tolerance capability, which uses wavelength-dependent optical switches (or mechanical switches). It uses a set of Tap-and-Continue Modules (TCMs) on the right side of Fig. 4(a). In a TCM1 shown in Fig. 4(b), an extremely small fraction of the input signal P_{in} (e.g., (1/1000) P_{in} [21]) is tapped and forwarded to the local station. The remaining power of the order of 99.9% is switched to any one of the other $(N-1)$ outputs. The tapping device used is fully programmable so that the tapped signal power is determined by considering the signal to noise ratio (SNR). To switch the signal to any of the $(N-1)$ outputs, switching elements (SEs) are used. The SE can be realized using a photonic directional coupler with electronic control [22]. A 1×*N*TCM module has $\lceil \log_2 N \rceil$ stages where stage *i* has twice the number of SEs as stage *(i-1)*. Fig. 4(d) shows an example of a TCM8. An input signal is tapped in using a tapping device. A small fraction of the signal power is directed toward the local station, while the rest continues to a multistage network of the SEs. By controlling the voltage on these SEs, the input can be connected to any output port(s). Hence, the TCM8 module can support multicasting traffic in the optical domain by controlling the voltage on these SEs. On the other hand, the 2×1 SW element shown in Fig. 4(b) and the 3×1 SW element shown in the right side part of Fig. 4(a) are used to select the alternative port of switch for normal operation, fault tolerance or multicasting.

The TAPs shown in the left side of Fig. 4(a) are used to select switches in the MFOXC. A small fraction of the signal power is directed toward the fault tolerant switch, while the rest continues to a normal switch. The benefit of tap is that all the signal power is directed toward the normal switches in normal operation mode. Only little signal power is transferred to the fault tolerant switch. However, if the normal switch is faulty, by controlling the taps, the signal power is directed toward the fault tolerant switch.

In addition to routing and switching signals, the MFOXC also serves as a source and sink of traffic in the network by an array of multi-wavelength transmitters and an array of multi-wavelength receivers, respectively. The source (sink) are the start (end) of a connection. Each inbound link and outbound link has its associated receiver (*Rx*) and transmitter (*Tx*), respectively. The bottom of Fig. 4(a) shows that each *Tx* or *Rx* is realized by an array of multi-wavelength transmitters or an array of multi-wavelength receivers, respectively. Each optical switch is extended with 1 additional port to support inbound link and outbound link for multicast. The resulting MFOXC has *2N* optical/mechanical switches with *(N+1)* inputs and *(N+1)* outputs as compared to *N* inputs and *N* outputs in Fig. 2.

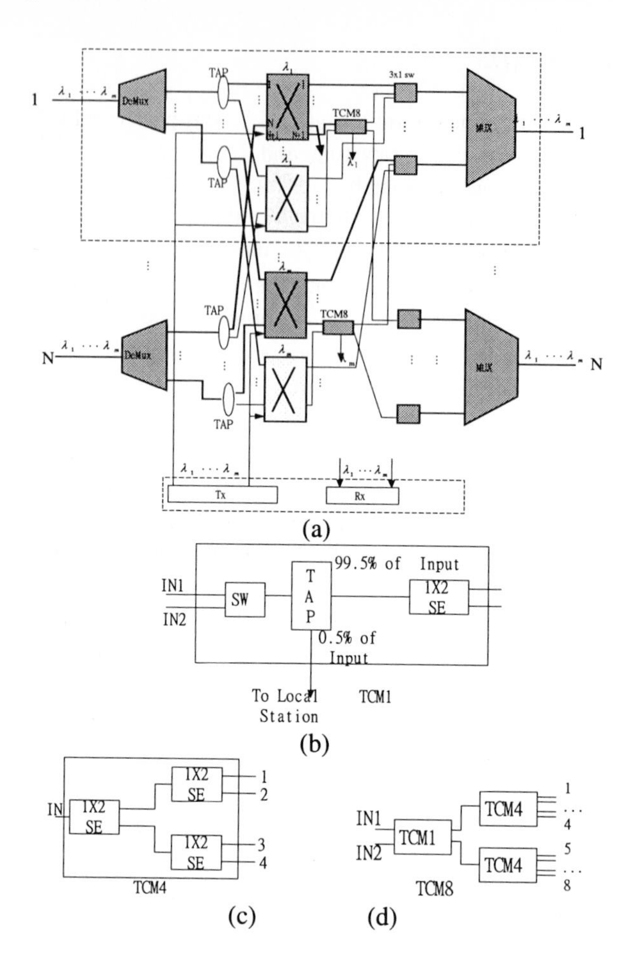

Fig. 4. (a) Structure of tap-based MFOXC, (b) a TCM1 module, (c) a TCM4, (d) a TCM8 (consists of $\lceil \log_2 N \rceil$ stages).

2.3 Splitter-Based MFOXC

The Splitter-and-Delivery (SaD) is a cross-connect with multicast capability that was proposed in [10]. A cross-connect consists of a set of SaD switches (see Fig. 5(a)) for each wavelength. A SaD switch consists of an interconnection of power splitters, optical gates (to reduce the excessive crosstalk), and photonic switches. Fig. 5(b) shows the organization of a cross-connect based on the SaD switch. In addition to the SaD switches, demultiplexers (multiplexers) are used to extract (combine) individual wavelengths. In the following we will propose two types of splitter-based MFOXC.

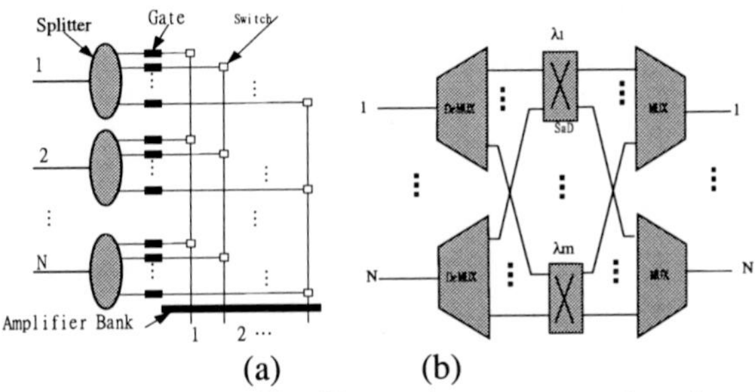

Fig. 5. (a) a SaD switch, (b) an $N \times N$ optical cross-connect based on the SaD.

Type I:

Fig. 6 shows an implementation of an $N \times N$ splitter-based MFOXC. Its architecture is based on the SaD switch. In order to achieve robustness on the splitter for reliability, we duplicate a splitter on each SaD switch. In Fig. 6(a) a fault tolerant and multicasting (FTM) module is proposed. The input lightbeam is initially transferred to one of the branches by controlling the SE. Each branch is split into n branches and connect to a switch. Hence, any input of the splitter can be connected to any output branch. Fig. 6(b) is a switch module equipped with N *FTM* modules and 2×1 switches. Each branch is switchable to an associated output by a 1×2 switch. Therefore, any input can be connected to none, one, several or all the output ports. This features a multicasting capability.

To implement the FTM module, four components, i.e., 2 splitters, optical gates, 2×1 SW, and 1×2 SE are integrated on a silicon board using planar silica waveguide technology [4-6].

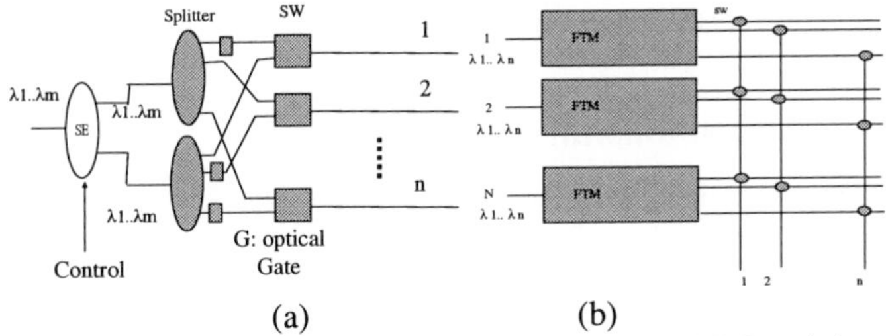

Fig. 6. (a) an FTM module based on SaD, (b) an FTM SaD switch.

TYPE II:

Fig. 7 shows an alternative implementation of an $N \times N$ splitter-based MFOXC. Its architecture is very similar to that of tap-based MFOXC. The main differences between these two architectures are the splitter-to-n (SPn) module shown in Fig. 7(b) and the SE element. The former is equipped with a $1 \times N$ splitter to be as a light-splitter in order to support multicasting in the optical domain. The latter is for selecting switches to operate for fault tolerance. Owing to the power spreading of light splitter, the optical loss is assumed to be compensated with optical amplifier (e.g. EDFA(Erbium Doped Fiber Amplifiers) or SOA (Semiconductor Optical Amplifiers)) properly located in the MFOXC(not shown in the figure). The SE can be realized using a photonic directional coupler with electronic control [22]. By controlling the SEs and the switches, the input can be connected to any output port(s). Hence, the SPn module can support multicasting traffic in the optical domain by controlling these SEs and switches.

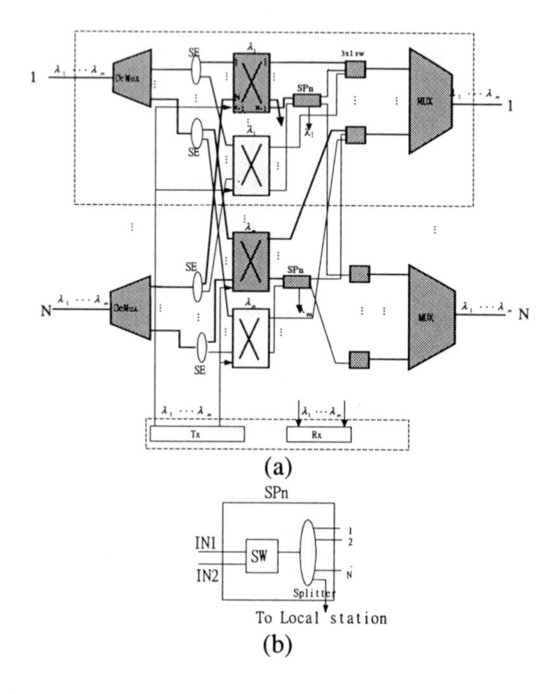

Fig. 7. (a) Structure of splitter-based MFOXC, (b) a $1 \times N$ splitter-to-n module.

3. Connection Establishment

Connection establishment in wavelength-routed networks is routing. We present a set of efficient routing algorithms for point-to-point requests, multicast, and multiple multicast, respectively.

3.1 Multiple Point-to-Point Routing

Point-to-Point routing is the most fundamental routing problem. A general form of Point-to-Point routing is routing for multiple requests: given r requests $R = \{(s_1,t_1),(s_2,t_2),...,(s_r,t_r)\}$, we are required to establish a communication path form s_i to t_i for each request (s_i,t_i), where $1 \le i \le r$. Being able to route for multiple requests is an essential ability of a communication network. We use a greedy approach to design our algorithm that employs a modified shortest-path algorithm to construct paths one by one in their length increasing order. Our modified shortest-path algorithm for path establishment takes into consideration edge weights for edges on the path under construction and uses source routing in which path establishment is initiated by the source node and a path is extended step-by-step from one node to the next toward the destination. The multiple Point-to-Point routing algorithm has the following structure:

```
Algorithm MPP( )    /* Multiple point-to-point algorithm */
{Construct a shortest path for each request (s_i, t_i), where
1 ≤ i ≤ r }
for i = 1 to r do
  Use Dijkstra's algorithm to compute the length L_i of the
shortest path from s_i to t_i ;
  Insert L_i into the sorted list of increasing order
  {L_π1, L_π2, ..., L_πi }
for i = 1 to r do
  Find the shortest path form s_π1 to t_πi using source routing;
```

3.2 Multicast Routing

Multicast requires transporting information from source s to a set of destinations $D = \{t_1, t_2, ..., t_g\}$. Multicast can be realized by first constructing a multicast tree *MT* rooted at *s*, including all nodes $\{t_1, t_2, ..., t_g\}$ in G, and then transmitting information form the root to all destinations along the tree edges using appropriate wavelengths. We are interested in finding an optical *MT* in which the total cost in terms of hops and wavelengths for multicast is minimum. It is clear that finding an optimal *MT* is equivalent to finding a minimum directed Steiner tree in *G* which is an NP-complete problem. To overcome the complexity of the problem, the Minimum Path Heuristic (MPH) [23] is proposed. In each iteration, the tree is extended by including a new destination with the smallest shortest-path to some node. The process is repeated until all destinations are included. The MPH algorithm is listed as follows.

```
Algorithm MPH( )  /* Minimum Path Heuristic */
Begin
     Input: 1.Directed Graph H=(V,E).
            2. Source s.
            3. Set of destinations X ⊆ V−{s}.
     Output: Directed Steiner tree for session (s,X).
     Let U be the nodes in the tree.
     U ← φ; T ← φ; Remaining ← |X|; U ← {s}; Let Path be an arbitrary
shortest path from s to any x ∈ X
     While Remaining > 0 do
     ∀v ∈ U do
       ∀ x ∈ X do
        if (cost(ShortestPath(v,x)) < cost(path)) then
```

```
                path ← ShortestPath(v,x)
        end if
      end ∀
    end ∀
    X ← X − {all dstination s in path}
    T ← T ∪ {all edges in path}
    U ← U ∪ {all nodes in path}
    Remaining ← |X|
    End while
    Return T
End
```

3.3 Multiple Multicast Routing

When several groups of multicast wish to take place concurrently, a more general communication pattern, the multiple multicast is formed. Given *r* groups of multicast $M_i = (s_i, D_i)$, where s_i is a source and $D_i = \{t_i^1, t_i^2, ..., t_i^{gi}\}$ are the destinations, $1 \le i \le r$. Assume M_i can be realized by a multicast tree MT_i. Let multicast forest $MF = \cup MT_i$. It is clear that several edges of different MT_i in *MF* may fall onto the same edge of *G* and hence, attempt to use the same wavelength at the same node in the network. This will possibly cause contention on a particular wavelength when these requests arrive simultaneously at a node. Hence an important task in implementing multiple multicast which is wavelength contention-free is to construct a minimum cost MF. To achieve this, we take a greedy approach to find an approximate optimal multicast tree for each multicast MT_i one-by-one employing Algorithm MPH, where all edges of MT_i are marked with an infinitely large weight as soon as MT_i is constructed, and therefore no wavelength contention can possibly occur. However, in such case, some routing requests may be rejected for insufficient network resource in terms of wavelength, port, and network connectivity, which is common in circuit-switched networks. Our algorithm for multiple multicast is described as follows:

```
Algorithm MM( )   /* Multiple Multicast algorithm */
Begin
    Input: Multiple multicast for M_1, M_2, ..., M_r, where M_i = (s_i, D_i)
    Output: Multicast forest MF = ∪MT_i
    Sort {M_1, M_2, ..., M_r} into increasing size order {M_π1, M_π2, ..., M_πr}
    for i =1 to r do
      Construct multicast tree M_πr for M_i using algorithm MPH
    For each e ∈ E(MT_i) mark weight(e) with weight ∞.
End
```

4. Fault Model and Failure Restoration

4.1 Fault Model

Many possible physical faults in a WRAON are depicted in [24]. Unlike previous researches we consider all potential failures in the active/passive components of the OXC and the fibers, and assume that some critical components (e.g. SE, tap) of the MFOXC are reliable. The fault model is in the following:

1) Channel fault: This fault can result from the failure of a designated transmitter or receiver for a wavelength. In this case, only the failed wavelength on a link between two OXCs can not be used for transmission and other wavelengths still work.
2) Link fault: This is typically caused by a fiber cut or an optical amplifier fault. The channel faults in a link can also be emulated as a link fault if the number of failed channels exceeds a threshold. Further, single link fault is assumed.
3) Switch fault: This is due to the failure of the power circuit or mechanical switch. The failed switch also results in the failure of the corresponding channel on all the associated input and output links of the MFOXC.
4) MFOXC node fault: This is the most severe fault, and may result from the power outage of the controller in the MFOXC. The optical switch faults in a MFOXC can also be considered as a node fault if the number of failed optical switches exceeds some threshold.

4.2 Failure Restoration

In this subsection, we present how the MFOXCs can be reconfigured to tolerate the modeled faults and give an example for each case. Assume that the fault detection mechanism in [24] is adopted in the proposed MFOXC because it can differentiate between a node fault and a link fault. The restoration scheme has two phases: one is link-based and the other is source-based.

a. Channel faults

The least cost solution for channel faults is to allocate some channels on each link between MFOXCs as spare channels. Upon detection of a channel failure, the controller in the MFOXC redirects the lightpath from the failed channel to the spare channel. This solution can be extended to tolerate multiple channel failures if there is more than one spare channel. If the number of spare channels is insufficient, the multiple channel faults can be divided into two classes. One class is handled by the link-based restoration, which still uses the spare channels. The other is by the source-based restoration that reestablishes the restoration path from the source node.

b. Link faults

We can use the line protection method to repair the lightpaths affected by a single link fault locally with the designated spare channels. Take for example the lightpath in Fig. 8. The original normal lightpath is from the source node to X and finally to the destination node Z. Assume there is a link fault between nodes X and Z.

After detecting the failure, this lightpath must be able to re-route around the failed link. In our link-based restoration mechanism, the controller in node *X* informs node *Y* the spare channel and the restoration path is *X-Y-Z,* and configures the optical switch in X to re-route the failed lightpath. The link-based restoration path is determined in advance based on the shortest path on the network without the failed link. The controller in node *Y* informs the successor node *Z* (*Z* is the other end node of the failed link) in the restoration path and the restoration path is *Y-Z*. It also informs node *X* that it has agreed on this restoration. If the restoration path stored in node *X* can not afford to restore the failed lightpath due to insufficient spare channels, the source-based restoration is performed and the original path is given up.

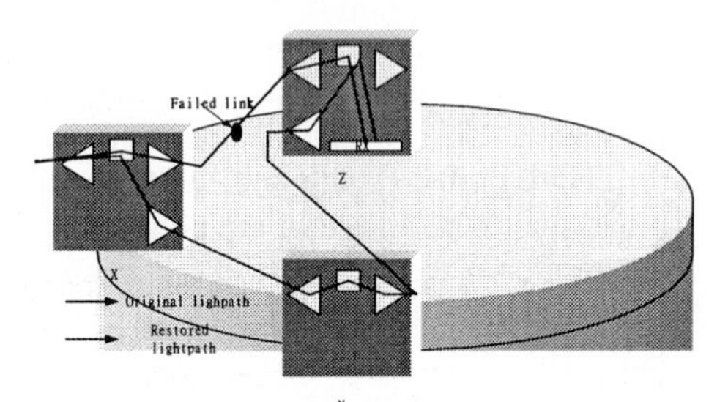

Fig. 8. Restoration from a link fault using a spare channel.

c. Switch fault

Upon detecting failure of the optical switch (mechanical switch), an example restoration route is shown in Fig. 9. This example shows that when an optical switch fails and there is an active lightpath passing through it, how the link-based restoration is performed. In Fig. 9, if the switch within node *Y* fails, node *Y* uses the spare switch S to operate. In this situation node Y needs not inform any other nodes in the path. When there is more than one active lightpath passing through the failed switch or more than one optical switch failure, the solution is also similar.

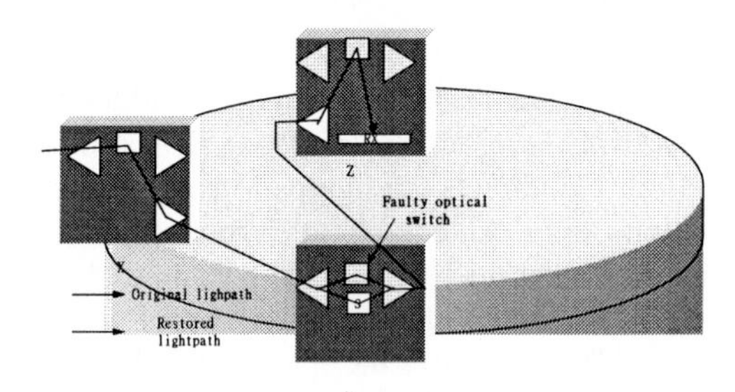

Fig. 9. Restoration from an optical switch fault.

d. MFOXC node fault

Due to the high cost of a complete spare node, we do not suggest sparing at the node level. In our fault model, the optical switch faults in a MFOXC can also be considered as a node fault if the number of failed switches exceeds some threshold.

However, in our proposed MFOXC, all the switches have spare ones to recover switch faults. The MFOXC node fault can be handled by the mechanism for switch fault.

5. Conclusions

We have proposed a series of multicasting and fault-tolerant optical crossconnect (MFOXC) architectures. First, a tap-based and two splitter-based MFOXC node architectures were presented for wavelength routed all-optical networks. The tap-based MFOXC is an OXC with multicasting and fault tolerance capability, which uses wavelength-dependent optical switches (or mechanical switches). It uses a set of Tap-and-Continue Modules (TCMs). The benefit of tap is that all the signal power is directed toward the normal switches in normal operation mode. Only little signal power is transferred to the fault tolerant switch. However, if the normal switch is faulty, by controlling the taps, the signal power is directed toward the fault tolerant switch. Second, a splitter-based MFOXC architecture based on the SaD switch was proposed. In order to achieve robustness on the splitter for reliability, the splitter on each SaD switch is duplicated. An alternative implementation of an $N \times N$ splitter-based MFOXC is very similar to that of tap-based MFOXC. The main differences between these two architectures are the splitter-to-n (SPn) module. Compared to the traditional optical crossconnect, the proposed MFOXC node not only has the advantage of multicast capability but also improves the capability of fault tolerance. It could be assigned to critical points in networks to improve reliability and multicast performance.

The communication patterns considered in our algorithms include three general types covering almost all current communication patterns. The MFOXC routing algorithms for point-to-point, multicast and multiple multicasts were presented, respectively. Furthermore, we propose a fault model which is more complete than the existing ones by considering both the active and passive faults. Unlike previous researches we consider all potential failures in the active/passive components of the OXC and the fibers, and assume that some critical components (e.g. SE, tap) of the MFOXC are reliable. In addition to the introduction of a fault model, the corresponding restoration mechanism was also proposed. The methods and architectures in this paper can be applied to any all-optical network.

References

1. R. Ramaswami, and K. N. Sivarajan, "Routing and Wavelength Assignment in All-Optical Networks," *IEEE/ACM Trans. on Networking*, 1995.
2. P. E. Green, "Optical Networking Update," *IEEE J. on Select. Areas in Commun.*, vol. 14, no. 5, pp. 764-779, June 1996.
3. S. Bannerjee and C. Chen, "Design of Wavelength-Routed Optical Networks for Circuit Switched Traffic," in *Proc. Globecom '96*, pp. 306-310, 1996.
4. Y. D. Jin and M. Kavehrad, "Optical Cross Connect based on WDM and Space-Division Multiplexing," *IEEE Photon. Technol. Lett.*, vol. 7, pp.1300 - 1303, Nov. 1995.

5. A. Watanabe, O. Okamoto, M. Koga, K. Sato, and M. Okuno, "8×16 Delivery and Coupling Switch Board for 320 Gbit/s Throughput Optical Path Cross-connect System," *Electron. Lett.*, vol. 33, no. 1, pp. 67 - 68, Jan. 1996.
6. S. Johansson, M. Lindblom, P. Granestrand, B. Lagerstrom, and L. Thylen, "Optical Cross-connect System in Broad-band Networks: System Concept and Demonstrators Description," *J. Lightwave Technol.*, vol. 11, pp. 688 - 694, May/June 1993.
7. S. Okamoto, A. Watanabe, and K-I. Sato, "Optical Path Cross-connect Node Architectures for Photonic Transport Network," *J. Lightwave Technol.*, vol. 14, pp. 1410 - 1422, June 1996.
8. E. Iannone and R. Sabella, "Optical Path Technologies: A Comparison among Different Cross-connect Architectures," *J. Lightwave Technol.*, vol. 14, pp. 2184 - 2196, Oct. 1996.
9. Y. D. Jin and M. Kavehrad, "Optical Cross Connect based on WDM and Space-Division Multiplexing," *IEEE Photon. Technol. Lett.*, vol. 7, pp. 1300 - 1303, Nov. 1995.
10. W. S. Hu and Q. J. Zeng, "Multicasting Optical Cross Connects Employing Splitter-and-Delivery Switch," *IEEE Photon. Technol. Lett.*, vol. 10, pp. 970 - 972, July 1998.
11. M. Ali and J. S. De, "Cost-Effective Implementation of Multicasting in Wavelength-Routed Networks," *IEEE J. Lightwave Technol.*, vol. 18, no. 12, pp. 1628-1638, Dec. 2000.
12. X. Zhang, J. Y. Wei, and C. Qiao, "Constrained Multicast Routing in WDM Networks with Sparse Light Splitting," *IEEE J. Lightwave Technol.*, vol. 18, no. 12, pp. 1917-1927, Dec. 2000
13. J. Armitage, et al., "Design of a Survivable WDM Photonic Network," in *Proc. Infocom' 97*, pp. 244-252, 1997.
14. S. Baroni, et al., "Link Failure Restoration in WDM Optical Transport Networks and the Effect of Wavelength Conversion," in *OFC' 97 Technical Digest*, pp. 123-124, 1997.
15. Y. Miyao and H. Saito, "Optimal Design and Evaluation of Survivable WDM Transport Networks," *IEEE J. on Select. Areas in Commun.*, vol. 16, no. 7, pp. 1190-1198, Sep. 1998.
16. B. Mikkelsen et al., "All-optical Noise Reduction Capability of Interferometric Wavelength Converters," *Electron. Lett.*, vol. 32, no. 6, pp. 566 - 567, 1996.
17. T. Gyselings, G. Morthier and R. Baets, "Strong Improvement in Optical Signal Regeneration and Noise Reduction through Asymmetric Biasing of Mach-Zehnder Interferometric All-optical Wavelength Converters," in *Proc. ECOC 1997*, pp. 188 - 191.
18. W. V. Parys, B. Van Caenegem, and B. Vandenberghe, "Meshed Wavelength-Division Multiplexed Networks Partially Equipped with Wavelength Converters," in *Proc. OFC 1998*, ThU1, pp. 359 - 360.
19. B. Mukherjee, Optical Communication Networks. New York: Mc-Graw-Hill, 1997.
20. Peter Ohlen, "Noise and Crosstalk Limitations in Optical Cross-Connects with Reshaping Wavelength Converters," *IEEE J. Lightwave Technol.*, vol. 17, no. 8, pp. 1294-1301, Aug., 1999.
21. P. Prucnal, E. Harstead, and S. Elby, "Low-loss, High-impedance Integrated Fiber-optic Tap," *Opt. Eng.*, vol. 29, pp. 1136 - 1142, Sept. 1990.
22. R. A. Spanke, "Architectures for Guided-wave Optical Space Switching Systems," *IEEE Commun. Mag.*, vol. 25, May 1987.
23. H. Takahashi and A. Matsuyama, "An Approximate Solution for the Steiner Problem in Graphs," *Math. Japonica*, vol. 24, no. 6, pp. 573 - 577, 1980.
24. C. Li and R. Ramaswami, "Automatic Fault Detection, Isolation, and Recovery in Transparent All-Optical Networks," *IEEE J. Lightwave Technol.*, vol. 15, no. 10, pp. 1784-1793, Oct. 1997.

Optical Backbones with Low Connectivity for IP-over-WDM Networks

Rui M.F. Coelho[1] and Mário M. Freire[2]

[1] Superior Scholl of Technology, Polytechnic Institute of Castelo Branco
Avenida do Empresário, 6000-000 Castelo Branco, Portugal
rmfcoelho@hotmail.com
[2] Department of Informatics, University of Beira Interior
Rua Marquês d'Ávila e Bolama, 6201-001 Covilhã, Portugal
mario@noe.ubi.pt

Abstract. This paper presents a performance assessment of wavelength routing optical networks with nodal degrees of 2 and 3. It is shown that the traffic performance of a network, with a random topology and an average nodal degree of 2, is better than ring (degree 2) and some chordal ring (degree 3) networks. It is also shown that the performance of a network, with a random topology and an average nodal degree of 3, is better than chordal ring networks with minimum diameter. This fact leads to the question of the existence of other degree three topologies that outperform chordal rings. We have investigated two new topology families with nodal degrees of 3 and an assessment of their performance is presented.

1 Introduction

IP (Internet Protocol) networks based on WDM (Wavelength Division Multiplexing) are expected to offer an infrastructure for the next generation Internet, since they provide network solutions to support the traffic growth driven by Internet-based services [1]-[6]. Actually, the worldwide deployment of WDM (Wavelength Division Multiplexing) transmission systems, to satisfy the bandwidth requirements imposed by the traffic growth, is seen as the first phase of optical networking. Recent technology developments, such as the advent of Optical Add/Drop Multiplexers (OADMs) and Optical Cross-Connects (OXCs), are enabling the evolution from that point-to-point WDM links to wavelength routing networks. In Europe, up to some time ago, research has mainly been directed to WDM ring architectures and their possible evolution to interconnected rings and mesh topologies, due to the simplicity of components and management functions required by the ring topology [7]. However, the restoration in the optical layer is more cost-effective in mesh networks than in rings, especially in long-haul environments with rapid demand growth [8], and, as a consequence, mesh networks are now being object of intense research, particularly in the United States. In [9], it is presented a study of the influence of nodal degree on the fiber length, capacity utilization, and average and maximum path lengths of wavelength routed mesh networks. It is shown that average nodal degrees varying between 3 and 4.5 are of particular interest.

I. Chong (Ed.): ICOIN 2002, LNCS 2343, pp. 327-336, 2002.

In this paper, we consider wavelength routing networks with low connectivity (topologies with nodal degrees of 2 and 3). A well-known family of regular topologies with degree 3 is the chordal ring topology. Chordal ring networks have been proposed by Arden and Lee [10], in the early eighties, for interconnection of multi-computer systems. Since then, some studies have been published concerning properties of chordal rings [10]-[13]. Recently, Freire and da Silva [14] have investigated the influence of the chord length on the traffic performance of wavelength routing chordal ring networks. They have shown that the best network performance is obtained with the chord length that leads to the smallest network diameter. In [15], the same authors have shown that the performance of a chordal ring network (which has a nodal degree of 3), with a chord length that leads to the smallest diameter, is similar to the performance of a mesh-torus network (which has a nodal degree of 4). Since a chordal ring network with N nodes has $3N$ links and a mesh-torus network with N nodes has $4N$ links, the choice of a chordal ring with minimum diameter, instead of a mesh-torus, reduces network links by 25%. Moreover, since chordal rings have lower nodal degree, they require in each switch, a smaller number of node-to-node interfacing (NNI) ports.

Here, we present an assessment of the traffic performance in wavelength routing networks with random topologies of average nodal degrees of 2 and 3. It is shown that the performance of a network, with a random topology and an average nodal degree of 2, is better than the performance of rings and some chordal rings. It is also shown that the performance of a network, with a random topology and an average nodal degree of 3, is better than chordal rings with smallest diameter. This fact, lead us to the following question: Is there a degree three topology that outperforms chordal rings? To study this subject, we have generalized the concept of chordal ring topology and we have investigated new degree three topologies.

The remainder of this paper is organized as follows. Section 2 describes degree three topologies, of which a chordal ring is a particular case. Section 3 briefly describes the analytical model used to compute the path blocking probability in wavelength routed optical networks. The performance assessment of wavelength routing networks, with nodal degrees of 2 and 3, is presented in Section 4. Main conclusions are presented in Section 5.

2 Degree Three Topologies

A chordal ring is basically a ring network, in which each node has an additional link, called a chord. The number of nodes in a chordal ring is assumed to be even, and nodes are indexed as 0, 1, 2, ..., N-1 around the N-node ring. It is also assumed that each odd-numbered node i (i=1, 3, ..., N-1) is connected to a node $(i+w) mod\ N$, where w is the chord length, which is assumed to be positive odd and, without loss of generality, we also assume that $w \leq N/2$, as in [10]. For a given number of nodes there is an optimal chord length that leads to the smallest network diameter. The network diameter is the largest among all of the shortest path lengths between all pairs of nodes, being the length of a path determined by the number of hopes.

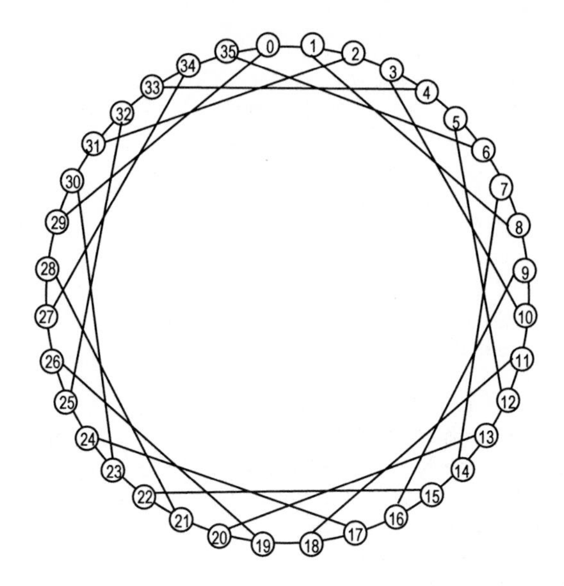

Fig. 1. Schematic representation of DTT(1,*N*-1,7) for *N*=36 nodes (chordal ring with a chord length of 7).

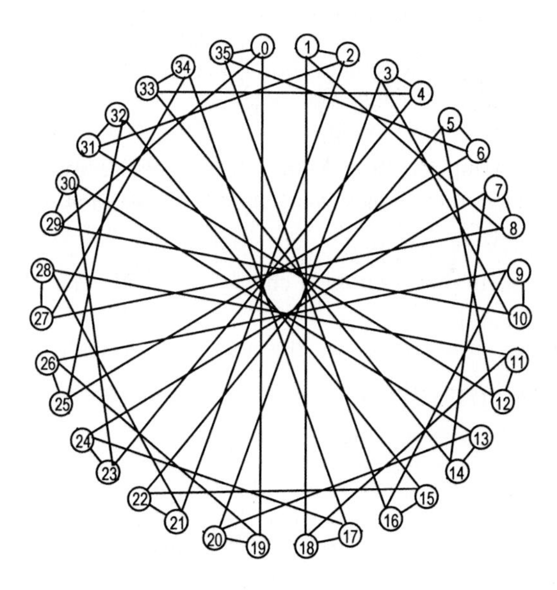

Fig. 2. Schematic representation of DTT(1,*N*/2-1, 7), for *N*=36 nodes.

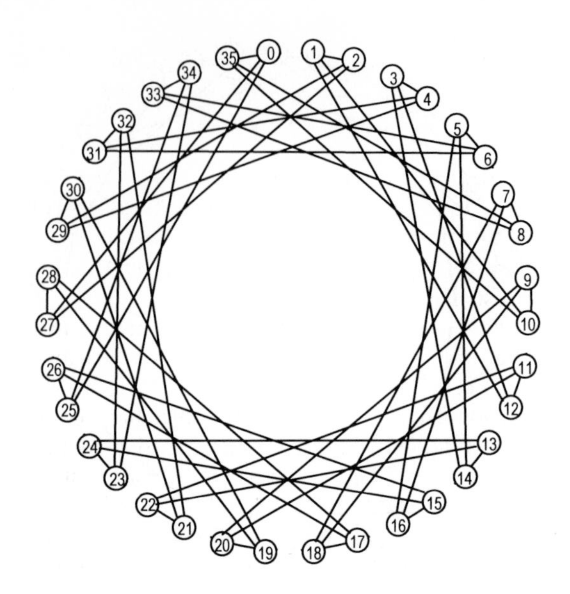

Fig. 3. Schematic representation of DTT(1, $\sqrt{N}$ +3, $\sqrt{N}$ +5), for N=36 nodes.

Here, we consider degree three topologies, of which a chordal ring is a particular case. In each node of a chordal ring, we have a link to the previous node, a link to the next node and a chord. Now, we assume that the links to the previous and to the next nodes are replaced by chords. Thus, each node has three chords, instead of one. Let w_1, w_2, and w_3 be the corresponding chord lengths. We represent a general degree three topology by DTT(w_1, w_2, w_3). We assume that each odd-numbered node i (i=1, 3, …, N-1) is connected to the nodes $(i+w_1) mod\ N$, $(i+w_2) mod\ N$, and $(i+w_3) mod\ N$, where the chord lengths, w_1, w_2, and w_3 are assumed to be positive odd with $w_1 \leq N-1$, $w_2 \leq N-1$, and $w_3 \leq N-1$. In this notation, a chordal ring with chord length w is simply represented by DTT(1,N-1,w).

In this paper we concentrate on the following topologies: DTT(1,N-1,w_3), DTT(1,N/2-1, w_3) and DTT(1, $\sqrt{N}$ +3, w_3). Such topologies are represented in Figs. 1, 2 and 3, for networks with 36 nodes.

3 Evaluation of Path Blocking Probability

To compute the blocking probability in optical networks with wavelength interchange, we have used the model given in [16], since it applies to topologies with low connectivity, has a moderate computational complexity, and takes into account dynamic traffic and the correlation between the wavelengths used on successive links of a multi-link path.

The following assumptions are used in the model [16]: 1) Session requests arrive at each node according to a Poisson process, with each session equally likely to be destined to any of the remaining nodes. 2) Session holding time is exponentially distributed. 3) The path used by a session is chosen according to a pre-specified criterion (e.g. random selection of a shortest path), and does not depend on the state of the links that make up a path; the session is blocked if the chosen path can not accommodate it; alternate path routing is not allowed. 4) The number of wavelengths per link, F, is the same on all links; each node is capable of transmitting and receiving on any of the F wavelengths; each session requires a full wavelength on each link it traverses. 5) Wavelengths are assigned to a session randomly from the set of free wavelengths on the associated path.

In addition to the above assumptions, it is assumed in [16] that, given the loads on links 1, 2, ..., i-1, the load on link i of a path depends only on the load on link i-1 (Markovian correlation model).

The analysis presented in [16] also assumes that the hop-length distribution is known, as well as the arrival rates of sessions at a link that continue, and those that do not, to the next link of a path. The session arrival rates at links have been estimated from the arrival rates of sessions to nodes, as in [16]. The hop-length distribution is a function of the network topology and the routing algorithm, and is easily determined for most regular topologies with the shortest-path algorithm. It is easy to find the hop-length distribution for the (bi-directional) N-node ring network. However, for the (bi-directional) N-node chordal ring with chord length w, it was not possible to obtain a general expression for the hop-length distribution. We have found analytical expressions for the hop-length distribution when the chord length is 3 (w=3), when the chord length is maximum ($w=N/2$ or $w=N/2-1$), when the chord length is as close as possible to the mean chord length ($w=N/4$), and for a chord length that leads to the smallest network diameter ($w=\sqrt{N}+3$). Concerning chordal rings, the analysis presented in this paper is focused on the latter chord length ($w=\sqrt{N}+3$), since it was shown in [14], that this chord length leads to the best performance of chordal rings. Concerning DTT(1, $N/2$-1, w_3) and DTT(1, $\sqrt{N}+3$, w_3) topologies, we analyzed the hop-length distributions of these topologies for 100 nodes. We have also found that, in the range $3 \le w_3 \le N/2-1$, the DTT(1, $N/2$-1, w_3) has two minimum diameters (for w_3=7 and w_3=43), while DTT(1, N-1, w_3) and DTT(1, $\sqrt{N}+3$, w_3) have only one minimum diameter.

Concerning the analysis of networks with random topologies, we follow [16]. Although the assumptions used for random topologies make easy the analysis of the hop-length distribution, the model does not ensure that the network is fully connected.

4 Assessment of Path Blocking Performance

In this section, we present a performance assessment of networks with nodal degrees of 2 and 3. For the case of degree three topologies, we concentrate the analysis on the following topologies: DTT(1, N-1, w_3), DTT(1, $N/2$-1, w_3) and DTT(1, $\sqrt{N}+3$, w_3). Results presented here are for networks without wavelength interchange, except the results of Fig. 7.

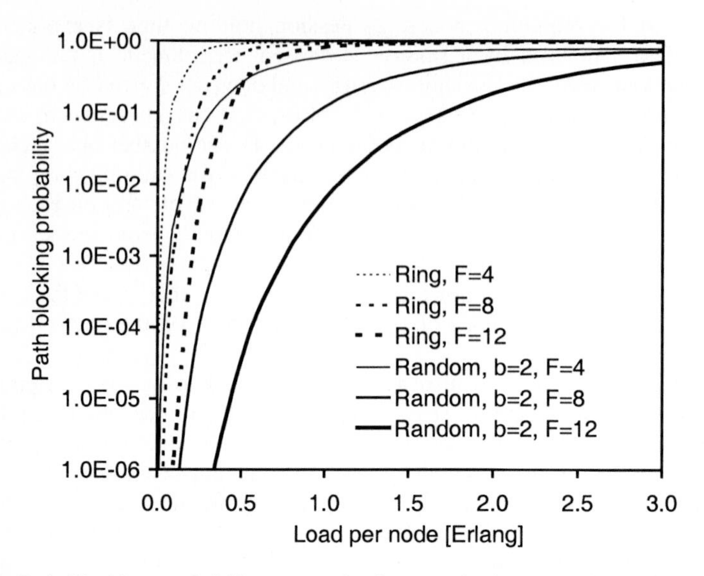

Fig. 4. Path blocking probability versus load per node for networks with random and ring topologies. N=100; F: number of wavelengths per link; b: average nodal degree.

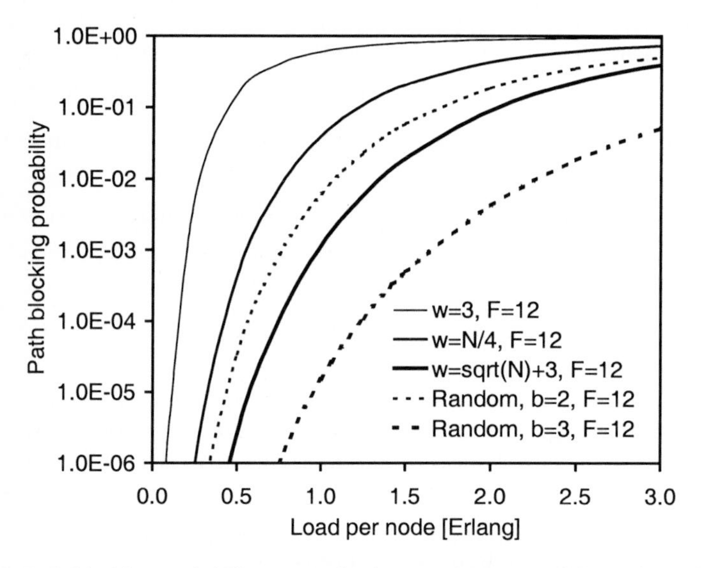

Fig. 5. Path blocking probability versus load per node for networks with random and chordal ring (DTT(1,N-1,w_3) topologies. For the chordal ring, chord lengths of 3, N/4 and $\sqrt{N}$ +3 have been considered. N=100; F: number of wavelengths per link, b: average nodal degree.

Fig. 4 shows the path blocking probability versus load per node for networks with random topology (which has average nodal degree of 2) and ring topology (which has nodal degree of 2). As can be seen in this figure, the network with the random topology has a better performance. If we compare the performance of networks with chordal ring (nodal degree of 3) and random (average nodal degree of 3) topologies, the best performance is obtained with the random topology (see Fig. 5), even for the chordal ring with smallest network diameter ($w_3=\sqrt{N}+3$). From Fig. 5, we may also see that the random topology with average nodal degree of 2 leads to a better performance than the chordal ring with chord lengths of 3 and $N/4$.

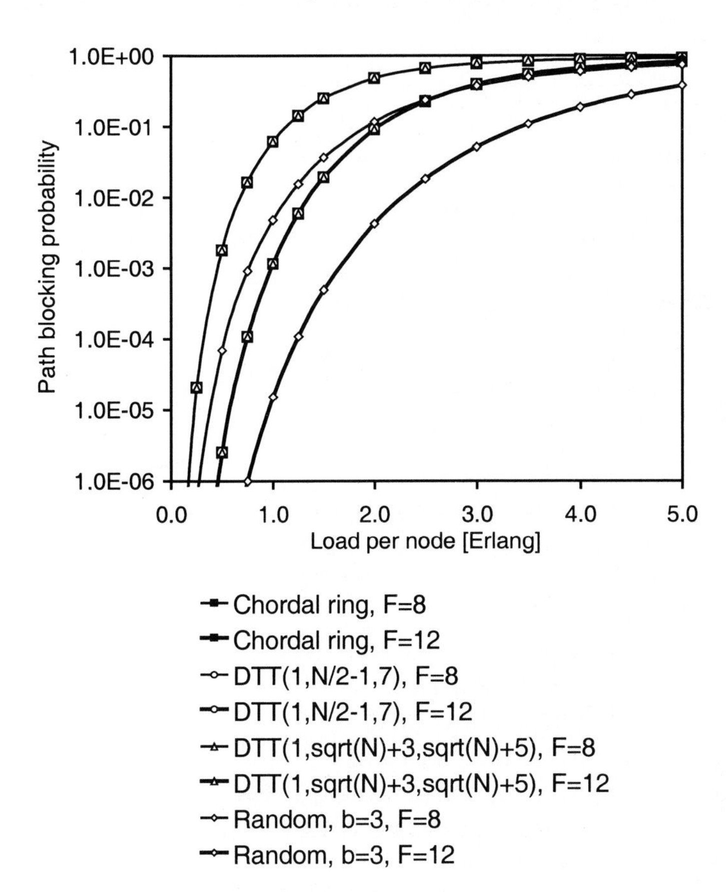

Fig. 6. Path blocking probability versus load per node for networks with DTT(1,N-1,$\sqrt{N}$+3) (chordal ring with smallest network diameter), DTT(1,N/2-1,7) and DTT(1,$\sqrt{N}$+3, $\sqrt{N}$+5). N=100; F: number of wavelengths per link, b: average nodal degree.

Fig. 6 shows the path blocking probability versus load per node for networks with DTT(1,N-1,$\sqrt{N}$+3) (chordal ring with smallest network diameter), DTT(1,N/2-1,7) and DTT(1, $\sqrt{N}$+3, $\sqrt{N}$+5). For these topologies, we have considered for w_3, the chord length that leads to the smallest network diameter. In the case of the DTT(1,N/2-1, w_3), two values of w_3 lead to smallest network diameter. However, in Fig. 6, only the results obtained with w_3=7 are shown, since the results obtained with

w_3=43 are exactly the same as the results obtained with w_3=7. Moreover, we have also observed that the performance of the three degree-three topologies with smallest network diameter is the same (see Fig. 6). These results are confirmed in Fig. 7, which shows the path blocking probability versus wavelength converter density. To explain this observation, we have compared the network diameter and the hop-length distribution. We have found that these network have i) the same minimum diameter and ii) the same hop-length distribution, when the network diameter is minimum.

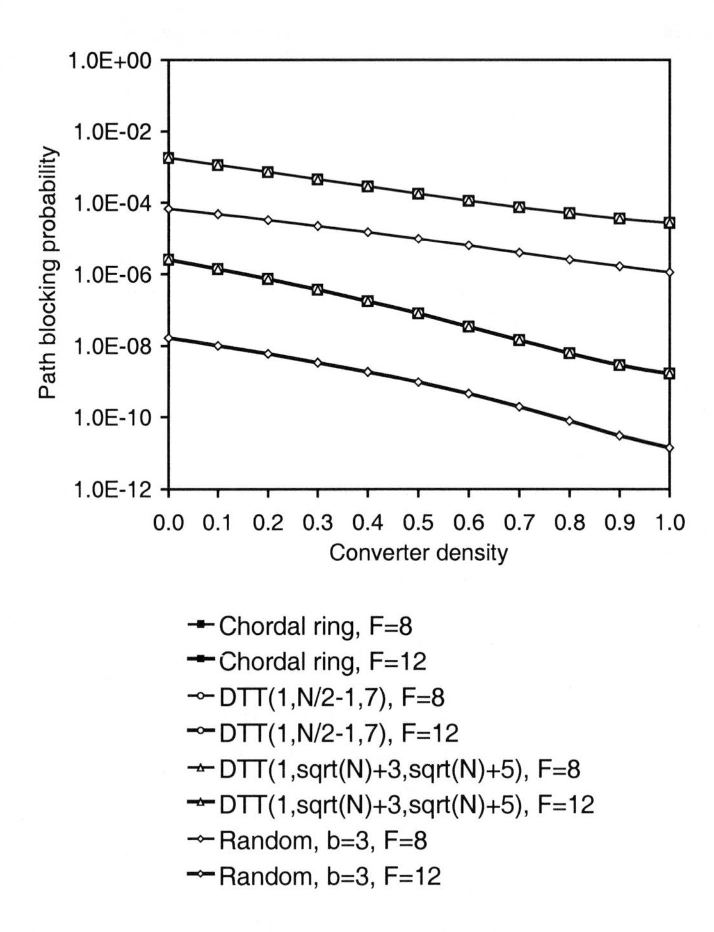

Fig. 7. Path blocking probability versus converter density for networks with the following topologies: DTT(1,N-1,$\sqrt{N}$ +3) (chordal ring with smallest network diameter), DTT(1,N/2-1,7) and DTT(1, $\sqrt{N}$ +3, $\sqrt{N}$ +5). N=100; load per node: 0.5 Erlang; F: number of wavelengths per link, b: average nodal degree.

In a further study, we have analysed diameters of each of the following families: DTT(1, 3, w_3), DTT(1, 5, w_3), ..., DTT(1, N-1, w_3), DTT(3, 1, w_3), DTT(3, 5, w_3), ..., DTT(3, N-1, w_3), ..., DTT(N-1, 1, w_3), DTT(N-1, 3, w_3), ..., DTT(N-1, N-3, w_3), for

N=100, and we found some interesting results. One of those results is concerned with the diameters of the DTT(w_1,w_2,w_3) families, for which w_2=(w_1+2)*mod N* or w_2=(w_1-2)*mod N*. The diameter of a family of this kind, i.e. DTT(w_1, (w_1+2)*mod N*, w_3) or DTT(w_1, (w_1-2)*mod N*, w_3), with $1 \le w1 \le 99$ and $w1 \ne w2 \ne w3$, is a shifted version (with respect to w_3) of the diameter of the chordal ring family DTT(1,N-1,w_3).

We also observed that each family, whose diameter is a shifted version of the diameter of the chordal ring, has four smallest diameters (which are 9, for N=100). Moreover, we found that, for a family of the type DTT(w_1, (w_1-2)*mod N*, w_3), those four minimum diameters occur at w_3=(13+w_1-1)*mod N*, w_3=(15+w_1-1)*mod N*, w_3=(85+w_1-1)*mod N*, and w_3=(87+w_1-1)*mod N*, while, for a family of the type DTT(w_1, (w_1+2)*mod N*, w_3), those four minimum diameters occur at w_3=(15+w_1-1)*mod N*, w_3=(17+w_1-1)*mod N*, w_3=(87+w_1-1)*mod N*, and w_3=(89+w_1-1)*mod N*.

We have also analysed the path blocking performance of wavelength routing networks with all DTT(w_1,w_2,w_3) of smallest diameter and we found that, for N=100 nodes, all topologies of smallest diameter have exactly the same path blocking probability of chordal ring with smallest diameter. We also found that all degree three topologies with smallest diameter have exactly the same hop-length distribution.

5 Conclusions

We presented an assessment of the traffic performance in wavelength routing optical networks with nodal degrees of 2 and 3. It was shown that the traffic performance of a network, with a random topology and an average nodal degree of 2, is better than ring (degree 2) networks and than chordal ring (degree 3) networks with chord lengths of 3 and $N/4$. It was also shown that the performance of a network, with a random topology and an average nodal degree of 3, is better than chordal ring networks with the chord length that leads to the best performance ($w_3=\sqrt{N}+3$). From these observations arose the question of the existence of degree three topologies that lead to a better performance than chordal rings. To study this subject, we considered a general family of regular degree three topologies, represented by DTT(w_1, w_2, w_3).

The performance analysis was focused on the following degree 3 topologies: DTT(1, N-1, w_3), DTT(1,N/2-1, w_3) and DTT(1, $\sqrt{N}$+3, w_3). It was observed that the performance of the three degree 3 topologies is the same when the network diameter is smallest. Moreover, we observed that all DTT(w_1,w_2,w_3) with smallest diameter have the same smallest diameter (which is 9 for N=100) and the same hop-length distribution, when the network diameter is smallest.

Acknowledgements. Part of this work has been supported by Fundação para a Ciência e Tecnologia, Portugal, in the framework of project TRANSPARENT (POSI/34559/CPS/2000), at IT-Coimbra.

References

1. Listanti, M., Sabella, R. (Eds.): Optical Networking Solutions for Next-Generation Internet Networks. IEEE Communications Magazine **38** (2000) 79-122.
2. Gerstel, O., Li, B., McGuire, A., Rouskas, G., Sivalingam, K. M., Zhang, Z. (Eds.): Protocols and Architectures for Next Generation Optical WDM Networks. IEEE Journal on Selected Areas in Communications **18** (2000).
3. Elsayed, K., Lerner, M. (Eds.): Topics in Internet Technology: IP in 2005 – Directions in Wireless and Optical Transport. IEEE Communications Magazine **39** (2001) 135-150.
4. Dixit, S. S., Lin, P. J. (Eds.): Advances in Packet Switching/Routing in Optical Networks. IEEE Communications Magazine **39** (2001) 79-113.
5. Hill, A., Neri, F. (Eds.): Optical Switching Networks: From Circuits to Packets. IEEE Communications Magazine **39** (2001) 107-148.
6. Elsayed, K., Lerner, M. (Eds.): Topics in Internet Technology: Directions in Optical and Wireless Transport II. IEEE Communications Magazine **39** (2001) 130-159.
7. Arijs, P., Caenegem, B., Demeester, P., Lagasse, P., Parys, W., Achten, P.: Design of Ring and Mesh Based WDM Transport Networks. Optical Networks Magazine **1** (2000) 25-40.
8. Doucette, J., Grover, W. D.: Influence of Modularity and Economy-of-Scale Effects on Design of Mesh-Restorable DWDM Networks. IEEE Journal on Selected Areas in Communications. **18** (2000) 1912-1923.
9. Hjelme, D. R.: Importance of Meshing Degree on Hardware Requirements and Capacity Utilization in Wavelength Routed Optical Networks. In M. Gagnaire and H. R. van As (Eds.): Proceedings of the 3rd IFIP Working Conference on Optical Network Design and Modeling (ONDM'99), Paris, France, February 8-9, 1999, 417-424.
10. Arden, B. W., Lee, H.: Analysis of Chordal Ring Network. IEEE Transactions on Computers **C-30** (1981) 291-295.
11. Morillo, P., Comellas, F., Fiol, M. A.: The Optimization of Chordal Ring Networks. In Q. Yasheng and W. Xiuying (Eds): Communication Technology, World Scientific Publishing (1987) 295-299.
12. Hu, X. D., Hwang, F. K.: Reliabilities of Chordal Rings. Networks **22** (1992) 487-501.
13. Hwang, F. K., Wright, P.E.: Survival Reliability of Some Double-Loop Networks and Chordal Rings. IEEE Transactions on Computers **44** (1995) 1468-1471.
14. Freire, M. M., da Silva, H. J. A.: Influence of Chord Length on the Blocking Performance of Wavelength Routed Chordal Ring Networks. In Jukan, A. (Ed.): Towards an Optical Internet: New Visions in Optical Network Design and Modelling, Kluwer Academic Publishers (2002) 79-88.
15. Freire, M. M., da Silva, H. J. A.: Performance Comparison of Wavelength Routing Optical Networks with Chordal Ring and Mesh-Torus Topologies. In Lorenz, P. (Ed.): Networking - ICN 2001, Lecture Notes in Computer Science, Vol. 2093. Springer-Verlag, Berlin Heidelberg (2001) 358-367.
16. Subramaniam, S., Azizoglu, M., Somani, A. K.: All-Optical Networks with Sparce Wavelength Conversion. IEEE/ACM Transactions on Networking **4** (1996) 544-557.

A Study of Blocking Probability in WDM Network

Kiyoung Jung, Jun Kyun Choi

Broadband Network Laboratory
Information and Communications University
Yuseong, Taejon, Korea, 305-732
{jjungki, jkchoi}@icu.ac.kr

Abstract. Calculating blocking probabilities in a network is hard work to do. Also transmission limitations and network upgrade strategies are all expected to mean that different links in a wavelength division multiplexed (WDM) network are expected to have different number of light channels per link. In this paper, we examine the new calculation method, Markov Random Field model, and compare it with other methods. However, the performance improvements depend on the number of wavelengths and fibers on each link regarding to the offered loads and the wavelength assignment scheme used.

1 Introduction

Recent advances in wavelength-division multiplexing (WDM) and optical switching technologies such as MPλS and OBS make it possible to contemplate the deployment of wavelength-routing networks that will provide backbone connectivity at very high data rates. In WDM networks, each node has a dynamically reconfigurable photonic switch. In wavelength-routed all optical WDM networks, a lightpath is an "optical communication path" between two nodes, established by allocating the same wavelength throughout the route of the transmitted data. The requirement that the same wavelength must be used on all the links along the selected path is known as a wavelength continuity constraint. Two lightpaths can share the same fiber link, only if they use different wavelengths. If each intermediate node on a path is equipped with a wavelength converter, the wavelength continuity is not required. However, the technology of all-optical wavelength conversion is not matured yet. [1]

Much research has been done in obtaining the call blocking performance of single-fiber WDM networks. A Markov chain (MC) model with the consideration of link-load correlation in [3] is accurate, and has a moderate complexity. As pointed out in [2], the MC model is an approximate model, because the arrival rate varies with the stat of the Markov chain. A simple analytical model is developed in [4] for two-hop

The research featured in this paper is sponsored by Korea Science and Engineering Foundation (KOSEF), and partially supported by Korean Ministry of Information and Communication (MIC).

I. Chong (Ed.): ICOIN 2002, LNCS 2343, pp. 337–345, 2002.

and mulithop paths. A more general model that can be used in any topology is proposed in [5]. Since link-load correlation is not considered, the analysis may not be accurate for sparse network topologies. There has also been considerable interest in analyzing the blocking performance of multifiber WDM networks. The independent wavelength load model in [6] is extended to multifiber networks in [7]. The results of this model are not numerically accurate for Poisson traffic because of the assumption that the load on one wavelength is independent of those on the other wavelength on a link. However these independent models are not accurate. It overestimates the blocking performance for $F = 1$ and underestimates for $F > 1$ in a mesh-torus network. The blocking models for first-fit wavelength assignment in [8] and [9] are also proposed to be applicable in multifiber networks. However, both of these models require intensive computation due to their iterative procedure to solve the Erlang fixed-point equation.

We study the effect on multiple fibers in circuit-switched all-optical WDM networks. Our study method is different from the ones used in [1], which assumes that traffic load of a network is approximated to a link network. We call it network correlation model. So we study the analytical model from network information matrices and use the Markov Random Field (MRF) model developed in [10]. In this model, calculation complexity of blocking probabilities in a network is partitions and approximated and we can use it with modest complexity.

2 Multifiber WDM Network Models

We consider the blocking probability models in WDM networks with two cases. First one is link correlation model and the other is network correlation model. This is based on the scope of calculation parameters; the former is on the line network and the later is network-wide one. The process can be divided into three parts. i) estimation of call state probabilities without wavelength constraint, ii) calculation of a route blocking probability with wavelength constraint. iii) last, network wide blocking can be estimated from the results in ii).

2.1 Link Correlation Models

We describe analytical model to compute the blocking performance of multifiber WDM networks in this session. In the analytical model, we assume call request is arrived on a route i as Poisson process with rate λ_i at each route and exponentially distributed call holding time with mean $1/\mu_i$ and then the utilization factor with $\rho_i = \lambda_i/\mu_i$. It is a big question that if the assumption of Poisson arrival is valid to the lightpath request process. But until more attractive process is found, we just use widely used assumptions for this paper. The arrival rate and service rate is based on the calls in specific route, so we let a route set R and its element r_i follows pre-assigned path, i.e. using fixed routing. A single fixed route is pre-selected for each source-destination ($s - d$) pair, and a wavelength assigned to a connection is uniformly randomly selected from the set of free wavelengths on that path. Let F_i be the number of fibers on a link i, and W_i be the number of wavelengths per fiber of link i. We assume that the W is same for all fiber in the network, but F can be different, vice versa. We also assume that an incoming request on one input port can be switched to any output port using OXC as long as the output port has the same wavelength free regardless of which fiber it is on. If the wavelength is not free on all of the F fibers at the output, the request is blocked on this wavelength. No wavelength converter is available at any node.

We define a light channel (LS) as a wavelength on a fiber on a link. A lightpath (LP) is a connection between a source-destination (s – d) pair using the same wavelength on all the links of a path. The LCs on a path may not be on the same fiber. Let a wavelength trunk (WT) w_i be a collection of the LCs/LPs using wavelength w_i as used in [1]. If there is F_1 fibers in a link, each lightpath requiring wavelength w_i can be use any fiber that have a free wavelength w_i; zero to F_1. This structure can be considered as the limited wavelength conversion capability.

In the line network case, we just consider about link correlations with the routes that use links in the network. For two-hop path, each link can be composed with different number of fibers, i.e., F_1 for first hop and F_2 for second hop. We can consider three bi-directional routes for each call that is on the two 1-hop and on 2-hop. The sum of every call on each link cannot exceed the WF, that is $C_{11} + C_{12} \leq WF_1$ and $C_{22} + C_{12} \leq WF_2$. C is the number of progressing calls in a route. Also we assume that the bi-directional link case, but we specify the direction of a link whenever needed.

Let's assume that in 2-hop links, each link is composed of different number of wavelengths due to the different number of fiber on each link. Then the steady state of number of progressing call in each route with the consideration of limited wavelength conversion capabilities is shown in (1).

$$\pi(c_{11},\, c_{12},\, c_{22}) = \frac{\dfrac{\left(\frac{\lambda_{11}}{\mu}\right)^{c_{11}}}{c_{11}!}\dfrac{\left(\frac{\lambda_{12}}{\mu}\right)^{c_{12}}}{c_{12}!}\dfrac{\left(\frac{\lambda_{22}}{\mu}\right)^{c_{22}}}{c_{22}!}}{\sum_{j=0}^{W\min(F_1,F_2)}\sum_{i=0}^{WF_1-j}\sum_{k=0}^{WF_2-j}\dfrac{\left(\frac{\lambda_{11}}{\mu}\right)^{i}}{i!}\dfrac{\left(\frac{\lambda_{12}}{\mu}\right)^{j}}{j!}\dfrac{\left(\frac{\lambda_{22}}{\mu}\right)^{k}}{k!}} \tag{1}$$

$$(0 \leq c_{11} + c_{12} \leq WF_1,\ 0 \leq c_{12} + c_{22} \leq WF_2)$$

Note that the number of fiber at each link is different than that in [1], so we consider that in (1) thereby c_{22} cannot exceed $W\min(F_1, F_2)$. In [2], the normalization factor contains the number of available wavelength through the two-hop network. But to calculate the normalization factor, another approximation is needed to satisfy the time reversibility in Markov chain. The choice of normalization factor and state variable depends on the application to use or the calculation complexity.

In (1) we can consider the link correlation in line network with 2-hop. So line network that have more than 2-hop is can be treated with the recursive form of (1) in [1], or decomposition technique that is described in [2]. For n-hop line networks, there can be limited wavelength conversion capability due to multifiber configurations. With the considerations of multifiber effect that give a limited conversion capabilities, we should find the number of ways to arrange x calls in F fibers link given W wavelengths. Among the available light channels, given F fibers and W wavelengths for each fiber, the possible way to arrange the x calls on WF light channel is, as given in [1],

$$f(x,F,W)=\begin{cases} 0 & x>(W-1)F \\ \binom{WF}{x} & x<W \\ \binom{WF}{x}-\sum_{i=1}^{\lfloor x/W \rfloor} f(x-iW,F-i,W) & \text{o/w} \end{cases} \tag{2}$$

then the probability that k wave trunks are available along an h-hop route is given recursively by

$$q^{(h)}(k)=\sum_{k'=0}^{|F_{\{1,\dots,h-1\}}|}\sum_{n_h=0}^{|F_h|} R(k\,|\,n_h,k',F_{\{1,\dots,h-1\}},F_h)\cdot q^{(h-1)}(k')\cdot p_h(|F_h|-n_h) \tag{3}$$

where $F_{\{1,\dots,h\}}$ is the set of wave trunks which have available wavelength λ_i exist on all links from hop 1 to hop h and $|F_{\{\cdot\}}|$ is the number of usable wave trunks through the hops, and $p_{h(x)}$ is the probability that last h-th hop have x wavelength trunks. To calculate $R(\cdot)$, we can use (2) and (3) with combinational wavelength matching following wavelength continuity constraint to calculate the conditional probability.

2.2 Network Correlation Model

In case of network-wide correlation, we can use the link-correlation model in (1) or the independent model. It affects on the accuracy of the equations. Also, some papers propose network partitioning that decomposes the mesh network into several line networks. But these modes use Erlang fixed point approximation (EPPA) to find the steady state probability in segmented 1-hop or 2-hops line network. This is the method reducing the load of route when the route is congested. As given in (1), The link correlation model as well as the network correlation model can use the exact distribution that is given by $G(C)$

$$\pi(\mathbf{n})=G(C)^{-1}\prod_{r\in R}\frac{\rho^{n_r}}{n_r!} \tag{4}$$

, where n is the vector of each link state and $G(C)$ is normalizing constant so that the distribution sums to unity. R is fixed (predefined) route set. Ideally the normalizing constant would be calculated by summing the distribution over every allowable state in S that possible value of n. However the state space for this process grows rapidly in the number of links and wavelengths, thereby calculating exact determination G a formidable task. [10] In the viewpoint of network, many routes are correlated with links. So we can't evaluate the exact Markov chain because there are too many state and normalization factors are hard to evaluate. To mention the network state, we will use some additional notations.

A network can be represented by simple finite non-directed graph $G = [V, J]$. Let the set V whose elements are called node (vertex) and set J whose elements are pair of nodes (edges), called edges. If $j = \{s, d\}$ is an edge of G, then nodes s and d are adjacent or neighboring. Also, we use R to denote the route set that is predetermined (fixed routing), respectively. A route of length l means a sequence of l edges $r = (u1, u2, \ldots, u_l)$ such that edge u_i has one common endpoint with $u_{i\text{-}1}$ and a second common endpoint with u_{i+1} and such that two consecutive edges of the sequence are different. The link capacity set $C = \{c_j \mid j = 1, 2, \ldots, J\}$ is matched with link j that means, in this case, the number of fiber on each link with number of wavelengths in each fiber (i.e. $c_j = W_j \cdot F_j$). The equivalence condition between two nodes is that a path between two nodes is exist with finite hops and we call the two nodes are accessible from each other. The equivalence class introduced on S from a partition of S_i is

$$\varphi_{R_{K'}}(\mathbf{n}_{R_K}) = \{n_{R_{K'}} \in S_{R_{K'}} : (\mathbf{n}_{R_K})_r, \forall r \in R_K\} \tag{5}$$

and all node are accessible from other nodes. This equivalence class concept is used later in this paper.

Actually, we want to know the blocking probabilities on arbitrary topology network. To do this, we can partition the network with line networks and star networks or just treat all partitions as star network. We shall use R_K denote the set of routes which use a link in $K \subseteq J$, and ∂R_K as the set out routes requiring links both within and outside of K. As described in section 2.1, calls requesting on route r_i arrive as a Poisson stream of rate λ_i independent of every other route and independent of the networks state. A newly arriving call is blocked and lost if there is no circuit for the call duration, or if there are circuits, there is no wavelength that is continuous through the required route. A call arrival rate and the call holding times are same as the line networks case.

Important thing we must consider is that, the process has spatial properties since it is the interactions between resources caused by competing calls on overlapping routes that makes the process difficult to analyze. In a recent paper, Zachary & Ziedins (1999) [10] describe a new approach, highlighting the application of Markov random field theory to the problem. They consider the equilibrium distribution as a finite random field on the set of resources (alternatively routes) and not that the process is Markovian with respect to various neighbor relations. The neighbor scheme between two nodes is that there exist the routes that contain two nodes in its path. And if two nodes, i and j, are neighbor, we simply write $i \sim j$. Then that induces the graph in which the links are the nodes and two resources, s and d, are connected if they are used by at least on common route. Then the graph $G = [J, \sim]$ can be described by Markov Random Field. We will re-introduce some preliminary results given in [10].

The normalizing constant θ_K for marginal distribution of $\mathbf{n}_{R_K}$ depends only on the number of calls active on the routes in the boundary set of K. Formally,

$$\pi_{R_K}(\mathbf{n}_{R_K}) = \theta_K(\mathbf{n}_{\partial R_K}) \prod_{r \in R_K} \frac{\rho_r^{n_r}}{n_r!} \tag{6}$$

This gives a clue the Markov random field method. It partitions the exact distribution equation in (4) into sub-partition K. And the computational complexity to get $\theta_K(\mathbf{n}_{\partial R_K})$ is easier than $G(C)$ in (4). Then when a sub-network, K, partitions the associated graphs external nodes (those not in K) into distinct groups, $G_1, G_2, \ldots, G_d$, those are equivalence class described in (5), the normalizing constant has a product from with each group contributing a factor

$$\theta_K(\mathbf{n}_{\partial R_K}) = \prod_{i=1}^{d} \theta_i\left(\mathbf{n}_{\partial R_K \cap R_K}\right) \tag{7}$$

Using (7), we can get the normalizing constant of sub-network K just regarding the routes that is in the set $R_K \cap R_{J \setminus K}$. This set means that only the routes across the K are considered in the normalizing constant.
Let the neighbor routes mapping function

$$\varphi_{R_{K'}}(\mathbf{n}_{R_K}) = \{n_{R_{K'}} \in S_{R_{K'}} : (\mathbf{n}_{R_K})_r, \forall r \in R_K\} \tag{8}$$

, then by saying $\mathbf{m}_{R_{K'}} \in \varphi_{R_{K'}}(\mathbf{n}_{R_K})$ we mean the components of $\mathbf{m}_{R_{K'}}$ and $\mathbf{n}_{R_K}$ agree for all the routes they have in common. Consider the subsets of J, $K' = \{j, k\}$, and $K = \{j\}$, for any two links such that $j \sim k$. Then, from (7)

$$\pi_{R_K}(\mathbf{n}_{R_K}) = \prod_{l:l \sim j} \theta_{jl}(\mathbf{n}_{R_j \cap R_l}) \prod_{r \in R_j} \frac{\rho_r^{n_r}}{n_r!} \tag{9}$$

and finally the coefficient is shown in (10).

$$\theta_{jk}(\mathbf{n}_{Rj \cap R_k}) = \sum_{\mathbf{m}_{R_{K'}} \in \varphi_{R_{K'}}(\mathbf{n}_{R_K})} \prod_{\substack{l:l \sim k \\ i \neq j}} \theta_{kl}(\mathbf{m}_{R_K \cap R_l}) \prod_{r \in R_{K'} \setminus R_K} \frac{\rho_r^{m_r}}{m_r!} \tag{10}$$

By using (9) and (10), we can calculate the normalization factor for the sub-network with route r that uses the links in K. It is the kind of network partitioning for the calculation of wavelength usage on each fiber, $\pi(\mathbf{n})$. It is worth to note that the formula in (10) gives exact solution if the subcomponent, K, partitions other networks to be isolated, like the root node of a tree network.

Another factor that effects on the blocking probability on WDM network is the wavelength continuity constraint. A route with several links may have available wavelength trunks. If there are some available wavelength trunks on the links along a route, we can check the existence of wavelength available for all links on the route r link by link and check them with between merged links and so on. Especially, in multifiber environment, the number of available wavelength trunks on each links can be different one by one. Then we just use the formula like (1) and (2) and the blocking probability, for a route r, can be described by the state probability of the number of usable wavelength in the fiber on the route and the probability of a lightpath establishment with wavelength constraint. To capture the wavelength trunk concept, we check the number of usable wavelength by using the equation in (2), and then the probability of k wavelength trunks are available through the route is obtained with (3). To count the number of wavelength through a route, some pre-calculation is needed to find the $F_{\{\cdot\}}$.

$$q^{(h)}(k) = \sum_{k'=0}^{|F_{\{1,\dots,h-1\}}|} \sum_{n_h=0}^{|F_h|} R(k \mid n_h, k', F_{\{1,\dots,h-1\}}, F_h) \cdot q^{(h-1)}(k') \cdot \pi(\mathbf{n}_{\{1,\dots,h\}}) \tag{11}$$

So we can claim that the blocking with using two probability formulas, (6) and (11), described in previous, can be calculated for a call on a route or the network wide.

3 Numerical Results

We applied the MRF methods with n nodes ring topology. In case of the ring networks, we can partition the network with n sub-networks, $K_1, K_2, \dots, K_n$, with respect to the links $\ell_1, \ell_2, \dots, \ell_n$. In a sub-network, we just consider the routes that use the link in the sub-network. In this case, a sub-network matches a link, one by one.

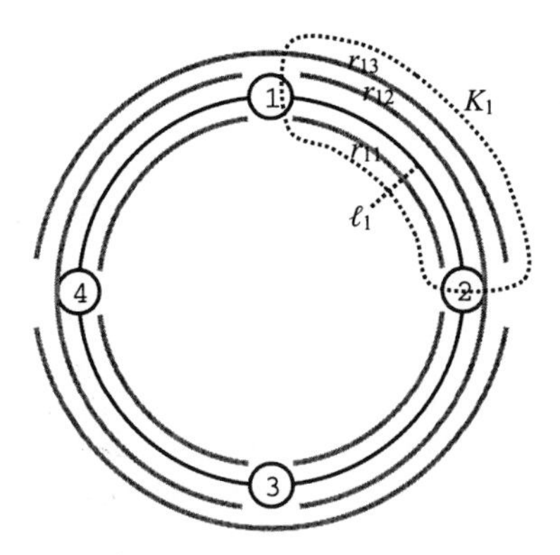

Fig. 1. Bidirectional Ring Network

In fig. 1, there are two types of route according to the number of hops they are using. Then we can calculate the blocking probabilities with regarding to the type of route.

Now we should compare our method with other methods those are proposed in the papers. We consider the multifiber case to apply the model, so we just compare the result given in [1]. A ring network is considered and we choose the parameters following the one given in [1]. Also, to get the result, the number of nodes are actually set to 16 and the route is choose according to the SPF algorithms with the maximum physical hop fixed to 4 for all routes. The result graph is given in fig. 2. The traffic load, ρ, is 0.4 for all routes.

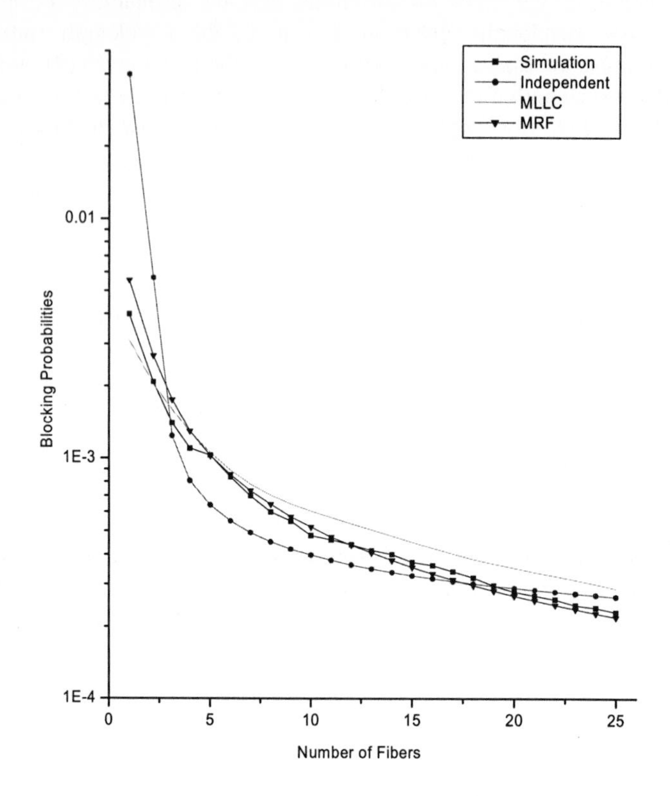

Fig. 2. Blocking probability for the ring network

The models considered in here are Independent mode given in [7], MLLC model given in [1] and our MRF model. As you can see in Fig. 2, MRF model is well done even the case that the number of fibers is increased. The gap between the simulation results and the MRF model is little larger in small number of fibers in each link. But as the number of fiber is increased, the gab is decreased. Also, the important factor, the computational complexity is given in [10].

We still evaluate our methods to other parameters such as number of wavelengths on the fiber, number of nodes in ring, network, other network topologies, and so on.

4 Conclusions

We've studied the effect of link correlation in arbitrary topology network. Also we consider the WDM network model with multifiber configurations. To evaluate the blocking performance of such networks, we have introduced the MRF theory to the network and so we taking the network wide correlation into account by decompose the network such as star networks and ring networks. We have shown that the model is accurate for variety network topologies by comparing the analytical results to the simulation results. Another consideration factor for us is the need of more accurate blocking probability model in WDM networks. We can guess that the more wavelength channel capacity is increased, the more we need to predict the network performance including the blocking probabilities. Also, as the number of fiber is increased in a network, we may want to get more accurate prediction to the better utilization of such huge network. The MRF model can be the one of the choice depending on the application.

References

1. Ling Li, A.K. Somani, "A New Analytical Model for Multifiber WDM Networks", IEEE Journal on Selected Areas in Comm. Vol. 18, 10, Oct, 2000, pp. 2138-2145
2. Yuhong Zhu, G.N.Rouskas, H.G. Perros, "Blocking in wavelength routing networks .I. The single path case", INFOCOM '99. Eighteenth Annual Joint Conference of the IEEE Computer and Communications Societies. Proceedings. IEEE , Volume: 1 , 1999, Page(s): 321 -328 vol.1
3. S. Subramaniam, M. Azizoglu, A.K. Somani, "All-optical networks with sparse wavelength conversion", Networking, IEEE/ACM Transactions on , Volume: 4 Issue: 4 , Aug. 1996, Page(s): 544 -557
4. J. Yates, J Lacey, D. Everitt, M. Summerfield, "Limited-range wavelength translation in all-optical networks", INFOCOM '96. Fifteenth Annual Joint Conference of the IEEE Computer Societies. Networking the Next Generation., Proceedings IEEE , Volume: 3 , 1996, Page(s): 954 -961 vol.3
5. T. Tripathi, K.N.Sivarajan, "Computing approximate blocking probabilities in wavelength routed all-optical networks with limited-range wavelength conversion", INFOCOM '99. Eighteenth Annual Joint Conference of the IEEE Computer and Communications Societies. Proceedings. IEEE , Volume: 1 , 1999, Page(s): 329 -336
6. R.A. Barry, P.A. Humblet, "Models of blocking probability in all-optical networks with and without wavelength changers", Selected Areas in Communications, IEEE Journal on , Volume: 14 Issue: 5 , June 1996, Page(s): 858 -867
7. G.Jeong, E.Ayanoglu, "Comparison of wavelength-interchanging and wavelength-selective cross-connects in multiwavelength all-optical networks", INFOCOM '96. Fifteenth Annual Joint Conference of the IEEE Computer Societies. Networking the Next Generation., Proceedings IEEE , Volume: 1 , 1996, Page(s): 156 -163 vol.1
8. A. Mokhtar, M. Azizoglu, "Adaptive wavelength routing in all-optical networks", Networking, IEEE/ACM Transactions on , Volume: 6 Issue: 2 , April 1998, Page(s): 197 -206
9. E. Karasan, E. Ayanoglu, "Effects of wavelength routing and selection algorithms on wavelength conversion gain in WDM optical networks", Networking, IEEE/ACM Transactions on , Volume: 6 Issue: 2 , April 1998, Page(s): 186 -196
10. S. Zachary, I. Ziedins, "Loss networks and Markov random fields", J. Appl. Probab. 36, 1999.

Instability Analysis for OBGP Routing Convergence in Optical Internet*

Sangjin Jeong and Chan-Hyun Youn

Information and Communications University, 58-4, Hwamam-Dong, Yuseong, Daejon, Korea, 305-732
{Sjjeong, Chyoun}@icu.ac.kr

Abstract. Optical Border Gateway Protocol (OBGP) is an extension to Border Gateway Protocol (BGP) for the Optical Cross Connects (OXCs) to be automatically setup and multiple direct optical lightpaths between many different autonomous domains. With OBGP, the routing component of a network may be distributed to the edge of the network while the packet classification and forwarding is done in the core. However, it is necessary to analyze the stable convergence functions of OBGP in case of lightpath failures. In this paper, we first describe the architecture of OBGP model and analyze the potential problems of OBGP, e.g. convergence behavior virtual BGP router in the lightpath failure. We then propose an OBGP convergence model derived from inter-AS relationship. The evaluation results show that the proposed model can be used for setting stable OBGP routing policy.

1 Introduction

Currently, optical networks are primarily used for the interconnection of large network domains such as enterprise networks, Internet Service Providers (ISPs), GigaPOPs and so on. Most of these networks already use Border Gateway Protocol (BGP) to manage the interconnection of their respective networks. More importantly, these large enterprise customers and ISPs are likely the first to use dark fiber and operate Dense Wavelength Division Multiplexing (DWDM) networks. Therefore, routing protocols for inter-domain networking might also be useful for interconnecting optical networks.

Instead, Optical BGP (OBGP) is a proposed extension to BGP for the manipulation of Optical Cross Connects (OXCs) to permit them to be automatically setup and configured as BGP speaking devices to support multiple direct optical lightpaths among many different autonomous systems (ASs). With OBGP, the routing component of a network may be distributed to the edge of the network, while the packet classification and forwarding is done in the core. OBGP also allows customers at the edge to control a subset of lightpaths within another network's wavelength cloud, so that customers can manage their own lightpath routing within that cloud. With the large number of adjacencies possible using OBGP lightpaths themselves may be used as a direct peering and transit mechanism between consenting ISPs. The

* This work was partially funded as part of KOSEF-Optical Internet Research Center.

I. Chong (Ed.): ICOIN 2002, LNCS 2343, pp. 346-356, 2002.

proposed protocol extensions allow carrier free networks, where the customers at the edge control and route lightpaths directly across an optical wavelength cloud [1][2].

These architectural approaches of customer empowered networks may require a fundamentally different architecture from the traditional ones. For example, caching and multi-homing can provide better reliability than fast restoration and protection on individual optical link for enterprise customers. Interconnection and direct peering also allow the enterprise or small ISP network to bypass the traditional hierarchical carriers and ISPs to establish direct peering with destination ISPs. One possible solution is to treat each OXC as a direct path between a pair of OBGP speakers. However, this significantly increases the session complexity of the OBGP, particularly for multiple parallel lightpaths. The alternative solution is to treat each OXC as an independent virtual BGP router with one input and output port, respectively. A virtual BGP router can then be set up for each OXC and separate OBGP sessions are initiated with its peers. This approach is much more scalable as each virtual BGP router configuration can be easily cloned from other virtual BGP routers [2].

The key for OBGP to scale to a very large Internet lies in the stability of inter-AS routing. If routes between ASs vary frequently — a phenomenon termed as flapping — then the virtual BGP routers spend a great deal of time to update their routing tables and to propagate the routing changes. Unstable inter-AS routing can cause unstable end-to-end routing [3]. The analysis of both the topology of routing system and the instability of routing system is important to evaluate network performance between end users. Since the customers of each network are connected to domains and provided various services, it is known that the instability of domains can severely affect the performance of their customers. In this paper, we discuss a routing convergence model to reduce the instability effectively in optical Internet.

2 Architecture of OBGP Model

OBGP is intended for the customers to control the routing of their lightpaths through another entity's optical wavelength cloud, as an overlay to an interior wavelength management protocol. OBGP allows the customer's topology to take precedence over the carrier's preferred topology as shown in Fig. 1. Large single domain wavelength clouds become unmanageable and are too difficult to optimize for traffic engineering purpose as a large single domain. The common solution is to break those single domains into many small domains that individually can be optimized. In every domain, as a more modest optimization mechanism, OBGP could be used [2].

OBGP requires a simple OXC switch as depicted in Fig. 2. OBGP routers with multiple paths in OXC paths are given preference over any path that goes through an electrical forwarding engine using standard BGP techniques for selecting shortest AS path, local preferences, and others. To multiplex and demultiplex wavelengths, Router B must use optical filters to separate out the individual wavelength. By using a simple optical switch, the individual light path can be treated in effect as an alternative path to router B. There are two ways to configure Router B. One is to treat each OXC as a direct path between a pair of BGP speakers. However, this significantly increases the complexity of any single BGP session, particularly, for many parallel lightpaths.

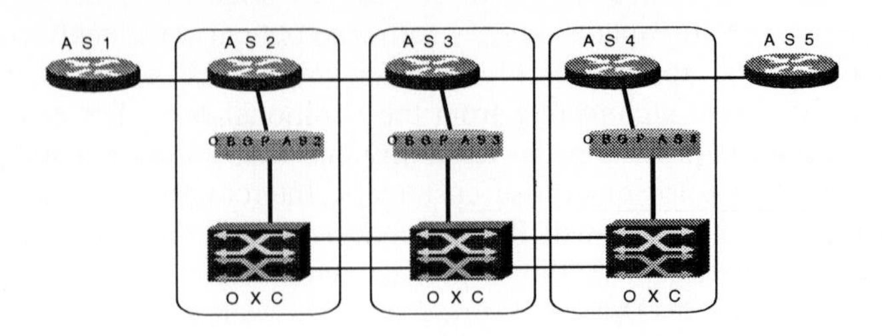

Fig. 1. OBGP configuration in optical Internet

Another is to treat each OXC as an independent virtual BGP router with only one input port and one output port. A virtual BGP router can then be set up for each OXC and separate BGP sessions initiated with its peers. This approach is much more scalable as each virtual BGP router configuration can be easily cloned from other virtual BGP routers [2]. To date there has been little effort in addressing the requirement for configuring, setup and managing wavelengths between domains, and allowing enterprises to manage their own wavelength configuration across a wavelength clouds. The conventional solution to date is for a carrier to operate a wavelength cloud and offer a managed lightpath service to the customers at the edge as shown in Fig. 2.

The main operations of OBGP consist of two phases. The first phase is the lightpath reachability phase. During this phase, sites advertise the availability of the optical lightpath to their sites through BGP. These announcements contain information on the OXC and the available lightpath through the OXC. The information is encoded using multi-protocol BGP extensions and extended community. This first phase allows sites to build up a lightpath Routing Information Base (RIB) that is used to determine if a lightpath is available across a number of OXC in different sites. The second phase is the lightpath establishment. This phase uses the information received from the lightpath reachability phase and then uses a BGP UPDATE message to communicate the lightpath establishment with the OXC sites on the path [1][2].

3 Potential Problems of OBGP with Virtual Router

The basic concept of virtual BGP routers is to treat each and individual OXC as a separate BGP router. The virtual router advertises itself independently of Router B in Fig. 2 with its own loopback address and its own set of IP address for its interfaces. Contrary to a normal BGP multi-router configuration, the virtual BGP router does not establish any Internal BGP (IBGP) connectivity even though it is within Router B's AS. It acts and behaves as an independent router by carrying its own set of routes, metrics, and others. The use of a virtual router for each OXC allows us to use standard BGP routing while no modifications are necessary to support optical lightpaths. In fact, the virtual BGP router assigns its own private (or public) AS such that AS path metrics are used for basic traffic engineering [2].

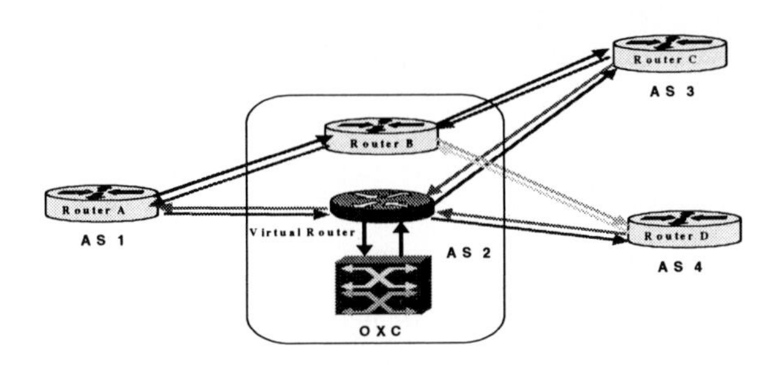

Fig. 2. Virtual router for OBGP

By instantiating a virtual BGP router, at first the owner of the OXC can establish OXC between neighbors that reduce the load on its electrical forwarding engine. Over time it can reconfigure the virtual BGP router to interconnect with other neighbors if traffic patterns change. More importantly, the owner of the OXC can establish OXCs between neighbors automatically. More intriguingly the virtual BGP router can also be easily reassigned into other router's AS domains. The main purpose of the OBGP OXC is to announce routes, perform route filtering, classify and provide standard BGP traffic network engineering capabilities to OBGP peers. As there is only one input and output port, there is no need to create a forwarding table within the OXC.

The loopback address for the virtual router used for OBGP connectivity is not same as the data forwarding address of the OXC. As such, under normal circumstances the virtual router does not aware any failures on the optical cross connect link between Routers A and C in Fig. 2. Therefore, if the interface card on either Router A or C detects a link failure, it immediately terminates the OBGP session with the neighbor – the virtual router. Either router A or C can terminate the session by sending an OBGP NOTIFICATION message to the virtual router. The virtual router, then updates its routing information database and sends Network Layer Reachability Information (NLRI) UPDATE messages to the other edge router indicating that the those addresses are unreachable across the OXC. Once the problem has been cleared, the router tries to re-establish the link across the OXC. To re-establish the link, three routers can re-initiate the OBGP sessions between the virtual router and routers A and C. The re-initiation of the BGP sessions start immediately after the receipt of the NOTIFICATION message even before the link failure has been cleared [2].

These processes demand OBGP update for topology information exchanged among domains. OBGP updates contain the reachability information to destination IP address prefix. Each OBGP UPDATE message can be classified into two types; route announcements and route withdrawal. Route announcement describes that a router has either learned of a new network attachment or has made a policy decision to prefer another route to a network destination. Route withdrawal is sent when a router makes a new local decision that a network is no longer reachable. Furthermore, each OBGP update message contains AS_PATH list to store traversed domains. Each AS_PATH is a list of identifiers for domains traversed by route [7]. Therefore the OBGP operation follows to keep the network reachability when the virtual router does not be aware any failures on the optical cross connect link. This can give rise to routing instability in OBGP by excessive update messages in case of the failure of virtual router or OXCs.

3.1 Instability Propagation

Internet routing instability is an important problem currently facing the Internet engineering community. High levels of network instability can lead to packet loss, increased network latency and time to convergence. At the extreme, high levels of routing instability have led to the loss of internal connectivity in wide-area networks [3]. Since OBGP is not deployed in the global Internet, it is hard to experimentally analyze the behaviors of OBGP. It is sufficient to evaluate BGP instead of analyzing OBGP behaviors. In order to analyze the propagation behavior of OBGP routing instability, we consider the instability propagation scheme based on inter-AS relationship suggested by Gao [4]. In [4], two ASs that exchange traffic was classified as one of customer-to-provider or peer-to-peer relationship.

Our analysis is based on BGP data collected in the global Internet for two years since 1998. These data are provided by NLANR [5]. Fig. 3 depicts the mapping of example inter-AS topology to leveled topology. Left side of figure and right side of figure indicates example inter-AS topology and leveled topology, respectively.

The level of each domain is defined as follows.

- Level 0 domain means the source of BGP routing instability. Level 1 domain represents the neighbor having peer-to-peer relationship with level 0. Level 2 domain depicts the neighbor having customer-to-provider relationship with level 0 or level 1 AS.

Fig. 4 shows Temporal Topology Variation (TTV) of one of the peer-to-peer AS of AS1239. TTV denotes absolute value of degree difference between consecutive days. AS1239 is identified most instable during 1998 [6]. According to our analysis, other neighbors of AS1239 show similar behavior. As we can see in the Fig. 4, level 1 ASs show rapid domain degree variation on 137th day (May 20 1998) same as AS1239. From the figure, we assume that the high instability of certain AS can be generated by the effect of peer-to-peer relation between AS pairs.

Table 1 summaries domain degree distribution of several neighbors ASs of AS1239. These ASs have peer-to-peer relationships with AS1239. Namely, peer-to-peer ASs of AS1239 shows rapid domain degree change. This behavior implies that high BGP routing instability of certain AS can impact on the domain degree of its peer-to-peer AS. Furthermore, with the results of Govindan and Reddy [9], we can classify certain domain as backbone nodes if their degree is over 28, and can understand that domains having peer-to-peer relationship with AS1239 are backbone nodes.

Table 2 summarizes the analysis of domain degree distribution of ASs that are two hop away (level 2) from AS1239. To analyze degree change of level 2 domains, we select one of peer-to-peer domains of AS1239, then investigate the BGP peer ASs of selected domain. In this paper, we choose AS2548 as an origin of level 2 AS. We classify BGP peers of AS2548 into peer-to-peer and customer-to-provider AS.

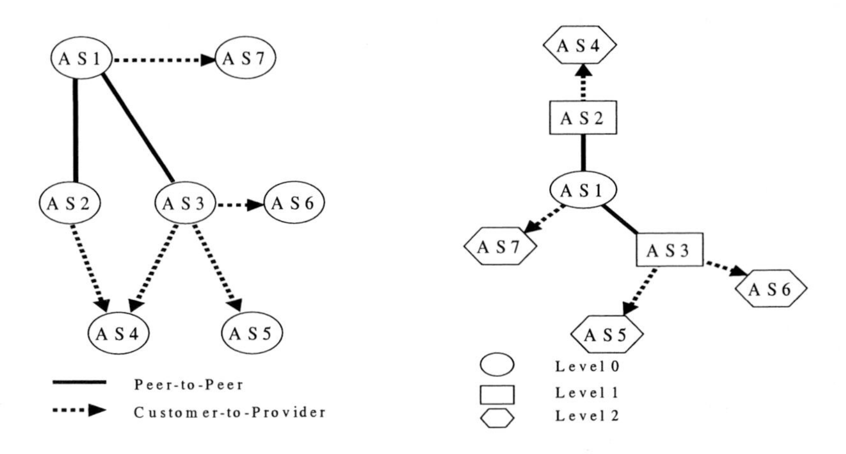

Fig. 3. Level architecture of inter-AS topology

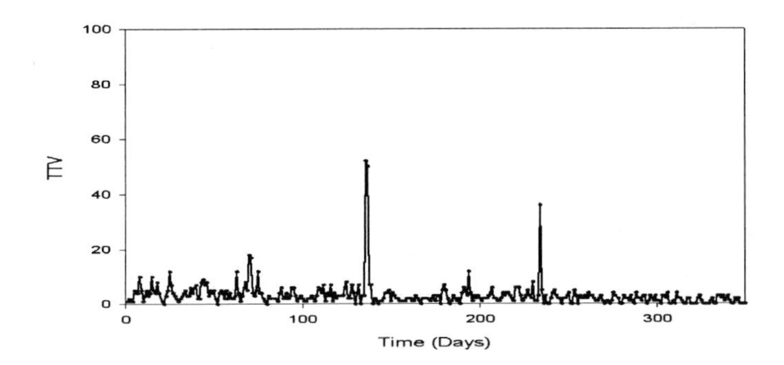

Fig. 4. Temporal topology variation of level 1 domain

According to analysis results, 92% of level 1 ASs show the change of TTV, i.e. the change of inter-AS topology, and 58% (or 92 customers) of level 2 ASs show the change of TTV. Since, as shown in the Table 2 and Table 3, the origin of rapid change of TTV on 137th day is the decrement of domain degree, we can interpret that 58% of level 2 AS experienced the decrement of domain degree on 137th day in 1998. Furthermore, according to [9], customer (or stub) domains have degree 1 or 2. Therefore, we can understand that the customer domains have lost the connection to their providers, and that the high value of TTV at level 1 ASs was due to the customers. In other words, the topology of backbone nodes having peer-to-peer relationship with each other does not change in the occurrence of BGP routing instability.

Table 1. Temporal topology variations in level 1 ASs of AS1239

AS Number	Domain degree (TTV)		
	May 19, 1998	May 20, 1998	May 21, 1998
1	205 (3)	153 (52)	201 (48)
701	802 (1)	514 (288)	806 (292)
1673	74 (0)	71 (3)	74 (3)
1740	101 (2)	92 (9)	103 (11)
2548	171 (7)	107 (64)	184 (77)
2914	137 (2)	125 (12)	135 (10)
3561	630 (6)	416 (214)	627 (211)
3847	48 (2)	35 (13)	51 (16)

Many researches on the topological characteristics of inter-AS routing system and the origins of Internet routing instability were reported in [3][8]. But there is few works about the propagation of Internet routing instability among domains and the development of systematic model that can represent the propagation of BGP routing instability. In this section, we analyze the propagation behavior of BGP routing instability by using inter-AS relationships suggested by [4] and propose BGP routing policy model to preserve stability in the inter-AS network during the occurrence of high level instability. It is important to analyze the propagation of instability, because it is possible to set up routing policy that can efficiently decrease the BGP instability and fast converge to stable network based on systematic propagation model of instability.

Table 2. Domain degree of level 2 ASs

Node Class	Number of domains (%)	Number of domains whose TTV is nonzero (%)	Number of domains whose TTV is zero (%)
Backbone (degree $\geq$ 30)	13 (7.6)	12 (92.3)	1 (7.7)
Customer	158 (92.4)	92 (58.2)	66 (41.8)

Arnaud et al. [2] discussed several possible schemes in the failure of lightpath. The following shows the approaches.

1. Static configuration at setup
2. Establish at configuration knowledge of the destination router
3. Use BGP UPDATE information such that Router A can make a dynamic decision

4 Model Description for OBGP Convergence Policy

For the time being, an AS may change the nature of its relationships with its neighbors. For example, a customer may grow large enough to renegotiate its relationship with a provider, and the AS pair may transit to a peer-to-peer relationship. As part of evolving to a new relationship, the two ASs may need to change their import and export policies. Ideally, these changes would occur simultaneously. However, in practice, each AS configures the routers independently with others. As a result, the BGP system may go through a transition period where one AS has changed its configuration and the other has not changed. Since these changes occur on a human time scale, it is important to carefully study the influence of the transition period on system stability.

In this section, we discuss OBGP convergence policy model based on state automaton for reducing Internet OBGP routing instability and verify our model with BGP data that are collected in the global Internet.

As proposed by [8], if AS_PATH in an OBGP routing information satisfies valley-free condition, it is possible to classify inter-AS topology into peer-to-peer and customer-to-provider relationships. On the other hands, these relationships translate into rules that determine whether or not an AS exports its best routes to a neighboring AS, e.g. normal export rules [4]. And the interaction of locally defined routing policies can have global ramifications for the stability of the OBGP system. Conflicting local policies among a collection of ASs can result in OBGP route oscillation. To avoid local oscillation, we consider policy model in the theorem.

4.1 Policy Management for Stable Convergence

The transition between states happens while traversing inter-AS topology. According to AS relationship between two ASs, there exist at most two possible transitions for each state. The output is generated during state transition. Output indicates the probability of pruning customer's logical connection to provider. Output consists of two values, 0 and p_{tr} , where 0 means that the provider does not prune the connection of customer that has customer-to-provider AS relation. p_{tr} represents the probability that the provider prunes connection of customer. Furthermore, while analyzing the state transitions in the automaton, the conceptual model of topology reconfiguration can be used to determine state transition probability p_{tr}. As proposed [4], the state transition probability p_{tr} can be determined by experimentally, by statistical analysis of large number of samples using inter-AS relationships.

Our proposed model can be applied to network topology as shown in Fig. 5. Let AS1 be the source of BGP routing instability. AS the instability propagates from AS1 to its neighbors, each AS sets its routing policy according to the AS relationship with its neighbor.

Namely, from the state automaton, the event α means that two ASs have peer-to-peer relationship and β means that two ASs have provider-to-customer relationship. In the above example, AS1 has one peer-to-peer AS (AS7) and two provider-to-customer ASs (AS2 and AS3). Since AS1 has peer-to-peer relationship with AS7, AS1 does not make a change of the edge for AS7. But in case of AS2 and AS3, the edge is provider-to-customer, so AS1 (provider AS) sets its routing policy to cut its customers according to the probability p_{tr}.

Therefore, by pruning its customers, AS1 can prevent customers from sending their traffic to itself, fast recover from high instability and control the propagation range of the instability. Furthermore, by choosing optimal p_{tr}, the BGP routing instability propagation range can be adjusted.

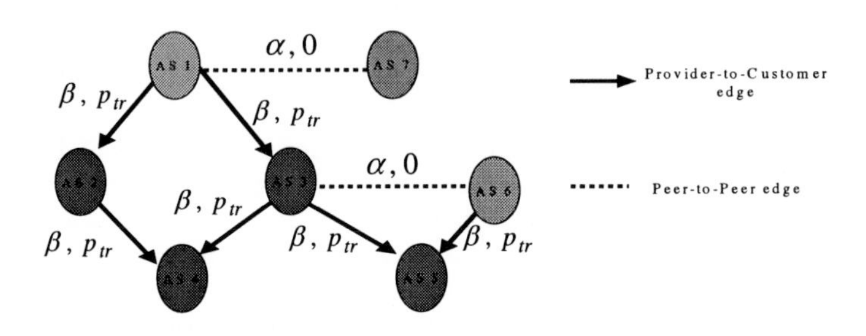

Fig. 5. Example of BGP routing instability propagation

Sang-Jin et al. [8] showed that the event probability for fast convergence policy is restricted by;

$$P(\textit{converge within Level n domain}) = 1 - \frac{(2 - p_{tr})^n}{2^n} \quad \textbf{(1)}$$

, where n is domain level and p_{tr} is given.

4.2 Experimental Analysis and Discussion

According to the characteristics of virtual BGP router, it is not easy to implement virtual BGP router. However, the virtual BGP router inherits the characteristics of BGP router, it is possible to analyze the behavior of BGP [1][2]. Therefore, before analyzing the proposed OBGP instability propagation model, we investigate the correlation of the number of BGP UPDATE messages by using BGP data measured at major IXPs in global Internet.

Since the data used in Fig. 6 and Fig. 7 were gathered once for every midnight in global Internet during two years, they show macroscopic behavior of BGP routing in global Internet. As we can see in the Fig. 6, the variation of domain degree follows regression line, i.e. it shows linearity in temporal domain.

Fig. 7 shows convergence rate according to various p_{tr}, the probability of instability absorption in provider-to-customer relationship, with respect to domain level.

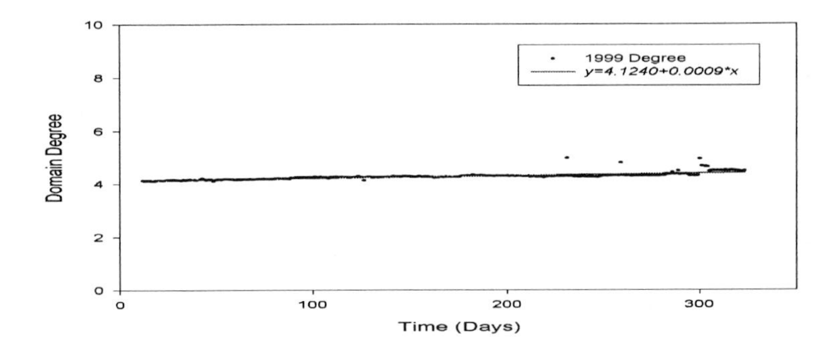

Fig. 6. Regression analysis of day-to-day distribution of domain degree in 1999

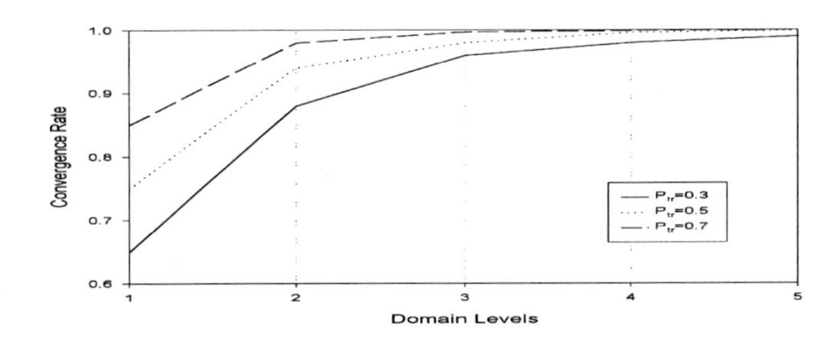

Fig. 7. Convergence rate versus domain level

From the figure, we can discover that the convergence rate of our proposed model increases with respect to the level of domains. This behavior implies that the routing instability decreases by propagating domains which have routing policy based on the proposed model.

Therefore, by Fig. 6, and Fig. 7, we claim that our proposed model can be used for setting up routing policy which converges to stable state of OBGP routing.

The propagation behavior of the instability can be used for setting up routing policy in domain. Without considering the propagation of the instability, high levels of routing instability in certain domain can cause the disconnection of its neighbor domains. However, our proposed model can be used for setting up routing policy in order to manage the propagation of routing instability.

The applicability of proposed model can be summarized as follows : when OBGP routing instability happens, by pruning some of customer connections, i.e. virtual BGP router peering, backbone domains reduce incoming customer traffic. Thus, backbone topology can be preserved, although OBGP routing instability occurs. In inter-AS routing system, it is possible to guarantee routing path between peer-to-peer ASs.

5 Conclusions

Internet routing instability is an important problem in the Internet engineering community. It is known that high levels of BGP routing instability can lead to packet loss, increased network latency and increased time to convergence. High levels of routing instability may lead to the loss of internal connectivity in wide-area networks at the extreme. However, there are few works on the propagation of the instability in inter-AS network and on the routing policy that can preserve the reliability despite of the occurrence of high level instability.

OBGP is an extension to BGP for the manipulation of OXCs to permit them to be automatically setup and configured as BGP speaking devices to support multiple direct optical lightpaths between many different autonomous domains. Since virtual BGP router which connects BGP router and OXC is a process which inherits the characteristics of BGP, the virtual BGP router is highly likely to show same behaviors i.e. routing instability.

In this paper, we discussed the instability considered in OBGP and proposed routing convergence model to reduce the instability in virtual BGP router. Moreover, Our instability propagation model and its analysis conclude that effective policy management scheme improves OBGP routing instability in case of lightpath failure.

References

1. Luciani, J., Rajagopalan, B., Awduche, D., Cain, B., Jamoussi, B., „IP over Optical Networks – A Framework", draft-ip-optical-framework-00.txt", Sep. 2000.
2. Bill St. Arnaud, Rene Hatem, Wade Hong, John Coulter, Marc Blanchet, Abdul Abdalla, Ian MacDonald, Florent Parent, Tom Tam, Mike Weir, „Optical BGP Networks", CA*Net3 News Archive, 2001.
3. C. Labovitz, A. Ahuja, A. Bose, and F. Jahanian, „Delayed Internet Routing Convergence", in Proc. ACM SIGCOMM, Aug. 2000.
4. Lixin Gao, „On Inferring Autonomous System Relationships in the Internet", in Proc. IEEE GLOBECOM, Nov. 2000.
5. http://www.nlanr.net/
6. http://www.merit.edu/
7. Y. Rekhter, „A Border Gateway Protocol 4", Request for Comments, 1771, Internet Engineering Task Force, March 1995.
8. Sang-Jin Jeong, Chan-Hyun Youn, Tae-Sang Choi, Tae-Soo Jeong, Daniel Lee, and Kyoung-Seon Min, „Policy Management for BGP Routing Convergence Using Inter-AS Relationship", Journal of Communications and Networks, Vol. 3, No. 4, Dec. 2001.
9. R. Govindan, A. Reddy, „An Analysis of Internet Inter-Domain Topology and Route Stability", in Proc. IEEE INFOCOM, pp. 850-857, 1997.

IV. Network Performance Issues

On Modeling Round-Trip Time Dynamics of the Internet Using System Identification

Hiroyuki Ohsaki[1], Mitsushige Morita[2], and Masayuki Murata[1]

[1] Cybermedia Center, Osaka University
1-30 Machikaneyama, Toyonaka, Osaka 567-0043, Japan
{oosaki, murata}@cmc.osaka-u.ac.jp

[2] Graduate School of Engineering Science, Osaka University
1-3 Machikaneyama, Toyonaka, Osaka 560-8531, Japan
m-morita@ics.es.osaka-u.ac.jp

Abstract. Understanding the end-to-end packet delay dynamics of the Internet is of crucial importance since it directly affects the QoS (Quality of Services) of realtime services, and it enables us to design an efficient congestion control mechanism. In this paper, we measure the round-trip time, and build a mathematical model representing its dynamics using system identification. We first measure, as the input and output data for system identification, the packet inter-departure time from a source host and the corresponding round-trip time measured by the source host. ICMP (Internet Control Message Protocol) is utilized to measure the round-trip time for each packet. We next model the network, seen by a specific source host, as a dynamic SISO (Single-Input and Single-Output) system. Using measurement results obtained from three different network configurations, we investigate how accurately the round-trip time dynamics of the Internet can be modeled with the system identification.

1 Introduction

Understanding the end-to-end packet delay dynamics of the Internet is of crucial importance since (1) it directly affects the QoS (Quality of Services) of realtime applications, and (2) it enables us to design an efficient congestion control mechanism for both realtime and non-realtime applications. For non-realtime applications, a delay-based approach for congestion control mechanisms, rather than a loss-based approach as used in TCP (Transmission Control Protocol), has been proposed (e.g., [1, 2]). The main advantage of such a delay-based approach is, if it is properly designed, packet losses can be prevented by anticipating impending congestion from increasing packet delays.

For a long time, queueing theory has been extensively used as a powerful tool to analyze packet-switched networks. In general, the queueing theory assumes stationarity of the network, and allows us to obtain several performance measures such as the average packet delay and the average packet loss probability. However, the stringent limitation of the queuing theory is its difficulty to analyze the *dynamic behavior* of the network. Several measurement-based studies suggest that the end-to-end packet behavior in the Internet is quite dynamic [3–5]. Another approach, being different from the queueing theory, should therefore be taken to investigate the packet delay dynamics of the Internet.

I. Chong (Ed.): ICOIN 2002, LNCS 2343, pp. 359–371, 2002.

In [6], the authors have proposed a novel approach to model the end-to-end packet delay dynamics of the Internet. The main idea of the approach is treating the network, seen by a specific source and destination pair, as a *black-box*, and modeling the end-to-end packet delay dynamics using *system identification* [7]. The end-to-end packet delay dynamics are modeled as a SISO (Single-Input and Single-Output) system based on the ARX (Auto-Regressive eXogenous) model. The input to the system is a packet inter-departure time from the source host, and the output is a (one-way) end-to-end packet delay variation measured by the destination host.

This paper is a direct extension of [6], and primarily focuses on an applicability of our approach to real networks. However, there is a major difference in the modeling approach. In [6], the network seen by a specific source and destination pair is modeled as a black-box. On the contrary, in this paper, the network seen by a specific source host is modeled; that is, the output is the round-trip time variation instead of the end-to-end packet delay variation. Although modeling the round-trip time dynamics suffers from more measurement noise than modeling the end-to-end packet delay, the modeling approach taken in this paper is easier to implement, so that desirable for practical purposes.

After discussing advantages and disadvantages of several measurement methods for the round-trip time, we present a measurement method using ICMP (Internet Control Message Protocol) to collect the input and output data for determining the coefficients of the ARX model. Since almost all hosts and routers respond to ICMP packets, this method can be used in various network environments. We then collect the input and output data from real networks. Three network configurations are used including both wired and wireless LANs. Using the input and output data obtained, coefficients of the ARX model are determined using the least-square method. We investigate how accurately the ARX model can represent the round-trip time dynamics of the Internet.

This paper is organized as follows. In Section 2, we summarizes related works in recent publications. In Section 3, a black-box approach for modeling the round-trip time dynamics of the Internet using the ARX model is explained. In Section 4, we discuss several measurement methods of the round-trip time, in particular, for collecting the input and output data for system identification. Section 5 shows several measurement and modeling results, and discuss how accurately the ARX model can capture the round-trip time dynamics. In Section 6, we discuss several possible applications of our approach, followed by conclusion of this paper.

2 Related Works

In the literature, there have been several measurement-based studies regarding the end-to-end packet delay [3, 4, 8, 9] and the end-to-end path characteristics [5, 10]. In [3], the authors have examined the end-to-end packet delay and loss behavior in the Internet using small UDP probe packets. In [4], the authors have examined the correlation between packet delay and packet loss experienced by a continuous-media traffic source, based on measurements of per-packet delays and packet loss. In [8], a large number of TCP measurements have been used to discuss two estimation problems: estimation of the retransmission timer (RTO) for a TCP connection, and estimation of the available bandwidth. In [9], the

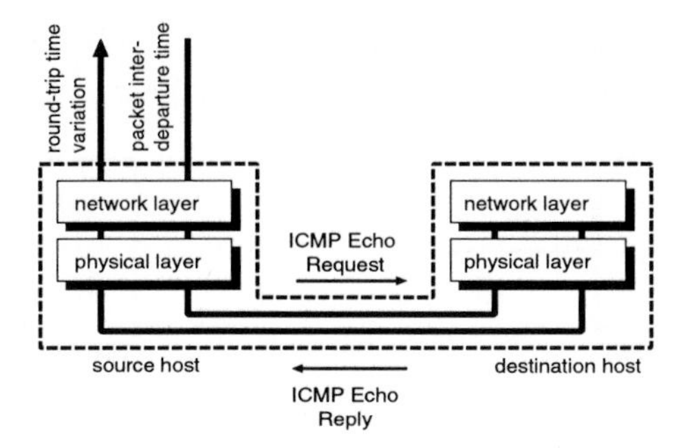

Fig. 1. Modeling round-trip time dynamics as SISO system

authors have presented an approach to characterize loss and delay characteristics of a transmission link based on end-to-end multicast measurements. In [5], the packet dynamics of the Internet have been analyzed based on measurements of about 20,000 TCP data transfers. In [10], the routing behavior of the Internet has been analyzed based on measurements of about 40,000 end-to-end traceroute results. However, those studies are limited to a statistical behavior of the end-to-end packet delays and/or path characteristics. In other words, the dynamics of the packet delay of the Internet, which is the main concern of this paper, has not been investigated.

Aside from analyses of the end-to-end packet delay, another area of measurement-based studies is regarding a black-box modeling of the network traffic [11–15]. In [11], the authors have proposed a traffic model for wide-area TCP traffic by characterizing several distributions of, for example, the packet inter-arrival time and the number of bytes transferred. In [12], the authors have proposed a fast algorithm to construct a CMRP (Circulant Modulated Rate Process) for traffic modeling. In [13], CMRP and ARMA (Auto-Regressive Moving Average) have been discussed as a traffic model. In [14, 15], a measurement-based tool for traffic modeling and queueing analysis has been developed, which uses CMPP (Circulant Modulated Poisson Process) for a traffic model. Those studies are closely related to our black-box modeling approach, but there is a significant difference. Those studies have focused on traffic modeling based only on outputs (i.e., observed amount of traffic). On the contrary, this paper focuses on modeling the round-trip time dynamics based on both inputs (i.e., packet inter-departure time) and outputs (i.e., round-trip time variation). In other words, this paper focuses on how the round-trip time of a packet sent from a source host is affected by its past packet transmission process.

3 Black-Box Modeling and System Identification

As depicted in Fig. 1, the network seen by a specific source host, including underlying protocol layers (e..g, physical, data-link, and network layers), is considered as a black-box. Our goal of this paper is to model a SISO system describing the round-trip time dynamics: i.e., the relation between a packet sending process from the source host and its resulting round-trip time observed at the source host. Effects of other traffic (i.e., packets coming from other hosts) are modeled as *noise*. As the input to the system, we use a *packet inter-departure time* from

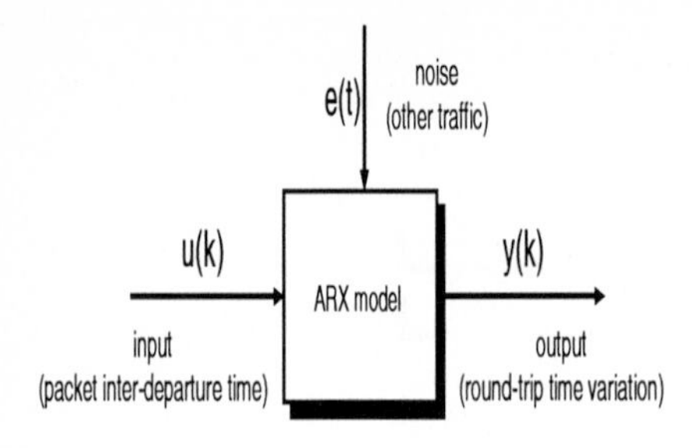

Fig. 2. ARX model for modeling round-trip time dynamics

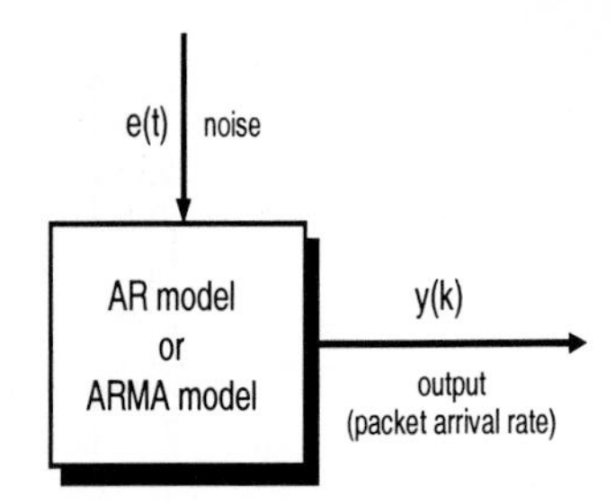

Fig. 3. AR model or ARMA model for modeling network traffic

the source host: i.e., the time interval between two consecutive packet transmissions from the source host. Use of the packet inter-departure time is straightforward since it directly affects the end-to-end packet delay. As the output from the system, we use a *round-trip time variation* measured by the source host: i.e., the difference in two consecutive round-trip times. We choose the round-trip time variation, instead of the round-trip time itself, as the output from the system. This choice is for reducing unstationarity of noise (i.e., effect of other traffic) on the measured round-trip time since the aggregated network traffic at a packet-level time scale is not stationary [16].

In this paper, the ARX model is used and its coefficients are determined using system identification [7]. Figure 2 illustrates a fundamental concept of using the ARX model for modeling the packet delay dynamics. The input to the ARX model is a packet inter-departure time from the source host, and the output from the ARX model is a round-trip time variation measured at the source host. Effects of other traffic (i.e., packets coming from other hosts) are modeled as the noise to the ARX model. Letting $u(k)$ and $y(k)$ be the input and the output data at slot k, the ARX model is defend as

$$A(q)\, y(k) = B(q)\, u(k - n_d) + e(k) \tag{1}$$

where $A(q)$ and $B(q)$ are given by

$$\begin{aligned} A(q) &= 1 + a_1 q^{-1} + \ldots + a_{n_a} q^{-n_a} \\ B(q) &= b_1 + b_2 q^{-1} + \ldots + b_{n_b} q^{-n_b+1} \end{aligned}$$

In the above equations, $e(k)$ is unmeasurable disturbance (i.e., noise), and q^{-1} is the delay operator; i.e., $q^{-1}u(k) \equiv u(k-1)$. The numbers n_a and n_b are the orders of polynomials. The number n_d corresponds to delays from the input to the output. For compact notation, ζ is introduced as

$$\zeta = [n_a, n_b, n_d] \tag{2}$$

In our case, $u(k)$ and $y(k)$ correspond to k-th packet inter-departure time and k-th round-trip time variation. All coefficients of the polynomials, a_n and b_n, are parameters of the ARX model, and are to be determined from input and output data using system identification. Refer to [7] for the detail of the ARX model.

Our approach of a black-box modeling using the ARX model is distinctive from other black-box approaches, which model network traffic using the

AR (Auto-Regressive) model or the ARMA (Auto-Regressive Moving Average) model [13, 17, 18]. Figure 3 illustrates a typical usage of the AR model or the ARMA model for modeling network traffic. Comparing Figs. 2 and 3, the ARX model has the input whereas either the AR model or the ARMA model does not. In other words, only the ARX model can represent the dynamics, i.e., how the past input data affects the future output data.

Note that the ARX model has a drawback in modeling the round-trip time dynamics; i.e., the ARX model is a linear time-invariant model, so it cannot rigorously capture non-linearity of the round-trip time dynamics. But it should be noted that the ARX model is applicable in various control engineering problems. This is because non-linear dynamical systems operating around the stable point can be well approximated by a linear system [7]. In Section 5, we will investigate how accurately the round-trip time dynamics can be described by the ARX model.

The system identification problem for the ARX model is formulated as a minimization problem, where the cost function is given by a loss function [7]. Because of space limitation, only the outline is shown in this paper, and interested readers should refer to [7] for more detail.

Let θ be a vector of all coefficients and $\psi(k)$ be a vector of all past n_a outputs and n_b inputs, respectively.

$$\theta = [a_1, \ldots, a_{n_a}, b_1, \ldots b_{n_b}]^T \tag{3}$$

$$\begin{aligned}\psi(k) = [&-y(k-1), \ldots, -y(k-n_a),\\ &u(k-n_d-1), \ldots, u(k-n_d-n_b)]^T\end{aligned} \tag{4}$$

Using Eq. (1), the output from the ARX model $\hat{y}(k|\theta)$ is given by

$$\hat{y}(k|\theta) = \psi^T(k)\,\theta \tag{5}$$

The loss function $V_N(\theta, Z^N)$ is defined as the sum of all squared prediction errors for N input and output data.

$$V_N(\theta, Z^N) = \frac{1}{N}\sum_{k=1}^{N}(y(k) - \hat{y}(k|\theta))^2 \tag{6}$$

where Z^n is the past input and output data defined as

$$Z^N = \{u(1), y(1), \ldots, u(N), y(N)\} \tag{7}$$

The solution $\hat{\theta}_N$ that minimizes the above loss function is easily obtained by the least squares method:

$$\hat{\theta}_N = \left[\sum_{k=1}^{N}\psi(k)\psi^T(k)\right]^{-1}\sum_{k=1}^{N}\psi(k)y(k) \tag{8}$$

4 Measurement Methods

For collecting the input and output data from real networks, it is necessary to send a series of probe packets into the network, and to measure their resulting round-trip times. For sending probe packets, one of the following protocols can be used.

- TCP (Transmission Control Protocol)
- UDP (User Datagram Protocol)
- ICMP (Internet Control Message Protocol)

In what follows, we discuss advantages and disadvantages of these protocols for sending probe packets to collect the input and output data, in particular, for system identification.

TCP has a feedback-based congestion control mechanism, which controls the packet sending process from a source host according to the congestion status of the network. Since it is an ACK-based protocol, it is easy for the source host to measure the round-trip time for each packet. However, TCP is not suitable for sending a probe packet because of the following reasons. First, for system identification purposes, the input data (i.e., the packet inter-departure time) should contain diverse frequencies. So the white noise, which equally contains all frequencies, is the ideal input data for system identification [7]. However, the packet inter-departure process of TCP would have limited frequencies. Second, most of system identification techniques assume an independence between the input and output data. However, because of a feedback-based nature of TCP, the packet inter-departure time is dependent on the past round-trip times, so the independence assumption cannot be satisfied with TCP.

On the contrary, UDP has no feedback-based control. The packet inter-departure time of UDP can be freely controlled. However, UDP is a one-way protocol. The destination host must perform some procedure to measure the round-trip time for each packet at the sender side. One possible way is to use *ICMP Destination Unreachable* message as in the *traceroute* program [19]. When the host receives a UDP packet to an unreachable port, it returns ICMP Destination Unreachable message to the source host. The source host can therefore measure the round-trip time by observing the elapsed time between the UDP packet transmission and the receipt of the corresponding ICMP packet. However, as specified in [20], generation of ICMP Destination Unreachable messages is limited to a low rate. Use of ICMP Destination Unreachable message is therefore not desirable to collect the input and output data for system identification.

ICMP is a protocol to exchange control messages such as routing information and node failures [21]. Since ICMP has no feedback-based control, the inter-departure time of ICMP packets can be freely controlled. Also it is easy to measure the round-trip time at the source host by using *ICMP Echo Request* and *ICMP Echo Reply* messages, as in the *ping* program. Although a part of network devices have a rate limitation for transmitting ICMP Echo messages [22], many network devices respond to ICMP Echo messages. So this method can be used in various network environments.

In this paper, we therefore choose ICMP Echo message as a probe packet. More specifically, the source host sends a series of ICMP Echo Request messages to the destination host, and the destination host returns ICMP Echo Reply messages. We have modified the ping program to dynamically change the packet inter-departure time (originally fixed at one second).

The detailed algorithm is described below.

Sender Algorithm (Source Host):

S1) Send ICMP Echo Request message of 1,500 bytes including IP and ICMP headers. The payload of the ICMP packet holds the timestamp of the packet transmission.

S2) Randomly choose the packet inter-departure time from the exponential distribution in order to schedule the next ICMP packet transmission.
S3) Go to S1.

Receiver Algorithm (Source Host):

R1) Wait for the receipt of ICMP Echo Reply message.
R2) Extract the timestamp from the payload.
R3) Calculate the round-trip time from the current time.
R4) Calculate the round-trip time variation from the previous round-trip time.
R5) Go to R1.

The destination host copies the payload of the received ICMP Echo Request message to the returning ICMP Echo Reply message. Thus, the ICMP Echo Reply packet contains the timestamp placed by the source host at its transmission time. This enables precise measurement of the round-trip time at the source host.

5 Modeling from Measured Data

Network Configurations

We have measured three sets of input and output data from the following three network configurations.

N1) 100 Mbps LAN without background traffic
N2) 100 Mbps LAN with background traffic
N3) 11 Mbps wireless LAN and 100 Mbps LAN

In the network configuration **N1**, both the source and destination hosts are directly connected to a single 100 Mbps switch. Because of a direct connection, there exists no background traffic, and the output data (i.e., the round-trip time variation) suffers from little observation noise. This network configuration enables us to investigate how accurately the round-trip time dynamics can be modeled in a high-speed and non-congested network.

The network configuration **N2** is a 100 Mbps LAN, which consists of five 100 Mbps switches connected in serial. The network configuration **N2** is a private LAN in our laboratory, where about 50 client computers and 10 server computers are connected. There are five switches between the source and destination hosts. Since intermediate switches process traffic from other computers, the round-trip time measured at the source host might be affected by existence of the background traffic. This network configuration is for investigating how the background traffic deteriorates the accuracy of the ARX model.

In the network configuration **N3**, the destination host is equipped with a 11 Mbps wireless LAN interface. The base station is connected to the network configuration **N2**. There are five 100 Mbps switches between the source host and the base station. In this case, the wireless LAN, which is much slower than 100 Mbps LAN, is the bottleneck. The round-trip time is expected to be significantly larger than other network configurations.

In each network configuration, we have collected both the packet inter-departure time $u(k)$ and the round-trip time variation $y(k)$ using the approach

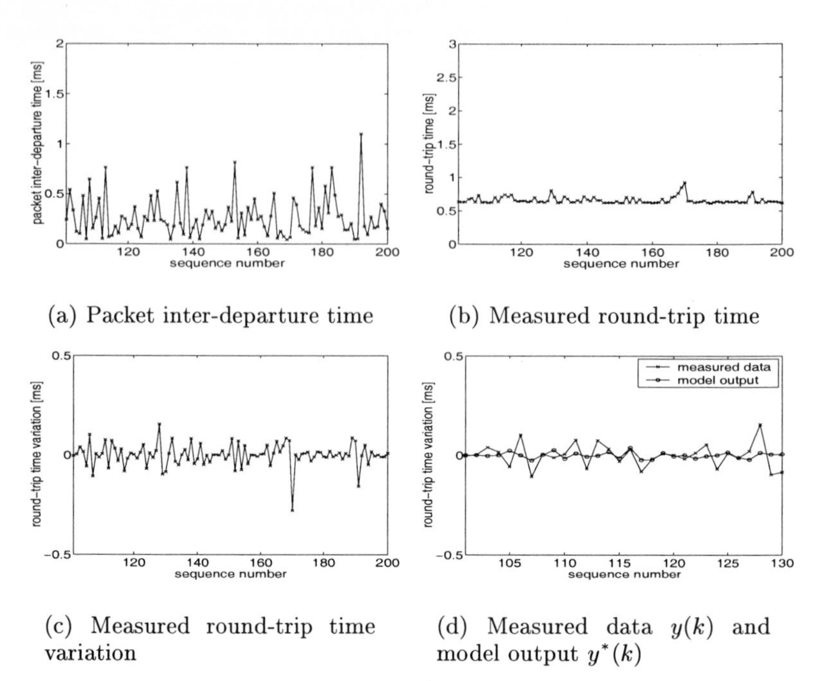

(a) Packet inter-departure time (b) Measured round-trip time

(c) Measured round-trip time variation (d) Measured data $y(k)$ and model output $y^*(k)$

Fig. 4. Results in network configuration N1 (100 Mbps LAN w/o Background Traffic)

described in Section 4. The source host sent 10,000 probe packets with an exponentially distributed inter-departure time. Note that lost packets are not included in the measured input and output data. Of all input and output data collected, we use the input and output data of 100 packet samples for coefficients determination and model validation of the ARX model. In what follows, we discuss how accurately the round-trip time dynamics can be modeled by the ARX model.

Network Configuration N1

In the network configuration **N1**, the mean packet inter-departure time has been set to 0.2 ms, resulting in the average packet transmission rate of 43.2 Mbps and the average round-trip time of 0.8 ms. Shown in Fig. 4 are results in the network configuration **N1** for $\zeta = [8, 8, 1]$. This figure shows: (a) the packet inter-departure time $u(k)$, (b) the measured round-trip time, (c) the measured round-trip time variation $y(k)$, and (d) comparison between the measured output data and the model output. More specifically, the "measured output data" is the measured round-trip time variation $y(k)$, and the "model output" is the simulated output from the ARX model, which is defined as

$$y^*(k|\theta) = \psi^{*T}(k|\theta)\,\theta \tag{9}$$

where

$$\begin{aligned}\psi^*(k|\theta) = [&-y^*(k-1|\theta), \ldots, -y^*(k-n_a|\theta),\\ &u(k-n_d-1), \ldots, u(k-n_d-n_b)]\end{aligned} \tag{10}$$

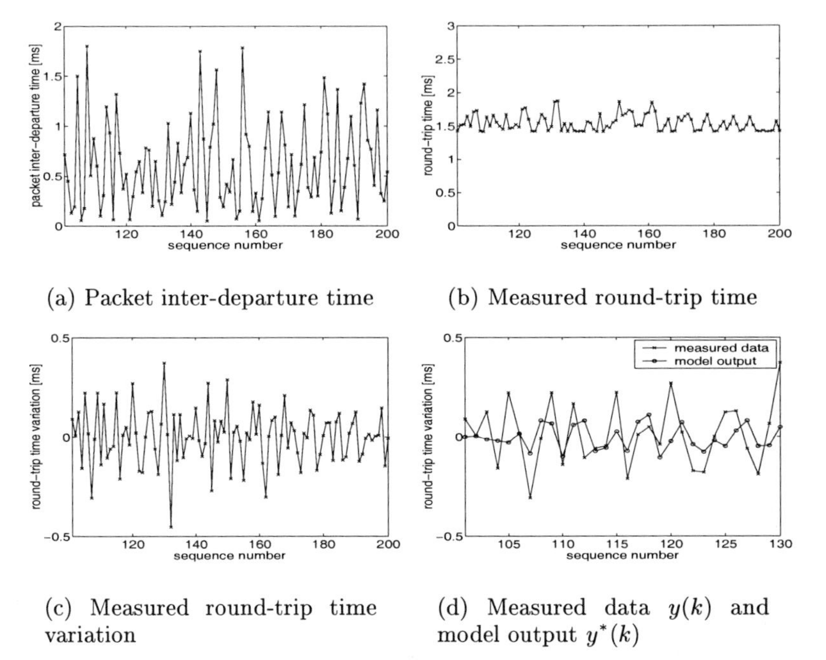

(a) Packet inter-departure time

(b) Measured round-trip time

(c) Measured round-trip time variation

(d) Measured data $y(k)$ and model output $y^*(k)$

Fig. 5. Results in network configuration N2 (100 Mbps LAN w/ Background Traffic)

Note the difference between $\hat{y}(k|\theta)$ and $y^*(k|\theta)$; i.e., $\hat{y}(k)$ is a 1-step ahead prediction from the measured inputs and outputs, whereas $y^*(k|\theta)$ is a simulated output only from the measured inputs assuming zero noise. There are several techniques for checking accuracy of the ARX model obtained by system identification [7]. Comparing the measured output data and the model output is one of the most intuitive approaches.

Figure 4(c) shows that the amplitude of the round-trip time variation is very small, whereas the packet inter-departure time dynamically changes. This is because there is no background traffic between the source and destination hosts. A slight change in the round-trip time would be caused by the processing delay variation at the host and/or by a timer granularity of the operating system, since the network is not a bottleneck in the network configuration **N1**. Figure 4(d) indicates that the ARX model cannot capture the round-trip time dynamics in the network configuration **N1**. Namely, the model output $y^*(k)$ is almost unchanged, although the measured round-trip time changes. This is caused by the weak correlation between the packet inter-departure time and the measured round-trip time; that is, in the network configuration **N1**, the round-trip time is almost independent of the packet inter-departure time.

Network Configuration N2

Figure 5 shows results in the network configuration **N2** for $\zeta = [8, 8, 1]$. In this case, the mean packet inter-departure time has been set to 0.6 ms, resulting the average packet transmission rate of 18.0 Mbps and the average round-trip time of 1.8 ms. Figure 5(c) shows that the amplitude of the round-trip time variation

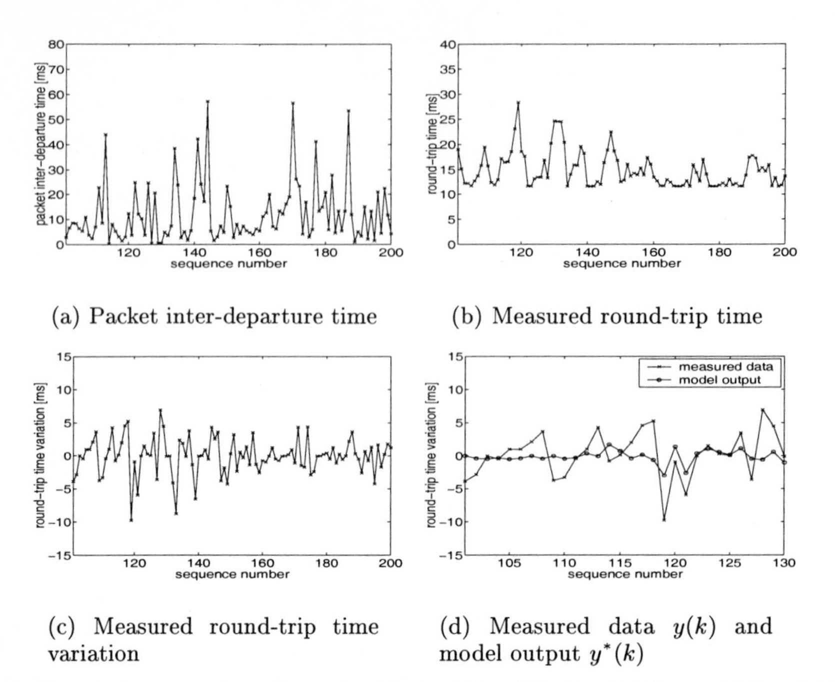

(a) Packet inter-departure time (b) Measured round-trip time

(c) Measured round-trip time variation (d) Measured data $y(k)$ and model output $y^*(k)$

Fig. 6. Results in network configuration N3 (11 Mbps Wireless LAN + 100 Mbps LAN)

is larger than that of the network configuration **N1**. The main reason for such a large amplitude would be the effect of the background traffic; that is, the round-trip time tends to become large when the network is congested. It can be found that the model output $y^*(k|\theta)$ and the measured output $y(k)$ roughly coincide but slightly differ. This is because the measured round-trip time variation is disturbed by other traffic, which is unknown so that not included in the model output $y^*(k)$.

Network Configuration N3

Results in the network configuration **N3** for $\zeta = [8, 8, 1]$ are shown in Fig. 6. In this case, the mean packet inter-departure time has been set to 12.0 ms, resulting the average packet transmission rate of 967 Kbps and the average round-trip time of 16.7 ms. Figure 6(c) shows that the amplitude of the round-trip time variation is much larger (about 10 ms) than the previous cases, **N1** and **N2**. Figure 6(d) indicates that the round-trip time dynamics is not correctly modeled by the ARX model. It is probably because the transmission delay at the wireless link is significantly changed, resulting in a large measurement noise. From these observations, we conclude that the round-trip time dynamics can be modeled by the ARX model when the network is moderately congested.

Choice of Model Orders and Number of Samples

In the above results, the orders and the delay of the ARX model is fixed at $\zeta = [8, 8, 1]$. In general, the accuracy of the ARX model is dependent on the

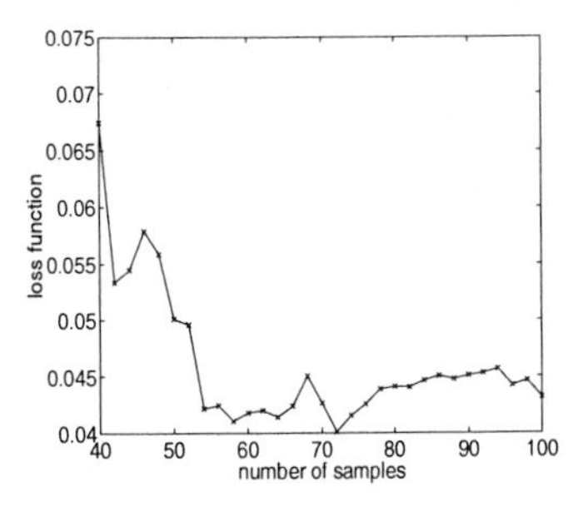

Fig. 7. Relation between loss function and the number of samples

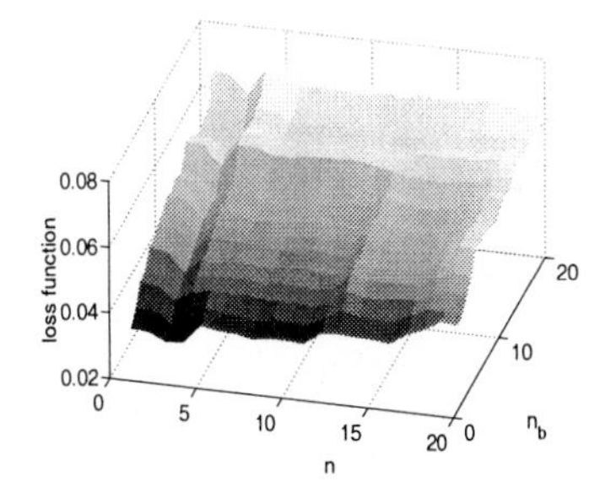

Fig. 8. Relation between loss function and the orders of the ARX model

choice of the orders and the delay of the ARX model, and the number of samples used for system identification. It is therefore desirable to carefully choose ζ and the number of samples to minimize the loss function $V_N(\theta, Z^N)$ (i.e., the sum of all squared prediction errors).

Figure 7 shows the relation between the loss function $J_N(\theta)$ and the number of samples used from the input and output data in the network configuration **N2**. In this figure, the orders and the delay of the ARX model is fixed at $\zeta = [8, 8, 1]$, while the number of samples is changed from 40 to 100. This figure shows a tendency that, as the number of samples increases, the loss function first decreases and then gradually increases. The similar tendencies are observed in other network configurations **N1** and **N3**, although the results are not included here.

We next show the relation between the orders of the ARX model and the loss function $V_N(\theta, Z^N)$ in Fig. 8. This figure uses 100 samples from the input and output data in the network configuration **N2**, and the orders of the ARX model, n_a and n_b, are changed from 1 to 20, respectively. This figure indicates that the loss function increases as the n_b increases. On the contrary, the choice of n_a has little effect on the loss function.

Another important factor in determining the orders of the ARX model is the highest frequency that should be captured by the ARX model. Namely, the ARX model is able to capture higher frequency of the output data (i.e., the round-trip time variation) with larger n_a and n_b. Moreover, the ARX model requires more computational burden and becomes less stable as the orders increase [7]. So the orders of the ARX model should be determined by taking account of a trade-off among accuracy, complexity, and stability.

6 Discussion and Conclusion

We discuss several possible applications of our approach — modeling the round-trip time dynamics of the Internet using the ARX model. Details of these topics will be discussed in the forthcoming paper, but it is worthwhile to discuss how our approach can be applied to various problems. The first and straightforward application would be to use our approach to *understand* the round-trip time dynamics of the Internet. We can analyze the round-trip time dynamics through the ARX model. Because the ARX model is one of LTI (Linear Time Invariant) models, various analysis techniques for LTI models in time- and frequency-domain can be utilized. The second application would be to *predict* the future round-trip time from the ARX model. As have shown in Section 5, the round-trip time of a packet is considerably disturbed by background traffic. Hence, it is difficult to predict the far future round-trip time. However, the ARX model can predict the near future round-trip time, which would be useful to, for example, QoS controlling mechanisms. As noted in Section 1, the third and possibly most important application would be to *design* a delay-based congestion control mechanism. Once the ARX model capturing the round-trip time is obtained, it would be possible to apply the optimal control theory to design an efficient delay-based congestion control mechanism. Congestion control of the Internet is a difficult problem because of its complexity such as heterogeneity of various network elements and non-negligible propagation delays. However, we believe that combination of the ARX model and the optimal control theory would help us to design a more efficient congestion control mechanism. We are currently working on designing a delay-based congestion control mechanism for stream video applications.

In this paper, we have proposed a novel approach to model the round-trip time dynamics of the Internet using system identification. The main idea is to model the network, seen by a specific source host, as a linear time-invariant ARX model. The input to the ARX model is the packet inter-departure time from the source host, and the output is the round-trip time variation measured at the source host. With the ICMP-based measurement method, we have collected three sets of the input and output data from real networks. Using the measurement results, we have determined coefficients of the ARX model, and have investigated how accurately the ARX model captures the round-trip time dynamics. We have found that the ARX model can capture the round-trip time dynamics when the network is moderately congested. We have also found that, when the network is not congested or the measured round-trip time is noisy, the ARX model fails to capture the round-trip time dynamics.

As a future work, it is important to validate effectiveness of our modeling approach for a through set of input and output data obtained from various network configurations. We are currentry measuring the input and output data in working LAN and WAN environments [23].

Acknowledgement

This work was supported in part by Research for the Future Program of Japan Society for the Promotion of Science under the Project "Integrated Network Architecture for Advanced Multimedia Application Systems"(JSPS-RFTF97R16301).

References

1. R. Jain, "A delay-based approach for congestion avoidance in interconnected heterogeneous computer networks," *ACM Computer Communication Review*, vol. 19, pp. 56–71, Oct. 1989.

2. L. S. Brakmo, S. W. O'Malley, and L. L. Peterson, "TCP Vegas: New techniques for congestion detection and avoidance," in *Proceedings of ACM SIGCOMM '94*, pp. 24–35, Oct. 1994.
3. J.-C. Bolot, "Characterizing end-to-end packet delay and loss in the Internet," *Journal of High-Speed Networks*, vol. 2, pp. 305–323, Dec. 1993.
4. S. B. Moon, J. Kurose, P. Skelly, and D. Towsley, "Correlation of packet delay and loss in the Internet," tech. rep., Department of Computer Science, University of Massachusetts, USA, Jan. 1998.
5. V. Paxson, "End-to-end Internet packet dynamics," in *Proceedings of ACM SIGCOMM*, pp. 139–152, Sept. 1997.
6. H. Ohsaki, M. Murata, and H. Miyahara, "Modeling end-to-end packet delay dynamics of the Internet using system identification," in *Proceedings of Seventeenth International Teletraffic Congress*, pp. 1027–1038, Dec. 2001.
7. L. Ljung, *System identification — theory for the user.* Englewood Cliffs, N.J.: Prentice Hall, 1987.
8. M. Allman and V. Paxon, "On estimating end-to-end network path properties," *ACM SIGCOMM '99*, pp. 263–274, Aug. 1999.
9. A. Adams, T. Bu, R. Caceres, N. Duffield, T. Friedman, J. Horowitz, F. L. Presti, S. B. Moon, V. Paxson, and D. Towsley, "The use of end-to-end multicast measurements for characterizing internal network behavior," *IEEE Communications*, pp. 152–159, May 2000.
10. V. Paxson, "End-to-end routing behavior in the Internet," in *ACM SIGCOMM '96*, pp. 25–38, May 1996.
11. R. Caceres, P. B. Danzig, S. Jamin, and D. J.Mitzel, "Characteristics of wide-area TCP/IP conversations," in *ACM SIGCOMM '91*, pp. 101–112, Sept. 1991.
12. H. Che and S.-Q. Li, "Fast algorithms for measurement-based traffic modeling," *IEEE Journal on Selected Areas in Communications*, vol. 16, pp. 612–625, June 1998.
13. L. A. Kulkarni and S.-Q. Li, "Measurement-based traffic modeling: capturing important statistics," *Journal of Stochastic Model*, vol. 14, no. 5, 1998.
14. S.-Q. Li, S. Park, and D. Arifler, "SMAQ: A measurement-based tool for traffic modeling and queueing analysis, Part I - design methodologies and software architecture," *IEEE Communications Magazine*, vol. 36, pp. 56–65, Aug. 1998.
15. S.-Q. Li, S. Park, and D. Arifler, "SMAQ: a measurement-based tool for traffic modeling and queueing analysis, Part II - network applications," *IEEE Communications Magazine*, vol. 36, pp. 66–77, Aug. 1998.
16. Y. Zhang, V. Paxon, and S. Schenker, "The stationarity of Internet path properties: routing, loss, and throughput," tech. rep., ACIRI, May 2000.
17. O. Ait-Hellal, E. Altman, and T. Basar, "A robust identification algorithm for traffic models in telecommunications," in *Proceedings of IEEE CDC '99*, Dec. 1998.
18. A. Sang and S.-Q. Li, "A predictability analysis of network traffic," in *Proceedings of IEEE INFOCOM 2000*, pp. 342–351, Mar. 2000.
19. S. Hares, "Essential tools for the OSI Internet," *Request for Comments (RFC) 1574*, Feb. 1994.
20. F. Baker, "Requirements for IP version 4 routers," *Request for Comments (RFC) 1812*, June 1995.
21. J. Postel, "Internet control message protocol," *Request for Comments (RFC) 792*, Sept. 1981.
22. S. Savage, "Sting: A TCP-based network measurement tool," in *Proceedings of USENIX Symposium on Internet Technologies and Systems*, pp. 71–79, Oct. 1999.
23. H. Ohsaki, M. Morita, and M. Murata, "Measurement-based modeling of Internet round-trip time dynamics using system identification," submitted to *the Second IFIP-TC6 Networking Conference (NETWORKING 2002)*, Nov. 2001.

Analyzing the Impact of TCP Connections Variation on Transient Behavior of RED Gateway

Motohisa Kisimoto[1], Hiroyuki Ohsaki[2], and Masayuki Murata[2]

[1] Graduate School of Engineering Science, Osaka University
1-3 Machikaneyama, Toyonaka, Osaka 560–8531, Japan
kisimoto@ics.es.osaka-u.ac.jp

[2] Cybermedia Center, Osaka University
1-30 Machikaneyama, Toyonaka, Osaka 567-0043, Japan
{oosaki, murata}@cmc.osaka-u.ac.jp

Abstract. Several gateway-based congestion control mechanisms have been proposed to support an end-to-end congestion control mechanism of TCP (Transmission Control Protocol). One of promising gateway-based congestion control mechanisms is a RED (Random Early Detection) gateway. In this paper, we analyze the transient behavior of the RED gateway when the number of TCP connections is changed. We model both TCP connections and the RED gateway as a single feedback system, and analyze the dynamics of the number of packets in the RED gateway's buffer when the number of TCP connections is increased or decreased. Through numerical examples, we quantitatively show that the transient performance of the RED gateway is quite sensitive to system parameters such as the total number of TCP connections, the processing speed of the RED gateway. We also show that control parameters of the RED gateway have little impact on the transient behavior of the RED gateway.

1 Introduction

Several gateway-based congestion control mechanisms have been recently proposed to support an end-to-end congestion control mechanism of TCP [1,2]. One of promising gateway-based congestion control mechanisms is a RED (Random Early Detection) gateway that randomly drops an incoming packet at the buffer [1]. A number of studies on the steady state performance of the RED gateway using simulation experiments have been performed [1,3,4]. Although effectiveness of the RED gateway is fully dependent on a choice of control parameters, it is difficult to configure them appropriately. For example, the authors of [1] have proposed a set of control parameters for the RED gateway, but this is only a guideline acquired empirically using simulation experiments. On the other hand, there are a few studies analyzing the characteristics of the RED gateway. Stability and transient behavior of the RED gateway in the steady state have been analyzed in [5,6,7,8] by assuming that the number of TCP connections is

I. Chong (Ed.): ICOIN 2002, LNCS 2343, pp. 372–383, 2002.

constant. It has not been cleared how the variation of the number of TCP connections affects the transient behavior of the RED gateway. In an actual network, the number of TCP connections changes frequently. When the number of TCP connections is increased or decreased, either buffer overflow or buffer underflow may occur, resulting in the performance degradation of the RED gateway. It is therefore important to evaluate the transient behavior of the RED gateway by taking account of the variation of the number of TCP connections.

In this paper, we analyze the transient behavior of the RED gateway by extending the analytic results obtained in [5]. More specifically, we analyze the dynamics of the number of packets in the RED gateway's buffer (i.e., the queue length) when one or more TCP connections newly start or terminate their data transmissions. Showing numerical results, we reveal how control parameters of the RED gateway affect its transient behavior.

This paper is organized as follows. In Section 2, we explain the algorithm of the RED gateway in short. In Section 3, the analytic model of the RED gateway is explained, which is used throughout this paper. In Section 4, we briefly present the derivation of the average state transition equations, which describe the dynamics of the RED gateway. In Section 5, using the average state transition equations, we analyze the transient behavior of the RED gateway when the number of TCP connections is changed. In Section 6, several numerical examples are presented to clearly show how control parameters of the RED gateway or system parameters affect the transient performance. In Section 7, we finally conclude this paper and discuss future works.

2 RED Algorithm

The RED gateway has four control parameters: min_{th}, max_{th}, max_p, and q_w. min_{th} is the minimum threshold and max_{th} is the maximum threshold. These thresholds are used to calculate a packet marking probability for every incoming packet. The RED gateway maintains its average queue length $\overline{q}$, which is calculated from the current queue length using EWMA (Exponential Weighted Moving Average) with a weight factor of q_w. The RED gateway calculates the packet marking probability p_b for every incoming packet from the average queue length. Namely, the RED gateway determines the packet marking probability p_b using the function shown in Fig. 1. In this figure, max_p is a control parameter that determines the maximum packet marking probability. The packet dropping mechanism of the RED gateway is not per-flow basis, so the same packet marking probability p_b is used for all the incoming packets.

3 Analytic Model

In this paper, we analyze the transient behavior of the RED gateway using the analytic results obtained in [5]. We show our analytic model in Fig. 2. The analytic model consists of a single RED gateway and multiple TCP connections. We assume that all TCP connections have an identical (round-trip) propagation delay (denoted by τ). We also assume that the processing speed of the RED

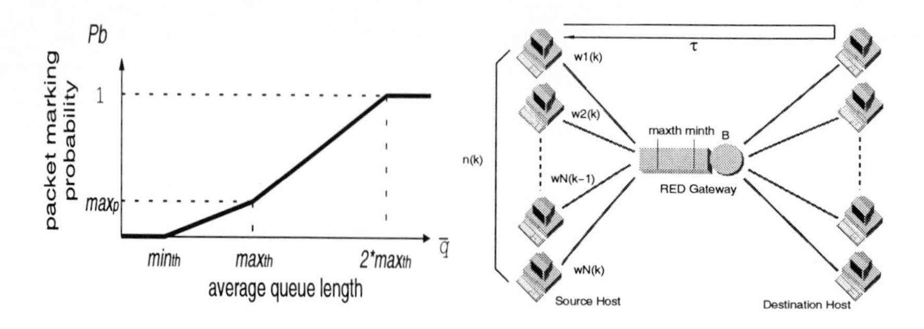

Fig. 1. Calculation of packet marking probability p_b.

Fig. 2. Analytic model.

gateway (denoted by B) is the bottleneck in the network. Namely, transmission speeds of all links are assumed to be sufficiently faster than the processing speed of the RED gateway.

We model the congestion control mechanism of TCP version Reno [9] at all source hosts. We further assume that all TCP connections change their window sizes (denoted by w) synchronously. Source hosts are allowed to send w packets without receipt of an ACK (ACKnowledgement) packet. Thus, the source host can send w packets during its RTT (Round Trip Time). In our analysis, we model the entire network as a discrete-time system, where a time slot of the system corresponds to an RTT of TCP connections. We define $w(k)$ as the window size of the source host at slot k. All source hosts are assumed to have enough data to transmit; that is, the source host is assumed to always send the number $w(k)$ of packets during slot k. We define $q(k)$ and $\overline{q}(k)$ as the current and the average queue lengths (i.e., the current and the average number of packets in the buffer of the RED gateway). We assume that both $q(k)$ and $\overline{q}(k)$ will not change during a slot [5]. For taking account of a TCP connections variation, the number of TCP connections at slot k is denoted by $n(k)$.

4 Derivation of Average State Transition Equations

In this section, we present the derivation of average state transition equations, which describe the dynamics of the RED gateway [5]. Refer to [5] for the detail of the analysis.

4.1 Derivation of State Transition Equations

Provided that the average queue length $\overline{q}(k)$ lies between min_{th} and max_{th}, and that the number $n(k)$ of TCP connections is constant, $p_b(k)$ is given by

$$p_b(k) = max_p \left(\frac{\overline{q}(k) - min_{th}}{max_{th} - min_{th}} \right)$$

The RED gateway discards an incoming packet with a probability $p_a(k)$:

$$p_a(k) = \frac{p_b(k)}{1 - count \cdot p_b(k)}$$

where *count* is the number of unmarked packets that have arrived since the last marked packet. The number of unmarked packets between two consecutive marked packets, X, can be represented by an uniform random variable in $\{1, 2, \cdots, 1/p_b(k)\}$. Namely,

$$P_k[X = n] = \begin{cases} p_b(k) & 1 \le n \le 1/p_b(k) \\ 0 & \text{otherwise} \end{cases}$$

Let $\overline{X}(k)$ be the expected number of unmarked packets between two consecutive marked packets at slot k. $\overline{X}_k$ is obtained as

$$\overline{X}_k = \sum_{n=1}^{\infty} n\, P_k[X = n] = \frac{1/p_b(k) + 1}{2}$$

The probability that at least one packet is discarded from $w(k)$ packets, $\overline{p}$, is given by

$$\overline{p} = \min\left(\frac{w(k)}{1/p_b(k)}, 1\right)$$

Therefore, by assuming that all TCP connections are in the congestion avoidance phase, the window size at slot $k+1$ is given by

$$w(k+1) = \begin{cases} \frac{w(k)}{2} & \text{with probability } \overline{p} \\ w(k) + 1 & \text{otherwise} \end{cases} \tag{1}$$

Note that in the above equation, it is assumed that all packet losses can be detected by duplicate ACKs [5]. The current queue length at slot $k+1$ is given by

$$\begin{aligned} q(k+1) &= q(k) + n(k+1)\, w(k+1) - B\left(\tau + \frac{q(k)}{B}\right) \\ &= n(k+1)\, w(k+1) - B\,\tau \end{aligned} \tag{2}$$

The average queue length at slot $k+1$ is given by

$$\overline{q}(k+1) = (1 - q_w)^{n(k)\,w(k)}\overline{q}(k) + \frac{q_w\{1 - (1 - q_w)^{n(k)\,w(k)}\}}{1 - (1 - q_w)}\, q(k)$$

4.2 Derivation of Average State Transition Equations

We derive average state transition equations that represent a typical behavior of TCP connections and the RED gateway [5]. We introduce a *sequence*, which

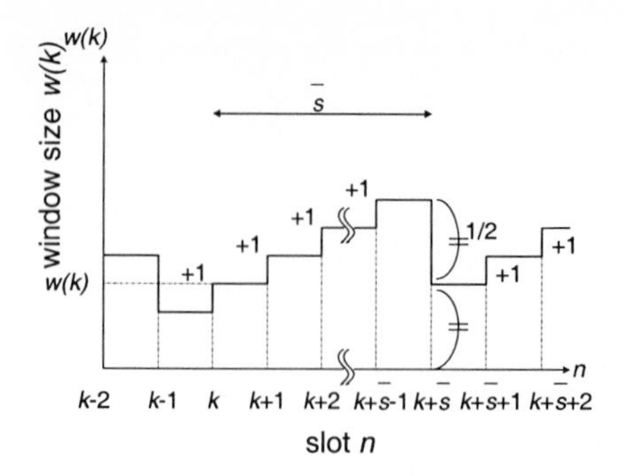

Fig. 3. Relationship between slot and sequence.

is a series of adjacent slots in which all packets from a source host have been unmarked by the RED gateway (Fig. 3). We then treat the entire network as a discrete-time system where a time slot corresponds to a sequence, instead of a slot. Let $\overline{s}(k)$ be the average number of slots that consists of a sequence that begins at slot k.

The average state transition equation from $w(k)$ to $w(k+\overline{s}(k))$ is obtained from Eq. (1) as

$$w(k+\overline{s}(k)) = \frac{w(k)+\overline{s}(k)-1}{2} \tag{3}$$

Note that $w(k)$ represents the expected value of the *minimum* window size. Similarly, the average state transition equations from $q(k)$ to $q(k+\overline{s}_k)$ is obtained from Eq. (2) as

$$q(k+\overline{s}(k)) = n(k+\overline{s}(k))\, w(k+\overline{s}(k)) - B\,\tau \tag{4}$$

The average state transition equation from $\overline{q}(k)$ to $\overline{q}(k+\overline{s}_k)$ is obtained as

$$\overline{q}(k+\overline{s}(k)) \simeq (1-q_w)^{\overline{X}(k)}\overline{q}(k) + \{1-(1-q_w)^{\overline{X}(k)}\}q(k) \tag{5}$$

Average state transition equations given by Eqs. (3), (4), and (5) describe the average behaviors of the window size, the current queue length, and the average queue length, respectively. An *average equilibrium value* is defined as the expected value in steady state. Let w^*, q^*, $\overline{q}^*$, and n^* be the average equilibrium values of the window size $w(k)$, the current queue length $q(k)$, the average queue length $\overline{q}(k)$, and the average number of TCP connections $n(k)$, respectively. Let us introduce $\delta\mathbf{x}(k)$ as the difference between the state vector $\mathbf{x}(k)$ and the average equilibrium point.

$$\delta\mathbf{x}(k) \equiv \begin{bmatrix} w(k)-w^* \\ q(k)-q^* \\ \overline{q}(k)-\overline{q}^* \\ n(k)-n^* \end{bmatrix}$$

By lineally approximating $w(k)$, $q(k)$, $\overline{q}(k)$, and $n(k)$ around their average equilibrium values, $\delta\mathbf{x}(k+\overline{s})$ can be written as

$$\delta\mathbf{x}(k+\overline{s}(k)) \simeq \mathbf{A}\delta\mathbf{x}(k) \tag{6}$$

where $\mathbf{A}$ is a state transition matrix.

5 Analysis of Transient Behavior

5.1 Types of TCP Connections Variation

We assume that N TCP connections exist in steady state. We also assume that all TCP connections are in the congestion avoidance phase. In this case, there are four types of changes in the number of TCP connections.

The first case is that ΔN ($\Delta N < N$) TCP connections of N TCP connections end their data transmissions (C1). In this case, $N - \Delta N$ TCP connections are in the congestion avoidance phase and will reach the steady state again. The second and the third cases (C2 and C3) are that ΔN TCP connections resume their data transmissions after an idle period. In these cases, the behavior of these ΔN TCP connections depends on the length of the idle period. When the idle period is short (C2), ΔN TCP connections operate in the congestion avoidance phase with using their previous window sizes. In this case, there exist totally $N + \Delta N$ TCP connections in the congestion avoidance phase.

On the other hand, when the idle period is long (in general, longer than the TCP's retransmission timer) (C3), ΔN TCP connections operate in the slow start phase with the initial window size. Moreover, the fourth case is that ΔN TCP connections newly start their data transmissions (C4). In this case, similar to the third case, there exist N TCP connections in the congestion avoidance phase and ΔN TCP connections in the slow start phase. In this paper, we analyze the transient behavior of the RED gateway in each case. We use two different approaches for the cases that all TCP connections are in the congestion avoidance phase (C1 and C2) and for the cases that some TCP connections are in the slow start phase (C3 and C4).

5.2 Cases C1 and C2: Congestion Avoidance Phase Only

We consider the cases that all TCP connections are in the congestion avoidance phase (C1 or C2). Let $u(k)$ ($\equiv n(k) - n(k-1)$) be the difference of the number of TCP connections from slot $k-1$ to slot k. For instance, when the number of TCP connections is increased by ΔN at slot i, $u(k)$ is given by

$$u(k) = \begin{cases} \Delta N & \text{if } k = i \\ 0 & \text{otherwise} \end{cases}$$

We analyze the effect of the TCP connections variation on the dynamics of the current queue length of the RED gateway. The main idea is to treat the TCP connections variation, $u(k)$, and the current queue length, $q(k)$, as the input to, and the output from the system formulated by Eq. (6), respectively. Namely, by adding both the input $u(k)$ and the output $q(k)$ to Eq. (6), we have

$$\delta\mathbf{x}(k+\overline{s}(k)) = \mathbf{A}\delta\mathbf{x}(k) + \mathbf{B}u(k)$$
$$q(k) = \mathbf{C}\delta\mathbf{x}(k)$$

where $\mathbf{B}$ and $\mathbf{C}$ are defined by the following equations.

$$\mathbf{B} = [\,0\;0\;0\;1\,]^T$$
$$\mathbf{C} = [\,0\;1\;0\;0\,]$$

Namely, the variation of the number of TCP connections, $u(k)$, is added to the number of active TCP connections, $n(k)$, by $\mathbf{B}$. And the current queue length of the RED gateway, $q(k)$, is extracted from the state vector by $\mathbf{C}$.

Using such a SISO (Single-Input Single-Output) model given by Eq. (7), the dynamics of the current queue length of the RED gateway can be precisely analyzed. For example, the evolution of the current queue length, $q(k)$, for a given TCP connections variation, $u(k)$, can be calculated by

$$q(k) = \sum_{i=0}^{k} u(i)\,\delta\mathbf{x}(k-i) \tag{7}$$

The great advantage of this approach is that various analytic techniques used in the control theory can be directly applied. For example, if the number of TCP connections is increased by ΔN at slot k, the input $u(k)$ becomes the impulse function [10]. Therefore, it is easy to analyze the dynamics of the current queue length, $q(k)$, by investigating the impulse response of the system. We can investigate the dynamics of the current queue length not only for an instantaneous TCP connections variation but also for an arbitrary TCP connections variation.

5.3 Cases C3 and C4: Congestion Avoidance Phase and Slow Start Phase

We next focus on the other two cases: when TCP connections resume their data transmissions (C3) and when TCP connections newly start their data transmissions (C4). Let $u'(k)$ be the difference in the total number of packets from slot $k-1$ to slot k, which are sent from all TCP connections in the slow start phase. More specifically, $u'(k)$ is defined by

$$u'(k) = \sum_{i=1}^{\Delta N} (w_i(k) - w_i(k-1))$$

where ΔN is the number of TCP connections operating in the slow start phase, and $w_i(k)$ is the window size of ith TCP connection. In the slow start phase, the window size is first initialized and then doubled every RTT. Thus, when

ΔN TCP connections newly start their data transmissions at slot i, $u'(k)$ is approximately given by

$$u'(k) \simeq \begin{cases} \frac{\Delta N}{n(k)} \times 2^{\overline{s}(k)(k-s_i-1)} & \text{if } k > s_i \\ 0 & \text{otherwise} \end{cases}$$

Similarly to the previous subsection, the dynamics of the current queue length of the RED gateway can be analyzed. Namely, the difference in the total number of packets, $u'(k)$, and the current queue length, $q(k)$, are added to the system given by Eq. (6) as the input and the output, respectively.

$$\delta \mathbf{x}(k + \overline{s}(k)) = \mathbf{A}\delta \mathbf{x}(k) + \mathbf{B}'u'(k)$$
$$q(k) = \mathbf{C}\delta \mathbf{x}(k)$$

where

$$\mathbf{B}' = \left[1\,0\,0\,0\right]^T$$

In the above equations, the window size of a TCP connection, $w(k)$, is increased by $u'(k)$ by $\mathbf{B}$.

6 Numerical Examples and Discussions

6.1 Performance Measures for Transient Behavior

Three performance measures called *overshoot*, *rise time* and *settling time* are widely used for evaluating the transient behavior of dynamic systems (Fig. 4) [11]. These are criteria for the damping performance (the overshoot), the response performance (the rise time), and both the response and the damping performance (the settling time). In this paper, we define the overshoot as the difference between the maximum and the equilibrium queue lengths. The rise time is defined as the time taken for the current queue length to reach the 90 % of the equilibrium queue length. The settling time is the time taken for the current queue length to converge within 5% of the equilibrium queue length. In general, all of these performance measures should be small for achieving better transient behavior. However, there is a tradeoff among the overshoot, the rise time, and the settling time. It is therefore important to balance these three performance measures according to the desired transient behavior.

These performance measures have the following implications to the RED gateway. A large overshoot means that the current queue length of the RED gateway grows excessively when the number of TCP connections is changed. Since the current queue length is limited by the buffer size, a large overshoot sometimes causes buffer overflow at the RED gateway. Otherwise, it results in a long queueing delay in the buffer. Hence, a small overshoot is desirable for preventing buffer overflow and minimizing the queueing delay. In addition, the rise time represents the convergence speed of the current queue length after

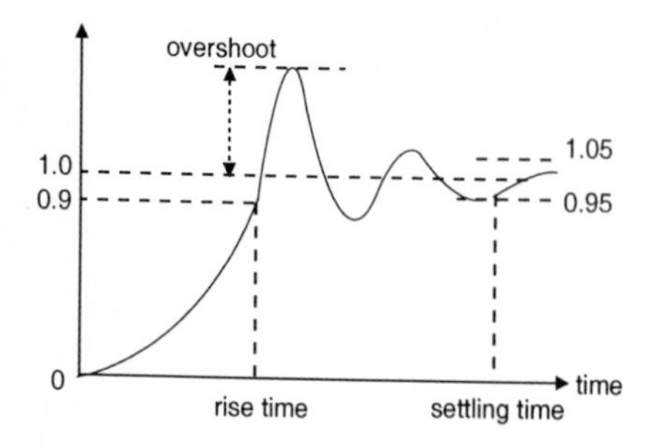

Fig. 4. Performance measures for transient behavior (overshoot, rise time, and settling time).

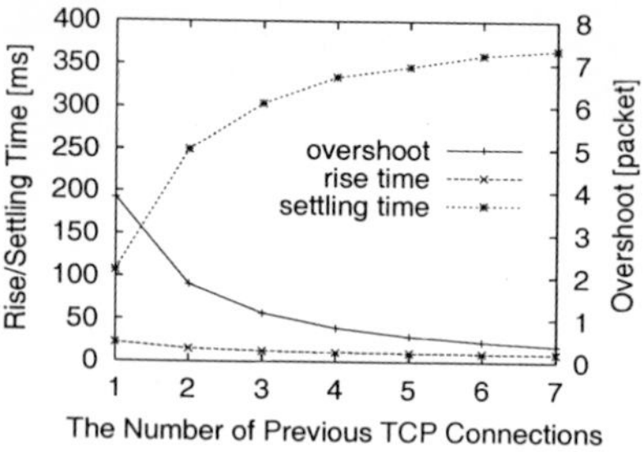

Fig. 5. Performance measures for transient behavior (the number of previous TCP connections N = 1–7).

a change of the number of TCP connections. As can be seen from Eq. (2), the current queue length, $q(k)$, directly reflects the window sizes $w(k)$. So it is possible to estimate the convergence speed of TCP connections from the rise time. The settling time implies the convergence speed of the current queue length to its equilibrium value after the number of TCP connections is changed.

6.2 Case C2: Congestion Avoidance Phase Only

Due to space limitation, we only show numerical examples for the case (C2): when ΔN TCP connections resume their data transmissions in the congestion avoidance phase after a short idle period. We use the equilibrium values, w^*, q^*, and $\overline{q}^*$, as the initial values for $w(k)$, $q(k)$, and $\overline{q}(k)$. We calculate the dynamics of the queue length $q(k)$ of the RED gateway using Eq. (7) when ΔN TCP connections resume at slot 0; i.e.,

$$n(k) = \begin{cases} N & \text{if } k < 0 \\ N + \Delta N & \text{if } k \geq 0 \end{cases}$$

Figure 5 shows performance measures for the transient behavior (the overshoot, the rise time, and the settling time) for different number N of TCP connections in steady state, n^*. In the following figures, unless explicitly stated, we use a set of control parameters of the RED gateway recommended by the authors of [1]. We also use the following system parameters: the processing speed of the RED gateway $B = 2$ [packet /ms], the propagation delay $\tau = 1$ [ms], and the number of resumed TCP connections $\Delta N = 1$. Figure 5 shows that the current queue length of the RED gateway changes more dynamically (i.e., a larger overshoot) when N is smaller. It is because when the number of TCP connections in steady state, N, is smaller, the impact of the resumed TCP connection becomes larger. The figure also shows that the overshoot is smaller than 1 [packet] when the number of TCP connections, N, is greater than 4. It suggests that the buffer

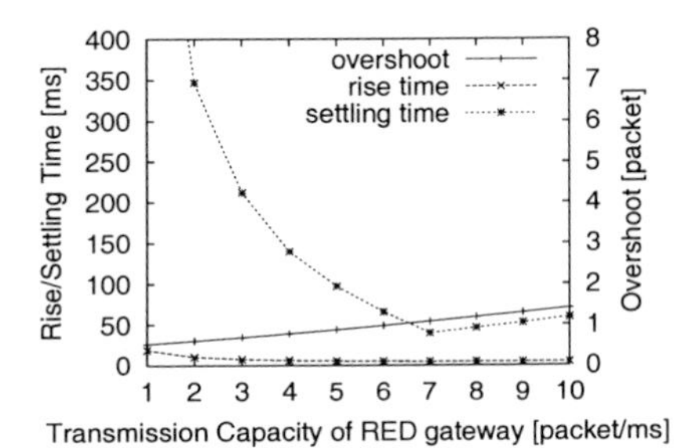

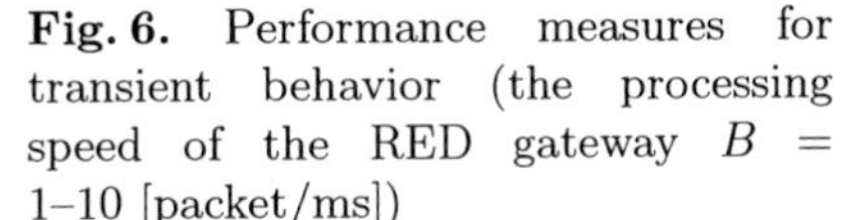
Fig. 6. Performance measures for transient behavior (the processing speed of the RED gateway B = 1–10 [packet/ms])

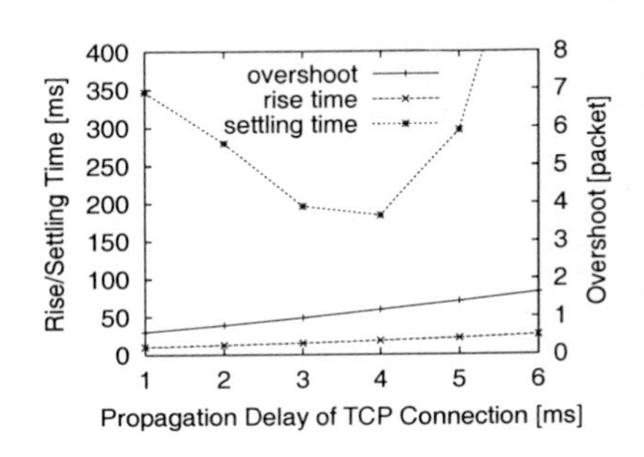

Fig. 7. Performance measures for transient behavior (the propagation delay of the TCP connection τ = 1–5 [ms])

overflow at the RED gateway is not likely to happen when the number of TCP connections is sufficiently large.

Shown in Fig. 6 is the case that the processing speed of the RED gateway, B, is changed from 1 to 10 [packet/ms]. One can find from this figure that as the processing speed of the RED gateway decreases, the overshoot and the settling time becomes small and long, respectively. This implies that the effect of TCP connections variation on the current queue length of the RED gateway sustains for a long period if the processing speed of the RED gateway is small.

Figure 7 illustrates the effect of the (round-trip) propagation delay of the TCP connection on the transient behavior of the RED gateway. In this figure, the propagation delay of the TCP connection, τ, is changed from 1 to 6 [ms]. This figure clearly shows that the transient behavior of the RED gateway is degraded when the propagation delay of the TCP connection is large. For example, as the propagation delay increases, both the overshoot and the rise time increase. This phenomenon can be understood by the fact that when both TCP connections and the RED gateway are considered as a single feedback system, a longer propagation delay corresponds to a longer feedback delay. In general, both the stability and the transient performance of a feedback system are degraded by a long feedback delay.

Figure 7 also shows that the settling time is minimized when the propagation delay of the TCP connection, τ, is about 4 [ms]. It can be conjectured from this phenomenon that the current queue length of the RED gateway will change slowly when the propagation delay of the TCP connection is short, and that the current queue length changes oscillatory when the propagation delay is long. From this observation, it is expected that the operation of the RED gateway becomes unstable if the propagation delay of the TCP connection is very long. In most feedback-based systems, a small feedback delay improves both the stability and the transient performance. However, in the congestion avoidance phase of TCP, the window size of the source host is increased every its RTT. In other

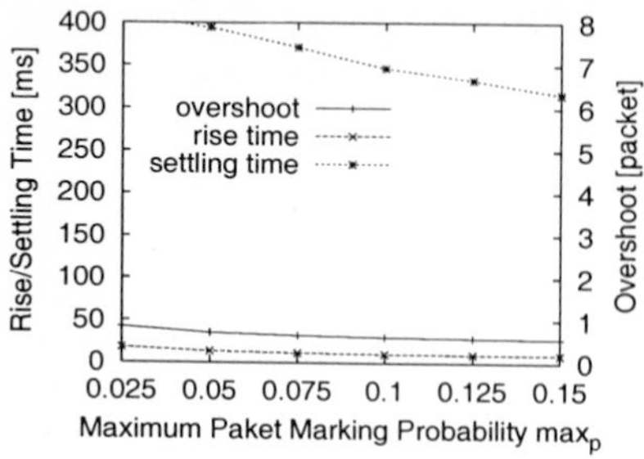

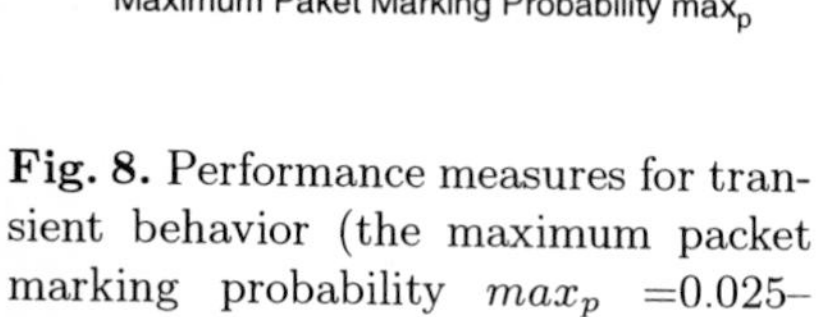

Fig. 8. Performance measures for transient behavior (the maximum packet marking probability max_p =0.025–0.15)

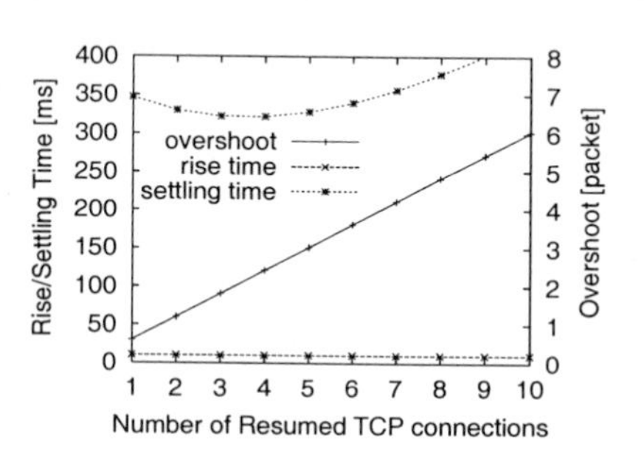

Fig. 9. Performance measures for transient behavior (the number of resumed TCP connections ΔN = 1–10)

words, the congestion avoidance phase of TCP has a feedback gain, which is dependent on the feedback delay.

We then investigate the effect of the maximum packet marking probability, max_p, on the transient behavior of the RED gateway. Figure 8 suggests that three performance measures — the overshoot, the rise time, and the settling time — are slightly increased as max_p increases. Namely, the maximum packet marking probability, max_p, has little impact on the transient behavior of the RED gateway. The maximum packet marking probability, max_p, should therefore be configured by taking account of the steady state performance of the RED gateway (e.g., the average throughput and the average queue length). Although we do not include results due to space limitation, we found that two threshold values, min_{th} and max_{th}, also have little impact on the transient behavior of the RED gateway.

We finally show the dynamics of the current queue length of the RED gateway for a different number of resumed TCP connections ΔN. In this figure, ΔN is changed from 1 to 10. It can be found from this figure that the current queue length of the RED gateway changes more excessively with a larger number of resumed TCP connections, ΔN. This phenomenon can be intuitively understood. Namely, when the number of resumed TCP connections is large, more packets arrive at the RED gateway. It gives a larger impact on the transient behavior of the RED gateway.

7 Conclusion and Future Work

In this paper, we have analyzed the impact of TCP connections variation on the transient behavior of the RED gateway by utilizing the average state transition equations obtained in [5]. We have modeled the entire network including both TCP connections and the RED gateway as a feedback system. We have investigated the transient behavior (in particular, the dynamics of the current queue length) of the RED gateway when the number of TCP connections is changed. We have quantitatively shown that the transient behavior of the RED gateway is sensitive to system parameters such as the number of TCP connections in the steady state, the capacity of the RED gateway, and the propagation delay of the TCP connection. We have also shown that the control parameters of the RED gateway have little influence on the transient behavior of the RED gateway.

As a future work, it is important to analyze the transient behavior of the RED gateway in various situations since our analytic approach enables us to investigate the transient behavior of the RED gateway for realistic TCP connection variation.

Acknowledgement. This work was supported in part by Research for the Future Program of Japan Society for the Promotion of Science under the Project "Integrated Network Architecture for Advanced Multimedia Application Systems" (JSPS-RFTF97R16301).

References

1. S. Floyd and V. Jacobson, "Random early detection gateways for congestion avoidance," *IEEE/ACM Transactions on Networking*, vol. 1, pp. 397–413, Aug. 1993.
2. B. Barden et al., "Recommendations on queue management and congestion avoidance in the Internet," *Request for Comments (RFC) 2309*, Apr. 1998.
3. D. Lin and R. Morris, "Dynamics of random early detection," in *Proceedings of ACM SIGCOMM '97*, pp. 127–137, Sep 1997.
4. M. May, J. Bolot, C. Diot, and B. Lyles, "Reasons not to deploy RED," in *Proceedings of IWQoS '99*, pp. 260–262, Mar. 1999.
5. H. Ohsaki, M. Murata, and H. Miyahara, "Steady state analysis of the RED gateway: stability, transient behavior, and parameter setting," to appear in *IEICE Transactions on Communications*, Jan. 2002.
6. M. May, T. Bonald, and J.-C. Bolot, "Analytic evaluation of RED performance," in *Proceedings of IEEE INFOCOM 2000*, pp. 1415–1424, Mar. 2000.
7. V. Sharma, J. Virtamo, and P. Lassila, "Performance analysis of the random early detection algorithm," available at `http://keskus.tct.hut.fi/tutkimus/com2/publ/redanalysis.ps`, Sept. 1999.
8. H. M. Alazemi, A. Mokhtar, and M. Azizoglu, "Stochastic approach for modeling random early detection gateways in TCP/IP networks," in *Proceedings of IEEE International Conference on Communications 2001*, pp. 2385–2390, June 2001.
9. W. R. Stevens, *TCP/IP Illustrated, Volume 1: The Protocols.* New York: Addison-Wesley, 1994.
10. R. Isermann, *Digital control systems, Volume 1: fundamentails, deterministic control.* Springer-Verlag Berlin Heidelberg, 1989.
11. N. S. Nise, *Control Systems Engineering.* New York: John Wiley & Sons, third ed., 2000

Static Document Scheduling with Improved Response Time in HTTP/1.1

Hojung Cha[1], Jiho Bang[2], and Rhan Ha[3]

[1] Dept. of Computer Science, Yonsei University, Seoul 120-749, Korea
hjcha@cs.yonsei.ac.kr
[2] Korea Information Security Agency, Seoul 138-803, Korea
jhbang@kisa.or.kr
[3] Dept. of Computer Engineering, Hongik University, 121-791, Korea
rhanha@cs.hongik.ac.kr

Abstract. This paper proposes a window-based connection scheduling technique for web server with static HTTP/1.1 objects. The method is based on the shortest connection first principle and schedules web documents dynamically according to the number of inlined-documents of the shortest connection. The simulation study shows that the proposed method greatly improves the server performance.

1 Introduction

A popular web site is concurrently accessed by many clients and the server should process the connections in an efficient way since a web server can only accept a limited number of connections simultaneously. There has been many researches on the efficient scheduling of web connections. Previous researches are typically classified into two categories, depending on the consideration of document sizes being requested. By using the size information of the requested documents, the average response time for connection can be improved as the technique favors small documents. The connection scheduling technique such as the shortest remaining processing time first (we call it SRPT-CS hereafter)[1] is an example of this technique. SRPT-CS schedules connections according to the amount of data to be served in the connection descriptor. It is a work conserving scheduling method which minimizes average response time. In the case where the document sizes are not known in advance, a typical approach is to use heuristic schedulings to improve the performance of short objects. Another category of connection scheduling is to assign priorities to clients or documents. [2,3] are the examples of this technique. Bender[2] suggested the Stretch-so-far EDF scheduling technique which assigns deadlines to the requested documents and processes the one with the earliest deadline first. Lu[3] proposed a connection scheduling technique with service delay feedback. With these techniques, however, there is a possibility of connection starvation unless the scheduling policy is carefully implemented.

I. Chong (Ed.): ICOIN 2002, LNCS 2343, pp. 384–393, 2002.

The problem of size-based web scheduling is that the overall response time of each connection is not guaranteed if a connection requests multiple inlined-documents at once. In this paper, we propose a window-based connection scheduling technique for HTTP/1.1 server, called WCS, which provides a better response time and fairness. The method is based on the shortest connection first principle and schedules web documents using a scheduling window which is dynamically determined by the number of inlined-documents of the shortest connection. For clarity, the paper only focuses on serving HTTP/1.1 requests for static contents whose sizes can be determined in advance.

The remainder of this paper is organized as follows. Section 2 describes the proposed window-based web content scheduling technique. Section 3 analyzes the performance of the proposed technique based on the simulation results. Section 4 concludes the paper.

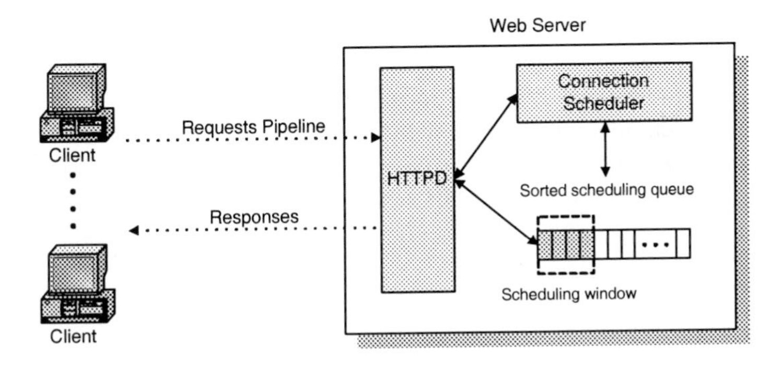

Fig. 1. The window-based connection scheduling model

2 Window-Based Connection Scheduling

Figure 1 illustrates the web server and connection scheduling model considered in our work. When a connection requesting multiple inlined-documents in a single document is created through pipelining, the connection scheduler stores the connection information, such as how many and which inlined-documents are requested, in the corresponding descriptor. The scheduler first sorts the connections by the total size of each connection and then apply scheduling rule based on the scheduling window. The scheduling window is a moving unit with which the scheduling is based on. It is important to decide the size of scheduling window, Δ. The scheduling window size can be determined in two ways. First, it can be decided by the number of requested inlined-documents of a connection in the head of the queue. We call this policy Δ_p. Another method is to use the average number of inlined-documents of the whole connections in the queues as the scheduling window size. We call this policy Δ_q. As will be shown in section 3,

Δ_p provides better response time than Δ_q and hence we use Δ_p as a scheduling policy in our work.

Now, we describe the WCS algorithm. The WCS scheduler sorts the connection information according to the total size of the required document in the queue (that is, the sum of the document size of multiple inlined-documents). The smallest connection is located in the queue head. The WCS scheduler then sets the scheduling window size Δ as the number of the requested inlined-documents of the connection in the queue head. The scope of connection scheduling is dependent on the scheduling window size. For example, if three inlined-documents are simultaneously requested in the connection in the queue head, Δ is 3 and the scheduler schedules the smallest three connections in the sorted queue. As the connections are sorted according to the connection size, the current scheduling scope is the first three connections. Within the scheduling scope the scheduler schedules the connections by round-robin fashion. While the connections are processed, they are not preempted by other connections which are out of the range. After finishing the processing of all inlined-documents of a connection in the queue head, the scheduler slides the scheduling window and adjusts its size to the number of requested inlined-documents remained in the next connection. If the connections in the scheduling window are finished prior to the connection in the queue head, the scheduler includes the next connection located outside of the scheduling window and starts the schedule. Table 1 shows the detailed WCS algorithm.

Table 1. The WCS algorithm

```
Procedure WCS(queue)
  Δ = c_des[qhead].total_req;
  while (c_des[qhead]->file is not NULL)
   for (i=0;i<Δ;i++)
     if (c_des[i]->file is not NULL)
      read (c_des[i]->file);
      c_des[i]->file = c_des[i]->file->next;
      c_des[i].total_req−−;
     else
      c_des[i-1]->next = c_des[i]->next;
      remove (c_des[i]);
      expand_window;
End Procedure
```

Lets explain the algorithm with an example. Table 2 details a sample web site with 5 HTML documents (*a.html*, *b.html*, *c.html*, *d.html* and *e.html*) where each document has multiple inlined-documents. Here, we refer the set of inlined-documents in the document *a.html* as *a.class*. Similarly, the set of inlined-documents in *b.html*, *c.html*, *d.html* and *e.html* are *b.class*, *c.class*, *d.class* and

Table 2. Size of HTML and inlined-documents

HTML(bytes)	Inlined-document(bytes)
a.html(3177)	a_1.gif(218) a_2.gif(2234) a_3.gif(20426) a_4.gif(39358)
b.html(102)	b_1.gif(2223) b_2.gif(13786)
c.html(1774)	c_1.gif(1956) c_2.gif(13786)
d.html(2153)	d_1.jpg(15671) d_2.gif(13786)
e.html(3638)	e_1.jpg(1356) e_2.gif(10703) e_3.jpg(135075)

e.class, respectively. 10 clients (A, B, C, D, E, F, G, H, I and J) are requesting web contents, where five of them (A, C, E, G, and I) request *a.html*, *b.html*, *c.html*, *d.html* and *e.html* documents and the rest (B, D, F, H, and J) request inlined-documents of *a.class*, *b.class*, *c.class*, *d.class*, and *e.class* in each HTML document. The sizes of each HTML document *a*, *b*, *c*, *d* and *e* classes are 62236, 16009, 15742, 29457, 147134 bytes, respectively. The ordered array of these 10 clients by their document size is (C, E, G, A, I, F, D, H, B, J).

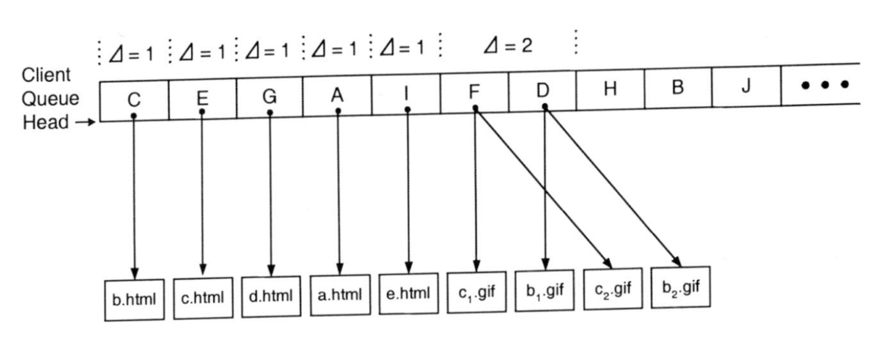

Fig. 2. Scheduling of HTML documents by WCS

The client C at the queue head requests one HTML document, and therefore Δ is set 1 by the scheduler and *b.html* document of C client is then processed. Similarly, E, G, A and I are scheduled one by one. Figure 2 illustrates the scheduling steps. The next clients F, D, H, B and J in the queue changes Δ as follows. F, in the head of the queue, requests 2 inlined-documents($c_1.gif$ and $c_2.gif$), so Δ is set 2 by the scheduler and the scope of scheduling ranges from F to D. Figure 2 shows the processing of inlined-documents requested by F and D clients with round-robin. When all the inlined-documents requested by F are processed, all the inlined-documents requested by D client are processed too. Thus, the scheduling starts from the next one in the queue, H. Since H requests 2 inlined-documents, Δ is set 2 and the current scheduling scope is up to B client. When all the documents requested by H are processed, two inlined-documents($a_1.gif$, $a_2.gif$) of total 4 inlined-documents requested by B client are processed and two inlined-documents($a_3.gif$, $a_4.gif$) remain to be processed. Thus, Δ is set 2 again. The next scheduling range is from B to J and the documents are processed by round-robin.

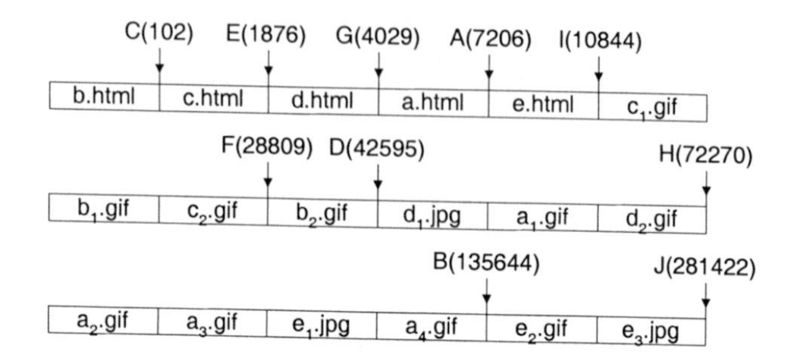

Fig. 3. WCS scheduling

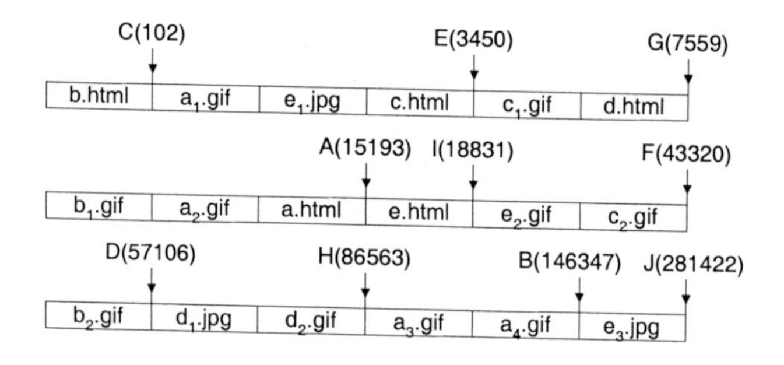

Fig. 4. SRTP-CS scheduling

Figure 3 illustrates the scheduling of 10 clients according to WCS in the above example. The arrows in the figure represent the time when the documents or all the inlined-documents requested by clients are processed completely. The alphabet and the number above each arrow indicate the corresponding finishing client and the amount of data processed by the web server until the time represented by the related arrow. Figure 4 shows the scheduling of the same clients by the SRPT-CS algorithm. As with WCS, SRPT-CS is a scheduling technique based on the document size. However, the difference is that WCS schedules documents according to the total size of documents or inlined-documents requested by the connection whereas SRPT-CS schedules them by the size of each single document or inlined-document. Furthermore, WCS uses the scheduling window, SRPT-CS does not. As predicted with the amount of data processed by the server, the connection completion times of all clients by WCS are faster than the completion times of corresponding clients by SRPT-CS, except the client C which is the case where both algorithms have the same completion time.

3 Analysis

We have compared WCS with two well-known techniques: SRPT-CS[1] and FIFO[1]. A modified version of Flash[5] web server is used to implement the proposed WCS algorithm. The clients use GETALL method[6] to request various web documents. When the client requests HTML documents, the server sends all inlined-documents to the client via a single TCP connection, which is a persistent connection. The documents used in the experiments are generated by SURGE[7]. That is, the document size follows the Pareto distribution and the document requests follows the Zipf-like law[7]. The number of inlined-documents also follows the Pareto distribution and therefore most of the HTML documents have small number of inlined-documents.

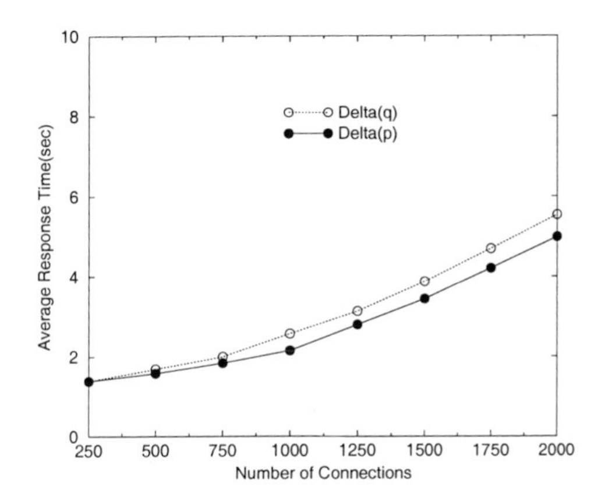

Fig. 5. Comparison of scheduling windows: Δ_p and Δ_q

Figure 5 shows the performance difference of two WCS scheduling policies, Δ_p and Δ_q. The documents used for simulations are generated by SURGE. As shown in the figure, the average response time of Δ_p is faster than that of Δ_q. Δ_p therefore provides better performance and is considered more efficient than Δ_q. With Δ_q the length of the sorted queue becomes longer as the connections are processed and hence the scope of the scheduling becomes wider. With Δ_p, however, the size of scheduling window is decided by the number of inlined-documents in the head of the queue. This means that the scheduling scope is

[1] FIFO is used in Apache[4].

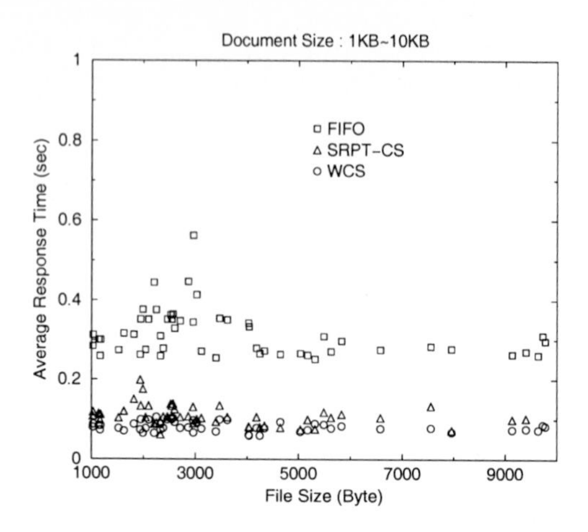

Fig. 6. Average response time vs. document size (1KB - 10KB)

independent of the connection queue length and therefore the average response time of Δ_p can be faster than that of Δ_q.

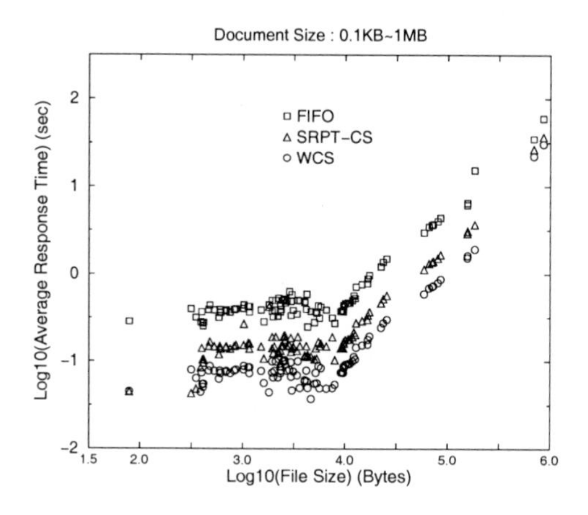

Fig. 7. Average response time vs. document size (0.1KB - 1MB)

Figure 6 and Figure 7 show the average response time of connections when the distribution of objects varies and the number of clients is 1000. Figure 6 and 7 show the results when the size of document is 1KB~10KB and 0.1KB~1MB, respectively. The document size and the average response time are, in Figure 7, plotted as *log-log* scales to present both the small and large documents together. The graphs show that the average response time of WCS is faster than SRPT-CS and FIFO. Also, it is shown that as the document size increases WCS and SRPT-CS have little increase in their average response times, while FIFO becomes slower in the average response time. In the case of SRPT-CS, since the scheduling scope includes all the connections which are requested at the time, the prolonged object scheduling may cause a delay in response time of each connection. FIFO processes the documents in the order they arrive regardless of their sizes. FIFO usually runs on round-robin principle and is affected by system scheduler in the server. The scheduling range of FIFO is extended to the whole connections and thus the processing of small documents can be delayed. The more the number of connections increases, the more the server is overloaded. As a result, the connection response time becomes longer.

Like other priority-driven scheduling algorithms, starvation of large connections can be the problem. Starvation is, however, not a concern when document sizes follow heavy-tailed distributions which are the cases for many web request distributions. Figure 6 and Figure 7 show that starvation even for the largest little one percent of all documents is far lower in WCS than in FIFO. We further experimented the algorithms with 2000 clients and noticed that the overall performance characteristics are similar to the case with 1000 clients, but the average response time of FIFO gets much worse than WCS and SRPT-CS.

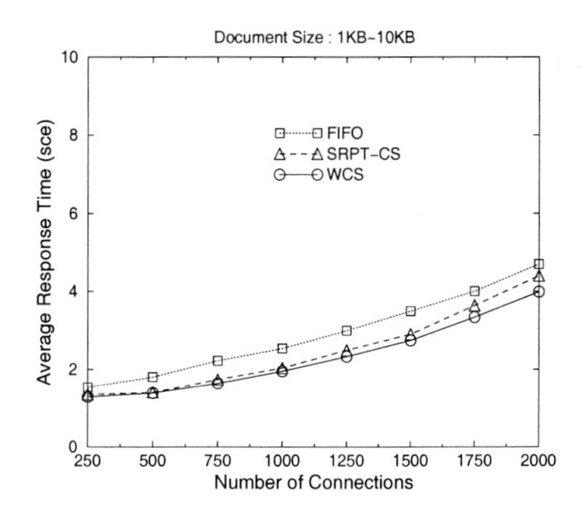

Fig. 8. Average response time vs. no. of connections (document size: 1KB - 10KB)

Figure 8 and 9 show the average transfer time when the number of connections increases. The figures show that WCS is generally faster in its average response time than SRPT-CS or FIFO. WCS shows a better performance especially in the case when the range of size distribution is bigger and more documents are requested per connection. Figure 9 shows the result with a wider range of document distribution. The performance improvement is distinctive in Figure 9 with wider range of document size. In the case of FIFO, the performance of web

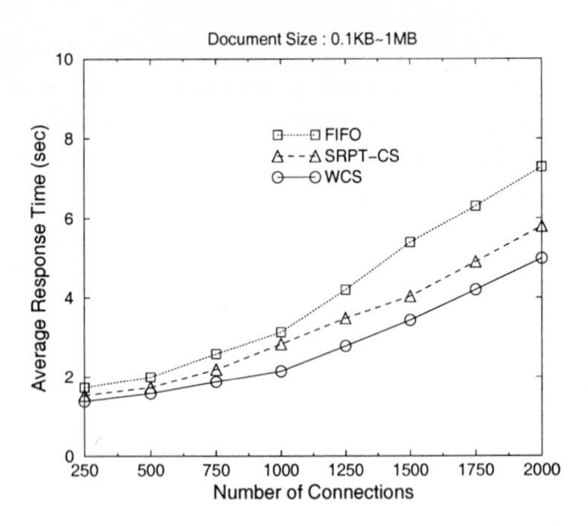

Fig. 9. Average response time vs. no. of connections (document size: 0.1KB - 1MB)

server is affected by the system scheduler as the system does not distinguish the web server process from other application processes. When the documents with different sizes are requested at the same time, SPRT-CS may cause a worst-case result as the algorithm considers the size of each single document and each connection is processed with round-robin fashion. With WCS, however, the worst result would not happen as the scheduling is performed based on the scheduling window. Based on these simulation results, we could conclude that WCS which makes use of a scheduling window performs better than SRPT-CS and FIFO.

4 Conclusion

This paper presented a window-based connection scheduling algorithm to improve the connection response time in HTTP/1.1. The priority in WCS is given to a connection where the total size of requested inlined-documents is the smallest. The scheduling scope of WCS is temporally restricted by the scheduling window. The simulation results show that WCS performs better than SRPT-CS and FIFO in terms of average connection response time. WCS especially shows better performance in the case when the range of size distribution is bigger and many documents are requested per connection. Future work includes dynamic web content scheduling and load balancing for web clustering systems.

Acknowledgement. This work is supported by the University Research Program of the Ministry of Information & Communication in republic of Korea (2001-076-3) and BK(Grant 991031001).

References

1. M. E. Crovella, R. Frangioso, M. Harchol-Balter: Connection Scheduling in Web Servers. Proc. of the 1999 USENIX Symposium on Internet Technologies and Systems. (1999) 243-254.
2. M. Bender, S. Chakravarti, S. Muthukrishnan: Flow and Stretch Metrics For Scheduling Continuous Job Streams. Proc. of the 9th Annual ACM-SIAM Symposium on Discrete Algorithms. (1998) 270-279.
3. C. Lu, T. Abdelzaher, J. Stankovic, S. Son: A Feedback Control Approach for Guaranteeing Relative Delays in Web Servers. IEEE Real-Time Technology and Applications Symposium. (2001).
4. The Apache Group: http://www.apache.org.
5. V. S. Pai, P. Druschel, W. Zwaenepoel: Flash: An Efficient and Portable Web Server. Proc. of the USENIX 1999 Annual Technical Conference. (1999) 199-212.
6. V. N. Padmanabhan, J. Mogul: Improving HTTP Latency. Computer Networks and ISDN Systems. **28** (1995) 25-35.
7. P. Barford, M. Crovella: Generating Representative Web Workloads for Network and Server Performance Evaluation. Proc. of Performance '98/ACM SIGMETRICS. (1998) 151-160.

TCP-Vegas Slow Start Performance in Large Bandwidth Delay Network*

Soo-hyeong Lee[1], Byung G. Kim[2], and Yanghee Choi[1]

[1] Department of Computer Science and Engineering,
Seoul National University, Seoul, Korea,
shlee@mmlab.snu.ac.kr
[2] Department of Computer Science,
University of Massachussetts at Lowell, Lowell, MA, USA

Abstract. With the rapid expansion of the Internet, it has become possible for end hosts that are separated long apart to be connected through high bandwidth links. This environment, called a Large Bandwidth Delay Network, poses a major challenge to the performance of the Internet. A long-delay connection usually suffers from being treated unfairly when competing with short-delay connections. A link, to avoid being under-utilized, has to be equipped with an a buffer as large as the bandwidth delay product of the longest connection.
TCP-Vegas is known as a potential solution to these problems. According to a number of previous studies, it is said to fairly treat connections with different propagation delays and avoid under-utilization even with a buffer that is independent of the bandwidth delay product.
In this paper we show, from simulation and analysis, that the current TCP-Vegas does NOT achieve high utilization in such a large bandwidth delay network, because of its slow-start phase. Moreover, to avoid loss, TCP-Vegas slow-start requires a buffer that is proportional to the square root of the bandwidth delay product. We propose a solution to these problems and analyze its performance.

1 Introduction

TCP is the protocol dominating the Internet[6]. TCP is a connection-oriented transport protocol that provides reliable delivery service using the unreliable IP network. There are several versions of TCP such as Tahoe, Reno, and Vegas. Every TCP version except Vegas adopts the congestion-avoidance scheme described in [1], whose basic idea is to slowly increase the number of packets in the network until a loss occurs, and to halve it when a loss does occur. Each version improves the performance of previous ones by revising behavior during loss recovery. However, Vegas tries to gain performance improvement more fundamentally, in that it does not incur loss because it does not wait for loss when reducing the number of packets in the network.

* This work was supported by Korea Science and Engineering Foundation under the contract number 20005-303-02-2. It was also supported by the Brain Korea 21 Project and National Research Laboratory Project of Korea.

I. Chong (Ed.): ICOIN 2002, LNCS 2343, pp. 394–406, 2002.

TCP has two phases: congestion-avoidance and slow-start. First, we need to define two terms: window of a connection is the number of packets belonging to the connection in the network; round-trip time is the time required for a packet to traverse the network from sender to receiver and then back from receiver to sender. Slow-start doubles the window every round-trip time, whereas congestion-avoidance increases the window by a packet every round-trip time. It can be shown that slow-start effectively opens the window exponentially, whereas congestion-avoidance opens the window with a speed increase slower than linear with time.

Vegas has a new congestion detection formula[1] known as dictating the per-connection queue length in the bottleneck link[3]. Vegas applies this formula to both congestion-avoidance and slow-start to avoid loss. Vegas congestion-avoidance keeps the value of the formula within two thresholds α and β ($\alpha < \beta$): it increases the window (by one packet per round-trip time) when the formula is below α, decreases the window (by one packet per round-trip time) when the formula is above β, and keeps the window unchanged if the formula lies between the two thresholds. Vegas slow-start stops its window doubling when the value of the formula exceeds a threshold γ. The window is doubled only every second round-trip time. In between, the window stays fixed. Let us define two terms: the period of time during which the window is doubled is a *doubling round*[2], and the period of time during which the window does not change is a *holding round.* The Vegas slow-start is characterized by periodic repetition of doubling and holding.

This paper studies Vegas because it is a potential solution to be used in the Large Bandwidth Delay Network. Vegas is known to achieve high link utilization and low loss [3] [4]. In addition, [8] and [9] observed that Vegas has the virtue of having a throughput independent of propagation delay. Moreover, [8] and [13] argued that such a good performance is obtained with buffer requirement independent of propagation delay. These characteristics are substantially different to those of Tahoe and Reno. In Reno, loss is inevitable, the link utilization is highly dependent on the buffer size, throughput is a decreasing function of propagation delay, and the buffer requirement should be at least proportional to the bandwidth delay product (BDP) to achieve the same link utilization[14].

This paper studies the Vegas slow-start because it is the most important part that affects the performance of Vegas, as shown by [12]. Although it is important to study Vegas slow-start, there has been no reported research on it. Every analytical study [8][13][9] favoring Vegas has only studied the steady state congestion-avoidance phase. Their lack of study on the slow-start may be the

[1] The inventor of Vegas used the term 'congestion detection mechanism' when referring to the whole process of calculating a formula and comparing it with thresholds to decide whether or not to increase the window. By 'congestion detection formula', we mean the formula used in the congestion detection mechanism. We use this term for the purpose of discussion in this paper.

[2] *Round* is defined to be the time interval whose length is equal to the round-trip time.

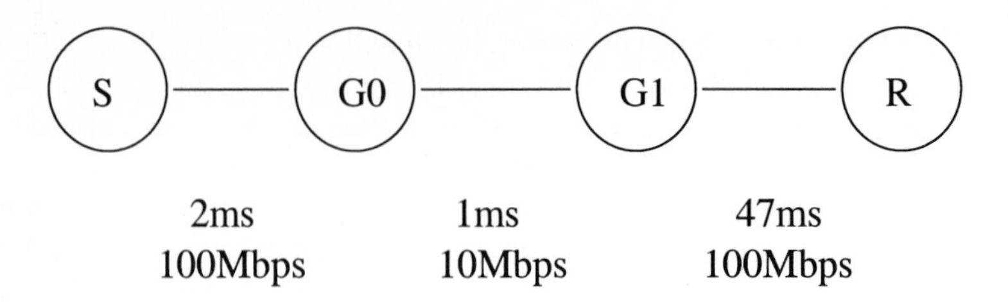

Fig. 1. Simulation Topology

reason for the discrepancy with some researchers' reports that Vegas performs poorly over either a gigabit network[5] or a long delay link[7].

This paper identifies that Vegas slow-start has two problems: 1) it is too conservative in utilizing the available bandwidth; 2) it can cause loss by overflowing the buffer. The first problem was reported recently in [12], whereas the second problem has been reported since its invention[3]. The first problem results in long convergence delay because congestion avoidance has to take the role of increasing the window. The second problem results in high maximum queue length (or equivalently, buffer requirement for no loss). This paper calculates both the convergence delay and the buffer requirement through analysis and confirms the analysis by simulation. It is worth noting that the two problems become worse as the network becomes a large bandwidth delay (LBD) network.

This paper proposes and evaluates a method to eliminate both problems. We show that a kind of pacing, named *Streak Doubling*, makes buffer requirement independent of the BDP of the connection and avoids the under-utilization.

The rest of this paper is organized as follows. We describe the problem and analyze it in Section 2. We describe our solution to the problem in Section 3. We show performance of our solution as the BDP changes in Section 4. A summary of our work and future plans is given in Section 5.

2 Symptom

The two problems, under-utilization of the link and a high buffer requirement, are apparent in the simulation. Figure 2 and Fig. 3 show the sending rate and queue length evolution of a connection under the simulation topology of Fig. 1 with a packet size of 1000B. The connection delivers data from host S to host R, through the bottleneck link G0-G1. The connection starting at 1 second under-utilizes the 10Mbps bottleneck link until it reaches the steady state at 13 seconds.

The queue length before 2.2 seconds is characterized by surges occurring every 0.2 second. Four surges are conspicuous, with heights of 2, 4, 8, and 16 packets. Given that each surge is twice as high as the previous one, the height of the largest surge would increase as the surge count increases, which is the case when the BDP increases as we show later in this section. Although the queue

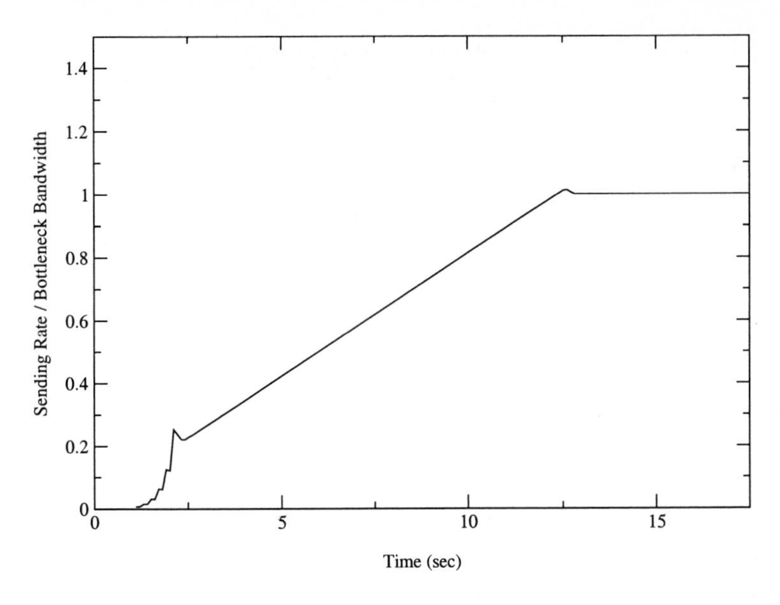

Fig. 2. Sending Rate of Normal Vegas

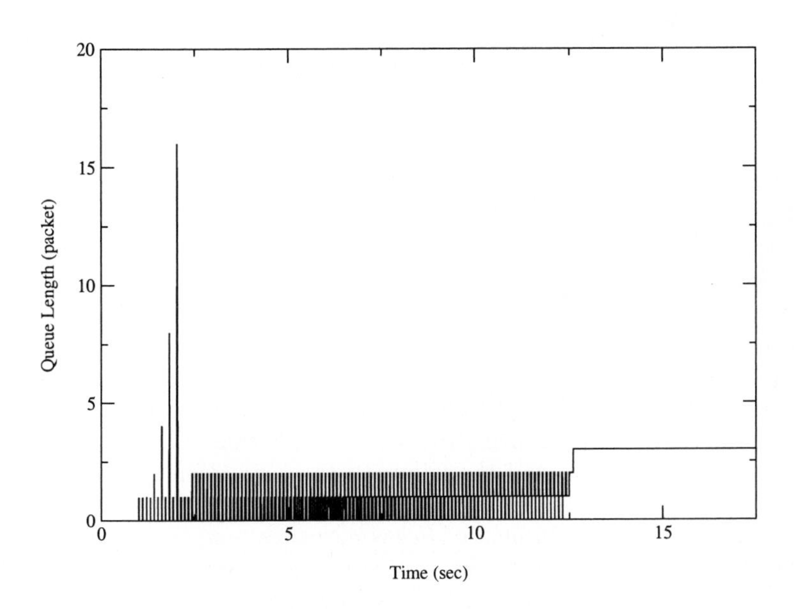

Fig. 3. Queue Length of Normal Vegas

buildup during slow-start may not seem very high in this specific simulation scenario, we can show that it is a monotonically increasing function of BDP, whereas steady state queue length is independent of the BDP. We call the queue buildup during slow-start as *temporary* because it disappears within a round-trip

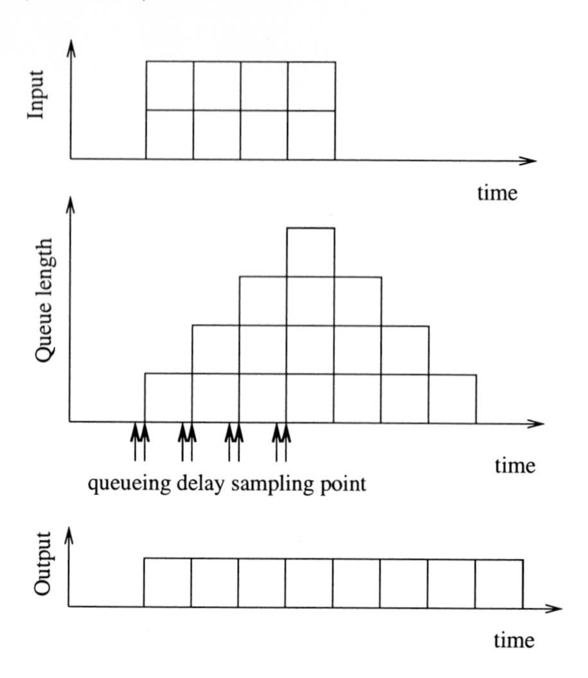

Fig. 4. Temporary Queue Buildup during Slow-Start

time. This is observed as every surge is followed by a one-packet queue buildup in a round-trip time of 0.1 second.

Temporary queue buildup during slow-start, which is the cause of the high buffer requirement, has long been identified by researchers. It is because slow-start sends two packets on receiving an ack[3]. $W/2$ acknowledgments create W packets with sending rate double the bottleneck bandwidth, which in turn create queue length of $W/2$. The packet stream departing the bottleneck link now has a rate equal to the bottleneck bandwidth. This is depicted in Fig. 4 where $W = 8$. The horizontal unit is the packet service time and the vertical unit is packets.

The problem of under-utilizing the bandwidth is caused by a combination of temporary queue buildup and the failure of the congestion detection formula. We develop this story in the following two subsections. First, we show that the congestion detection formula of Vegas measures temporary queue buildup. Second, we show that the under-utilization becomes even worse as the BDP of the connection increases, and that buffer requirement is no exception.

2.1 Failure of the Vegas Congestion Detection Formula

Vegas slow-start suffers from premature termination of slow-start, because it overestimates $Diff$ due to the overestimated RTT. In this subsection, we show the reason of the RTT overestimation.

[3] We use ack as an abbreviation of acknowledgment.

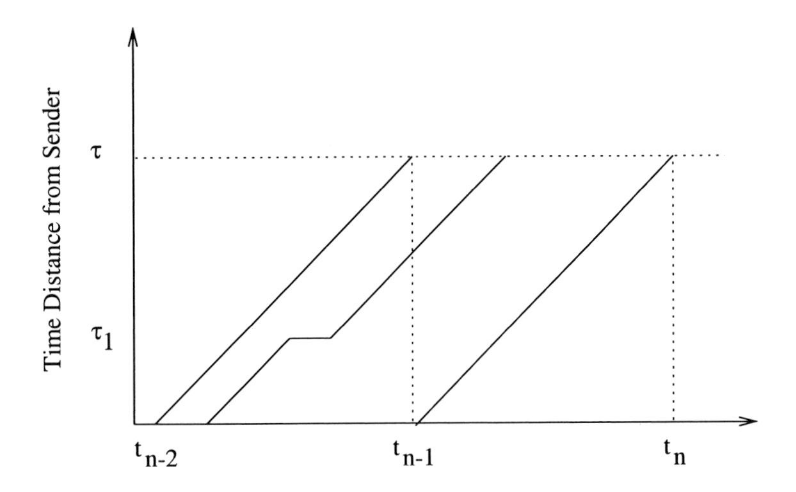

Fig. 5. Vegas Sees Queueing Delay of One and a Half RTT Before

Suppose t_n is the beginning of the n-th round. The Vegas *congestion detection formula* at time t_n is as follows:

$$Diff \triangleq \left\{1 - \frac{baseRTT}{RTT(t_n)}\right\} W(t_n), \tag{1}$$

where W(t) is window size at time t and *baseRTT* is the least of Round-Trip Times (RTT) until then. We assume $baseRTT = \tau$, where τ is the round-trip propagation delay.

Vegas slow-start checks whether $Diff > \gamma + 0.5$ only at the beginning of a doubling round. Thus t_n and t_{n-2} are beginnings of doubling rounds, and t_{n-1} is the beginning of a holding round.

$RTT(t_n)$ is the average of the round-trip times experienced by the acknowledgments received during $[t_{n-1}, t_n)$. Figure 5 shows that these are sampled by packets sent during $[t_{n-2}, t_{n-1})$. The solid line is the trajectory of a packet from when it departs the sender (the vertical axis value is 0) until it returns to the sender as an ack (vertical axis value is round-trip propagation delay τ). Through its journey, the packet cannot stop except in the bottleneck link, located τ_1 away from the sender, where it waits for service in the queue.

$RTT(t_n)$ suffers from a temporary queue buildup, because t_{n-2} is the beginning of a doubling round. Small arrows in Fig. 4 show the sampling times. Suppose C is the bottleneck link bandwidth in packets per second, then i-th packet (i begins at 1) in a round experiences a queueing delay of $\lfloor \frac{i}{2} \rfloor C^{-1}$. The average round-trip time the Vegas sender sees is

$$RTT(t_n) = baseRTT + \frac{W(t_{n-1})}{4} C^{-1}, \tag{2}$$

which is higher than the correct round-trip time, *baseRTT*.

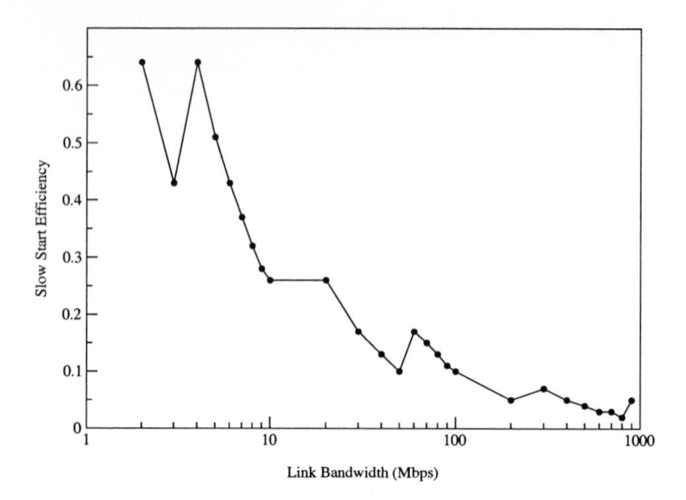

Fig. 6. Slow-Start Efficiency for Various Bottleneck Bandwidths where Round-trip Propagation Delay is 100ms and Packet Size is 1000B

2.2 Large Bandwidth Delay

The problem is that the termination of the Vegas slow-start does not extend in proportion to the increase of the BDP of the connection. Moreover, the rate of window increase during congestion-avoidance is indirectly proportional to BDP. Hence, as BDP increases, under-utilization of the bottleneck link becomes aggravated because the convergence delay increases.

Suppose $W^* \triangleq C\tau$ is the ideal window size that fully utilizes the bottleneck link, and the *slow-start efficiency* η is defined to be the ratio of the slow-start termination window to the ideal window. The simulation result in Fig. 6 shows that η decreases rapidly as bottleneck bandwidth increases. Increasing the propagation delay or decreasing the packet size has the same effect. In general, as we apply (2) to (1), while keeping in mind that window size assumes only integer powers of 2, we have:

$$\frac{f\left(\gamma_N\right)}{2} \leqslant \eta < f\left(\gamma_N\right) \tag{3}$$

where $\gamma_N \triangleq \frac{\gamma+0.5}{W^*}$ and $f\left(\gamma_N\right) \triangleq \gamma_N + \sqrt{16\gamma_N + \gamma_N^2}$. Note that $f\left(\gamma_N\right)$ is a strictly decreasing [4] function of W^*.

We have a long convergence delay if the slow-start efficiency is low and the propagation delay is high. Let us define *rise time* T_r as the portion of convergence delay other than the slow-start duration. Rise time is derived[5] as

[4] η is not a strictly decreasing function of W^*. Occasional surges in Fig. 6 reflect that the window size assumes only integer powers of 2.

[5] This equation is a formulation of the fact that the window increases one packet a round-trip time during the congestion-avoidance phase.

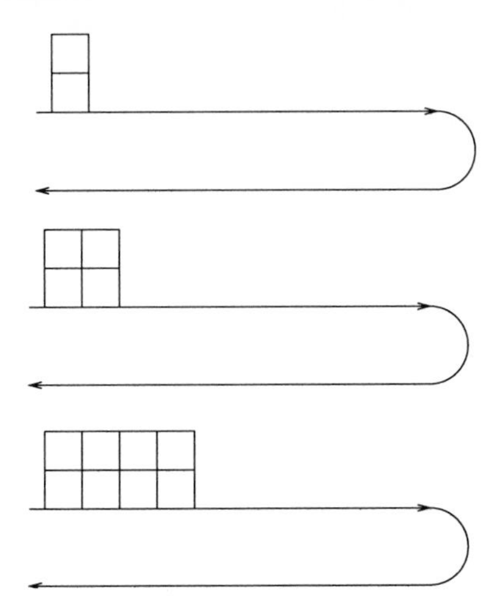

Fig. 7. Packet Spacing of Vegas slow-start

$T_r = ((1-\eta)\, W^* + \alpha)\, \tau$ if the slow-start efficiency is below unity. We can see that the rise time increases as the square of the propagation delay as we apply (3) here.

We also have a high buffer requirement even if the slow-start efficiency is low. The buffer requirement is given[6] as $Q^{max} = \frac{\eta W^*}{2}$, which increases as the square root of BDP.

One would think that controlling γ might solve some of the problems. In general, a higher γ results in a higher η, which means a lower convergence delay. However, one cannot make η fixed to a desired value by controlling γ, because the window can only assume integer powers of 2. The exact value of η is predictable but not controllable, especially when γ is relatively high compared to BDP.

There are other ways to make η closer to unity, which we show in the next section. However, high η always results in an enormous buffer requirement if two packets are sent on the moment of the receipt of an ack. In general, we face the following trade-off:

$$\frac{T_r}{\tau} + 2Q^{max} = W^* + \alpha. \tag{4}$$

As BDP increases, either round count for convergence ($\frac{T_r}{\tau}$), or the buffer requirement (Q^{max}) should be increased.

3 Solution

We consider a kind of pacing, named Streak Doubling, that strictly requires packet spacing equal to ack spacing.

[6] This is a formulation that W packets create queue length of $W/2$ if the sending rate is double the bottleneck bandwidth. This was depicted in Fig. 4.

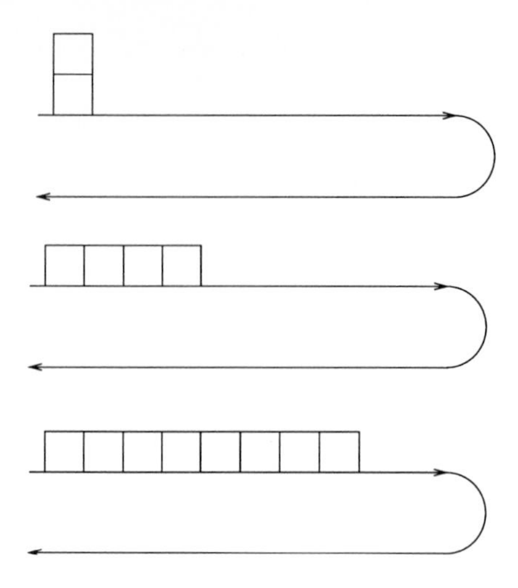

Fig. 8. Packet Spacing of Streak Doubling

We depict the operation of this algorithm and that of normal Vegas in Figs. 7 and 8. We show three consecutive doubling rounds with windows of 2, 4, and 8. The square is a packet, whose area is the amount of data that can be served by the bottleneck link during the horizontal side length of the square. Hence, the bottleneck link can serve a single row of back-to-back squares without queueing. A single column of double row packets means that the two packets are sent at the speed of the access link, which is assumed to be at least twice as fast as the bottleneck link in the figure. Normal Vegas slow-start, in Fig. 7, always sends a double row of back-to-back squares, which results in a temporary queue buildup as shown in Fig. 4. Conversely, pacing never sends a double row of back-to-back squares, as shown in Fig. 8, thus avoiding a temporary queue buildup which would lead to premature termination of the Vegas slow-start.

Streak doubling, in Fig. 8, has two periods: an active period and an inactive period. The active period has a sending rate equal to the available bandwidth, whereas the inactive period has a sending rate of zero. During the first half of the active period, each packet is sent on receipt of an ack without increasing the window. The latter half of the active period is a replica of the first half except that a packet is sent by a timer-scheduled event of the window increase instead of on receipt of an ack. The period of the scheduled event is the average packet service time of the bottleneck link. In the beginning, it has to send two packets as fast as possible in order to estimate the average packet service time. Streak doubling has a maximum temporary queue length of two packets because of this.

Streak doubling provides constant maximum queue length and constant round counts for rise time without regard to BDP: $Q^{\max} = 2$ and $T_r = \alpha\tau_f$. For comparison with normal Vegas slow-start, we show an equation that looks similar to (4) and is different only in the right hand side: $\frac{T_r}{\tau_f} + 2Q^{max} = 4 + \alpha$.

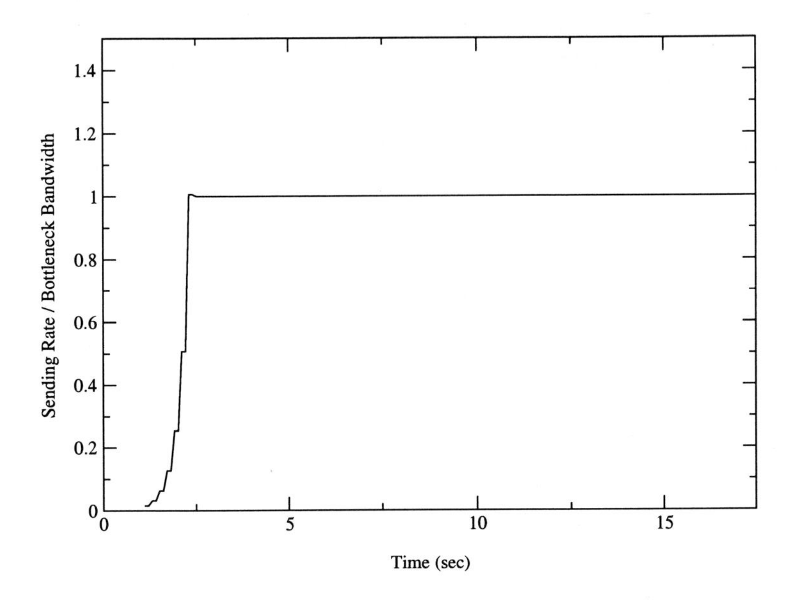

Fig. 9. Sending Rate of Streak Doubling

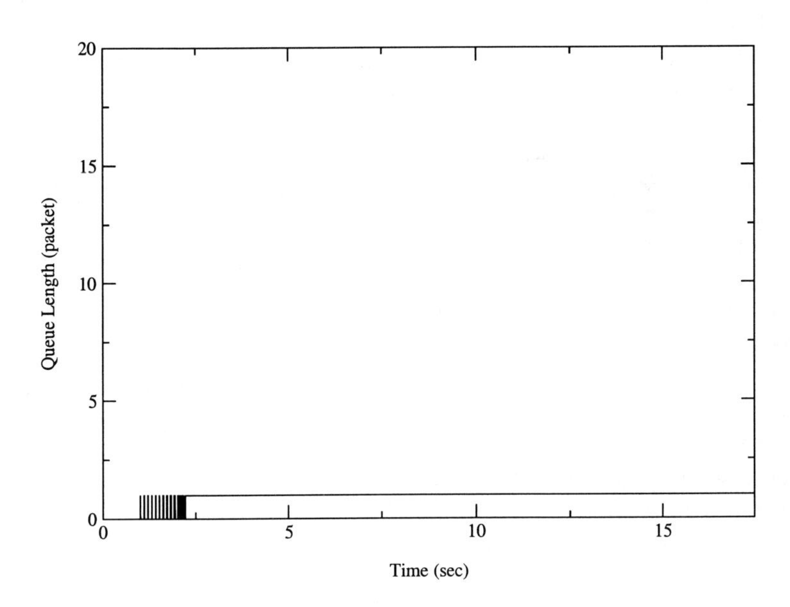

Fig. 10. Queue Length of Streak Doubling

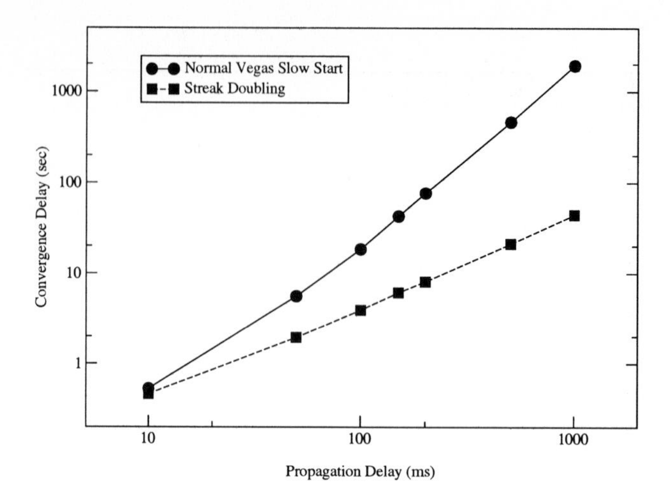

Fig. 11. Convergence Delay

4 Performance

We see from Fig. 11 that the convergence delay T_c [7]of streak doubling is proportional to the propagation delay τ, whereas the convergence delay of normal Vegas increases almost as the square of the propagation delay. For a propagation delay of 1 second, streak doubling has only a one-hundredth of the convergence delay of normal Vegas. Using a least squared error fit, assuming a good fit to $T_c = A(\tau)^B$, we obtain the exponent of $B = 0.996$. [8] Applying the same fitting to the normal Vegas results in the exponent of $B = 1.80$.[9]

We see from Fig. 12 that the maximum queue length during slow-start is a constant of two packets for streak doubling, whereas that of normal Vegas is a monotonic increasing function of propagation delay.

5 Conclusions

Some researchers have argued that Vegas is suited for large bandwidth delay network because of its two merits in steady state: being fair irrespective of propagation delay, and highly utilizing the bandwidth with buffer requirement irrespective of bandwidth-delay product.

[7] The convergence delay T_c for the single connection case is the sum of the rise time T_r and the slow-start duration T_{ss}. However, it is acceptable to equate convergence delay with rise time because the simulation result shows that the slow-start duration is negligible.

[8] τ is in *msec* and T_c is in *sec*. $A = 0.040$, $\sigma[\log A] = 0.040$, and $\sigma[B] = 0.018$.

[9] $A = 0.0061$, $\sigma[\log A] = 0.150$, and $\sigma[B] = 0.0677$.

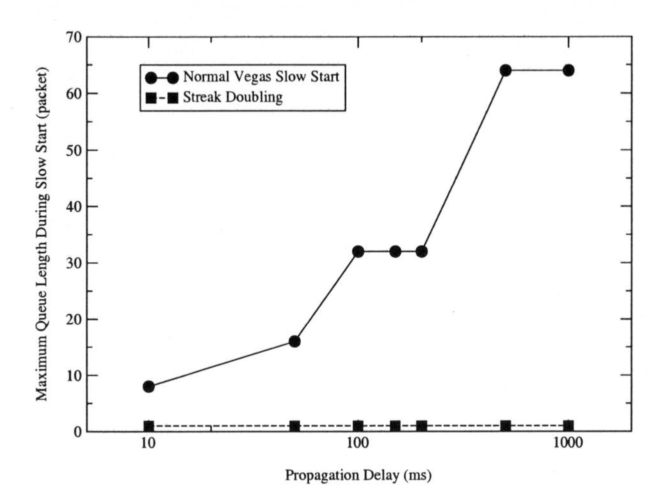

Fig. 12. Maximum Queue Length During Slow-Start

We have shown that the current Vegas shows poor performance in a large bandwidth delay network because of its slow-start phase. We identified two problems of Vegas slow-start: under-utilization and temporary queue buildup. The under-utilization can lead to a lower throughput of Vegas in LBD. The temporary queue buildup requires buffer proportional to the square root of the BDP. We have proposed a kind of pacing as the solution to solve the under-utilization problem with a constant buffer requirement.

We are planning two areas of further study. First, we plan to further explore whether Vegas is suitable for a large bandwidth-delay network. There are many issues left unsolved, such as convergence delay in reaching steady state in the multiple connections case. Second, we plan to implant the slow-start of Vegas into the more popular Reno. We believe that it will improve the performance because the initial slow-start of Vegas does not incur the coarse retransmission timeout caused by burst loss. The lossless slow-start of Vegas is beneficial to its employer as well as to the network, unlike the congestion-avoidance mechanism of Vegas which is detrimental to its employer when competing with the more aggressive connections such as Reno.

References

1. V. Jacobson, *Congestion Avoidance and Control*, Proceedings of the ACM SIGCOMM'88, August 1988.
2. S. Keshav, *A Control-Theoretic Approach to Flow Control*, Proceedings of the ACM SIGCOMM '91, pp. 3-15, September 1991.
3. Lawrence S. Brakmo and Larry L. Peterson, *TCP Vegas : End to End Congestion Avoidance on a Global Internet*, IEEE J. of Selected Areas in Communication, pp.1465-1480, October 1995.

4. J. S. Ahn, P. B. Danzig, Z. Liu, and L. Yan, *Evaluation of TCP Vegas: Emulation and Experiment*, Proceedings of the ACM SIGCOMM'95, Cambridge, MA, August 1995.
5. J. S. Ahn and P. B. Danzig, *Packet Network Simulation: Speedup and Accuracy Versus Timing Granularity*, IEEE Transactions on Networking, 4(5):743-757, October 1996.
6. K. Thompson, G. Miller, and R. Wilder, *Wide-Area Internet Traffic Patterns and Characteristics*, IEEE Network, 11(6):10-23, November/December 1997.
7. Y. Zhang, E. Yan, and S.K. Dao, *A Measurement of TCP over Long-Delay Network*, Proceedings of 6th Intl. Conf. on Telecommunication Systems, pages 498-504, March 1998.
8. T. Bonald, *Comparison of TCP Reno and TCP Vegas via Fluid Approximation*, Workshop on TCP, 1998.
9. Jeonghoon Mo, Richard J. La, Venkat Anantharam, and Jean Walrand, *Analysis and Comparison of TCP Reno and Vegas*, IEEE INFOCOM'99, pp.1556-1563, 1999.
10. S. McCanne and S. Floyd, *Ns(network simulator)*, http://www-mash.cs.berkeley.edu/ns
11. G. Hasegawa, M. Murata, and H. Miyahara, *Fairness and Stability of Congestion Control Mechanisms of TCP*, IEEE INFOCOM'99, 1999.
12. U. Hengartner, J. Boliger, and T. Gross, *TCP Vegas Revisited*, IEEE INFOCOM 2000, 2000.
13. O. Ait Hellal and E. Altman, *Analysis of TCP Vegas and TCP Reno*, Telecommunications Systems, 2000.
14. H. Lee, S. Lee, and Y. Choi, *The Influence of the Large Bandwidth-Delay Product on TCP Reno, NewReno, and SACK*, the 15th International Conference on Information Networking, Beppu, Japan, January 2001.

Performance Analysis of a Generic GMPLS Switching Architecture under ON–OFF Traffic Sources*

Ling-Chih Kao and Zsehong Tsai

Department of Electrical Engineering, National Taiwan University, Taipei, TAIWAN
d6942004@yahoo.com.tw, ztsai@cc.ee.ntu.edu.tw

Abstract. This paper proposes a queueing model including the control plane and the switching buffer mechanism of a GMPLS switch for evaluating the performance of a GMPLS switching architecture. With the proposed model, one can select appropriate parameters for the label-setup policy and the label-release policy to match the traffic load and network environment. Key performance metrics, including the throughput, the label-setup rate, and the fast path bandwidth utilization can be obtained via the analytical results. Numerical results and simulations are used to verify the accuracy of our proposed queueing model. The trade-off among these performance metrics can be observed as well.

1 Introduction

In recent years, it has been a trend to provide a wide range of data services over the same backbone network via newly adopted technologies such as Multi-Protocol Label Switching (MPLS) [1] and Multi-Protocol Lambda Switching (MPλS) [2] to overcome the scalability and complexity issues. Many other layer-3 switching technologies [3] have also been specially designed to solve the dilemma of routing table scalability and overload of routing processing. But these proposed technologies may not be compatible with one another. In order to integrate these techniques, the Internet Engineering Task Force (IETF) proposes MPLS. The basic concept of MPLS is that packet forwarding is based on a fixed short length label instead of longest matching search, which can shorten packet transit time. There are two most popular approaches for connection setup in MPLS. The traffic-driven method is to trigger label-setup according to traffic demand, while the topology-driven system is based on routing information. In addition, by way of Constraint-based Routed Label Distribution Protocol (CR-LDP) [4] or Resource ReSerVation Protocol (RSVP) [5], it is possible to include QoS mechanism in MPLS. When it is necessary to combine traffic engineering aspect of MPLS and bandwidth provision capability of DWDM, MPλS [2] is found to play a major role. Meanwhile, considering that there are many different underlying data-link and physical layer technologies, Generalized Multi-Protocol Label Switching (GMPLS) [6,7] is thus suggested to extend MPLS to encompass time-division, wavelength (lambdas) and spatial switching.

In order to control different switching operations under GMPLS, the label defined for various switches is required to be of different formats [8], and related signaling and

* This work was supported by National Science Council, R.O.C., under Grant NSC 90-2213-E-002-076, and by Ministry of Education, R.O.C., under Grant 89E-FA06-2-4-7.

I. Chong (Ed.): ICOIN 2002, LNCS 2343, pp. 407–418, 2002.

routing protocols also need modifying [6,9]. However, the key operations of the control plane of these various switching protocol suites are found to be similar. Although basic functions of the GMPLS control plane have been discussed or defined in the literature, operation procedures for efficient resource control have still being defined and their impact on performance have still being investigated. Some papers [10]–[12] proposed performance queueing models for MPLS or GMPLS, but most detailed operations of the GMPLS control plane are not well covered. At the same time, it is often found that a sophisticated queueing model which can evaluate the performance of the GMPLS switching network is not easy to build. Therefore, a model embracing detailed operations of the GMPLS control plane is thus strongly needed.

In this paper, we develop a queueing model to characterize the behavior of most operations of a GMPLS switch. The aggregation of IP streams can save usage of labels (or lambdas) and thus alleviate the processing load of the GMPLS controller. The label-setup policy is based on the accumulated packets in the default path buffer. The label-release policy is controlled by an adjustable label-release timer. Efficient resource allocation mechanism is thus achieved by fine tuning the flexible label-setup policy and the adjustable label-release timer. Although our queueing model is traffic-driven oriented, the behavior of the topology-driven system can be approximately obtained via extreme case of this traffic-driven model. Some key performance measures, such as the throughput, the label-setup rate, and the path bandwidth utilization, can all be derived in the proposed model.

The remainder of the paper is organized in the following. In Section II, the queueing model for a GMPLS switch is described. In Section III, the analysis procedure is proposed. In Section IV, three performance measures are derived. Numerical experiments and simulation results are discussed in Section V. Conclusions are drawn in Section VI.

2 Queueing Model

In this section, a queueing model characterizing the behavior of an aggregated IP stream passing through a GMPLS switch is proposed. The number of labels is assumed to be enough for all incoming flows. The bandwidth allocated to each label (or a flow) is fixed. Therefore, we can focus on investigating the steady-state performance of a GMPLS switch without label contentions. For simplicity, we focus on the case that only one single flow is included in this queueing model. The results can then be easily extended to the general case. Regarding the traffic source, an aggregated stream (equivalently a flow) is assumed to be consisted of N homogeneous IPPs (Interrupted Poisson Process), where each IPP has an exponentially distributed *on* (*off*) duration with mean equals $1/\alpha$ ($1/\beta$) and λ is the arrival rate in *on* state. Note that this traffic source model includes four parameters to match most Markovian traffic patterns[1]

The queueing model for a GMPLS switch, GMPLS queueing model, is shown in Fig. 1. The solid lines in Fig. 1 denote the paths that packets go through and the dotted lines are the signaling paths. There are three major functional blocks in this model: the GMPLS controller, the default route module, and the fast route module. The functions

[1] Self-similar process is not considered in this paper.

of the control plane are included in the GMPLS controller. The default route module stands for the IP-layer (layer-3) and data-link layer (layer-2) on the default path. The fast route data-link is represented by the fast route module. In this GMPLS architecture, the *label* is used as a generic term. When GMPLS is used to control TDM such as SONET, time slots are labels. Each frequency (or λ) corresponds to a label when FDM such as WDM is taken as the underlying switching technology. When the switching mechanism is space-division multiplexing based, labels are referred to as ports. Six queueing nodes are included in this model: *Default_Route, Fast_Route, Label_Pool, Fast_Route_Setup, Label_Release, and Label_Release_Timer.* Traffic served by traditional routing protocol will be served by the *Default_Route* node, whose buffer stores the packets which cannot be processed in time by the IP processor on the default path. Meanwhile, the *Fast_Route* node serves packets whose stream has been assigned a label. The fast path buffer stores the packets which can not be processed in time by the fast route. The *Label_Pool* stores the labels which represent the availability of the fast path and the fast path is available if there is a label in the *Label_Pool.* The *Fast_Route_Setup* node represents the time required to set up an LSP (Label Switched Path) for an aggregated stream. The *Label_Release* node represents the time required to release a label. The *Label_Release_Timer* node represents a label-release timer. This timer indicates the maximum length of idle period of an aggregated stream before its label is released for other use. Once an aggregated stream is granted a label, it is served with its own *Fast_Route* node and uses its own label-release mechanism. As a result, this model is used to examine the protocol efficiency instead of label competitions. The assumed details of label operations over this queueing model are described as follows.

When an aggregated IP stream arrives, two possible operations may occur. In the first case, incoming traffic has been assigned a fast path. Then its packets will be directly sent to the *Fast_Route* node. In the second case, incoming traffic has not been assigned a fast path. All the packets are continuously served by the *Default_Route* node (via the *default path*) during the label-setup operations under this situation. If the accumulated packets in the buffer of the *Default_Route* node have not reached the triggering threshold (m), it is served by the *Default_Route* node through traditional IP routing protocol. However, if the accumulated packets in the buffer of the *Default_Route* node reach the triggering threshold, the flow classifier & label_setup_policy manager will trigger the default route module to send a setup request packet through the *Fast_Route_Setup* node to its downstream LSR (Label Switch Router) until the egress LSR for negotiating an appropriate LSP according to the current network resources. The GMPLS controller will set up a path called the *fast path* for this stream and assign it a label.

The label manager maintains an activity timer to control the label-release operation of the flow. The label is released only if the activity timer indicating that the maximum allowed inactive duration has been reached. Incoming packets will be blocked if the accumulated packets in the *Default_Route* node exceed the buffer size of the *Default_Route* node but the stream has not been assigned a fast path, or if the accumulated packets in the *Fast_Route* node exceed the buffer size of the *Fast_Route* node when the stream has been assigned a fast path.

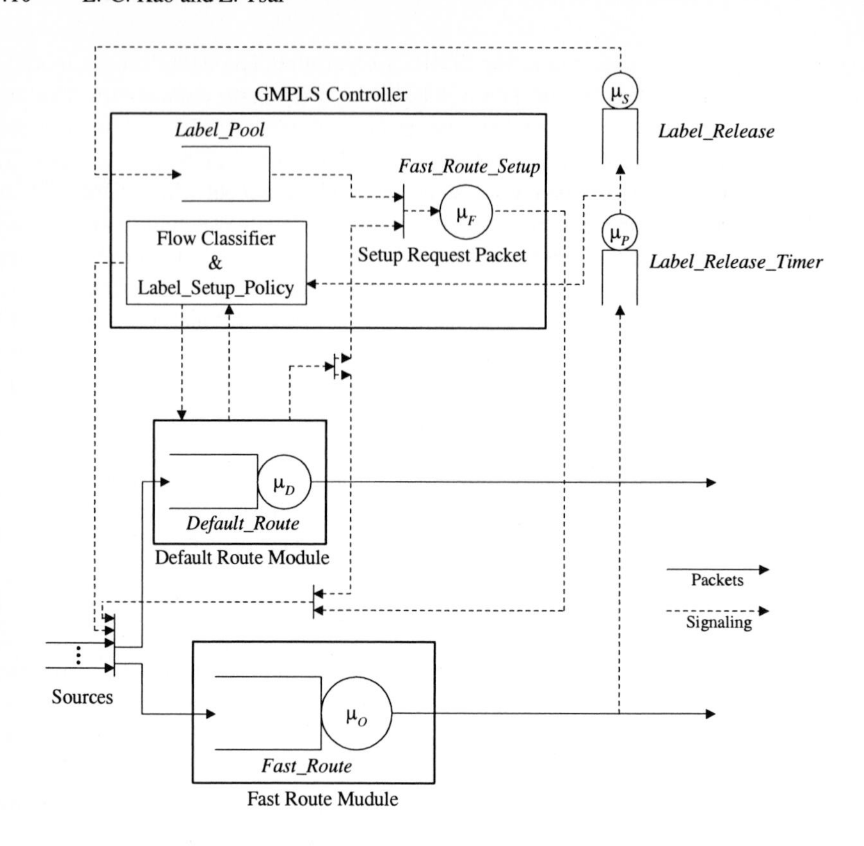

Fig. 1. GMPLS queueing model.

3 Steady-State Analysis

We here propose a procedure to calculate steady-state distribution of a GMPLS switch model as shown in Fig. 1. We adopt the following notations:

$1/\mu$: packet length (bit per packet).
C_D: default path capacity (bps).
C_O: fast path capacity (bps).
$\mu_O = \mu C_O$: service rate of the *Fast_Route* node.
$T_{rel} = \frac{1}{\mu_P}$: the average sojourn time of the *Label_Release_Timer* node, where μ_P is the service rate of the *Label_Release_Timer* node.
μ_S: service rate of the *Label_Release* node.
μ_F: service rate of the *Fast_Route_Setup* node.
T_{LSP}: label-setup latency.
$\mu_D = \mu C_D$: service rate of the *Default_Route* node.
n: buffer size of *Default_Route* node.
t: buffer size of *Fast_Route* node.
n_I: the number of packets in the *Default_Route* node.

n_S: the number of packets in the *Fast_Route* node.
n_T: the number of labels in the *Label_Pool* ($n_T = 1$, if the fast path is available; $n_T = 0$, otherwise).
n_O: the number of IPPs in *on* state.
π_T: the state of the *Label_Release_Timer* node ($\pi_T = 0$, if the *Label_Release_Timer* node is idle; $\pi_T = 1$, otherwise).
π_R: the state of the *Label_Release* node ($\pi_R = 0$, if the *Label_Release* node is idle; $\pi_R = 1$, otherwise).
m: the triggering threshold which represents the minimum number of accumulated packets that will trigger label-setup operation.

In order to avoid the state size explosion problem, we employ the technique of state-aggregation approximation. The aggregated system state of the GMPLS queueing model is defined as the number of IPPs in *on* state, and we use π_k to denote the steady-state probability, where k (ranging from $0 \sim N$) is the aggregated state. The aggregated transition diagram is shown in Fig. 2. We then employ the state vector (a, b, c, d, e, f) to represent the internal state of the aggregated state k of the GMPLS queueing model when $n_I = a, n_S = b, n_O = c, \pi_T = d, \pi_R = e$, and $n_T = f$. The behavior of the GMPLS queueing model can then be predicted by a Markov chain.

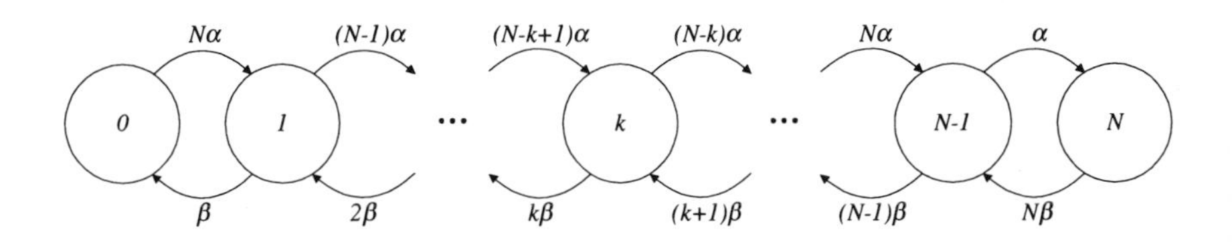

Fig. 2. Aggregated state-transition diagram of the GMPLS queueing model.

In this model, the service time is assumed to be exponentially distributed in all nodes. At the same time, silence interval, burst length and IP packet size are also assumed to be exponentially distributed. According to the above definitions, the global balance equations in state k of aggregated state-transition diagram are listed as follows.

$$P_{n,t,k,0,0,0} = (1 - \mu_O - \mu_D)P_{n,t,k,0,0,0} + k\lambda P_{n,t-1,k,0,0,0} \quad (1)$$

$$P_{n-i,t,k,0,0,0} = (1 - \mu_O - \mu_D)P_{n-i,t,k,0,0,0} + \mu_D P_{n-i+1,t,k,0,0,0} + k\lambda P_{n-i,t-1,k,0,0,0}\ , 1 \leq i \leq n-1 \quad (2)$$

$$P_{0,t,k,0,0,0} = (1 - \mu_O)P_{0,t,k,0,0,0} + \mu_D P_{1,t,k,0,0,0} + k\lambda P_{0,t-1,k,0,0,0} \quad (3)$$

$$P_{n-i,t-j,k,0,0,0} = (1 - \mu_O - \mu_D - k\lambda)P_{n-i,t-j,k,0,0,0} + \mu_D P_{n-i+1,t-j,k,0,0,0} + k\lambda P_{n-i,t-j-1,k,0,0,0} + \mu_O P_{n-i,t-j+1,k,0,0,0}\ , 1 \leq i \leq n-1,\ 1 \leq j \leq t-2 \quad (4)$$

$$P_{n,t-j,k,0,0,0} = (1 - \mu_O - \mu_D - k\lambda)P_{n,t-j,k,0,0,0} + \mu_O P_{n,t-j+1,k,0,0,0} + k\lambda P_{n,t-j-1,k,0,0,0}\ , 1 \leq j \leq t-2 \quad (5)$$

$$P_{0,t-j,k,0,0,0} = (1 - \mu_O - k\lambda)P_{0,t-j,k,0,0,0} + \mu_O P_{0,t-j+1,k,0,0,0} + k\lambda P_{0,t-j-1,k,0,0,0} + \mu_D P_{1,t-j,k,0,0,0}\ , 1 \leq j \leq t-2 \quad (6)$$

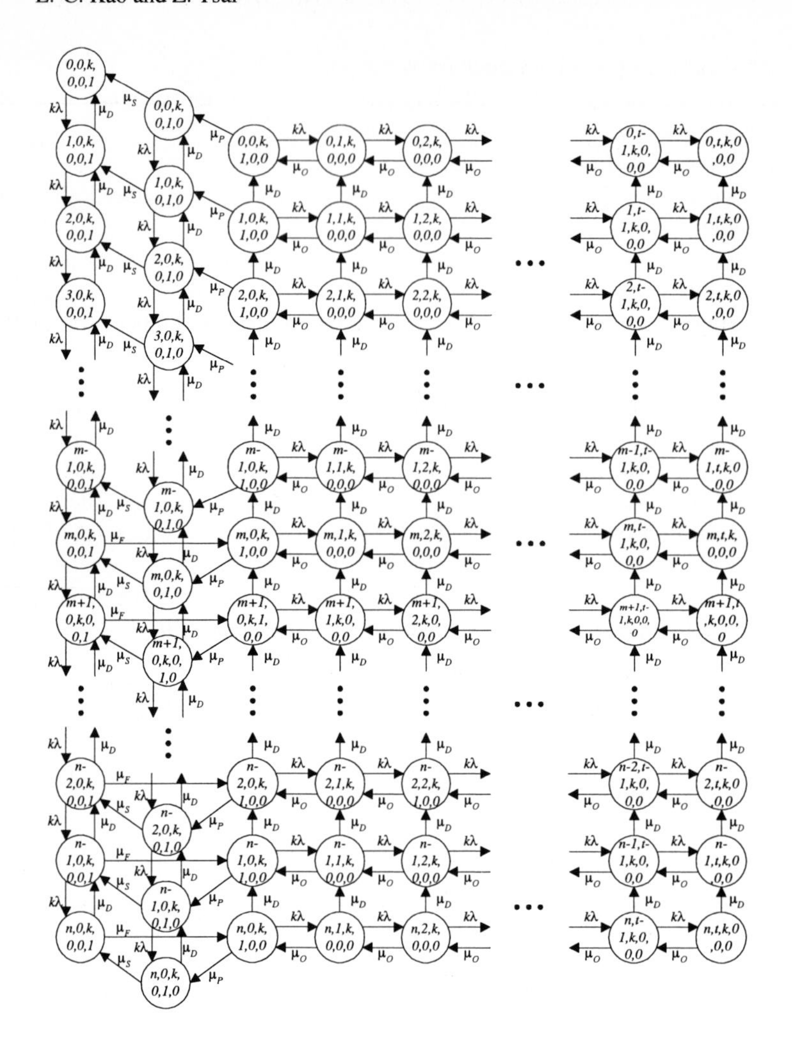

Fig. 3. Detailed state-transition diagram in aggregated state k.

$$P_{n,1,k,0,0,0} = (1 - \mu_O - \mu_D - k\lambda)P_{n,1,k,0,0,0} + \mu_O P_{n,2,k,0,0,0} + k\lambda P_{n,0,k,1,0,0} \tag{7}$$

$$P_{n-i,1,k,0,0,0} = (1 - \mu_O - \mu_D - k\lambda)P_{n-i,1,k,0,0,0} + \mu_O P_{n-i,2,k,0,0,0} + k\lambda P_{n-i,0,k,1,0,0} + \mu_D P_{n-i+1,1,k,0,0,0}\ , 1 \le i \le n-1 \tag{8}$$

$$P_{0,1,k,0,0,0} = (1 - \mu_O - k\lambda)P_{0,1,k,0,0,0} + \mu_O P_{0,2,k,0,0,0} + k\lambda P_{0,0,k,1,0,0} + \mu_D P_{1,1,k,0,0,0} \tag{9}$$

$$P_{n,0,k,1,0,0} = (1 - \mu_D - \mu_P - k\lambda)P_{n,0,k,1,0,0} + \mu_O P_{n,1,k,0,0,0} + \mu_F P_{n,0,k,0,0,1} \tag{10}$$

$$P_{n-i,0,k,1,0,0} = (1 - \mu_D - \mu_P - k\lambda)P_{n-i,0,k,1,0,0} + \mu_O P_{n-i,1,k,0,0,0} + \mu_D P_{n-i+1,0,k,1,0,0} + \mu_F P_{n-i,0,k,0,0,1}\ , 1 \le i \le n-m \tag{11}$$

$$P_{n-i,0,k,1,0,0} = (1-\mu_D-\mu_P-k\lambda)P_{n-i,0,k,1,0,0} + \mu_O P_{n-i,1,k,0,0,0} + \mu_D P_{n-i+1,0,k,1,0,0}\ , n-m+1 \le i \le n-1 \quad (12)$$

$$P_{0,0,k,1,0,0} = (1-\mu_P-k\lambda)P_{0,0,k,1,0,0} + \mu_O P_{0,1,k,0,0,0} + \mu_D P_{1,0,k,1,0,0} \quad (13)$$

$$P_{n,0,k,0,1,0} = (1-\mu_S-\mu_D)P_{n,0,k,0,1,0} + \mu_P P_{n,0,k,1,0,0} + k\lambda P_{n-1,0,k,0,1,0} \quad (14)$$

$$P_{n-i,0,k,0,1,0} = (1-\mu_S-\mu_D-k\lambda)P_{n-i,0,k,0,1,0} + \mu_P P_{n-i,0,k,1,0,0} + k\lambda P_{n-i-1,0,k,0,1,0} + \mu_D P_{n-i+1,0,k,0,1,0}\ , 1 \le i \le n-1 \quad (15)$$

$$P_{0,0,k,0,1,0} = (1-\mu_S-k\lambda)P_{0,0,k,0,1,0} + \mu_P P_{0,0,k,1,0,0} + \mu_D P_{1,0,k,0,1,0} \quad (16)$$

$$P_{n,0,k,0,0,1} = (1-\mu_F-\mu_D)P_{n,0,k,0,0,1} + \mu_S P_{n,0,k,0,1,0} + k\lambda P_{n-1,0,k,0,0,1} \quad (17)$$

$$P_{n-i,0,k,0,0,1} = (1-\mu_F-k\lambda-\mu_D)P_{n-i,0,k,0,0,1} + \mu_S P_{n-i,0,k,0,1,0} + k\lambda P_{n-i-1,0,k,0,0,1} + \mu_D P_{n-i+1,0,k,0,0,1}\ , 1 \le i \le n-m \quad (18)$$

$$P_{n-i,0,k,0,0,1} = (1-\mu_D-k\lambda)P_{n-i,0,k,0,0,1} + \mu_S P_{n-i,0,k,0,1,0} + k\lambda P_{n-i-1,0,k,0,0,1} + \mu_D P_{n-i+1,0,k,0,0,1}\ , n-m+1 \le i \le n-1 \quad (19)$$

$$P_{0,0,k,0,0,1} = (1-k\lambda)P_{0,0,k,0,0,1} + \mu_S P_{0,0,k,0,1,0} + \mu_D P_{1,0,k,0,0,1} \quad (20)$$

where $P_{a,b,c,d,e,f}$ is the steady-state probability of the state vector (a,b,c,d,e,f). The detailed state-transition diagram corresponding to equations (1)–(20) is shown in Fig. 3.

4 Performance Measures

One key performance metric is the throughput. We define T_d and T_f as the average throughput at the *Default_Route* node and *Fast_Route* node respectively. $T_{total} = T_d + T_f$ is the total throughput. Their formulas are given by

$$T_d = \mu_D\{\sum_{k=0}^{N}\sum_{i=1}^{n}\sum_{j=1}^{t} P_{i,j,k,0,0,0}\pi_k + \sum_{k=0}^{N}\sum_{i=1}^{n}(P_{i,0,k,1,0,0} + P_{i,0,k,0,1,0} + P_{i,0,k,0,0,1})\pi_k\} \quad (21)$$

$$T_f = \mu_O \sum_{k=0}^{N}\sum_{i=0}^{n}\sum_{j=1}^{t} P_{i,j,k,0,0,0}\pi_k \quad (22)$$

Since the label-setup rate is proportional to the required label processing load, it is included as another key metric. The label-setup rate S_R is defined as the average number of label-setup operations in the *Fast_Route_Setup* node per unit time and given by

$$S_R = \mu_F \sum_{k=0}^{N}\sum_{i=m}^{n} P_{i,0,k,0,0,1}\pi_k \quad (23)$$

Regarding the path bandwidth utilization we focus on the prediction of the ratio of wasted bandwidth on the fast path. For the fast path, the time periods considered to be "reserved" by an aggregated stream include the packet transmission time by the *Fast_Route* node (with time ratio B_f), the idle period waiting for label-release timeout (with time ratio B_t) and the time required to release a label (with time ratio B_r). However, only the period that the packets are transmitted by the *Fast_Route* node is considered effectively utilized. Hence, the fast path bandwidth utilization U_F is given by $U_F = \frac{B_f}{B_f+B_t+B_r}$, where $B_f = \sum_{k=0}^{N}\sum_{i=0}^{n}\sum_{j=1}^{t} P_{i,j,k,0,0,0}\pi_k$, $B_t = \sum_{k=0}^{N}\sum_{i=0}^{n} P_{i,0,k,1,0,0}\pi_k$, and $B_r = \sum_{k=0}^{N}\sum_{i=0}^{n} P_{i,0,k,0,1,0}\pi_k$.

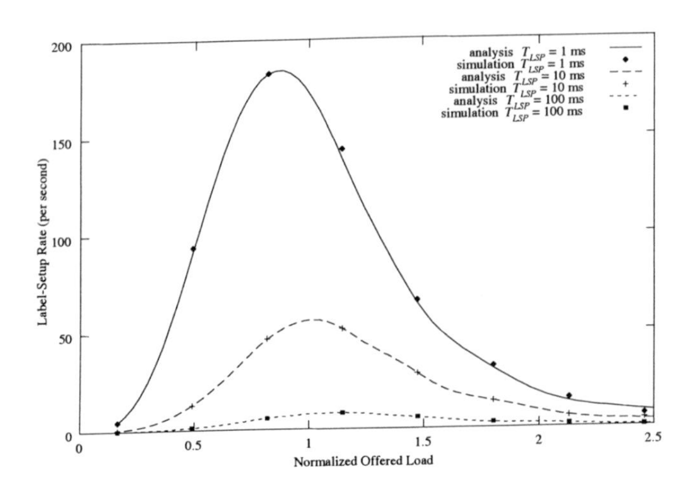

Fig. 4. Label-setup rate as a function of normalized offered load with $T_{rel} = 1$ ms and $m = 3$ under different T_{LSP}.

5 Numerical Examples

In this section, we demonstrate the applicability of the queueing model and discuss analytical and simulation results of the proposed generic GMPLS switch. We also illustrate the trade-off among key system parameters. Throughout this section, we set the number of IPPs (N) to 5, the average silence interval ($1/\alpha$) to 0.2 sec, the average burst length ($1/\beta$) to 0.8 sec, the average IP packet size to 512 bytes, the fast path capacity to 150 Mbps, the default path capacity to 100 Mbps, the average label-release latency ($\frac{1}{\mu_S}$) to 0.2 ms, the buffer size of *Default_Route* node (n) to 50 packets, and the buffer size of *Fast_Route* node (t) to 30 packets. The normalized offered load is defined as $N\lambda\left(\frac{\alpha}{\alpha+\beta}\right)/\mu_D$.

From Fig.s 4 and 5, one can observe that the longer the label-setup latency (T_{LSP}), the lower the label-setup rate. In other words, when it takes time to set up an LSP due to long T_{LSP}, more traffic will go through the default path. Hence, a switch with very large T_{LSP} should not be considered a topology-driven system because almost all traffic still goes through its default path. When traffic load becomes large, we also notice the increase of LSP lifetime. As a result, the label-setup rate decreases as traffic increases.

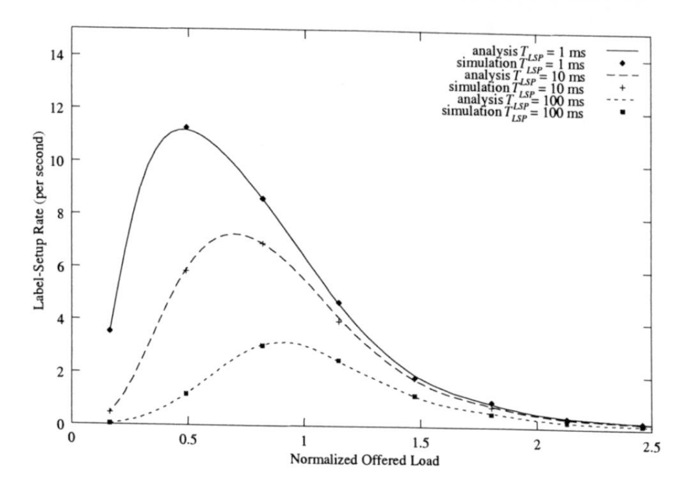

Fig. 5. Label-setup rate as a function of normalized offered load with $T_{rel} = 50$ ms and $m = 3$ under different T_{LSP}.

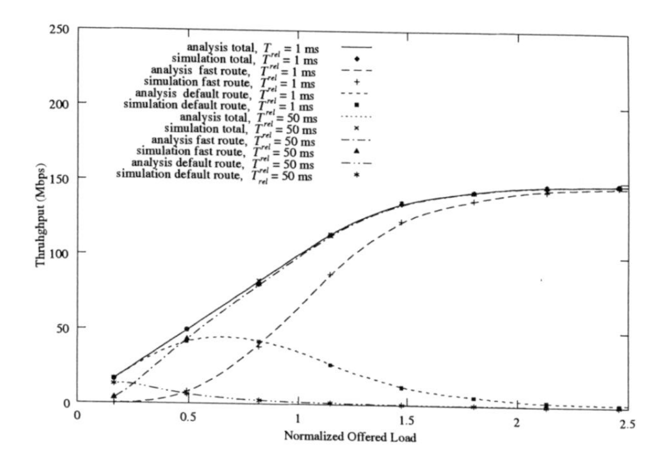

Fig. 6. Throughput as a function of normalized offered load with $T_{LSP} = 1$ ms and $m = 3$ under different T_{rel}.

Although the total throughput is almost the same under different T_{LSP} and label release timer (T_{rel}), the difference exists in the behavior of default path and fast path. With our model, one can determine how much traffic is served by the fast path. We plot the throughput as a function of normalized offered load with $m = 3$ under different T_{LSP} and T_{rel} in Fig. 6 and Fig. 7. From these two figures, one can find that the default path throughput will increase with the increase of traffic load if total traffic load is light. When most traffic starts to be switched to the fast path, the default path throughput decreases. Additionally, one can observe that most traffic will go through the fast path with larger T_{rel}, even under different T_{LSP}. Another phenomenon is that most traffic

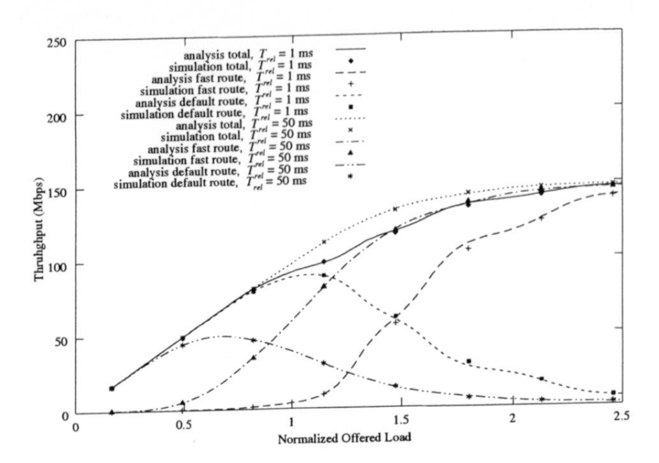

Fig. 7. Throughput as a function of normalized offered load with $T_{LSP} = 100$ ms and $m = 3$ under different T_{rel}.

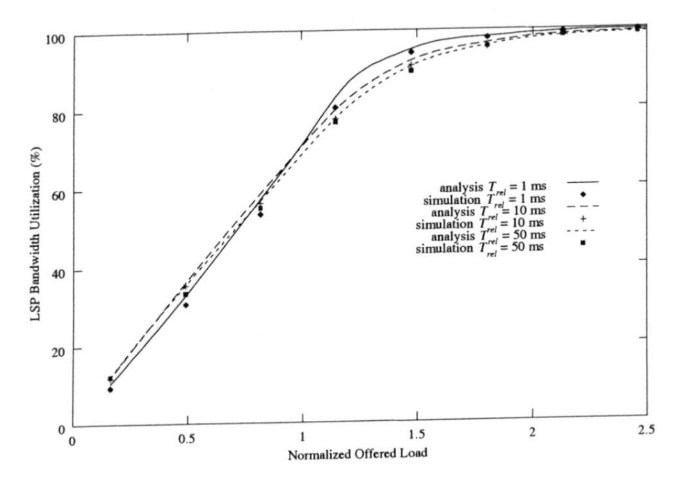

Fig. 8. Fast path bandwidth utilization as a function of normalized offered load with $T_{LSP} = 10$ ms and $m = 3$ under different T_{rel}.

goes through the default path with small T_{rel} and large T_{LSP} in the range of small to medium traffic condition.

The ratio of wasted bandwidth can be predicted by the fast path bandwidth utilization. From Fig. 8, one can know that the fast path bandwidth utilization for smaller T_{rel} is always higher than that for larger T_{rel} under arbitrary traffic load as long as T_{rel} is set sufficiently long. However, one can observe that the fast path bandwidth utilization with small T_{rel} (such as 1 ms) is lower than that with larger T_{rel} (such as 10 ms and 50 ms) when traffic load is light or medium. The reason is that the time ratio waiting for label-release timeout (B_t) under smaller value of T_{rel} will increase under such load. When traffic is heavy, this phenomenon will diminish because B_t becomes small.

From the above results, one can know that when T_{rel} is small, the system behavior is traffic-driven oriented. However, in the case that T_{rel} is extremely large, the system behavior approaches a topology-driven GMPLS switch.

6 Conclusions

The queueing model for a generic GMPLS switching architecture is proposed. On the basis of the approximated analysis and simulation results, one can effectively fine tune the resource utilization level or label processing load. Furthermore, the trade-off between the fast path bandwidth utilization and the label-setup rate can be observed. Hence, an appropriate value of label release timer T_{rel} can be carefully selected to meet the requirement of both. For a network with large round-trip time and sufficient resources in the fast path, if one uses a small value of T_{rel}, most traffic will go through the default path instead of the fast path. Therefore, choosing large value of T_{rel} is preferred. For a network with a small round-trip delay and insufficient resources in the fast path, it is adequate to use the system with a small value of T_{rel}.

Our study shows that the best performance of a GMPLS switch can be achieved only when its control plane parameters are appropriately tuned. In the future, we will investigate a mechanism to reduce the out-of-sequence problem due to dynamic path changes in GMPLS.

References

1. E. Rosen, A. Viswanathan, and R. Callon, "Multiprotocol Label Switching Architecture," *RFC 3031,* Jan. 2001.
2. D. Awduche, Y. Rekhter, J. Darke, and R. Colton, "Multi-Protocol Lambda Switching: Combining MPLS Traffic Engineering Control with Optical Crossconnects," *IETF Internet draft-awduche-mpls-te-optical-02.txt,* July 2000.
3. C. Y. Metz, *IP Switching: Protocols and Architextures,* MacGraw-Hill, 1999.
4. O. Aboul-Magd, et al., "Constraint-Based LSP Setup using LDP," *IETF Internet draft-ietf-mpls-cr-ldp-05.txt,* Feb. 2001.
5. D. O. Awduche, et al., "RSVP-TE: Extensions to RSVP for LSP Tunnels," *IETF Internet draft-ietf-mpls-rsvp-lsp-tunnel-09.txt,* Aug. 2001.
6. P. Ashwood-Smith, et al., "Generalized Multi-Protocol Label Switching (GMPLS) Architecture," *IETF Internet draft-ietf-ccamp-gmpls-architecture-00.txt,* June 2001.
7. A. Banerjee, et al., "Generalized Multiprotocol Label Switching: An Overview of Routing and Management Enhancements," *IEEE Commun., Mag.,* pp. 2–8, Jan. 2001.
8. J. Sadler, et al., "Generalized Switch Management Protocol (gsmp)," *IETF Internet draft-sadler-gsmp-tdm-labels-00.txt,* Feb. 2001.
9. P. Ashwood-Smith, et al., "Generalized MPLS–Signaling Functional Description," *IETF Internet draft-ietf-mpls-generalized-signaling-05.txt,* July 2001.
10. S. Nakazawa, K. Kawahara, S. Yamaguchi, and Y. OIE, "Performance Comparasion with Layer 3 Switches in Case of Flow-And Topology-Driven Connection Setup," *IEEE GLOBECOM'99,* pp. 79–86, Rio de Janeiro, Brazil.

11. L.-C. Kao and Z. Tsai, "Performance Analysis of Flow-Based Label Switching: the Single IP Flow Model," *IEICE Trans. Commun.*, vol. E83-B, no. 7, pp. 1417–1425, July 2000.
12. L.-C. Kao and Z. Tsai, "Steady-State Performance Analysis of MPLS Label Switching," *IEICE Trans. Commun.* vol. E84-B, no.8, pp. 2279–2291, Aug. 2001.

Performance Analysis of Multicast MAC Protocol for Multichannel Dual Bus Networks

Shiann-Tsong Sheu[1] and Jenhui Chen[2]

[1] Department of Electrical Engineering,
Tamkang University, Tamsui, Taipei, Taiwan 25137, R.O.C.
stsheu@ee.tku.edu.tw
[2] Department of Computer Science and Information Engineering,
Tamkang University, Tamsui, Taipei, Taiwan 25137, R.O.C.
jenhui@cs.tku.edu.tw

Abstract. The WDM-based multichannel dual bus networks (MCDBN) for high-speed LAN/MAN was proposed in the few years. When a station requests to connect to another, the transmitter of source station and the receiver of destination station must listen to a same channel (wavelength). The problem of using minimal wavelengths for a set of requests on MCDBN had been proved as NP-hard. Intuitively, the problem of providing the multicast services over MCDBN is more complicated. In this paper, we will analyze the random approach (RAND) and the Best Effort Multicasting Approach (BEMA) which was proposed for multicast packet transmissions on MCDBN. The derived analysis results are very close to the simulation results.

1 Introduction

The dual bus topology for high-speed LAN/MAN was proposed in the few years [11,12]. One of the major characteristics of such multi-access networks is that it provides a low access delay even when the number of connecting nodes is large. Several distributed protocols on optical bus networks, such as distributed queue dual bus (DQDB) [11], had been proposed. However, the maximum network capacity is still bounded due to only one channel is available on the optical fiber [3]. For solving this problem, the Wavelength Division Multiplexing (WDM) [2, 14,15] is used to support a large number of transmission channels in parallel. Several multi-channel photonic star/bus/ring networks with unicasting protocols had been reported in which either node has tunable-transmitter(s) (TT) and fixed-receiver(s) (FR) [1,3,5,6] or node has tunable-transmitter(s) (TT) and tunable-receiver(s) (TR) [8,9,13]. In paper [10], the wavelength/receiver assignment problem (WRAP) of unicast services is defined on a multi-channel dual bus network (MCDBN) to assign a transmission wavelength and a receiver for each of the request such that the network throughput is maximized and the number of assigned wavelengths is also minimized. The paper also proves the WRAP is NP-hard. Obviously, the problem of providing multicast services over MCDBN is more complicated than that of supporting unicast services. For example, when

I. Chong (Ed.): ICOIN 2002, LNCS 2343, pp. 419–430, 2002.

a node requests to transmit a packet to several multicast members, if it can not find a channel which is listen by all these members, it must transmit the packets over several different channels to complete this transmission. Therefore, how to minimize the number of transmission times of a same packet is also a complicated problem. So far, only few protocols have been proposed for multicast services on MCDBN. In paper [16], we had proposed a simple and efficient protocol, named as the Best Effort Multicasting Approach (BEMA), for multicast services. The BEMA tries to minimize the transmission times and maximize the channel utilization. In this paper, we will precisely analyze the average transmission times of the proposed BEMA.

The rest of this paper is organized as follows. The architecture of WDM-based multi-channel dual bus network (MCDBN) is introduced in Section 2. In Section 3, we present the BEMA multicasting protocol. The detailed analyses of the BEMA and random approach are addressed in Section 4. The simulation models and simulation results are reported in section 5. Finally, some conclusion remarks are given in Section 6.

2 The Architecture of MCDBN

The architecture of a MCDBN consists of two contra-directional optical buses, which allow full duplex communications between any two nodes as shown in Fig. 1 The network contains N stations. Each station has m TRs and n TRs. The two buses are called Bus A and Bus B, respectively. For each bus, the entire bandwidth is divided into $C+1$ channels $\{\lambda_0, \lambda_1, \lambda_2, \cdots, \lambda_C\}$, where channel λ_0 is the dedicated control channel and others are data channels. For each channel, fixed length slots are generated by the headend station (also called slot generator) periodically, passed to the downstream stations and is terminated by the terminator.

3 The BEMA Multicasting Protocol

In this section, we will briefly introduce the proposed BEMA multicasting protocol for the MCDBN.

The data slot format of the MCDBN is the same as that of the DQDB. The control signals are arranged in the payload field. Fig. 2 shows the format of a control slot. A control slot includes three major fields: the multicast request field REQ_MULTICAST, the BROADCAST field and the RESERVATION field. The REQ_MULTICAST field contains four subfields: SRC, MDST, LEN and SWAVE. The BROADCAST field contains SRC and UWAVE subfields. It is used for each station to inform all stations the wavelength usage information. The C-bit RESERVATION field, one bit for each wavelength, carries the requests of slots issued from the downstream stations. Based on this protocol, within a cycle time, each station can maintain the newly and correctly traffic information of all stations in a *traffic status matrix* (TSM). The TSM is defined as follows:

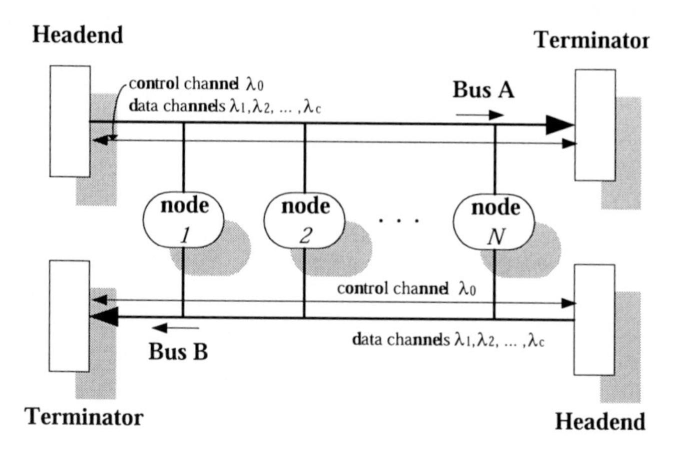

Fig. 1. The construction of a WDM-based dual bus networks.

- **Traffic Status Matrix:** $M = \{t(i,j)\}_{N\times N}, i,j = 1, 2,\ldots, N$, where $t(i,j) = k, k \in \{1,\ldots,C\}$. Each element $t(i,j) = k(k > 0)$ indicates that wavelength λ_k is used for station i to transmit packet to station j, and $t(i,j) = 0$ otherwise.

For illustration, consider the example shown in Fig. 3. The Traffic Status Matrix of stations 8 is :

$$M_8 = \begin{pmatrix} 0 & 2 & 0 & 0 & 0 & 0 & 0 & 0 \\ 0 & 0 & 1 & 2 & 0 & 0 & 0 & 0 \\ 0 & 0 & 0 & 1 & 0 & 3 & 0 & 0 \\ 0 & 0 & 0 & 0 & 2 & 0 & 0 & 1 \\ 0 & 0 & 0 & 0 & 0 & 2 & 0 & 0 \\ 0 & 0 & 0 & 0 & 0 & 0 & 2 & 3 \\ 0 & 0 & 0 & 0 & 0 & 0 & 0 & 0 \\ 0 & 0 & 0 & 0 & 0 & 0 & 0 & 0 \end{pmatrix}.$$

In [10], an efficient *load balancing wavelength assignment algorithm* (LB-WAA) had been proposed for the unicast traffic on multi-channel dual bus network. However, for a multicast packet, the number of transmission times for it may be larger than one because that the number of tunable receivers equipped in each station is small. In detail, it is possible that all TRs of multicast members are assigned to listen on different channels. As a result, a multicast packet may be required to transmit to all multicast members by different wavelengths. For example, if there are any two multicast members do not tune their receivers into a same channel at that time, the multicast packet must be transmitted twice (one for each station). This causes the bandwidth wasted seriously. In the next paragraphs, we will briefly describe the BEMA protocol.

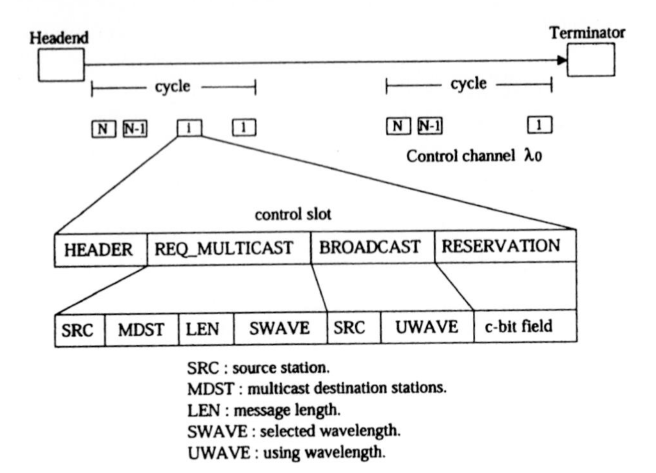

Fig. 2. The slot format on control channel.

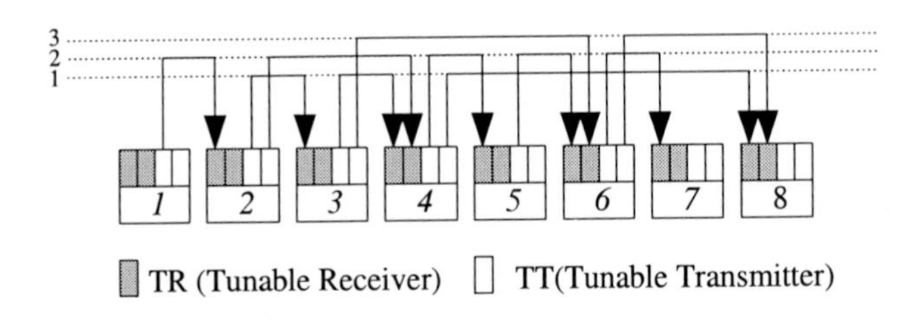

Fig. 3. An example of MCDBN for C=3, T=2, R=2 and N=8.

Assume a multicast packet arrives in station s and let $\mathcal{D} = \{d_1, d_2, \ldots, d_k\}$ $(d_1 < d_2 < \ldots < d_k)$ denote as the multicast members of this multicast packet. For simplicity, the number of the multicast members is denoted as $|\mathcal{D}|$ $(|\mathcal{D}| = k)$. In BEMA, the way of selecting wavelengths for a multicast packet is divided into three major steps: (1) The BEMA first finds the most shared wavelength (say w) among these multicast members in set $\mathcal{D}$, which is listened by the most multicast members. The number of multicast members (NM_i) which listen the wavelength i can be calculated by the following equation

$$NM_i = \sum_{j=1}^{|\mathcal{D}|} \sum_{k=1}^{N} f_s^i(k, d_j), \tag{1}$$

where

$$f_k^j(p, q) = \begin{cases} 1, & \text{if } t_j(p, q) = k, \\ 0, & \text{otherwise.} \end{cases} \tag{2}$$

Therefore, the wavelength w can be easily determined by $w = \{k \mid NM_k \geq NM_i, \forall\, i, k = 1, 2, \dots, C\}$. (2) For each multicast member (say d_i), the BEMA checks whether the selected wavelength w is being used by any of receiver in it. If it does, station d_i will be removed from set $\mathcal{D}$ (ie., $\mathcal{D} = \mathcal{D} - \{d_i\}$). Otherwise, if there exists any idle receiver in it, an idle receiver will be selected and tuned to this wavelength w. Similarly, d_i will be removed from set $\mathcal{D}$. (3) If any member is still left in $\mathcal{D}$ ($|\mathcal{D}| > 0$), it will repeat steps (1) and (2) until all multicast member are considered. As mentioned before, if set $\mathcal{D}$ only contains one member, the LBWAA will be applied to further improve the network throughput. This is performed in step (1) to find a proper wavelength instead of finding a random wavelength.

4 Analysis of BEMA Protocol

4.1 Analysis of Random Approach

Before analyzing the BEMA, we first discuss and analyze the number of transmission times in general cases which using random method to select wavelength/receiver for each multicast packet. In detail, a source station will first select a random wavelength to transmit. If there is any multicast member does not listen to this wavelength, the source station will random select another random wavelength (excluding the wavelength already selected) to cover the remaining stations. This process will be repeated until all multicast members have received the multicast packet successfully. For simplicity, Let w and k denote the total wavelengths and numbers of multicast members of stations in the network, respectively. Moreover, let tt and tr represent the numbers of TTs and TRs equipped in each station, respectively. For the sake of practice, the value of tr is often less than or equal to the total wavelengths w.

We analyze all the different *combinations* and permutations that take just T times to finish a multicast transmission. Now, the number of combination which k multicast members with tr TRs connect in a network with w wavelengths is

$$\left(C^{w}_{tr}\right)^{k} \tag{3}$$

There are two cases to consider for sender taking just T times to finish a multicast transmission. At first, we consider a situation that a multicast member cannot receive any packet before sender transmits the T-th transmission. In other word, the receiver did not receive any information at the first $T-1$ transmissions. Let $S(w, tr, T)$ denote the combinations of a multicast member receives packet at the T-th transmission successfully. We have

$$S(w, tr, T) = C^{w-T}_{tr-1}. \tag{4}$$

The second is that one multicast member may receive the packet before the T-th transmission. Let $F(w, tr, T)$ denote the combinations of a multicast member receives packet before the T-th transmission successfully. This case can be considered as a multicast member has TR(s) which connect to these $(T-1)$ wavelengths that were selected by sender before. Thus, we have

$$F(w,tr,T) = C_{tr}^{w} - C_{tr}^{w-(T-1)}. \tag{5}$$

For all k multicast members, the number of total combinations of sender transmits T times to complete the transmission can be summed as follows:

$$\begin{aligned} C_0^k S(w,tr,T)^0 F(w,tr,T)^k + C_1^k S(w,tr,T)^1 F(w,tr,T)^{k-1} \\ + \ldots + C_k^k S(w,tr,T)^k F(w,tr,T)^0. \end{aligned} \tag{6}$$

Substituting Eq. (4) and Eq. (5) into Eq. (6) leads to:

$$\begin{aligned} C_0^k (C_{tr-1}^{w-T})^0 (C_{tr}^{w} - C_{tr}^{w-(T-1)})^k + C_1^k (C_{tr-1}^{w-T})^1 (C_{tr}^{w} - C_{tr}^{w-(T-1)})^{k-1} \\ + \ldots + C_k^k (C_{tr-1}^{w-T})^k (C_{tr}^{w} - C_{tr}^{w-(T-1)})^0, \end{aligned} \tag{7}$$

which simplifies to

$$\begin{aligned} &\sum_{i=0}^{k} C_i^k (C_{tr-1}^{w-T})^i (C_{tr}^{w} - C_{tr}^{w-(T-1)})^{k-i} \\ &\qquad = (C_{tr-1}^{w-T} + C_{tr}^{w-T} - C_{tr}^{w-(T-1)})^k - (C_{tr}^{w} - C_{tr}^{w-(T-1)})^k. \end{aligned} \tag{8}$$

Thus, from Eq. (3) and Eq. (8), the probability $P_{random}(T)$ of all multicast members receive the packets after sender transmitted T times (with random selecting wavelength) can be computed as follows

$$P_{random}(T) = \frac{\left(C_{tr-1}^{w-T} + C_{tr}^{w-T} - C_{tr}^{w-(T-1)}\right)^k - \left(C_{tr}^{w} - C_{tr}^{w-(T-1)}\right)^k}{\left(C_{tr}^{w}\right)^k}. \tag{9}$$

Since $C_r^n = C_{r-1}^{n-1} - C_r^{n-1}$, we have $C_{tr}^{w-(T-1)} = C_{tr-1}^{w-T} + C_{tr}^{w-T}$. Thus, the probability $P_{random}(T)$ can be further reduced as

$$P_{random}(T) = \frac{\left(C_{tr}^{w} - C_{tr}^{w-T}\right)^k - \left(C_{tr}^{w} - C_{tr}^{w-(T-1)}\right)^k}{\left(C_{tr}^{w}\right)^k}. \tag{10}$$

Notice that when $T \le w - tr + 1$, the maximum number of transmission times is $w - tr + 1$ at the worst case. Now, we can derive the average number of transmission times of a sender to complete a multicast service. That is,

$$\begin{aligned} EXP_{random} &= 1 \times P_{random}(1) + 2 \times P_{random}(2) + \ldots + w \times P_{random}(w) \\ &= \sum_{i=1}^{w} i \times P_{random}(i). \end{aligned} \tag{11}$$

4.2 Analysis of BEMA Approach

To analysis the BEMA technique, we have to consider the number of TRs. We note that the derived combination cases of a station equipped with one TR and several TRs are quite different. Thus, we divide the conditions into two parts.

A Single Tunable Receiver. The transmission times in BEMA can be considered a Stirling number problem. To arrange n tunable receivers into r wavelengths, we can use $S_2(n, r)$ function to calculate the result.

The way to derive all possible onto mapping functions, say H, from distribute k distinct stations into w distinct wavelengths can be formalized as the following generating functions. Since no wavelength is empty (by *onto* definition), each wavelength is listening by at least one station. Therefore, according to the following two equations,

$$\begin{aligned}(\frac{x}{1!}+\frac{x^2}{2!}+\frac{x^3}{3!}+\frac{x^4}{4!}+\ldots)^w &= (e^x-1)^w \\ &= \sum_{i=0}^{w} C_i^w(-1)^i e^{(w-1)x}\end{aligned} \tag{12}$$

and

$$\sum_{i=0}^{w} C_i^w(-1)^i \sum_{k=0}^{\infty} \frac{1}{k!}(w-i)^k x^k = \sum_{k=0}^{\infty} \frac{x^k}{k!}\Big(\sum_{i=0}^{w}(-1)^i C_i^w (w-i)^k\Big). \tag{13}$$

We can easily derive H by the coefficient of $x^k/k!$ from above equations. Thus, we have

$$H = \frac{1}{w!}\Big(\sum_{i=0}^{w}(-1)^i C_i^w (w-i)^k\Big). \tag{14}$$

Since $S_2(k, w) = \frac{1}{w!}\big(\sum_{i=0}^{w}(-1)^i C_i^w (w-i)^k\big)$, we have

$$H = S_2(k, w). \tag{15}$$

From Eq. (15), the probability $P^1_{BEMA}(T)$ of all multicast members with one TR receives the multicast packet just after sender transmitted T times can be derived as following:

$$P^1_{BEMA}(T) = \frac{C_T^w T! S_2(k, T)}{w^k}. \tag{16}$$

Then we can get the expectation EXP_{BEMA} by summering form above $P^1_{BEMA}(T)$:

$$EXP^1_{BEMA}(T) = \frac{1C^w_1 1!S_2(k,1) + \ldots + wC^w_w S_w(k,w)w!}{w^k} = \frac{\sum_{i=1}^{w} iC^w_i i!S_2(k,i)}{w^k}. \quad (17)$$

In the case of $k < w$, the maximum transmission times of a multicast packet should be less than or equal to k. Therefore, we obtain the generalized equation:

$$EXP^1_{BEMA}(T) = \frac{\sum_{i=1}^{\min(k,w)} iC^w_i i!S_2(k,i)}{w^k}. \quad (18)$$

The Multi-tunable Receivers. In a multi-tunable receivers model, a station has a chance to connect several channels at a time. Therefore, increasing the number of available receivers will reduce the transmission times. Let $\mathcal{R} = k \cdot tr$ be the total participation multicast members' tunable receivers. And let ψ denote the maximum transmission times of a multicast packet in the case of multiple tunable receivers. Considering the relation between $\mathcal{R}$ and w, there are two different cases:

- If $\mathcal{R} \leq w \Rightarrow$ the maximum transmission times will be no more than $\psi = k$ times;
- Otherwise, $\mathcal{R} > w \Rightarrow$ the maximum transmission times will be no more than $\psi = \lfloor \frac{w}{tr} \rfloor$ times.

First, assume sender only transmits the packet over one wavelength to finish the transmission, all multicast members must listen to a same wavelength. So we can get the combinations as follows:

$$C^w_1 (C^{w-1}_{tr-1})^k. \quad (19)$$

From Eq. (19), we want to finish the transmission time at the T-th time, the total number of combinations is

$$C^w_1 (C^{w-1}_{tr-1})^k - C^w_2 (C^{w-2}_{tr-2})^k + \cdots + (-1)^{tr-1} C_t r^w (C^{w-tr}_{tr-tr})^k = \sum_{i=1}^{tr} (-1)^{i-1} C^w_i (C^{w-i}_{tr-i})^k. \quad (20)$$

Thus, the probability of exact T times is

$$P^{tr}_{BEMA}(T) = \frac{\sum_{j=T}^{tr} (-1)^{j-T} C^w_j (C^{w-j}_{tr-j})^k T! S_2(k,T)}{(C^w_{tr})^k}. \quad (21)$$

Obviously, from Eq. (21), the expectation value EXP^{tr}_{BEMA} can be obtained by the following equation:

$$\begin{aligned} EXP^{tr}_{BEMA} &= 1 \times P^{tr}_{BEMA}(1) + 2 \times P^{tr}_{BEMA}(2) + \ldots + \psi \times P^{tr}_{BEMA}(\psi) \\ &= \sum_{i=1}^{\psi} i P^{tr}_{BEMA}(i) \\ &= \frac{\sum_{i=1}^{\psi} \sum_{j=i}^{tr} i(-1)^{j-i} C^{w}_{j} (C^{w-j}_{tr-j})^{k} i! S_2(k,i)}{(C^{w}_{tr})^{k}}. \end{aligned} \tag{22}$$

5 Simulation Models and Results

For simplicity, we assume there are N stations are equally spaced on Bus A. Each station is equipped with a TT and tr TRs. We assume that the distributions of the multicast size and destination stations for all packets are uniform.

To measure the precision of analysis, for each simulation run (100,000 time slots), the average of transmission times is calculated. In the first simulation shown in Fig. 4, we compare the derived analysis result and the simulation result of random approach (RAND) under different number of tunable receivers, different channels and multicast sizes. We can see that the mathematical analyses are very close to the simulation results of RAND approach. Fig. 5 shows the comparisons of the derived analysis result and the simulation result of BEMA approach under different number of tunable receivers, different channels and multicast sizes. We also can see that the analyses are still very close to the simulation results of BEMA approach.

The second simulation is shown in Fig. 6. In this simulation, we investigate how the number of transmission times is affected by the network size. We note that the multicast size (k) is equal to the network size (N). We can see that a larger number of wavelengths or multicast size is, a larger transmission times will obtain. Moreover, we can also find that the analyses are almost matching the derived simulation results.

6 Conclusion

In this paper, we analyzed the average transmission times of both random approach (RAND) and best effort multicast approach (BEMA) for providing multicast service over WDM-based multi-channel dual bus networks (MCDBN). Simulations show that the derived analytical results are very close to the simulation results.

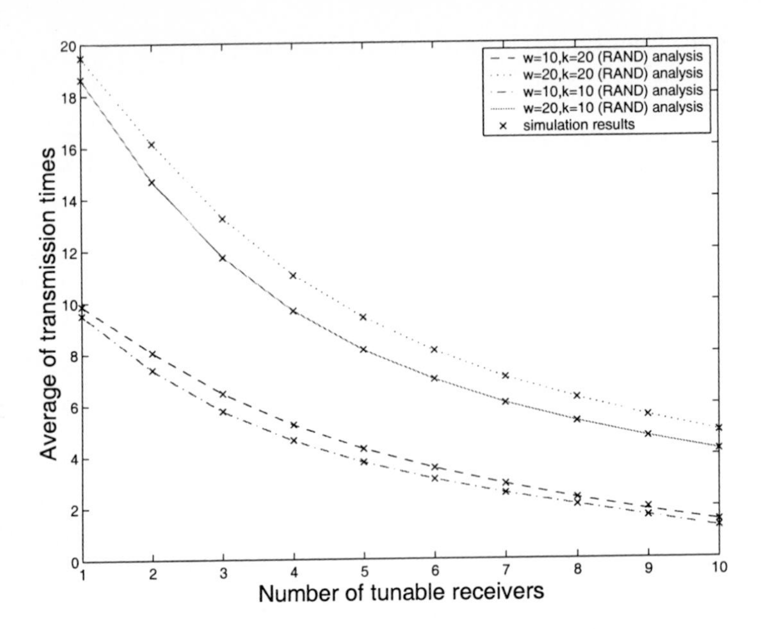

Fig. 4. Comparisons of analysis and simulation results obtained by RAND approach under different number of channels (w), number of tunable receivers (tr) and multicast size (k).

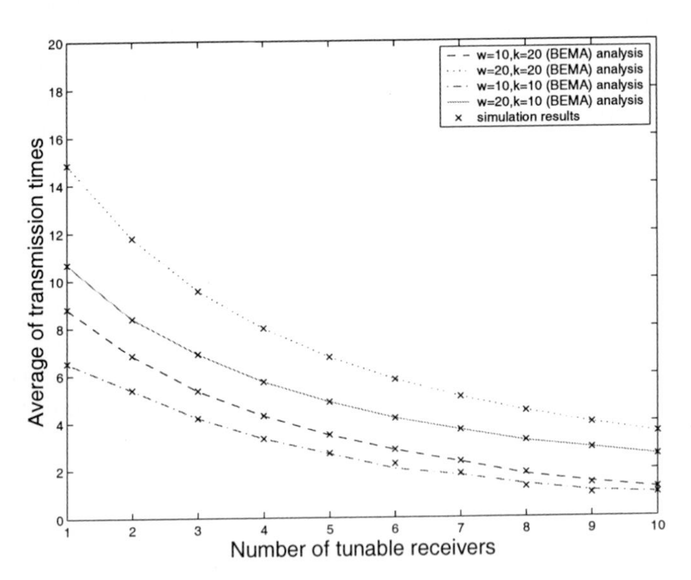

Fig. 5. Comparisons of analysis and simulation results obtained by BEMA approach under different number of channels (w), number of tunable receivers (tr) and multicast size (k).

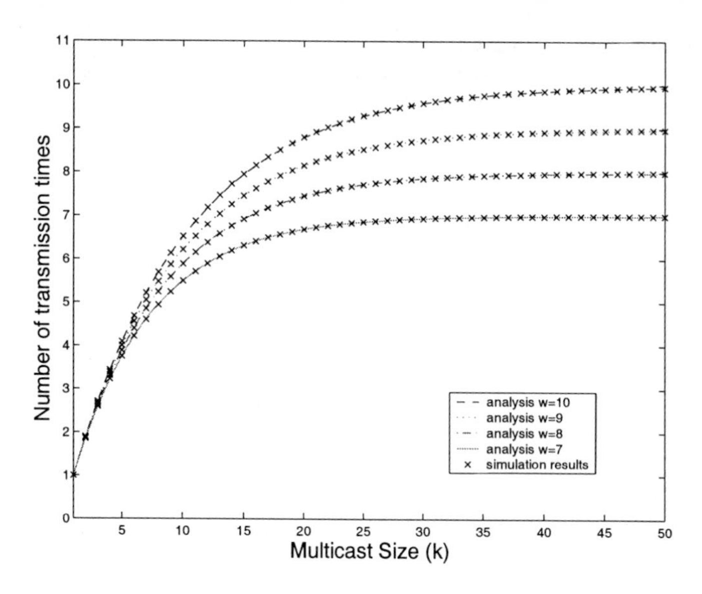

Fig. 6. Comparisons of analysis and simulation results obtained by BEMA approach under different number of channels (w) and multicast size (k) when $tr = 1$.

References

1. S. Banerjee and B. Mukherjee, "An Efficient and Fair Probabilistic Scheduling Protocol for Multi-channel Lightwave Networks," *Proc. IEEE SUPERCOMM/ICC'92*, Chicago, IL, pp. 1115–1119, May. 1992.
2. C. A. Brackett, "Dense Wavelength Division Multiplexing Networks: Principles and Applications," *IEEE J. Select. Areas Commun.*, vol. 6, no. 8, pp. 948–964, Aug. 1990.
3. K. W. Cheung, "A Subcarrier-Multiplexed Photonic Dual Bus with Tunable Channel Slotted Access for Local/Metropolitan Area Network Applications," in *Proc. OFC'92*, San Jose, California, pp. FD5:300, Jan. 1992.
4. K. W. Cheung, "Design and Implementation Considerations for Wavelength-Division Multiplexed (WDM) Photonic Dual Bus," *Proc. IEEE SUPERCOMM/ICC'92*, Chicago, IL, pp. 848–854, May 1992.
5. K. W. Cheung, "EQEB-A Multi-channel Extension of DQDB Protocol with Tunable Channel Access," in *Proc. GLOBECOM'92*, Orlando, Florida, pp. 1610–1617, Dec. 1992.
6. K. W. Cheung, "Adaptive-Cycle Tunable-Access (ACTA) Protocol: A Simple, High Performance Protocol for Tunable-Channel Multi-Access (TCMA) Networks," *Proc. IEEE ICC'93*, Geneva, Switzerland, pp. 848–854, Jun. 1993.
7. N. F. Huang and S. T. Sheu, "A slot interleave multiple access scheme for DQDB metropolitan area networks," *Proc. IEEE INFOCOM'93*, San Francisco, CA, pp. 1075–1082, Mar. 1993.
8. N. F. Huang and S. T. Sheu, "On the Wavelength Assignment Problem of Multichannel Photonic Dual Bus Networks," *Proc. IEEE GLOBECOM'94*, San Francisco, CA, pp. 1925–1929, Nov. 1994.

9. N. F. Huang and S. T. Sheu, "DTCAP-A Distributed Tunable-Channel Access Protocol for Multi-Channel Photonic Dual Bus Networks," *Proc. IEEE INFOCOM'95*, Boston, MASS, pp. 908–915, Apr. 1995.
10. N. F. Huang and S. T. Sheu, "A Wavelength Reusing/Sharing Access Protocol for Multi-Channel Photonic Dual Bus Networks," *IEEE J. Lightwave Tech.*, vol. 14, no. 5, pp. 678–692, May 1996.
11. IEEE P802.6 Working Group, "Proposed standard: Distributed Queue Dual Bus (DQDB) Metropolitan Area Network," approved draft D12, Feb. 1990.
12. Y. M. Lin, D. R. Spears, and M. Yin, "Fiber-based local access network architectures," *IEEE Commun. Mag.*, pp. 64–73, Oct. 1989.
13. J. C. Lu and L. Kleinrock, "A WDMA Protocol for Multichannel DQDB Networks," *Proc. IEEE GLOBECOM'93*, Houston, TEXAS, pp. 149–153, Nov. 1993.
14. B. Mukherjee, "WDM-based Local Lightwave Networks Part I: Single-Hop Systems," *IEEE Network*, pp. 12–27, May 1992.
15. B. Mukherjee, "WDM-based Local Local Lightwave Networks Part II: Multi-Hop Systems," *IEEE Network*, pp. 20–32, Jul. 1992.
16. S. T. Sheu and J. D. Fang, "An Efficient Multicasting Protocol for Multi-channel Dual Bus Networks," *Proc. 15th IASTED Int'l Conf. Applied Informatics'97*, Innsbruck Austria, pp. 375–378, Feb. 1997.

A New Packet Marking Strategy for Explicit Congestion Notification in the Internet

Byung-Chul Kim, Young-Soo Choi, and You-Ze Cho

School of Electrical Engineering and Computer Science
Kyungpook National University, Korea
{bckim, guru0109}@palgong.knu.ac.kr, yzcho@ee.knu.ac.kr

Abstract. A window-based flow control is a sort of feedback-based congestion control mechanisms, and has been widely used in current TCP/IP networks. In the feedback-based congestion control, feedback delay of the congestion information from the bottleneck node is one of the important elements that affect the performance. Recently, the use of an explicit congestion notification (ECN) mechanism as congestion indication from the network to source hosts has been actively discussed in the IETF. In this paper, we propose an enhanced marking strategy for ECN that can reduce the transfer delay of congestion information of a router in multiple-hop network environments. At a transit router, the proposed method relays the congestion experienced (CE)-bit contained in an incoming IP packet to the head-of-line packet with a corresponding flow which is waiting for transmission in the output buffer. Simulation results showed that ECN with the proposed mark-relay strategy produced a superior performance in terms of throughput and fairness between TCP flows.

1 Introduction

The TCP protocol is extensively used on the Internet as a reliable data transfer protocol. In TCP congestion control mechanisms, packet loss is used for feedback information of network status since packet loss implies the occurrence of congestion in the network [1]-[2]. Therefore, until a packet loss occurs in the network, TCP gradually increases its congestion window size. Namely, TCP begins in slow-start phase, and its congestion window is doubled every round trip time (RTT) until the window size reaches the slow-start threshold. After this threshold, the window is increased linearly at a rate of one segment each RTT. When the congestion window size exceeds the available bandwidth, excess packets are queued at the buffer of the bottleneck router for some period. If the congestion window size increases further, then the packets at the buffer of the router overflow, thereby leading to packet loss. If the source host detects the occurrence of packet loss in the network using a time-out mechanism, then reduces its window size to one and enters the slow-start phase. TCP Reno has another mechanism called fast retransmission, which uses the receipt of duplicate ACK packets to detect occurrence of packet loss. It reacts to the packet loss by halving its congestion window and retransmitting the missing segment, without waiting for the

I. Chong (Ed.): ICOIN 2002, LNCS 2343, pp. 431–442, 2002.

timer to expire. After the congestion window size is reduced, the congestion in the network is relieved and the source host then increases its window size again. Consequently, since the occurrence of packet loss in network is used as the signal for congestion by the TCP congestion control mechanism, packet loss cannot be avoided. And, packet losses introduce performance degradation, such as an increment of traffic in the network and the addition of a significant transfer delay as a result of packet retransmissions.

Accordingly, active queue management and the use of an explicit congestion notification (ECN) have been proposed and discussed in the IETF to improve the performance of TCP congestion control [3]-[6]. Active queue management schemes attempt to detect congestion before the queue overflows, and then notify the host node of incipient congestion status. As such, active queue management can provide several advantages, including avoiding the undesirable synchronization of packet loss across multiple flows and the reduction of unnecessary queueing delays and packet losses. ECN is a mechanism for explicitly notifying source hosts of congestion in a network using marked packets. A packet is marked at a router, which uses active queue management such as a random early detection (RED) mechanism, by setting a congestion experienced (CE)-bit in the IP header instead of dropping it when impending congestion is detected. The mark gets echoed by the TCP receiver through setting an ECN echo flag in the header of ACK and the TCP source host then reacts to the mark as it would do to a dropped packet. The advantage of the ECN mechanism is that unnecessary packet losses can be prevented if the source hosts respond to the ECN message properly. In addition, ECN has been proven to be a better way of delivering congestion information to the source host and improves the performance of TCP [7].

The performance of feedback-based congestion control is generally affected by the feedback delay involved in the congestion information reaching the source host form the bottleneck node. Therefore, ECN inherits the bias against connections with long RTTs. Hamann and Walrand proposed a new fair window algorithm for ECN capable source hosts that correct the inherited bias of ECN-TCP against connections with long RTT [8]. Liu and Jain suggested the use of the mark-front strategy which marks the packet at the head of the output queue in stead of marking the incoming packet [9].

The feedback-based congestion control requires the cooperation of three parts: (1) congestion identification, (2) congestion notification, and (3) response to congestion notification [10]. In the context of ECN, the first part may be regarded as finding an answer to the question of when to mark a packet, the second part as identifying which packet to mark, and the third part addressing the question how a source should react to a congestion indication from the network. In general, the RED mechanism is used to decide when to mark a packet. Regarding as which packet is marked, Floyd suggested marking the incoming packet [5], while Liu and Jain presented the mark-front strategy [9]. For the third question, Floyd presented a mechanism using the standard TCP congestion response [5], while Hamann and Walrand suggested a new fair-window algorithm [8].

In this paper, we concentrate on the first and second parts to resolve the performance degradation of connections with long RTTs. Most ECN implementations use the mark-tail strategy. That is, when congestion is detected, the router marks the incoming packets that have just entered the buffer of each router. However, with this strategy, an incoming marked packet can experience a queueing delay until all previously buffered packets will have been transmitted. Thus, to reduce this queueing delay, the mark-front strategy was proposed where a packet is marked at the time it is sent, thereby providing the faster congestion information delivery and reflecting the up-to-date congestion information [9]. However, since a packet can still experience a queueing delay in the buffer at each transit router, the congestion information contained in an incoming packet from an upstream node cannot be immediately transferred to its destination. Therefore, in existing marking strategies for implementing ECN, this additional queueing delay is unavoidable and increases proportionally with the number of routers the flow passes through.

Accordingly, this paper proposes a mark-relay strategy for ECN implementation that can significantly reduce the transfer delay of congestion information from a router to its source host in a multiple-hop network environment. At each transit router, the proposed method relays the CE-bit contained in an incoming IP packet to the head-of-line packet with a corresponding flow that is waiting for transmission in the output buffer.

This paper is organized as follows. Section 2 presents some background on ECN mechanisms and motivation for our research. In Section 3, we introduce the proposed mark-relay strategy for ECN-capable router implementation. Section 4 compares the performance of ECN in multiple-hop network environments using simulation according to the marking strategy. Finally, Section 5 presents conclusions and discusses future work.

2 Related Works

ECN enhances active queue management by detecting an impending congestion status before a buffer overflows and drops packets probabilistically [3]-[4]. When ECN is implemented, packets are marked rather than dropped. As such, congestion information can be quickly transferred to the source host where the transfer rate can be immediately adjusted without additional delay in waiting for a duplicated ACK or timeout. As a result, unnecessary packet drops can be avoided. To support ECN, both the router and the end hosts are changed and operate as shown in Fig. 1.

In the connection setup phase, the source and destination TCPs have to negotiate their ECN capability. This is done by the sender setting both an ECN-echo and congestion window reduced (CWR)-flag in the SYN packet and by the receiver setting the ECN-echo flag in the SYN ACK packet. When this agreement has been reached, the sender then sets an ECN capable transport (ECT)-bit in the IP header of the data packets for that flow, to indicate to the network the capability and willingness of the sender to participate in ECN.

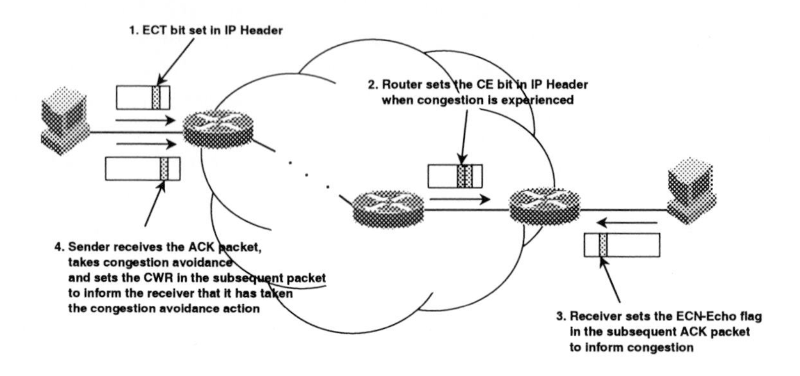

Fig. 1. Summary of ECN operation

When a router has detected congestion using active queue management, it first checks the ECT-bit in the IP packet header. Then, it sets a CE-bit in the IP header if an ECT-bit has been set. When the receiver receives a packet with a CE-bit set, it sets an ECN-echo flag in the next outgoing ACK for the flow. The receiver continues to do this in subsequent ACKs until it receives a data packet from the sender with a CWR-flag set.

An ECN-capable sender reacts to an ACK with an ECN-echo flag set as a lost packet, i.e. reduces its congestion window by half, yet without the need to retransmit the marked packet. If a sender receives multiple congestion indications, including a timeout, duplicate ACKs, and ECN-echo ACK, it only needs to react to the congestion once per RTT. After a sender has responded to a congestion notification, it sets a CWR-flag in the next outgoing data packet to tell the receiver that it has reacted to a congestion notification. An ECN-capable router usually uses RED gateways with a modified algorithm to set the CE-bit in the IP header. The difference is that the router marks packets instead of dropping them. The RED algorithm detects incipient congestion before the buffer overflows, based on an average queue length, and then marks packets probabilistically.

Several researches presented the advantages of ECN over packet drops for congestion control [5]-[7]. The advantages of ECN include reducing unnecessary packet drops, and faster delivery of congestion notification instead of relying on packet drops and the coarse granularity of the timer, which may add significant delay in the source recognizing congestion in the network.

However, ECN still inherits the bias against connections with long RTTs. Hamann and Walrand presented a new fair window algorithm for ECN capable source hosts that correct the inherited bias of ECN-TCP against connections with long RTT [8]. Their method prevents the connections with small RTT from opening their congestion window too quickly, and allows connections with large RTT to be more aggressive in opening their congestion windows. This is achieved by weighting the congestion window increment $1/cwnd$ by a window slope factor Swnd which increases with the RTT of the connection.

Another important issue to consider is whether different packet marking strategies will have an impact on ECN performance. The standard ECN uses the mark-tail strategy. In ECN using the mark-tail strategy, the CE-bit of an incoming packet is set according to the congestion status at the time it enters the buffer of each router. Although this method is easy to implement, imposes an additional queueing delay on the feedback of the congestion information to the source host because after setting the CE-bit the incoming packet must wait until all previously buffered packets will have been transmitted. Liu and Jain suggested the use of the mark-front strategy which mark the packet at the head of the queue in stead of marking the incoming packet [9]. As a result, mark-front strategy avoids the additional queueing delay that can occur at a router in the ECN with mark-tail strategy as explained above, and can reflect the up-to-date congestion status of the router.

However, when using ECN with the mark-tail and mark-front strategies in a multiple-hop network environment, an incoming packet with a CE-bit set can still experience longer queueing delays at each downstream node. These delays are particularly serious when the network becomes more congested. Theses additional queueing delays at each transit node then cause a slower response to the congestion status of the bottleneck link, resulting in packet loss and performance degradation.

3 A New Mark-Relay Strategy for ECN

In ECN using the proposed mark-relay strategy, the transfer delay of the CE-bit information contained in an incoming packet can be reduced by relaying the information to the head-of-line packet with a corresponding flow waiting for transmission in the output buffer of each transit router. The mark-relay strategy can be easily implemented by adding a CE-bit relay function to the mark-front strategy.

In the proposed mark-relay strategy, each router maintains a table in order to store the CE-bit information relayed from the upstream node, as shown in Fig. 2. This table then maintains a relayed CE-bit on a per-flow basis. The Flow ID can be identified by its source and destination IP address tuple, or can be defined according to the flow aggregation levels. As such, a table entry for a particular flow is created upon the arrival of the first packet and then maintained only as long as there is at least one packet waiting for transmission for that flow in the buffer. The entry for a particular flow in the relayed CE-bit table is removed when there is no longer a packet for that flow in the router buffer. Meanwhile, when the router receives each incoming packet, it stores the CE-bit information on the incoming packet in the relayed CE-flag in the table related to the corresponding flow. If the CE-bit of an incoming packet is one, the value is stored in the relayed CE flag in the table, then the CE-bit of the incoming packet is reset. Thereafter, the CE-bit of the next outgoing packet is

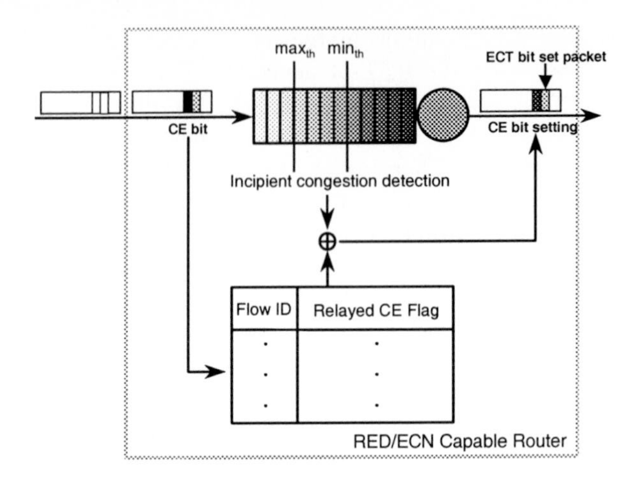

Fig. 2. ECN with the proposed mark-relay strategy

then set according to the congestion status, as determined by an active queue management mechanism and the relayed CE flag for the corresponding flow. Namely, the CE-bit of the outgoing packet is set to the value of the logical OR between the congestion status of the current node and the relayed CE flag in the table. As a result, the feedback delay of the congestion information to its source host can be reduced considerably.

4 Performance Evaluation

In this section, we compare the performance of the different marking strategies for ECN implementation in a router. To compare the marking strategies, simulations are performed using the OPNET simulation tool [11].

4.1 Simulation Configuration

Fig. 3 illustrates the multiple-hop network model used for the simulation [12]. In this model, S_M is the TCP source host for the multiple-hop flows and D_M is its corresponding destination. In addition, the single-hop transit flows are also considered as back-ground traffic at each hop. S_1 - S_N are the TCP source hosts for the single-hop flows and D_1 - D_N are their corresponding destinations. The dotted lines indicate the connection paths. It is assumed that S_M and S_1 consisted of two bulk file transfer sources, respectively. All bulk file transfer sources are considered as infinite greedy sources. In addition, TCP source hosts for the single-hop flows, S_1 - S_N, also consist of small-sized file transfer sources that were used for generating bursty background traffic at each hop. Average file sizes used for small-sized file transfer source are assumed to be 10KB and 100KB with the exponential distributions. And, we change the offered load of background traffic form 20% to 80% of link capacity by changing the average

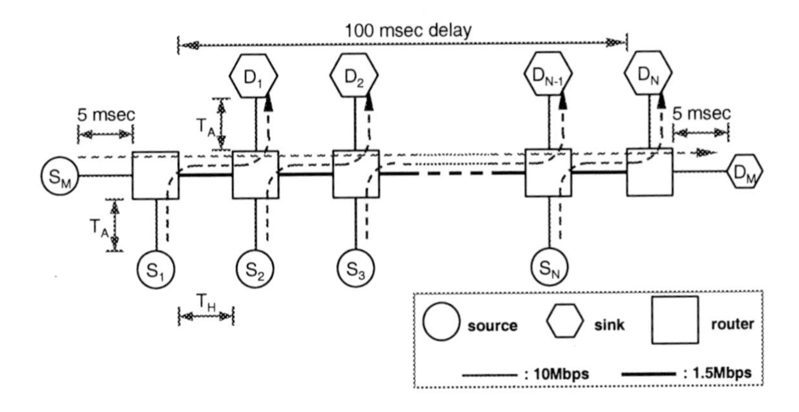

Fig. 3. Simulation network model with N hops

transfer rate of files (files/hours). We evaluate the performance of the marking strategies under the multiple-hop environment by comparing the throughput of bulk file transfer flows.

The number of hops is varied to investigate the effect of the number of transit routers. The delays at each link are assumed as shown in Table 1, to remove the effect of any propagation delay difference between the single-hop and multiple-hop flows, and the effect of the queueing delay on the transferring the congestion information is investigated. In the Table 1, T_A is the access link delay of each sender and receiver host, and T_H is the link delay between routers.

In the simulation, TCP-Reno is used for the TCP flows and each flow was fed using an FTP traffic generator. The TCP timer granularity is set to 100msec, and the maximum segment size (MSS) is set to 1024Bytes. The simulations are continued for about 10 minutes in each case. The router uses the RED as the active queue management mechanism and the parameters for each router are set as follows:

- $B = 100$ [packets] : router buffer size
- $P_{max} = 0.1$: maximum marking probability
- $max_{th} = 60$ [packets] : maximum threshold level
- $min_{th} = 30$ [packets] : minimum threshold level
- $W_q = 0.002$: weighting factor for the queue length average

Table 1. Link delay values for simulation

Number of Hops	2	4	6	8	10
T_H[msec]	50.00	25.00	16.67	12.50	10.00
T_A[msec]	30.00	42.50	46.67	48.75	50.00

4.2 Simulation Results

In this section, we investigate the performances of the mark-tail, mark-front and mark-relay strategies for ECN-capable router implementation in terms of the average throughput. The average throughput is defined as follows: The average bits per second forwarded to the application layer by the TCP layer in the receiver. This result was to compare the performance of the bulk file transfer flows.

4.2.1 Performance Comparison Relative to the Number of Hops

First, we fix the offered load of background traffic to 60% of the link capacity, and evaluate the performance of marking strategies according to changing the number of hops.

Fig. 4 compares the average throughput of the bulk file transfer flows between the various marking strategies relative to an increase in the number of hops. The average throughputs decrease for all the marking strategies as the number of transit hops increases. This is because the additional queueing delay involved in transferring the congestion information to the sender, which is experienced in the downstream routers, produces slower response to congestion. The slower response to congestion then result in packet drops, packet retransmissions, and link underflows. Accordingly, The feedback delay involved in returning congestion information from a congested router to the sender causes degradation in the throughput.

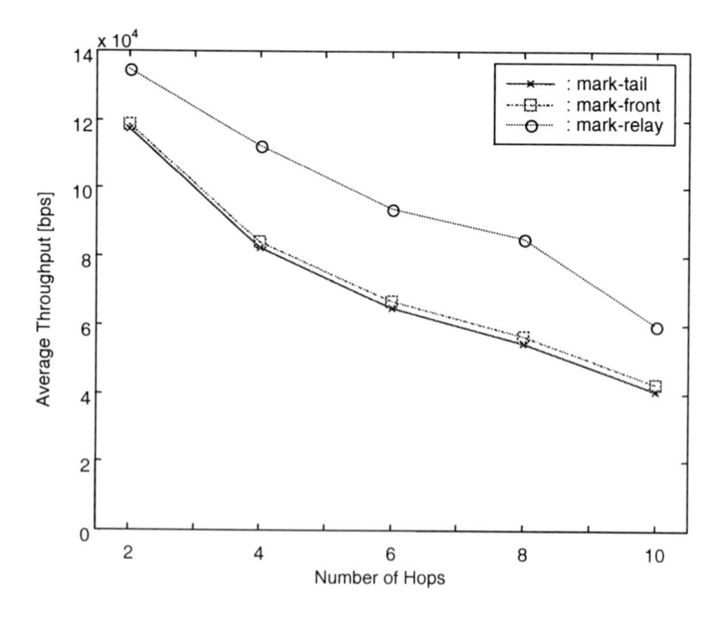

Fig. 4. Average throughput of bulk file transfer flows relative to number of hops

Compared with the mark-tail strategy, although the mark-front strategy reduce the marked packet transfer delay at those nodes where marking occurred, the additional queueing delay experienced at the downstream nodes can not be eliminated. As a result, very little enhancements in performance are observed with the mark-front strategy in multiple-hop network environments. In contrast, the mark-relay strategy provides a much faster congestion response than the other two marking strategies in multiple-hop network environments. That is, the mark-relay strategy is able to reduce the additional queueing delay for transferring the CE information at each transit router by relaying the CE-bit contained in an incoming packet to the head-of-line packet with a corresponding flow. As such, the mark-relay strategy is able to reduce the throughput degradation, as the number of hops increases, and produces a relatively high throughput compared with the mark-tail and mark-front strategies.

Fig. 5 compares the average throughput of the multiple-hop flows with that of the single-hop flows connected to the first router. That is, the performance degradation of the multiple-hop flows is measured using the average throughput ratio of the multiple-hop flows to the single-hop flows. The results show that the multiple-hop flows generally experience a higher performance degradation compared with the single-hop flows, even though the end-to-end link delay of the multiple-hop flows is set to be the same as that of the single-hop flows. That is because of some more marking and the increased queueing delay for transferring congestion information at the downstream nodes. And, performance

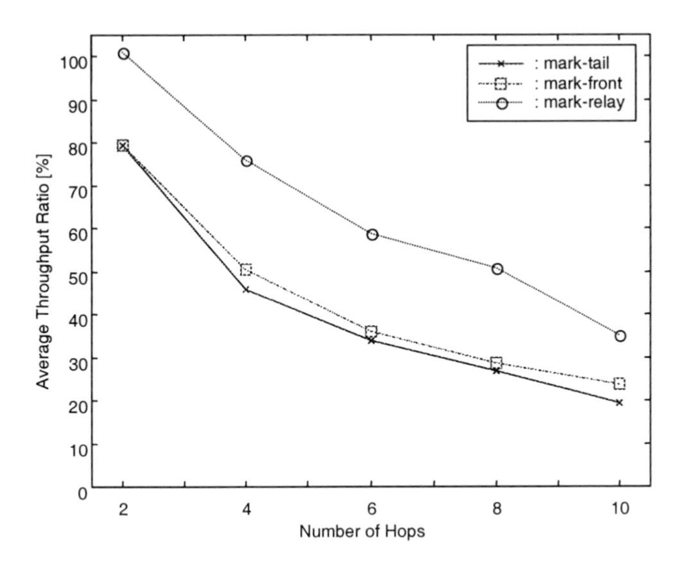

Fig. 5. Performance degradations of multiple-hop flows compared with single-hop flows relative to number of hops

difference among the marking strategies mainly causes by increased queueing delay for transferring congestion information at the downstream nodes. Also, more performance degradation can be seen relative to an increase in the number of transit hops. However, the performance degradation ratio of the mark-relay strategy is lower than that of the other two marking strategies, thereby indicating an improved fairness between the multiple-hop flows and the single-hop flows when the mark-relay strategy is used in the ECN-capable routers.

4.2.2 Performance Comparison Relative to the Offered Load

Next, we fix the number of hops to 8, and evaluate the performance according to changing the offered load of background traffic at each transit hops.

Fig. 6 compares the average throughput of the multiple-hop flows among the marking strategies relative to an increase in the offered load of background traffic. As we can expect, increased offered load can introduce the more queue build-ups and queueing delay at each transit router. Accordingly, the relative throughput degradation of multiple-hop flows increases as the offered load of the background traffic increases, even though the number of transit hops is equivalent. And, the very little enhancements in performance of mark-front strategy over mark-tail strategy can be found in the multiple-hop environments as offered load increases. However, mark-relay strategy relatively increases the throughput compared with the other two marking strategies in any offered load conditions.

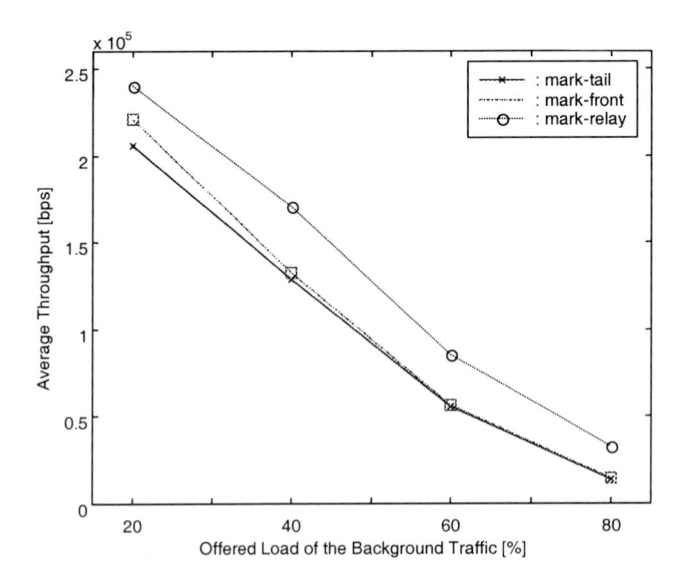

Fig. 6. Average throughput of bulk file transfer flows relative to offered load of background traffic

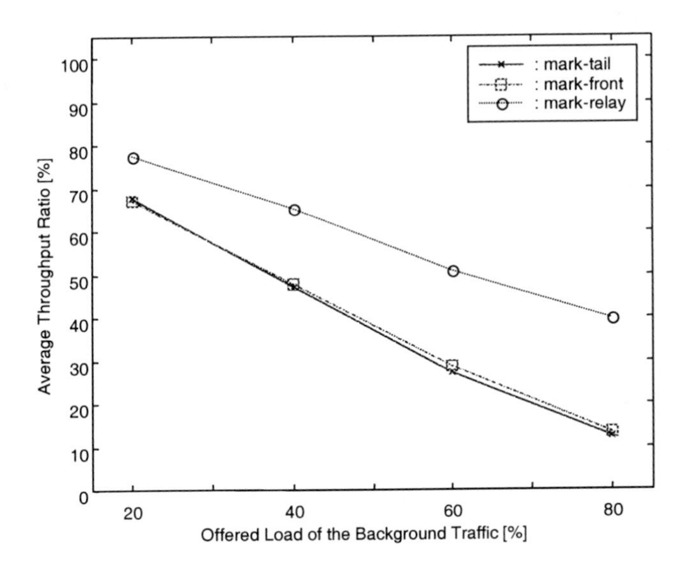

Fig. 7. Performance degradations of multiple-hop flows compared with single-hop flows relative to offered load of background traffic

In the Fig. 7, we compare the average throughput of the multiple-hop flows and that of the single-hop flows connected to the first router. Similar to the previous section, we can see that the performance degradation ratio of the mark-relay strategy is lower than that of the other strategies. And, relative advantage of the mark-relay strategy increases as the offered load of background traffic increases. Accordingly, the fairness of mark-relay strategy between multiple-hop flows and single-hop flows is relatively increased compared with that of the other marking strategies, as the offered load of background traffic increases.

5 Conclusions

ECN with the mark-relay strategy was proposed, and its performance compared with existing marking strategies for ECN-capable router using simulation. The mark-relay strategy can effectively reduce the transfer delay of congestion information through transit routers in multiple-hop network environments. At a transit router, the proposed method relays the CE-bit contained in an incoming IP packet to the head-of-line packet with a corresponding flow that is waiting for transmission in the output buffer. Simulation results showed that ECN with mark-relay strategy produced a better performance than the existing mark-tail and mark-front strategies in terms of throughput and fairness between TCP flows.

In future work, the performance of marking strategies will be investigated under more complicated scenarios with realistic traffic models, such as self-similar traffic model, compared to the simplified traffic source model used in this paper. And, the effect of the congestion detection mechanisms of active queue management will be examined because fast and efficient congestion detections are another key issues in feedback based congestion control.

Acknowledgement. This work was partially supported by the Brain Korea 21 project and the Ministry of Information and Communication, Korea.

References

1. V. Jacobson, "Congestion Avoidance and Control," in Proc. of SIGCOMM 88, pp. 314-329, Aug. 1988.
2. W. R. Stevens, TCP/IP Illustrated, Volume 1: The Protocols, New York: Addison-Wesley, 1994.
3. S. Floyd and V. Jacobson, "Random Early Detection Gateways for Congestion Avoidance," IEEE/ACM Transactions on Networking, vol. 1, no. 4, pp. 397-413, Aug. 1993.
4. B. Braden et al., "Recommendations on Queue Management and Congestion Avoidance in the Internet," RFC2309, Apr. 1998.
5. S. Floyd, "TCP and Explicit Congestion Notification," ACM Computer Communication Review, vol. 24, no. 5, pp. 8-23, Oct. 1994.
6. S. Floyd and K. Ramakrishnan, "A Proposal to add Explicit Congestion Notification (ECN) to IP," RFC2481, Jan. 1999.
7. J. H. Salim and U. Ahmed, "Performance Evaluation of Explicit Congestion Notification (ECN) in IP Networks," Internet Draft draft-hadi-jhsua-ecnperf-01.txt, Mar. 2000.
8. T. Hamann and J. Walrand, "A New Fair Window Algorithm for ECN Capable TCP (New-ECN)," in Proc. of INFOCOM 2000, pp. 1528-1536, Mar. 2000.
9. C. Liu and R. Jain, "Improving Explicit Congestion Notification with the Mark-Front Strategy," Computer Networks, vol. 35, no. 2-3, pp. 185-201, Jan. 2001.
10. I. Leung and J. Muppala, "Effect of Different Marking Strategies on Explicit congestion Notification (ECN) Performance," in Proc. of IEEE ICC 2001, pp. 1812-1816, June 2001.
11. "OPNET Modeler 6.0", http://www.opnet.com.
12. S. Floyd, "Connection with Multiple Congested Gateways in Packet Switched Networks, Part 1: One-Way Traffic," ACM Computer Communication Review, vol. 21, no. 5, pp. 30-47, Oct. 1991.

Improving the Fairness and the Response Time of TCP-Vegas*

Soo-hyeong Lee[1], Byung G. Kim[2], and Yanghee Choi[1]

[1] Department of Computer Science and Engineering,
Seoul National University, Seoul, Korea,
shlee@mmlab.snu.ac.kr
[2] Department of Computer Science,
University of Massachussetts at Lowell, Lowell, MA, USA

Abstract. Unfairness of the Internet has galvanized numerous studies toward fair allocation of bandwidth. Study of TCP-Vegas is one of them. TCP-Vegas, although not perfect, at least enables bandwidth allocation independent of propagation delay, which is radically different behavior from that of current Internet. In the current Internet, a long-delay connection usually receives less throughput than short-delay connections. Until now, two necessary conditions have been identified to make TCP-Vegas achieve fair allocation of bandwidth: correct estimation of propagation delay and finer control of window by adopting single threshold rather than two.

In this paper, we propose three more fixes to achieve fair bandwidth allocation. First, we provide a fix for packet size independence. Second, we provide a fix regarding the reference value in the control. Third, we provide a fix for reducing both the oscillation and the convergence delay. We argue that fixes of ours and those of previous researchers constitute the necessary and sufficient condition for fair allocation of bandwidth.

1 Introduction

TCP is one of the most widespread protocols in the Internet. TCP is a connection-oriented transport protocol that provides reliable delivery service using unreliable IP network. There are several versions of TCP such as Tahoe, Reno, and Vegas. Every TCP version except TCP-Vegas adopts congestion avoidance scheme described in [1] whose basic idea is slowly increasing the number of packets in the network until loss occurs and halving it down when loss occurs. Each version improves performance of previous ones by revising behavior during loss recovery. In contrast, TCP-Vegas tries to gain performance improvement more fundamentally in that it does not incur loss because it does not wait for loss in reducing the number of packets in the network.

* This work was supported by Korea Science and Engineering Foundation under the contract number 20005-303-02-2. It was also supported by the Brain Korea 21 Project and National Research Laboratory Project of Korea.

I. Chong (Ed.): ICOIN 2002, LNCS 2343, pp. 443–454, 2002.

The principle of TCP-Vegas is to have per-connection queue length constant. This allows fair bandwidth allocation as well as loss prevention. It is well known that connections get the same throughput if all the connection have the same per-connection queue length.[1] Loss is prevented with a finite amount of buffer if the number of connection is bounded. If the connections are too many or the buffer is not large enough, TCP-Vegas resorts to TCP-Reno-like lossy behavior, hence TCP-Vegas is always no worse than TCP-Reno.

We focus on the fairness of TCP-Vegas because its fairness is superior to that of Tahoe and Reno. Tahoe and Reno are versions of TCP currently dominating the Internet. Under TCP-Reno or Tahoe, a long-delay connection usually receives less throughput than short-delay connections.

However, even TCP-Vegas has been known to be unfair for two reasons [13][11]. First reason is that TCP-Vegas relies on correct measurement of propagation delay, which is usually not met. A connection that incorrectly measures the propagation delay steals more bandwidth. A connection gets the less correct measurement of propagation delay as it shares the link with the more other connections. There has been no effective way[2] to enforce correct measurement of propagation delay. In this paper, we assume that every connection has correct propagation delay measurement. This may be implemented by giving SYN packets higher queueing priority than other types of TCP packets.

Second reason is that TCP-Vegas actually does not try to make per-connection queue length constant. Instead, it tries to bound per-connection queue length within a region marked by two thresholds, α and β, which are usually set as 1 packet and 3 packets respectively. The known solution of this artifact is to have both thresholds have the same value, which is adopted in this paper.

This paper shows that Vegas is still unfair even when the above two problems are solved, because of its dependence on packet size[3]. In the current Vegas, the connection with bigger packet gets more bandwidth. We propose three fixes as a solution.

First, we propose to redefine the thresholds, α and β, which are set equal to each other in this paper, as a byte count instead of as a packet count. We call the byte count threshold as the *reference*. This fix constitutes the core of our solution. The remaining fixes deal with the performance of this first fix.

Our second fix is about choosing the value of the reference. The first fix alone exhibits a significant departure from the perfect fairness, if the reference is not large enough. We show that a higher reference improves the fairness. We can always make the bandwidth allocation fairer by increasing the reference.

[1] This is the single bottleneck case. For the multiple bottleneck case, this type of allocation can provide a proportional fairness, instead of max-min fairness.

[2] [11] argued that using single threshold instead of two may solve this problem by incurring oscillation, which is later disproved by [13]. [9] mentioned the possibility of expiring fixed-delay measurement without any further development.

[3] Precisely, we mean Path MTU(Maximum Transmission Unit) size when speaking of packet size. MTU is the largest allowable payload size for link level frame. Path MTU is the least MTU along the forward path. MTU ranges from 296B of SLIP to 65536B of Hyperchannel.

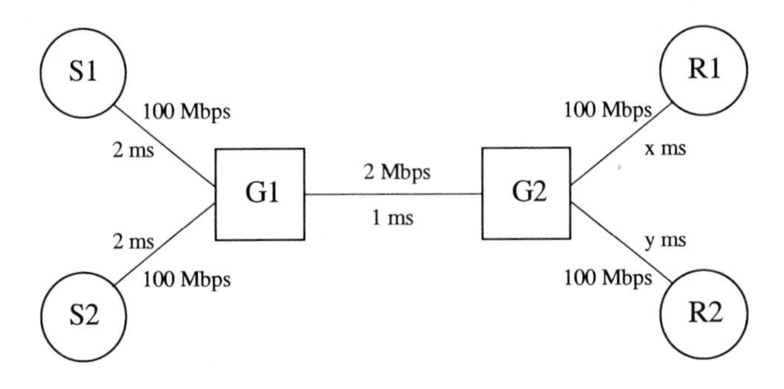

Fig. 1. Simulation Topology

However, this comes at the price of an increase in both the convergence delay and the buffer requirement.

Our third fix is to reduce the needed reference value in achieving the same level of fairness. Specifically, we propose to upgrade the congestion detection formula by adding a half packet. This improves the fairness when used under the same reference value, in that it eliminates the preference of larger-packet connections in the above two fixes.

Our last fix is to reduce the convergence delay when the reference value is high. This is a departure from the traditional window control in that the window change speed is not necessarily one packet a round. The window may change either by multiple packets a round or by less than one packet a round. We argue that this does not cause any harm on the stability of the congestion control, by providing a proof.

The rest of this paper is organized as follows. We describe the problem in Section 2. We describe the main fix in Section 3. We show the effect of the value of the reference in Section 4. We show how to fully eliminate the packet size dependence by upgrading the congestion detection formula in Section 5. In Section 6, we present a faster window control, which shortens the transient state without losing the stability. A summary of our work and future plans is given in Section 7.

2 Vegas Throughput Is Dependent on Packet Size

We show that TCP-Vegas is unfair because connections with different packet sizes receive different throughput in the steady state.

Fig. 2 shows that connection with bigger packet gets more bandwidth. There are two connections sharing a bottleneck link under the topology of Fig. 1. One connection has Path MTU of 1500B(limited by Ethernet or PPP), while the other has Path MTU of 296B(limited by SLIP). For many graphs in this paper, the legend of the graph means the packet size, and the vertical axis has no unit because it is a ratio of two values of the same dimension.

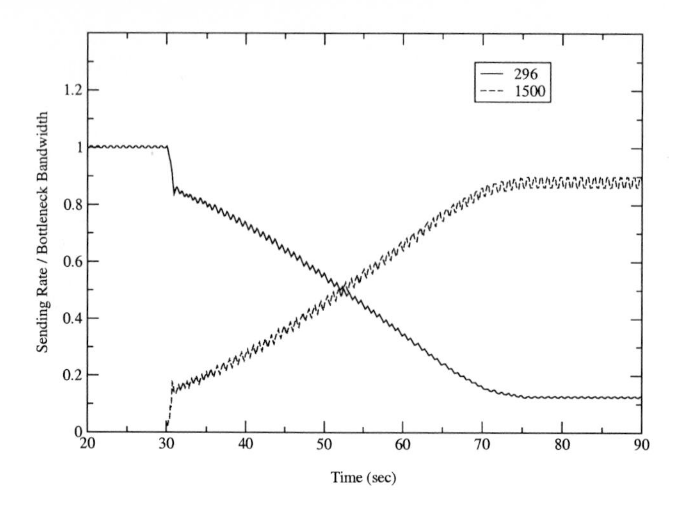

Fig. 2. Bigger Packet gets More Bandwidth

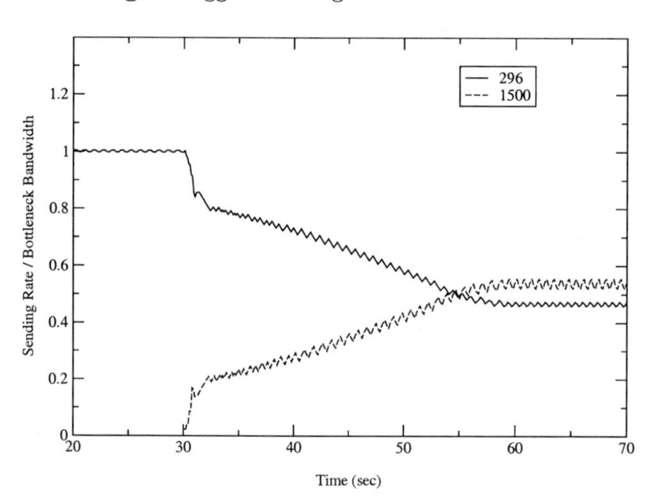

Fig. 3. Equalizing the Formula in Byte

To show the effect of the current subject, we eliminated all the known causes of unfairness. We have made them know their correct propagation delay. We set two thresholds, α and β, equal to two packets.

3 Packet Size Independence

We can obtain much improved fairness by changing the semantic of the threshold from a packet count to a byte count. Fig. 3 is when the reference byte count is set

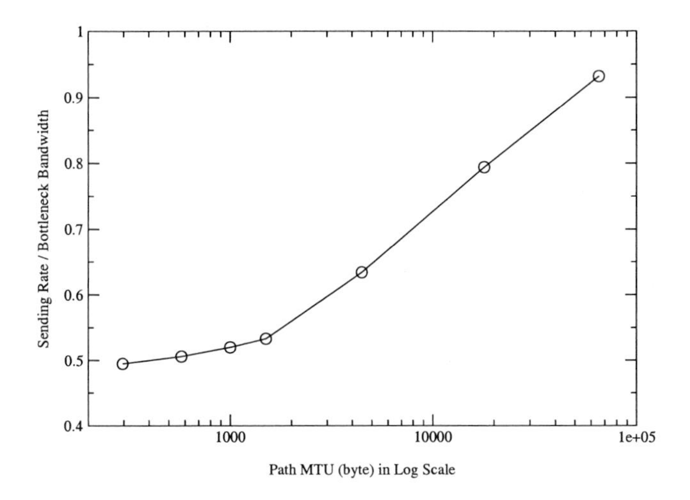

Fig. 4. Connection with Larger Packet Gets More Throughput When Competing with Connection with 296B Packet

to 3kB, which is chosen to be twice as high as the largest MTU in this simulation setting.

However, this improvement is still not perfect. We can observe that connection with bigger packet still gets slightly more bandwidth. We conducted a simulation of two connections, the Path MTU of one of them being changed, while that of the other having 296B Path MTU. The reference byte count was 3000 bytes. Figure 4 shows that the error is too high to be ignored and is increasing as the difference in packet size widens.

4 Reducing the Error by Raising the Thresholds

In general, the fairness can be made better by raising the threshold.

When the reference byte count is raised to 131070 B, which is chosen to be twice as the largest possible MTU, varying the difference in packet size does not cause difference of more than six percent in throughput.

However, there are two problems remaining in raising the reference.

The first problem is that the better treatment to connections with the larger MTU is all the same, though somewhat mitigated.

The second problem is that raising the reference count increases both the buffer requirement and the convergence delay. The increase in queue length is inferred by considering the reference byte count as the mean queue length. The increase in the convergence delay is shown in Fig. 6, where the steady state is reached only after 420 seconds.

The first problem is dealt with in the next section, and the second problem is dealt with in the section after the next one.

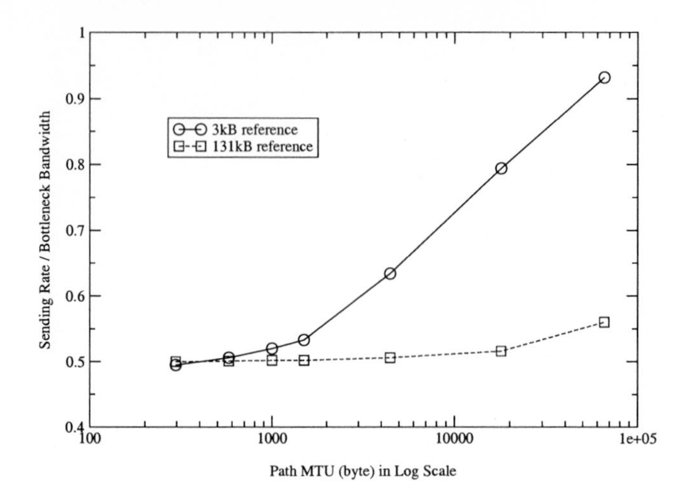

Fig. 5. Raising the Reference Byte Count Improves the Fairness

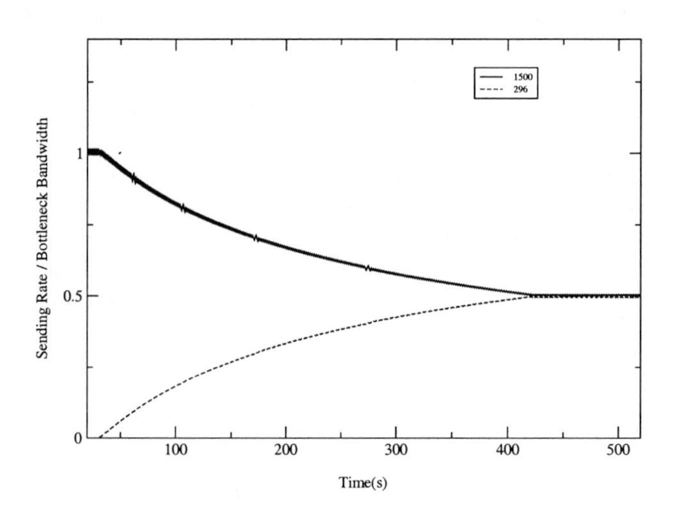

Fig. 6. Reference Byte Count Raised

5 Making the Error Independent of Packet Size

The results of the previous two sections suggest that the traditional interpretation of the congestion detection formula may be incorrect. The congestion detection formula has been known to mean the per-connection queue length. If the per-connection queue length is all the same throughout the connections, the connections should receive the same throughput. However, this is inconsistent with what the previous two sections have shown.

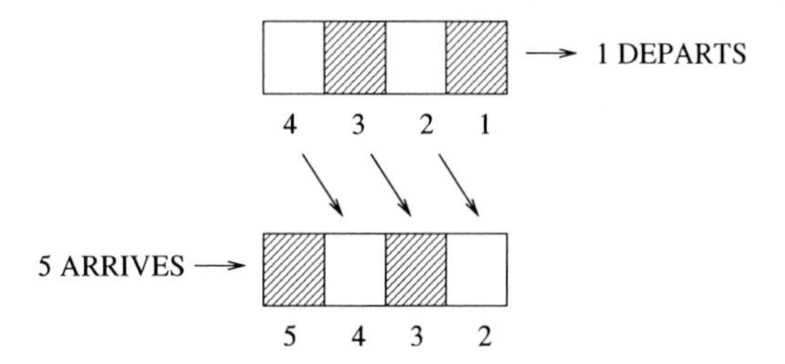

Fig. 7. Failure of Per-connection Queue Length Computation

In this section, we show that the known interpretation of the congestion detection formula is indeed incorrect. Our proposed fix is simple and is shown to improve the fairness under the same reference byte count and to eliminate the preference of larger-packet connections, thus possibly enabling lowered byte reference count for the same level of fairness.

The fix is to add one half packet to the congestion detection formula. We argue that the resulting value means the per-connection queue length. This is because the packet arriving at the bottleneck queue fails to count itself when measuring queueing delay.

Figure 7 depicts this phenomenon when per-connection queue length is two. Just before a packet arrives, another packet of the same connection departs the bottleneck queue. So every packet sees only one packet in the queue ahead of itself, although there are always two packets in the queue.

There is yet another aspect remaining to be clarified: the figure says that we should add one packet, whereas our fix is to add a half packet. In fact, as shown by the proof in the appendix, the ideal fix is to add $\frac{P}{n}$, where P denotes the packet size, and n is the number of connections. The reason for not using this ideal fix is that it requires knowing the number of connections. Choosing a fixed n results in both the preference of small-packet connections when more than n connections share the bottleneck link and the preference of large-packet connections when less than n connections share it. We chose $n = 2$ because of the following two reasons: we want to make it impossible for the larger-packet connection to get more throughput; we want to make the queue length error as small as possible. The first reason suggests using either 1 or 2 as n, and the second reason dictates that $n = 1$ makes the error too large when the number of connections is large.

Fig. 8 shows that we can improve the fairness from Fig. 3 by adding a half packet even when using a low reference of 3kB.

This fix can also improve the fairness when using a higher reference count of 131kB, as shown in Fig. 9. This figure clearly shows that the fix makes the dependency on the packet size unobservable when the number of connections is two.

However, there is a shortcoming in this fix. It requires the reference count to be greater than half the largest possible MTU. Otherwise, the connection with

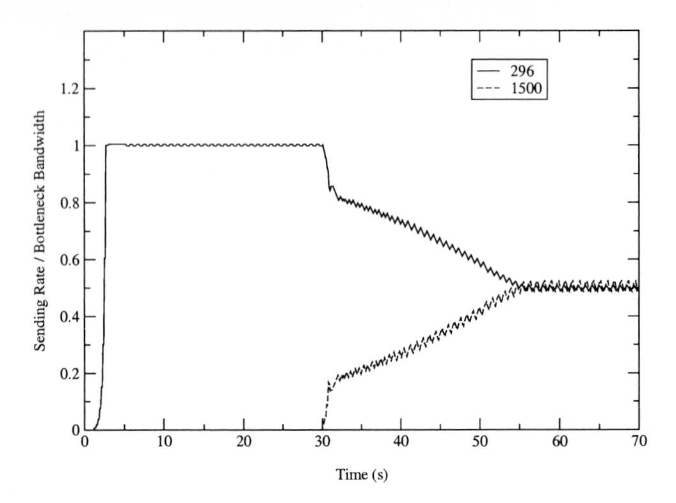

Fig. 8. Equalizing the Formula Plus One Half Packet

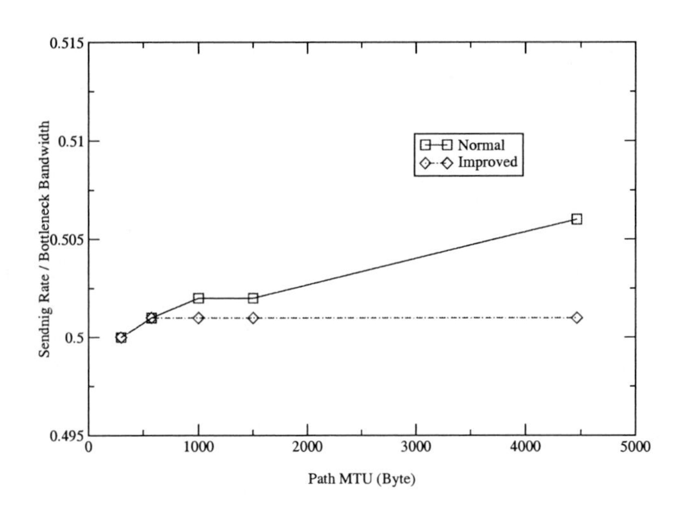

Fig. 9. Fairness Improvement by Adding a Half Packet(The Reference Count is 131kB)

the packet greater than twice the reference count would receive poor bandwidth, because adding a half packet would always result in a value above the reference, which in turn forces to decrease the window.

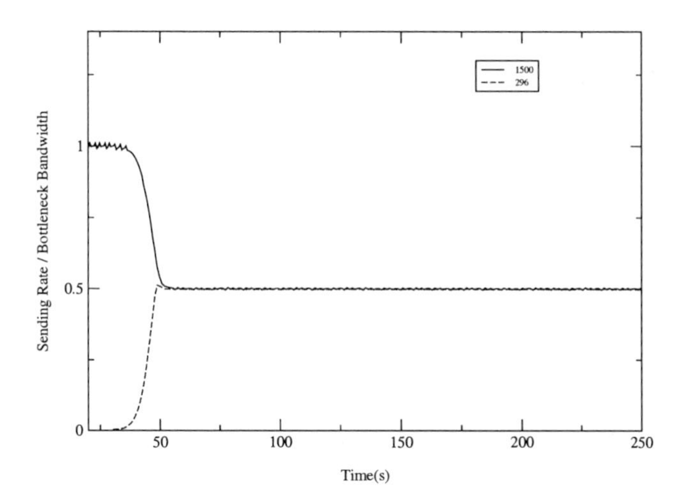

Fig. 10. After Using the Fast Control from Fig. 6

6 Reducing Both the Oscillation and the Convergence Delay

In this section, we present a fast window control scheme designed to solve the problem of long convergence delay, which is a consequence of raising the byte reference count.

Our fast window control is to make the rate of window change proportional to the error, which is defined as the difference between the reference and the estimate of the per-connection queue length. The normal Vegas also compares the estimate of the per-connection queue length with the reference. However, this is only used in determining the direction of change, and has no influence in the rate of the window change. The rate of the window change of the normal Vegas can assume one of the following three: plus one packet a round, minus one packet a round, or no window change during a round.

Our fast window control is sometimes slower than the normal Vegas, thus eliminating the oscillation.

Figure 10 shows the application of our fast control on Fig. 6. The fast control reduces the convergence delay from 390 seconds to 30 seconds.

Figure 11 shows how much the convergence delay is reduced by our fast control. The simulation was carried out with two connections, where a connection with a controlled packet size competes with the other whose packet size is 65535B. The result indicates that, without fast control, the convergence delay is indirectly proportional to the smallest MTU, and that the fast control makes the convergence delay almost independent of the MTU size.

We argue that our fast control scheme is stable, based on the following argument: the fast control is stable if the normal Vegas is stable regardless of the MTU size. A small-packet connection changing multiple packets of window

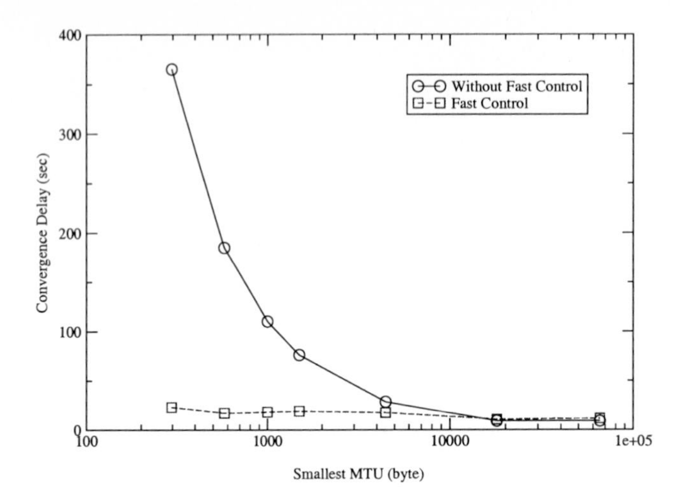

Fig. 11. The Fast Control Reduces the Convergence Delay

a round is not different from a large-packet connection changing one packet a round.

7 Conclusions

Vegas shows much improved fairness than Reno or Tahoe. Its bandwidth allocation is independent of the propagation delay, whereas Reno or Tahoe give less throughput to long delay connections.

This improved fairness results from its principle of congestion control: it tries to keep constant number of packets in the bottleneck queue. In theory, this principle implies the perfect fairness in throughput.

However, Vegas does not show perfect fairness, because of some implementation problems. Two such problems have been identified by previous studies: it does not always obtain correct measurement of the propagation delay; it does not actually equate the per-connection queue length among all the connections.

This paper explains another problem resulting from different path MTU sizes among connections. To solve this problem, this paper proposed one main fix and three subsidiary fixes. The main fix is that the Vegas should try to keep constant number of bytes, rather than packets, in the bottleneck queue. The subsidiary fixes answer the following three questions: how many bytes should be kept in the bottleneck queue; whether the main fix is sufficient; what the side effect is and how to solve it.

In doing this, we found that the Vegas implementation has yet another discrepancy from its principle, in an aspect other than the data unit being packet or byte. We found that the measurement of the per-connection queue length is incorrect.

We also proposed an improved window control, which allows the rate of the window change to be much faster or much slower than the conventional rate of one packet a round. A formal proof and a thorough validation on its stability is our future work.

References

1. V. Jacobson, *Congestion Avoidance and Control*, Proceedings of ACM SIGCOMM'88, August 1988.
2. S. Keshav, *A Control-Theoretic Approach to Flow Control*, Proceedings of the SIGCOMM '91, pp. 3-15, September 1991.
3. Lawrence S. Brakmo and Larry L. Peterson, *TCP Vegas : End to End Congestion Avoidance on a Global Internet*, IEEE J. of Selected Areas in Communication, pp.1465-1480, October 1995.
4. J. S. Ahn, P. B. Danzig, Z. Liu, and L. Yan, *Evaluation of TCP Vegas: Emulation and Experiment*, ACM SIGCOMM '95, Cambridge, MA, August 1995.
5. J. S. Ahn and P. B. Danzig, *Packet Network Simulation: Speedup and Accuracy Versus Timing Granularity*, IEEE Transactions on Networking, 4(5):743-757, October 1996.
6. J.C. Hoe, *Improving the Start-up Behavior of a Congestion Control Scheme for TCP*, Proceedings of ACM SIGCOMM'96, August 1996.
7. Y. Zhang, E. Yan, and S.K. Dao, *A Measurement of TCP over Long-Delay Network*, Proceedings of 6th Intl. Conf. on Telecommunication Systems, pages 498-504, March 1998.
8. T. Bonald, *Comparison of TCP Reno and TCP Vegas via Fluid Approximation*, Workshop on TCP, 1998.
9. Jeonghoon Mo, Richard J. La, Venkat Anantharam, and Jean Walrand, *Analysis and Comparison of TCP Reno and Vegas*, IEEE INFOCOM'99, pp.1556-1563, 1999.
10. S. McCanne and S. Floyd, *Ns(network simulator)*, http://www-mash.cs.berkeley.edu/ns
11. G. Hasegawa, M. Murata, and H. Miyahara, *Fairness and Stability of Congestion Control Mechanisms of TCP*, IEEE Infocom 1999.
12. A. Aggarwal, S. Savage, and T. Anderson, *Understanding the Performance of TCP Pacing*, IEEE INFOCOM 2000, 2000.
13. C. Boutremans and J. Le Bourdec, *A Note on the Fairness of TCP Vegas*, International Zurich Seminar on Broadband Communications, 2000.
14. Jin-Ru Chen and Yaw-Chung Chen, *Vegas Plus: Improving the Service Fairness*, IEEE Communication Letters, Vol. 4, No. 5, May 2000.
15. U. Hengartner, J. Boliger, and T. Gross, *TCP Vegas Revisited*, IEEE INFOCOM 2000.

A The Congestion Detection Formula Does Not Mean the Per-connection Queue Length

In this section, we explain why the sender under-estimates its per-connection queue length, and the motivation of our solution. The reason of the under-estimation is that the queueing delay measurement is under-estimated and that the queue length is only inferred by the measured queueing delay.

First, we need to show how the per-connection queue length is inferred from the queueing delay. This is done by the Vegas *congestion detection formula*. Suppose t_n is the beginning of the n-th round. At time t_n, the congestion detection formula is defined as follows:

$$Diff \triangleq \tau_q * \lambda, \tag{1}$$

where the queueing delay is $\tau_q \triangleq RTT(t_n) - baseRTT$, and the throughput is $\lambda \triangleq \frac{W(t_n)}{RTT(t_n)}$, W(t) is window size at time t, and $baseRTT$ is the least of Round-Trip Times (RTT) until then. We assume $baseRTT = \tau$, where τ is the round-trip propagation delay. $RTT(t_n)$ is the average of the round-trip times experienced by the acknowledgments received during $[t_{n-1}, t_n)$. The unit of queue length is that of window used in this formula.

Second, we show a formal argument about the inevitable under-estimation of the queueing delay. We have already tried to give an intuitive explanation of this phenomenon, in Fig. 7. However, readers may not feel assured. This is driven by the steady-state assumption. Under steady-state, the aggregate queue length is the same and the per-connection queue length is the same. Thus if a packet of size P arrives, a packet of the same size must have left just before. Hence the larger the packet, the less bytes of queue are ahead of it in the queue. From this reason, we can say that the queueing delay is more under-estimated with larger packet size. Formally, the measured queueing delay is a decreasing function of the connection i's packet size P_i: $\frac{Q_A - P_i}{C}$, where the aggregate queue length is $Q_A \triangleq \sum Q_i$, Q_i is the actual per-connection queue length of connection i, and the bottleneck bandwidth is C. Thus, the per-connection queue length of connection i is under-estimated by $\frac{\lambda}{C} P_i$, if we combine this and the last paragraphs. If the bandwidth is fairly allocated and there are n connections, this amount equals $\frac{P_i}{n}$, because $\lambda = \frac{C}{n}$.

Using the above argument, we can infer that each connection should consider its amount of under-estimation in choosing the target per-connection queue length, in order to have the same per-connection queue length for every connection. When the per-connection queue lengths are actually the same, it is natural that the connection with bigger packet sees lower estimated queue length. However, the current Vegas implementation tries to equate the estimated queue length rather than the actual queue length. Our proposal is that a connection with bigger packet should set its target queue length lower. More specifically, when there are n connections, connection i should set its target as $alpha - \frac{P_i}{n}$. However, the number of connections n is not available, thus a conservative approach is to set the target as $alpha - \frac{P_i}{2}$. With this scheme, no connection with bigger packet can steal the bandwidth.

This phenomenon is problematic when the packet size is heterogeneous. Note that our proposed scheme does not cause any harm when every packet size is the same.

V. Quality of Services

Radio Resource Allocation in GSM/GPRS Networks

Jean-Lien C. Wu[1], Wei-Yeh Chen[2], and Hung-Huan Liu[1]

[1] Department of Electronic Engineering, National Taiwan University of Science and Technology, 43, Keelung Road, Section 4, Taipei 106, Taiwan, R.O.C.
{jcw, vincent}@nlhyper.et.ntust.edu.tw
[2] Department of Information Management, National Taiwan University of Science and Technology, 43, Keelung Road, Section 4, Taipei 106, Taiwan, R.O.C.
D8609002@mail.ntust.edu.tw

Abstract. GPRS is a packet switched access mode for GSM system to improve wireless access to the Internet. In this paper, we study the design of radio resource allocation for GPRS and GSM services by allowing guard channels to be temporarily allocated to GPRS users to increase channel utilization. The call admission controller and channel allocation controller are employed to achieve good channel utilization and preserve the QoS of GSM services. Simulation results show that at low voice traffic load, there is no need to apply admission control to GPRS connections. While at high voice traffic load, applying call admission control to GPRS connections can guarantee the performance of voice service, but result in high GPRS connection blocking and low channel utilization. Furthermore, the QoS of voice service not being affected by the introduction of GPRS can be obtained by allowing voice arrivals to preempt the ongoing GPRS connections.

1 Introduction

General Packet Radio Service (GPRS) [1], initiated in 1994, is an European Telecommunications Standard Institute (ETSI) standard for packet data transmission using the core GSM (Global System for Mobile Communications) radio access network. Consequently, GPRS shares the GSM frequency bands with telephone and circuit-switched data traffic, and makes use of many properties of the physical layer of the original GSM system. Since radio resources of a cell are shared by both the GPRS and GSM voice services, how to efficiently allocate radio resources between these two services and at the same time not degrading the QoS of voice service is an important issue.

Brasche *et al.* [2] first introduced GPRS, described the GPRS protocol and demonstrated its performance. Different scheduling strategies were proposed by Sau *et al.* [3,4] to guarantee the QoS in GPRS environment. The performance analysis of radio link control and medium access control (RLC/MAC) protocol of GPRS was investigated by Ludwig *et al.* [5]. The performance of integrated voice and data for GPRS was analyzed in [6,7]. However, the above researches focused on the performance of GPRS traffic, none has discussed the impact of accommodating GPRS traffic on the performance of voice services. In this paper, we will study the

I. Chong (Ed.): ICOIN 2002, LNCS 2343, pp. 457-468, 2002.

interaction of resource allocation between voice traffic and GPRS traffic and the impact on the system performance.

Static guard channel scheme [8] is commonly used to prioritize GSM voice handoff calls because of its low implementation complexity. These guard channels can be temporarily allocated to GPRS to increase channel utilization, and will be de-allocated to handoff calls when necessary.

The rest of this paper is organized as follows: In section 2, we will briefly introduce the radio interface of GPRS. The channel de-allocation and call admission control mechanisms are described in section 3. Section 4 provides the simulation results of the proposed scheme and section 5 concludes this paper.

2 Radio Interface

GPRS uses the same TDMA/FDMA structure as that of GSM to form physical channels. Each physical channel can be assigned to either GPRS or GSM service. The physical channel dedicated to packet data traffic is called the packet data channel (PDCH). The basic transmission unit of a PDCH is called a radio block. To transmit a radio block, four time slots in four consecutive TDMA frames are used [9]. Four different coding schemes, CS-1 to CS-4, are defined for the radio blocks [10] and are shown in Table 1.

Table 1. GPRS coding schemes

Coding scheme	Code rate	Payload	Data rate (kbits/s)
CS-1	1/2	181	9.05
CS-2	~2/3	268	13.4
CS-3	~3/4	312	15.6
CS-4	1	428	21.4

Radio blocks can be sent on different PDCHs simultaneously, thus reducing the packet delay for transmission across the air interface. The allocated channels may vary by allocating one to eight time slots in each TDMA frame depending on the number of available PDCHs, the multi-slot capabilities of the mobile station, and the current system load [11]. With coding scheme and multi-slot allocation, higher date rate can be achieved.

To support the packet-switched operation of GPRS, PDCHs are assigned temporarily to mobile stations. The base station controller (BSC) controls the resources in both the uplink and downlink directions. We will focus on the uplink data transfer to investigate the radio resource allocation. To avoid access conflicts in the uplink direction, the BSC transmits in each downlink radio block header an uplink state flag indicating which mobile station is allowed to transmit on the corresponding uplink PDCH.

3 The Proposed Radio Resource Allocation Scheme

The proposed scheme employs the channel allocation and admission control mechanism to guarantee the QoS and improve the channel utilization. Each GPRS connection request can be associated with two bandwidth parameters: the requested bandwidth (*b_req* Kbps) and the minimum required bandwidth (*b_min* Kbps). Each GPRS connection request demands for a bandwidth of *b_req* Kbps, and the minimum bandwidth to be guaranteed is *b_min* Kbps once this connection request is admitted. The bandwidth allocated to each GPRS connection can vary between *b_req* and *b_min* Kbps.

Upon the arrival of a GPRS connection request, the call admission controller has to figure out the number of channels required. Let *c_req* denote the number of channels allocated for GPRS to offer a bandwidth of *b_req* Kbps if it is admitted, and *c_min* denote the minimum number of channels required to offer a bandwidth of *b_min* Kbps for an admitted GPRS connection. Assume each PDCH can provide a bandwidth of I Kbps. Then *c_req* and *c_min* can be obtained as follows:

$$c\_req = \left\lceil \frac{b\_req}{I} \right\rceil, \text{ and } c\_min = \left\lceil \frac{b\_\min}{I} \right\rceil, \tag{1}$$

where $\lceil x \rceil$ is the ceiling function of x.

The channel allocation model is depicted in Fig. 1 where GSM voice service and GPRS share the same common pool of the physical channels. A number of guard channels are reserved for prioritized voice handoff calls. C denotes the total number of channels of the common pool, C_G denotes the number of guard channels reserved for voice handoff calls which can be temporarily allocated for GPRS. C_{voice} denotes

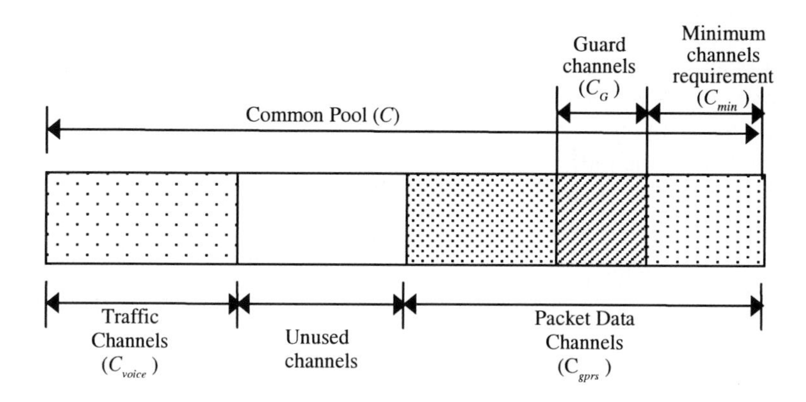

Fig. 1. The channel allocation model

the number of channels used by voice calls. C_{gprs} denotes the number of channels used by GPRS connections. C_{min} denotes the number of channels guaranteed for admitted GPRS connections. The number of available channels for voice service denoted as C_{avail} can be expressed as $C_{avail} = C - C_{min}$.

3.1 The Channel Allocation Controller

The channel allocation controller is employed to dynamically adjust channels allocated for both services to achieve better channel utilization. When the network is congested, the channel allocation controller is responsible for de-allocating some channels of the existing GPRS connections to fulfill the minimum bandwidth requirement for the admitted GPRS connections and voice calls.

When a new voice call is admitted, the channel allocation controller will allocate one channel to this voice call from the unused channels. If there are no unused channels, it will try to de-allocate one PDCH from the existing GPRS connections whose allocated bandwidth is larger than its minimum bandwidth requirement. The remaining PDCHs should still provide bandwidth for the ongoing GPRS connections to maintain their minimum bandwidth requirement. If a handoff call arrives and all the guard channels are used up by voice handoff calls and GPRS connections, the guard channels temporarily allocated to GPRS as PDCH must then be de-allocated for voice handoff calls.

3.2 The Call Admission Controller

The call admission controller is employed to control the number of GPRS to guarantee the QoS of voice service and admitted GRPS connections. A GPRS connection request will be admitted under two conditions. Firstly, the admission of a GPRS connection can still maintain the blocking probability of new and handoff calls below P_{tnb} and P_{thd}, where P_{tnb} is the target blocking probability of new calls, and P_{thd} is the target blocking probability of handoff calls. Secondly, the network should have enough bandwidth to guarantee a bandwidth of *b_min* Kbps for this request, that is, $c\_min \leq C - C_{voice} - C_{min}$.

To find the blocking probability of new and handoff calls after having admitted a GPRS connection, the traffic model for personal communication system [12] is used. Fig. 2 shows the state-transition diagram for the static guard channel scheme. The mean arrival rate of new call requests and handoff call requests are denoted as λ_n and λ_h, respectively. The mean residence time of a mobile unit in a cell is denoted by $1/\mu$. Having admitted a GPRS connection request, the system needs to allocate *c_min* channels to guarantee its minimum QoS requirement. Then the number of available channels for voice service, $\tilde{C}_{avail}$, can be expressed as $\tilde{C}_{avail} = C - (C_{min} + c\_min)$.

Let i be the system state corresponding to the number of voice calls in the system. $P(i)$ denotes the steady-state probability of a total of i voice calls in the system, and the probability can be easily obtained from the M/M/c/c queueing model as

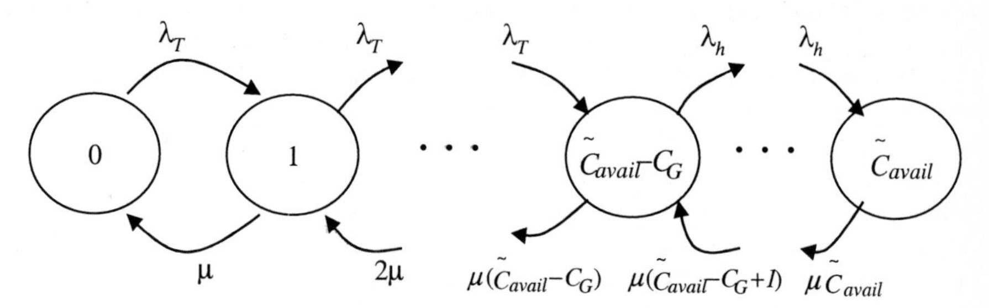

Fig. 2. The state-transition diagram for the static guard channel scheme

$$P(i)=\begin{cases}(\frac{\lambda_T}{\mu})^i\frac{P(0)}{i!}, & \text{if } i\le\tilde{C}_{avail}-C_G\\(\frac{\lambda_T}{\mu})^{\tilde{C}_{avail}-C_G}(\frac{\lambda_h}{\mu})^{i-(\tilde{C}_{avail}-C_G)}\frac{P(0)}{i!}, & \text{if } \tilde{C}_{avail}-C_G<i\le\tilde{C}_{avail}\end{cases}\tag{2}$$

$$P(0)=\left[\sum_{i=0}^{\tilde{C}_{avail}-C_G}\frac{(\frac{\lambda_T}{\mu})^i}{i!}+(\frac{\lambda_T}{\mu})^{\tilde{C}_{avail}-C_G}\sum_{i=\tilde{C}_{avail}-C_G+1}^{\tilde{C}_{avail}}\frac{(\frac{\lambda_h}{\mu})^{i-(\tilde{C}_{avail}-C_G)}}{i!}\right]^{-1}\tag{3}$$

where $\lambda_T = \lambda_n + \lambda_h$. The new call blocking probability P_{nb} and handoff call blocking probability P_{hb} can be expressed respectively as

$$P_{nb}=\sum_{i=\tilde{C}_{avail}-C_G}^{\tilde{C}_{avail}}P(i)\text{, and } P_{hb}=P(\tilde{C}_{avail})\tag{4}$$

The call admission controller then compares these two values with the target values P_{tnb} and P_{thb}, respectively. If both P_{nb} and P_{hb} are smaller than the target values respectively, and the available channels are enough to guarantee the minimum bandwidth requirement, the GPRS connection requests will be accepted. On the other hand, GSM new call requests will be accepted if $C_{avail} - C_G > 0$, and handoff call requests will be accepted if $C_{avail} > 0$.

4 Simulation Assumptions and Results

The total number of channels, C, is assumed to be 100. For simplicity, we assume the arrival of new and handoff calls form a Poisson process with rate λ_n and λ_h, respectively, and let $\lambda_n = \lambda_h = \lambda$. According to the study of the effect of different number of guard channels to voice call blocking probability for C =100, the number of guard channel being 2 is chosen. Let the new call arrival rate be 0.20 calls/sec for low voice traffic load and 0.23 calls/sec for high voice traffic load. The call holding time, new or handoff, is assumed to be exponentially distributed with a mean of 180 seconds.

The arrivals of GPRS connection requests are assumed to form a Poisson process with rate λ_{gprs}. In the simulation, CS-2 coding scheme is used and its corresponding transmission rate is 13.4 *Kbps* per PDCH. Assume the packet length of each GPRS connection is exponentially distributed with a mean of 2×13.4 *Kbits*, corresponding to the mean service time of 2 seconds if one PDCH is allocated.

We also assume that the mobile station has the multi-slot capability and the maximum number of PDCHs that can be allocated to one mobile station is 4. In other words, 1 to 4 time slots per TDMA frame can be allocated to one mobile station. For simplicity, the allocated channels are not restricted to be in the same frame. Referring to the call blocking probabilities given in [13], the target new call blocking probability P_{tnb} is chosen to be 0.05 and the target handoff call blocking probability P_{thb} is 0.005.

To investigate the performance of the proposed scheme, three scenarios are considered:

Scenario-1 : GPRS traffic shares the radio resources with GSM voice traffic without resource management.

Scenario-2 : GPRS traffic shares the radio resources with GSM voice traffic with channel de-allocation mechanism, *i.e.*, the channels of existing GPRS connections will be de-allocated to voice calls when no resources are available in the system.

Scenario-3 : GPRS traffic shares the radio resources with GSM voice traffic, and both channel de-allocation and call admission control mechanism are employed.

Performance of the three scenarios with increasing GPRS mean arrival rate under different voice traffic load are studied. Performance measures of interest are blocking probability of new and handoff voice calls, GPRS connection rejection ratio, and channel utilization. Fig. 3 and Fig. 4 show the comparison of the blocking probability of new voice call and handoff call respectively with GPRS mean packet size being 2×13.4 *Kbits*. It can be seen that accommodating GPRS without any resource management, *i.e.,* scenario-1, would severely degrade the performance of voice service. Besides, at low voice traffic load, admission control on GPRS arrivals is not necessary. Scenario-2 gives almost the same performance as scenario-3. At high voice traffic load, the blocking probability of voice service for scenario-2 becomes worse with increasing GPRS traffic load. The blocking probability of voice calls, new and handoff call, for scenario-3 still maintains below certain value in despite of the increasing GPRS traffic load. The reason for handoff call blocking probability exceeds P_{thb} at high voice traffic load is that although the guard channels temporarily used by GPRS connections can be de-allocated to handoff calls, the amount of bandwidth of the de-allocated GPRS connection must still be greater than or equal to its minimum bandwidth requirement. If all the GPRS connections are admitted with minimum required bandwidth and the guard channels have been used up by handoff calls or GPRS connections, handoff arrivals would be blocked.

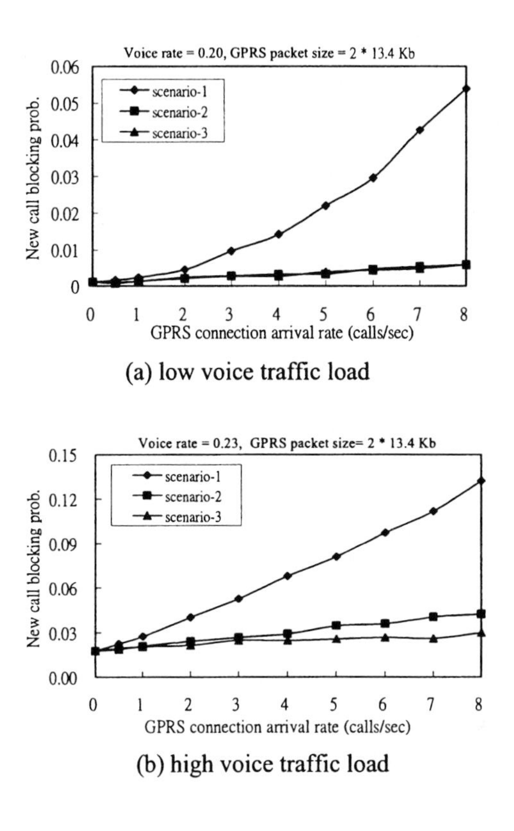

Fig. 3. Comparison of new call blocking

Fig. 5 shows the comparison of GPRS connection rejection. It shows that scenario-3 will suffer large connection rejection ratio, especially at high traffic load, compared with the other two scenarios. This is because when the traffic load increases, a GPRS connection request will most probably fail the admission control test, causing high rejection ratio. On the other hand, scenario-1 gives the lowest GPRS rejection ratio among the three scenarios.

Fig. 6 compares the channel utilization. At low voice traffic load, the channel utilization of all three scenarios are almost the same when the GPRS arrival rate is less than 3 calls/sec. With increasing GPRS arrival rate, scenario-2 will have the largest channel utilization. At high voice traffic load shown in Fig. 6 (b), the channel utilization has similar trend and characteristics as the low voice traffic load case. Scenario-3 has the lowest channel utilization. This is because a large portion of GPRS connection requests are rejected by the call admission controller at high traffic load.

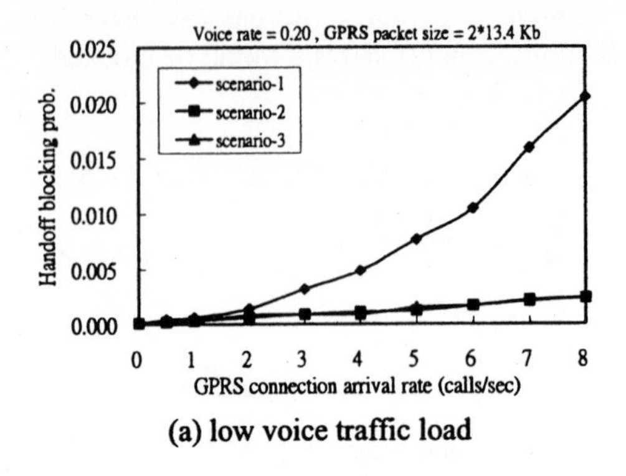

(a) low voice traffic load

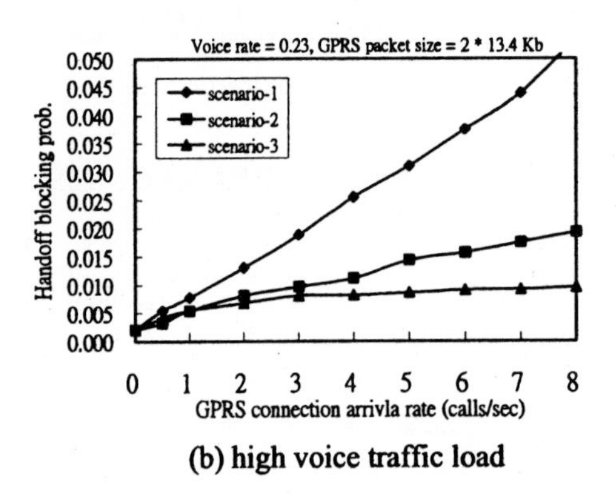

(b) high voice traffic load

Fig. 4. Comparison of handoff call blocking

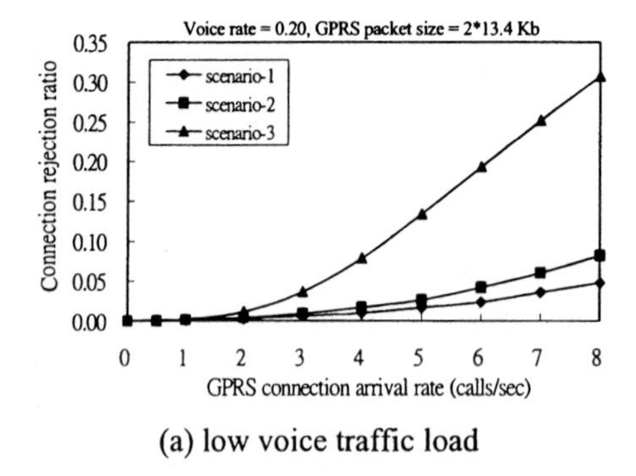

(a) low voice traffic load

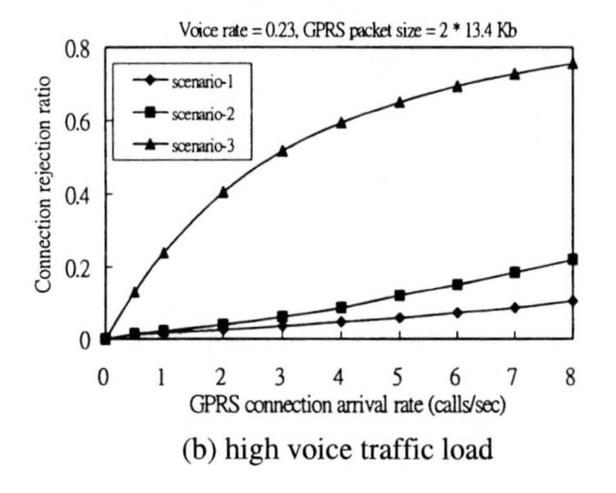

(b) high voice traffic load

Fig. 5. Comparison of GPRS connection rejection

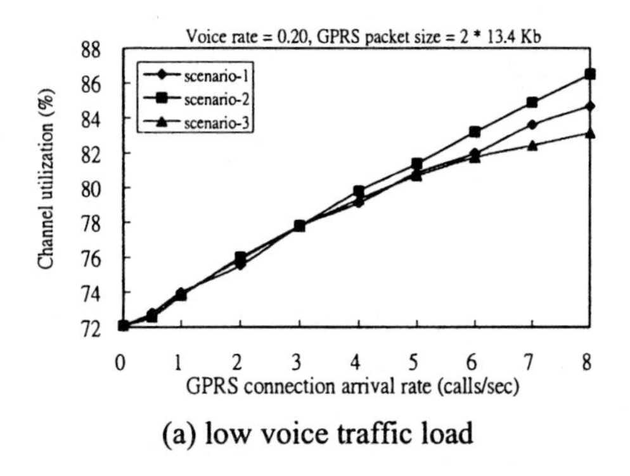

(a) low voice traffic load

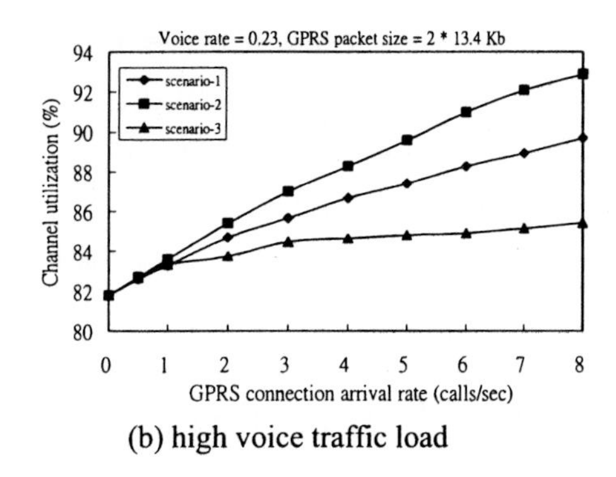

(b) high voice traffic load

Fig. 6. Comparison of channel utilization

From Fig. 3 and Fig. 4, it can be seen that the voice blocking probability will be increased with increasing GPRS traffic load even if applying channel de-allocation and call admission control mechanisms to GPRS traffic. Therefore we modified the previous scheme to guarantee the voice blocking probability not to be affected by the increasing GPRS traffic load. The modification is described as follows. When channel de-allocation mechanism can not provide channels for the arriving voice call, new or handoff, the network preempts an ongoing GPRS connection to service the arriving voice call. In addition, when a GPRS connection request arrives and there are no unused channels, the connection request is queued in the buffer. The preempted GPRS connections will also be queued in the buffer and are given higher priority to resume their services whenever there are channels available. Both kinds of connections are served in a first come first served (FCFS) manner. In this part of simulation, the buffer size is assumed to be infinite to avoid the GPRS connection request being blocked due to buffer overflow. We will investigate the mean packet delay of GPRS traffic with different multi-slot capability under different traffic load.

Fig. 7 shows the blocking probability of new call and handoff call with voice traffic load being 0.23 calls/sec and GPRS mean packet size being 2×13.4 *Kbits*. It can be seen that with voice preemption, the blocking probability of voice service is well below the target blocking probability (P_{tnb} and P_{thb}) and is independent of GPRS traffic load.

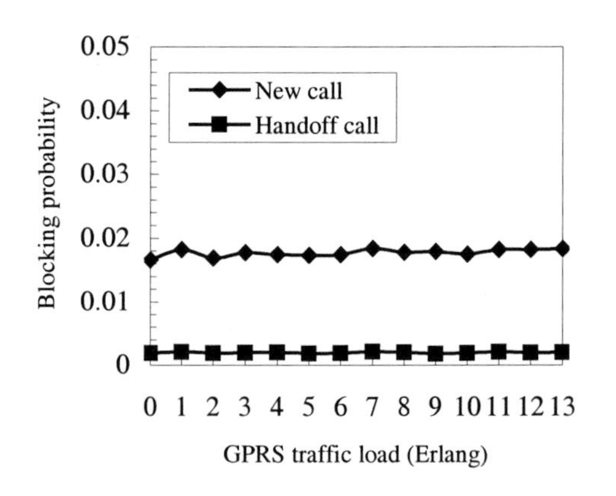

Fig. 7. Voice blocking probability of new call and handoff call

The effect of different multi-slot allocation to mean packet delay of GPRS traffic is shown in Fig. 8. In the figure, slot = 1 (or 2, 4) means that the maximum number of PDCHs that can be allocated to one mobile station is 1 (or 2, 4). It can be seen that at low GPRS traffic load, the mean packet delay can be effectively reduced with multi-slot allocation. While at high GPRS traffic load, the improvement is not obvious. The reason is that at high traffic load, a large portion of GPRS connections are allocated only one channel despite of multi-slot capability.

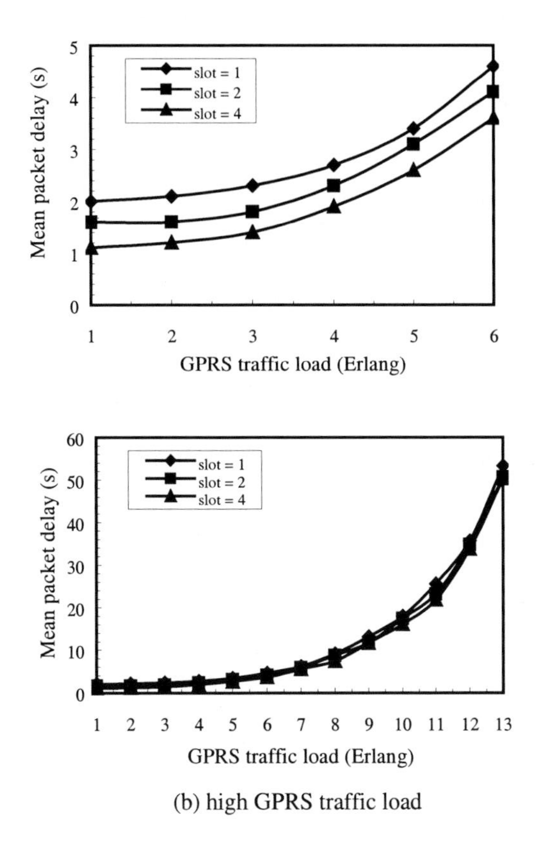

(b) high GPRS traffic load

Fig. 8. The effect of multi-slot allocation

5 Conclusions

Since GPRS shares radio resources with voice service, how to allocate bandwidth between the two services is an important issue. The introduction of GPRS service should not degrade the QoS of existing voice services. Guard channels can be temporarily allocated to GPRS connections to improve channel utilization. As voice traffic load increases, the channels of some ongoing GPRS connections are de-allocated to arriving voice calls. The de-allocation must still maintain the minimum required QoS of the de-allocated connections.

Simulation results show that at low voice traffic load, there is no need to apply admission control to GPRS connections. At high voice traffic load, the call admission control guarantees the blocking probability of new and handoff calls to be below certain value. But this will result in high GPRS rejection and low channel utilization.

To guarantee the QoS of voice service not to be affected by the introduction of GPRS, voice arrivals are allowed to preempt the ongoing GPRS connections. The mean packet delay of GPRS traffic can be effectively reduced with multi-slot allocation at low GPRS traffic load.

References

1. ETSI, "GSM 03.60 General packet radio service (GPRS) : Service description, Stage 2," v. 5.2.0, Jan. 1998
2. G. Brasche and B. Walke, "Concepts, services, and protocols of the new GSM phase 2+ general packet radio service," *IEEE Commun. Mag.* ,vol. 35, no. 8, pp. 94-104, Aug. 1997
3. J. Sau and C. Scholefield, "Scheduling and quality of service in the general packet radio service," *Proceedings of IEEE ICUPC'98*, vol. 2, pp. 1067-1071, Florence, Italy, Oct. 1998
4. J. S. Yang, C. C. Tseng, and R. G. Cheng, "Dynamic scheduling framework on RLC/MAC layer for general packet radio service," *Proceedings of IEEE ICDCS'2001*, pp.441-447, Phoenix, Arizona, April 2001
5. R. Ludwig and D. Turina, "Link layer analysis of the general packet radio service for GSM," *Proceedings of IEEE ICUPC'97*, vol. 2, pp. 525-530, San Diego, USA, Oct. 1997
6. M. Mahvadi and R. Tafazolli, "Analysis of integrated voice and data for GPRS," *International Conference on 3G Mobile Communication Technology*, pp.436-440, March 2000
7. S. Ni and S. G. Haggman, "GPRS performance estimation in GSM circuit switched services and GPRS shared resource systems," *Proceedings of IEEE WCNC'99*, vol. 3, pp. 1417-1421, New Orleans, USA, Sep. 1999
8. D. Hong and S. S. Rappaport, "Traffic model and performance analysis for cellular mobile radio telephone systems with prioritized and no-protection handoff procedure," *IEEE Trans. Veh. Technol.*, vol. 35, no. 3, pp. 77-92, Aug. 1986
9. ETSI, "GSM 03.64 General packet radio service (GPRS) : Overall description of the GPRS radio interface, Stage 2," v. 7.0.0, July 1999
10. ETSI, "GSM 05.03 General packet radio service (GPRS): Channel coding, Stage 2," v.6.0.0, Jan. 1998
11. J. Cai and D. Goodman, "General packet radio service in GSM," *IEEE Commun. Mag.* , vol. 35, no. 10, pp. 122-131, Oct. 1997
12. G. C. Chen and S. Y. Lee, "Modeling the static and dynamic guard channel schemes for mobile transactions," International Conference on Parallel and Distributed Computing and Systems, pp. 258-265, Las Vegas, Nevada, Oct. 1998
13. T. W. Yu and C. M. Leung, "Adaptive resource allocation for prioritized call admission over an ATM-based wireless PCN," *IEEE J. Select. Areas Common.,* vol. 15, no. 7, pp. 1208-1225, July 1997

A New Active RED Algorithm for Congestion Control in IP Networks

Jahon Koo[1], Kwangsue Chung[1], Hwasung Kim[1], and Hyukjoon Lee[2]

[1] School of Electronics Engineering, Kwangwoon University, Korea
jhkoo@adams.gwu.ac.kr, kchung@daisy.gwu.ac.kr, hwkim@daisy.gwu.ac.kr
[2] Department of Computer Engineering, Kwangwoon University, Korea
hlee@daisy.gwu.ac.kr

Abstract. In order to reduce the increasing packet loss rates caused by an exponential increase in network traffic, the IETF (Internet Engineering Task Force) is considering the deployment of active queue management techniques such as RED (Random Early Detection). While active queue management in routers and gateways can potentially reduce packet loss rates in the Internet, this paper has demonstrated the inherent weakness of current techniques and shows that they are ineffective in preventing high loss rates. The inherent problem with these queue management algorithms is that they all use static parameter setting. So, in case where these parameters do not match the requirement of the network load, the performance of these algorithms can approach that of a traditional Drop-tail.
In this paper, in order to solve this problem, a new active queue management algorithm called ARED (Active RED) is proposed. ARED computes the parameter based on our heuristic method. This algorithm can effectively reduce packet loss while maintaining high link utilizations.

1 Introduction

It is important to avoid high packet-loss rate in the Internet. When a packet is dropped before its reaches its destination, all of the resources it has consumed in transit have been wasted. In extreme cases, this situation can lead to congestion collapse. Over the last decade, TCP and its congestion control mechanisms have been instrumental in controlling packet loss and in preventing congestion collapse across the Internet. Optimizing the congestion control mechanisms used in TCP has been one of the most active areas of research in the past few years. The problem of congestion collapse encountered by early TCP/IP protocols has prompted the study of end-to-end congestion control algorithms in the late 80's and proposal such as congestion avoidance, which forms the basis for the TCP congestion control in current implementations. The essence of this congestion control scheme is that a TCP sender adjusts its sending rate according to the rate (probability) of packets being dropped in the network (which is considered a measure of network congestion). This algorithm is relatively well understood and several models have been proposed and verified with increasing degrees of

I. Chong (Ed.): ICOIN 2002, LNCS 2343, pp. 469–479, 2002.

accuracy through simulation and Internet measurements. But end-to-end congestion control mechanisms such as those in TCP are not enough to prevent congestion collapse in the Internet (for starters, not all applications might be willing to use them), and they must be supplemented by control mechanisms inside the network. This has led the IETF to recommend the use of active queue management in Internet router queue management [1,2,3,4].
In traditional implementations of router queue management, the packets are dropped when a buffer becomes full in which case the mechanism is called Drop-tail. More recently, other queue management mechanisms have been proposed. The IETF has singled out Random Early Detection (RED) as one queue management scheme recommended for rapid deployment throughout the Internet. RED algorithm alleviate many of the problem found in other active queue management algorithms such as random drop, early packet discard, and early random drop in that they can prevent global synchronization, reduce packet loss rates, and minimize biases against bursty sources using simple,low-overhead algorithm [5].
But, while RED can potentially improve packet loss rates, we show that its effectiveness is highly dependent upon its operating parameters. In fact, in cases where these parameters do not match the requirements of the network load, the performance of the RED algorithm can approach that of a traditional Drop-tail algorithm.
While RED algorithm can certainly outperform traditional Drop-tail algorithm, our experiments show that it is difficult to parameterize RED algorithm to perform well under different congestion scenarios. In this paper, we propose a new active RED algorithm to solve this problem. Section 2.1 first discusses active queue management mechanisms such as RED for controlling the average queue size and reducing unnecessary packet drops. Section 2.2 discusses for steady-state operation of RED as a feedback control. Section 3.1 shows that the effectiveness of RED is critically dependent upon the rate at which congestion notification is provided to the source and the network load. In Section 3.2 we make recommendations for RED configuration. Section 4 describes the design of an Active RED algorithm that determines suitable operating parameters for a RED depending on the traffic load. The simulation results are presented algorithmin Section 5. Our conclusions are presented in Section 6.

2 Background

2.1 Active Queue Management

It has long been known that Drop-tail queue management can result in pathological packet-dropping patterns, particularly in simple simulation scenarios with long-lived connections, one-way traffic, and fixed packet sizes. A more relevant issue for actual networks is that with small-scale statistical multiplexing, Drop-tail queue management can result in global synchronization among multiple TCP connections, with underutilization of the congested link resulting from several

connections halving their congestion window at the same time. This global synchronization is less likely to be a problem with large-scale statistical multiplexing. However, there is a fundamental tradeoff between high throughput and low delay with any queue management, whether it is Active Queue Management such as RED (Random Early Detection) algorithm or simple queue management such as Drop-tail algorithm. Maintaining a low average delay with Drop-tail queue management means that the queue will have little capacity to accommodate transient bursts, and can result in an unnecessarily- high packet drop rate.
The main motivation for Active Queue Management is to control the average queueing delay while at the same time preventing transient fluctuations in the queue size from causing unnecessary packet drops. For environments where low per-packet delay and high aggregate throughput are both important performance metric, active queue management can allow a queue tuned for low average per packet delay while reducing the penalty in unnecessary packet drops that might be necessary with Drop-tail queue management can sometimes come at the cost of a higher packet drop rate. In environments with highly burst packet arrivals, Drop-tail queue management can result in an unnecessarily large number of packet drops, as compared to Active Queue Management, particularly with similar average queueing delays. Even if there is full link utilization, a higher packet drop rate can have two consequences, wasted bandwidth on congested links before the points of loss, and a higher variance in transfer times for the individual flows [5,6].
One might ask if unnecessary packet drops really matter, if full link utilization can be maintained. Unnecessary packet losses result in wasted bandwidth to the point of loss only if there are multiple congested links, where other traffic could have made more effective use of the available bandwidth upstream of the point of congestion. Paths with multiple congested links might seem unlikely, given the lack of congestion reported within many backbone networks. However, even with uncongested backbone networks, a path with a congested link to the home, a congested link at an Internet exchange point, and a congested transoceanic link would still be characterized by multiple congested links.
The second possible consequence of unnecessary packet losses even with full link utilization can be a higher variance in transfer times. For example, small flows with an unnecessary packet drop of the last packet in a transfer will have a long wait for a retransmit timeout, while other active flows might have their total transfer time shortened by one packet transmission time.

2.2 Steady-State Operation of RED

In this section, we combine TCP congestion control model with RED (Drop module) control element and derive the state behavior of the resulting feedback control system. We consider a system of n TCP flows passing through a common link l with capacity c as in Figure 1.
We can view this system as a feedback control system with the controlled systems being the TCP senders, the controlling element being the drop module, the feedback signal – the drop probability, and the controlled variables – the TCP sending rates.

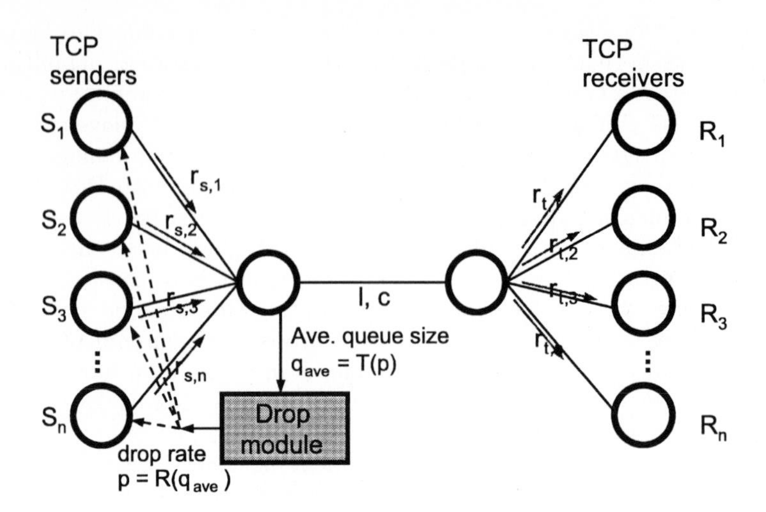

Fig. 1. An n-flow feedback control system

As discussed in [7], we have derived an expression for the long-term (steady-state) average queue size as a function of packet drop probability denoted by $\boldsymbol{q_{ave} = T\ (p)}$. If we assume that the drop module has a RED feedback control function denoted by $\boldsymbol{p = R\ (q_{ave})}$, where q_{ave} is an estimate of the long-term average of the queue size, and if the following system of equations has a unique solution (p_s, q_s), then the feedback system in Figure 1 has an equilibrium state (p_s, q_s). Moreover, the system operates on average at (p_s, q_s), i.e., its long-term average of packet drop probability is p_s and average queue size is q_s.
Figure 2 illustrates this result: the equilibrium point (p_s, q_s) is at the intersection of the curves $\boldsymbol{q_{ave} = T\ (p)}$ and $\boldsymbol{p = R\ (q_{ave})}$. This system may or may not be stable around the equilibrium point, depending $\boldsymbol{T}$ and $\boldsymbol{R}$.
Therefore, the feedback system we are about to model has a time lag of about one RTT between the moment a signal is sent by the RED module (by dropping or ECN marking) and the moment the controlled system (TCP sender) reacts to this signal. The increase or decrease in the TCP sending rate produces an increase or decrease in the instantaneous queue length at bottleneck link l, which prompts the RED module to again change its drop rate, and the process repeats. In this case, equilibrium state (p_s, q_s) converges into equilibrium point.

3 The Analysis of RED

3.1 The Action of RED

The RED algorithm uses a weighted average of the total queue length (size) to determine when to drop packets. When a packet arrives at the queue, if the weighted average queue length ($avg_q < -(1 - w_q)\bullet avg_q + w_q \bullet q$, where w_q is a fixed parameter and q is the instantaneous queue length) is less than a

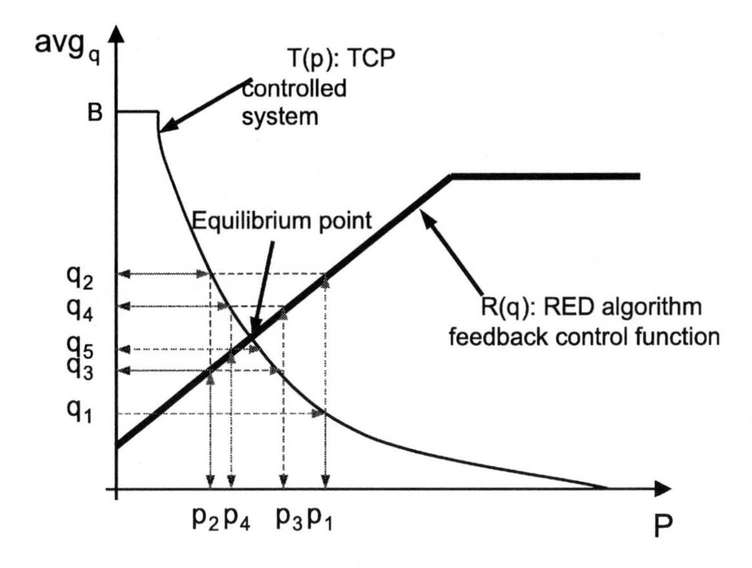

Fig. 2. Equilibrium point for a RED feedback system

minimum threshold value, min_{th}, no drop action will be taken and the packet will simply be enqueued. If the average is greater than min_{th} but less than a maximum threshold, max_{th}, an early drop test will be performed as described below. An average queue length in the range between the thresholds indicates some congestion has begun and flows should be notified via packet drop. If the average is greater than the maximum threshold value, a forced drop operation will occur. An average queue length in this rang indicates persistent congestion and packets must be dropped to avoid a persistently full queue. Note that by using a weighted average, RED avoids overreaction to bursts and instead reacts to longer-term trends. Furthermore, because the thresholds are compared to weighted average, it is possible that no forced drops will take place even when the instantaneous queue length is quite large. For example, Figure 3 illustrates the queue length dynamics in a RED router (gateway) in simulation network [5,8].

The early drop action in the RED algorithm probabilistically drops the incoming packet when the weighted average queue length is between the min_{th} and max_{th} thresholds.

In contrast, the forced drop action in the RED algorithm is guaranteed to drop the incoming packet. In the case of early drops, the probability that the packet will be dropped is dependent on several other parameters of algorithms. An initial drop probability $P_b = max_p\ (avg_q\text{-}min_{th})/(max_{th}\text{-}min_{th})$, is computed, where max_p is the maximum drop probability and avg_q is the weighted average queue length (average queue size). Note that given a weighted average queue length, the impact of min_{th} is dependent on both max_p and max_{th}. This mean that one may find a value for min_{th} that results in good performance, but it may only be

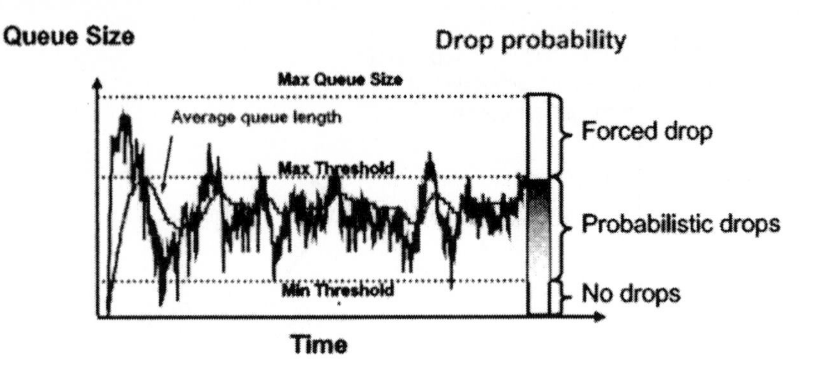

Fig. 3. The queue length dynamics in a RED router

in combination with certain values of max_p and max_{th}. In principle, this is the case for all the parameters. The main control parameters for RED algorithm are summarized in Table 1.

Table 1. RED parameters

qlen	The maximum number of packet that can be enqueued
min_{th}	Threshold which the average queue length must exceed before any dropping or marking is done
max_{th}	Threshold which the average queue length must exceed before all packets are dropped or marked
w_q	Weight for updating average queue length. Large values increase the weight of more recent queue length data
max_p	Maximum marking probability, Determines how aggressively the queue drops or marks packets when congestion occurs

3.2 Recommendations for Configuring RED

When support for ECN (explicit congestion notification) is not available, RED must resort to packet drops to signal congestion [9]. This leads to an interesting optimization problem where the RED queue must pick a max_p value that minimizes the overall packet loss, including drops due to early detection and packet drops due to buffer overflow. When extremely large values of max_p are used, packet loss rates are dominated by drops due to early detection, while with extremely small values of max_p packet loss is mostly due to queue overflow [10].

In order to examine the performance of RED using packet drops to signal congestion, we performed a set of experiments using the ns simulator [11]. Figure 4 shows the network topology used in these experiments. Note that each link is full duplex, so acknowledgements flowing on the reverse path do not interfere with data packets flowing on the forward path. In these experiments, the RED algorithm is controlled using max_p.

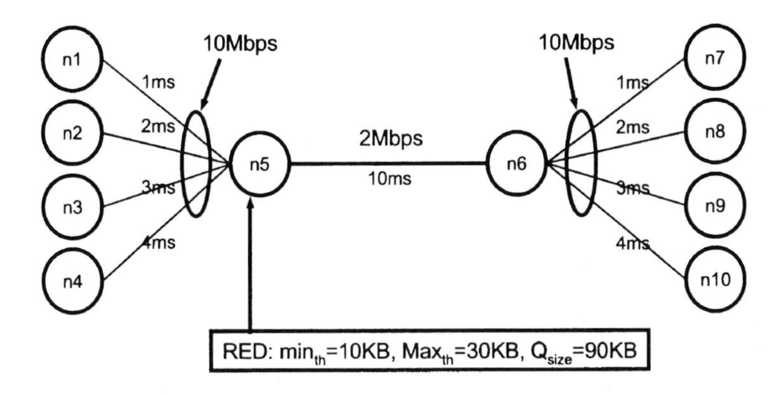

Fig. 4. Network Topology

Figure 5 shows the loss rates for 32 and 64 connections in a Drop-tail and in a RED queue for a range of max_p values. As the figure shows, RED algorithm has only a marginal impact on the packet loss rates observed. For small values of max_p, early detection is ineffective and the loss rates in the RED algorithm approach loss rates in the Drop-tail algorithm. As max_p is increased, loss rates decrease slightly since the RED queue is able to send congestion notification back to the sources in time to prevent continual buffer overflow.
Finally, as max_p becomes large, the RED algorithm causes a slight increase in packet loss rates over Drop-tail algorithm. Note that the value of max_p that minimizes the loss rates is different in Figure 5. As more connections are added, the optimal value of max_p increases.

4 Active RED Algorithm

From the discussion in the previous sections, it is clear that setting RED parameters based on traffic load. If max_p doesn't match up to optimal value, the queue length oscillates between 0 and full buffer size (B). In this case, equilibrium state (p_s, q_s) doesn't converge into the equilibrium point.
ARED algorithm proposed to solve the problem of existing RED algorithm, which adjusts the value of $max_p$to converge into equilibrium state reducing change rate of queue size depending on quantity of packets if possible. Figure 6 shows the proposed ARED algorithm.

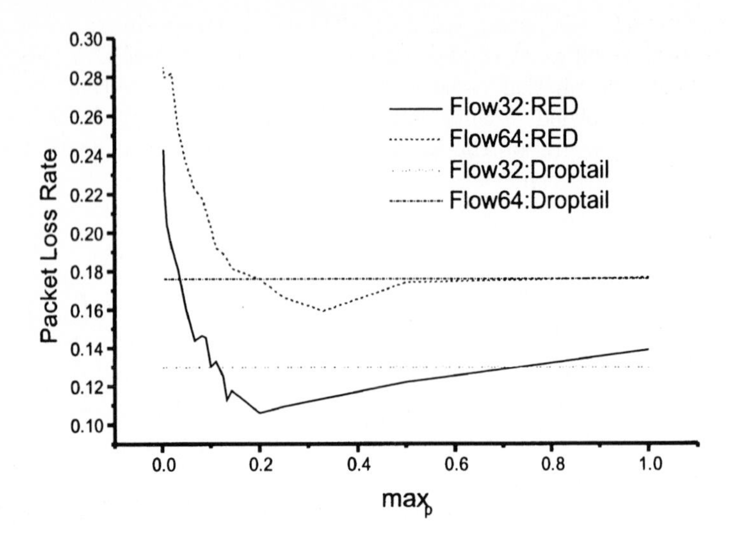

Fig. 5. Impact of max_p on RED algorithm

```
if (The change of Average Queue size is large
  && (ΔW old_gain < margin1 * ΔW current_gain))
  max_p increased

if (Average Queue Size during ΔT is reduced
  && (Average Queue Size < margin2 * max th ))
  max_p decreased

if (max_p increased)
  max_p = max_p + d_1;

else if (max_p decreased)
  max_p = max_p - d_2;
```

Fig. 6. Active RED algorithm

This paper proposes the activity of ARED algorithm adjusts the value of max_p to reduce range of change supervising ΔW(width), which is the value between maximum length and minimum length of queue and the degrees, which is the change of average queue length during any time ΔT. For example, if max_p is not suitable for present network congestion situation, the change of queue length is oscillated extremely. However, ARED is algorithm that adjusts the value of

max_p based on observed queue length dynamics in the situation of network congestion. So, ARED algorithm reduces the entire packet loss rate as reducing the degrees of change of queue length.

5 Simulation and Evaluation for ARED

In order to evaluate the performance of our ARED algorithm, a number of experiments have been performed on the basis of the ns (Network Simulator) of LBNL (Lawrence Berkely National Laboratory) [11].

5.1 Simulation for Change of Queue Length

We first evaluated the performance of queue length oscillation on RED and ARED algorithm. In the first experiment, each of 32 TCP flows sends on a simple network configuration as shown in Figure 4.
Figure 7 shows the comparison of queue dynamics in the all-TCP experiment with RED ($max_p = 0.01$) and ARED algorithm. From Figure 7, we can see that the queue length is RED and ARED both fluctuate greatly, and the queue length oscillation of ARED is smaller than that of RED.

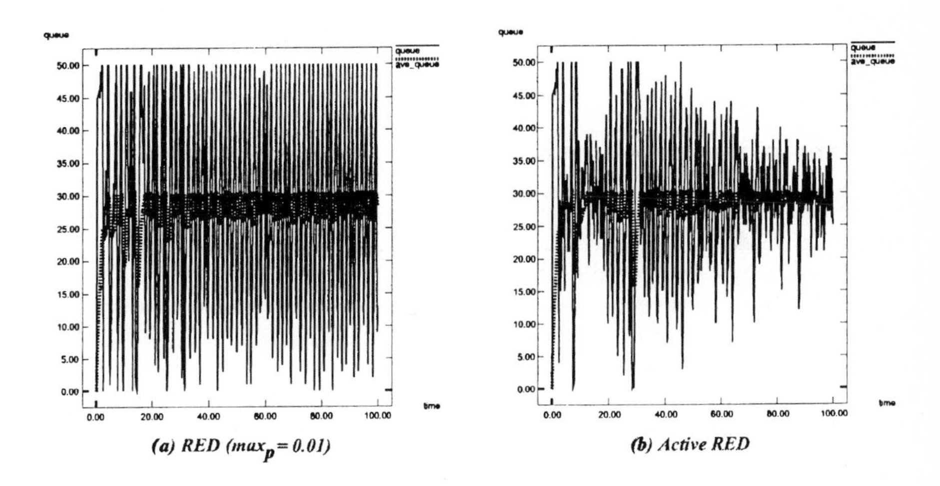

(a) RED ($max_p = 0.01$) *(b) Active RED*

Fig. 7. Queue length oscillation

5.2 Simulation for Packet Loss Rate

In order to explore the feasibility of the ARED algorithm, we ran another set of experiments using the same network as shown in Figure 4. In this experiment, we change the number of active connection. Figure 8 shows the loss rates observed for 4, 8, 16, 32, and 64 connections. The results in Figure 8 show that when the

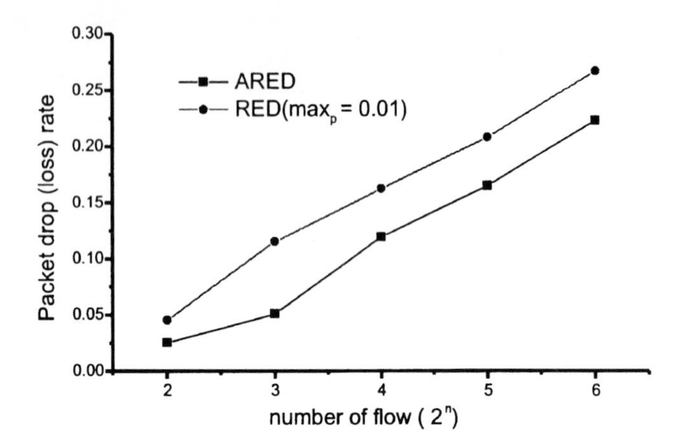

Fig. 8. Simulation result for packet loss rate

number of active connection increases, the packet loss rate increases. And we can see that packet loss rate of ARED is smaller than that of RED.
The results show that ARED algorithm consistently outperforms RED algorithm in packet loss rate. This is because the ARED is algorithm that adjusts the value of max_p based on observed queue length dynamics in the situation of network congestion.

6 Conclusion and Further Work

In this paper, the new active queue management algorithm, called ARED algorithm, is proposed to improve the existing RED algorithm. Congestion is controlled effectively using our algorithm. ARED is algorithm that adjusts the value of max_p based on observed queue length dynamics in the situation of network congestion. The simulation results show that ARED algorithm is able to effectively reduce packet loss in congested networks.
Further work involves studying the performance of the algorithm under a wider range of parameters, network topologies and real traffic traces, and considering method of improving fairness about unresponsive flow.

References

1. B. Braden, D. Clark, J. Crowcroft, B. Davie, S. Deering, D. Estrin, S. Floyd, V. Jacobson, G. Minshall, C. Partridge, L. Peterson, K. Ramakrishnan, S. Shenker, J. Wroclawski, and L. Zhang, "Recommendations on Queue Man-agement and Congestion Avoidance in the Internet," RFC 2309, April 1998.
2. V. Jacobson and M. Karels, "Congestion Avoidance and Control," Proc. SIGCOMM '88, August 1988.
3. S. Floyd, and K. Fall, "Router Mechanisms to Support End-to-End Congestion Control", LBL Technical report, February 1997.

4. Stevens, W., "TCP Slow Start, Congestion Avoidance, Fast Retransmit, and Fast Recovery Algorithms", IETF RFC 2001, January 1997.
5. S. Floyd and V. Jacobson, "Random Early Detection Gateways for Congestion Avoidance," IEEE/ACM Transactions on Networking, 1(4): 397-413, August 1993.
6. W. Feng, D. Kandlur, D. Saha, and K. Shin, "Blue: A New Class of Active Queue Management Algorithms," University of Michigan Technical Report CSE-TR-387-99, Apr. 1999.
7. V. Firoiu, and M. Borden, "A Study of Active Queue Management for Congestion Control," Proc. IEEE INFOCOM '2000, March 2000.
8. M. Christiansen, K. Jeffay, D. Ott, and F.D. Smith "Tuning RED for Web Traffic" IEEE/ACM Transactions, June 2001.
9. K. Ramakrishnan and S. Floyd, "A Proposal to add Explicit Congestion Notification (ECN) to IP", RFC 2481, January 1999.
10. W. Feng, D. Kandlur, D. Saha, and K. Shin, "A Self-Configuring RED Gateway," Proc. IEEE INFOCOM '99, March 1999.
11. UCB LBNL VINT, "Network Simulator ns (Version 2)", http://www-mash.cs.berkeley.edu/ns/.

Flow Admission Control for MPLS Support of DiffServ

Ji-Young Lim and Ki-Joon Chae

Dept. of Computer Science and Engineering,
Ewha Womans University, Korea
{jylim,kjchae}@ewha.ac.kr

Abstract. In this thesis, an efficient traffic flow admission control mechanism supporting DiffServ is proposed to provide QoS in MPLS networks. We propose a dynamic and flexible flow admission control for MPLS support of DiffServ to prevent the waste of resources. Ingress LSRs may collect information to find out the congested area using control messages exchanged in QoS routing. We exclude the congested LSR from routing decision and LSP aggregation. Thus, the proposed model realizes improved traffic engineering by efficient resource utilization and traffic flow admission control while satisfying the MPLS property of separating control and forwarding.

1 Introduction

MPLS(MultiProtocol Label Switching)[1], which is regarded as a core technology for migrating to next generation Internet can support various high-speed and valuable services by high-speed switching and traffic engineering. The goal of MPLS QoS has been to establish parity between the QoS features of IP and MPLS, not to make MPLS QoS somehow superior to IP QoS. One of the main reasons that MPLS supports the IP QoS model is that MPLS, unlike IP, is not an end-to-end protocol[2].

IntServ(Integrated Services)[3] and DiffServ (Differentiated Services)[4] are the methods that provide QoS in IP networks. Since both IntServ and MPLS use RSVP(Resource ReSerVation Protocol)[5,6,7] as the signaling protocol to reserve resources, supporting IntServ over MPLS is not difficult[8]. DiffServ dose not reserve resources for each flow but classifies flows depending on their properties. It determines the Behavior Aggregation (BA) for each class at edge routers and forwards or drops the packets according to their BAs at core routers.

The DiffServ model is appropriate for MPLS, which makes routing decisions at edge routers and forwards packets at core router. DiffServ defines traffic profiles that have to be satisfies by packets being forwarded. For EF(Expected Forwarding)[9], resources for the peak data rate are reserved whereas resources for the committed data rate are reserved for AF(Assured Forwarding)[10]. In this paper, a traffic flow admission control model that controls the amount of AF flows without reserving resources for AF is proposed to prevent the wasting of resource that occur when excessive resources are allocated for EF flows. Also, it saves the number of labels used by aggregating LSPs and cuts down on the number of the dropped packets in heavily-congested networks.

I. Chong (Ed.): ICOIN 2002, LNCS 2343, pp. 480-493, 2002.

Remainder of the paper is organized as follows: MPLS support of DiffServ is presented in section 2 as related works. Section 3 describes the proposed traffic flow admission control model. In section 4, performance evaluation is shown. Finally, section 5 concludes the paper.

2 MPLS Support of DiffServ

First issue in supporting DiffServ over an MPLS network is to provide some way to ensure that packets marked with various DSCPs receive the appreciate QoS treatment at each LSRs in the network. However, the DSCPs are carried in IP headers, and LSRs do not examine them when they forward packets. There has to be some way to determine appreciate PHBs(Per Hop Behaviors) from the label headers.

The shim header for MPLS has a 3-bit field, EXP(see Figure 1), that is reserved for an experimental use. The problem with this field is that it is only 3-bit long and therefore, can only express eight different values. The DiffServ standards allow for up to 64 DSCPs. In a draft from IETF Working Group, two methods are proposed to solve this problem[11]:

Label	EXP	S	TTL

Label: Label value (20bits)
EXP: Experiment (3bits)
S: Bottom of Stack (1bit)
TTL: Time to Live (8bit)

Fig. 1. MPLS Header

- E-LSP (EXP-Inferred PSC (PHB Scheduling Class) LSPs)
  In figure 2, a single E-LSP is in use and may carry packets requiring up to eight different PHBs. The PHB to be applied to any packet is determined by examining the EXP bits.
- L-LSP (Label-Only-Inferred PSC LSPs)
  In figure 2, two L-LSPs have been established. The first one carries only default packets. LSRs R2 and R3 know that packets that arrive on the lower LSP must be default packets and should be queued accordingly since they were told at LSP establishment time. The upper LSP carries packets that may be AF11, AF12, or AF13. LSRs R2 and R3 use the label to determine that the packets are AF1*y* packets and they examine the EXP bits to find the drop preference, that is, the value of *y*.

E-LSPs offer several advantages over L-LSPs. Since more PHBs can be supported on a single E-LSP, E-LSPs reduce the total number of LSPs, which may be helpful if label space is a limited resource. The E-LSP model is most similar to the standard DiffServ model: an LSR just looks at bits in the header to decide what PHB to apply to a packet. However, L-LSPs also offer advantages, including the ability to support an arbitrarily large

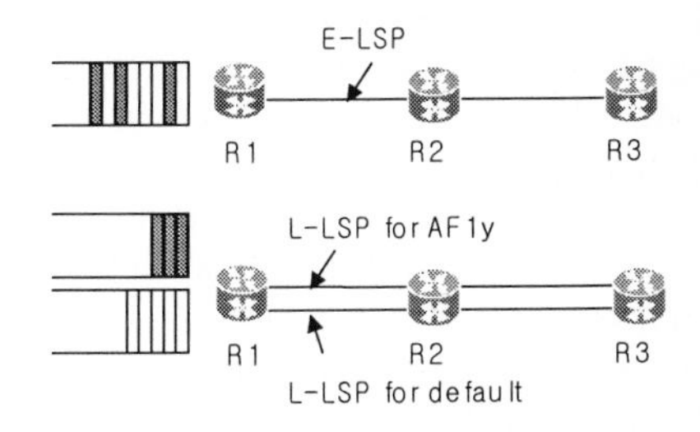

Fig. 2. E-LSP and L-LSP

number of PHBs, not just eight. In addition, because the L-LSPs provide one LSP for each LHB or PHB scheduling class, they provide the possibility of engineering different paths for different PHBs.

3 Traffic Flow Admission Control Model

Flow admission control and traffic control are essential requirements to reserve resources and establish the connection in networks providing connection-oriented services. Since Internet services were best effort service, Internet did not need flow admission control and traffic control. But both of them became necessary as requests for guaranteed QoS services increased.

IntServ provides the traffic flow admission control by accepting a new flow when the required resources are reserved. Also, it provides traffic control in order to observe its traffic profiles when packets are forwarded.

DiffServ forwards, drops or delays the packets of a flow according to the traffic profiles of the class, which is a kind of traffic control. As flow admission control, it reserves peak-data- rate resources for EF and committed-data-rate resources for AF. Excessive resource reservation for EF is a critical drawback for DiffServ causing low resource utilization, since MPLS, unlike IP, can provide the complete resource reservation over the connection.

Therefore, we propose a dynamic and flexible traffic flow admission control in order to efficiently provide DiffServ over MPLS while, at the same time, avoid congestion. It adjusts the number of AF flows that exist in MPLS networks by forwarding AF packets using left-over resources by EF. Using this mechanism, reserving resources for AF id unnecessary.

Due to the class configuration difference between E-LSP and L-LSP over one LSP, we apply our proposed mode to both E-LSP and L-LSP. Also, since the established LSP allows aggregation of LSPs through its modification, which is proposed in IETF[12], an efficient LSP aggregation scheme is proposed.

3.1 EF Flow Admission Control

EF is the delay and loss-sensitive premium service and allocates the bandwidth as the peak rate. Figure 3 shows the proposed flow admission control for EF class. It consists of 3 modules as follows: the first module looks up the LSP that enables LSP aggregation with a new flow. The second module decides whether more bandwidth should be allocated for the selected LSP. According to the decision made, the third model either requests a label for the new LSP or modifies the LSP with more resource.

First of all, we are going to describe the LSP candidate selection procedure for LSP aggregation, when a new EF flow arrives at a DiffServ-aware edge LSR. It examines the LSP candidates whose properties are as same as those of the new LSP. The properties that are considered in this process are the destination in MPLS networks, DiffServ DSCPs and transferable classes over the LSP. As for the LSP aggregation decision module, it is applied to the selected LSP candidates. It determines whether more bandwidth for the new EF flow should be allocated. If it is

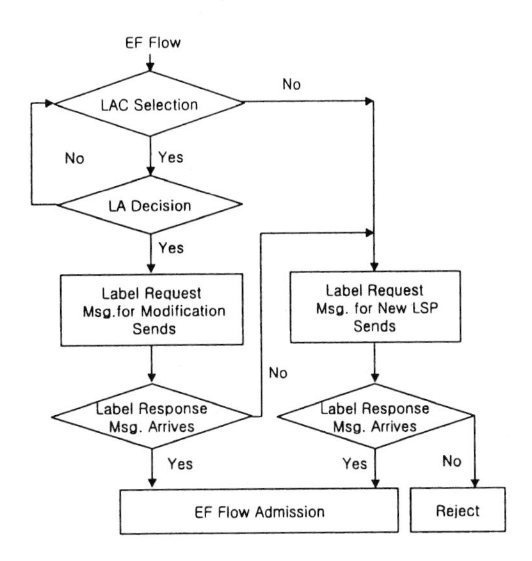

Fig. 3. EF Flow Admission Control

decided so, it selects one LSP and forwards the Label Request message to the selected LSP in order to modify its property. If the Label Response message is positive, the new flow is forwarded with the existing flows that are being forwarded over the LSP. If no LSP candidates are capable of allocating any more resources, a new LSP is established via LDP(Label Distribution Protocol)[13] and CR-LDP(Constrained-based Routing LDP)[14]. In the case no new LSPs can be established, the EF service is rejected.

3.2 AF Flow Admission Control

AF is a delay-insensitive class that forwards packets according to drop preference and the drop rates of the network that are shown in pre-defined BAs. Therefore, the bandwidth for AF can be reserved for the committed data rate. As described earlier, we use the unused portion of the allocated EF bandwidth when forwarding packets marked for one of AF*xy* PHBs.

In figure 4, the proposed flow admission control for AF class is shown. It is less complicated than that for EF because it has no LSP modification procedure to reserve more resources. Firstly, it selects the LSP Aggregation candidates when an AF flow arrives at a DiffServ-aware edge LSR, which is similar to the EF flow admission control. Then the LSP aggregation decision module determines available LSPs among the selected candidates. If the available LSP is determined, packets of the new AF flow are forwarded through it. If no LSP is determined, Label Request message is forwarded to establish a new LSP connection. When a new LSP establishment is not possible, this AF flow is rejected.

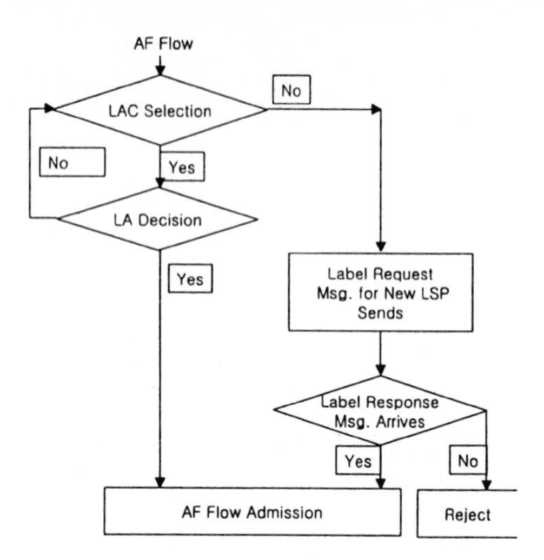

Fig. 4. AF Traffic Flow Admission Control

3.3 LAC (LSP Aggregation Candidate) Selection

In this section, we describe in detail LAC selection that is a part of flow admission control. The goal of LSP aggregation is to increase resource utilization and to reduce the number of entries in FTN(FEC to NHLFE)[1] table as well as the number of routing decision processing. The success of LAC selection depends on how mapping is done between EXP field values of the LSP and DSCP values of DiffServ. We describe the LAC selection for L-LSP and E-LSP respectively.

- **LAC Selection for L-LSP**

L-LSP can distinguish between EF and AF as label values and PHBs have different drop preference among AF*xy*s as EXP field values. When trying to forward flows for all the PHBs, one LSP for EF and best effort classes and 4 LSPs for the four AF*x*s become necessary. Therefore, the LAC selection for EF is given a flow-id, some destination information and traffic profile for an incoming flow and it is made to look for the LSPs that have the same PHB and destination. LAC selection for AF looks for the LSPs that have been the same AF*x* and destination information. Its results are LSPs that are applicable to LSP aggregation decision module described in following section. Table 1 shows the mapping between EXP field values and PHBs.

Table 1. The Mapping between L-LSP EXP Field Values and PHBs

EXP Field Values	PHB
000	AFx1
001	AFx2
010	AFx3

- **LAC Selection for E-LSP**

E-LSP distinguishes PHBs as EXP field values. The mapping between PHBs and EXP filed values is pre-defined. If all PHBs for AF are considered, flows for EF, best effort and two of AF*xs* may be forwarded over the same LSP. In this case, we confront a very difficult problem of forwarding two AF*xys* over the same LSP. We forward the flows for EF, best effort and one AFx over one LSP. Using the LAC selection for EF, the LSPs that have the same destination information is found the LSP that has the same AFx is found using LAC selection for AF. Table 2 shows the mapping between E-LSP EXP filed values and PHBs used in this paper.

Table 2. The Mapping between E-LSP EXP Field Values and PHB

EXP Field Values	PHB
000	EF
001	AFx1
010	AFx2
011	AFx3
111	DF

3.4 LA (LSP Aggregation) Decision

In this section, we describe the LA decision that determines whether the LSP selected via LAC selection model is able to aggregate with the new flow. It analyzes the post-aggregation effects and selects one LSP that has the best effect.

3.4.1 Decision A: Decision According to the Sum of Traffic Transferred

Decision A is proposed to reduce the waste of resources for EF that is a weakness of DiffServ. Since DiffServ allocates resource as PDR(Peak Data Rate), and not CDR (Committed Data Rate), when the gap between PDR and CDR becomes great, the unused portion of the resources grows as well. To prevent this situation, BB(Bandwidth Broker)[15,16] is proposed to manage resources efficiently. A BB in each subnet manages and allocates resources for the members of the subnet and exchanges information between BBs of other subnets. That is, a BB is a kind of a resource management server that manages resources for EF especially in connectionless networks.

Decision A is based on the fact that amount of traffic transferred over an outgoing-link is not larger than its total capacity. LSPs that do not satisfy the following condition is removed from the list of the selected LSP aggregation candidates.

$$b_{tr\_o_i} + b_{con\_o_i} + b_{new\_o_i} < B_{o_i}$$

B_{o_i} is total link capacity of outgoing-link of the LSR i, o_i, $b_{tr\_o_i}$ is the sum of the bandwidth used by traffic transferred toward the outgoing-link, o_i via LSR i and $b_{con\_o_i}$ is the sum of bandwidth used by traffic transferred toward the outgoing-link, $o_i \cdot b_{new\_o_i}$ is the bandwidth required by traffic of new flow that goes toward the outgoing-link, $o_i \cdot b_{tr\_o_i}$ is a measurement value for any flows that traverse outgoing-link, $o_i \cdot b_{new\_o_i}$ is determined as follows.

$$b_{con\_o_i} = b_{conEF\_o_i} + b_{conAF\_o_i}$$

$b_{conAF\_o_i}$ is the sum of CDR for AF flows transferred toward outgoing-link, o_i. $b_{conEF\_o_i}$ is also the sum of CDR for EF flows transferred toward outgoing-link, o_i. If AF traffic may be full in the queue of LSR *i*, it is limited by multiplying $b_{conEF\_o_i}$ by $\alpha(>1)$.

3.4.2 Decision B: Reflection of Congested Link Information at Ingress LSR

Decision B uses the control information exchanged among network routers for making routing decision such as LSA(Link State Advertisement) messages of OSPF(Open Shortest Path First)[17]. The procedure of Decision B is as follows: if the queue of LSR *i* is full, it sets the congestion bit in the control message and advertises the fact at the next message exchange time.

LSRs that get the message classify LSR *i* as a congestion area and exclude the path, including LSR *i* when making routing decisions. LSRs do not release LSR *i* until they are given the next control message. In this paper, because we use the routing algorithm proposed in [18], which does not exchange control messages frequently, we prevent congestion by setting the congestion bit before congestion occurs before the queue becomes full.

4 Performance Evaluation

This section evaluates the performance of the proposed flow admission control. It consists of LAC selection modules according to L-LSP and E-LSP and LA Decision modules according to Decision-A and Decision-B. Table 3 shows the simulation list with their combination.

4.1 Simulation Evaluation

Figure 5 shows the topology of an ISP backbone network used in our study (also used in [18]). For simplicity, all the links are assumed to be bidirectional and of the same capacity, with *C* units of bandwidth in each direction. Flows arrive at a source node according to a Poisson process with the rate λ. The destination node of a flow is chosen randomly from all nodes except the source node. The holding time of a flow is exponentially distributed with the mean $1/\mu$.

Following [19], the offered load is given by $\rho = \lambda N h / \mu L C$, where *N* is the number of source nodes, *L* is the number of links and *h* is the mean number of hops per flow, averaged across all source-destination pairs. Consider the case of traffic *k* types of flows, each flow of type *i* having a mean holding time $1/\mu_i$ and requesting bandwidth

Table 3. Simulation List

	Resource Reservation		LSP Types	LA
	EF	AF		
llsp+res	O	O	L-LSP	If resource is reserved
elsp+res	O	O	E-LSP	
llsp+nolmt	O	×	L-LSP	unlimited
elsp+nolmt	O	×	E-LSP	
llsp+A	O	×	L-LSP	Decision-A
elsp+A	O	×	E-LSP	
llsp+A&B	O	×	L-LSP	Decision-A + Decision-B
elsp+A&B	O	×	E-LSP	

B_i. Let ρ_i be the offered load on the network due to flows of type i, where the total offered load, $\rho = \sum_{i=1}^{k} \rho_i$. We consider the case of traffic with 2 types of flows in table 4. The fraction of flows for EF, ρ_1/ρ is equal to that of flows for AF, ρ_2/ρ. The parameters used in simulation are C=20, N=18, L=60, h=2.58.

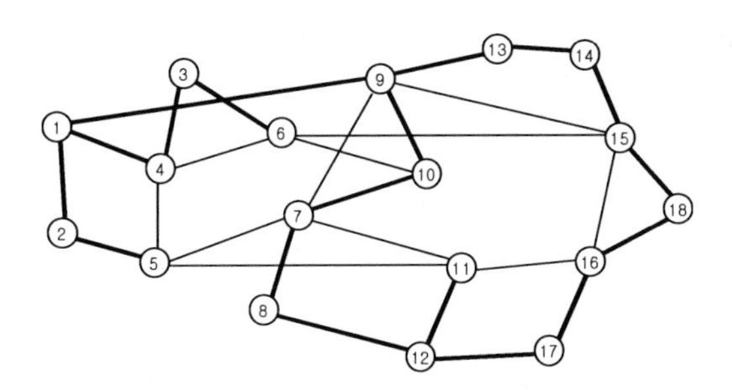

Fig. 5. ISP Backbone Network

4.2 Blocking Probabilities and Bandwidth Utilization

In this paper, we omit some graphs because the results of *llsp+nolmt* and *elsp+nolmt* and of *llsp+res* and *elsp+res* are very similar respectively.

First of all, we describe the blocking probability in figure 6 and 7. In figure 6 and 7, *llsp+res* that reserves the required resources for flows of EF and AF has the highest blocking probabilities of EF and AF respectively. The blocking probabilities of EF and AF in *elsp+res* are similar to those in *llsp+res*. In Figure 8, the bandwidth utilization values of *llsp+res* and *elsp+res* are the lowest among others. These show that the required resource reservation of EF and AF gives the bad effect to support DiffServ in MPLS.

In figure 7, the blocking probabilities of *llsp+A* and *elsp+A* using Decision-A are higher than those of *llsp+A&B* and *elsp+A&B* using Decision-A and Dicision-B. In figure 8, the bandwidth utilization of *llsp+A* is lower than that of *llsp+A&B* and bandwidth utilization of *elsp+A* is also lower than that of *elsp+A&B*. It shows that Decision-B that excludes the path including the congested area gives a good performance.

In figure 7 and 8, the results of *llsp+A* and *llsp+A&B* are higher than those of *elsp+A* and *elsp+A&B* respectively while results of *llsps* and *elsps* have little difference in figure 6. In figure 7 and 8, the results of *llsp+nolmt* and *elsp+nolmt* that do not limit forwarding of AF flows are the best until now.

Table 4. Traffic Model

	EF	AF1*y*	AF2*y*	AF3*y*
Required bandwidth	3	1		
Flow duration	60	30		
PDR	3	1		
CDR	1	1		

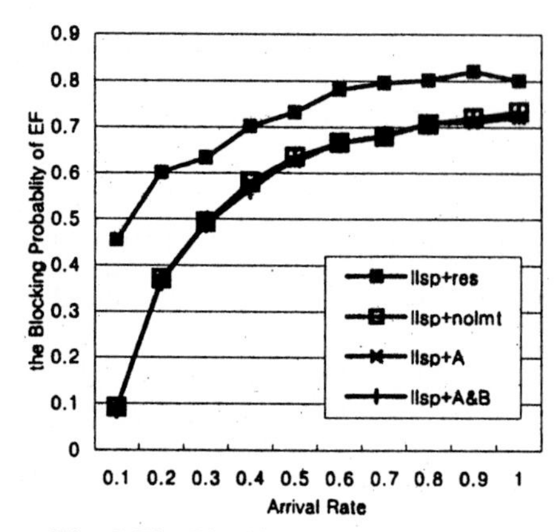

Fig. 6. The Blocking Probability of EF

4.3 LSP Aggregation

Figure 9 and 10 show the percentages of the LSP aggregation for EF and AF flows respectively. In figure 9, the result of *llsp+res* is the lowest and the results of *llsp+nolmt*, *llsp+A* and *llspA&B* are higher than those of *elsp+A* and *elsp+A&B*. In figure 10, *llsp+nolmt* is the highest and the results of *elsp+A*, *elsp+A&B* is higher than those of llsp+A, llsp+A&B and *llsp+res*. The results of *elsps*(*elsp+A and elsp+A&B*) and *llsps*(*llsp+A and llsp+A&B*) are a little contrary to each other. The proposed models show a little better results than the original scheme(*llsp+res* and *elsp+res*).

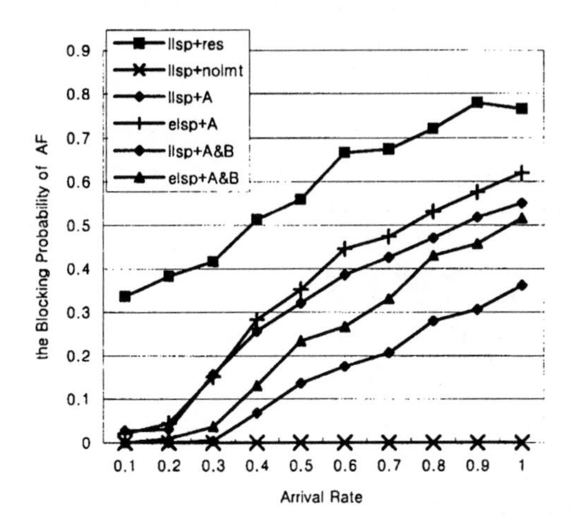

Fig. 7. The Blocking Probability of AF

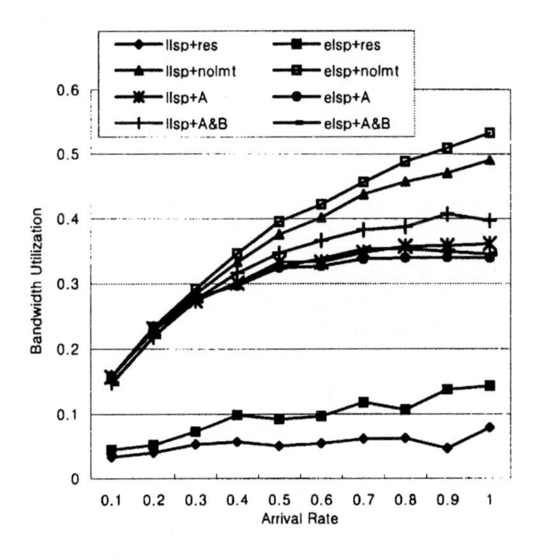

Fig. 8. Bandwidth Utilization

4.4 Delays and Drop Rates

First, we describe the average delays of packets for EF and AF flows. In figure 11, the delays of packets for EF are 0 in *res*. However, the delays of packets for EF are 0 in *A&C* while they are about 1 in *A* and *nolmt*. These results do not related to L-LSP and E-LSP. Figure 12 shows the delays of packets for AF. *llsp+A&B* and *llsp+res* do not

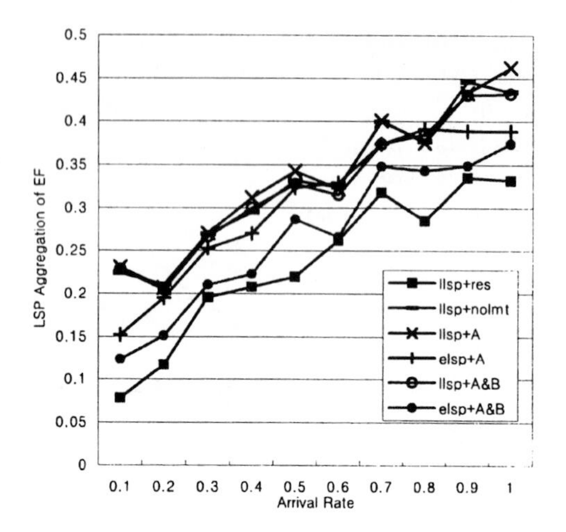

Fig. 9. The Percentage of LSP Aggregation for EF

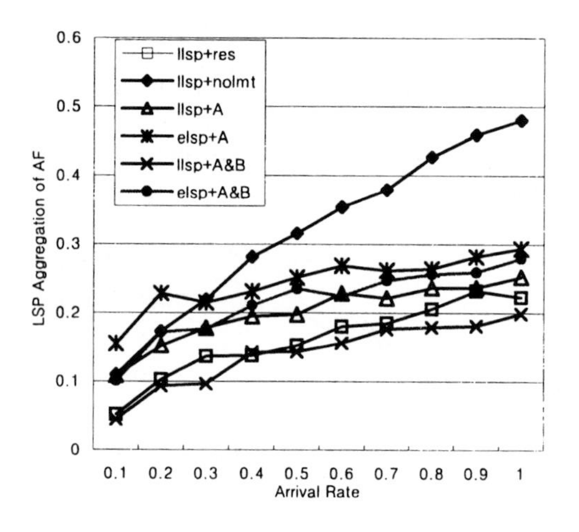

Fig. 10. The Percentage of LSP Aggregation for AF

cause additional delay while others cause very long additional delays. The Results show that our proposed model using *A&B* satisfies the requirements of EF and AF.

Since drop rates of EF is 0, the graph is omitted. In figure 13, drop rates of *llsp+res* is 0 and those of *llsp+A&B* are about 0, while those of nolmt are the highest. We can explain that delays of *lsp+A* are longer than those of *llsp+nolmt* in figure 12 because many packets in llsp+nolmt are lost. The Results show that our proposed model using *A&B* satisfies the requirements of EF and AF.

5 Conclusion

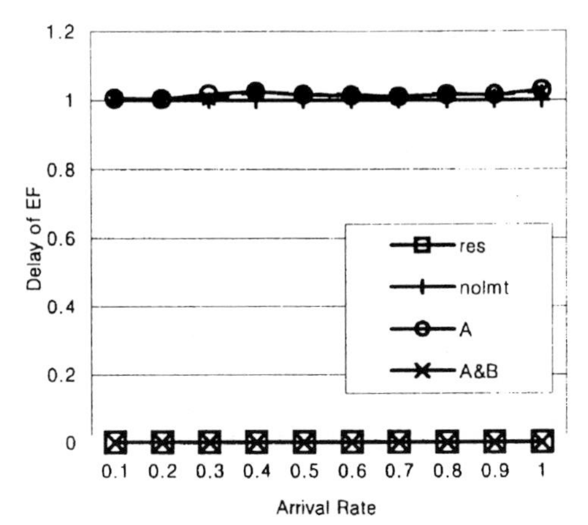

Fig. 11. Delay of EF

Research and standardization activities on policy-based resource allocation and flow admission are yet inactive in MPLS. Since they require centralized servers in the network such as IPOA (IP over ATM), they result in overhead. As a solution to the problem, a mechanism that performs flow admission for DiffServ traffic at the edge of the MPLS network without any help of the server is proposed in this paper. It dynamically adjusts the amount of admissible traffic based on transmittable capacity over one outgoing port(Decision A). It then transmits the traffic while avoiding congested area resulting traffic loss(Decision B). Ingress LSR may collect information to find out congested area if it uses control messages exchanged that reflect QoS link states in QoS routing. Our proposed modes using Decision A and B together shows good performance that the requirements of EF and AF are satisfied and the weakness of resource reservation in DiffServ is overcome. Thus, the proposed model realizes improved traffic engineering by efficient resource utilization and traffic flow admission control.

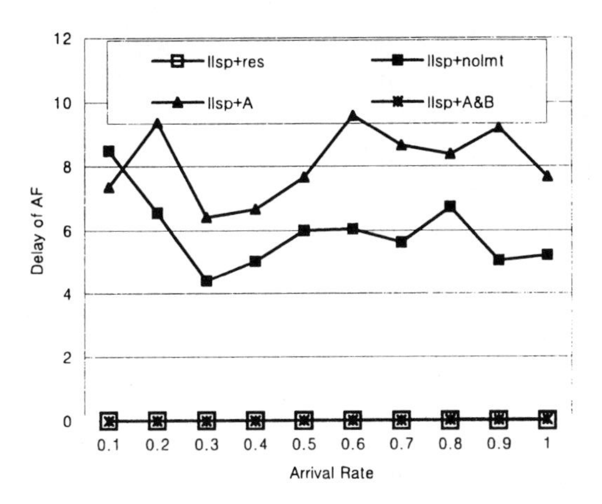

Fig. 12. Delay of AF

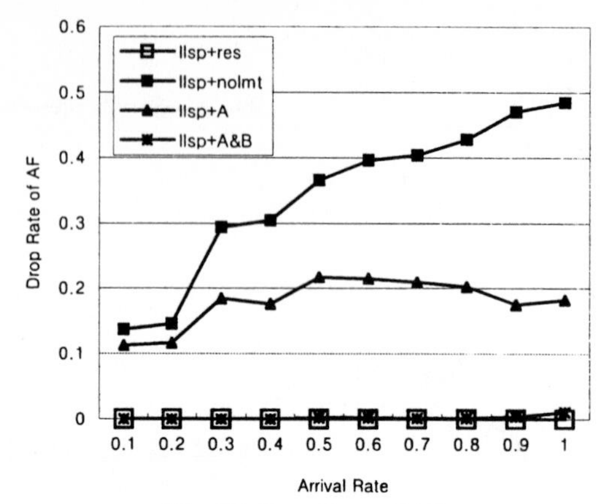

Fig. 13. Drop Rate of AF

References

1. E. Rosen, A. Viswanathan, R. Callon: Multiprotocol Label Switching Architecture, RFC 3031, Jan. 2001.
2. Bruce Davie, Yakov Rekhter: MPLS - Technology and Applications, Morgan Kaufmann Publishers, 2000.
3. R. Braden, et al.: Integrated Services in the Internet Architecture: an Overview, RFC 1633, Jun. 1994.
4. S. Blake, et al.: An Architecture for Differentiated Services, RFC 2475, Dec. 1998.
5. A. Mankin, et al.: Resource ReSerVation Protocol (RSVP) Version 1 Applicability Statement Some Guidelines on Deployment, RFC 2208, Sep. 1997.
6. R. Branden et al.: Resource ReSerVation Protocol (RSVP) Version 1 Functional Specification, RFC 2205, Sep. 1997.
7. J. Wroclaswski: The Use of RSVP with IETF Integrated Services, RFC 2210, Sep. 1997.
8. D. O. Awduche, L. Berger, D. Gan, T. Li, V. Srinivasan, G. Swallow: RSVP-TE: Extensions to RSVP for LSP Tunnels, Internet Draft (work in progress) draft-ietf-mpls-rsvp-lsp-tunnel-08.txt, Feb. 2001.
9. V. Jacobson, K. Nichols, K. Poduri: An Expected Forwarding PHB, RFC 2598, Jun. 1999.
10. J. Heinanen, F. Baker, W. Weiss, K. Wroclawski: Assured Forwarding PHB Group, RFC 2597, Jun. 1999.
11. F. L. Faucheur, L. Wu, B. Davie, S. Davari, P. Vaananen, R. Krishnan, P. Cheval, J. Heinanen: MPLS Support of Differentiated Services, Internet Draft (work in progress) draft-ietf-mpls-diff-ext-08.txt, Feb. 2001.
12. J. Ash, Y. Lee, P. Ashwood, Smith, B. Jamoussi, D. Fedyk, D. Skalecki, L. Li: LSP Modification Using CR-LDP, Internet Draft (work in progress) draft-ietf-mpls-crlsp-modify-03.txt, Mar. 2001.
13. L. Andersson, P. Doolan, N. Feldman, A. Fredette, B. Thomas: LDP Specification, RFC 3036, Jan. 2001.
14. B. Jamoussi: Constraint-Based LSP Setup using LDP, Internet Draft (work in progress) draft-ietf-mpls-cr-ldp-05.txt, Feb. 2001.
15. R. Neilson, J. Wheeler, F. Reichmeyer, S. Hares: A Discussion of Bandwidth Broker Requirements for Internet2 Qbone Deployment, Internet2 Qbone BB Advisory Council, Ver. 0.7, Aug. 1999.

16. Ca*net II Differentiated Services Bandwidth Broker System Specification, British Columbia Institute of Technology, Technology Center, Group for Advanced Information Technology, Oct. 1998.
17. J. Moy: OSPF Version 2, RFC 2328, Apr. 1998.
18. Ji-Young Lim, Ki-Joon Chae: Differentiated Link based QoS Routing Algorithms for Multimedia Traffic in MPLS Networks, 15th International Conference on Information Networking(ICOIN-15), pp. 587-592, Japan, Feb. 2001.
19. A. Shaikh, J. Rexford, K. Shin: Load-Sensitive Routing of Long-Lived IP Flows, *ACM SIGCOMM*, p pp. 215-226, Sep. 1999.

An ACK-Based Redundancy Control Algorithm in Periodic FEC

So-Hyun Lee, Myoung-Kyoung Ji, Tae-Uk Choi, and Ki-Dong Chung

Department of Computer Science
Pusan National University
Kumjeoung-Ku, Pusan 609-735, Korea
{shlee, bluesky, tuchoi, kdchung}@melon.cs.pusan.ac.kr

Abstract. Error propagation is the main problem in transmitting a real-time interactive video over the Internet. A lot of work and research has been done to solve this problem. Periodic FEC[4], proposed by Choi et al., is a scheme that prevents error propagation with small redundant information. However, this scheme does not have an effective redundancy control mechanism to adapt to network conditions. In this paper, an ACK-based redundancy control mechanism is proposed to reduce the amount of redundancy based on current network conditions. Also, we propose a TCP-friendly redundancy control algorithm of Periodic FEC to have a similar bandwidth with the TCP control mechanism. Through simulations, we confirm that the proposed algorithm provides a good PSNR when the network conditions are good and shares a similar bandwidth with TCP.

1 Introduction

Development of technology of computers and the Internet has lead to many interactive video applications such as video phones, video conferences, etc. But, the transmission of realtime interactive video over the network is not easy or challenging because of variable packet loss, delay and bandwidth of the Internet. Most of video compression standards (MPEG I, MPEG II, H.261, etc) have high compression efficiency by reducing spatial and temporal redundancy of frames. Because these standards use motion estimation and compensation, a small loss in a frame can severely affect quality. As well, the error can be propagated to the subsequent frame and amplified.

Error propagation techniques have been proposed in many research papers[1]. The simplest approach is to transmit I-frame periodically, but this requires large bandwidth. Another approach is the conditional replenishment method used in NV[2] and VIC[3], which intra-codes and transmits only macro-blocks including errors in a frame. Choi at al. classified these techniques for error propagation prevention into codec-level and network-level schemes. The codec-level schemes

[0] This work was supported by grant No. R05-2002-000-00345-0 from the Basic Research Program of the Korea Science & Engineering Foundation.

I. Chong (Ed.): ICOIN 2002, LNCS 2343, pp. 494–503, 2002.

such as ET(Error Tracking) and RPS(Reference Picture Selection) prevent error propagation in encoding and decoding time. The network-level schemes such as FEC and Retransmission control the error during transmitting and receiving packets. These two schemes can be combined with each other to improve the performance of the video applications. Periodic FEC[4] and RESCU[5] are examples of the combined schemes.

The Periodic FEC scheme reduces redundant information and prevents error propagation effectively. However, because of its poor redundancy control mechanism, Periodic FEC needs excessive bandwidth. We have to analyze network conditions to prevent transmitting useless redundant information. When network conditions are good, much redundant information is not required. However, when the network is overload, redundancy has to increase to protect the periodic frames. A RTCP-based feedback mechanism can be used for redundancy control of Periodic FEC. Since Periodic FEC uses the FEC mechanism to protect only periodic frames, it is difficult and takes a long time for the RTCP-based redundancy control method to collect feedback information of the receiver. In this paper, we propose an ACK-based redundancy control scheme that adapts to network states and determines the optimal amount of redundancy.

When the network loss rate is high, sending much redundancy to protect the periodic frames may give rise to the problem of network congestion. TCP controls a transmission rate using its AIMD(Additive Increase Multiplicative Decrease) mechanism while UDP has no control mechanism. As UDP traffic shares bandwidth with TCP traffic on the same path of the network, it can have a higher transmission rate than TCP traffic[8]. To prevent unfairness of bandwidth sharing like this, we consider TCP-friendliness to control redundant information. In the simulations, the proposed scheme shows a fair bandwidth share and a better performance when the network conditions are not good. The rest of this paper is organized as follows. Section 2 presents the overview of Periodic FEC. Section 3 describes the ACK-based redundancy control algorithm, and the TCP-friendly redundancy control algorithm is shown in section 4. And next, section 5 shows the performance of the proposed schemes, and section 6 concludes this paper.

2 Periodic FEC

The purpose of Periodic FEC is to restrict error propagation within a period. It divides video frames into periodic frames and non-periodic frames and proposes two methods to prevent error propagation[4]. The first is the frame reference method in the codec level. The other is the transmission method in the network level.

2.1 The Basic Ideas of Periodic FEC

Fig.1 shows the references of the periodic frame and non-periodic frame. A periodic frame makes reference to its previous periodic frame, and a non-periodic frame depends only on its previous periodic frame. It means that the effect of loss of non-periodic frame is limited to the period only if the periodic frame has no error in the period. However, if the periodic frame is lost, the error propagates to the next period. Therefore, a mechanism to protect the periodic frame is required.

Periodic FEC transmits a periodic frame with redundant information to protect the periodic frame. That is to say, the nth periodic frame is transmitted with redundant information of the $(n-i)^{th}$ periodic frames. The maximum value of i is referred to as the order of the scheme[6]. The receiver recovers it using the redundant information, when the packet of the periodic frame is lost. Fig.2 shows an example of a media-dependant FEC scheme to protect the periodic frame, in which the order is 2. The redundant information of the $(n-2)^{th}$ periodic frame is encoded using a lower quantization parameter than that of the $(n-1)^{th}$ periodic frame.

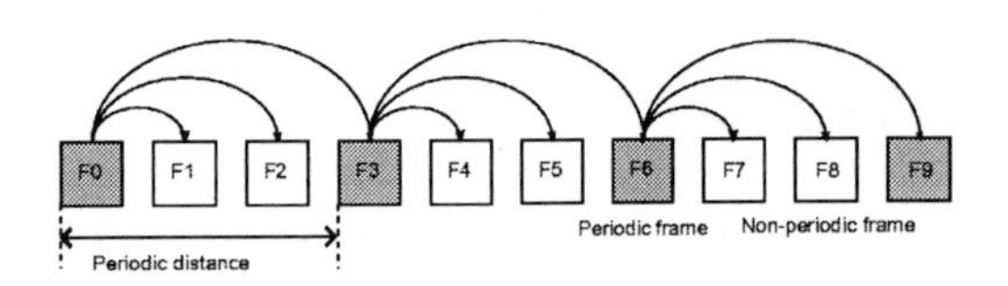

Fig. 1. The PSNRs of each scheme depending on network loss rates

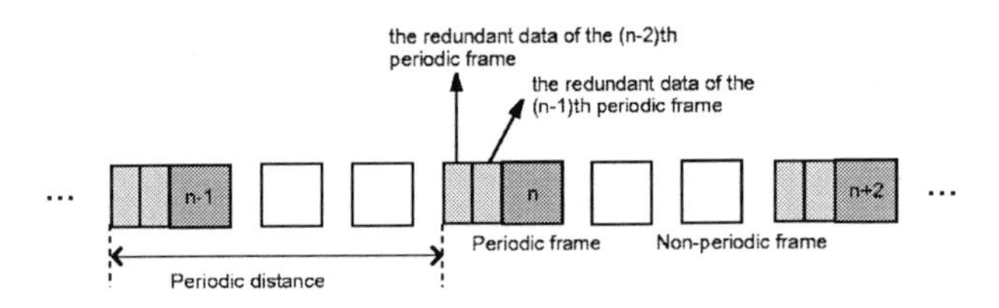

Fig. 2. The redundant information of Periodic FEC

2.2 The Problem of Redundancy Control

Bolot et al. proposed an adaptive FEC scheme in which the order value can be controlled dynamically depending on the packet loss rate of the network[6]. In this scheme, as the network is overloaded, the order value becomes larger and the amount of redundancy increases so that the packet loss rate after recovery can be decreased. But, because of redundant information for all the frames, the scheme requires a large amount of additional bandwidth. To save the bandwidth, the Periodic FEC scheme sends redundant information only for periodic frames, and needs an effective feedback mechanism to control the order values. The RTCP-based feedback information contains the average loss rate of all the frames not just the periodic frames. When it uses a RTCP-based feedback mechanism, it may get undesirable performance. The most important feedback to control the redundancy is the loss information of periodic frames over the network. The NACK-based feedback information offers a fast response but few feedback packets in the case of low loss rate of periodic frames. It may result in a lack of feedback about network conditions. The other is the ACK-based feedback mechanism that provides a fast response and enough feedback regardless of network conditions. We propose an ACK-based feedback mechanism to control the redundancy.

3 ACK-Based Redundancy Control

3.1 ACK-Based Feedback

Rejaie proposed RAP that controls the transmission rate according to ACK-based feedback information[7]. The ACK packet contains valuable information such as A_{curr}, N and A_{last} to show the current network state fast and effectively. A_{curr} is the sequence number of the packet being acknowledged. N is the sequence number of the last packet before A_{curr} that was still missing, or 0 if no packet was missing. A_{last} is the sequence number of the last packet before N that was receive, or 0 if was the first packet.

Similarly, A_{curr}, N and A_{last} of periodic frames are calculated and feedbacked to the sender in the Periodic FEC scheme. Table 1 shows a series of ACK messages depending on loss patterns. In loss patterns, the digits are a number of a periodic frame transmitted to the receiver successfully, and "_" denotes a missing periodic frame. In the last ACK of the figure, A_{curr} is 11, N is 8 and A_{last} is 6. These mean that the receiver has received frame 11, the packets between frame 6 and frame 8 have lost, and packet 6 was received before the consecutive packet losses. The length of the last consecutive losses L_{loss_burst} is calculated by $(N - A_{curr})$, and the length of the last consecutive successes $L_{success\_burst}$, which it means safe arrival of a packet, is calculated by $(A_{last} - N)$.

3.2 The ACK-Based Redundancy Control Algorithm

Periodic FEC controls the redundancy by increasing or decreasing the order value. When the order value is 2, it sends the main information along with the redundant information of two previous periodic frames. It means that the error recovery in the receiver is possible accept when three or more consecutive periodic frames are lost. So, if three periodic frames were lost continuously, they cannot be recovered. The higher the order value is, the larger loss burst can be recovered. The most important factor of the network to control the order value is not the packet loss rate but the length of the loss burst of the periodic frames.

Table 1. A sequence of ACK messages depending on loss patterns

loss patterns	A_{last}	N	A_{curr}	L_{loss_burst}	$L_{success\_burst}$
1 _	0	0	1	0	1
1 _ _ 4	1	3	4	2	1
1 _ _ 4 _ 6	4	5	6	1	1
1 _ _ 4 _ 6 _ 9	6	8	9	2	1
_ 6 _ _ 9 10	6	8	10	2	2
6 _ _ 9 10 11	6	8	11	2	3

Procedure ACK_based_RC()
Begin Procedure
  For (A new ACK message is arrived)
    calculate L_{loss_burst} and $L_{success\_burst}$.
    If ($L_{loss\_burst} \geq order$) **then** $order \leftarrow L_{loss\_burst}$.
    Else If ($L_{success\_burst} > T_i$) **then** $order \leftarrow (order - 1)$.
    End If
  End For
End Procedure

Fig. 3. The ACK-based redundancy control algorithm

If the order is always set to the length of the loss burst or more, the periodic frame will be protected.

In this paper, we propose an ACK-based redundancy control algorithm to adjust the order value whenever an ACK message is arrived. It calculates the size of a loss burst L_{loss_burst} and the size of a success burst $L_{success\_burst}$. As L_{loss_burst} becomes larger, the algorithm increases the order value larger than or equal to the last L_{loss_burst}. As L_{loss_burst} becomes smaller or $L_{success\_burst}$ grows larger, it decreases the order value to reduce the useless redundant information. T_i is the threshold of order i to decrease the order value. If there are few losses of periodic frames, the consecutive ACK messages occur, and $L_{success\_burst}$ will increase. When $L_{success\_burst}$ is larger than the threshold, the algorithm decreases the order value by 1. If there are few losses after reducing the order,$L_{success\_burst}$ will be larger because it has accumulated. Therefore, the lower the order is, the higher the threshold value is needed. Fig.3 shows the ACK-based redundancy control algorithm.

4 TCP-Friendly Redundancy Control

When network congestion occurs, TCP controls the transmission rate using its AIMD(Additive Increase Multiplicative Decrease) mechanism. However, UDP or RTP doesn't have any rate control mechanism. So, applications using UDP or RTP should control the transmission rate by themselves, That is to say, RTP flows should be controlled in a method similar to the rate control mechanism of TCP in order to share the network bandwidth fairly with competing TCP traffic in a same network path[8]. Mathis et al. proposed a model that estimates the transmission bandwidth of TCP using an average loss rate and delay of the network[9]. Equ.(1) shows the model. M is the maximum packet length, $t_{RTT}$is the round-trip delay, l means the average loss rate and C denotes a constant that is 1.22 or 1.36 in general.

$$r_{tcp} = \frac{C \times M}{t_{RTT} \times \sqrt{l}} \tag{1}$$

The algorithm of Fig.3 does not consider the competing TCP traffic when controling the amount of redundancy. In the algorithm, as the packet loss rate is increased because of network overload, the order value increases and the redundant information becomes larger and larger, while TCP traffic reduces the transmission rate by itself. This results in unfair bandwidth share. The TCP-friendly algorithm estimates TCP bandwidth using Equ.(2) and control that the amount of original and redundant data should be less than the estimated bandwidth of TCP. It is necessary that the additional bandwidth due to the increase of the order should be estimated. Table 2 shows the average additional bandwidths AB_i and quantization parameter Q_i of the order i, which are measured through many experiments. In order to increase the order, Equ.(2) should be satisfied. When the current transmission rate is r_{curr} and TCP bandwidth is r_{tcp}, the TCPfriendly redundancy control algorithm is shown in Fig.4.

$$r_{curr} + \sum_{i=order}^{L_{loss\_burst}} AB_i > r_{tcp} \tag{2}$$

Table 2. The average amount of redundancy according to the order values

order	Q_i	AB_i(bytes)
0	10	-
1	13	315
2	16	225
3	19	150
4	22	123
5	25	105

5 Experiments

In order to evaluate the proposed scheme, it was compared to RTCP-based, ACK-based and TCP-friendly redundancy control schemes. The transmission

---

Procedure TCP_friendly_RC()
Begin Procedure
  For (A new ACK message is arrived)
    calculate L_{loss_burst} and $L_{success\_burst}$.
    estimate r_{tcp}.
    $R \leftarrow \sum_{i=order}^{L_{loss\_burst}} AB_i$.
    If ($L_{loss\_burst} \geq order$) and ($r_{curr} + R < r_{tcp}$) **then**
      $order \leftarrow L_{loss\_burst}$ and $r_{curr} \leftarrow (r_{curr} + R)$.
    Else If ($L_{success\_burst} > T_i$) **then** $order \leftarrow (order - 1)$.
    End If
  End For
End Procedure

Fig. 4. The TCP-friendly redundancy control algorithm

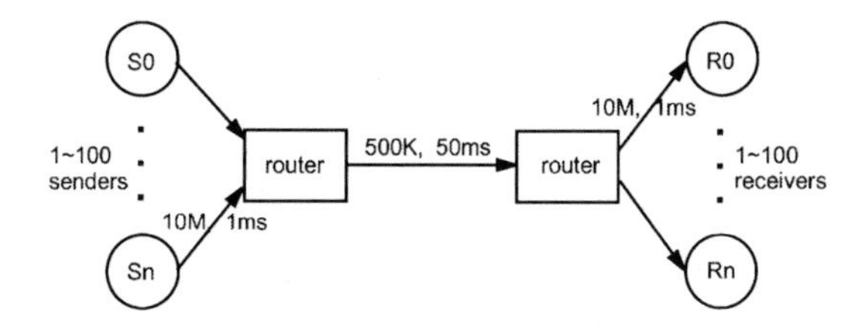

Fig. 5. The topology of the NS-2 simulator

rate was measured to show the amount of redundancy, and PSNR was measured to denote the quality of video. We used the NS-2 simulator to generate various network traffics[10]. Fig.5 shows the topology of NS-2 for the simulation. We used a simple type that all end users pass through the single bottleneck link of 500Kbps, while all other local links of 10Mbps are congestion-free. Propagation delay of the bottle link is set to 50ms. Variable losses and delays are generated as the number of TCP flows is changed from 1 to 100. We modified the tenelor H.263 source to implement the schemes, and the packet size was 1024 bytes.

Fig.6 and Fig.7 shows the transmission rates and PSNRs of the schemes according to various loss rates. The RTCP-based scheme doesn't control the order value fast and effectively. As shown in the figures, while the packet loss rate is low, its transmission rate and PSNR are small. As the packet loss rate is high, its amount of transmission increases. The ACK-based scheme shows a fast reaction to network conditions. However, it may work poorly when the network is underloaded. Its transmission rate overall is higher than the RTCP-based scheme, and so is PSNR. However, it sends much more data than TCP in overloaded situations. The TCP-friendly algorithm shows the similar amount of redundancy to and the higher PSNR than the other schemes when the network loss rate is around 10%. And, transmission data and PSNR decreases in the overloaded situations (the loss rate is about 25%) because it controls the amount of re-

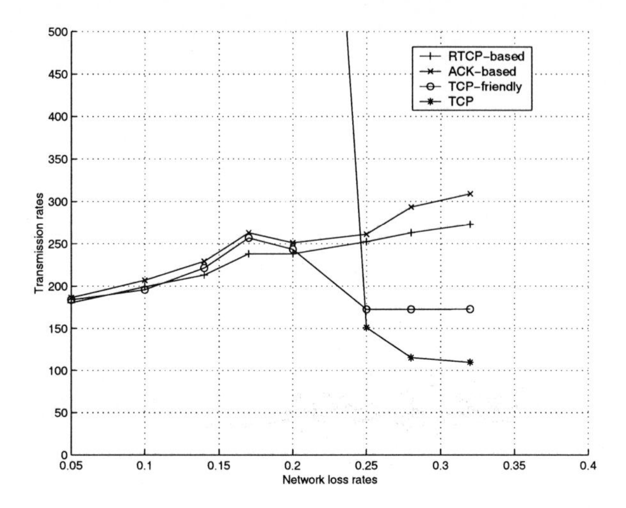

Fig. 6. The average transmission rates of each scheme depending on network loss rates

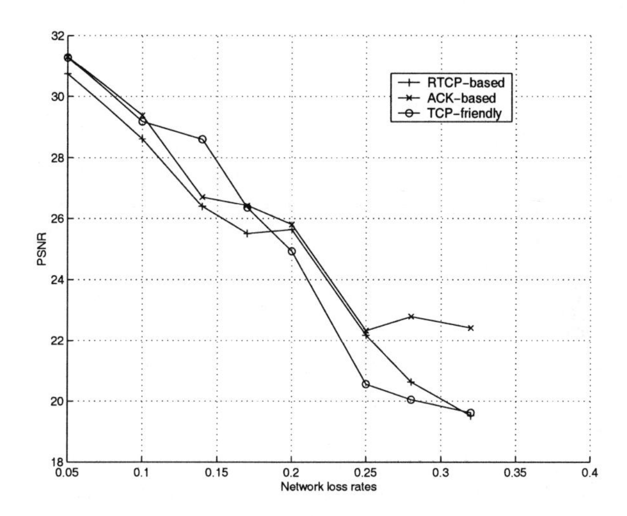

Fig. 7. The PSNRs of each scheme depending on network loss rates

dundancy by considering competing TCP flows. Fig.8 shows TCP-friendliness when the average loss rate is 25%. The RTCP-based scheme and the ACK-based scheme send much more data than the estimated TCP bandwidth. The TCP-friendly scheme has a similar transmission rate to the TCP bandwidth. Consequently, the proposed scheme shows a fair bandwidth share and a better performance when network conditions are good.

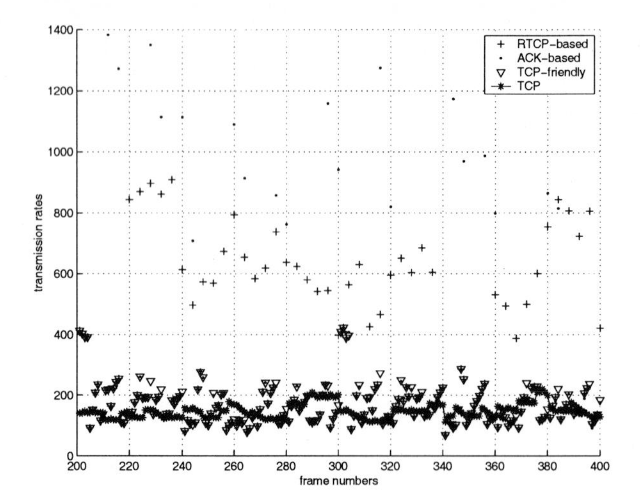

Fig. 8. The transmission rates of each scheme when the average loss rate is 25%

6 Conclusion

The Periodic FEC scheme reduces the amount of redundancy and prevents error propagation effectively. It needs an effective redundancy control mechanism that can adapt to network conditions and minimize the additional bandwidth of redundancy. In this paper, we propose the TCP-friendly redundancy control scheme, which utilizes the ACK-based feedback and control the redundancy using TCP-friendliness. Through the simulation, we confirm that the proposed scheme provides a better performance as well as a fair bandwidth share with TCP.

References

1. J-C. Bolot, T. Turletti, "Adaptive error control for packet video in the Internet", in Proceedings of International Conference on Internet Protocols, Lausanne, September, 1996.
2. R. Frederick, "Experiences with real-time software video compression", in Proceedings of 6th Packet Video Workshop, Protland, OR, Sept. 1990.
3. S. McCanne and V. Jacobson, "Vic: A flexible framework for packet video", in Proceedings of ACM SIGCOMM, Philadelphia, PA, Sept. 1990.
4. Tae-Uk Choi, Myoung-Kyoung Ji, Seong-Ho Park, Ki-dong Chung, "An Adaptive Periodic FEC Scheme for Internet Video Applications", in Proceedings of Tyrrhenian International Workshop on Digital Communications, Taormina, Italy, Sept. 2001
5. I. Rhee, S.R. Joshi, "Error recovery for interactive video transmission over the Internet", IEEE journal on selected on areas in communications, VOL.18, NO6, June 2000.
6. J-C. Bolot, T. Turletti, "Experience with Control Mechanisms for Packet Video in the Internet", Computer Communication Review, January 1998.

7. Reza Rejaie, "An End-to-End Architecture for Quality Adaptive Streaming Applications in the Internet", Ph.D. Dissertation, Computer Science Department, USC, Sept. 1999.
8. Dorgham Sisalem, "TCP-friendly Congestion Control for Multimedia Communication in the Internet", Ph.D Dissertation, Computer Science, University Berlin, 2000.
9. M. Mathis, J. Semke, J. Mahdavi, and T. Ott, "The macroscopic behavior of the TCP congestion avoidance algorithm, IEEE Network, 11(6), November/December 1997.
10. ns - Network Simulator. See http://www-mash.cs.berkeley.edu/ns/ns.html

Real-Time Multimedia Data Transmission Module Based on Linux

Nam-Sup Park, Sang-Jun Nam, and Tai-Yun Kim

Dept. of Computer Science & Engineering, Korea University, 1, 5-ga, Anam-dong, Seongbuk-ku, Seoul, 136-701, Korea
{nspark, sjnam, tykim}@netlab.korea.ac.kr

Abstract. Recently the demand for multimedia services on the Internet has increased. But, server based systems that offer multimedia data services are mostly unable to satisfy the expectations of the users. In this paper we propose SRTPIO (Special RTP Input/Output) module, that processes RTP (Real-time Transport Protocol) data in the kernel with SIO (Special Input/Output) Mechanism, as a solution to transport the multimedia data in the server based system more efficiently. SIO mechanism improves the transfer speed because it reduces the overheads which are generated in the process of data copying and context-switching between the user mode and the kernel mode, taking place in the kernel-level of a general server based system. SRTPIO module which integrates SIO mechanism and RTP data processing in the kernel supports efficient multimedia data transfer architecture.

1 Introduction

Large scale MOD(multimedia-on-demand) servers, which are capable of providing independent and interactive access to a vast amount of multimedia information to a large number of concurrent clients, will be required to enable a wide spread deployment of exciting multimedia applications[1].

Large scale storage servers that can provide location transparent, interactive concurrent access to hundreds of thousands of independent clients will be an important component of the future information superhighway infrastructure[2].

The RTP (Real-Time Transport Protocol)[3] is used for these requirements, but the RTP does not control the subsystem which is made up of storage device and network data link layer. Most of the RTP based multimedia applications need the elimination of data and control passing overheads in the I/O process[4].

Most of the previous work in this area was concentrated on reducing data-copying for network protocol processing[5], but did not provide Real-Time processing.

In this paper, we propose a scheme to realize SIO (Special Input/Output) mechanism as a method to reduce the overhead to allow more efficient real-time multimedia data transfer. SRTPIO (Special RTP Input/Output) module has been implemented on

I. Chong (Ed.): ICOIN 2002, LNCS 2343, pp. 504-515, 2002.

the Linux operating system as an instance of the scheme. We got performance enhancement of 12.5% compared with the transmission speed on the same platform used in the existing Linux operating system. This paper is categorized into following sections: section 2 discusses related works about the RTP and the I/O subsystem. In section 3 we propose a method to eliminate all the data-copying presented by user applications and to provide for real-time requirements. The implementation of SRTPIO is described in section 4 and the basic performance using SRTPIO module is measured and compared with that of the existing Linux system in section 5. Finally, section 6 presents our conclusions.

2 Related Work

This section is made up of RTP[3] and MARS system. RTP has gained widespread acceptance as the transport protocol for voice and video on the Internet[6]. MARS[8] project has rectified I/O sub system problems.

2.1 Multimedia Data Transmission Stack Using RTP

There has recently been a flood of interest in the delivery of multimedia services on the Internet. The growing popularity of Internet telephony, streaming audio and video services (such as those provided by Real Audio) and the Mbone are all indicators of this trend. To support these applications, standards are being developed to insure interoperability[6]. RTP is used by H.323 terminals as the transport protocol for multimedia. But it doesn't handle about resource reservation, timely delivery, QoS guarantee and reversed transmission prevention.

Multimedia applications transfer data between UDP and RTP at the very high rate. Therefore, Data copying and context switching has been identified as sources of I/O inefficiency. Recently there has been developed the researches on RTP transmission speed enhancement and QoS guarantee [6, 7]. But existing RTP doesn't consider about shifting RTP protocol's transmission stack to another Protocol layer. In this paper, we propose the mechanism which shifts RTP protocol stacks to kernel areas.

2.2 Existing File I/O for Networked Multimedia and MARS

Fig. 1 shows the layered architecture used in the storage devices and the network I/O systems of current LINUX operating systems. As shown in Fig. 1, existing file I/O and network I/O has following limitations. We summarize the limitations to support networked multimedia data in existing LINUX systems. Unnecessary data copying is a performance penalty: Clearly, in a MOD server implemented in an user area, the data transfer path from a disk to the network interface involves two memory copies: the first copy is made by a *read()* call, which moves data from the kernel buffer cache to an user area buffer. The second copy is made by a *send()* call, which moves the data from an user area buffer into the mbuf chain in the kernel, on the socket layer. This approach works well for small sized accesses in general purpose I/O, such as traditional text and binary file accesses. However, multimedia data such as audio, video and animations do not possess any caching properties. That is, extra data copying results in performance penalty. In the case of multimedia data files, the buffer cache blocks can be reused only if several sessions are phase-locked each other in a small (few tens of seconds) time interval in order to read the same file. Such behavior among interactive clients is rare. Therefore, retrieving multimedia data through a buffer cache does not provide any performance benefits. Also, any application initiated data transfer from a disk file to the network requires the use of two different buffer systems. These buffer systems cause excessive extra data copying and system call overheads. Data copying from memory to memory not only requires processor time but also consumes memory and system bus bandwidth[8].

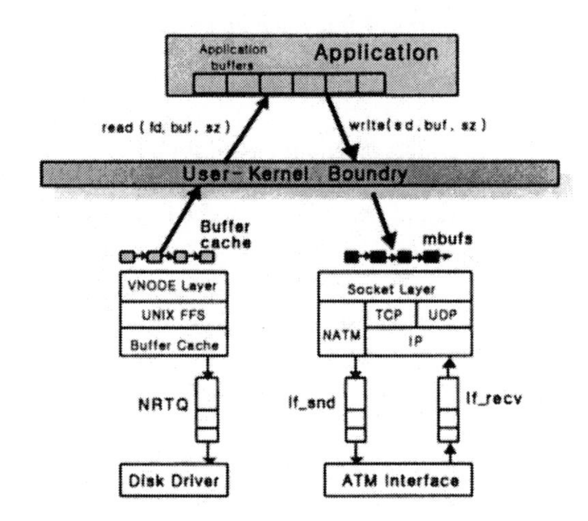

Fig. 1. General Data Transmission Mechanism of Server System

MARS[7] rectifies these limitations and proposes a new kernel buffer management system called mmbuf (Multimedia Memory Buffer) which shortens the data path from a storage device to network interface.

MARS presented the design of a new mmbuf buffering system[8]. But MARS has following problems.

1. mmbuf is configured with two pointers' list. Many pointer operations have to be happened. These result in the inefficiency.
2. The memory allocation in the kernel using the new mmbuf's structure is somewhat excessive.
3. The coexistence with another device is not flexible because of building a new mmbuf's structure.

In this paper, to transport multimedia data in the server system more efficiently, we propose that SRTPIO (Special RTP Input/Output) module processes RTP data in the kernel with SIO(Special Input/Output) Mechanism.

3 The SRTPIO Module That Processes the RTP Data

As described in section 2, RTP has some problems resulted from data copying and context switching. MARS doesn't provide coexistence with other NICs. It has a little inefficiency from the standpoint of the pointer operation. This section proposes the efficient transmission of multimedia data, the SIO mechanism and the SRTPIO module in the kernel.

3.1. SIO(Special Input/Output) Mechanism

In this section, we propose the SIO mechanism, the interface between the kernel area and the user area. The SIO mechanism provides the enhancements in the data transmission.

Multimedia data such as audio or video are transmitted in the serial stream. When the system transmits the multimedia data, the garbage copying and the overhead by context switching arises in a general server system. If a system call would be in the kernel area instead of the user area, the processing time on a system could be shortened.

Especially, these overheads are more serious in case of protocols that receive massive in-coming calls at the same time like UDP. Part (a) of Fig. 2 shows the UDP transmission path on LINUX. A transmission path from the application to the network interface via an UDP protocol is created as you can see in (a) of Fig. 2. The *sendto* function is called by the Application of the user mode. An extra data copy is made by the *sendto* function of the user mode.

The general process of UDP is: After establishing a socket to bind an UDP transmission, the *recvfrom* function waits for the client's connection. The *sys_sendto* called by the *sendto* function calls the *udp_sendto*. This attaches a UDP header and calls the network device interface through the *ip_build_xmit* function [11,12,13,14].

In case of using the transmission mechanism which is embedded the SIO mechanism, the data copying is reduced. As shown in (b) of Fig. 2, the transmission process is as follows. After establishing a socket and binding it to an UDP transmission, the *recvfrom* waits for the client's connection. The *sendto* calls the *sys_sio* of SIO system in the kernel area. After the *sys_sio* creates data from the client's information and parameter passed over by SIO system call, it is passed to the Network Interface Card through the *sys_sendto* function in the kernel area. A UDP header is attached through the *udp_sendto* function. Then the packets are transmitted by network device interface through the *ip_build_xmit* function.

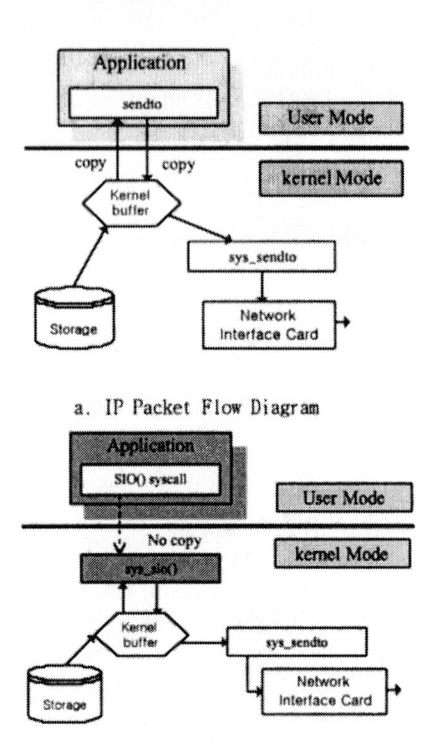

a. IP Packet Flow Diagram

b. SIO System Call Mechanism

Fig. 2. The comparison of IP Packet Flow

Keeping this model compatible to all kinds of kernel mechanisms with minimum modification, we decreases the amount of data copying. The following merits can be concluded from comparing the SIO mechanism with the MARS mechanism.

1. The mmbuf manages two pointers' list in MARS mechanism. So many pointer operation is required. The SIO mechanism enhances the transmission speed by reducing the procedures of pointer operation.
2. Compared with the MARS mechanism, high adaptability is guaranteed through using existing buffer cache so that no excessive mmbuf's assignments is required.
3. SIO mechanism is more compatible with other devices than MARS mechanism.

Using advantages of this SIO mechanism, we propose SRTPIO module used by the network interface in section 3-2.

3.2. SRTPIO Module Action Principle

SRTPIO module that we propose in this paper utilizes the SIO mechanism. RTP protocol is compatible with UDP based SIO transmission mechanism. Fig. 3 shows the structure of SRTPIO integration module we propose in this paper.

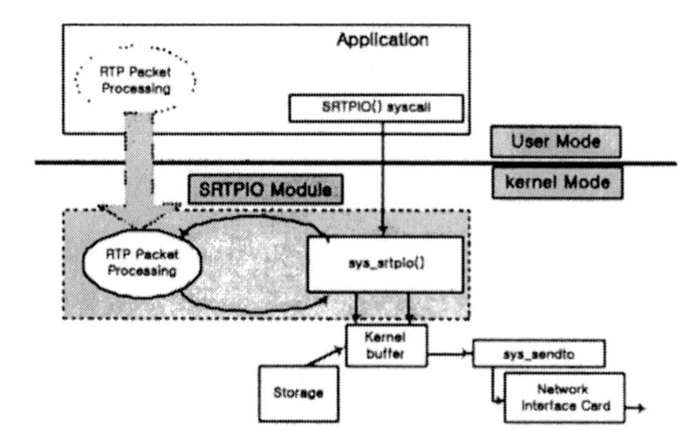

Fig. 3. SRTPIO Module Configuration

The transmission process of a general RTP based application is as follows. The server system that receives a client's request on a specific multimedia service searches and locates the position of a multimedia data in response of the specific request for service. The data from this allocated position are transferred from the buffer of the kernel area to that of the application. A RTP packet is created from this received data and passed as the UDP class. Iterative works are processed to service such requests. That consequently results in continual data copying and context switching between the user area and the kernel area.

But, the SRTPIO module informs the position and information of data, which is requested by client services and received by the server system, of the kernel using a SIO call. The *sys_sio* function creates RTP packets. It sends information and data to the user area. The RTP header is attached in the kernel area. These packets are passed into network interface. These continual processes are repeated from next packet to last packet.

As shown in Fig. 3, We moved RTP protocol into kernel area. So, We reduced much transaction between the kernel area and the user area. Also, the interface and architecture of a video or audio based application can be simplified much more by including RTP protocol in the kernel area. SRTPIO module offers a general performance improvement by using SIO system call in UDP based processing and in processing of RTP packets. Criteria of performance improvements are as following.

1. Improvement of input/output: We Reduced garbage copying and context switching by using SIO system call between kernel and application area.
2. Enhancement of RTP packet process: In case of creating a RTP packet, the processing speed can be improved by having the module in the kernel area as stated in this paper. Also, the interface of RTP application can be simplified.
3. Adaptation of kernel mechanism: Service of efficient multimedia data can be supported by correction of minimum kernel mechanism.

4 The Implementation of SRTPIO Module

Presently there are no server systems that have a module such as that in Fig. 3. We added and embodied a new SRTPIO transmission module in the kernel area of a LINUX system that do not have suitable transmission mechanisms in multimedia data transmission.

We loaded the RTP protocol in the kernel area of the LINUX system. As shown in part (a) of Fig. 2 in section 3.1, In case of UDP there exists data copying and context switching between the kernel area and the user area. We registered the SIO system call that takes charge of UDP based RTP packet transmission in the kernel area to avoid these disadvantages. The pseudo code of *sys_sio* system call is as follows.

```
Asmlinkage int sys_srtpio(fd, RTP_INFO)
{
...
sioread(fd, buf, len);
rtp_packet=
   rtp_hdr_build(RTP_INFO, buf, len);
server_len= siosendto(rtp_packet);
...
if(success)flag == 1;
else flag == 0;
return flag;
}
```

The algorithm can be basically grouped into 3 parts. We coded/designed the multimedia data transmission mechanism calling/embedding 3 functions; *sioread*, *rtp_hdr_build* and *siosendto*, which are given the information of RTP header that is passed to the user area application. The flow of the process of SRTPIO module can be shortly described as:

1. A request message from the client to server is passed.
2. Communication is established in server side.
3. Server application passes information of multimedia data to the kernel area using SIO system call.
4. The *sioread* function searches the position pointer of data.
5. The *rtp_hdr_build* function builds RTP packet header in RTP header field.
6. After *siosendto* function receives RTP packet call by SIO mechanism in a way presented in section 3.1, RTP packet is sent to the network interface.

The following algorithm shows how SRTPIO module of a server application can be embodied.

```
_syscall2(setting parameter)
int main(void)
{
RTP_INFO setting;
socket setup;
protocol, address, port establishment;
...
```

```
bind establishment;
start time;
n = sio(fd, RTP_INFO);
end time;
if(n)RTP Packet Service Success;
else RTP Packet Service Failed;
}
```

SRTPIO module that is kept in the kernel area calls the SIO system call. And SIO system call sends the information of RTP packet with a pointer of a file that is to be transmitted. RTP_INFO's contents contain the data about the header that represents important information of RTP packet. Version information and Sequence Number in the module of SRTPIO is sent to the kernel area. Application is on standby mode while the values are returned. SRTPIO module could reduce the number of times of data copying between the user area and the kernel area.

5 Experimental Results

We compared SRTPIO module with general LINUX system. We estimated the performance of only SIO mechanism that reduces route of general transmission step of UDP base in section 5.1. Section 5-2 compares the performance of SRTPIO module that has RTP protocol in the kernel area and SIO mechanism with a general LINUX server system.

5.1. SIO Module Performance Evaluation

SRTPIO module that is kept in the kernel area calls a SIO system call. Linux System-I transmits multimedia data packets as a server system and Linux system-II receives these packets as a client system. Hardware specifications of the two systems are stated in Table 1.

Table 1. Testbed Hardware Specification

	Linux System I (Server)	Linux System II (Client)
Hostname	netlab1.korea.ac.kr	mullab3.korea.ac.kr
CPU	PentiumIII 600MHz	PentiumIII 600MHz
RAM	128M	128M
Kernel Version	2.2.12-20	2.2.5-15
GCC Version	2.91.66	2.91.66
Libc6 Version	2.1.1	2.1.1

Table 2 is showing data that is used in the SIO mechanism performance estimation.

Table 2. SIO Experimental Environment

Packet Total Size	Transmission Type	Data Type	Data Generation
32 byte	Continuous	General Data	Random

The detail about an experiment and achievement scenario is as following. Because the existing LINUX system does not support RTP, RTP header file is built-in into the kernel area according to the advice of RFC 1889[3].

We compared the transmission speed of a general LINUX with that of SIO mechanism proposed in section 3.1. The processing outputs are as follows. We compared the packet transmission time of a general LINUX with that of SIO mechanism LINUX.

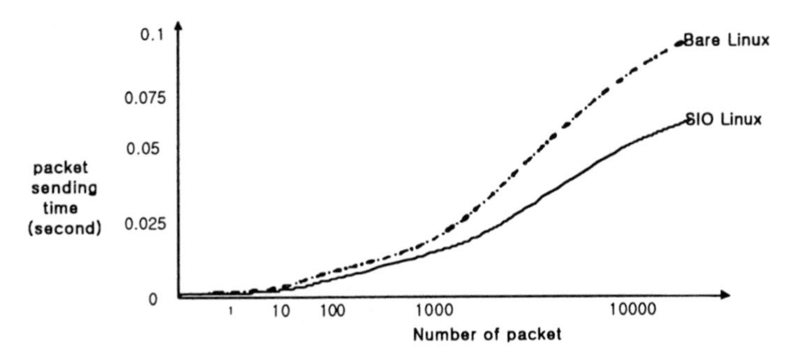

Fig. 4. Transmission speed comparison

The server system that receives client's requests transmitted packets in a period of 1, 10, 100, 1000, 10000. Afterwards, the outputs of this experiment were compared to the transmission completion time in server side is shown in Fig. 4.

Fig. 4 shows the arithmetic mean of 100 times experiment. As shown in Fig. 4, the transmission time of the LINUX of SIO mechanism is improved in the view of speed. The number of transmission packet is much less(6% ~ 23%) than that of general LINUX about each packet sink.

5.2. SRTPIO Module Performance Evaluation

We embedded RTP protocol into the LINUX kernel area and thereafter made an introduction about how to improve SRTPIO's performance by using the advantage of SIO mechanism. The hardware specification of the performance estimation is same as Table 1.

We composed a more generalized transmission scenario than the experiment in section 5.1. In section 5.1 SIO mechanism did not take any considerations on packet damage on the side of the recipient because it did not need to support real-time demands. In this experiment we transmitted RTP packets in an interval of 30ms to imitate the flow of audio stream. Also, we let 10 clients connect to the server system at the same time. Table 3 arranges the environment of a performance experiment of SRTPIO module.

Table 3. SRTPIO Experimental Environment

	Packet Size	Transmission Duration	Concurrent Connection Session Count	Packet Count by session
Result	32 bytes	30 ms	10	1000

We compared the time per session of a general LINUX with that of SRTPIO mechanism proposed in section 3.2. The processing outputs are as follows. We compared the time per session of a general LINUX with that of SRTPIO mechanism LINUX.

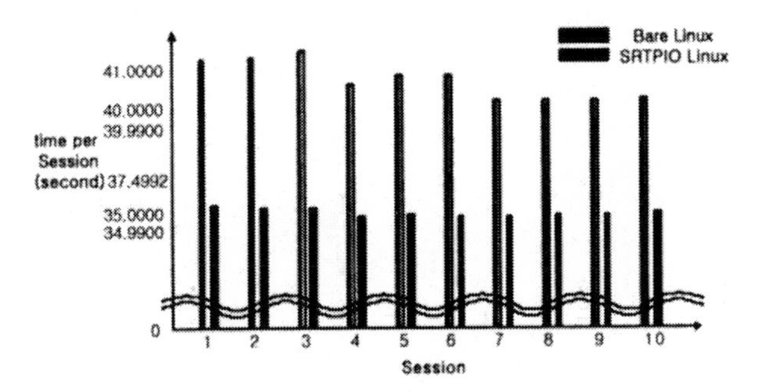

Fig. 5. Transmission speed comparison per session

Fig. 5 shows the arithmetic mean of 100 times experiment. We could see that each group of 10 sessions transmit 1000 packets at similar time rate. There were a performance elevation of about 12.5% when SRTPIO LINUX system handled packet transmission with an interval of 30 ms compared to a general LINUX system.

We transmitted multimedia data in the kernel area of the LINUX environment server system and showed performance elevation. Finally table 4 compares SRTPIO module LINUX with others.

Table 4. The comparison of Quality

	Bare LINUX	MARS	SIO LINUX	SRTPIO LINUX
Read/Write Merit	×	○	○	○
Transmission Buffer Modification	×	○	△	△
Enhancement of Transmission Speed	×	○	○	○
RTP Packet Processing	×	×	×	○
Kernel Memory Spending	×	○	△	△
Flexibility with another NIC device	○	×	○	○

(○:high, △:Low, ×:None)

6 Conclusion and Future Directions

Both data and user's requests are increasing at a rate beyond the capability of a server system. For example there are video-conferences, remote education and etc.

In this paper we proposed SRTPIO module for transmission support of multimedia data to satisfy these requests. There is no need of data copying from user area to the kernel area. And there is no necessity for scheduling the user area application. SRTPIO shows a performance elevation of about 12.5% compared with that of bare LINUX in the aspect of transmission speed. These approaches can be applied well in the server environment that provides multimedia data service.

There are following things to be researched hereafter. SRTPIO module server could not stop the transmission in the middle of streaming service. This problem can be solved as follows. First, Clients' application would be able to have the streaming data information. Second, we could put identifier into the stream.

We would design the priority scheduler for SRTPIO module. If we would combine SRTPIO with the scheduler as mentioned, we could apply these technologies to multimedia server systems.

References

1. Millind Buddhikot and Guru Parulkar, "Efficient Data Layout, Scheduling and Playout Control in MARS", ACM/Springer Multimedia Systems Journal, pp. 199-211, Volume 5, Number 3, 1997.
2. Millind Buddhikot, Guru Parulkar and Gopalakrishnan, R., "Scalable Multimedia-On-Demand via World-Wide-Web (WWW) with QOS Guarantees", Sixth International Workshop on Network and Operating System Support for Digital Audio and Video, NOSSDAV96, Zushi, Japan, April 23-26, 1996.
3. H. Schulzrinne and S. Casner, "RTP: the Real-time Transport Protocol", Audio-Video Transport Working Group, RFC 1889, January 1996.
4. Jose Carlos Brustoloni, "Effects of Data Passing Semantics and Operating System Structure on Network I/O Performance", Ph.D. Dissertation, Technical Report CMU-CS-97-176, School of Computer Science, Carnegie Mellon University, September 1997.
5. Moti N. Thadani and Yousef A. Khalidi, "An Efficient Zero-Copy I/O Framework for UNIX", Technical Report, SMLI TR-95-39, Sun Microsystems Lab, Inc., May 1995.
6. Rosenberg, J. and Schulzrinne, H., "Timer Reconsideraion for Enhanced RTP Scahability", INFOCOM '98, Seventeenth Annual Joint Conference of the IEEE Computer and Communications Societies, Proceedings IEEE, pp. 233-241, Volume 1, 1998.
7. Apache HTTP server project, Puneet Sharma, Deborah Estrin, Sally Floyd and Van Jacobson, "Scalable Timers for Soft State Protocols", INFOCOM '97, Sixteenth Annual Joint Conference of the IEEE Computer and Communications Societies, Driving the Information Revolution, Proceedings IEEE, pp. 222-229, Volume 1, 1997.
8. Milind M. B., Dakang W., Guru M.P. and Xin J.C., "Enhancements to 4.4 BSD UNIX for Efficient Networked Multimedia in Project MARS", Multimedia Computing and Systems, Proceedings IEEE International Conference on, pp. 326-337, 1998.

9. P. Druschel and L. L. Peterson. "Fbufs: A highbandwidth cross-domain transfer facility", In Proceedings of the Fourteenth ACM Symposium on Operating System Principles, pp. 189-202, Dec. 1993.
10. Kevin Fall and Joseph Pasquale, "Improving Continuous-Media Playback Performance with In-kernel Data Paths", Proceedings of the International Conference on Multimedia Computing and Systems, May 14-19, Boston, Massachusetts. IEEE-CS, pp. 100-109, 1994.
11. M. Beck, H. Bohme, M. Dziadzka, U. Junitz, R. Magnus and D. Verworner, "Linux Kernel Internals", 2nd Edition, Addison-Wesley, 1999.
12. Richard M. S., Roland M. and Andrew O., "The GNU C Library Reference Manual", Edition 0.05, DRAFT last updated, 3 1993 for version 1.07 Beta.
13. W. Richard Stevens, "TCP/IP Illustrated", Volume 3, Addison Wesley, April 1996.
14. Daniel P. Bovet and Marco Cesati, "Understanding the Linux Kernel", O'Reilly, January 2001.

An Adaptive TCP for Enhancing the Performance of TCP in Mobile Environments

Ying Xia Niu[1], Choong Seon Hong [1] , and Byung-Deok Chung [2]

[1] School of Electronics and Information, Kyung Hee University, 449-701 Korea
niuyx@networking.kyunghee.ac.kr cshong@khu.ac.kr

[2] Access Network Laboratory, KT, 305-811 Korea
bdchung@kt.co.kr

Abstract. Transmission control protocol (TCP) has been designed and tuned as a reliable transfer protocol for wired links. However, it incurs end-to-end performance degradation in mobile environments. Recent years, many protocols have been proposed to enhance the performance of TCP in mobile environments. Although these methods simulate better than original TCP, but they either need intermediaries (such as base station) to modify TCP, can not handle the end-to-end encrypted traffic or do not perform well in both high bit error rates and disconnections. In this paper, we propose a protocol named adaptive TCP, which is a combination of TCP HACK (Header Checksum Option) and Freeze-TCP. By using the adaptive TCP, we can get a true end-to-end TCP and improve the performance of TCP over mobile environments in both high bit error rates and disconnections. We use OPNET to simulate our proposal, the results have shown that our proposal performs substantially better than original TCP in cases where there are bursty corruptions and long or frequent disconnections.

1 Introduction

The integration of Internet and wireless communications is a trend to provide multimedia services for both fixed and mobile users. Many applications are built on top of TCP, and will continue to be in the foreseeable future. So the performance of TCP in mobile environments has received much attention in recent years. TCP has been designed, improved and tuned to work efficiently in wired network where the packet loss is very small. Whenever a packet is lost, it is reasonable to assume that congestion has occurred on the connection path. Hence, TCP triggers congestion recovery algorithms when packet loss is detected. These algorithms work reasonably well as the assumption on packet losses remains valid in most situations. However, in mobile environments (e.g. Fig.1.), the bit error rate is much higher and a wireless connection might be temporally broken due to a handoff (e.g. in Fig.1, when the mobile host moves from cell 1 to cell 2) or other temporal link impairment. As a result, the assumption that packet loss is (mainly) due to congestion is no longer valid and the original TCP cannot work well in a heterogeneous network with both wired and wireless links.

I. Chong (Ed.): ICOIN 2002, LNCS 2343, pp. 516-526, 2002.

The challenges for TCP over mobile environments are:

(1) high bit error rates (BER) (2) long time disconnections (3) frequent disconnections.

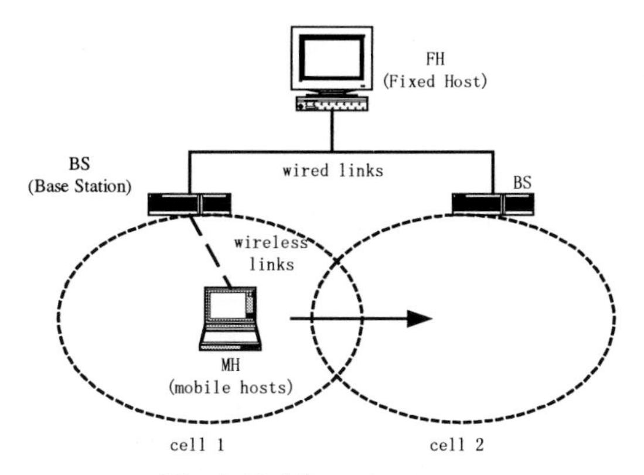

Fig. 1. Mobile environment

In the following, we first summarize the existing proposed solutions, indicating their strengths and weaknesses. Then see why we propose the combination of TCP HACK and Freeze-TCP mechanisms to enhance TCP in mobile environments and identify its advantages and drawbacks.

2 Strengths and Drawbacks of Existing Solutions

Recently, many protocols have been proposed to alleviate the poor end-to-end TCP performance in the heterogeneous network environments, these mechanisms include Indirect TCP (I-TCP), Snoop TCP and New Snoop TCP, Mobile TCP (M-TCP), Explicit bad-state notification (EBSN), TCP HACK and Freeze-TCP etc.

In the following, we will describe the mechanisms of these proposals and from Table 1 we can see the strengths and drawbacks of each solution.

I-TCP [5] suggests that any interaction from a mobile host (MH) to a host on the fixed network (FH) should be split into two separate interactions - one between the MH and the Base Station (BS) over the wireless medium and the other between the BS and the FH over the fixed network. The data sent to MH is received, buffered and ACKed by BS. It is then the responsibility of BS to deliver the data to MH. The BS communicates with the MH on a separate connection using a variation of TCP that is tuned for wireless links and is also aware of mobility. By using I-TCP it can achieve better throughputs than standard TCP, but it does not preserve end-to-end semantics of TCP because the BS partitions the TCP connection.

The Snoop protocol [6] introduces a module, called snoop agent, at the base station. The agent monitors every packet that passes through the TCP connection in both directions and maintains a cache of TCP packets sent across the link that has not yet been acknowledged by the receiver. The snoop agent retransmits the lost packet if it has cached and suppresses the duplicate acknowledgments (ACKs). The Snoop module performs extremely well in high BER environments and maintains end-to-end

TCP semantics. But it does not perform as well either in the presence of lengthy disconnections or in environment where there are frequent disconnections.

Table 1. Characteristics of various TCP proposed for heterogeneous network environments

	I-TCP	M-TCP	Snoop	New Snoop	EBSN	Freeze-TCP	TCP HACK
Maintains end-to-end TCP semantics?	N	Y	Y	Y	N	Y	Y
Requires the intermediate node to modify TCP?	Y	Y	Y	Y	Y	N	N
Handle encrypted traffic?	N	N	N	N	N	Y	Y
Need symmetric routing?	Y	Y	Y	Y	Y	N	N
Handle long Disconnections ?	Y	Y	N	Y	Y	Y	N
Handle frequent disconnections?	Y	Y	N	Y	Y	Y	N
Handle high bit error rate?	Y	Y	Y	Y	N	N	Y

The New Snoop protocol [7] was proposed to overcome the shortcomings of Snoop protocol. It uses a two-layer hierarchical cache architecture cache scheme. The main idea is to cache the unacknowledged packets at both Mobile Switch Center (MSC) and Base Station (BS), thus forming a two-layer cache hierarchy. If a packet is lost due to transmission errors in wireless link, the BS takes the responsibility to recover the loss. If the loss/interruption is due to a handoff, the MSC performs the necessary recovery. With this proposed hierarchical cache architecture, New Snoop protocol can effectively handle the packet losses caused by both handoffs and link impairments. But both Snoop protocol and New Snoop protocol need the intermediary (Such as a Base Station) to do TCP modifications and New Snoop even needs the MSC's participation.

The M-TCP [8] has the same goals as I-TCP and snoop TCP: to prevent the sender window from shrinking if bit errors or disconnection but not congestion cause current problems. M-TCP wants to improve overall throughput, to lower the delay, to maintain end-to-end semantics of TCP, and to provide a more efficient handover. It splits the TCP connection into two parts as I-TCP does. An unmodified TCP is used on the standard FH-BS connection, while an optimized TCP is used on the BS-MH connection. The BS monitors all packets sent to the MH and ACKs returned from the MH. And it retains the last ACK. If the BS does not receive an ACK for some time, it assumes that the MH is disconnected. It then shut down the TCP sender's window by sending the last ACK with a window set to zero. Thus, the TCP sender will go into persist mode. The M-TCP approach does not perform caching/retransmission of data via the BS. If a packet is lost on the wireless link, it has to be retransmitted by the original sender. This maintains the TCP end-to-end semantics. But it still requires a substantial base station involvement.

An explicit bad-state notification (EBSN) scheme [9] does not split the connection in two connections, it uses two types of acknowledgments: one is a partial acknowledgment informing the sender that the packet had been received by the base station and the other is a complete acknowledgment which has the same semantics as the normal TCP acknowledgment, i.e. the receiver (MH) received the packet. So it can distinguish the losses on the wired portion from the losses on the wireless link. Now the base station is responsible for retransmissions on the wireless link, while it delays timeout at the sender by sending a partial acknowledgement. This idea is that these explicit notifications prevent the sender from dropping congestion window. It also requires an intermediate node to modify TCP.

Freeze-TCP [10] is to move the onus of signaling an impending disconnection to the client. A mobile node can certainly monitor signal strengths in the wireless antennas and detect an impending handoff; and in certain cases, might even be able to predict a temporary disconnection (if the signal strength is fading, for instance). In such a case, it can advertise a zero window size, to force the sender into the zero window probe mode and prevent it from dropping its congestion window. It suggests the receiver to start advertising a window size of zero just before a round-trip-time (RTT) period. When re-connectivity detects, the TCP will operate at exactly the same point where it had been forced to stop. It offers a way to resume TCP connections even after longer interruptions of the connectivity. But it is the insufficient isolation of packet losses.

TCP HACK (Header Checksum Option) [12] is a solution based on the premise that when packet corruption occurs, it is more likely that the packet corruption occurs in the data and not the header portion of the packet. This is because the data portion of a packet is usually much larger than the header portion for many applications over typical MTUs. It introduced two TCP options: the first option is for data packets and contains the 1's-complement 16-bit checksum of the TCP header (and pseudo-IP header) while the second is for ACKs and contains the sequence number of the TCP segment that was corrupted. These "special" ACKs do not indicate congestion in the network. Hence, the TCP sender does not halve its congestion window if it receives multiple "special" ACKs with the same value in the ACK field. With this scheme, TCP is able to recover these uncorrupted headers and thus determine that packet corruption and not congestion has taken place in the network. TCP HACK performs substantially better than both TCP SACK and NewReno in cases where burst corruptions are frequent. But it is not able to deal with the problems caused by lengthy disconnections or by frequent disconnections.

3 Our Proposed Solution: Adaptive TCP

In this section, we will first analyze the problems of TCP in mobile environments, describe the requirements for enhancing the performance of TCP, then give the proposal of our solve.

3.1 Problems of TCP in Mobile Environments

The TCP Tahoe detects missing ACKs via three duplicate acknowledgements or time-outs. When high bit error rates or disconnections occur, it will be considered that congestions have happened, it just could not distinguish between the congestion and packet loss or disconnections. Then the TCP sender will trigger slow start mechanism: it drop the congestion window down to 1, and first grows it by a factor of 2 each time an ACK is received, until it reaches half of the threshold of the congestion window. The wrong assumption drastically decreases the performance of TCP in those cases.

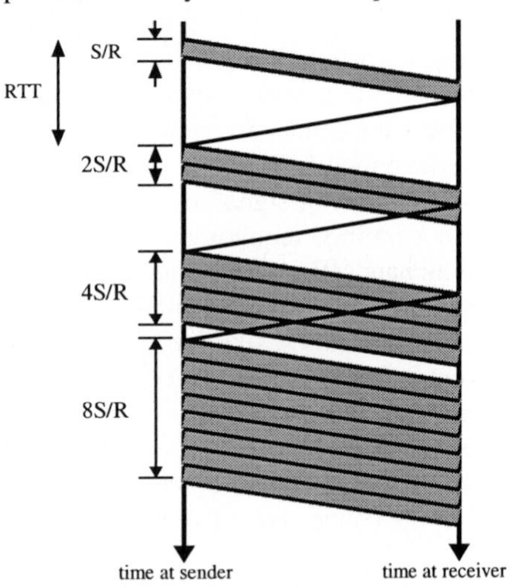

Fig. 2. TCP timing during slow-start

Fig.2. shows the TCP timing during slow start. S means the MSS (maximum size segment) is S bits; RTT is the round-trip time and R denotes the transmission rate of the link from the sender to the receiver. So the idle time (approximate) of the sender in slow start stage is given by:

Idle time=P[RTT]-(2^P-1)S/R

In which P=min {Q, K-1}, K is the number of windows that cover the object and Q is the number of times the sender would stall [15].

If the slow-start scheme is not triggered, then the number of extra packets can be transferred is approximate given by [10]:

Extra Segments=W^2/8+WlgW-5W/4+1;

W is the unACKed packets can be sent in a congestion window.

It should be noted that the upper two expressions are under the assumptions that no corruptions in the traffic and ignoring all protocol header overheads.

3.2 Requirements for Enhancing the Performance of TCP

Our goal in developing a new TCP protocol is to provide a general solution to the problem of improving TCP's efficiency for mobile applications. Specifically, we want to design a protocol that has the following characteristics:

- It should preserve end-to-end TCP semantics. I-TCP and EBSN are not end-to-end-semantics. So, if a sender receives an acknowledgement, it assumes that the receiver got the packet. Receiving an acknowledgement now only means (for the mobile host and a correspondent host) that the foreign agent received the packet. The correspondent node does not know anything about the partitioning, thus a crashing access node may also crash applications running on the correspondent node assuming reliable end-to-end delivery.

- It should not require the intermediate node to do TCP modifications. I-TCP, Snoop, New Snoop, M-TCP and EBSN all need the intermediate node to do TCP modifications. So the intermediary will become the bottleneck and add the third point of failure besides the endpoints themselves.

- It can handle encrypted traffic. As network security is taken more and more seriously, encryption is likely to be adopted very widely. Finally, all efforts for snooping and buffering data in the intermediate nodes may be useless if certain encryption schemes are applied end-to-end between the correspondent host and mobile host. Using IP encapsulation security payload the TCP protocol header will be encrypted, so that the intermediate nodes may not even know that the traffic being carried in the payload is TCP. Furthermore, retransmitting data from the foreign agent may not work any longer because many security schemes prevent replay attacks and retransmitting data from the foreign agent may be misinterpreted as replay. Encrypting end-to-end is the way many applications go, therefore, it is not clear how these schemes (I-TCP, Snoop, New Snoop, M-TCP and EBSN) could be used in the future.

- It doesn't need a symmetric routing. The protocols that need to modify TCP in an intermediate node usually require that traffic to and from the end mobile host is routed through the same intermediate node. But in some networks, data and ACKs can take different paths, so these schemes based on intermediary involvement cannot be accomplished and may result in non-optimal routing.

- It can handle long time disconnections. I-TCP, M-TCP, New Snoop, EBSN and Freeze-TCP all can handle long time disconnections.

- It can handle frequent disconnections. I-TCP, M-TCP, New Snoop, EBSN and Freeze-TCP all can handle frequent disconnections.

- It can handle high BER. I-TCP, M-TCP, Snoop, New Snoop, and TCP HACK all can handle high BER.

From Table1 and the analysis above we can get a conclusion that none of the existing mechanisms can satisfy all these characteristics. Then how about we combine two of them? The protocols to be combined must satisfy the first four characteristics above respectively. We can see that only the TCP HACK and Freeze-TCP can accord with all the first four conditions. The Freeze-TCP can handle long time disconnections and frequent disconnections while the TCP HACK can handle high BER. So it will be a good solve to combine Freeze-TCP and TCP HACK. With the combination of these two mechanisms, we can reach our goal perfectly.

3.3 Our Solve: Adaptive TCP

Fig.3. shows the work flow of our proposal, We add TCP HACK to the original TCP, it can work well when the BER is high. Once disconnections happen, the Freeze-TCP will be triggered. Until the disconnections are all recovered, it will continue work as TCP HACK.

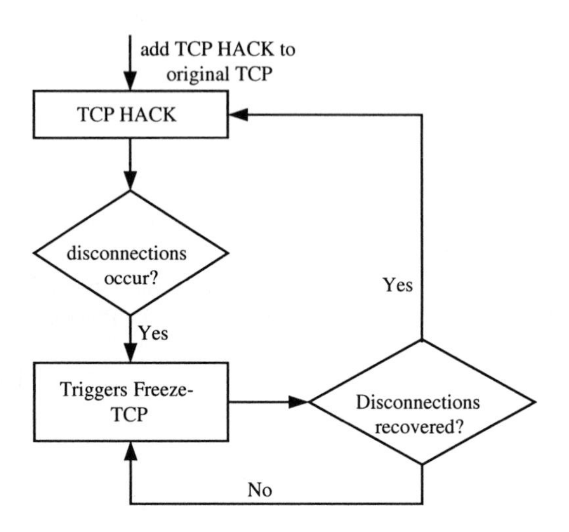

Fig. 3. Work Flow of the proposed adaptive TCP

By using the proposed adaptive protocol, we don't need to have any change in the intermediate nodes, but we should make some modifications on the TCP sender and receiver.

The main idea behind our adaptive protocol is to prevent the sender from dropping its congestion window as in the mobile environment of bursty error or disconnections. By our proposed protocol, if bursty error occurs, TCP HACK will find that packet corruption but not congestion has taken place in the network in most conditions by checking the TCP header, then the TCP receiver will send special ACK to the sender, so the TCP sender can react appropriately, retransmitting the corrupted packets, not dropping its congestion windows. While in long or frequent disconnections, while the mobile host predicts an impending disconnection before it happens, Freeze-TCP will be triggered. Then the receiver can send some zero window acknowledgements to the sender in advance of the disconnection. When the sender receives the zero window acknowledgements, it will stop sending and "freeze" the current state of its congestion window and further timers. As the disconnection is recovered, the sender can begin just at the same point where it had been forced to stop. Because the TCP time simply does not advance, so there is no timers expire.

According to TCP HACK, we extend TCP by including two additional TCP options. The first is Header Checksum option, contains the 16bit 1's complement checksum of the TCP header and the pseudo-IP header. The second option is the

Header Checksum ACK option that is included in "special" ACKs generated in response to packet corruption.

According to Freeze-TCP, if the receiver (Mobile Host) can sense an impending disconnection, it should try to send out a few (at least one) acknowledgements, wherein its window size is advertised as zero. The waning period prior to disconnection is the round-trip-time (RTT). The author of [10] claimed that Experimental data corroborates this: warning periods longer or shorter than RTT led to worse average performance in most cases.

4 Performance Evaluation

We carried our experiments by using OPNET modeler 8.0 which was installed on the computer with Celeron 800 CPU and 128M RAM. The simulation model is shown as Fig.4. We ran our experiments by sending bulk data from the server to the client. The connect link between the two routers is a wired link but has been configured by different error bit rates to simulate the lossy wireless link and we deliberately stopped the transmission to simulate the long and frequent disconnections. The experiments over bursty errors and disconnections are run separately.

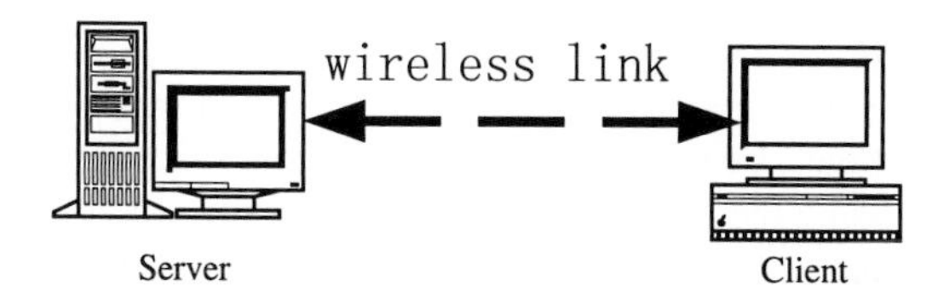

Fig. 4. Simulation model

4.1 Bursty Errors

First, we ran experiments to compare the performance of our adaptive TCP and original TCP in bursty error conditions. 5E-05 ~ 5E-01 packet error rate were considered. The burst packets length is configured to 5 packets.

We assume that errors are only on the forward path; packets on the reverse path (the ACK packets) are not corrupted.

From Fig.5. it can be seen that our proposal performs substantially better than original TCP in the presence of bursty errors. Especially when the packet loss rate is high, when the packet loss rate is less than 5.0E-05, the throughput of the adaptive TCP and original TCP will be almost the same.

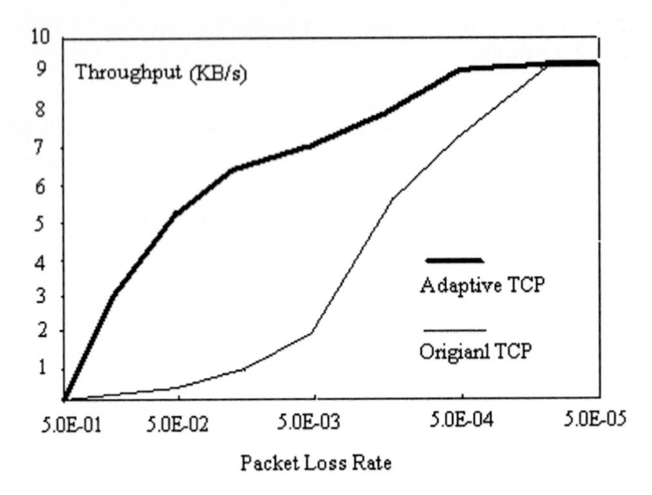

Fig. 5. Throughput for various packet loss rates

4.2 Disconnections

We next performed experiments to compare the performance of our adaptive TCP and the original TCP in disconnections. The server is on the "sender side" and the client runs on the "receiver side" which emulates a mobile node implementing the Freeze-TCP scheme. Disconnection time is ranging from 0.5ms to 3s. We do simulations to get the transfer time of each protocol over different disconnection time. From Fig.6. we can see that using the adaptive protocol can improve the performance of TCP over disconnections.

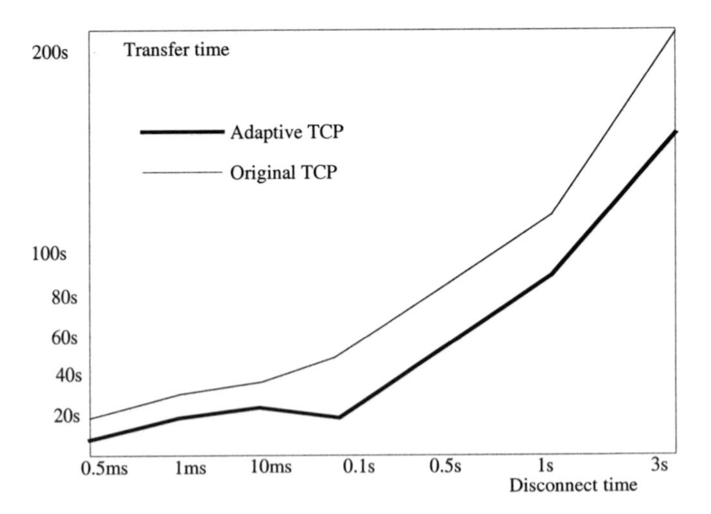

Fig. 6. Transfer time for various disconnect time

5 Drawbacks

There are some problems should be considered by using our proposed adaptive TCP:

- The software overload: Using adaptive TCP, more complex software will be on the sender and receiver sides, but while memory sizes and CPU performance permanently increase, the bandwidth of the air interface remains almost the same. Therefore the higher complexity is no real disadvantage any longer as it was in the early days of TCP.

- The Freeze-TCP needs the receiver to predict an impending disconnection. However, if a disconnection cannot be predicted, the behavior and performance of the adaptive protocol will be exactly that of TCP HACK only.

6 Conclusions and Future Work

In this paper, we summarized the existing protocols, indicating their strengths and weaknesses, and analyzed the current status of TCP enhancements being considered for adoption in future versions of TCP. We proposed an adaptive TCP, it is the combination of TCP HACK and Freeze-TCP. By using this protocol we can get the benefits of the original two mechanisms, and recover the TCP from both packet loss due to corruption in lossy environment and long or frequent disconnections in mobile computing.

Simulations have been done to test our adaptive protocol by using OPNET modeler 8.0. The results proved that our proposed protocol performs much better than original TCP in the cases that corruptions or disconnections are frequent.

In [10] the author suggested to use Freeze-TCP with the protocol suggested in [14], and the performance of TCP over disconnections will be improved much more; also in [12] the author suggested to run TCP HACK with the TCP SACK suggested in RFC 2018[11] together, it will get a better result over bursty errors. But in our simulations, we didn't add these two protocols. Simulations will be done adding these protocols to our adaptive TCP.

The TCP HACK can work well when the ACKs are not corrupted much. If the ACKs are lost often, the efficient of TCP HACK will not be so prominent. So we want to improve it by modifying the special ACK in TCP HACK. The main idea is in the special ACK, it not only contains the sequence number of the packet that is just received and data corrupted; but also it contains the sequence numbers of the other packets in the same window that have been received but are data corrupted. It should be mentioned that these sequence numbers of packets have been sent in the special ACKs already, but the corresponding retransmitted packets have not been received yet.

Works are being done to test the effectiveness of our proposal in situations where ACKs are also susceptible to packet corruption. And we'd like to run the experiments under the condition that congestion, corruptions and disconnections occur at the same time.

Acknowledgements. This work was supported by Kyung Hee University.

References

[1] Behronz A.Forouzan, "TCP/IP protocol suite" international editions 2000,271-311
[2] Jochen Schiller, "Mobile Communications" Pearson Education Limited 2000, 290-307
[3] M.Allman,V.Paxson,and W.Stevens, "TCP congestion control", IETF RFC 2581,1999
[4] Jiangping Pan,Jon W.Mark and Xuemin Shen, "TCP performance and its improvement over wireless links" GlobeCom2000
[5] Ajay Bakre and B.R. Badrinath, "I-TCP: Indirect TCP for mobile hosts", Tech. Rep., Rutgers University, May 1995.
[6] H.Balakrishnan, S.Seshan, Eamir, and R.H. Katz, "Improving TCP/IP performance over Wireless Networks", In Proc.1st ACM Conf. on Mobile computing and Networking, November 1995.
[7] Jian-Hao Hu,Kwan L.Yeung,Wiew Chee Kheong and Gang Feng, "Hierarchical Cache Design for Enhancing TCP over Heterogeneous Networks with Wired and Wireless Links" GlobeCom2000
[8] K.brown and S.Singh, "M-TCP: TCP for Mobile Cellular Networks", ACM computer Communications Review (CCR), vol.27, no.5, 1997
[9] N.Vaidya, Overview of work in mobile-computing (transparencies).
[10] Tom Goff, James Moronski, D.S.Phatak, "Freeze-TCP: a true end-to-end TCP enhancement mechanism for mobile environment" InfoCom2000
[11] M.Mathis, J.Mahdavi, S.Floyd, A.Romanow, "TCP selective acknowledgment options" IETF RFC 2018, 1996
[12] R.K.Balan, B.P.Lee,K.R.R.Kumar, L.Jacob,W.K.G.Seah, A.L.Ananda, " TCP HACK:TCP header checksum option to improve performance over lossy links", InfoCom2001
[13] Hari Balakrishnan, Venkata N. Padmanabhan, Srinvasan Seshan, Mark Stemm, Elan Amir, Randy H.Katz, "TCP Improvements for Heterogeneous Networks: The Daedalus Approach".
[14] Ramon Caceres and Liviu Iftode,"Improving the performance of reliable transport protocols in mobile computing environments" IEEE JSAC Special Issue on Mobile Computing Network,1994.
[15] James F. Kurose and Keith W. Ross, "Computer Networking" 2001 by Addison Wesley Longman, Inc. 167-260.
[16] Modeler & Radio powered by OPNET Simulation Technology. SIMUS Technologies. Inc.
[17] IT Decision Guru powered by OPNET Simulation Technology. SIMUS Technologies.Inc.
[18] IETF PILC WG homepage, http://www.ietf.org/html.charters/plic-charter.html

A Novel Scheduler for the Proportional Delay Differentiation Model by Considering Packet Transmission Time

Yuan-Cheng Lai[1], Wei-Hsi Li[2], and Arthur Chang[1]

[1]Deptartment of Information Management
National Taiwan University of Science and Technology
laiyc@cs.ntust.edu.tw
[2]Deptartment of Computer Science and Information Engineering
National Cheng Kung University
leews@locust.csie.ncku.edu.tw

Abstract. The *proportional delay differentiation model* provides a consistent packet delay differentiation between various classes of service. The *waiting time priority* (WTP) scheduler is a priority scheduler in which the priority of a packet increases in proportion to its waiting time, and it is known as the best scheduler to achieve the proportional delay differentiation model. This paper proposes an *advanced WTP* (AWTP) scheduler, modified from WTP, that accounts for the packet transmission time. Simulation results reveal that when the link utilization is moderate (60%~80%), this scheduler not only obtains more accurate delay proportion than the WTP scheduler, no matter in short or long timescales, but also reduces the average queuing delay (waiting time). The effects of mean packet size and variance of packet size, on both schedulers' performance are also examined. AWTP always outperforms WTP.

1 Introduction

Conventional Internet applications such as ftp, email, and telnet, all employ best-effort service. However, many multimedia applications require some guarantee of the quality of services (QoS). The best-effort service is not suited to such applications because it can not promise any guarantee of packet loss, delay, or jitter.

The Internet Engineering Task Force (IETF) has developed *Integrated Services* (IntServ) [1] and *Differentiated Services* (DiffServ) [2] to meet the QoS requirements of multimedia applications. The IntServ approach employs the *Resource Reservation Protocol* (RSVP) [1] to reserve the bandwidth in the router along the flow path by setting up the flow state. The Diffserv approach guarantees the performance for each service class by marking the *Differentiated Service Code Point* (DSCP) in the Type of Service (ToS) field of the IPv4 or IPv6 packet header [3].

Diffserv has proceeded in two directions - *absolute service differentiation* and *relative service differentiation*. The absolute differentiated service ensures that an admitted user can enjoy certain and steady performance, while the relative differentiated service ensures that users with a higher service class experiences better performance than the users with a lower service class. The relative service

I. Chong (Ed.): ICOIN 2002, LNCS 2343, pp. 527-538, 2002.

differentiation can be classified into four models, namely, *strict prioritization*, *price differentiation*, *capacity differentiation*, and *proportional differentiation* [4]. The first three models suffer some shortcomings in achieving relative service differentiation, and only the proportional differentiation model can perform the controllable and predictable relative service differentiation [5]. That is, the proportional differentiation model offers the network manager a means of varying quality spacing between service classes according to the given pricing or policy criteria, and ensures that the differentiation between classes is consistent in any measured timescale.

A packet scheduler, called the WTP (waiting time priority) scheduler, has been proposed to yield the proportional delay differentiation model [4-9]. This scheduler uses the waiting time of the HOL (Head-of-Line) packets to determine their priorities, before one is selected and serviced. In this paper, we propose an AWTP (advanced WTP) scheduler which uses the waiting times and transmission times of the HOL packets to determine their priorities, supposing that one of them has been selected and serviced.

The rest of the paper is organized as follows. Section 2 outlines the background, including the proportional differentiation model and the waiting time priority scheduler. Section 3 gives the newly proposed scheme – the AWTP scheduler. In Section 4, some simulation results are shown and their implications are addressed. Finally, some conclusions are given in Section 5.

2 Background

2.1 Proportional Differentiation Model

The main concept of the proportional differentiation model is that the performance metric of each class should be proportional to its corresponding differentiation parameter, which can be defined by a network manager.

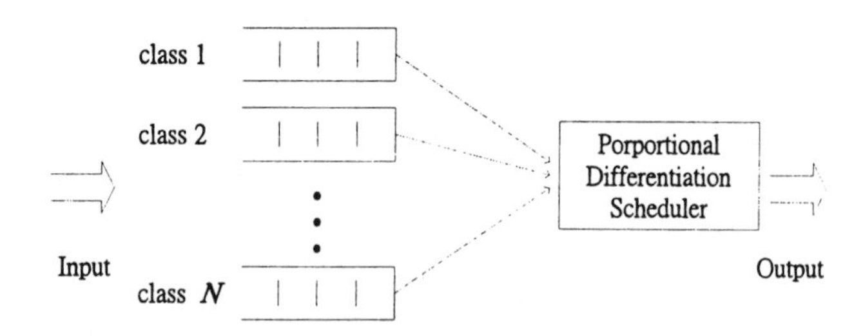

Fig. 1. Proportional differentiation model

Figure 1 depicts the proportional differentiation model by considering N classes of service. A queue is associated with each class of service, and a packet scheduler is responsible for deciding which packet should be served next. Let $Q_i(t,t+\tau)$ be the measured performance of class i, over time interval $(t,t+\tau)$, where τ is the measured timescale. For any arbitrary pair of classes, i and j, the following relationship is maintained.

$$\frac{Q_i(t,t+\tau)}{Q_j(t,t+\tau)} = \frac{C_i}{C_j}, \quad (1 \leq i, j \leq N) \tag{1}$$

where $C_1 < C_2 < \ldots < C_N$, are the *quality differentiation parameters* (QDPs) [5].

The queuing delay, packet loss, or jitter could be as the performance metric in this model. Adopting queueing delay and packet loss, this model is called the proportional delay differentiation model and proportional loss differentiation model, respectively. This paper focuses on the proportional delay differentiation model. Let $\overline{D}_i(t,t+\tau)$ be the average queuing delay of the class-i HOL packet during the time interval $(t,t+\tau)$ and let δ_i be the *delay differentiation parameters* (DDP) [5] in the order $\delta_1 < \delta_2 < \ldots < \delta_N$. For all pairs of service classes, i and j, the proportional delay differentiation model is specified by

$$\frac{\overline{D}_i(t,t+\tau)}{\overline{D}_j(t,t+\tau)} = \frac{\delta_i}{\delta_j}. \quad (1 \leq i, j \leq N) \tag{2}$$

2.2 WTP Scheduler

The WTP scheduler is based on Kleinrock's Time-Dependent-Priorities algorithm, which is a priority scheduler in which the priority of a packet increases in proportion to its waiting time [7]. For all classes of service, the WTP scheduler delivers the HOL packet with the highest priority.

C. Dovrolis et al. found that the WTP scheduler can yield the proportional delay differentiation when the network traffic load is heavy [4,5]. Let $W_i(t)$ be the waiting time of the HOL packet that is waiting in queue i, at time t. Accordingly, the WTP scheduler can achieve proportional delay differentiation if the priority $P_i(t)$ of a packet at time t is set according to,

$$P_i(t) = \frac{W_i(t)}{\delta_i}. \quad (1 \leq i \leq N) \tag{3}$$

Hiroshi Saito et al. proposed a local optimal proportional delay packet scheduler to maintain the delay ratio and reach an optimal decision when no further packets are arriving [6]. Their proposed scheduler can more accurately approximate the delay proportion than can the WTP scheduler, when the link utilization is light-load. K. H. Leung et al. [9] theoretically calculated the achieved proportion using the link utilization and DDPs. Thus these authors can ensure the consistency of delay ratio between all pairs of classes, independently of link utilization.

3 AWTP Scheduler

Packet sizes over the Internet are not identical. Y. Matsumoto et al. have stated that the effect of the packet size cannot be neglected and have examined the effects of packet size distribution on the system's performance [8]. The AWTP presented here accounts for the size of all HOL packets, rather than ignoring them, like the WTP algorithm.

3.1 Algorithm

The difference between WTP and AWTP is that the former uses the waiting times of the HOL packets to determine their priorities before one of them is selected and serviced, while the latter uses the waiting times and transmission times of the HOL packets to determine their priorities, under the assumption that one of them has been selected and serviced.

In order to ascertain the condition after one HOL packet has been transmitted, the proposed AWTP scheduler employs a pseudo-service. Let $W_j(t)$ be the waiting time of the HOL packet of class j and let T_j be its transmission time. When the pseudo-served packet belongs to class i, our scheduler calculates its corresponding maximum proportion (MP_i) as,

$$MP_i = \max_{1 \le j \le N} \{\frac{W_j(t) + X_i}{\delta_j}\}, \quad (X_i = 0, \text{ if } j = i;\ X_i = T_i, \text{ if } j \ne i), \tag{4}$$

where X_i is the extra waiting time caused by transmitting the HOL packet of class i. For any class i, its corresponding MP_i is calculated. Then, the minimum value of all MP_i, and the corresponding index are given by,

$$MMP = \min_{1 \le i \le N} \{MP_i\};$$
$$C = \operatorname{argmin}_{1 \le i \le N} \{MP_i\}. \tag{5}$$

The AWTP scheduler chooses the HOL packet of class C and actually transmits it. When more than one MP_i equals MMP, the AWTP algorithm randomly selects one of them.

3.2 An Example

Consider three classes (N = 3) and let the current queuing delays of three HOL packets be $W_1(t)$=0.6, $W_2(t)$=0.8, and $W_3(t)$=1.0 time units. Their packet transmission times are T_1=0.6, T_2=0.05, and T_3=0.5 time units. The DDPs for the three classes are δ_1=1, δ_2=2, and δ_3=4. From Eq. (3), for WTP, the priorities of the HOL packets at time t are $P_1(t)$=0.6, $P_2(t)$=0.4, and $P_3(t)$=0.25; consequently, the WTP algorithm delivers the HOL packet of class 1. By employing Eqs. (4) and (5), given in Table 1, the AWTP algorithm calculates MP_1=0.7, MP_2=0.65, and MP_3=1.1, represented by a circle, and gives MMP=0.65 and C=2, marked by a double circle. Thus AWTP selects the HOL packet of class 2 to transmit. The selections of the WTP scheduler (class 1) and the AWTP scheduler (class 2) are clearly different.

Table 1. Example for the AWTP algorithm

	j=1	j=2	j=3
Pseudo-service of class-1 HOL packet	0.6	0.7	0.4
Pseudo-service of class-2 HOL packet	0.65	0.4	0.2625
Pseudo-service of class-3 HOL packet	1.1	0.65	0.25

3.3 Characteristics of AWTP

The AWTP algorithm exhibits the following characteristics.

- *Delay proportion is maintained.* For each HOL packet of class i, the calculated MP_i is the maximum proportion, that is, the most over-proportional value. To keep the average waiting times for all classes as proportional to the given DDPs, the AWTP algorithm selects the packet which generates the minimum MP. Thus, the AWTP algorithm can meet the target ratio as close as possible.
- *Small packet is preferred.* A large packet often leads to a large maximum proportion. Thus, it is less likely to be selected and give the opportunity to a small packet. A small packet only causes few differences between the pseudo waiting time and the current waiting time, so its MP is typically small and it is more likely to be selected as MMP. Therefore, the AWTP algorithm prefers small packets. Reviewing the above example, the packet transmission time $T_2(0.05)$ is shorter than $T_1(0.6)$ and $T_3(0.5)$, so ATWP selects the packet of service class 2, rather than the packet of class 1 or class 3.

  As well known, the shortest job first (SJF) algorithm yields the optimal queueing delay. The preference for small packets causes ATWP to have a much shorter delay than WTP. The most striking advantage of the AWTP scheduler is that its efficiency in achieving proportional delay differentiation is drastically improved.

  The packet of class 1 is preferred. WTP and AWTP all prefer the class-1 packet because of its lower DDP. However, AWTP more strongly prefers the class-1 packet than does WTP. For AWTP, when the pseudo-served packet belongs to class 1, its transmission time is divided by a larger *DDP*, giving a smaller MP_1. Thus, the class-1 packet is more likely to be selected. For example, when the current waiting times of all HOL packets are in the ratio, 1:2:4, WTP randomly selects a packet while AWTP selects the class-1 packet when all HOL packets are of equal size.
- *AWTP differs from WTP, even when all packet sizes are equal.* AWTP does not reduce to WTP when all packet sizes are equal. For AWTP, the extra transmission time is divided by its individual *DDP*, and which result is then added to the corresponding waiting time. Thus, the equal packet size does not ensure that both schedulers make the same selections.

4 Simulation Results and Their Implications

The simulations presented here seek to compare the performance of the AWTP algorithm with that of the WTP algorithm. These simulations are conducted in two parts. First, the queuing delay ratio, the improved delay ratio, and the queuing delay distribution are considered in a comparison of the performance of both schedulers. Also, the delay ratios of AWTP under various different timescales are observed, to understand its short and long-term behaviors. Second, the average packet size and its variance are changed to determine their effects on the proposed AWTP scheduler.

In all simulations, three service classes (N=3) are assumed and the corresponding DDPs are set as $\delta_1=1$, $\delta_2=2$, and $\delta_3=4$. The arrival of the packets in each class follows a Pareto distribution with shape parameter 1.9, and the packet size distribution for all classes is such that 40% of packets are 50 bytes, 50% packets are 500 bytes, and 10% packets are 1500 bytes, corresponding to real network traffic. Thus, the mean packet size is 420 bytes. The total bandwidth is 2520 bytes/sec and the buffer size for each class is infinite, i.e., no packet loss. The traffic shares of all classes are equal (1/3). In each simulation, at least 150,000 packets are generated for stability.

4.1 Case 1:

4.1.1 Queuing Delay Ratio

Figure 2 reveals the achieved delay ratios of class 3 over class 1 (R_{31}) and class 2 over class 1 (R_{21}) under various link utilizations. According to DDPs, the delay ratio, R_{31} should be four and R_{21} should be two.

When the link utilization is heavy-load (over 85%), the AWTP and WTP schedulers can reach the desired delay ratio in a long timescale. In such a case, the delay ratio achieved by the AWTP algorithm is the same as that achieved by the WTP algorithm. Moreover, both schedulers behave in the same way, that is, they select the same packet to transmit, because the queuing delay of the HOL packet greatly exceeds the transmission time. Thus, the original selection of the WTP scheduler is still not changed by considering the transmission time of the HOL packet.

Under moderate-loaded link utilization (60%~80%), neither the AWTP nor the WTP scheduler reaches the desired delay proportion. However, the delay ratio achieved by the former is closer to the desired ratio than that achieved by the latter. For example, when the traffic load is 70%, R_{31} is 3.14 by AWTP and 2.69 by WTP. When the traffic load is moderate, the queuing delay is close to the packet transmission time. Thus, the effect of AWTP is very obvious; that is, its selection differs greatly from the selection of WTP. The mean queuing delay of class-1 packets is greatly reduced due to the preference of AWTP for class-1 packets, resulting in higher R_{31} and R_{21}.

4.1.2 Improved Delay Ratio

To exhibit the queuing delay saved by AWTP, the improved delay ratio is defined as follows,

$$\frac{W-A}{W}, \tag{6}$$

where W and A are the queuing delays made by the WTP and AWTP schedulers, respectively. Figure 3 plots the improved delay ratios for three classes under a range of link utilization.

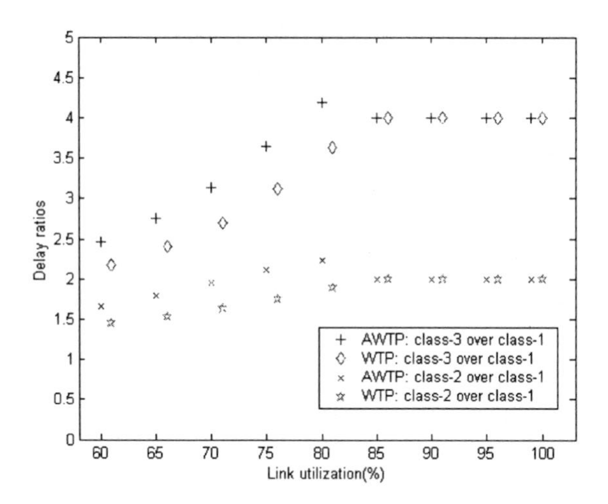

Fig. 2. Achieved delay ratios of AWTP and WTP

When the link utilization is heavy-load, the improvement of AWTP is zero because both schedulers make the same selections, as stated above. In a moderate-loaded link, the AWTP algorithm prefers small packets, so the queuing delays for the three classes are reduced. Moreover, the AWTP algorithm prefers class 1, so the improved delay ratio of class 1 is better than those of classes 2 and 3. For instance, when the traffic load is 70%, the improved delay ratio of class 1 is around 24%, but those of classes 2 and 3 are approximately 12% and 9%, respectively.

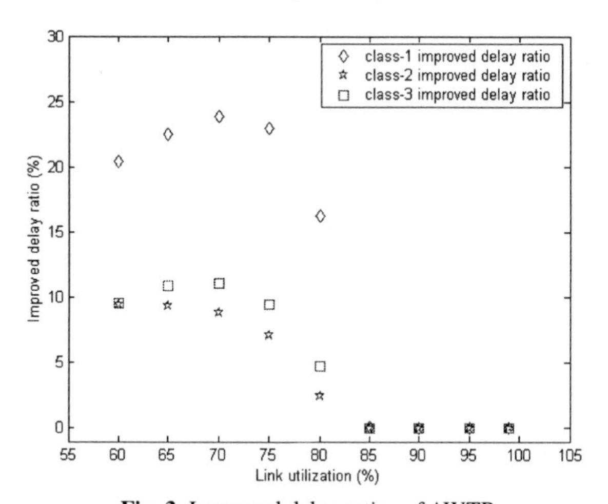

Fig. 3. Improved delay ratios of AWTP

4.1.3 Queuing Delay Distribution

Figure 4 plots the queuing delay distributions of all classes for AWTP and WTP under 70% utilization. The figure shows that AWTP has a more low-delay packets and a fewer high-delay packets, than WTP. This result also explains the reduction in the mean queuing delay when the AWTP algorithm is adopted. More importantly, this finding demonstrates that AWTP does not cause the starvation.

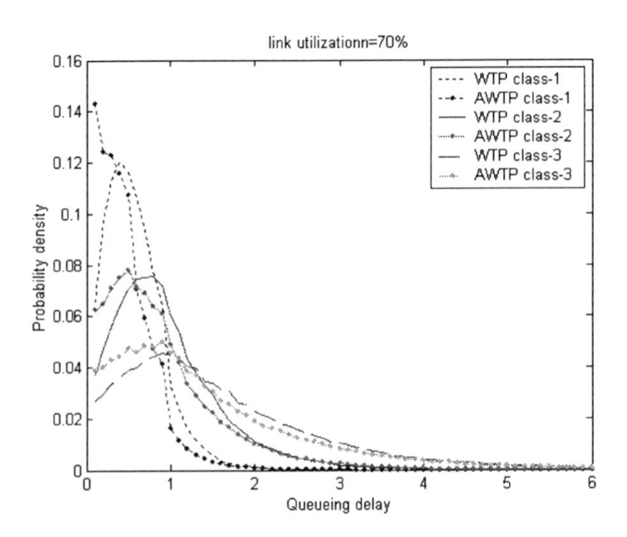

Fig. 4. Queuing delay distributions of AWTP and WTP

4.1.4 Timescale

The previous three investigations are based on long timescales. The achievement of the desired delay ratio by the AWTP algorithm in short timescales remains an important issue. In this simulation, the delay ratios between successive classes are measured using five time intervals τ, i.e., 100, 500, 1000, 5000, and 10000 p-units, where a p-unit is the average packet transmission time. After the delay ratios between all pairs of successive classes are measured, all data for each time interval are averaged to obtain the mean delay ratio.

Figure 5 shows five percentiles, 50%(median), 5%, 25%, 75%, and 95%, of the average delay ratio when the link utilization is 70%. Both schedulers have broad ranges in a short timescale and condensed ranges in a long timescale. Also, WTP has the more concentrated ranges than AWTP, yielding more predictable behavior. However, these concentrated ranges are far away from the target ratio. The AWTP scheduler has broader ranges, but nearer the target, and this situation is preferred.

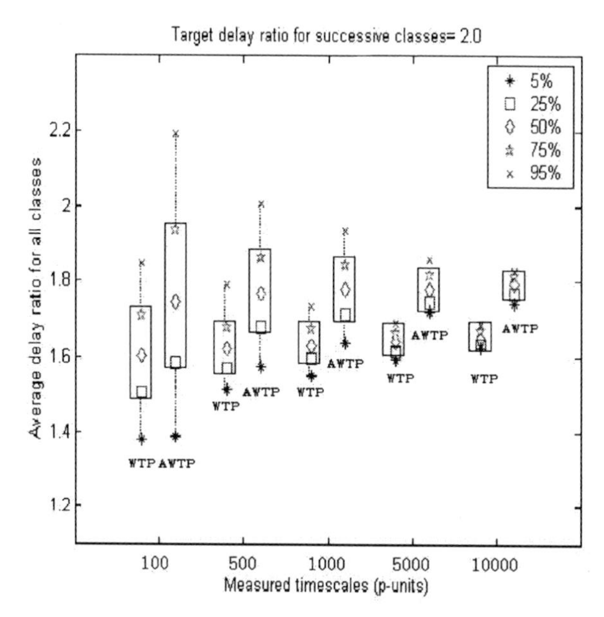

Fig. 5. Five percentiles of the average delay ratios under various measured timescales

4.2 Case 2:

The effects of the average packet size and the variance of the packet sizes on AWTP must be evaluated since the AWTP algorithm accomplishes proportional delay differentiation by considering the packet transmission time.

4.2.1 Average Packet Size

The variance of the packet sizes is fixed to 100,000 and the average packet size ranges from 200 to 550 bytes. In Fig. 6, the AWTP scheduler increases R_{31} and R_{21} with packet size while WTP keeps them stable. Thus, the gap between AWTP and WTP becomes larger. When the average packet size increases, the effect of considering transmission time is more obvious for the AWTP scheduler, yielding a more correct proportion.

As mentioned above, AWTP prefers class-1 packets. AWTP will more strongly prefer class-1 packets as packet size increases. The preference causes the queueing delay of class-1 packets to fall and that of class-3 packets to rise. For class 3, the improved queuing delay ratio becomes negative when the gain from the small packet preference is less than the loss due to the class-1 packet preference. As the packet size becomes larger, although the improved delay ratio of class 1 increases, the improved delay ratios of classes 2 and 3 decrease, resulting in a decline in the overall improved delay ratio, as shown in Fig. 7.

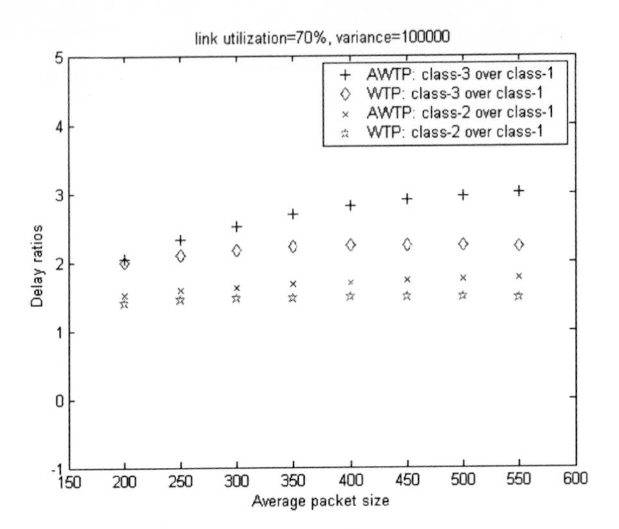

Fig. 6. Delay ratios for various average packet sizes

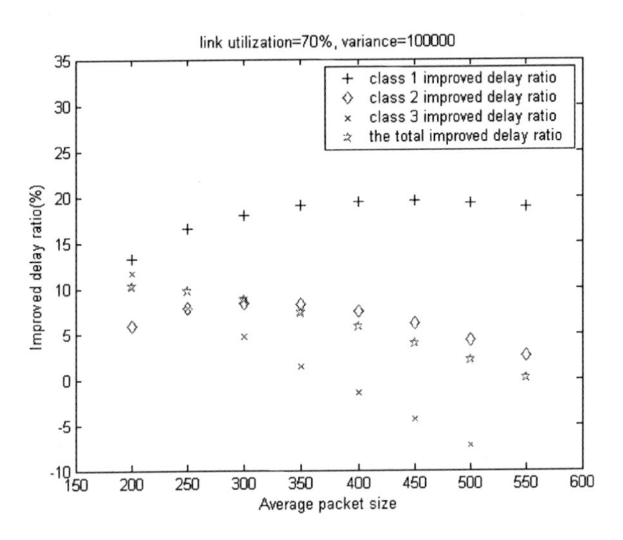

Fig. 7. Improved delay ratios for various average packet sizes

The simulation reveals that AWTP can give a more accurate delay ratio when the packet size increases, but it can not increase the total improved delay ratio without limit.

4.2.2 Variance of Packet Sizes

The average packet size is fixed at 600 bytes and its variance ranges from 100,000 to 350,000. In Fig. 8, the total improved delay ratio increases with the variance of packet size. When the variance of the packet size is increasing, the difference between the packet sizes grows. The preference for small packets is an important characteristic of the AWTP scheduler. Hence, a larger difference in packet size corresponds to a greater benefit in using the AWTP. Normally, the individual improved delay ratio for each class also increases.

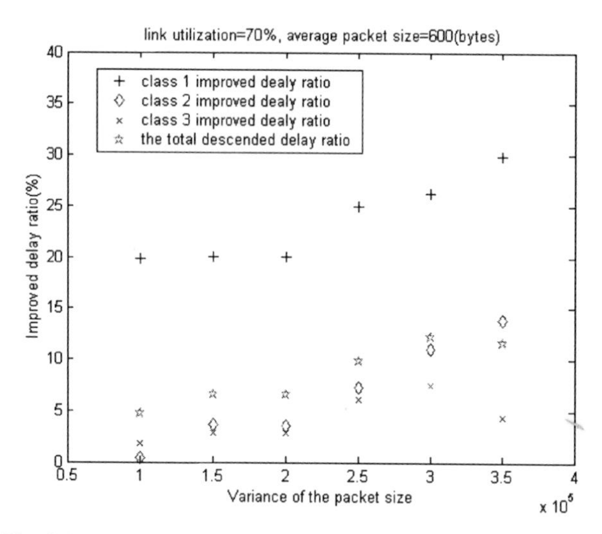

Fig. 8. Improved delay ratios for various variances of packet sizes

5 Conclusions

This research presented the AWTP scheduler to achieve proportional delay differentiation. The AWTP algorithm can maintain the delay proportion and reduce the average queuing delay by considering the packet transmission time. The simulation demonstrates that AWTP is the same as WTP when the traffic load is heavy and outperforms WTP when link utilization is moderate. AWTP actually achieves delay ratio near the target in long and short timescales, and reduces the average queuing delay. Finally, the effects of the mean packet size and the variance of the packet sizes, are examined to compare the performance of both schedulers. The AWTP scheduler always outperforms the WTP scheduler.

The only shortcoming of AWTP is that its computation complexity is $O(N^2)$, greater than $O(N)$, the complexity for WTP. However, this problem is not serious for two reasons. (1) In Differentiated Services, the flows with the same service degree are multiplexed to a queue by marking packets with the same Differentiated Services Code Point (DSCP). The limitation on the number of DSCP constrains the number of service classes. Consequently, N is usually small and $O(N^2)$ does not far exceed $O(N)$. (2) Currently, the bottleneck of the network is typically caused by the limitation of the link capacity, rather than the lack of the router's computation capacity. Accordingly, accelerating the transmission with some increase in computation time overhead is still profitable and worthwhile.

References

1. R. Braden, D. Clark, S. Shenker: Integrated Services in the Internet Architecture: an Overview. RFC 1633 (1994)
2. S. Blake, D. Black, M. Carlson, E. Davies, Z. Wang, and W. Weiss: An Architecture for Differentiated Services. RFC 2475 (1998)
3. K. Nichols, S. Blake, F. Baker, and D. L. Black: Definition of the Differentiated Services Field (DS Field) in the IPv4 and IPv6 Headers. IETF RFC 2474 (1998)
4. C. Dovrolis, D. Stiliadis, and P. Ramanathan: Proportional Differentiated Services: Delay Differentiation and Packet Scheduling. ACM SIGCOMM (1999)
5. C. Dovrolis, and P. Ramanathan: A Case for Relative Differentiated Services andthe Proportional Differentiation Model. IEEE network (1999)
6. H. Saito, C. Lukovszki, I. Moldovan: Local Optimal Proportional Differentiation Scheduler for Relative Differentiated Services. Computer Communications and Networks (2000)
7. L. Kleinrock: Queueing Systems, Volume 2: Computer Applications. Wiley-Interscience (1976)
8. Y. Matsumoto, Y. Takahashi, and T. Hasegawa: The effects on Packet Size Distributions on Output and Delay Processes of CSMA/CD. IEEE Transactions on Communications, Vol. 38, NO. 2 (1990)
9. K. H. Leung, John C. S. Lui, and David K. Y. Yau: Characterization and Performance Evaluation for Proportional Delay Differentiated Service. Network Protocols (2000)

Load Balancing Using Modified Cost Function for Constraint-Based Routing over MPLS Networks*

Sung-Dae Kim[1], Hahnearl Jeon[1], Jai-Yong Lee[1],
Yong-Joon Kim[2], and Hyung-Taek Kim[2]

[1] Department of Electric & Electronic Engineering, Yonsei University,
134 Shinchon-dong Seodaemun-gu Seoul, Korea
{sungdk, bird, jyl}@nasla.yonsei.ac.kr

[2] Mobile Telecommunication Network Group, LG Eletronics,
533 Hogye-dong Dongan-gu Anyang, Korea
{yjkim, htkim}@lge.com

Abstract. As the traffic grows rapidly in the Internet, the QoS (Quality of Service) guarantee of the flows and the traffic engineering problems have become very important issues. MPLS (MultiProtocol Label Switching) has more advantages to solve the problems than existing IP routing because of more free using of multiple paths between the source and destination pairs. Particularly the availability of Constraint-based Routing and explicit route in MPLS made the problems referred above solved efficiently. But the CSPF algorithm used in Constraint-based Routing has the characteristic that it selects the shortest path of all paths which meet the traffics' QoS constraints. So, even though Best-effort traffic which is only routed to the shortest path can be routed to alternative path, it is dropped by the intermediate nodes during congestion period. Therefore, if we reroute the best-effort traffic to the alternative path when congestion happens, the network resource could be used efficiently and the load on intermediate router in the path can be less. In this paper, we present the modified cost function used in path computation for best-effort traffic and verify that the using of that cost function make traffic engineering more effective through simulation.

1 Introduction

Traffic Engineering (TE) is the technique of controlling how traffic flows through network in order to optimize network resource (for example, bandwidth) utilization and network performance.

* This work was supported by grant No. 1999-2-303-005-3 from the interdisciplinary Research program of the KOSEF and LG, Co. Ltd.

I. Chong (Ed.): ICOIN 2002, LNCS 2343, pp. 539-548, 2002.

A major goal of Internet Traffic Engineering is to facilitate efficient and reliable network operations while simultaneously optimizing network resource utilization and traffic performance [1].

Traffic Engineering is needed in the Internet mainly because in today's Internet, existing IP routing is generally based on the destination address of the traffic and simple metrics such as hop-count or delay, and Shortest Path First (SPF) algorithm such as dijkstra or bellman-Ford algorithm. Though the simplicity of this approach allows existing IP routing to scale to very large networks, the utilization of network resources can be usually poor. Existing destination-based IP routing sometimes can result in unbalanced traffic distribution by using shortest path between source and destination pairs. And it made the satisfaction to QoS of traffic very difficult. Therefore, Traffic Engineering has become an essential requirement to optimize the utilization of network resources and, at the same time, to satisfy a desired QoS requirement of traffic.

In order to do Traffic Engineering effectively, the Internet Engineering Task Force (IETF) introduces Multi-Protocol Label Switching (MPLS).

Basic MPLS concepts can be explained as follows :

(1) MPLS uses fixed length label to represent some information about packets without using IP address

(2) A label is used to locally identify data packets that require equivalent forwarding

(3) The fixed-length label representation of forwarding information simplifies decisions based on using an exact-match algorithm. This simplified forwarding paradigm is called label switching.

(4) A router that takes part in label switching is a label switching router (LSR).

(5) Labels are added at an MPLS ingress node, swapped at intermediate LSRs (Label swapping), and removed at an MPLS egress node. The path along which this occurs for any particular starting label is called a label-switched path (LSP)

With MPLS, one can apply the traditional traffic engineering technique to compute and set up label switched paths, and one can then assign traffic to these paths based on the combination of destination address, source address, and possibly other information such as port numbers by using extended Link-state routing protocol, Constraint-Based Routing (CBR), explicit route, label distribution protocol [2][3][4][5]. The packets entering the ingress node will be distributed toward the various paths according to the specified patterns, leading to better network utilization.

In IP routing there is only one routing path computation algorithm (SPF) for traffic. On the other hand, In MPLS, roughly speaking, there are two kinds of routing path computation algorithms : Shortest Path First algorithm which is the same method in IP routing and Constraint-based Shortest Path First (CSPF) algorithm. In this paper we call this routing which use two different path computation algorithms "Constraint-based Routing (CBR)"

The SPF is for best-effort traffic and the other is for QoS traffic in MPLS.

The rest of this paper is organized as follow. Section 2 overview the Constraint-Based Routing which includes SPF algorithm and CSPF algorithm. Section 3 displays problem statement in CBR and Section 4 shows our proposal to distribute the traffic when the congestion arises. Simulation and the result are illustrated in Section 5. And the conclusion is given in Section 6.

2 The Overview of SPF and CSPF Algorithm in Constraint-Based Routing (CBR)

Formally, Constraint-based Routing (CBR) is defined as follows. Consider a network that is represented by a graph (V,E), where V is the set of nodes and E is the set of links that interconnect these nodes. Associated with each link is a collection of attributes. For each pair of nodes in the graph, there is a set of constraints that have to be satisfied by the path from the first node (ingress node) in the pair to the second one (egress node). This set of constraints is expressed in terms of the attributes of the links and is usually known only by the node at the head end of the path (the first node in the pair, namely ingress node). The goal of constraint-based routing is to compute a path from one given ingress node to a egress node, such that the path doesn't violate the constraints and is optimal with respect to some scalar metric.

Once the path is computed, constraint-based routing is responsible for establishing and maintaining forwarding state along such a path using signaling protocol such as CR-LDP or RSVP.

The main difference between conventional IP routing and constraint-based routing is as follows.

Conventional IP routing algorithms aim to find a path that optimizes a certain scalar metric (for example, minimizes the number of hops), while constraint-based routing algorithms set out to find a path that optimizes a certain scalar metric and at the same time does not violate a set of constraints. It is precisely the ability to find a path that doesn't violate a set of constraints that distinguishes constraint-based routing from plain IP routing.

As we mentioned above, constraint-based routing requires the ability to compute a path, such that the path :

(1) Is optimal with respect to some scalar metric (for example, hop count)

(2) Does not violate a set of constraints such as bandwidth, administrative policy and so on.

Constraint-based Routing (CBR) computes routes that are subject to constraints such as bandwidth and administrative policy. Because Constraint-based Routing considers more than network topology in computing routes, it may find a longer but lightly loaded path better than the heavily loaded shortest path. In the long run network traffic can be distributed more evenly.

In general, the constraint-based routing problem is known to be intractable for most realistic constraints. However, in practice, a very simple well known algorithm (in this paper, we call this algorithm "Constraint-based Shortest Path First (CSPF) algorithm") can be used to find a feasible path if one exists:

Briefly speaking, CSPF algorithm consists of two steps.

- First eliminate resources that do not satisfy the requirements of the traffic trunk attributes.

- Next, run a shortest path algorithm on the residual graph.

Let's compare CSPF algorithm with Shortest Path First (SPF) algorithm used in the traditional IP routing. For example in Figure 1, in case we use SPF algorithm, the shortest path between node 1 and node3 is through link 1-3 with 1 hop. But when the reserved bandwidth of link 1-3 is 100Mbps, and that of link 1-2 is 20Mbps, finally that of link 2-3 is 30Mbps, we assume that a traffic enters Node 1 requiring the 60Mbps bandwidth. In that case, that traffic was routed via Node2 by the CSPF algorithm which was computed by ingress node Node1, because the shortest path (Node1-Node3) does not meet the bandwidth constraint.

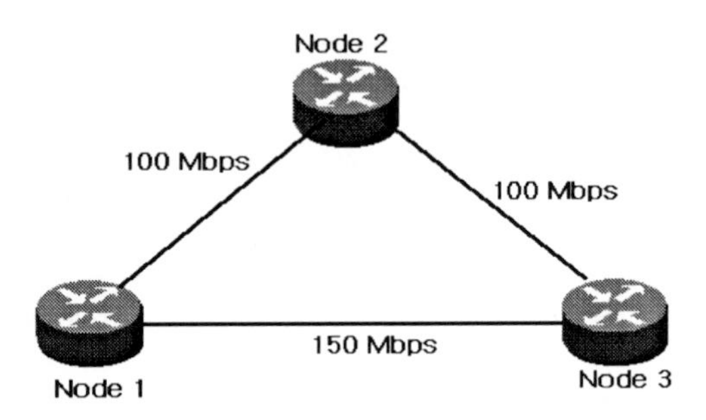

Fig. 1. The comparison of existing Routing using SPF algorithm and Constraint-based Routing using CSPF algorithm

It should be noted that the reservable bandwidth of a link is equal to the maximum reservable bandwidth set by network administrators minus the total bandwidth reserved by LSPs traversing the link. It does not depend on the actual amount of available bandwidth on that link. For example, if the maximum reservable bandwidth of a link is 150Mbps, and the total bandwidth reserved by LSPs is 100Mbps, then the reservable bandwidth of the link is 50 Mbps, regardless of whether the link is actually carrying 100 Mbps of traffic or more or less. In other words, CSPF algorithm does not compute LSP paths based on instantaneous residual bandwidth of links.

3 Problem Statement

As explained earlier, Constraint-base Shortest Path First (CSPF) algorithm in MPLS network is used to find a path satisfying constraint of the incoming traffic to Ingress node. In this paper, we assume that the constraint is only bandwidth for simplifying simulation environment.

Each Label Switched Router (LSR) within MPLS network uses SPF algorithm to find a shortest path to each destination of the incoming traffics with only one metric (ex, delay) cost function. When the shortest path between one source and destination pair is congested by the increasing traffic, QoS traffic and simple best-effort traffic is intermixed in that path, the best-effort traffic have no choice but to be dropped on the way traversing the path. The reason is why Ingress node reserves the resource of the links along the path taken by using CSPF algorithm and the intermediate LSRs in that path guarantee the bandwidth requirement of QoS traffic. The LSRs in the path taken by using SPF algorithm guarantee nothing for the best -effort traffic. When the shortest path is uncongested, no problems happen. But, when the shortest path grows congested and it was shared by best-effort traffic and QoS traffic, the performance of the best-effort traffic was degraded even though there are alternative non-shortest paths to the destination. And because CSPF algorithm is also a kind of SPF algorithm, even though there are paths having more capacity than the shortest path, the shortest path was used as long as it meets the constraints of the traffic.

4 Proposed Method to Get a Efficient Load Balancing Effect

Ingress node performs two kinds of algorithm - CSPF algorithm for QoS traffic and SPF algorithm using Modified cost function for best-effort traffic. And LSRs in the MPLS network performs only one algorithm - SPF algorithm using Modified cost function for best-effort traffic. The modified cost function is as follows :

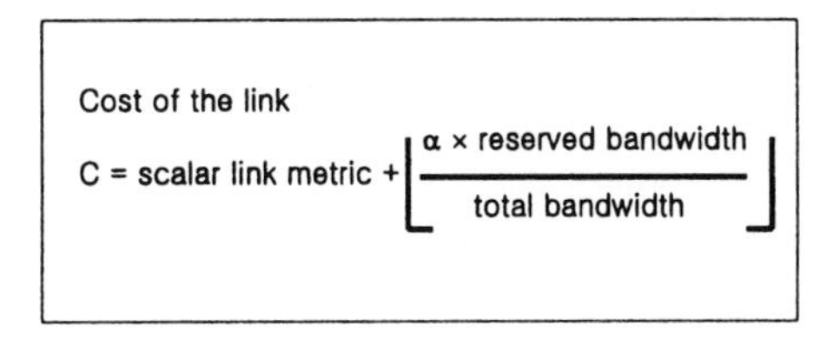

Fig. 2. Modified Cost Function for best-effort traffic suggested in this paper : where α is the constant value (bias value) which made $\lfloor\ \rfloor$ part value be 0 or positive value but α can be selected randomly by the network administrator and $\lfloor X \rfloor$ = the greatest integer less than or equal to X. ie, $\lfloor 1.2 \rfloor = 1$.

The second part of the cost formula considers the network congestion.

As the link bandwidth is reserved by QoS traffic, the link cost grows to increase and each router finds the next hop different to the conventional SPF using cost function considering only one metric such as delay.

If the second part ⌊ ⌋ of the above cost function is omitted, cost function is equal to that of the existing SPF.

Figure 3 shows the flow diagram of switching sequence of traffic incoming to MPLS network in this paper.

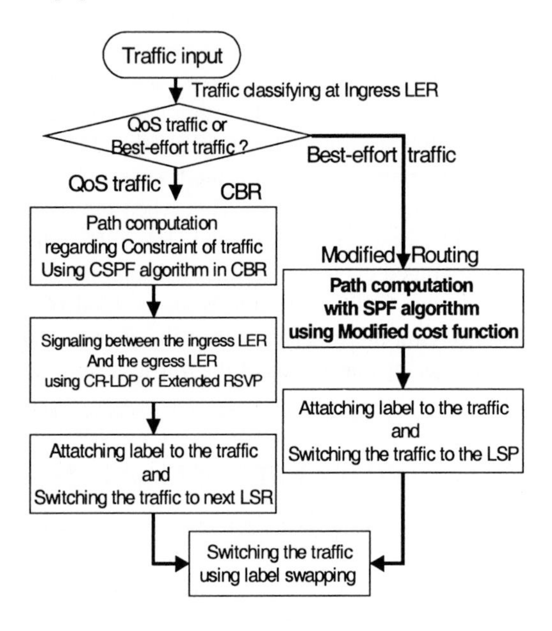

Fig. 3. The switching sequence of incoming traffic in MPLS Networks

When the shortest path is congested, because the LSRs and Ingress node know the reserved bandwidth information of the path, each router recomputes the shortest path using SPF with modified cost function and reroutes the best-effort traffic to new computed shortest path having lower cost.

Finally, we will get a load balancing effect and network optimization effect.

5 Simulation Result

We used Network Simulator version 2 (NS2) [6] and the MPLS module (MNS) introduced in reference [7].

The network topology used in our simulation was shown in Figure 4. It consists of seven MPLS routers (from LSR1 to LSR7) and three IP nodes.

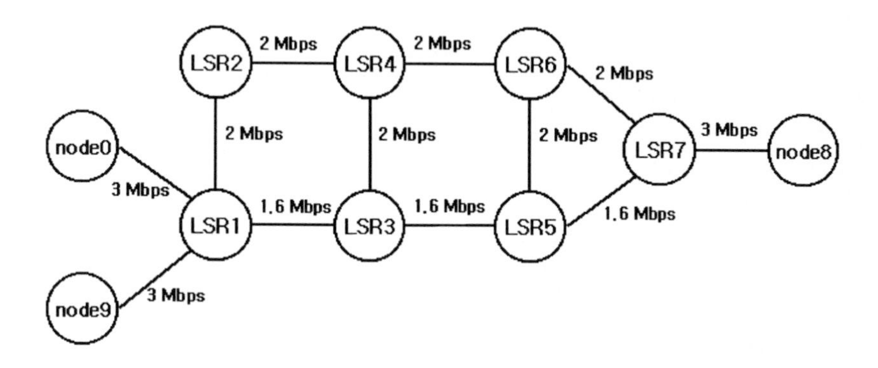

Fig. 4. Network Topology used for simulation

All traffic is the size equal to 200bytes/packet.

In this simulation, we regard the situation that a link was filled with 80% load as "congestion" and use the delay metric of the cost function as hop count. In the other words, the cost function of link is like this : cost of the link C = delay = hop count.

In the first simulation, Node 9 sends simple best-effort UDP traffic (350Kbps) destined for node 10 at 0.3sec. At 0.7sec, Node 9 also send best-effort UDP traffic (350Kbps). The two best-effort traffics are routed to the path node9 – LSR1 – LSR3 – LSR5 – LSR7 – node 8. And after, node0 sends src0 (bandwidth requirement : 800Kbps) at 3.5sec and src1 (bandwidth requirement : 700Kbps) at 4.8sec and finally src2 (bandwidth requirement : 600Kbps) at 6.0 sec. We assume that all traffics have the same destination node (node 8).

The simple best-effort traffic is routed by simple SPF algorithm and the resulting routing path is node9 - LSR1 - LSR3 - LSR5 - LSR7 – node8 because it is a shortest path between node9 and node8. But when the QoS traffic src0 was entered the Ingress node LSR1 (at 0.3sec), it is routed by CSPF algorithm and the resulting routing path is the same path as the path taken by the simple best-effort traffic and the second QoS traffic src 2 is routed to node8 via LSR2 – LSR4 – LSR6 –LSR7 – node 8 as a result of the CSPF computation at ingress node LSR1. The last QoS traffic src 3 is routed to the same path of the first QoS traffic (src0) because that path has the sufficient bandwidth to route (this situation can be called "congestion"). During this period, the simple best-effort traffics are dropped by the intermediate LSRs which only guarantee the QoS traffics (src0, src3). After the time interval 6.0sec and 17.5sec when the src3 stop, the bandwidth of the best-effort traffics increases to 350Kbps.

The result is shown in Figure 5.

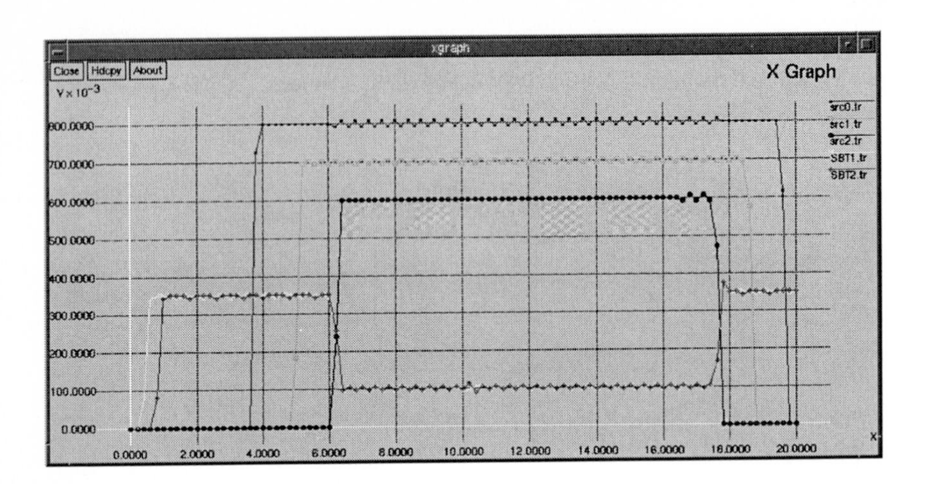

Fig. 5. Results using Normal CSPF (for QoS traffic) and SPF (for best-effort traffic) algorithm

As described earlier, Figure 5 shows that the bandwidth of the simple best-effort traffics (UDP source 350Kbps) decrease from 350Kbps to about 100Kbps which is the residual bandwidth after guaranteeing the bandwidth of traffic src0.

In the second simulation, the sequence of the simulation is the same as that of the previous one. The only difference is that all the LSRs in the MPLS network use SPF algorithm with modified cost function.

In this simulation, we also regard the situation that a link was filled with 80% load as "congestion" and use α value as 1.25 to make $\lfloor\ \rfloor$ value of the cost function be greater than 1.

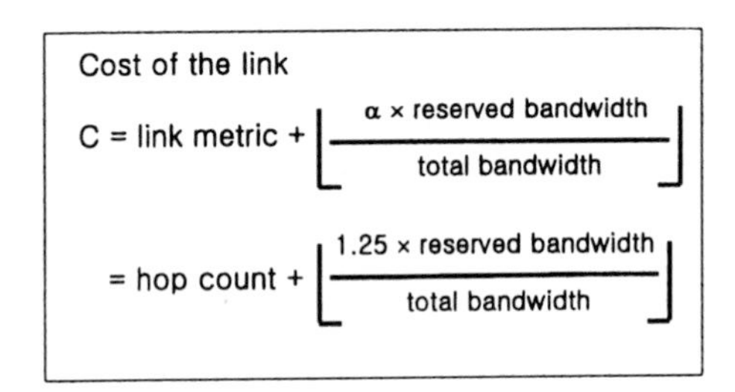

For example, when the load of one link is 70%, the cost function value is equal to delay value because $\lfloor 1.25 \times 0.7 \rfloor = \lfloor 0.875 \rfloor = 0$, and when the load of one link is 80%, the cost function value is equal to delay value + $\lfloor 1.25 \times 0.8 \rfloor$ = delay value + $\lfloor 1 \rfloor$ = delay value +1 and so on.

And we use the delay metric of the cost function as hop count.

As the QoS traffic's bandwidth reservation rate reaches to over 80% of the link, each LSRs recomputes the shortest path according to the modified cost function. when congestion happens, the shortest path becomes to be differ from the path computed by simple SPF algorithm in the previous simulation.

First, each simple best-effort traffic (350Kbps) at node 0 is sent to the node 8 at 0.3sec and 0.7 and QoS traffics (src0, src1, src2) is sent sequentially at 3.5sec, 4.8sec, and 6.0. The best-effort traffics, src0, and src2 traffic are mixed in the path LSR1 - LSR3 - LSR5 - LSR7 – node8 after 6.0sec. The other QoS traffic (src1) are routed to the same path as the path (node0 – LSR1 – LSR2 – LSR4 – LSR6 – LSR7 – node8) satisfied with their QoS constraints in the first simulation.

In this second simulation, between the time interval 6.0sec-17.5sec ("congestion period"), the LSRs compute new shortest path (LSR1 – LSR2 – LSR4 – LSR6 – LSR7 – node8) with the modified cost function and reroute best-effort traffics to that path.

The result is shown in Figure 6.

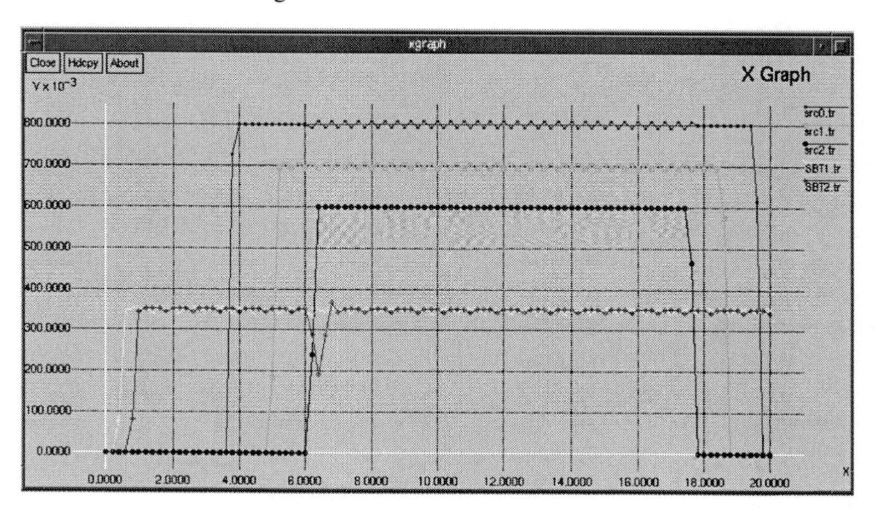

Fig. 6. Results using Normal CSPF (for QoS traffic) and Modified cost function in SPF (for best-effort) when congestion happens

After src2 traffic (600Kbps) entered the ingress node LSR1(at 6.0sec), the bandwidth of the best-effort traffic is transiently decreased owing to recomputation of shortest path and after that short time interval, the bandwidth of each best-effort traffic grow to original bandwidth and the other QoS traffics are guaranteed as earlier simulation. The table below is the received packet number at node8.

Table 1. The comparison of the received packet number at node8

	Existing SPF	Modified SPF
Src 0	7968 packets	7968 packets
Src 1	5957 packets	5957 packets
Src 2	4288 packets	4288 packets
SBT1	**2514 packets**	**4156 packets**
SBT2	**2429 packets**	**4125 packets**

As shown in the table, we know that in the second simulation destination node8 has more packet than in the first simulation.

Finally, comparing the first simulation results with the second simulation results, in the second simulation best-effort traffic has more bandwidth and the network resource (link utilization) was used more efficiently than the first simulation.

6 Conclusion

In this paper we suggested the idea to reroute the best-effort traffic to optimize the network capacity in MPLS network by using modified cost function. CSPF algorithm works well for satisfying the QoS requirement of traffic in MPLS. But there exists congestion situation mixed with QoS traffic and simple best-effort traffic. When there are residual bandwidth after reserving bandwidth for QoS traffic in the Shortest path, the bandwidth of the shortest path is always used for QoS traffic. In that case, though there are alternative paths to route best-effort traffic, it can't be routed to alternative paths only by using Shortest Path First (SPF) algorithm such as Dijkstra algorithm. But, By using modified cost function proposed by us we can route or reroute simple best-effort traffic (load-balancing) when congestion happens and thereby get the optimization effect.

References

1. Awduche, D., et al., "Requirements for Traffic Engineering over MPLS" RFC 2702, Sep 1999
2. Awduche, D., "MPLS and Traffic Engineering in IP Networks." IEEE Communications Magazine (vol. 37, no.12) Dec 1999
3. Ghanwani, A., B. Jamoussi, D.Fedyk, P. Ashwood-Smith, L.Li, and N. Feldman. "Traffic Engineering Standards in IP Networks Using MPLS." IEEE Communications Magazine (vol. 37, no.12) Dec 1999
4. Swallow, G. "MPLS Advantages for Traffic Engineering." IEEE Communications Magazine (vol. 37, no. 12) Dec 1999
5. Bruce Davie and Yakov Rekhter : "MPLS Technology and Applications" , Morgan Kaufmann Publishers, 2000
6. URL : http://www-mash.cs.berkeley.edu/ns/
7. URL : http://www.raonet.com

Performance Analysis of AAL2 Trunking/Switching for Digital Telephone Switching Systems

Seong-Ho Jeong[1] and John Copeland[2]

[1] Department of Information and Communications Engineering
Hankuk University of F. S., Korea
shjeong@hufs.ac.kr

[2] School of Electrical and Computer Engineering
Georgia Institute of Technology, Atlanta, GA, USA
copeland@ece.gatech.edu

Abstract. To satisfy the market need for voice over ATM (VoATM), ATM Adaptation Layer Type 2 (AAL2) has been defined in the ITU-T Recommendation I.363.2. AAL2 is designed with the capabilities to support multiple voice users per ATM connection and to transport compressed/silence-suppressed variable-bit-rate (VBR) voice traffic. Recently, there is a growing interest in the impact of AAL2 trunking/switching on the digital telephone switching systems. This paper analyzes the performance and the feasibility of AAL2 trunking/switching for VBR voice transport when the traditional digital telephone switching system is replaced by the AAL2 trunking/switching system.

1 Introduction

ATM technology has gained popularity as a key technology for implementing high-speed backbone network that can efficiently transport data with different traffic characteristics and service quality requirements. One of the questions that should be answered for public network operators is the actual impact of ATM technology on voice transport.

Basically, there are three standards-based methods that can be used for the transport of voice traffic over ATM networks: unstructured and structured circuit emulation using AAL1, and variable-bit-rate (VBR) voice transport using AAL2.

Unstructured circuit emulation allows the user to establish an ATM connection to emulate a circuit, such as a full T1 (1.544Mbps) or E1 (2.048Mbps), over the ATM backbone. Structured circuit emulation establishes an ATM connection to emulate N x 64 Kbps circuits, such as a fractional T1 or E1, over the ATM backbone. Voice transport using AAL2 allows multiple users to share one ATM connection to transport VBR voice traffic by taking advantage of statistical multiplexing.

In the development of ATM standards, AAL1 found its niche as a way to allow ATM to replace Time Division Multiplexing (TDM) circuits at fixed rates such as T1 or E1. The use of AAL1 was subsequently extended to allow replacement of 64 Kbps

I. Chong (Ed.): ICOIN 2002, LNCS 2343, pp. 549-560, 2002.

circuits (or traditional digital voice circuits), providing a means to convey voice on ATM backbones instead of TDM infrastructures. Although AAL1 was designed to support time-critical constant-bit-rate (CBR) circuits used to transport voice applications, AAL1 was not developed to optimize bandwidth efficiency for voice applications. On the other hand, the cooperation between the ATM Forum, which identified market needs, and the protocol experts at the ITU-T resulted in a new AAL that is much better suited for voice-over-ATM (VoATM) applications: AAL2.

Recently, there is a growing interest in the impact of AAL2 trunking/switching on the digital telephone switching systems. In this paper, we evaluate the performance and the feasibility of AAL2 trunking/switching when the traditional digital telephone switching system is replaced by the AAL2 trunking/switching equipment. The rest of the paper is organized as follows. Section 2 briefly describes the key characteristics of AAL2, and then Sections 3 and 4 present the performance and the feasibility of AAL2 trunking/switching in detail. Finally, Section 5 gives a brief summary of the paper.

2 ATM Adaptation Layer Type 2 for VBR Voice Transport

The basic functions of the AAL2 protocol are consistent with AAL1, in that both protocols enhance the service provided by the ATM layer to support the next higher layer user. However, AAL2 goes beyond AAL1 by defining a structure that includes functions supporting higher layer requirements neither considered nor possible within the structure of AAL1.

AAL2 provides for the bandwidth-efficient transmission of low-bit-rate, short, and variable packets in delay sensitive applications [4].

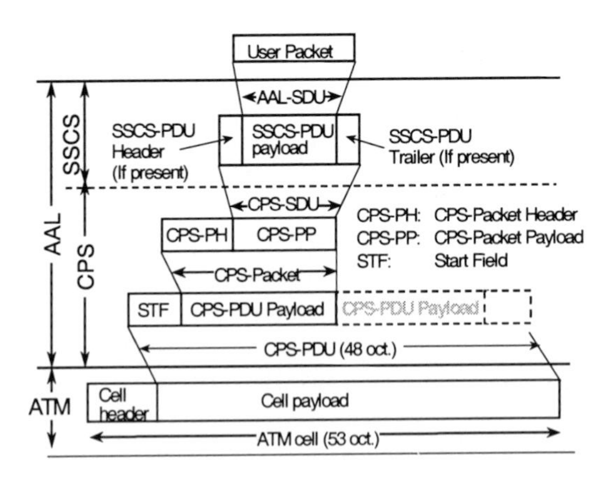

Fig. 1. Data unit naming conventions for AAL2

Figure 1 shows the naming conventions of AAL2 data units. AAL2 is subdivided into the Common Part Sublayer (CPS) and Service Specific Convergence Sublayer (SSCS). Different SSCS protocols may be defined to support specific AAL2 user services, or group of services. The SSCS may also be null, merely providing for the mapping of the equivalent AAL primitives to the AAL 2 CPS primitives. Each user information is mapped into a CPS-Packet at the CPS level. A CPS-Packet consists of a 3-octet CPS-Packet Header (CPS-PH) followed by a CPS-Packet Payload (CPS-PP).

Multiple user channels can be carried on a single ATM virtual circuit using AAL2 with varying traffic conditions for each individual user, or channel. The structure of AAL2 provides for the packing of short length packets into one (or more) ATM cells, and the mechanisms to recover from transmission errors. AAL2 allows a variable payload within the CPS-packet, which makes it possible to support various voice users. This functionality provides a dramatic improvement in bandwidth efficiency over either structured or unstructured circuit emulation using AAL1. Furthermore, VBR voice services using compression and silence detection/suppression suggest great potential for highly efficient voice transport over ATM networks.

In summary, compared with AAL1, AAL2 provides the following advantages:

- Support for compression and silence suppression
- Support for VBR voice services and high bandwidth efficiency
- Support for multiple user channels with varying bandwidth on a single ATM connection

3 Performance Analysis of AAL2 Trunking

In this section, we analyze the performance and the feasibility of AAL2 trunking for the digital telephone switching system.

3.1 Trunking Performance – Average Analysis

Figure 2 shows the system model for AAL2 trunking where compressed and silence-suppressed CPS-Packets originated from multiple users are fed into a FIFO queue for transmission via a T1 ATM trunk.

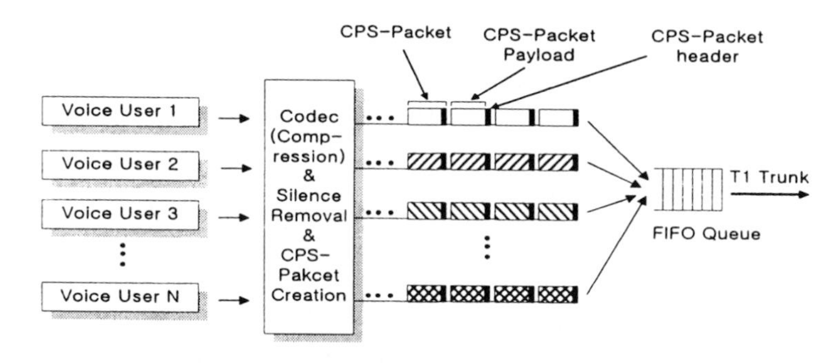

Fig. 2. System model for AAL2 trunking

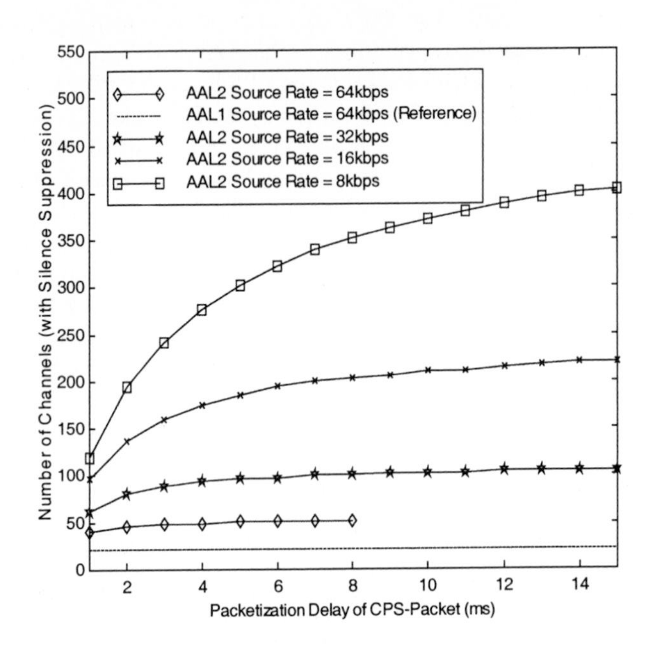

Fig. 3. AAL2 trunking performance (effect of compression and silence suppression)

Figure 3 illustrates the effect of applying both compression and silence suppression to voice sources. The voice activity factor was assumed to be 0.35. The data rate of voice source varies from 64 Kbps to 8 Kbps. As shown in the figure, AAL2 can realize at least 90% more efficiency than its counterpart AAL1 due to silence suppression. Note that the efficiency is given in terms of the number of channels which can be supported over a T1 ATM trunk. Compression scheme makes the efficiency even better. Therefore, AAL2 which can support compression and silence suppression is definitely an attractive alternative to AAL1. In addition, we observe, from Figure 3, that the more the source rate is reduced via compression, the more the utilization is dependant on the packetization delay.

3.2 Trunking Efficiency – Uniform Arrival/Service Model

In this section, the number of voice sources that can be supported on a given ATM trunk is analyzed using the "uniform arrival and service" model [1]. The multiplexing system model used for our analysis is based on Figure 2, where there are N independent voice sources. Our source model is based on the assumption of the two-state, on-off process. All voice sources are homogeneous, which implies that every source has the same traffic distribution characteristics. Each voice source generates information when it is 'on' at a rate of 1 unit of information per unit of time. The unit of time is defined to be the average duration of a talkspurt. The duration of the average silence period is denoted by $1/\lambda$. Thus, the voice activity factor can be calculated as $\lambda/(1+\lambda)$. The information generated by active sources is fed into the finite common queue or buffer where it is served at a rate of c, the link capacity. The lengths of both on and off periods are assumed to be exponentially distributed.

Let $P_i(t,b)$, $0 \le i \le N$, $t \ge 0$, $b \ge 0$, be the probability that i sources are active and buffer content does not exceed b at time t, where N is the number of voice sources multiplexed into a fixed-bandwidth trunk. The probability at time "$t+\Delta t$", $P_i(t+\Delta t,b)$, can be expressed as follows:

$$P_i(t+\Delta t,b) = \{N-(i-1)\}\lambda\Delta t P_{i-1}(t,b) + (i+1)\Delta t P_{i+1}(t,b) + [1-\{(N-i)\lambda + i\}\Delta t]P_i\{t,b-(i-c)\Delta t\} + 0(\Delta t^2)$$

where c is the trunk capacity. The first three terms on the right hand side correspond to the contributions from the states of "on" and "off" periods, and no change. By taking limit $\Delta t \rightarrow 0$, and setting $\delta P_i / \delta t = 0$, we obtain:

$$(i-c)\frac{dF_i(b)}{db} = (N-i+1)\lambda F_{i-1}(b) - \{(N-i)\lambda + i\}F_i(b) + (i+1)F_{i+1}(b)$$

where $F_i(b)$ is the steady state probability that i sources are on and the buffer content does not exceed "b". We can solve the set of differential equations to obtain each $F_i(b)$, and further derive the buffer overflow probability as:

$$G(b) = Pr(\textit{buffer content} > b) = 1 - \mathbf{1}'\mathbf{F}(\mathbf{b}), \quad b \ge 0$$

where $\mathbf{1}$ denotes the vector with unity for all its components and prime denotes transposition. Based on this formula, we can obtain CPS-Packet loss probability by approximating it as the buffer overflow probability $G(b)$.

Figure 4-(a) shows the impact of voice activity factor on the trunking performance given the computed packet loss probability. The number of voice sources supported over a T1 ATM trunk ranges from about 33 to 100, and the voice activity varies from 0.2 to 0.4. CPS packetization delay is assumed to be 4 ms. An increase in the voice activity factor indicates either an increase in the mean talkspurt length or a decrease in the mean silence length. As shown in the figure, as the voice activity factor increases, the packet loss probability increases. Therefore, given a maximum packet loss probability, the maximum number of voice sources carried over a T1 ATM trunk decreases as the voice activity increases. In other words, the increase in the voice activity factor causes a lower trunking efficiency.

Figure 4-(b) shows the relationship between the packet loss probability and packetization delay of CPS-packet. In the figure, the data rate of voice sources is 32 Kbps (ADPCM). As shown in the figure, given a packet loss probability bound, the maximum number of voice sources supported over a T1 ATM trunk increases as the packetization delay increases. Therefore, it is important to specify an appropriate user requirement for the packetization delay to maximize the trunking efficiency while meeting the desired packet loss objective.

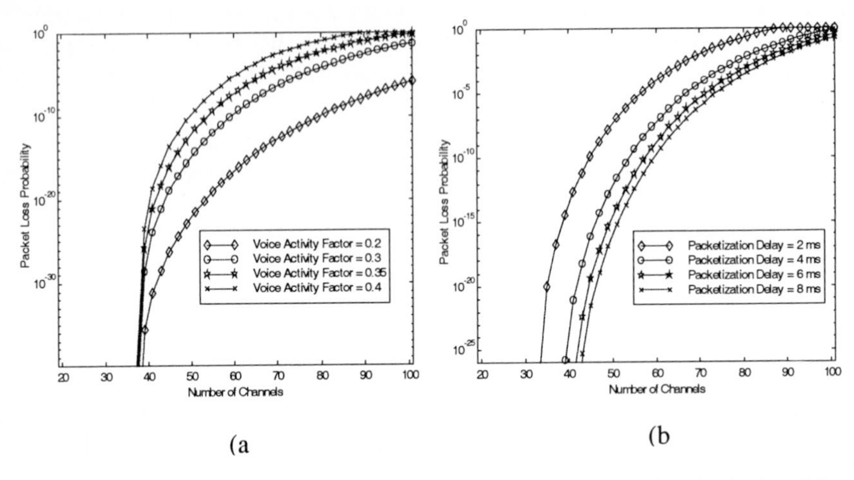

(a) Effect of voice activity factor and (b) Effect of packetization delay

Fig. 4. Packet loss performance

4 Performance Analysis of AAL2 Switching

In this section, we analyze the performance of AAL2 switching. The desire to provide seamless connectivity and to avoid double transcoder insertion is driving the AAL2 switching. Figure 5 shows AAL2 switching operations within an ATM (AAL2) switching node. AAL2 switching is performed at the AAL level where packets that are encapsulated within an ATM cell are switched based on their VPI/VCI/CID (Channel Identifier) values. The traditional VPI/VCI table used for ATM cell switching is extended one more level by introducing CID entries to identify AAL type 2 connections. An ATM cell received at an AAL2 switch is demultiplexed by retrieving CPS-Packets

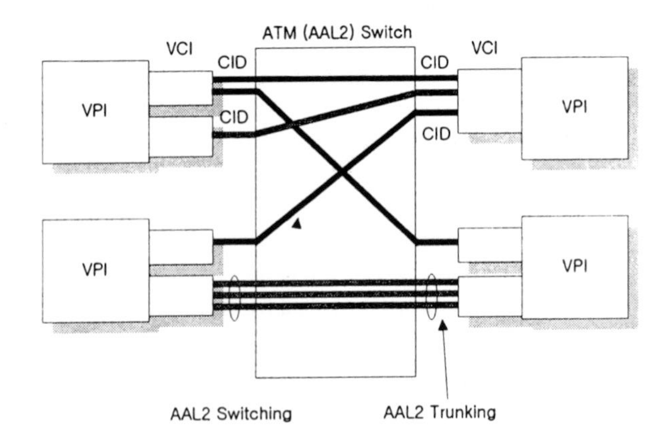

Fig. 5. AAL2 switching operations

(partial and full) that are encapsulated within the ATM cell. Any complete CPS-Packet is then assembled into an outgoing ATM cell according to the entries found in the VPI/VCI/CID table. The VPI/VCI/CID table entries are updated during the time of AAL2 connection setup (AAL2 signaling). If an AAL2 connection is routed through ATM switches, which do not support AAL2 switching then it is considered to be AAL2 trunking and those switches support only ATM level switching.

4.1 Digital Switching System Hierarchy

In this section, we describe general digital switching system hierarchy which is used as a reference model for our simulation in the paper. Calls through the North American network follow a hierarchical path. The search for a path through the network for a long-distance call follows a hierarchy similar to that in Figure 6. After a call leaves a class 5 switch, a path is hunted through the class 4 office followed by class 3, class 2, and class 1. In addition, there are international gateway offices (extension of class 1) which a central office calls to complete international-destination calls through cables, satellite, or microwaves. Figure 6 also shows the different classes of switching system in the North American switching network.

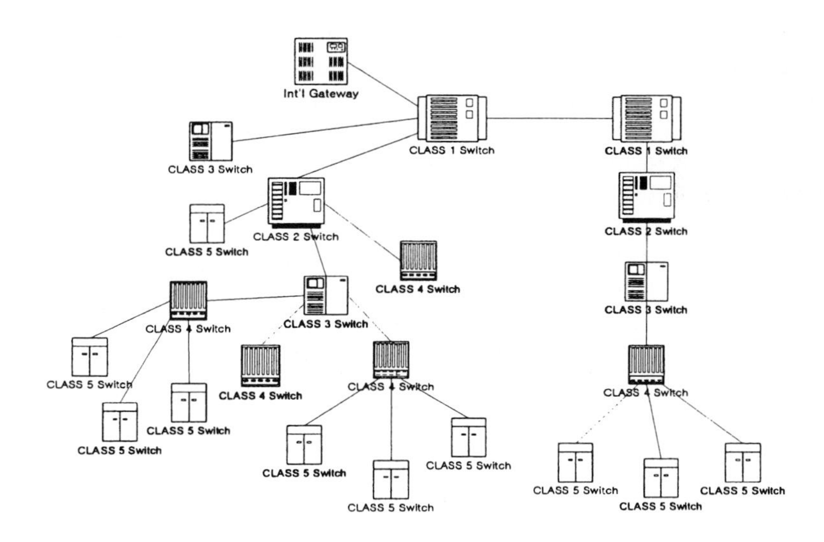

Fig. 6. Digital switching hierarchy

The following describes the key characteristics of each switch which is replaced by ATM/AAL2 switching equipment in the paper.

- Local exchange (class 5). It is also referred to as the end office (EO). It interfaces with subscribers directly and connects to tall centers via trunks. It records subscriber billing information.
- Tandem and toll office (class 4). Most class 5 Central Offices interface with the tandem offices. The tandem offices primarily switch trunk traffic between class 5 offices; they also interface with higher-level toll offices. Toll operator services can be provided by these offices.
- Primary toll center (class 3). The class 3 toll center can be directly served by class 4 or class 5 offices, depending upon the trunk deployment. In other words, if the normal number of trunks in these offices are exhausted, then traffic from lower-hierarchy offices can home into a class 3 office. Class 3 offices have the capability of storing, modifying, prefixing, translating, or code-converting received digits as well as finding the most efficient routing to higher-level toll offices.
- Sectional toll center (class 2). It functions as a toll center and can home into class 1 offices.
- Regional toll center (class 1). It functions as a toll center and can home into international gateway offices.
- International gateway. These offices have direct access to international gateway offices in other countries. They also provide international operator assistance.

The advantage of the hierarchical network is that it provides an efficient way of searching for a path through the network. The disadvantage is that if the primary, sectional, or regional toll center goes down, then large areas of North America may become inaccessible. The above hierarchy is used as a reference network model for our simulation in the next section.

4.2 Simulation Results

We performed simulations (we used OPNET) to analyze the performance of AAL2 switching. The simulation model is based on the results from the performance analysis in Section 3. The key parameters for simulation and values of each parameter are as follows. The number of voice sources is 50, and the data rate of each voice source is 64 Kbps. Voice activity factor is assumed to be 0.35, and silence is suppressed at the AAL2 CPS level. The default service rate of class 3 switch is T3 (45Mbps), and the default service rate of class 4 switch is T3 as well. Finally, the default service rate of class 5 switch is T1 (1.544 Mbps), and the default value of Timer_CU (Composite User) is 0.5 ms. At the CPS transmitter, Timer_CU is used to guarantee the maximum waiting time for CPS-Packets which have been already packed but not transmitted because the ATM cell has not been completely filled up.

Figure 7 shows a process model for each voice source. A voice source is 'on' when the talker is actually speaking. Initially, voice source is 'on', and during 'on' periods, the voice source generates user packets at regular intervals ('on' state). During silence periods, the voice source is inactive ('off' state) and generates no CPS-Packets.

Figure 8 shows the basic process model for ATM (AAL2) switches. Whenever any user packet (AAL2 SDU (service data unit)) arrives at the ATM (AAL2) switch, it creates CPS-Packets based on the user requirements, and the CPS-Packets are buffered into CPS-PDU buffer ('arrival' state). When CPS-PDU buffer is full or Timer_CU expires, the CPS-PDU is sent to ATM layer to create an ATM cell and the ATM cell is transmitted over a physical link ('svc_sta' state and 'svc_com' state).

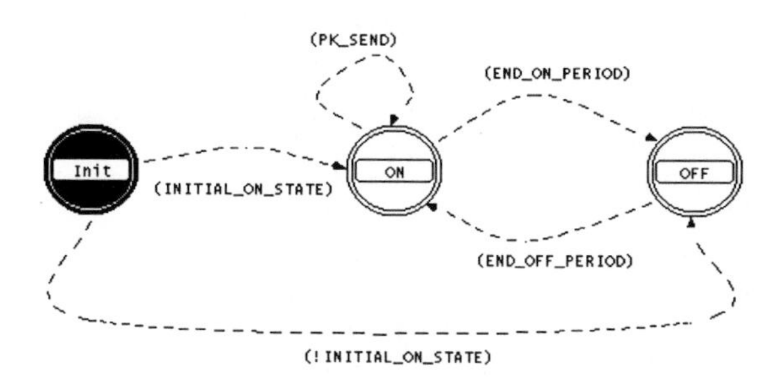

Fig. 7. Voice source model

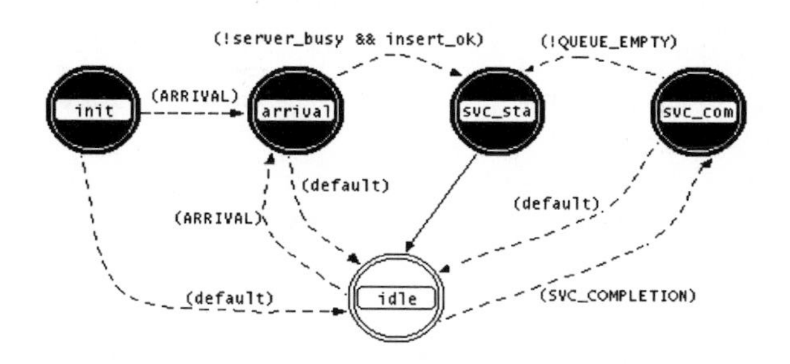

Fig. 8. Process model for AAL2 switching nodes

Figures 9 and 10 show the basic node model for each class 5 (AAL2) switching node and class 3/4 (AAL2) switching nodes, respectively.

Figure 11 shows our simulated network model for the North American switching network, which consists of class 5 (Local exchange), class 4 (Tandem and toll office), and class 3 (Primary toll center) AAL2 switches. As shown in the figure, there are one class 3 switch, four class 4 switches, and 120 class 5 switches. Each class 4 switch can accommodate 30 class 5 switches, and each class 5 switch can accept 50 voice sources. The number of voice sources and switching nodes is based on the analysis results in Section 3.

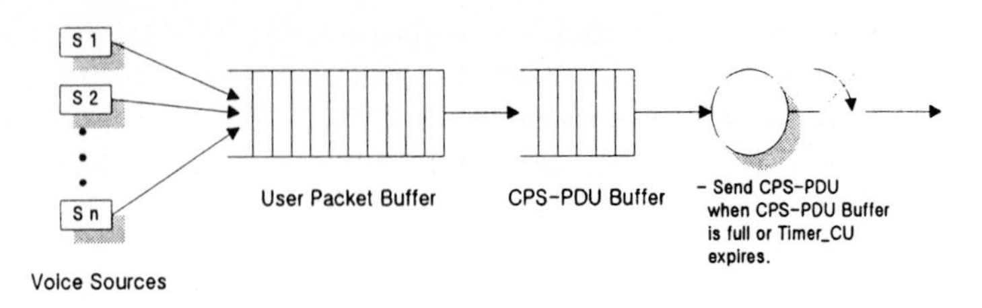

Fig. 9. Node model for Class 5 (AAL2) switching nodes

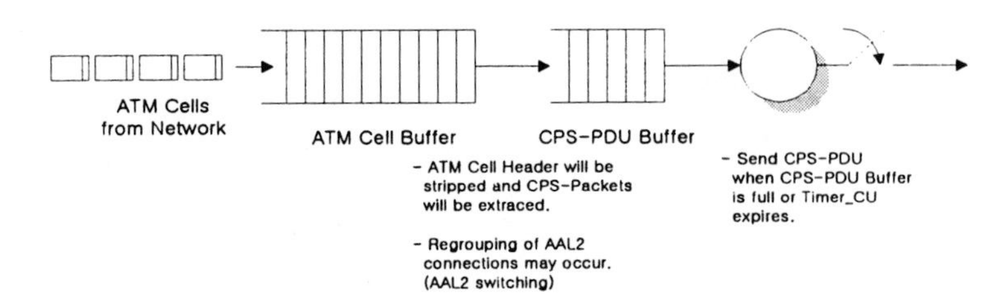

Fig. 10. Node model for Class 3 and 4 (AAL2) switching nodes

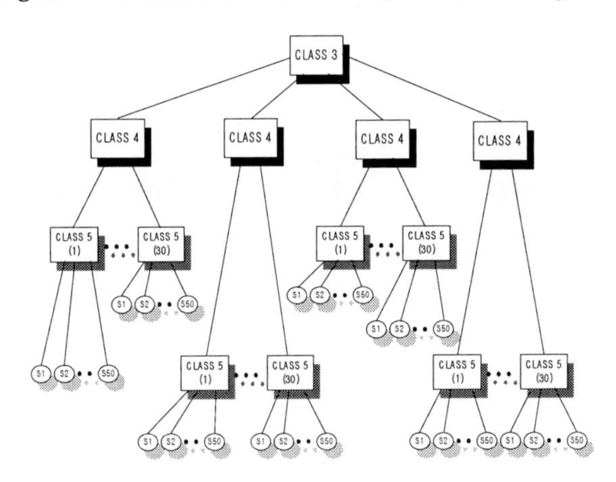

Fig. 11. Network model

Figure 12 shows the end-to-end delay of CPS-Packets when the link capacity between class 3 and 4 switches varies. Even if there are 30 T1 (~45 Mbps) links between class 4 and 5 switches, approximately 26 Mbps for each class 4 switch is enough to support 30 class 5 switches.

Figure 13 shows the end-to-end delay of CPS-Packets when the Timer_CU value which determines the CPS-Packet transmission varies. Lower Timer_CU value gives shorter delay, but very low Timer-CU value may degrade ATM cell utilization. If the Timer_CU value is set to a large value, it takes long to fill up an ATM cell. Accordingly, this causes a long waiting time until an ATM cell is sent to the ATM layer. However, note that with the Timer_CU value which is longer than or equal to about

0.5 ms, the end-to-end delay is not affected. This is because each traffic source generates CPS-Packets at the peak rate. In other words, even if the Timer_CU value increases, incoming CPS-Packets will fill up an ATM cell before the Timer_CU expires. The Timer_CU value for the saturation will be different if voice source rates are different.

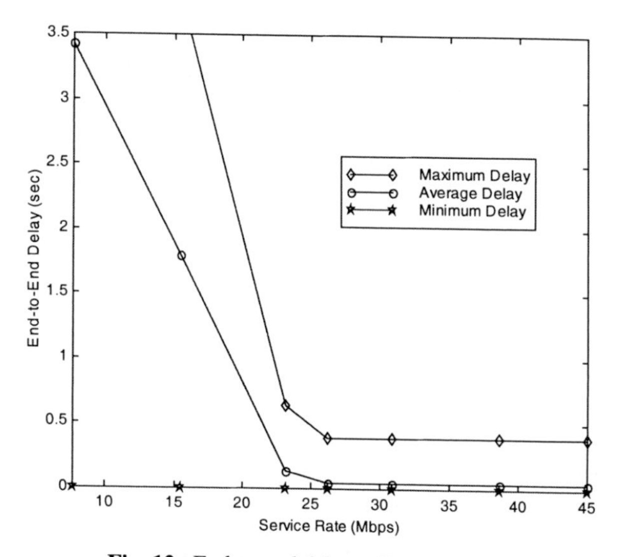

Fig. 12. End-to-end delay: effect of link rate

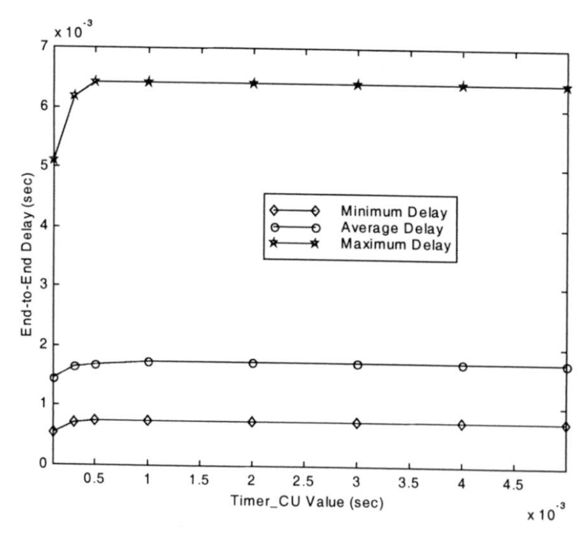

Fig. 13. End-to-end delay performance: effect of Timer_CU value

5 Concluding Remarks

In this paper, we analyzed the performance and the feasibility of AAL2 trunking/switching for the digital telephone switching systems. Specifically, we first analyzed the AAL2 trunking performance using uniform arrival/service model, and we then presented simulation results on the performance of AAL2 switching, in particular, when the traditional digital telephone switching system is replaced by the AAL2 switching system. The analysis results showed that AAL2 can provide high trunking/switching efficiency using the key features of AAL2 including support for compressed/silence-suppressed VBR voice and support for multiple voice users per ATM connection.

References

[1] Anick, D., Mitra, D., Sondhi, M.: Stochastic Theory of a Data-Handling System with Multiple Sources, The Bell System Technical Journal, Vol. 61, No.8, October 1982.

[2] Iyer, J., Jain, R., Dixit, S.: Performance of VBR Voice over ATM: Effect of Scheduling and Drop Policies, ATM Forum 97-0608, July 1997.

[3] Ali, S.: Digital Switching Systems - System Reliability and Analysis, McGraw-Hill, 1997.

[4] ITU-T Recommendation I.363.2: B-ISDN ATM Adaptation Layer (AAL) Specification: Type 2 AAL, 1997.

[5] Jeong, S.-H., Copeland, J.: Cell Loss Ratio and Multiplexing Gain of an ATM Multiplexer for VBR Voice Sources, IEEE LCN'98, pp. 384-389, October 1998.

[6] Sriram, K., McKinney, S.: Voice Packetization and Compression in Broadband ATM Network, IEEE JSAC, Vol. 9, No.3, pp. 294-304, April 1991.

[7] Kang, S., Sung, D.: Real-Time Cell Loss Estimation for ATM Multiplexers with Heterogeneous ON/OFF Sources, ICC'97, 1997.

[8] Feng, J., Lo, K.-T., Mehrpour, H.: Simulation Analysis on Statistical Multiplexing of MPEG Video Sources, ICC'97, 1997.

[9] Li, S.-Q.: A New Performance Measurement for Voice Transmission in Burst and Packet Switching, IEEE Transactions on Communications, Vol. COM-35, No. 10, October 1987.

[10] Daigle, J.: Models for Analysis of Packet Voice Communications Systems, IEEE JSAC, Vol. SAC-4, No. 6, September 1986.

[11] Saito, H.: Bandwidth Management for AAL2 in UBR VCs, ATM Workshop'99, 1999.

VI. Home Networking and Local Access Protocols

A Study on the Optimization of Class File for Java Card Platform*

Do-Woo Kim and Min-Soo Jung

Dept. of Computer Engineering, Kyungnam University, Masan, South KOREA
{dwkim, msjung}@eros.kyungnam.ac.kr

Abstract. The Java Card technology enables programs written in the Java programming language to run on smart cards and other resource-constrained devices. Smart cards represent one of the smallest platforms in use today. The memory configuration of a smart card is limited. The greatest challenge of Java Card technology design is to fit Java system software in smart card while conserving enough space for applications. The solution is to support a subset of the features of the Java language and to apply a split model to implement the Java virtual machine and to optimize Java bytecode. In this paper, we describe in detail the optimization algorithm of Java bytecode.

Keyterms : Java bytecode, Java Card, optimization, Java virtual machine

1 Introduction

The explosion of the Internet and of wireless digital communication has rapidly changed the way we connect with other people. Businesses, the government and healthcare organizations continue to move towards storing and releasing via networks, Intranets, Extranets, and the Internet. These Organizations are turning to smart card to make this information readily available to those who need it, while at the same time protecting the privacy of individuals and keeping their informational assets safe from hacking and unwanted intrusions.

The interest in smart cards is a result of the benefits they provide. One benefit is their built-in computational power. Security, portability, and ease of use are the other key advantage of smart cards. But, because smart card applications were developed to run on proprietary platforms, applications from different service providers cannot coexist and run on single card. Lack of interoperability and limited card functions prevent a broader deployment of smart card applications.

---

* This work was supported by grant No.2001-149-2 from the basic research program of Information Technology Assessment

I. Chong (Ed.): ICOIN 2002, LNCS 2343, pp. 563-570, 2002.

Java Card technology offers a way to overcome obstacles hindering smart card acceptance. It allows smart cards and other memory-constrained devices to run applications (called applets) written in the Java programming language. Essentially, Java Card technology defines a secure, portable, and multiapplication smart card platform that incorporates many main advantages of the Java language. But, because of its small memory footprint, the Java Card platform supports only carefully chosen, customized subset of the features of the Java language. Also, for reasons of small memory footprint, it is essential to optimize Java bytecode[2,3,10].

The organization of the rest of this paper is as follows. In section 2, we simply describe the characteristics of Java Card technology and optimization techniques. In section 3, we describe the optimization algorithm of Java bytecode. In section 4, we describe the result of test using the optimization algorithm. Finally, we conclude with a summary in section 5.

2 Related Works

2.1 The Java Card Virtual Machine

The configuration of the Java Card virtual machine is related to production and execution of the Java Card bytecode. A primary difference between the Java Card virtual machine(JCVM) and the virtual machine(JVM) is that the JCVM is implemented as two separate, as depicted in Figure 1[2,3].

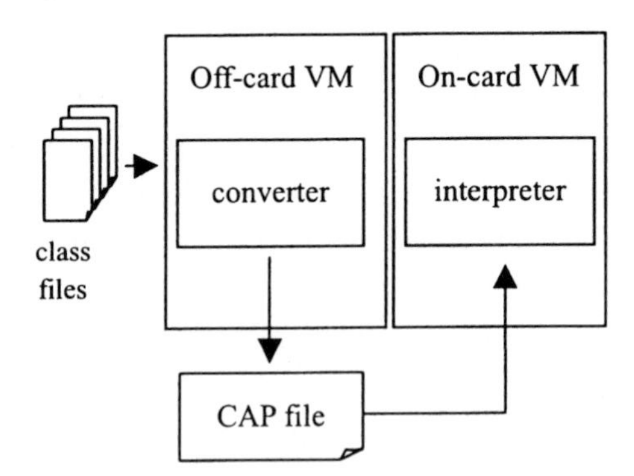

Fig. 1. Java Card virtual machine

The on-card portion of the Java Card virtual machine includes the Java Card virtual machine includes the Java Card bytecode interpreter. The Java Card converter runs on a PC or workstation. The converter is the off-card of the virtual machine. Taken together, they implement all the virtual machine functions-loading Java class files and executing them with a particular set of semantics. The converter loads and preprocesses the class files that make up a Java package and outputs a CAP(converted applet) file. The CAP file is then loaded on a Java smart card and executed by the interpreter. In addition to creating a CAP file, the converter generates an export file representing the public APIs of the package being converted.

2.2 The CAP File Format

A Java CAP file consists of eleven component, such as Header, Directory, Applet, Import, ConstantPool, Class, Method, StaticField, ReferenceLocation, Export, Descriptor. Each component describes an aspect of the CAP file contents, such as class information, executable bytecode, linking information, verification information, and so forth. The JAR file format is used as the container format for CAP files, A CAP file is a JAR file that contains a set of components, each stored as an individual file in the JAR file[2,3,5].

The content of each component defined in a CAP file must conform to the corresponding format specified. All components have the following general format:

```
component {
   u1 tag
   u2 size
   u1 info[ ]
}
```

2.3 Optimization Techniques

Loop Optimization

This technique is focused on identifying loop invariant and moving them out of the loop to improve the performance of the Java programs. In the following example, the loop condition *k < A[0] + 3.2 – (2 * A[n-1])* and the assignment statement *z = x – 4 * n* are invariants and should be moved out of the for-loop[9].

```
void foo(float[] A) {
  float x = 3.5f;
  int n = A.length;
  float z;
   for( int k = 0; k < A[0] + 3.2 - (2 * A[n-1]); k++) {
               float y = k * 3;
               z = x - 4 * n;
               System.out.println(k + " " + y);
   }
}
```

Compression of Java Bytecode

This technique proposes to factorize recurring instruction sequences into new instructions, yielding a more concise program that is executable on a Java virtual machine with an extended instruction set. By expressing the new instructions as macros over existing instruction, the Java virtual machine need only be extended to support such general macro instructions, not instructions specific to any one program[1,10].

3 Optimization Algorithm of Java Bytecode

Our approach is to make map file that represents the location information of "00" in CAP file, because there are padding element such as "00" in CAP file to satisfy CAP file format.

3.1 Map File Generation

To create map file that stores the location information of "00", Our Mapfile generator read CAP file as input. The basic process is as follows:

```
Loop ( End of CAP file )
    read 8byte in CAP file
     8 byte is assigned to Temp[ ] variable
     for each byte in Temp[ ]
         if ( each byte == 0 )
           write location information in map file
```

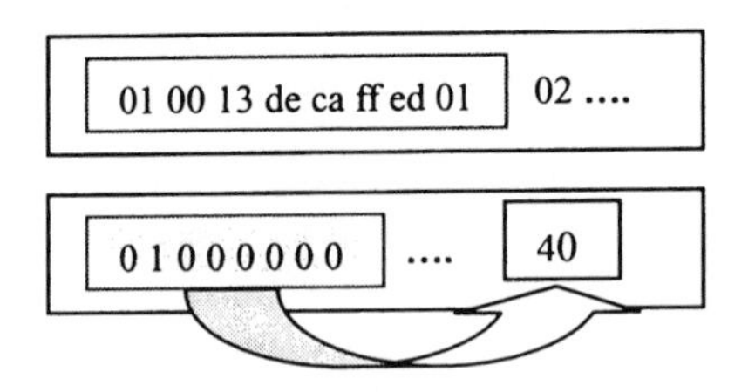

After this process is done, two files are created. One file is output file that removes "00" stream in CAP file. The other is map file that contains location information about "00".

01 **00** 13 de ca ff ed 01 02 02 **00** 01 09 a0 **00 00**
00 62 03 01 0c 01
02 **00** 1f 00 13 **00** 1f **00 00 00** 0b **00** 36 **00** 0c **00**
67 **00** 0a **00** 13 **00** 09 **00** 6c **00 00 00 00 00 00** 01
00 00
04 **00** 0b 01 **00** 01 07 a0 **00 00 00** 62 01 01
05 **00** 36 **00** 0d 02 **00 00 00** 06 80 03 **00** 03 80 03
01 01 **00 00 00** 06 **00 00** 01 03 80 0a 01 03 80 0a
:
0b **00** 6c 01 **00** 01 **00 00 00 00** 01 **00** 03 **00** 02 **00**
00 00 00 1c **00** 84 **00** 01 **00** 1e **00** 11 **00 00 00 00**
01 09 **00** 14 **00** 30 **00** 09 **00 00 00 00** 07 01 **00** 1f
00 33 **00** 46 **00 00 00 00 00** 0d **00** 1c **00** 1e **00** 1e
ff ff **00** 1e **00** 1c **00** 20 **00** 20 **00** 22 **00** 24 **00** 27

Fig. 2. HelloWorld.CAP

```
01 13 de ca ff ed 01 02 02 01
09 a0 62 03 01 0c 01 02 1f 00
13 1f 0b 36 0c 67 0a 13 09 6c
01 04 0b 01 01 07 a0 62 01 01
05 36 0d 02 06 80 03 03 80 03
01 01 06 01 03 80 0a 01 03 80
0a06 03 80 0a 07 03 80 0a 09
03 80 0a 04 03 80 0a 05 06 80
10 02 03 80 0a 03
:
:
:
0b 6c 01 01 01 03 02 1c 84 01
1e 11 01 09 14 30 09 07 01
1f 33 46 0d 1c 1e 1e ff ff 1e 1c
20 20 22 24 27 2a 2e 01 b0 01
10 01 40 02 41 03 44 10 04b4
41 06 b4 b4 44 02 44 04 b4 31
06 68 a1
```

```
40 23 81 55 55
5d f9 43 81 24
70 a7 10 76 00
00 00 00 51 1c
a1 08 90 44 04
20 04 20 02 02
01 04 42 2f fd
42 02 5e af 55
79 57 95 7d 51
55 55 00 00 00
30
```

Fig. 3. HelloWorld.out and HelloWorld.map

4 Evaluation

To test the effectiveness of our optimization algorithm, we applied many class files and CAP files. Table 1 shows the original size of class files and the difference of the original file size from generated file that output file removing "00" stream in class file adds to map file containing the location information of "00" stream. We can observe that, on average, the reduction is closer to 85% of the original size of each class file.

Table 1. Classic class file

No.	File Name	File Size			
	(Cap files)	Original	Delete Zero Byte	Zero Map	Difference Size
		*.cap	*.out	*.map	*.cap-(*.out+*.map)
1	ithmeticException	220	155	28	37
2	rayStoreException	220	155	28	37
3	Bitmap	745	545	94	106
4	Boolean	625	421	79	125
5	Button	1861	1348	233	280
6	Byte	982	696	123	163
7	teArrayInputStream	1335	924	167	244
8	teArrayOutputStream	1584	1165	199	220
9	Calendar	8849	6448	1107	1294
10	CalendarImpl	2036	1529	255	252
11	Caret	1136	782	143	211
12	Character	2062	1539	258	265
13	CheckBox	1491	1090	187	214
14	Class	1560	1317	207	126
15	ClassCastException	219	154	28	37
16	ClassNotFoundException	216	151	28	37
:	:	:	:	:	:
116	Writer	1230	856	154	220
	Average Byte	1451.71	1065.53	182.41	215.34
	Size Effective Ratio				15% ↓

Also, we can observe that, on average, the reduction is closer to 85% of the original size of each CAP file in Table 2.

Therefore, we can confirm that this algorithm can be applied to optimize Java bytecodes for devices with memory-constrained. But class loader have overhead to compound generated output file and map file. It may bring about the delay of run-time.

Table 2. CAP file

No.	File Name	File Size			
	(Cap files)	Original	Delete Zero Byte	Zero Map	Difference Size
		*.cap	*.out	*.map	*.cap-(*.out+*.map)
1	HelloWorld.cap	2337	1779	293	265
2	JavaLoyalty.cap	2575	1948	322	305
3	JavaPurse.cap	6431	5244	804	383
4	NullApp.cap	2155	1616	270	269
5	SampleLibrary.cap	2138	1629	268	241
6	wallet.cap	2763	2098	346	319
	Average Byte	3066.5	2385.67	383.83	297
	Size Effective Ratio				10% ↓

5 Discussion

We think that Run-time overhead is negligible because of following linear increment to input file size. Also, in case of finally generated mask file to install on card, our optimization algorithm is no effect. The reason is that resolution is performed not by index but by token-based linking. Therefore, finally generated file, masking file, has not unnecessary elements.

Table 3. Masking file

No.	File Name	File Size			
	(Masking files)	Original	Delete Zero Byte	Zero Map	Difference Size
		*.c	*.out	*.map	*.cap-(*.out+*.map)
1	Mask.c	166K	165K	20K	185K
	Average Byte	166K	165K	20K	185K
	Size Effective Ratio				10% ↑

6 Conclusion

We have described optimization algorithm for Java bytecode. It reduces the overall memory footprint of bytecode for memory-constrained devices to about 85% of their original size. But, to apply this algorithm may cause the execution time overhead. Also, to execute generated files the existing Java virtual machine is modified.

We are studying to apply our optimization algorithm to the Java virtual machine. We will design and implement the class loader subsystem of the Java virtual machine. In case of the Java Card, we will modify installation program. Also, we currently studying the issue related to Java bytecode optimization such as dependence analysis of Java bytecode and compression.

Reference

1. T. Lindholm and F. Yellin, *The Java™ Virtual Machine Specification* ADDISON-WESLEY, 1997.
2. Sun Microsystems, Inc., The *Java Card™ 2.1.1Virtual Machine Specification,* SUN, 2000
3. Sun Microsystems, Inc., The *Java Card™ 2.1.1Runtimel Environmen(JCRE)t Specification,* SUN, 2000
4. A. Taivalsaari, *Implementation a Java Virtual Machine in the java programming Language*, SUN Lab, 1997.
5. B. Venners, *Inside the Java Virtual Machine*, McGraw-Hill, 1997.
6. Min-Soo Jung, Jong-Dong Lee, "Design and Implementation of Call Graph Viewer for Java", The 25th KISS Spring Conference, pp74~76, 1998.
7. Dong-Hang Ryu, Min-Soo Jung, "Design and Implementation of Java ByteCode analyzer", The 25th KISS Spring Conference, pp77~79, 1998.
8. Do-Woo Kim, Min-Soo Jung, Dong-Hang Ryu, Min Jin, "Design and Implementation of Java Virtual Machine Simulator", The 25th KISS Fall Conference, pp 422~424,1998.
9. Kuznetsov, E. "Optimizing performance execution with Java", Java Report, pp. 49-51, 1997
10. Lars R. Clausen, Ulrik Pagh Schultz, Charles Consel, and Gilles Muller, "Java Bytecode Compression for Low-End Embeded Systems", ACM Transactions on Programming Languages and Systems, Vol. 22, No. 3, 2000
11. A. Adl-Tabatabai and M. Cierniak and Huei-Yuan Lueh and V. M. Parikh and J. M. Stichnoth, "Fast, Effective Code Generation in a Just-In-Time Java Compiler", *Proceedings of the ACM SIGPLAN*, pp. 280~290, 1998
12. Bradly. Q., Horspool. R. N., and Vitek. J. "An efficient compressed format for Java archive files" In Proceedings of CASCON'98, pp.294-302, 1998
13. Clausen. L. R, "A Java bytecode optimizer using side-effect analysis", Concurrency: Practice and Experience 9, pp1031-1045, 1997
14. *http://www.artima.com/,* ARTIMA SOFTWARE COMPANY
15. *http://java.sun.com/*, Sun Microsystem, Java Home Page

Design and Implementation of Small-Sized Java Virtual Machine on Java Platform Jini*

Jun-Young Jung and Min-Soo Jung

Dept. of Computer Engineering, Kyungnam University, Masan, South KOREA
mrj@hawk.com.kyungnam.ac.kr, msjung@eros.kyungnam.ac.kr

Abstract. With the opening of mobile internet age, concern of internet about mobile device, which can access network anytime and anywhere, is improving. Jini HomeNetwork in mobile based on Java platform is computing environment using feature of Java. To support Jini, Java virtual machine has to be executed and support RMI, but Java virtual machine of Java2ME CLDC couldn't support. In this paper, Java virtual machine which can access to HomeNetwork was designed using Java2ME CLDC. Small-sized VM is not only a program to each mobile device, but also support standard Java class file format, CLDC class and environment of Accessing Jini. Consequently, This Java virtual machine can support real networking of mobile device.

Keyterms : Java virtual machine, Jini, RMI

1 Introduction

Today, computing market has an interest in internet mobile device. For example, mobile device such as cellular phone or PDA is being attracted by reason of portability and mobility and makes a great change of computing environment. Java and Jini are representative technologies which lead such change. Java is remarkable by reason of dynamic application download, compatibility of cross platform, advanced experience of user, non-connectivity and security problem[1,6,7]. And, Jini expanding features of java into environment of the whole network, is a technology to realize all the components connected on the Internet[2,3,9,10,11]. But java technology performed on desktop platform can't be applied to mobile device such as a portable digital device[12,13,14].

Because Java needs enough of resource in current system, the device with a limited resource such as cellular phone or PDA can't allow all the technologies of Java[12,13,14].

* This was work supported by grant from the Basic Research Program of the Korea Science & Engineering Foundation.

I. Chong (Ed.): ICOIN 2002, LNCS 2343, pp. 571-580, 2002.

Sun Microsystems published K-Java Virtual Machine. As VM designed to be suitable to a little memory, processing ability of CPU and limited wireless network, it can't connect to Jini because the VM eliminated RMI(Remote Method Invocation) from JVM[12,13,16].

The organization of the rest of the paper is as follows. We simply describe the architecture and execution order of small-sized VM in section 2, the structure and role of small-sized VM connecting to Jini in section 3 and the implementation of small-sized VM in section 4. Finally, we conclude with a summary in section 5.

2 Related Works

2.1 The Architecture of the K-Java Virtual Machine

Figure 1 represents Architecture of the K-Java Virtual Machine(K-JVM). The K-Java Virtual Machine consists of class loader system, execution engine and runtime data areas. The K-JVM is basically similar to existing JVM[1,6,7,12,13].

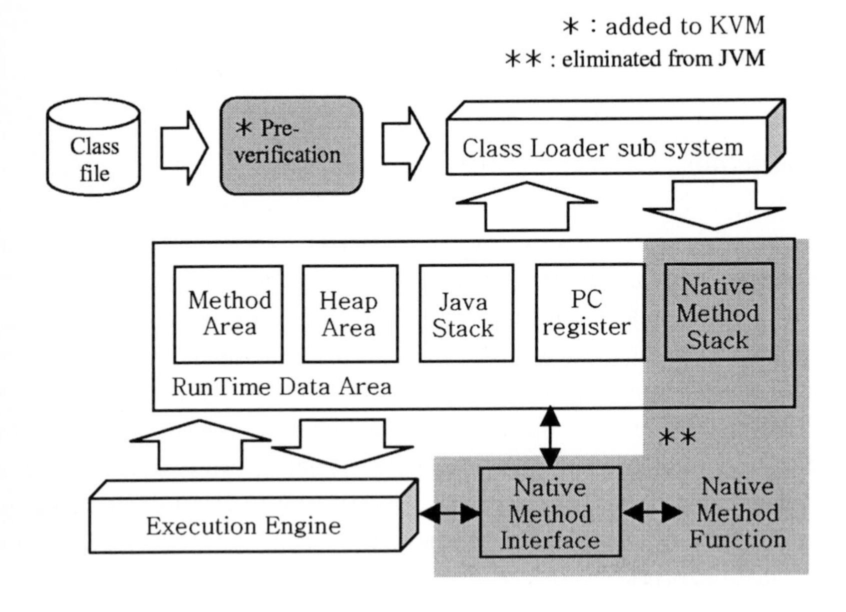

Fig. 1. Internal Structure of K-Java Virtual Machine

But K-JVM validates classfile through the pre-verification step before loading classfile into class loader subsystem. Runtime data area in the K-JVM doesn't have native method stack to support native method function.[12,13,14]

2.2 Verification of the K-Java Virtual Machine

The verification is executed in the class loader subsystem to load classfile. Verification checks whether loaded data type obeys semantics of the Java language or is satisfied with the internal format of classfile[1,6,7].

K-JVM and JVM have the same execution process but are somewhat different from execution method[12,13].

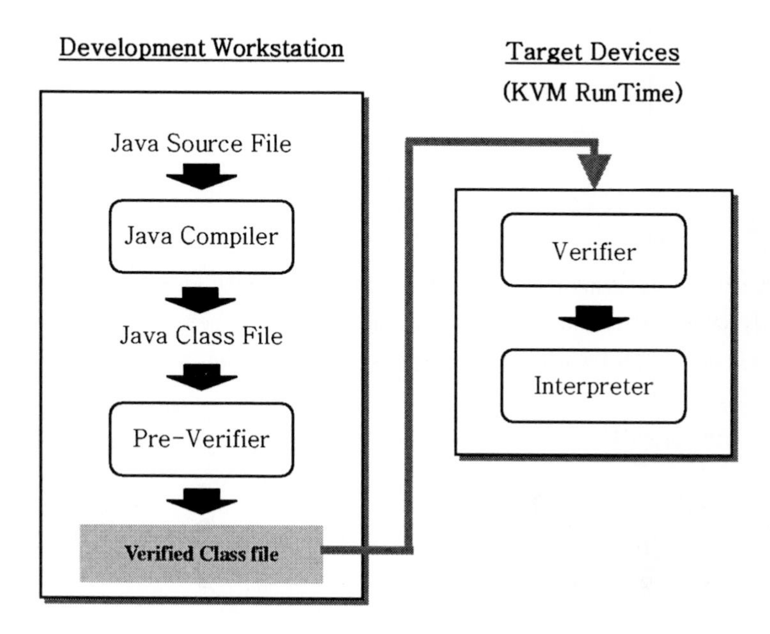

Fig. 2. Verification step of K-Java Virtual Machine

2.3 Reflection

Reflection is a technology that java program gets the information about object's member or class in execution through class, object, method, field, thread, the number of execution stack, contents of the stack and other execution structure in internal VM, then allows to work on the member of class using the information. It is applied to Object Serialization or RMI[5,8,9].

A VM supporting CLDC doesn't have the reflection function and can't support RMI and object serialization. Therefore, K-JVM can't support lookup service and join service that Jini supports[2,3,12,13,14].

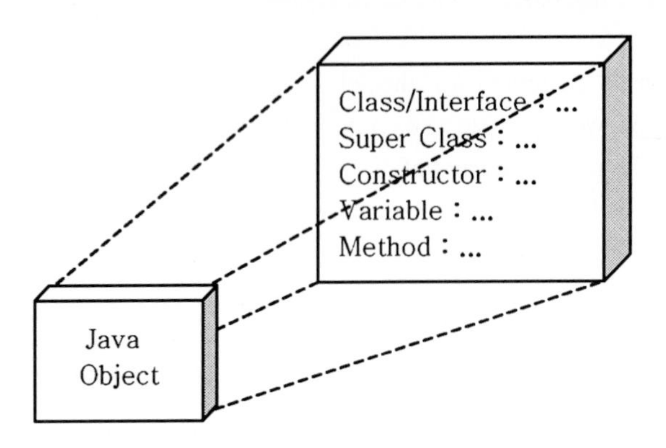

Fig. 3. Java reflection environment

2.4 Jini

Jini based on java technology is object oriented-distributed network architecture that expands java application environment into the whole network and connects all electronic devices, software and service[2,3].

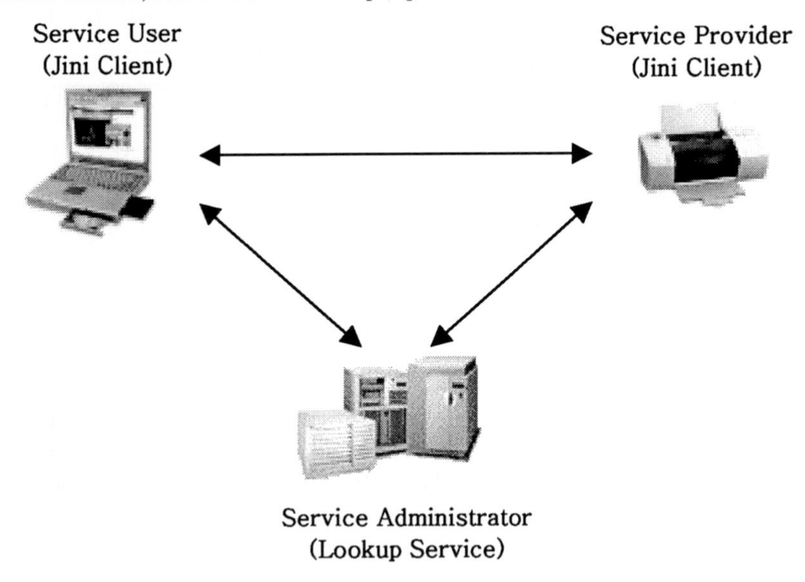

Fig. 4. Jini System Component

Figure 4 shows Jini system components. This system consists of service user(Jini Client), service provider(Jini Service) and service administrator(Lookup Service). Each component communicates with each other using Discovery, Join and Lookup protocol. Also, communication between service user and service provider take advantage of Java RMI(Remote Method Invocation)[2,3].

2.5 Jini HomeNetwork and K-Java Virtual Machine

Jini HomeNetwork is a kind of distributed system based on Java. It needs at least Java2SE 1.2.2 version to support a right performance because all the services in Jini use RMI. Jini is able to invoke method of remote object using lookup and join service. So if user wants various services, Jini obtains and uses the necessary object flexibly in remote system. Java platform supporting RMI connects to Jini and uses service. But, KVM being used in branch of mobile removed the reflection function because of restricted resource(e.g. power, memory, cpu, and so on). So KVM doesn't support RMI that allows to connect to Jini and to receive the service[2,3,5,8].

3 Small-Sized JVM Connecting to Jini

3.1 System Organization Applied to Jini and Small-Sized JVM

Jini, one of the distributed system architectures based on Java, consists of service, client and lookup service. Figure 5 shows system organization. The mobile device must embed small-sized VM for a client a to interact with Jini system.

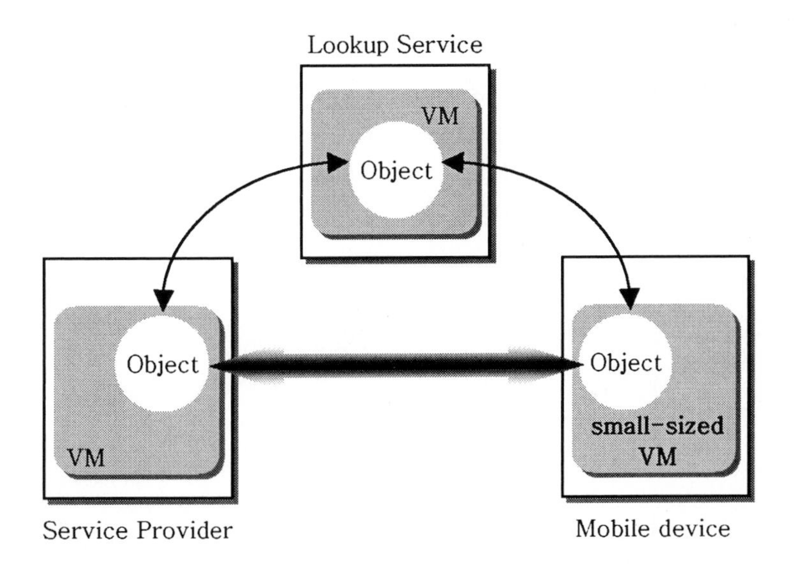

Fig. 5. System Organization

3.2 System Component

Figure 6 shows components of the small-sized VM. The source code is organized into loader step, initialization step and execution step like KVM of Java2ME.

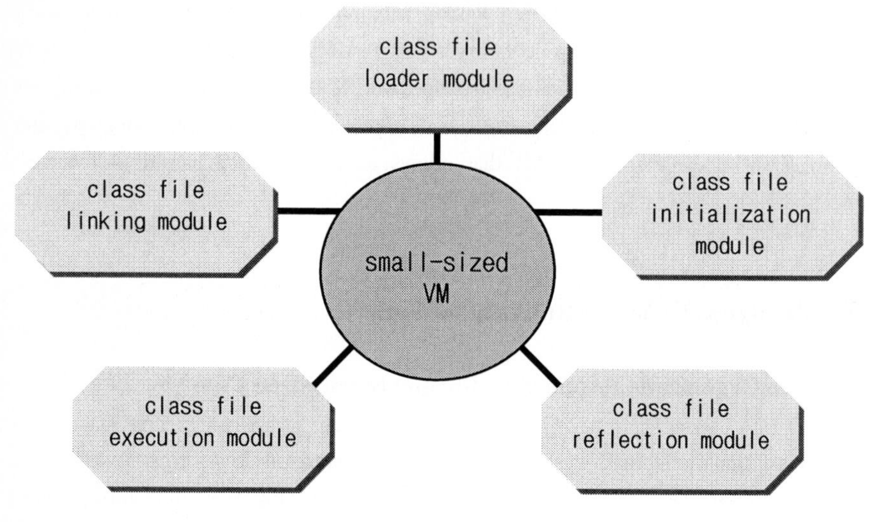

Fig. 6. System Components

3.3 Verification at the Linking Step

Shown as Figure 7, verification of small-sized VM checks if information of pre-verification is in the loaded classfile. If the information exists, VM performs verification step based on Java2ME specification. Otherwise, verification is omitted in execution of vm because the loaded class file format differs from Java2ME.

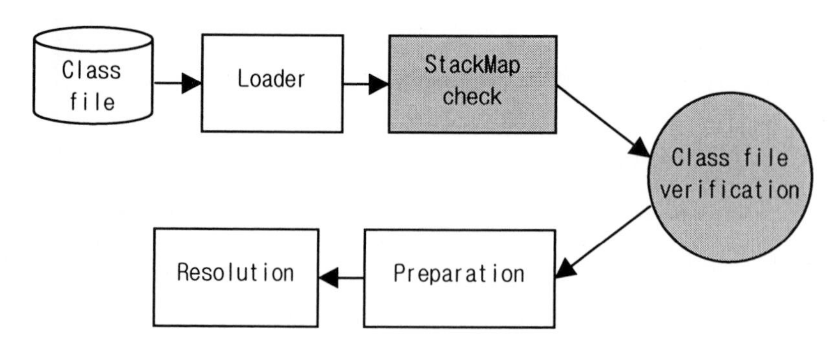

Fig. 7. Verification at the linking step

3.4 Class File Reflection

Shown as Figure 8, reflection consists of message analysis module and object creation module of class/interface. The message analysis module analyzes data received from lookup service. The object creation module of class/interface creates object to invoke object of service and class method of service.

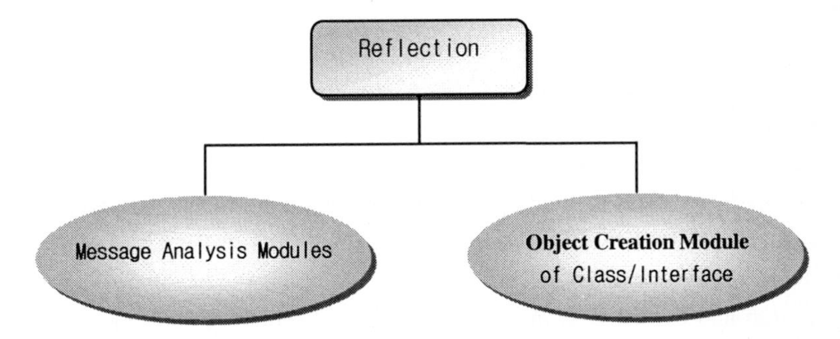

Fig. 8. Reflection component

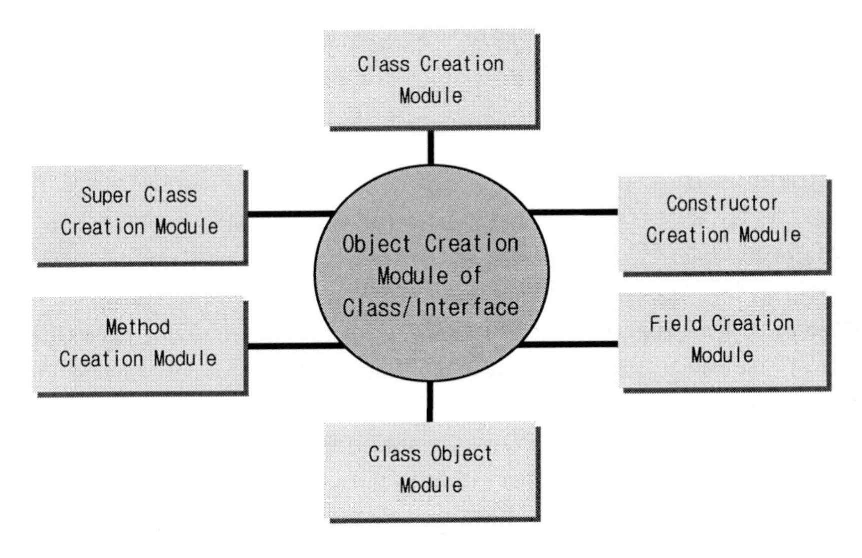

Fig. 9. Object Creation Module of Client/Interface

Shown as Figure 9, this module gets information about class of object, constructor, field and method. Then it allows small-sized VM of client to create object and invoke method of object in the JVM of server.

3.5 Reflection Processor Module

The processReflection module creates object to support RMI in the VM.

- Class Creation Module
  Used to create object after obtaining class information from object's class. Object is created by VM.
- Super Class Creation Module
  Get super class information from class information about object.
- Constructor Creation Module
  Get constructor information from class information about object
- Method Creation Module
  Get name for class method and information about attribute(return type, argument type) related with the class method.
- Field Creation Module
  Search and extract type information for specific field of class or interface from class information about object.
- Class Object Creation Module
  Create object which can invoke method using the above information about class at the execution time.

4 Implementation Issue

4.1 Major Module

Instead of implementing JVM, it is implemented with released source code based on KVM of Sun Microsystems. The execution sequence of KVM and standard JVM is similar.

- InitializeClassPath : Initialize classpath needed for class loading
- InitializeThreading : Create first execution thread and initialize VM registers.
- InitializeJavaSystemClasses : Load the standard Java system classes needed by the VM
- InitializeVerifier : Initialize the bytecode verifier.
- checkStackMap : After loading a class, it checks the existence and non-existence of verification information through class file. If verification information exists, vm performs verification. Otherwise, it does not perform verification.
- getClass : Find a class with the given name, loading the class if necessary.
- getSpecialMethod : Find a specific special method (<clinit>, main) using a simple linear lookup strategy.

- initializeClass : After loading a class(java.lang.System, java.lang.String, java.lang.Thread, java.lang.Class), it must be initialized by executing the possible internal static constructor '<clinit>'. This will initialize the necessary static structures. This function sets up the necessary Class.runClinit frame and returns to interpreter.
- Interpret : Translate and execute the bytecode of class file

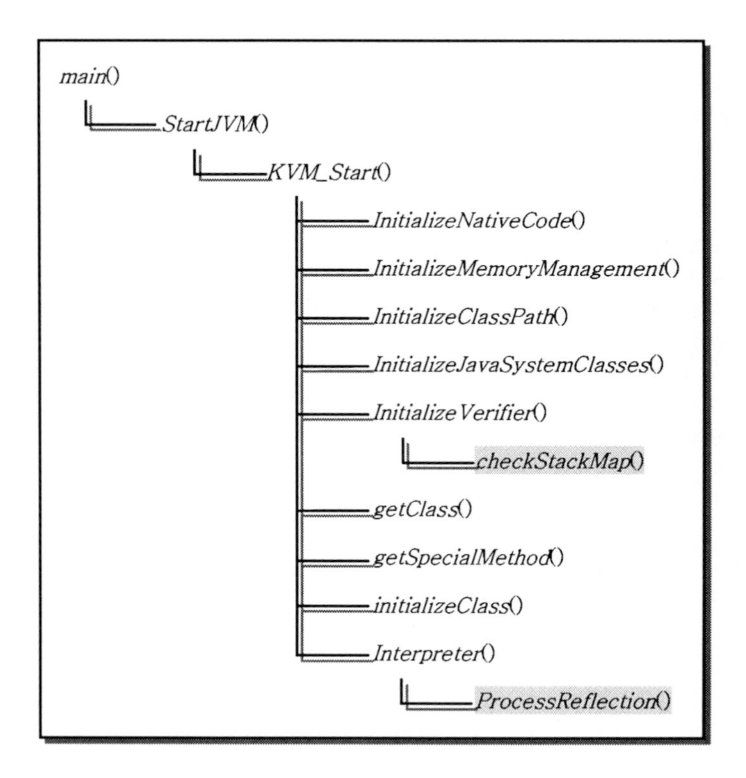

Fig. 10. Major module list of small-sized VM

5 Conclusion

In this paper, small-sized Virtual Machine checks format of class file and executes. It is able to interoperate between Java2 platform by supporting execution of Java2ME CLDC class file and Java2SE standard classfile. Also, it has the necessary RMI function to interact with Jini HomeNetwork based on java by adding reflection. If mobile device or digital device embedding small-sized VM connects to network without procedure such as installation or handiwork, it is possible to process various tasks over network

If program embedded in mobile device or digital device is written from Machine language or C language into Java language, there is no need for programmer to write program which is fitted in each mobile device. Also it provides a little more convenient program environment for programmer.

References

1. B. Venners, *Inside Java Virtual Machine*, McGraw-Hill, (1997)
2. W. KEITH EDWARDS, *Core Jini*, Sun MicroSystems(1999)
3. Sing Li, *Professional Jini*, Wrox(2000)
4. http://java.sun.com/, Sun Microsystems, Java Home Page
5. http://java.sun.com/products/jdk/1.1/docs/guide/reflection/spec/, Java Core Reflection
6. Yong-Jae Kwag, Java Virtual Machine Programming, Infobook, 1999
7. Tim Lindholm, Frank Yellin The Java Virtual Machine Specification,
8. http://java.sun.com/j2se/1.3/docs/guide/rmi/, RMI Specification
9. http://artima.com/, a resource for Java and Jini developers
10. http://www.jini.org/, Jini HomePage
11. http://www.jiniworld.net/, Jini and HomeNetwork
12. http://java.sun.com/j2me/, Java2 Platform Micro Edition Specification, White Paper
13. KVM WhitePaper, http://java.sun.com/products/ kvm
14. Eric Giuere, Java 2 Micro Edition, WILEY, 2000

Development of Internet Home Server to Control Information Appliances Remotely

Jonghwan Lee[1], Yongho Kim[2], and Kyongsok Kim[1]

[1] Dept. of Computer Science, Pusan National University, Keumjeong-gu, Pusan 609-735, Republic of Korea
{jhwlee, gimgs}@asadal.pnu.edu

[2] Dept. of ICOMM, KIMM (Korea Institute of Machinery & Materials), 66 Sangnam-dong, Changwon, Kyoungnam 601-010, Republic of Korea
yhkim@icomm.re.kr

Abstract. As the Internet and home networking technologies are developing rapidly, more users want to control their home appliances from a remote location, i.e. from outside the home via the Internet. However, some problems hinder them from doing so. First, an IP address is allocated dynamically to a subscriber to utilize the limited resources by the network service providers. As a result, it is difficult for users to make connections from outside the home to the home network. Secondly, for security, firewalls or proxies may be installed on the home network. This causes similar problem, too. In this paper, we propose a new approach to solve these problems and present our implementation. We built an Internet Home Server (IHS) outside the home, not inside of it, and it provides similar functions to those of the home server. Through IHS, users can be guaranteed to connect to home appliances from a remote location.

1 Introduction

The future computing environment will be ubiquitous [1]. We can recognize this from the fact that information appliances are replacing traditional appliances. These information appliances communicate with each other by connection to a home network. For the physical connection, many kinds of standards, such as HomePNA, HomeRF, PLC and Wi-Fi, have been issued [6, 7]. Also, some standards, such as Jini, Havi, UPnP and etc., have been developed to provide the inter-operability between them [8]. Here, inter-operability means the ability of an application running on an in-home appliance to detect and use the functionality of other appliances that are connected to the same network [12].

So far, a good amount of research about home networking has been concentrating on cooperation among multimedia devices such as PC, TV, VCR, etc. [2, 3]. However, few studies have been done on typical home appliances, especially kitchen appliances. Nowadays, technical improvements grant them new facilities in information appliances. For instance, kitchen appliances provide not only their traditional functions but also new services like displaying messages received from the network or being controlled from a remote location.

I. Chong (Ed.): ICOIN 2002, LNCS 2343, pp. 581–588, 2002.

Generally, a user accesses the home server located on a PC and then makes use of GUIs similar to the real control panel. A home server is an application which maintains and manages the status of the information appliances connected to network as well as provides the interface between them. People can easily access the Internet thanks to the development of network technologies. Also, they have been increasingly requesting a way of accessing the various information appliances which are link to the home network from outside the home through the Internet. However, they have some difficulties in doing that using PC, PDA and WAP-enabled mobile phones.

First, the network service providers allocate available IP addresses to subscribers in a way as to use the limited number of IP addresses efficiently. This means that a user from outside the home should access the information appliances which have dynamic IP addresses.

Secondly, if the home network has a firewall for security, a user may not be able to access the home network neither.

To overcome these problems, we developed a new application called by the Internet Home Server (IHS). This has a similar functionality to that of a home server but is located outside the home to guarantee accessibility among them. The home gateway is a residential application which makes a connection to the IHS and a home appliance .

After making a connection from the home gateway to the IHS, the latter permits a user to control his/her home appliances using web and/or mobile interfaces. When he/she accesses the IHS, a logical communication channel is created between him/her and the home appliance via the home gateway. Finally, he/she can control the appliance and receive the results of actions through the channel. In the view of accessibility, the user can always control his/her appliances using the IHS as he/she does at home.

We construct the IHS in the development of a new LG Electronics's washing machine, which is already being sold on the market. The Remote Control Server (RCS) is the name of LGE's Internet Home Server. And LG Ln-gate (LG Living network Gateway) is the name of LGE's Home Gateway [10]. Details are explained in Section 5.

In the following sections, we will describe our approach in detail. We will also present the background of our approach and the reasons for developing the Internet Home Server.

2 Home Network : Home Server and Home Gateway

2.1 Home Server

A home server which resides on the home network has the ability to control home appliances connected to the home network. There are digitalized and network-enabled appliances such as washing machines, refrigerators, microwave ovens and air conditioners. A user can obtain the information on appliances from the home server since it also manages information about those appliances.

In addition, it offers graphical user interfaces (GUIs) similar to those of real appliances in order for a user to control them. These user interfaces of appliances should be put together at the home server. A user can control home appliances through these interfaces at the home server.

In the information appliances middleware area, although many de facto standards are being spread without a dominating one, many are based on the Java platform. In May, 2000, OSGi (Open Services Gateway initiative) standard was announced which can be used for controlling home appliances. OSGi is independent of platforms and applications, supports multi service and many home network techniques, and can coexist with other standards [9].

2.2 Home Gateway

As shown in Fig. 1., a home gateway can be defined as a device that connects the home network to the Internet. It has a network interface card (NIC) which connects to the Internet outside the home and has a routing ability to access home appliances on the home network. Therefore, a home gateway can transfer a received message from the Internet to a designated appliance.

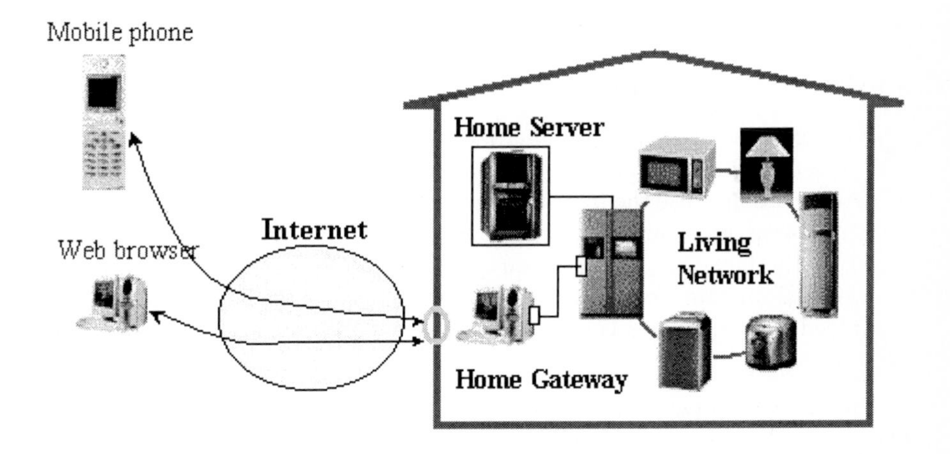

Fig. 1. A traditional home network including both a home server and a home gateway

As you see in Fig. 1., a traditional home network for residential information appliances, especially kitchen appliances, includes both a home server and a home gateway. A user can control and/or monitor his/her appliances through this home server. Whenever a user tries to control an appliance from outside the home using a web browser or an Internet-accessible mobile phone via the Internet, he/she has to access the home server through the home gateway. To guarantee his/her accessibility to the desired appliance via the Internet, the home gateway preserves the connection between the home network and the Internet. If the connection is not guaranteed, it is hard to control the appliance via the Internet.

3 Difficulties in Communication between the Internet and the Home Network

A few information appliances have built-in NICs which enable them to connect to the Internet directly. In this case, they need the TCP/IP protocol and application-level software which allow a user to control them. If they had the ability to directly connect to the Internet with the NIC, the home network could be constructed more easily. However, this causes some problems such as increasing the cost because of the NIC and developing a new application to control the product [4, 5].

All devices which connect to the Internet must have an individual Internet Protocol (IP) address. Currently, the assignment of IP addresses to Internet-accessible devices follows the Internet Protocol version 4 or IPv4 which uses 32-bit system results in the limited number of IP addresses. The total number of IP addresses are not enough to cover all devices connected to Internet. To avoid this limitation, network service providers assign an available IP address to a subscriber dynamically. A subscriber who wants to use a static IP address pays more than a subscriber who uses a dynamic IP address.

In order to use the Internet at a low cost, most households subscribe to a network service provider and use a dynamic IP service. An IP address changes whenever customers use the Internet. As a result, it is difficult for a user to access a home network from outside the home through an IP address. This problem is caused by the fact that a user doesn't know what the IP address of the home gateway is. If making a connection from the Internet to the home gateway is not guaranteed, the access via the Internet using a web browser or a mobile phone has no meaning.

The IETF has announced a new standard, the Internet Protocol version 6 or IPv6, in which an IP address is constructed with 128 bits. There is less of a limitation on IP addresses than that on IPv4. Therefore, if the home gateway has a device which follows the IPv6 standard, it can have a static IP address. However, only few devices follow the IPv6 standard in the experimental environment. It is hard to predict when IPv6 is going to spread. Even if IPv6 starts spreading, it will still be difficult to connect to the home network from the Internet when firewalls or proxies are on the home network.

In addition, if the home server resides on the home network, the cost of managing the home network will increase. Furthermore, it is hard for most housekeepers to operate the home server easily because they are not familiar with.

4 Internet Home Server (Virtual Home Approach)

An Internet Home Server (IHS) is a server through which a user can control an appliance at home. The IHS plays a similar role as home server in the view of allowing a graphic user interface to access and control the home appliances. The difference between the IHS and the home server is that the IHS resides on outside the home, not inside of it. The home gateway resides on the home network and it connects the IHS with an appliance on the home network. The scenario is as follows: a user accesses to the IHS to control an appliance.

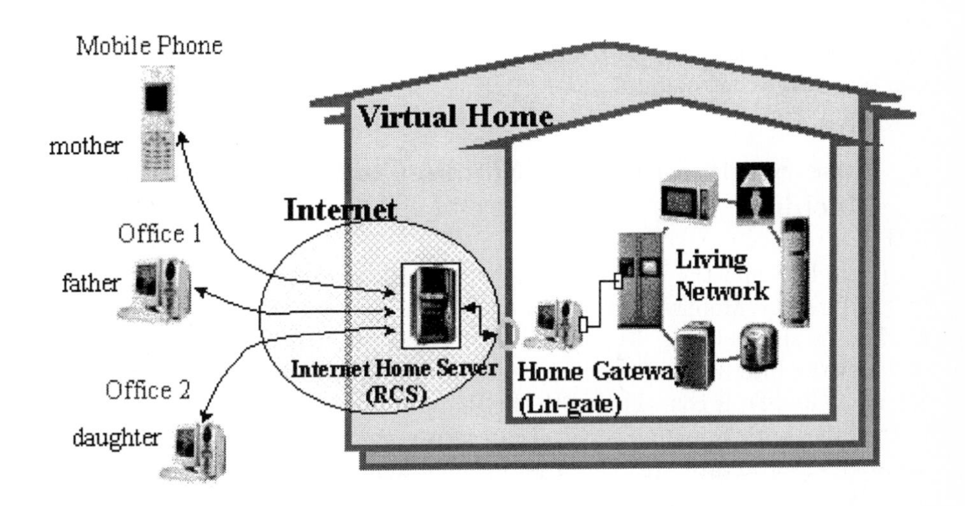

Fig. 2. Virtual Home using Internet Home Server

A user connects to the IHS and sees the interface looks like a real appliance's control panel. When he/she chooses what to do and clicks a start button, a message is sent to the IHS and the IHS transfers it to the home gateway. Next, the home gateway transforms it into a form which an appliance understands and sends a transformed message to the appliance. The appliance operates according to the message and sends the result of its operation back to the IHS. Finally, the IHS displays the reply from the appliance to the user. In this scenario, the user can treat it as if he/she is controlling a real appliance at home. As shown in Fig. 2., we call this Virtual Home including the IHS, the home gateway, the home network and the information appliances.

As mentioned previously, due to the dynamic IP address problem, it is very difficult to access an appliance at home from the Internet. Our approach has the same problem since a user cannot access the home gateway from the Internet. To overcome this problem, we choose the method that the home gateway makes a connection to the IHS and preserves the connection. In this way, a user can control a home appliance in spite of the dynamic IP address problem, firewalls and proxies. Even if the connection is broken, the home gateway automatically reconnects to the IHS. As a result, the connection is always preserved.

The IHS provides web and mobile user-interface. Therefore, the members of a household may access one appliance at the same time. The situation as seen in Fig. 2., which describes how a father controls a washing machine using a web interface while a mother controls the same appliance using a mobile interface simultaneously, may cause unpredictable status.

This would not be an exceptional situation, but a usual one. To manage multi-user access, the IHS creates a logical communication channel through which only one user can control an appliance. Until the user finishes his/her operations, the IHS manages this channel and prevents another user from controlling the same appliance. Furthermore, the IHS informs the user who is prevented that another user is already controlling it.

5 Case Study : Internet/Mobile-Controlled Washing Machine

Both the IHS and the home gateway were realized during the development of LGE's Internet/Mobile-controlled washing machine, which is being sold on the market. As shown in Fig. 3., the IHS has 5 representative modules. A web server provides the Internet and mobile interface to the user. A Remote Control Server Daemon (RCSD) is connected with the Home Gateway, LG Ln-gate. In addition to this, the Authentication server, Data server and SMS module cooperate with the web server and the RCSD for performing functions of IHS as follows.

- preserving a connection between the RCSD and the Ln-gate
- authenticating the Home Gateway's connection
- authenticating Internet or mobile access
- providing web control interface
- providing mobile control interface
- managing appliances' information
- notifying the user of what has happened
- storing the messsages for future analysis

There are two access points at the IHS. (A) is where a user connects using a web browser or mobile phone. (B) is where the Home Gateway connects. When both access points are connected to, a logical communication channel can be made. A user can control a washing machine through the control panel interface on a Web page which looks like a real one. When a user chooses a washing course and clicks the start button on the Web control panel to operate the real washing machine, a message is sent to the RCSD. The RCSD can transfer this message to the Ln-gate, since the RCSD already has a communication channel with the Ln-gate. The Ln-gate then transforms the received message into LnCP-based order codes and sends it to the washing machine [11]. Now the washing

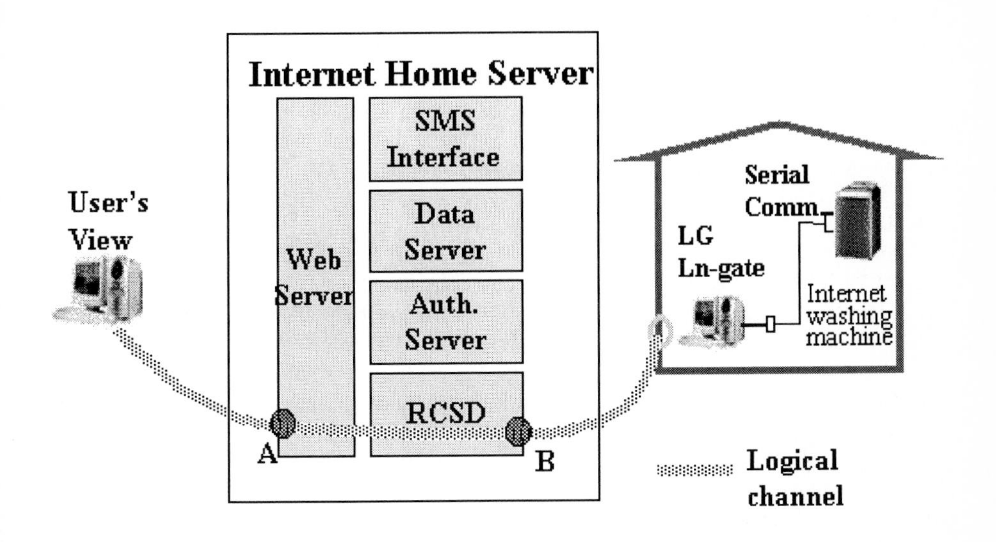

Fig. 3. Internet Home Server with a logical communication channel

machine actually runs according to the user's commands. When the washing 7 machine starts its operation, it informs the Ln-gate of its status. The result of its operation is sent to the Web server via the Ln-gate and the RCSD. Finally, a user recognizes it on the Web control panel.

At present, the IHS is able to manage appliances, i.e. a washing machine. We expect it to be adapted to various kinds of appliances in the near future.

6 Conclusion

In this paper, we proposed a new approach to communicate between a user and a home appliance via the Internet. To overcome the difficulties of accessing a Home Server such as a dynamic IP address problem, firewalls, and proxies, we developed the Internet Home Server outside home. We adapted our approach to LGE's Internet washing machine, which is being sold on the market. We are also convince that a user can always connect to a washing machine through our Internet Home Server.

In the future, we will extend the IHS in order to manage various kinds of appliances made by different vendors. In advance of doing this, we need to make the standards of the protocols dealing with remote control of home appliances and messages between the IHS and the home gateway.

References

1. M. Weiser: Some Computer Science Issues in Ubiquitous Computing. Communication of the ACM, 36(7), pp. 74-84, 1993.
2. T. Nakajima: System Supports for Networked Home Appliances on Commodity Operating Systems. International Conference on Multimedia and Expo (ICME2001), Tokyo, Japan, 2001.
3. T. Pham, G. Schneider, S. Goose: A Situated Computing Framework for Mobile and Ubiquitous Multimedia Access using Small Screen and Composite Devices. ACM Multimedia 2000 Los Angeles CA USA, 2000.
4. Margherita Washing Machine, http://www.magarita2000.com/index/index.html
5. Carrier, http://www.myappliance.carrier.com/
6. Bluetooth, http://www.bluetooth.org
7. HomePNA, Home Phoneline Networking Alliance: Simple, High-speed Ethernet Technology for Home, A white paper, Jun 1998
8. Universal Plug and Play Forum, http://www.upnp.org
9. Open Service Gateway Initiative, http://www.osgi.org
10. DreamLG, http://www.dreamlg.com
11. Koon-Seok Lee, Haon-Jong Choi, Seung-Myun Baek: A New Control Protocol for Home Appliances-LnCP. 2001 IEEE International Symposium on Industria Electronics, Vol. I. Pp286 291, June 2001.
12. G. O'Driscoll: The Essential Guide to Home Networking Technologies. Prentice-Hall, inc., ISBN:0-13-019846-3, (2000)

Design of HG-iPhone Service Based on H.323 in the Home Gateway System

YeunJoo Oh[1], KwangRoh Park[1], Kyungshik Lim[2], and KyoungRok Cho[3]

[1]Home Network Team, Network Technology Laboratory, ETRI,
161 Gajeong-Dong, Yuseong-Gu, Daejeon, Korea.
{oyj63246, krpark}@etri.re.kr

[2] Kyungpook National University,
1370 Sangyuk-Dong, Puk-Gu, Daegu, Korea.
kslim@knu.ac.kr

[3] Chungbuk National University,
48 Gaesin-Dong, Hyungduk-Gu, Cheongju, Korea.
krcho@cbucc.chungbuk.ac.kr

Abstract. In this paper, we represent the HG-iPhone Service, based on H.323 protocol, which is one of the service functionalities of the Home Gateway system. Internet Phone services are currently capable of offering inexpensive telephone services over IP networks, although with a lower quality of service in comparison to that available through public switched telephone networks (PSTNs). IP telephone services have other limitations as well: for example, end-users should make the phone calls manually, and the service equipment (i.e., the user's personal computer) should be turned on. To overcome these limitations, our approach focuses on the minimization of additional procedures. Our approach benefits the user financially as well. This paper offers a model of the HG-iPhone Service that could be provided as an embedded system in our Home Gateway system.

1 Introduction

In the last few years, there are an increasing number of uses for services utilizing VoIP (Voice over IP) technologies. The most well-known application amongst them is Internet Telephony, so-called "IP Telephony", which provides a voice service through packet switch systems. Thus, both voice and data can be transmitted using an integrated mechanism as well as media [1],[2],[3].

The remarkable development of IP telephony started in 1995. VocalTec was a pioneer in the IP telephony market with PC-software which provides a voice connection between PCs over IP-based networks. The product was ideally suited for the Internet. Shortly after that, several other competing software packages were launched consecutively. In 1996, early inter-networking trials between IP networks and PSTNs were conducted. In 1997, the Delta Three launched the first phone-to-phone service for commercial use.

I. Chong (Ed.): ICOIN 2002, LNCS 2343, pp. 589-597, 2002.

Currently IP telephony offers inexpensive rates. However, IP telephony calls lack certain qualities when compared to calls over PSTNs. People are not willing to tolerate a reduction in quality due to packet losses, delay and jitters, deteriorated voice quality, etc. for a reduction in price.

In this paper, we propose the HG-iPhone Service, which provides both a cost effective service and satisfactory playback quality for end-users. To provide the functionality of Internet Phone service, we use the Home Gateway system as the base system. The reasons are as follows:

First, we can make use of the characteristics that the Home Gateway system [4] is always on. Second, because the HG-iPhone Service functionality is provided as one of the basic services in the Home Gateway system, user does not need to purchase any additional equipment such as an IP Phone or a PC system.

2 Internet Telephony

2.1 Internet Telephony Services

Internet Telephony provides a telephone service, which means a transmission of packetized voices over an IP network such as the Internet. Namely, it transforms analog voice signals into 64Kbps sampling and compresses the digitized data through an encoding algorithm (e.g., G.723.1, GSM6.10). The encoded data is transmitted over the IP network, using a standard protocol such as H.323 [1],[2].

There are several types of Internet Telephony including PC-to-PC, PC-to-Phone, and Phone-to-Phone according to the terminals used or connection schemes [2]. Most recently, the PC-to-fax, fax-to-fax and Unified Messaging System (UMS) services have also emerged as new areas of application for Internet Telephony services.

2.2 Trends in Internet Telephony Standards

The leading signaling protocol 'standard' for IP telephony network is based on H.323. The network architecture of the H.323 consists of four types of network elements: terminals, gatekeepers, gateways, and multipoint control units (MCU). The minimal configuration consists of two terminals connected to a local area network (LAN) [5],[6],[7].

H.323 defines the system, control procedures, media descriptions, and call signaling. H.225.0 defines media packetization, stream synchronization, control message packetization, and control message formats. H.245 defines negotiation of channel usage and capabilities exchange. It is used for opening and closing of channels for audio, video, data, and camera mode requests. RAS (registration, admission, status) is used for communication between a terminal and a gatekeeper. An optional data channel can be supported with the T.120 series of ITU standards. Codecs needed for speech and video coding are defined by G.711, G.722, H.728, H.723.1, G.729, H.261, and H.263, respectively.

The venders (ex. MicroSoft, Netscape, and Intel, etc.) which lead in the field of PC Client Applications for personal computers, adopted H.323 as the Internet Telephony standard [6]. On the other hand, VoIP Forum and IMTC (International Multimedia Teleconferencing Consortium) use the G.723.1 algorithm as the codec algorithm, because the G.723.1 algorithm has the higher compression rate although it is more complex than others. Therefore, considering compatibility with other products, the proposed service in this paper also uses the H.323 protocol and G.723.1 codec.

3 HG-iPhone Service

3.1 Features

When customers use the Internet Phone service within the Home Gateway system, the HG-iPhone Service can provide the follow benefits to them:

• *Fast and simple connection*

In order to use Internet Phone services generally, users may have to turn on the proper software program or connect the system to the Internet. These procedures can cause user inconveniences. However, HG-iPhone Service uses the Home Gateway system which is always powered on. In other words, when users want the Internet Phone service, it can provide the service to them as soon without delay. It also allows uses to follow a simplified procedure to connect to the service.

• *The convenient UI(user interface)*

The VoIP service applications such as DialPad, Wowcall, and Net2phone require users to turn on the PC and then execute the proper program or connect the site. However, HG-iPhone Service uses a telephone unit, so it can provide a convenient and easy way of connecting to the Phone service communication.

• *Cost effective*

When making Long Distance Calls, the service can benefit users by reducing telephone charges as it transmits the calls over the Internet instead of commercial PSTNs.

• *QoS(Quality of Service) support*

This service reduces the bandwidth by transmitting compressed audio data using G.723.1 codec. Furthermore, to reduce delays caused by encoding and decoding, it supports the hardware codec algorithm.

• *Point-to-point connection model*

This proposed service has a point-to-point model that directly connects a caller and a callee.

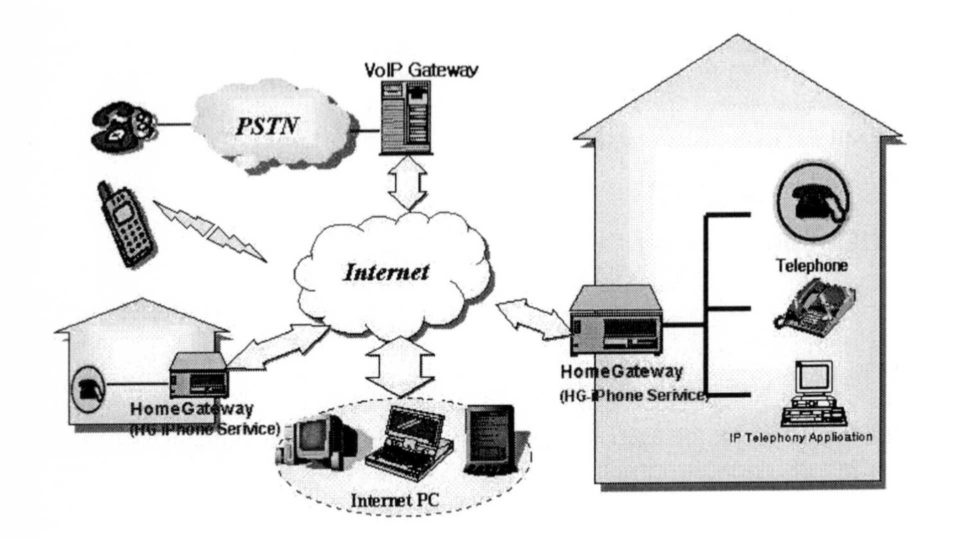

Fig. 1. IP Telephony service model using the HG-iPhone Service of Home Gateway system

3.2 Functionalities

The HG-iPhone Service allows service between a subscriber using a computer (IP-voice) and another using a phone on a PSTN by using the VoIP gateway located between the Private Branch exchange (PBX) and a TCP/IP network. It also allows service between two users on an IP network.

Figure 1 shows an IP Telephony service model providing the HG-iPhone Service within Home Gateway system.

Figure 2 illustrates the architecture of HG-iPhone Service. The functionalities of each module in the architecture are outlined as follows:

• *System Control module*

controls the entire procedure of the HG-iPhone Service modules such as H.323 Protocol module, Audio Data processing module, and UI module.

• *User Interaction module*

detects On/Off hook event and DTMF signal from telephone unit. And, when a user connects or disconnects the peer user, it sends dialing and ring or ringback tone to the user.

• *H.323 Protocol module*
consists of several protocol modules, including H.225.0-Q.931 for call control, H.245 for bearer controls, and RTP/RTCP for voice data transmission.

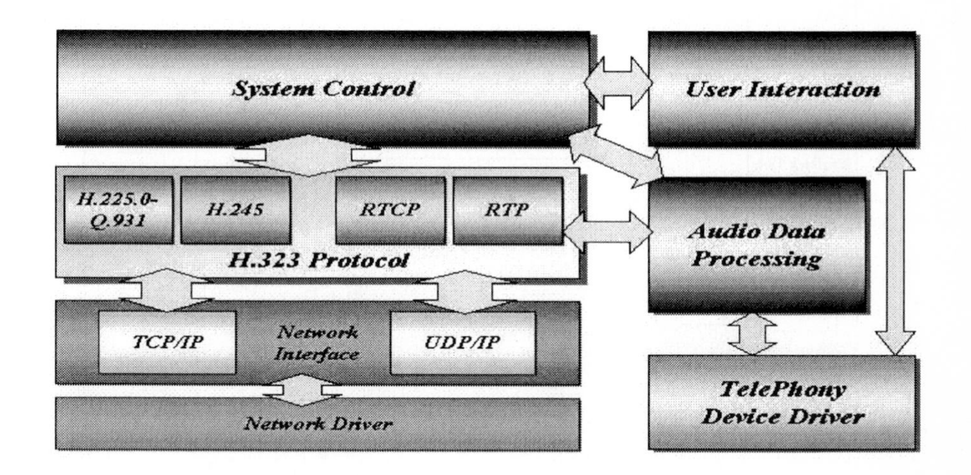

Fig. 2. The architecture of HG-iPhone Service

• *Telephony Device Driver module*
generates dial, ring, and ringback tones. Also, it transmits encoding audio using G.723.1 codec to Audio Data Processing module, and decodes audio data received from Audio Data Processing, then sends the decoded audio to the Telephone device.

• *Audio Data Processing module*
receives the encoded data from Telephony Device Driver module. Also, it rearranges the audio data received from RTP module in the playback buffer, and then sends the data to Telephony Device Driver.

3.3 Call Flows

This subsection describes the call procedure flows of the HG-iPhone Service proposed in this paper.

3.3.1 Call Connection Procedure

Figure 3 illustrates the flow of the call connection procedure between User1 and User2. The action begins with User1.

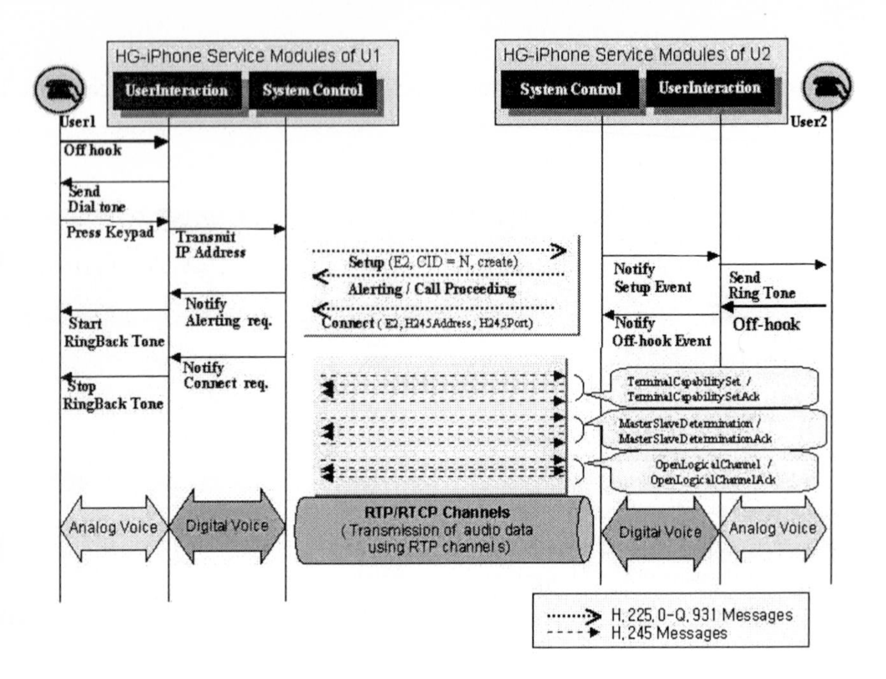

Fig. 3. Sequence flow for connecting a call

1) At the first, when User1 picks up the phone in order to make a call to User2, the UserInteraction module in the HG-iPhone Service of U1 detects the off-hook event. Subsequently, the module sends a dial tone to User1's handset.

2) User presses the IP address of User2 using the keypad.

3) The UserInteraction module of U1 transforms the DTMF signals inputted from the keypad into an IP address and then, delivers it to the SystemControl module of U1. Using the IP address, the SystemControl module performs TCP connection processing for exchanging an H.225.0-Q.931 message.

4) After TCP connection processing, an H.225.0-Q.931 signaling procedure is performed [5],[7]. The SystemControl module of U1 sends a Setup message which contains information (CID=N, conferenceGoal=create). The SystemControl module of U2, when it receives the message, sends Alerting or Call Proceeding messages. In this case, the telephone of User2 is ringing and then, if User2 takes it off the hook, the SystemControl module of U2 sends a Connect message to the HG-iPhone Service of U1.

5) When it receives the Connect message, the HG-iPhone Service of U1 performs a H.245 control procedure using the IP address and Port number included in the Connect message. At the first, the HG-iPhone Services of U1 and U2 exchange the TerminalCapabilitySet and TerminalCapabilitySetAck messages in order to negotiate the terminal's capability. And then, both HG-iPhone Services perform a H.245 master-slave determination procedure. Lastly, both HG-iPhone Services open RTP/RTCP channels

for the audio transmission from exchanging OpenLogicalChannel and OpenLogicalChannelAck messages [5],[7].

6) When the connection processing is complete, the users can talk with each other through RTP channels that transmit voice packets.

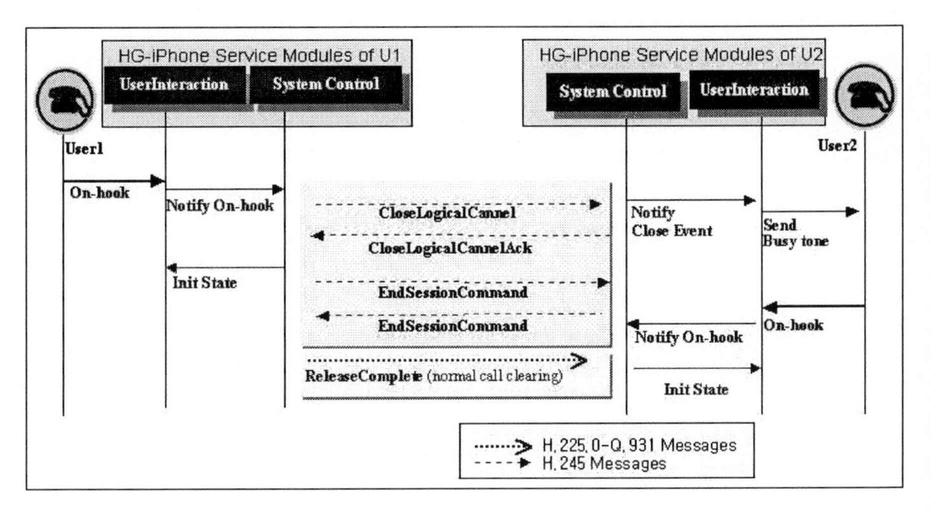

Fig. 4. The procedure for releasing the call

3.3.2 Call Release Procedure

When one of the users hangs up, the SystemControl module detects on-hook state and closes all connected sessions, then it enters into its initial state. Figure 4 illustrates the procedure for the disconnection after User1 and User2 hang up.

3.3.3 Call Rejecting Procedure

In case a callee is busy or rejects the call, after receiving a Setup message, the callee sends the ReleaseComplete message including the information of the call-rejected. In this case, the User1's phone gives a busy tone signal.

3.4 QoS Support

To achieve a perceived quality and to compensate the realtime sensitive factors such as jitter and end-to-end delay and packet loss, our system uses buffering and redundant data transmission mechanisms.

• *Audio codec*

The HG-iPhone Service uses the G.723.1 codec (6.3 kbps). This codec is more complex than other codecs. However, because it has higher compression rates, it has been used on several IP Telephony standards. Therefore, it is used in this proposed service and also, it is provided as a hardware codec in order to minimize the processing delay for encoding/decoding and the required bandwidth for the network.

• *Buffering mechanism*
The HG-iPhone Service sets the play buffer size to be 480 bytes. So, the play buffer can include 20 frames, in other words, the initial playback latency is about 0.6 seconds. The determination for buffer size is subjective, and needs more objective analysis. With the play buffer, the proposed service achieves both the compensation for jitter and delay and the rearrangement for out of sequence packets.

• *Redundant data transmission mechanism*
In the real time sensitive applications, the packets that don't arrive within a restricted time boundary are considered as lost packets. These cases can occur in the Packet Network such as the Internet. So, in this case, an appropriate recovery mechanism is required because the buffering is impossible for recovery the lost packets. Therefore, the HG-iPhone Service used the redundant data transmission mechanism [8].

4 Simulation and Conclusion

The HG-iPhone Service proposed in this paper was developed over Linux operating system (OS) (kernel ver2.2.16). The InternetPhoneJack PCI Card produced by Quicknet Co. was used as a Telephony Interface. And linux-ixj-1.0 source code was used as a Device Driver. H.323 protocol module developed by our laboratory has H.225.0-Q.931, H.245, and RTP/RTCP protocols, except for the RAS function. It used the G.723.1 codec.

To test the compatibility of HG-iPhone Service, we used the experimental set up as shown in Figure 5. In these experiments all the machines were connected via Ethernets. One of them consists of the HG-iPhone Service Application and Linux OS. Another consists of the MS-NetMeeting Application and Win98 OS. We could observe the test that was performed very well. Furthermore, the audio quality between the caller and the callee was satisfactory. However, further objective experimental results are necessary. Therefore, we are now experimenting with several well-know tools designed for voice transmission over the Internet, namely DialPad [9], and OpenPhone [10].

In this paper we proposed the HG-iPhone Service in order to provide an Internet Phone service functionality in the Home Gateway system. The HG-iPhone Service is composed of G.723.1 codec, H.323 module, and Phone Device Interface module, and also has the appropriate mechanism [8] to recover packet losses and to compensate for limitations such as jitters and delays. We observed the compatibility by testing several Internet Phone Applications such as MS-NetMeeting and Dialpad. In order to apply this service to the market, the critical factor is QoS. Therefore, continuous study in the

QoS is necessary. In particular, more objective and various experimental results are needed for the performance of the HG-iPhone Service.

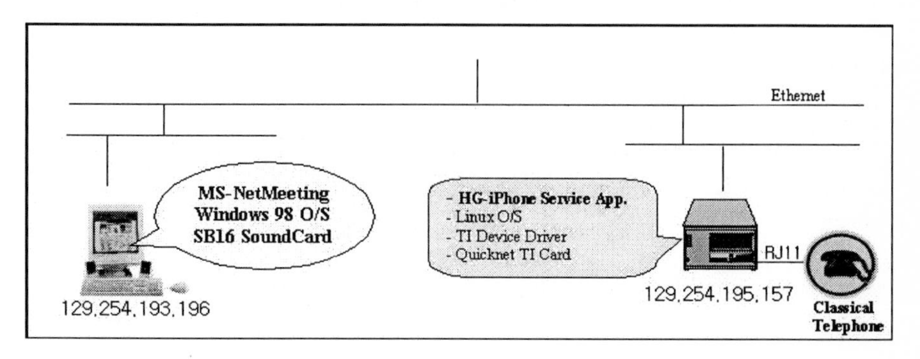

Fig. 5. Experimental set up for compatibility test

References

[1] J. Park, D. Ko, "The Design and implementation of Real Phone," Telecommunications Review, V.9, No.1, pp.106-113, Jan. 1999.

[2] O. Kwon, "Telecommunication Service using VoIP," Report of ITPS, Vol.12, No.23, Dec. 2000.

[3] Lasse Huovinen, Shuanghong Niu, "IP Telephony," http://www.hut.fi/~lhuovine/study/iwork99 /voip.html

[4] K. Park, J. Kim, J. Yang, "Home Gateway Technology," KICS, V.17. No.11, p.101-110.

[5] ITU-T Recommendation H.323(1999), Packet-based multimedia communications systems.

[6] ETRI, "Technology &Market Reports of 30 Strategic IT Fields," Dec. 1999.

[7] S. Jang, "Implementation of H.323 Internet Phone based PC-to-PC," CEIC2000, pp.154-156, Nov. 2000.

[8] Y. Oh, N. Baek, K. Park, H. Jung, K. Lim, "An Adaptive Packet Loss Recovery Scheme for Realtime Data in Mobile Computing Environment", KISS Journal V.28, No.3, pp.389-405, Sept. 2001.

[9] SEROME tech. Inc., "DialPad," http://www.dialpad.co.kr

[10] Open H.323 Project, "OpenPhone," http://www.openh323.org/code.html

An Efficient Access Scheme for Variable Length Packets in Slotted ALOHA Channel

Nor Diana Yaakob, Tanabe Hidehiko, and Umeda Hiroyuki

Department of Electrical and Electronics Engineering
Fukui University, Japan.
Zip-Code 910-8507, Fukui-Ken,
Fukui-Shi, Bunkyo 3-Chome, 9-1.
Tel/Fax: (81)778-27-8570
umeda@dignet.fuee.fukui-u.ac.jp

Abstract. The slot division access method is suggested as an efficient method of transmitting variable length packets through Slotted ALOHA channel. Variable length packets are divided into two groups according to division parameter p, and synchronized their transmission at the beginning or at the end of a coming slot. By this method, the maximum throughput reached a value of 0.26. In addition, variable length packets are divided into four groups according to three division parameters α, β and γ, and synchronized their transmission to two different slots. This method increased the throughput value to 0.35, and also showed an improvement in average delay performance. Finally, we proposed a contention access scheme by combining a reservation-slot to the aloha-slot, where we can expect an increase in throughput value and improved delay performance.

Keywords: *Variable length packets, division parameter, reservation ALOHA access scheme.*

1 Introduction

In Slotted ALOHA system, transmitted packets are assumed to be fixed-length and synchronized their transmission at the beginning of a coming slot. Collision in S-ALOHA channel happened when more than one packet tried to access a slot, but for variable length packets, collision also occurred as the slot's length is set to the packet's maximum length. This led to the decline in transmission efficiency. Here we proposed a model where packets are divided into certain groups according to the slot's division parameter, and expect an increase in throughput value and improvement in delay performance. We also examine the effect on throughput and delay value when reservation ALOHA access scheme is added to this method. In addition, we will evaluate the performance from both simulation analysis and theoretical interpretation.

I. Chong (Ed.): ICOIN 2002, LNCS 2343, pp. 598-608, 2002.

2 An Efficient Access Method for Variable Length Packets

2.1 Division in Slot Access Scheme

During transmission, variable length packets are divided into two groups, one synchronized its transmission at the beginning of a coming slot, while the other at the end of a slot. It is later shown that the throughput value can be derived from the packets maximum collision length. Let the packet length be $\tau_i = \tau \times i (i = 1,2,..,m)$ and assume $\tau = 1$. As to maintain packet's generality let the packet's maximum length m where $m = 2^k (k = 1,2,.., l)$. Let all assumptions be dimensionless in time as to simplify calculations.

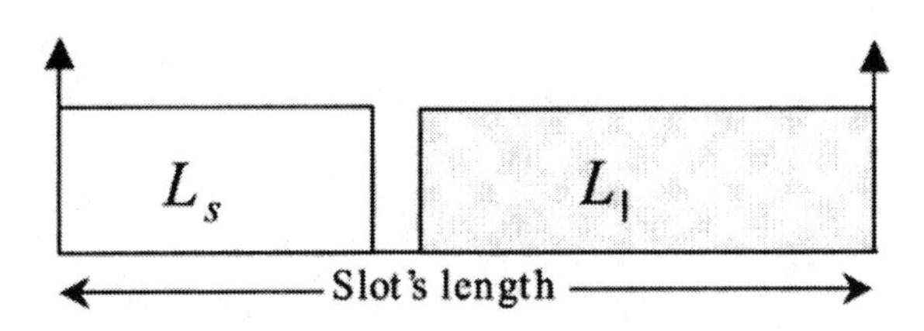

Fig. 1. Division in slot access model

Fig.1 is the model of division in slot access scheme. Packet of length 1 to $[p \times 2^k]$ is defined as short packet L_s while packet of length $[(p \times 2^k)+1]$ to 2^k is defined as long packet L_l . As shown in Fig.1, L_s packet is synchronized at the beginning of a coming slot while L_l is synchronized at the end of a slot. The same argument continues even when L_s and L_l replaced each other.

Here we define p as the parameter of dividing the packets. By calculating the packet length combination patterns for a successful transmission, p is the value when the total cumulative packet length for a successful transmission reached its maximum value. (Refer to Appendix A).

2.2 Two Slots for Four Groups of Variable Length Packets Access Scheme

Here we suggested that variable length packets are divided into four groups according to three division parameters α, β and γ. As shown in Fig.2, we prepared a frame containing two slots, named Slot A and Slot B. Packet of length 1 to $[\alpha \times 2^k]$ is defined as packet L_s, $[(\alpha \times 2^k)+1] \sim [\beta \times 2^k]$ as L_s', $[(\beta \times 2^k)+1] \sim [\gamma \times 2^k]$ as L_l' and $[(\gamma \times 2^k)+1] \sim [2^k]$ as L_l. L_s packet and L_l packet synchronized their transmission at the front and end of Slot A, while L_s' and L_l' packet synchronized their transmission at the front and end of Slot B. As to enhance transmission performance, we suggested that when packets for Slot A are not there to be transmitted during a time, Slot A can be used as Slot B and vice versa. By using this method we can expect an increase in throughput value and improved delay performance. The division parameters α, β and γ can be derived in a similar way as in Appendix A. Results showed that the throughput value for two slots for four groups access method reached its maximum when $\alpha = 0.09$, $\beta = 0.39$ and $\gamma = 0.70$.

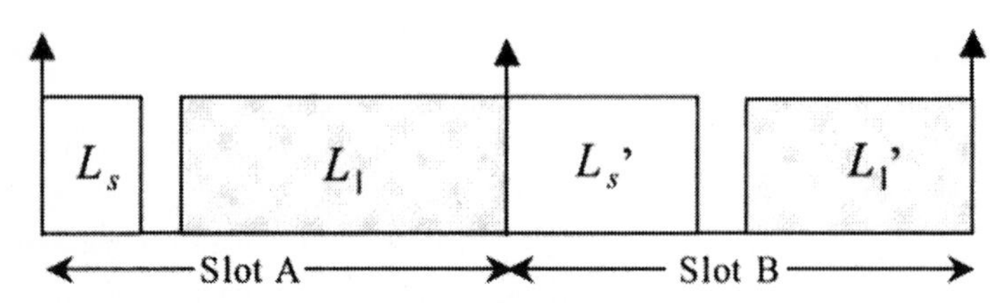

Fig. 2. Two slots for four groups access model

2.3 Deriving the Theoretical Values

The theoretical values of division in slot access and two slots for four groups access method can be derived from packet's maximum collision length. Packets with length i arrived with uniform distribution $f(\tau_i) = 1/m$ which followed the independent and identical Poisson process at rate σ. The probabilities that packet of length i is successfully transmitted (i.e. no collision occurred) can be obtained by calculating the maximum collision length T_{ij} as given below;

$$\prod_{j=1}^{m}\left(1-\sum_{\delta=1}^{\infty}\frac{(\lambda_i T_{i,j})^{\delta}}{\delta!}\exp(-\lambda_i T_{i,j})\right)=\exp\left(-\sum_{j=1}^{m}\lambda_i T_{i,j}\right) \tag{1}$$

Here we write $\lambda_i = \sigma f(\tau_i) = \sigma / m$, therefore the value of λ_i depends on m. The maximum collision length T_{ij} can be defined as the same value of the slot's length when packet τ_i collided with packet τ_j. Below are the examples of successful and unsuccessful transmission in division slot access scheme when m=1024. Fig.3(a) is an example of successful transmission where no collision occurred between packets and the sum of τ_i and τ_j do not exceed the slot's length. Therefore in this case T_{ij} becomes 0. But for Fig.3(b), Fig.3(c) and Fig.3(d), when collision occurred in the slot, the value of T_{ij} equals to the slot's length T_{ij}=1024.

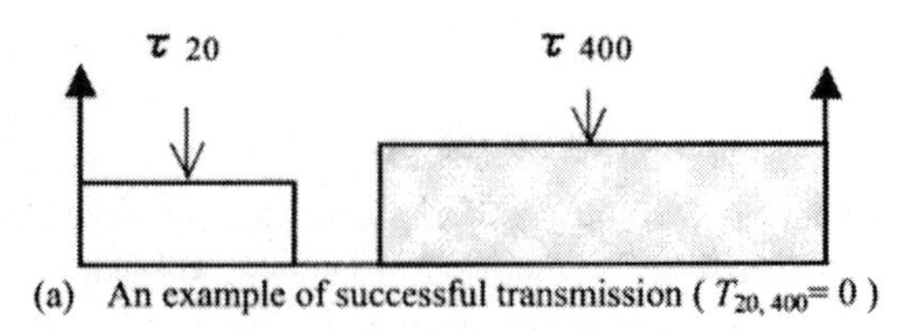

(a) An example of successful transmission ($T_{20,400}$= 0)

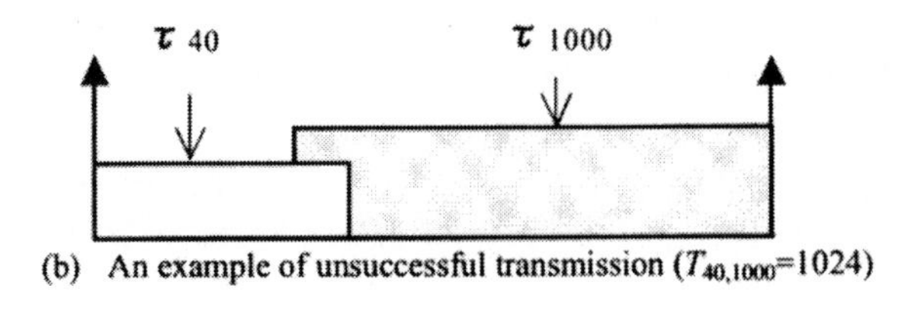

(b) An example of unsuccessful transmission ($T_{40,1000}$=1024)

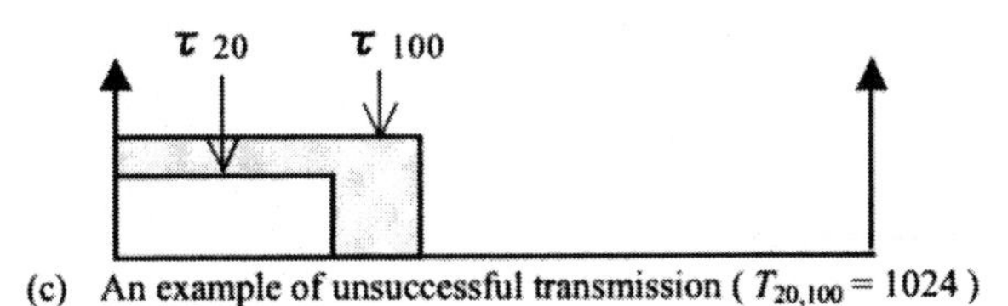

(c) An example of unsuccessful transmission ($T_{20,100}$ = 1024)

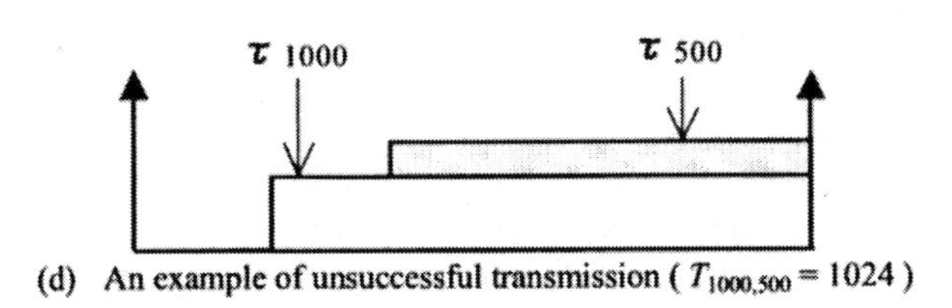

(d) An example of unsuccessful transmission ($T_{1000,500}$ = 1024)

Fig. 3. Maximum collision length in ALOHA slot

The average number of packets transmitted to the channel can be calculated as below;

$$G = \sum_{i=1}^{m} \tau_i \lambda_i \tag{2}$$

While the average throughput value is

$$S = \sum_{i=1}^{m} \tau_i \lambda_i \exp\left(- \sum_{j=1}^{m} \lambda_i T_{i,j} \right) \tag{3}$$

Transmission delay can be calculated as

$$D = \frac{1}{m} \sum_{i=1}^{m} D_i = \frac{1}{m} \left(0.5 + L_i + \left(\exp\left(\sum_{j=1}^{m} \lambda_i T_{i,j} \right) - 1.0 \right) \left(\frac{K + 1.0}{2.0} \right) \right) \tag{4}$$

K : Average number of slots needed for retransmission
L_i : Slot's length required for i length packet

As for two slots for four groups access method, the average traffic can be calculated the same way as in Eq.(2). But the average throughput is the total value of throughput in Slot A and Slot B, leading the equation to be

$$S=\sum_{i=1}^{m}S_i=S_{AA}+S_{BB}+S_{BA}+S_{AB}$$
$$=\sum_{i=1}^{\alpha}\tau_i\lambda_i\left(\exp\left(-\sum_{j=1}^{m}2\lambda_iT_{i,j}\right)(1-p_Bp_b)+\exp\left(-\sum_{j=1}^{m}\lambda_iT_{i,j}\right)p_Bp_b\right)$$
$$+\sum_{i=\alpha+1}^{\beta}\tau_i\lambda_i\left(\exp\left(-\sum_{j=1}^{m}2\lambda_iT_{i,j}\right)(1-p_Ap_a)+\exp\left(-\sum_{j=1}^{m}\lambda_iT_{i,j}\right)p_Ap_a\right)$$
$$+\sum_{i=\beta+1}^{\gamma}\tau_i\lambda_i\left(\exp\left(-\sum_{j=1}^{m}2\lambda_iT_{i,j}\right)(1-p_Ap_a)+\exp\left(-\sum_{j=1}^{m}\lambda_iT_{i,j}\right)p_Ap_a\right)$$
$$+\sum_{i=\gamma+1}^{m}\tau_i\lambda_i\left(\exp\left(-\sum_{j=1}^{m}2\lambda_iT_{i,j}\right)(1-p_Bp_b)+\exp\left(-\sum_{j=1}^{m}\lambda_iT_{i,j}\right)p_Bp_b\right) \quad \textbf{(5)}$$

S_{AA} : Throughput of Slot A
S_{BB} : Throughput of Slot B
S_{AB} : Throughput of Slot A when used as Slot B
S_{BA} : Throughput of Slot B when used as Slot A
p_a : Probabilities of no packets for Slot A are transmitted
p_b : Probabilities of no packets for Slot B are transmitted
p_A : Probabilities of no packets for Slot A is waiting in front of Slot B
p_B : Probabilities of no packets for Slot B is waiting in front of Slot A

The transmission delay for two slots for four groups access method can be calculated as below;

$$D=\frac{1}{m}\sum_{i=1}^{m}D_i$$
$$=\frac{1}{m}\sum_{i=1}^{\alpha}\left(0.5+0.5p_Bp_b+(1.0-p_Bp_b)+L_i+\left(\frac{G_i}{S_i}-1.0\right)\left(\frac{K+1.0}{2.0}\right)\right)$$
$$+\frac{1}{m}\sum_{i=\alpha+1}^{\beta}\left(0.5+0.5p_Ap_a+(1.0-p_Ap_a)+L_i+\left(\frac{G_i}{S_i}-1.0\right)\left(\frac{K+1.0}{2.0}\right)\right)$$
$$+\frac{1}{m}\sum_{i=\beta+1}^{\gamma}\left(0.5+0.5p_Ap_a+(1.0-p_Ap_a)+L_i+\left(\frac{G_i}{S_i}-1.0\right)\left(\frac{K+1.0}{2.0}\right)\right)$$
$$+\frac{1}{m}\sum_{i=\gamma+1}^{m}\left(0.5+0.5p_Bp_b+(1.0-p_Bp_b)+L_i+\left(\frac{G_i}{S_i}-1.0\right)\left(\frac{K+1.0}{2.0}\right)\right) \quad \textbf{(5)}$$

From Eq.(2) to Eq.(6), we can plot the calculated result and compare it with the simulation result. In simulations, random sample numbers are used to produce packets and fixed back-off algorithm is applied during transmission of unsuccessful packets.

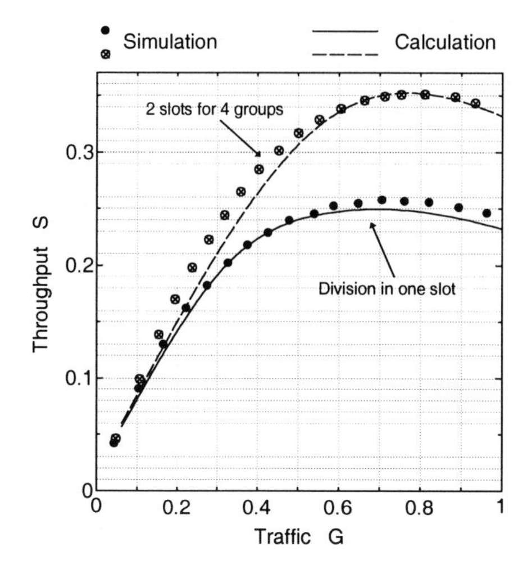

Fig. 4. Traffic-throughput characteristics for ALOHA access scheme

From Fig.4 we can see that the simulation plot draw near to the calculation plot. For division in slot access method, the maximum throughput value is 0.26,while for two slots for four groups access method the maximum throughput value is 0.35.

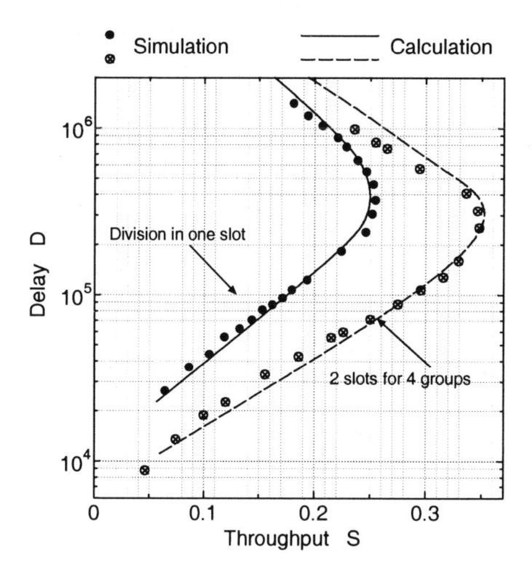

Fig. 5. Throughput- delay characteristics for ALOHA access scheme

Fig.5 showed the result for transmission delay for both access scheme. The calculation plot draw near to the simulation plot, and two slots for four groups access showed an improved value compared to division in slot access method.

3 Reservation ALOHA Access Scheme

As shown in Fig.6, we prepared a frame consisted of an ALOHA slot and reservation slot. ALOHA slot is for variable length packets, while reservation slot is for fixed length packet. Packets length varies from 1 to 2^{k+1} [bit]. Packet of length $1\sim2^k$ is called short packet G_S while packet of length $(2^k+1)\sim2^{k+1}$ is called long packet G_ℓ. Furthermore packet G_ℓ is cut off into fixed length packet 2^k (which equals to the half of the frame's length) and to another short packet group (we called this group as G_S', as to distinguish it from G_S).

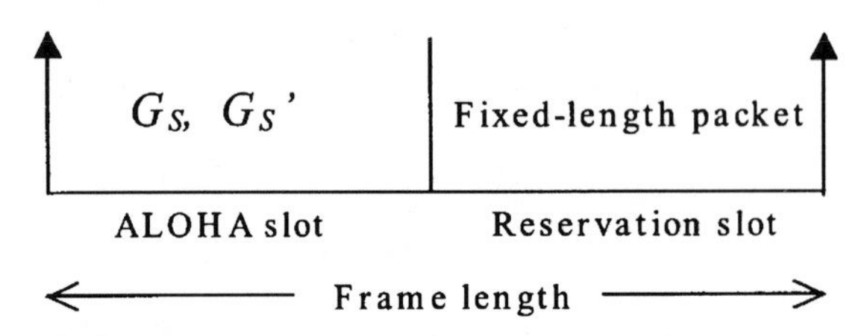

Fig. 6. Reservation ALOHA access scheme model

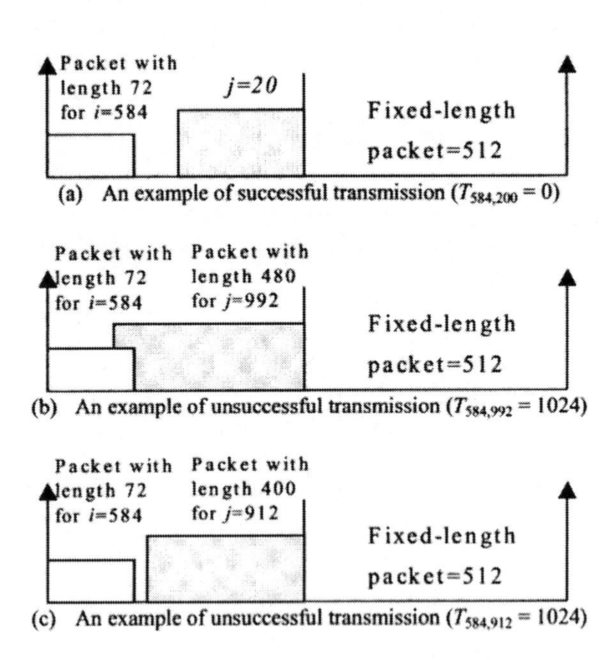

Fig. 7. Maximum collision length for reservation ALOHA access scheme

Fig.7 showed the maximum collision length patterns occurred in reservation ALOHA access scheme. Fig.7(a) is an example of a successful transmission, where there is no collision between G_S and G_S' packets in ALOHA slot. As for Fig.7(b),the sum of G_S' packets are larger than 512, and the fixed length packet can only be transmitted to the reservation slot if G_S' packets are successfully transmitted, therefore $T_{584,992} = 1024$. But in Fig.7(c), G_S' packets are successfully transmitted and produced two fixed length packets. One fixed-length packet is to be transmitted to the reservation slot, while the other one begin its transmission to the next coming frame. Therefore the maximum collision length correspond with the reservation slot's length, $T_{584,912} = 512$.

Now we can calculate the average throughput for ALOHA packets as below;

$$S_A = \sum_{i=1}^{512} S_{A_i} = 2\sum_{i=1}^{512} \tau_i \lambda_i \exp\left(-\sum_{j=1}^{1024} \lambda_i T_{i,j} \right) \tag{6}$$

While the throughput of fixed length packets can be calculated as below;

$$S_R = \sum_{i=513}^{1024} S_{R_i} = 512\sum_{i=513}^{1024} \lambda_i \exp\left(-\sum_{j=1}^{1024} \lambda_i T_{i,j} \right) \tag{7}$$

As to enhance transmission performance, if the reservation slot is vacant,(i.e. no fixed length packet is transmitted) the reservation slot can be used as an ALOHA slot. As we observe the packets from an infinite number of transmissions, the throughput of variable length packets accessing a vacant reservation slot can be calculated as below;

$$S_A' = \sum_{i=1}^{512} S_{A_i}' = \frac{1-2S_R}{1+2S_R} \sum_{i=1}^{512} S_{A_i} \tag{8}$$

When packet G_S or G_S' are transmitted successfully in the vacant reservation slot, we can expect that the transmission of fixed length packets are increased at the same time. The amounts of increased fixed length packets are;

$$S_R' = \sum_{i=513}^{1024} S_{R_i}' = \frac{1-2S_R}{1+2S_R} \sum_{i=513}^{1024} S_{R_i} \tag{9}$$

Therefore the overall throughput can be calculated as

$$S = \sum_{i=1}^{1024} S_i = \sum_{i=1}^{1024} (S_{A_i} + S_{R_i} + S_{A_i}' + S_{R_i}') \tag{10}$$

The overall traffic when reservation slot is used as an ALOHA slot is

$$G = \sum_{i=1}^{1024} G_i = \frac{2}{1+2S_R} \sum_{i=1}^{1024} \lambda_i \tau_i \tag{11}$$

Lastly, since we have known the traffic and throughput value, we can apply it to calculate the theoretical value of transmission delay D. Since a vacant reservation slot can be used as an ALOHA slot, we can expect a decrease in transmission delay value.

$$D = \frac{1}{1024}\sum_{i=1}^{1024}\left(0.5 + L_i + \left(\exp\left(\sum_{j=1}^{1024}\lambda_i T_{i,j}\right) - 1.0\right)\left(\frac{K+1.0}{2.0}\right)\right) \quad \textbf{(12)}$$

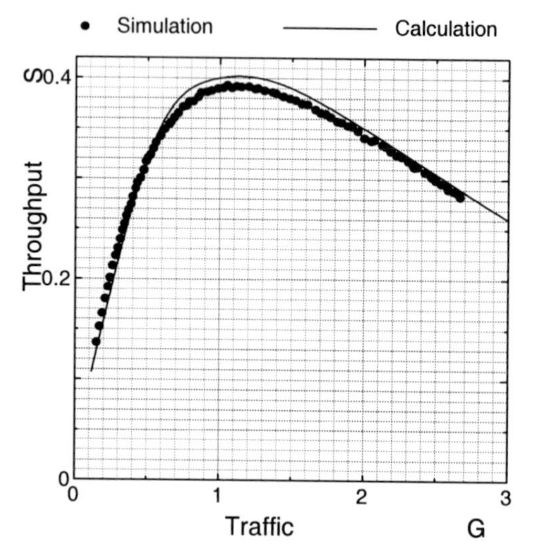

Fig. 8. Traffic- throughput characteristics in reservation ALOHA access scheme

Fig.8 showed that for traffic and throughput value, the simulation plot draw near to the calculation plot. We can also see that the reservation ALOHA access scheme increased the division in slot access method's maximum throughput value from 0.26 to 0.40.

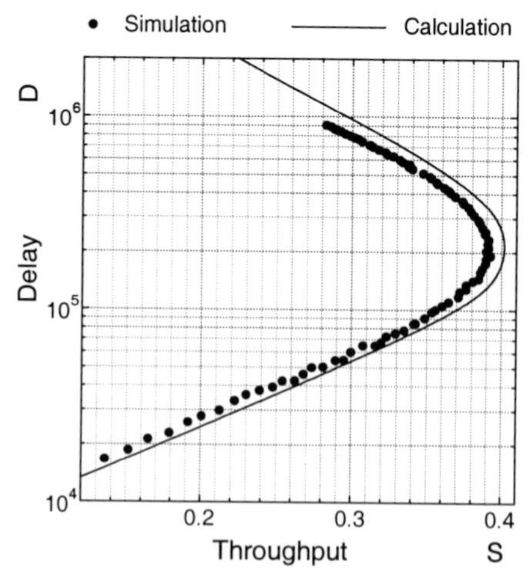

Fig. 9. Throughput-delay characteristics of reservation ALOHA access method

As for transmission delay, the simulation plot draw near to the calculation plot. We can also see that the reservation ALOHA access scheme improved the transmission of fixed length packets.

4 Conclusions

We have proposed the idea of grouping the variable length packets in slotted ALOHA channel with division parameters and gained an improvement in both throughput value and delay performance. Furthermore, by adding a reservation slot to an ALOHA slot, we could expect an increase of fixed length packets throughput and reduced transmission delay. To analyze the basic characteristics, end-to-end propagation delay is disregarded as to simplify the network model. The throughput value and transmission delay's validity is checked from both and theoretical assumptions.

References

1. Abramson N.,"The Throughput of Packet Broadcasting Channels", *IEEE Trans.Comm.Com-25*.pp117-128 (Jan 1977)
2. Ferguson M., "An Approximate Analysis of Delay for Fixed and Variable Length Packets in Unslotted ALOHA Channel", *IEEE Trans.Comm.Com-25*,pp.644-654(1977)
3. Hiroyuki Umeda, "Throughput Performance of Reservation ALOHA Access Scheme with Variable Length Packets", *IEICE Trans.Comm.(B)*, Vol.J82-B,No.8,pp.1608-1612,August 1999.

Appendix

Derivation of the optimum division parameter in slotted ALOHA channel for variable length packets.

As shown in Fig.1, variable length packets are divided into two groups L_S and L_ℓ and their length varied as below;

$$L_\ell = (m_1 + 1 \le i \le m) \qquad \textbf{(A.1)}$$
$$L_S = (1 \le i \le m_1)$$

L_S packet synchronized its transmission at the front of a coming slot, while L_ℓ packet synchronized its transmission at the end of a coming slot. The slot's length is equal to the packet's maximum length m. Let L_S's length be x and L_ℓ's length be y. When the total of $x + y \le m$, it means no collision occurred between packets. For example, when packet' maximum length m = 256, the number of combination patterns which x

and y can take when the rate of packets occupying a slot $z = (x + y)/m$, is shown in Fig.A.1.

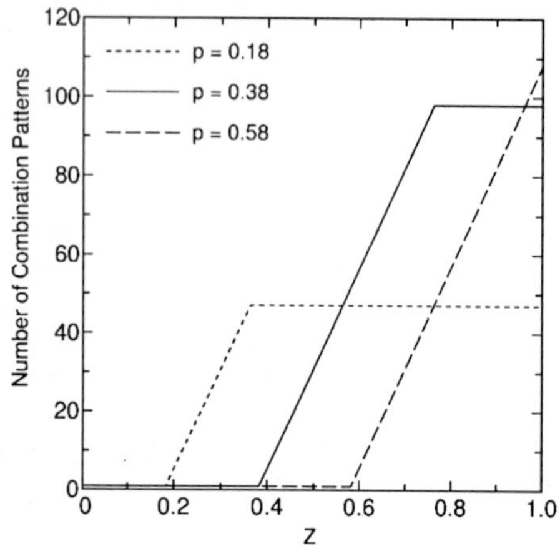

Fig. A.1. Number of combination patterns for a pair of packets vs. Z

Moreover, the total cumulative packet length T calculated from Fig.A.1 is expressed in Fig.A.2

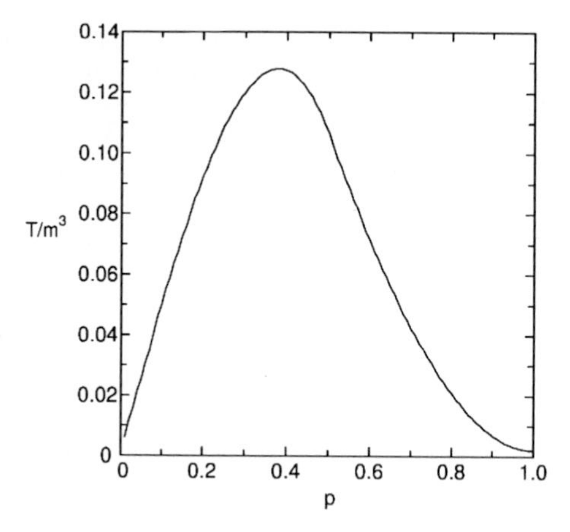

Fig. A.2. Total cumulative packet length of successful packets vs. division parameter p

We can see from Fig.A.2, when $p = m_1 / m = 0.38$, it showed the largest value of total cumulative packet length for successful transmission.

The division parameters for two slots for four groups access method can be derived similarly as in Appendix A. The total cumulative packet length for successful transmission showed the largest value when division parameters $\alpha = 0.09$, $\beta = 0.39$, $\gamma = 0.70$

A Channel Assignment Scheme for Handoff in Wireless Mobile Networks

Kil-Woong Jang and Ki-Jun Han

Department of Computer Engineering, Kyungpook National University, Daegu, Korea
jangkw@netopia.knu.ac.kr and kjhan@bh.knu.ac.kr

Abstract. In this paper, we propose a new channel assignment scheme, which is a hybrid of the channel reservation scheme and the channel carrying scheme. It is designed to efficiently carry out handoffs in wireless mobile networks. We evaluated our scheme with a two-cell model using both Markov analysis and computer simulation. The analytical and simulation results indicate our scheme may offer better performance than the conventional reservation scheme and channel carrying scheme in terms of handoff blocking probability and channel utilization.

1 Introduction

Handoff is one of the most challenging technical issues the evolving and next-generation wireless networks pose. In general, to be successful, the handoff should be completed while the mobile station is in the overlap region. Another important issue to consider is the minimum reuse distance among cells for which the use of the same channel produces an acceptable co-channel interference level. With some approaches, a channel can be allocated to a cell even if co-channel interference is present [2]. A handoff call may be blocked if there is no free channel for the next cell. Since blocking a handoff call is less desirable than blocking a new call, specific schemes have been developed to prioritize handoff calls [1]. Two schemes are briefly described below.

1.1 Channel Reservation Scheme

The channel reservation scheme offers a generic means of improving the probability of successful handoff by simply reserving a number of channels exclusively for handoffs [7]. In each cell, the channels are divided into two subsets which are reserved for the new calls and the handoff calls. Some channels are allocated for the new and handoff calls while others are allocated for the handoff calls only. Specially, in each cell a threshold is set, and if the number of channels currently being used in the cell is below that threshold, both new and handoff calls are accepted. However, if the number of channels being used exceeds this threshold, incoming new calls are blocked and only handoff calls are admitted.

I. Chong (Ed.): ICOIN 2002, LNCS 2343, pp. 609-617, 2002.

The advantage of this scheme is its simplicity in implementation. However, since fewer channels are granted to the new calls, it runs the risk of inefficient spectrum utilization and reduction of the total carried traffic.

1.2 Channel Carrying Scheme

This scheme is to allow mobile terminal to carry its current channels into the next cell not to be blocked the handoff request. Due to channel movement, however, it shortens the reuse distance and may violate the minimum requirement of the reuse distance. We let δ be the minimum reuse distance. This scheme uses $(\delta+1)$-channel assignment scheme to ensure the minimum reuse distance requirement is not violated [1]. In addition, in order to ensure the same channels do not get closer than δ, this scheme restricts channel movement in the following way: if the configuration of cellular system is assumed to be hexagonal, channels in a cell is divided into six groups. Channels of each group can be carried into a particular cell. Detailed procedures of the channel carrying scheme can be found in [1]. The disadvantage of this scheme basically requires fewer channels than other schemes because its minimum reuse distance is $\delta+1$.

In this paper, we propose a new channel assignment scheme for handoffs, which stems from the channel reservation scheme and the channel carrying scheme to improve performance of the existing approaches discussed thus far.

2 Our Scheme

In our scheme, we first divide channels into two types as shown in Fig. 1. The normal channels are assigned for both new calls and handoff calls. However, the handoff channels are exclusively allocated to handoff calls. The way handoff channels are allocated depends on whether the next cell has a channel for the incoming handoff call. If the next cell doesn't have a channel, the handoff call is assigned an idle channel of N_C channels in the current cell. If there is no available channel in the next cell during handoff, the mobile station is allowed to carry its handoff channel in the current cell.

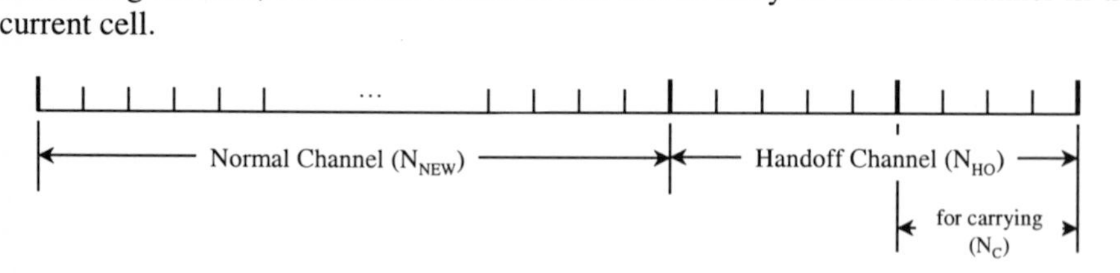

Fig. 1. Allocation of Channels in Our Scheme

In our scheme, when a new call arrives at the cell C_i, an idle normal channel is assigned if a free normal channel is available. Otherwise, the new call is blocked. When a handoff call request is received from the adjacent cell C_{i-1}, the normal channel is first scanned into C_i and assigned if available. Otherwise, the handoff channel is then scanned to find a free handoff channel in C_i. Finally, the handoff call is accepted into C_i if a free handoff channel is available. If there is yet no available handoff channel in C_i, the subscriber requesting the handoff checks for the possibility of being able to carry his currently occupying channel to C_{i-1}. If the subscriber has been using a normal channel in C_{i-1} and there is an idle handoff channel of N_C channels in C_{i-1} at the same time, then he exchanges his current normal channel with the handoff channel and carries it to C_i. On the other hand, if the subscriber has been using a handoff channel in C_{i-1} and there is an idle handoff channel of N_C channels in this cell, then he exchanges his current channel with it, and carries it to C_i. Finally, if the subscriber fails to find a channel, then the handoff cannot be accomplished.

When a call using a normal channel is terminated in C_i, the normal channel is replaced by the carried handoff channel if there is a carried handoff channel being used by other subscribers in C_i. Following this, the released normal channel is returned to the cell that originally assigned it.

It should be noted that there could be co-channel interference in our scheme. Suppose that there are two cells, C_A and C_B, and both cells have the same channel set. If call X and call Y are ongoing in cells C_A and C_B, respectively, then co-channel interference occurs if call X moves towards cell C_B and call Y has been already using the same channel occupied by call X. This happens when the following two events simultaneously occur:

1. The handoff call carrying a channel occurs in more than one cell.
2. The same channels of N_C channels are used in cells C_A and C_B

3 Performance Evaluation

In this section, we first present a Markov model to obtain the blocking ratio of the new and handoff calls, P_N and P_H, respectively. These are the most important QoSs (Quality of Services) in which we are interested.

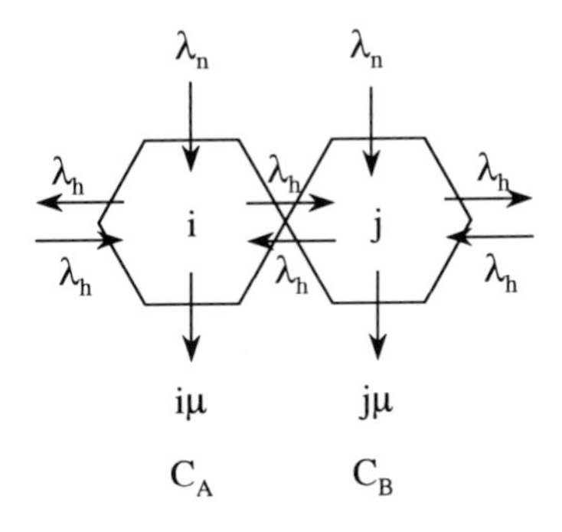

Fig. 2. Network Model

It is impossible to solve the analysis of the entire planar system. In general, performance analysis is done using one cell [6]. However, the one-cell model does not accurately capture the features of this scheme. To reduce disadvantages with a one-cell model, we consider a two-cell model [1], as shown in Fig. 2. We modeled data traffic using an ON/OFF process [5,6] where the durations of the ON and OFF are exponentially distributed; voice packets during the ON period are generated periodically. We made some assumptions for analysis:

1. Traffic is assumed to be symmetrically distributed over all the cells.
2. New call requests arrive according to the Poisson process with rate $\lambda_{n,}$ and are homogeneous among all cells.
3. Handoff calls arrive according to the Poisson process with rate λ_h.
4. A call duration time is assumed to be exponentially distributed with a mean $1/\mu$.

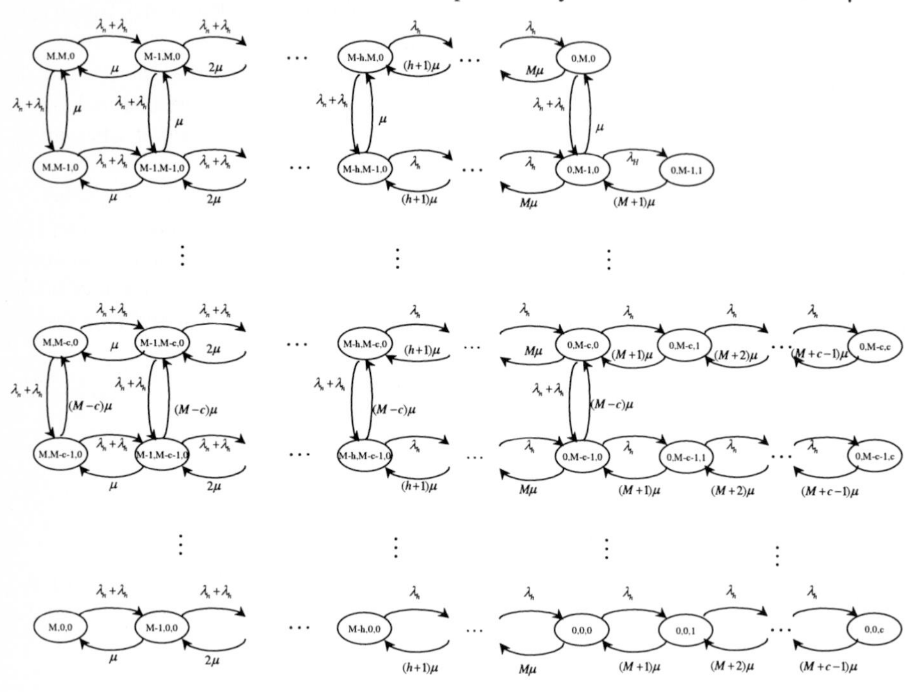

Fig. 3. Markov chain for our scheme using the two-cell model

Let N denote the total number of distinct channels available. We assume that the values of N_{NEW} and N_C in a cell are h and c, respectively. Let $N_A \in \{0, \ldots, M\}$ and $N_B \in \{0, \ldots, M\}$ be the number of available channels in cell C_A and cell C_B, respectively. In addition, let $N_{A\rightarrow B}$ be the number of carried channels whose value ranges from $\{0, \ldots, c\}$. Now, the factor $(N_A, N_B, N_{A\rightarrow B})$ is the state of the Markov chain, as shown in Fig. 3, whose description is similar to [1]. Let $P_{i,j,k}$ denote the steady state probability of the state $\{N_A=i, N_B=j, N_{A\rightarrow B}=k\}$. We obtain these probabilities by exploiting the symmetrical property and by applying standard numerical Markov chain techniques. The equilibrium equation for the state occupancy probabilities, $P_{i,j,k}$, can be written

$$\begin{aligned}
&P(i,j,k)(\alpha_1(M-j)\mu+\alpha_2(\beta_1\lambda_n+\lambda_h)\\
&+\alpha_3(\beta_2\lambda_n+\lambda_h)+\alpha_4(M-i+1+k)\mu)\\
&=P(i,j+1,k)\alpha_1(\beta_2\lambda_n+\lambda_h)+P(i-1,j,k)\alpha_2(M-i+1+k)\mu\\
&+P(i,j-1,k)\alpha_3(M-i+1)\mu+P(i+1,j,k)\alpha_4(\beta_1\lambda_n+\lambda_h)
\end{aligned} \tag{1}$$

We also define some inclusion functions as follows:

$$\alpha_1=\begin{cases}0 & (j=M)\vee(k>0)\\ 1 & (j<M)\wedge(k=0)\end{cases} \tag{1a}$$

$$\alpha_2=\begin{cases}0 & k=c\\ 1 & k\neq c\end{cases} \tag{1b}$$

$$\alpha_3=\begin{cases}0 & j=0\\ 1 & j\neq 0\end{cases} \tag{1c}$$

$$\alpha_4=\begin{cases}0 & i=M\\ 1 & i\neq M\end{cases} \tag{1d}$$

$$\beta_1=\begin{cases}0 & i<M-h\\ 1 & i\geq M-h\end{cases} \tag{1e}$$

$$\beta_2=\begin{cases}0 & j<M-h\\ 1 & j\geq M-h\end{cases} \tag{1f}$$

The handoff rate λ_h in the two-cell model is actually dependent of new call rates. Over all the states, λ_h is given by

$$\lambda_h=\omega\lambda_n \tag{2}$$

Since $P_{i,j,k}$ depends on λ_h, we iteratively solve the Markov chain. Having determined $P_{i,j,k}$, we can obtain P_N and P_H by summing over the appropriate states, which are defined as the probability that all channels are active. Therefore, P_N and P_H are given by

$$P_N=\sum_{i=0}^{M-h}\sum_{j=0}^{M}\sum_{k=0}^{c}P_{i,j,k} \tag{3}$$

$$P_H=\sum_{k=0}^{M-c-1}P_{0,M-k,k}+\sum_{j=0}^{M-c}P_{0,j,c} \tag{4}$$

Similarly, we developed a Markov chain model of the channel reservation scheme published in [1]. Since no channel movement is allowed in this scheme, the pair (N_A, N_B), $N_A \in \{0, \ldots, M\}$, $N_B \in \{0, \ldots, M\}$, suffices to characterize the state of the two-cell model. Since $P_{i,j}$ depends on λ_h, we iteratively solve the Markov chain. P_N and P_H can be obtained by

$$P_N = \sum_{i=0}^{M-h} \sum_{j=0}^{M} P_{i,j} \tag{5}$$

$$P_H = \sum_{j=0}^{M} P_{0,j} \tag{6}$$

Similarly, we developed a Markov chain model of the channel carrying scheme published in [1]. In this scheme, the channel reuse distance is given (δ+1) to ensure the minimum reuse distance requirement is not violated. Therefore, the number of channels in a cell is smaller than other schemes. P_N and P_H can be obtained by

$$P_N = \sum_{k=0}^{M/6} \sum_{j=0}^{M-k} P_{0,j,k} \tag{7}$$

$$P_H = \sum_{k=0}^{M-M/6-1} P_{0,M-k,k} + \sum_{j=0}^{M-M/6} P_{0,j,M/6} \tag{8}$$

Next, when we let M_S denote the average number of mobile stations in a cell. The channel utilization, U, is defined by

$$U = \frac{M_s}{M} \tag{9}$$

Now, we can obtain using the state occupancy probabilities by

$$U = \frac{1}{2M} \sum_{i=0}^{M} (M-i) \sum_{j=0}^{M} (M-j) \sum_{k=0}^{c} P_{i,j,k} \quad \text{for our scheme} \tag{10}$$

$$U = \frac{1}{2M} \sum_{i=0}^{M} (M-i) \sum_{j=0}^{M} (M-j) P_{i,j} \quad \text{for the channel reservation scheme} \tag{11}$$

$$U = \frac{1}{2M} \sum_{i=0}^{M} (M-i) \sum_{j=0}^{M} (M-j) \sum_{k=0}^{M/6} P_{i,j,k} \quad \text{for the channel carrying scheme} \tag{12}$$

The co-channel interference probability, P_c, is defined by

$$P_c = \frac{N_{coll}}{N_{call}} \tag{13}$$

where N_{call} represents the average number of channels for the new calls and the handoff calls, and N_{coll} represents the average number of collided channels in a cell.

Now, we can typically obtain P_H and U by

$$P_H = \sum_{k=0}^{M-c-1} P_{0,M-k,k} + \sum_{j=0}^{M-c} P_{0,j,c} + P_c \tag{14}$$

$$U = \frac{(1-P_c)}{2M} \sum_{i=0}^{M} (M-i) \sum_{j=0}^{M} (M-j) \sum_{k=0}^{c} P_{i,j,k} \tag{15}$$

We carried out simulation to verify our analytical Markov model. The same assumptions were also used in the simulation. Simulation was done for a 10*10 cell cellular system. Average call duration time was assumed to be 180 s. N, δ, ω and μ were assigned 600, 4, 0.3 and 1, respectively. Two different values, 0.8 and 0.7, were used for N_{NEW} / M, which is hereafter denoted by T.

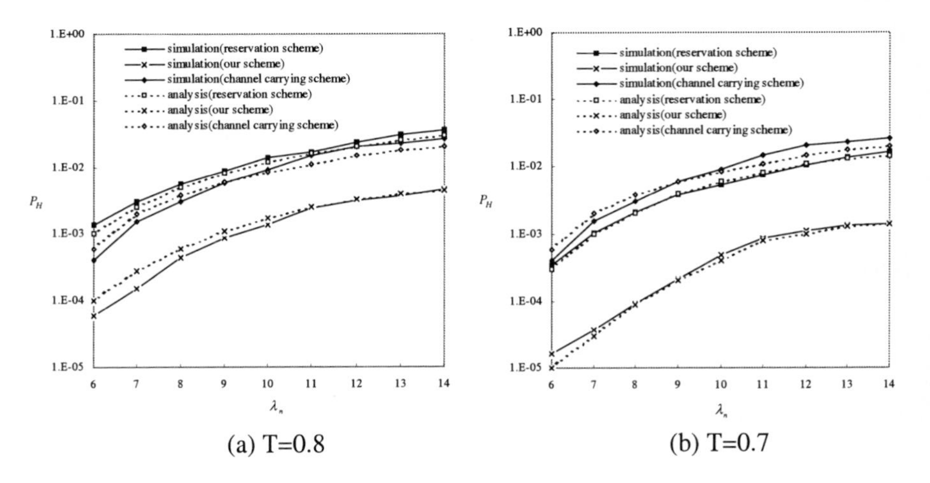

(a) T=0.8 (b) T=0.7

Fig. 4. Blocking Ratio of Handoff calls

Fig. 4 shows the blocking ratio of handoff calls. In this figure, we can see our scheme favors handoff requests over new calls by giving more opportunities for channel access. In addition, we can clearly see our scheme performs much better than the channel reservation scheme and the channel carrying scheme over all traffic loads. Our scheme offers a lower handoff blocking ratio than the channel carrying scheme. This is because the number of occupied channels in a cell is greater, compared to the channel carrying scheme, since its minimum reuse distance is δ instead of (δ+1). In addition, our scheme may offer a lower handoff blocking ratio than the channel reservation scheme due to a channel carrying.

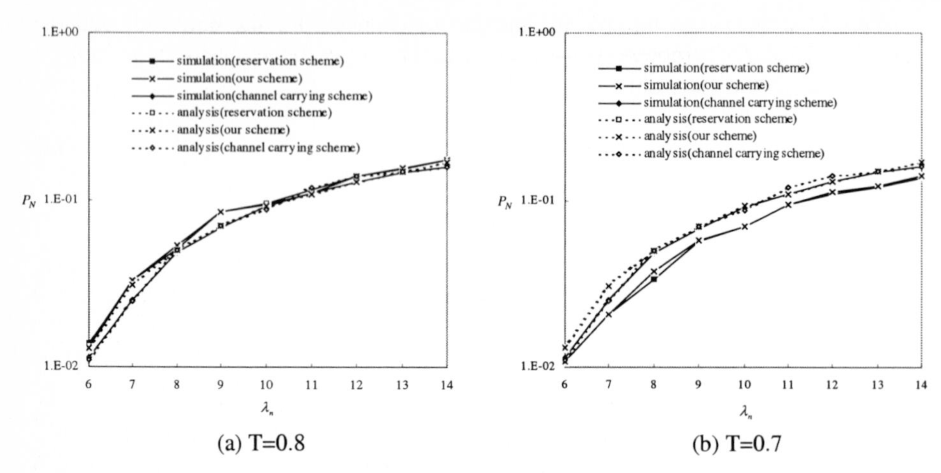

(a) T=0.8 (b) T=0.7

Fig. 5. Blocking Ratio of New Calls

Fig. 5 shows the blocking ratio of the new calls. In this figure, we can see the new call blocking ratio of our scheme is very close to the other two schemes. This is because if the number of occupied channels in a cell in our scheme reaches N_{NEW}, the new calls are blocked.

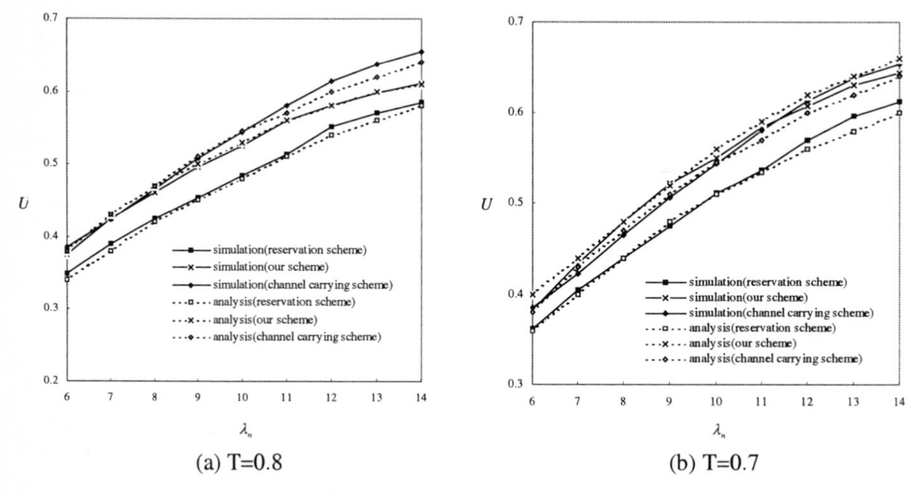

(a) T=0.8 (b) T=0.7

Fig. 6. Channel Utilization

In Fig. 6, we plotted the values of the channel utilization as the call arrival rate varies. As shown in this figure, our scheme offers a higher channel utilization than the channel reservation scheme under all traffic loads. The channel exchange of our scheme may improve the channel utilization due to a higher channel reuse. The channel reservation scheme requires a small value of T to maintain the constraint on the handoff blocking ratio, thus resulting in low channel utilization.

In our scheme, on the other hand, the handoff channels need not be reduced much to maintain the constraint on the handoff blocking ratio since the channels can be carried to the next cell. Finally, from Figs. 4, 5 and 6, we can see the simulation results are in sound agreement with the numerical results obtained from Markov analysis.

4 Conclusions

In this paper, we presented a new channel assignment scheme for efficient handoffs in wireless networks. Our scheme uses two types of channels: normal channels for both new and handoff calls and handoff channels exclusively for handoff calls. If the number of channels currently being used in the cell is below the number of normal channels, both new and handoff calls are accepted. However, if the number of channels being used exceeds the number of normal channels, new calls are blocked and only handoff calls are admitted. The way handoff channels are allocated depends on whether the next cell has a channel for the incoming handoff call. If the next cell doesn't have a channel for the incoming handoff call, the handoff call is assigned an idle channel of handoff channels for carrying in the current cell. Thus, if there is no available channel in the next cell during handoff, the mobile station is allowed to carry its handoff channel in the current cell.
We evaluated our scheme with a two-cell model using Markov analysis and computer simulation. The results indicate our scheme may offer a higher channel utilization as well as a lower handoff blocking ratio than the channel reservation scheme and the channel carrying scheme under all traffic loads.

References

1. J. Li, Ness B. Shroff and E. K. P. Chong.: Channel Carrying: A Novel Handoff Scheme for Mobile Cellular Networks. IEEE/ACM Trans. on Networking, vol. 7, no. 1, Feb. (1999) 38-50
2. L. J. Cimini, G. J. Foschini, C. L. I and Z. Miljanic.: Call blocking performance of distributed algorithms for dynamic channel assignment. IEEE Trans. Commun., vol. 42, no. 8, Aug. (1994) 2600-2607
3. W. S. Jeon and D. G. Jeong.: Comparison of Time Slot Allocation Strategies for CDMA/TDD Systems. IEEE J. Select. Areas Commun., vol. 18, no. 7, July (2000) 1271-1278
4. S. Choi and G. Shin.: Predictive and adaptive bandwidth reservation for hand-offs in QoS–sensitive cellular networks. Proc. ACM SIGCOMM'98, (1998) 155-166
5. S. Deng.: Empirical Model of WWW Document Arrivals at Access Link. Proc. ICC'96, (1996) 1797-1802
6. G. Anastasi, L. Lenzini and E. Mingozzi.: MAC Protocols for wideband Wireless Local Access: Evolution Toward Wireless ATM. IEEE Personal Commu., Oct. (1998) 53-64
7. R. Ramjee, R. Nagarajan and D. Towsley.: On Optimal call admission control in cellular networks. Proc. IEEE INFOCOM, San Francisco, CA, (1996) 35-42

A Relationship between Densities and Numbers of Hidden Stations of Packet Radio Network

Tetsuya Shigeyasu[1], Hideyuki Ishinaka[2], and Hiroshi Matsuno[3]

[1] Graduate School of Science and Engineering,
Yamaguchi University
1677-1 Yoshida, Yamaguchi 753-8512, Japan
sigeyasu@ib.sci.yamaguchi-u.ac.jp
[2] Yokogawa Electric Corporation
2-9-32 Naka-machi Musashino, Japan
Hideyuki_Ishinaka@yokogawa.co.jp
[3] Faculty of Science, Yamaguchi University
1677-1 Yoshida, Yamaguchi 753-8512, Japan
matsuno@sci.yamaguchi-u.ac.jp

Abstract. An interesting relationship between the throughput of CSMA and a connection ratio which reflects a density of a PRN, which is shown in this paper, motivate us to derive equations giving a number of hidden stations as a function of connection ratio. A new method of computational experiment through this equation is also presented which allows us to evaluate effects of hidden terminals on PRNs easily.

1 Introduction

It is obvious that, if we consider a contention based media access control (MAC) protocol, a packet radio network (PRN) contains hidden terminals [1] unless it is fully-connected. In other words, if we let density of a fully-connected graph be 1.0, graphs with densities below 1.0 always have hidden terminals. Note that, we do not have to conclude that throughput of contention based MAC protocol on a PRN with density below 1.0 is always less than the one on a PRN with density 1.0, because, as is pointed out in [2], it has a possibility to make parallel transmissions[1] in a PRN with density below 1.0. This implies that, when we intend to consider the throughput of MAC protocol on PRNs with densities below 1.0, we should analyze combination of two factors, (i) hidden terminal and (ii) parallel transmission. It follows that the throughput performance evaluation of MAC protocol is not an easy task, then, we are motivated to investigate the relationship between density of a PRN and throughput of an MAC protocol on the PRN.

As far as we have investigated, many arguments about hidden terminals have been made without giving the precise definition of hidden terminal. Then, in this paper, we will give the formal definitions of "hidden station" and "hidden station ratio" for developing discussions in an analytic way.

[1] In [2], "simultaneous transmission" is used for representing the same meaning as "parallel transmissions" in this paper.

I. Chong (Ed.): ICOIN 2002, LNCS 2343, pp. 618–627, 2002.

In Section 2, we firstly introduce the definition of *connection ratio* of a PRN and present an experimental result which shows an interesting relationship between the throughput of CSMA and connection ratio of a PRN. This result brings out our motivation to investigate the relationship between the connection ratio and the number of hidden stations.

Section 3 exposes the relationship between the existence of hidden stations and connection ratio by deriving the equation which gives the number of hidden stations as a function of connection ratio of a PRN. Furthermore, in this section, the result of computational experiment is also presented, which supports the correctness of the derived equation.

In Section 4, we discuss about the contribution of this paper to the evaluation method of MAC protocols. Concretely, a new computational experiment on MAC protocol evaluation becomes possible, since, by using the formula in this paper, we can save much time for producing PRNs with fixed ratio of hidden stations.

2 Effect of Connection Ratio on Throughput of CSMA

This section considers a simple question; "How do structural properties of PRNs reflect to the throughput of CSMA ?" In this section, we introduce *connection ratio* as a parameter of structural properties of PRN which reflects a density of PRN. Next, we present a result which implies an interesting relationship between the throughput of CSMA and the connection ratio of a PRN.

Definition 1. (connection ratio) For any node v in the node set of a graph $G = (V, E)$, if the number of nodes which are adjacent to the node v is $\alpha(|V|-1)$ $(0 \leq \alpha \leq 1)$, it is said to be that the graph G has a *connection ratio* α.

An example of graph whose connection ratio is 0.4 is presented in Figure 1

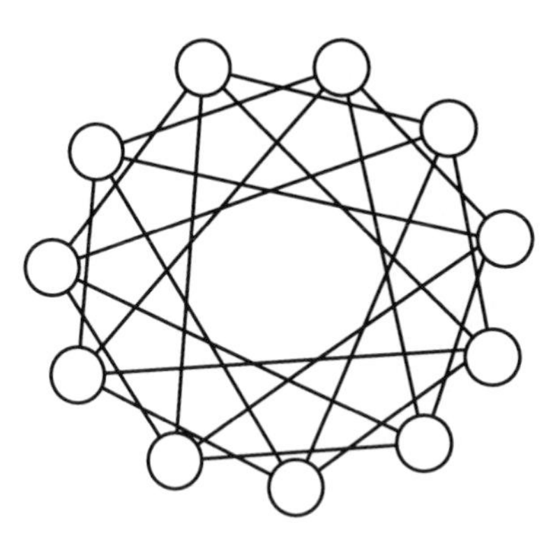

Fig. 1. Example of a graph with connection ratio 0.4

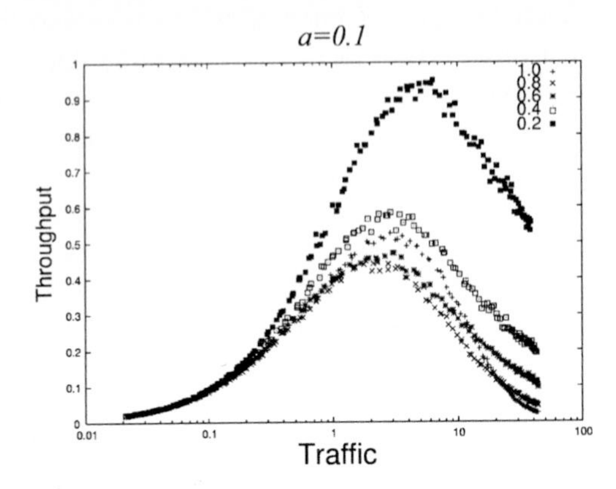

Fig. 2. Throughputs of CSMA with various connection ratios(a=0.1)

Figure 2 shows a result of computer simulation where the effect of connection ratio on throughput performance of CSMA is presented. This simulation is performed on the assumption that PRN consists of 50 terminals which randomly send packets to neighbor terminals. The value a in Figure 2 represents the ratio of propagation delay to packet transmission time. If for some reasons such as packet collision, a station fails to transmit a packet, it sets a random value to the timer and begins to decrement the value, retransmitting the packet when the timer expires.

In Figure 2, we can observe that throughputs of connection ratios 0.8 and 0.6 are lower than the throughput of connection ratio 1.0, however, interestingly, throughput of connection ratios 0.4 and 0.2 are higher than the throughput of connection ratio 1.0. From this experiment, we can conclude that throughput of CSMA is much affected by the density of the PRN.

3 Hidden Stations and Connection Ratio

It can be considered that there are two effects on the throughput of CSMA caused by connection ratio, one is positive effect and the other is negative effect;

Two effects can be considered as causes of effecting connection ratio on the throughput of CSMA, one is positive effect and the other is negative effect;

Positive. In a PRN which is not fully connected, parallel transmission is available (Figure 3). Several investigations have been made for proving positive effect of parallel transmission, [3,4,5,2].

Negative. Two terminals are within transmitting/receiving range of some terminals, but they are out-of-range each other, two such terminals are said to be "hidden". It is well known that existence of hidden terminals degrades the performance of contention based media access control protocols such as CSMA and MACA rapidly [6].

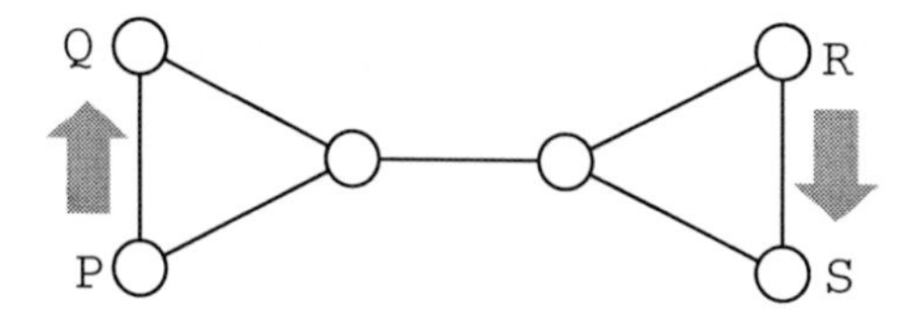

Fig. 3. Parallel transmissions

We focus attention on the negative effect of hidden terminals on the throughput of CSMA.The main purpose of this paper is to expose the relationship between the existence of hidden stations and connection ratio of PRN.

In many papers, discussions on hidden terminals were developed with ambiguous definition of hidden terminal, for example,

- Two terminals can be within range of the stations but out-of-range of each other; or they can be separated by some physical obstacle opaque to UHF signals. Two such terminals are said to be "hidden" from each other [1], and
- Two vertices (stations) i and j can generate a hidden terminal interference if and only if they are two hops away, namely when $d_{ij} = 2$. [7].

These descriptions about hidden terminal are not satisfactory for our discussion, since, according to these descriptions, for any arbitrary terminal T, we can not identify hidden terminals of the terminal T.

On the other hand, although the term "hidden station" is used instead of the term "hidden terminal", a precise definition about hidden terminal is given in [8]. We follow this description in [8] and give the formal definition of *hidden station* and the definition of *hidden station ratio* which are required in the following discussion.

Definition 2. (hidden station) For any terminal T in a PRN, a terminal at a distance of 2 hops from the terminal T is called *hidden stations of* T.

Definition 3. (hidden station ratio) For any graph $G = (V, E)$, *hidden station ratio* $H(V)$ is defined by the following formula;

$$H(V) = \frac{1}{|V|} \sum_{v \in V} \frac{h_v}{|V| - 1}$$

where
$h_v = \{v \in V \mid \text{for all } u \in V \text{ such as } v \neq u$, the distance of the shortest path between v and u is $2\}$.

3.1 Derive the Number of Hidden Stations as a Function of Connection Ratio

The number of edges $|E|$ of the graph consisting of n nodes and whose connection ratio is α is given by

$$(n-1)\alpha n \times \frac{1}{2}.$$

This formula can be also expressed as follows;

$$\alpha = \frac{|E|}{\frac{(n-1)n}{2}} = \frac{|E|}{{}_nC_2}. \tag{1}$$

Since ${}_nC_2$ represents the number of edges of a fully connected graph with n nodes, the connection ratio α can be considered as the density of edges of the graph G.

For any node $v \in V$, let $D_i(v)(\subseteq V)$ be a set of nodes whose distance from the node v are $i(i \geq 1)$, and let $D_{i\leq}(v)(\subseteq V)$ be a set of nodes whose distance from the node v are greater than or equal to $i(i \geq 1)$. Furthermore, for any set $S \subseteq V$, we define the sets $D_i(S)(\subseteq V)$ and $D_{i\leq}(S)(\subseteq V)$ as follows (See Figure 4.);

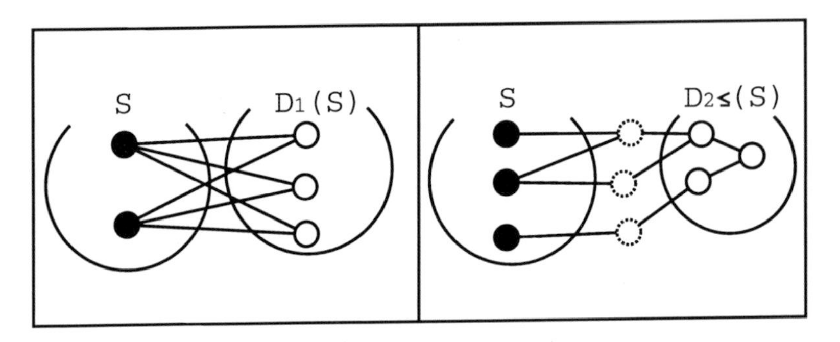

Fig. 4. Examples of sets $D_1(S)$ and $D_{2\leq}(S)$

$D_i(S) = \{v \in V \mid$ for all $u \in S$, the distance between nodes v and u is i $\}$ and
$D_{i\leq}(S) = \{v \in V \mid$ for all $u \in S$, the distance between nodes v and u is greater than or equal to i $\}$.
Then, the following lemma holds.

Lemma 1. Consider a graph $G = (V, E)$ of the connection ratio $\alpha(0 \leq \alpha \leq 1)$. Then, for any two disjoint sets $P, S \subseteq V$ $(P \cap S = \phi)$, the following formula holds;

$$|D_{2\leq}(S) \cap P| = |P|(1-\alpha)^{|S|}. \tag{2}$$

Proof. For any node $v \in S$, the number of nodes in the set P whose distances from the node v are greater than or equal to 2 can be expressed as follows;

$$|D_{2\leq}(v) \cap P| = |P|(1-\alpha).$$

By noting that the above formula holds for any node in S independently, it is easy to see that the lemma holds.

□

Theorem 1. Consider a graph $G = (V, E)$ of the connection ratio $\alpha(0 \leq \alpha \leq 1)$. Then, for any node $v \in V$, the number of nodes whose distances from the node v is 2 can be expressed as follows;

$$|D_2(v)| = (|V| - 1)(1 - \alpha)[1 - (1 - \alpha)^{\alpha(|V|-1)}]. \tag{3}$$

Proof. For any node $v \in V$, it is obvious that $|D_2(v)|$ is given by the following formula;

$$|D_2(v)| = (|V| - 1) - |D_1(v)| - |D_{3\leq}(v)|. \tag{4}$$

By noting that the node v or a node in $D_1(v)$ should not be included in the set $D_{3\leq}(v)$, the set $D_{3\leq}(v)$ can be expanded as the following formula;

$$D_{3\leq}(v) = D_{2\leq}(D_1(v)) \cap D_{2\leq}(v)$$

Since the sets $D_1(v)(\subseteq V)$ and $D_{2\leq}(v)(\subseteq V)$ are disjoint, from Lemma 1, we can get

$$\begin{aligned} D_{3\leq}(v) &= D_{2\leq}(D_1(v)) \cap D_{2\leq}(v) \\ &= (|V| - 1)(1 - \alpha)(1 - \alpha)^{(|V|-1)\alpha}. \end{aligned}$$

where $|D_1(v))| = (|V| - 1)\alpha$ and $|D_{2\leq}| = (|V| - 1)(1 - \alpha)$,

Thus,

$$\begin{aligned} |D_{3\leq}(v)| &= |D_{2\leq}(D_1(v)) \cap D_{2\leq}(v)| \\ &= (|V| - 1)(1 - \alpha)(1 - \alpha)^{(|V|-1)\alpha} \end{aligned}$$

This completes the theorem. □

Note that, if nodes in a PRN correspond nodes of a graph $G = (V, E)$, from Definition 1, $|D_2(v)|$ in Theorem 1 can be considered as the number of hidden stations of the node v.

Figure 5 shows behaviors of $H(V)$ with changing connection ratio of V in case of the graphs of 16, 32, 64, and 128 nodes. From this figure, we can observe the following interesting facts;

1. The ratio of hidden stations are inversely as the connection ratio from some fixed connection ratio which varies according to the number of nodes of a graph.
2. Let β be a ratio of hidden stations. Then, the formula $\alpha + \beta = 1$ holds while the relation of inverse proportion is kept between α and β.
3. Let α_{min} be the lowest point of connection ratio at which $\alpha_{min} + \beta = 1$ holds. Then, the value of α_{min} becomes smaller in proportion as the number of nodes of PRN increases. In other words, the range, in which inverse proportion between α and β holds, becomes wider in proportion as the number of nodes of PRN increases.

The last item listed above implies that $\alpha+\beta = 1$ holds if $|V|$ is infinite. The following corollary proves that this conjecture is true.

Corollary 1. Consider a graph $G = (V, E)$ with no isolated node of connection ratio α. For any node $v \in V$, suppose that (1) $\alpha|V|$ nodes in the node set V are adjacent to the node v, and (2) the value β $(0 \leq \beta < 1)$ exists such that $\beta|V|$ nodes are located at a distance of 2 from the node v. Then, $\alpha + \beta = 1$ holds if $|V| \to \infty$.

Proof. From the equation (3) in Theorem 1, β can be described as follows;

$$\beta = \frac{|D_2(v)|}{|V| - 1} = (1 - \alpha)[1 - (1 - \alpha)^{\alpha(|V|-1)}]$$

From the assumption that we only consider a graph with no isolated node, it is obvious that the case of $\alpha = 0$ never happens. Then, the following calculation leads to the completion of the theorem.

$$\begin{aligned}\beta &= \lim_{|V| \to \infty} (1 - \alpha)[1 - (1 - \alpha)^{\alpha(|V|-1)}] \\ &= (1 - \alpha)\end{aligned}$$

□

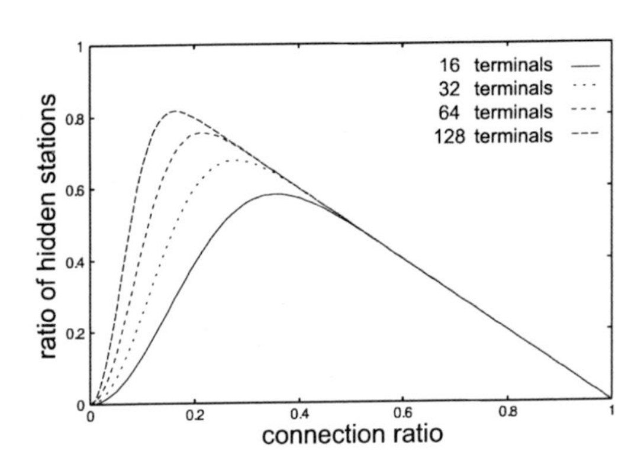

Fig. 5. Relationship between ratio of hidden station and connection ratio

3.2 Computational Experiment

For obtaining the conviction that our derivation described above is well performed, we made the following computational experiment on computing the number of hidden stations in graphs with 16, 32, 64, and 128 terminals.

1. Fix a connection ratio, produce randomly 10 graphs with that connection ratio, and perform the following procedure for these 10 graphs.
2. For a node v in the graph, count the number of hidden stations of the node v. Repeat this for all nodes in the graph, and take an average of hidden stations.

Figure 6 shows a result of computational experiment in which the above procedure is executed with connection ratios in decreasing order at intervals of 0.05, that is, 1, 0.95, 0.9 ,0.85, 0.8,... The experiment is terminated when connection ratio reaches the value with which no graph can be produced. We can see that the graphs in Figure 6 agree well with the graphs in Figure 5. Rough behavior of the graph of 16 terminal due to an insufficient number of nodes for computational experiment.

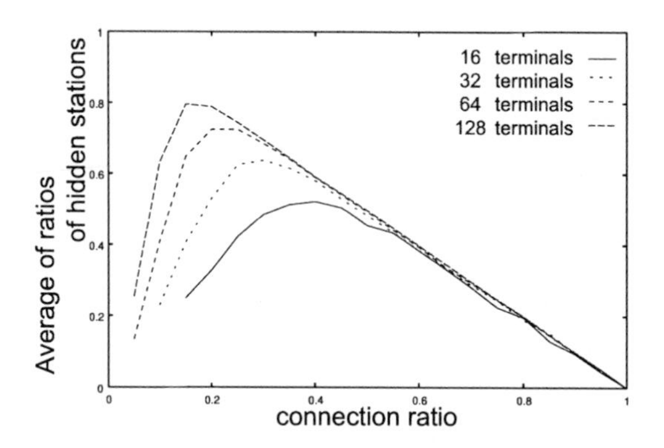

Fig. 6. Result of computational experiment

4 Easy Computational Experiment Method Evaluating Hidden Stations Effect

Careful consideration have been paid to hidden terminal problem in constructing contention based media access control (MAC) protocol such as CSMA [9] and MACA [10].

Many protocols have been proposed for coping with the hidden terminal problem in the literature [11] [4] [3] [12] [13] [8] and effectiveness of these proposed protocols were verified through computer simulations. With a viewpoint of using graphs appropriate to performing the aimed simulation, we can classify the methods of these computer simulations into the following two types ;

1. Designing graphs adequate to the evaluation of proposed protocol [12] [13] [8], and
2. Producing graphs randomly with some constraint such as the number of nodes [11] [4] [3].

Investigating an influence of hidden terminals on a PRN is an important issue of evaluating the proposed MAC protocol.

If we use the first method for this purpose, we can make such evaluation by designing graphs with various arrangements of hidden terminals on a given PRN. However, the problem of this method is that much number of graphs should be designed for making a good experimentation.

On the other hand, it is needless to say that much number of graphs can be produced quickly with the second method, but arrangements of hidden terminal in these produced graphs should be identified later. The problem is that algorithms for obtaining such arrangements seems to require plenty of computational time.

In Theorem 1 of this paper, we presented a formula which gives the number of hidden stations as a function of connection ratio of a PRN. By using this formula, we can easily obtain graphs with various hidden station ratios by the following procedure.

1. Fix a hidden station ratio H with which evaluation of aimed protocol will be performed,
2. Determine enough number of terminals for making experimentation according to graphs depicted in Figure 5. (Note that sufficient number of terminals becomes larger as the experiment requires larger ratio of hidden stations, because only the linear part of graphs in Figure 5 can be used for this experiment.)
3. Obtain the connection ratio C for the intended experiment by the corresponding graph of terminals determined in above step (Note that any graph with any number of terminals can be depicted through the formula (3) in Theorem 1), and
4. Produce graphs of connection ratio C as many as necessary.

This procedure allows us to make computational experiments for investigating effect of hidden terminals on a PRN easily.

5 Conclusions

In this paper, we discussed about the relationship between a connection ratio and a ratio of hidden terminals in a PRN. Our main future task is to investigate throughput performance of various MAC protocols in terms of hidden station ratio based on the computational experiment newly introduced in this paper.

References

1. F. A. Tobagi and L. Kleinrock, "Packet Switching in Radio Channels: Part II - The Hidden Terminal Problem in CSMA and Busy-Tone Solution," *IEEE Trans. on Communications*, COM-23, pp. 1417–1433, 1975.
2. T. Minami, M. Ichimura, and H. Matsuno, "An algorithm for optimal simultaneous transmission scheduling on packet radio network," *Trans. IEICE*, Vol.J82-B, No.6, pp.1151–1158, 1999.
3. T. S. Yum and K. W. Hung, "Design algorithms for multihop packet radio networks with multiple directional antennas stations," *IEEE Trans. Commun.*, Vol.40, No.11, pp.1716–1724, 1992.
4. H. Shigeno and Y. Matsushita, "Design and performance evaluation of new peer-to-peer MAC scheme STDA/MA for wireless LAN using directional antennas," *Trans. IEICE*, Vol.J79-B-I, No.2, pp.41–50, 1996.
5. O. Akizuki, S. Suzuki, K. Mutsuura, and S. Ooshita, "TDMA with parallel transmission: TDMA/PT," *Trans. IEICE*, Vol.J78-B-I, No.12, pp.846–854, 1995.

6. A. S. Tanenbaum: Computer networks (third edition), *Prentice-Hall Inc.*,1996.
7. A. Alan Bertossi and A. Maurizio Bonuccelli, "Code assignment for hiddenterminal interference avoidance in multihop packet radio network," *IEEE/ACM Trans. on Networking 3*, pp. 441–449, 1995.
8. T. Ozugur and M. Naghshineh and P. Kermani and C. M. Olsen and B. Rezvani and J. A. Copeland, "Balanced Media Access Methods for Wireless Networks," *Proc. of ACM/IEEE MobiCom*, pp. 21–32, 1998.
9. L. Kleinrock and F. A. Tobagi, "Packet Switching in Radio Channels: Part I - Carrier Sense Multiple Access Modes and Their Throughput Delay Characteristics," *IEEE Trans. on Communications*, COM-23, pp.1400–1416, 1975.
10. P. Karn, "MACA - a new channel access method for packet radio," *ARRL/CRRL Amateur Radio 9th*, Computer Networking Conference , pp.134–140, 1990.
11. H. Matsuno, T. Ebisui, and H. Ando, "Effect of an extra ability to central stations in CTMA," *Trans. IPS Japan*, Vol.39, No.4, pp.1049–1057, 1998.
12. C. L. Fullmer and J. J. Garcia-Luna-Aceves, "Solutions to Hidden Terminal Problems in Wireless Networks," *Proc. of ACM SIGCOMM '97*, pp. 39–49, 1997.
13. V. Bharghavan, A. Demers, S. Shenker, and L. Zhang, "MACAW: A Media Access Protocol for Wireless LAN's," *Proc of ACM SIGCOMM '94*, pp. 212–225, 1994.

VII. Network Management

A CORBA-Based Performance Management Framework for Multi-layer Network Environment

Daniel Won-Kyu Hong[1], Choong Seon Hong[2], Yoo Jae Hyoung[1]
Dong-Sik Yun[1], and Woo-Sung Kim[1]

[1] Access Network Lab., R&D Group, KT
463-1 Junmin-Dong Yusung-Gu, Daejeon 305-811 South Korea
{wkhong,styoo,dsyun,kwsun}@kt.co.kr
[2] School of Electronics and Information, Kyung Hee Univerity
1 Seocheon-Ri Kiheung, Yongin, Kyungki-Do 449-701 South Korea
cshong@khu.ac.kr

Abstract. Many researchers have addressed issues of performance management based on network elements, however, little attention has so far been given to issues of the scalable performance management framework uniformly applicable to heterogeneous network environment. To address this deficiency, this paper proposes a CORBA-based performance management framework for multi-layer network environments, such as ATM, Frame Relay (FR) and Internet. The framework includes a generic network model that can uniformly be applicable to such environments and performance metrics for each network. In addition, this paper describes the inter-relationship among heterogeneous networks in terms of performance management and network resource management scheme as a result of performance data collection and analysis. A system architecture is also proposed, including pro-active and re-active performance management, which is implemented through CORBA objects and provides IDL interfaces. It orchestrates network resource reconfiguration through the methods of performance measurement, policy management, and analysis.

1 Introduction

The performance management of telecommunications networks has undertaken on a network element (NE) basis in an uncoordinated manner, which requires sophisticated human expertise for relatively trivial tasks. Different performance management schemes are needed for each heterogeneous network element because the management interfaces or protocols provided by network elements for controlling the performance vary from vendor to vendor. The networks become increasingly diverse and complex, and the demands upon them likewise increase. To address this situation, the authors propose a distributed hierarchical performance management framework that can commonly be applicable for heterogeneous networks, not at network element level but at network management level.

I. Chong (Ed.): ICOIN 2002, LNCS 2343, pp. 631–640, 2002.

Attention is focused on the heterogeneous network environment composed of ATM, Frame Relay (FR) and Internet. ATM and FR networks take the role of Internet access by providing QoS guaranteed path between subscriber routers and network service provider's backbone routers. The scalable network model that can also be applicable to heterogeneous network without modification or adaptation is described. The network is designed through the technology of Rumbaugh's object modeling technology (OMT) [7] for maximizing modularity, which corresponds to the managed information base (MIB) in terms of TMN management concept [13,14].

On the other hand, it is very difficult for network service providers to manage the heterogeneous network elements because these have their own proprietary management protocols and managed objects. To solve this problem, the unified element management interface using CORBA IDL [6] is proposed. It gives a unified view of the management protocols and managed objects of heterogeneous network elements to the proposing performance management framework.

In the proposed multi-layer network model, there is client/server relationship between Internet and ATM or FR because the routers in Internet are connected with ATM VP PVCs, FR PVCs or ATM/FR inter-working PVCs. Therefore, the PVCs in ATM or FR serve as links in Internet for connecting routers. The reliable performance management in ATM and FR network affects the reliable Internet service provision. There are two levels of performance management. One is in terms of client/server relationship between Internet and ATM or FR, and the other is in terms of federation between ATM and FR for providing reliable data transfer path with ATM/FR inter-working [9,10,11]. The performance metric for each network is identified and the inter-relationship to orchestrate the overall performance management of the heterogeneous networks is defined.

There are two kinds of performance management approaches: pro-active and re-active. The pro-active approach adjusts resources of server network before the request or complaint of subscriber or client network is received. The re-active approach adjusts resources of server network after receiving the request or complaint of subscriber or client network. Both approaches are implemented because they are significant.

Under the multi-layer network environment, the functional module of hierarchical performance management framework is designed and implemented using CORBA [8]. CORBA is a very useful middleware in deploying the geographically distributed large-scale network management system by providing object interaction, location and migration transparencies. Following is a description of the detailed implementation model composed of several CORBA objects for performance management under multi-layer networks, and the scenarios for performance data collection, analysis and metric correlation method and network resource reconfiguration based on the two approaches.

The next section presents the hierarchical network model that can uniformly be applicable to multi-layer network without any modification and can guarantee scalability. Section 3 tackles the performance management framework, including metrics and their relationship among multi-layer networks; performance data

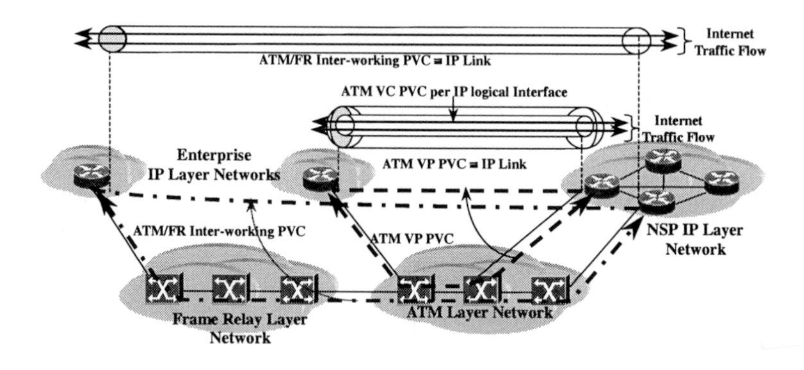

Fig. 1. Multi-layer network environment composed of ATM, FR and Internet

collection and analysis scheme; two approaches of performance management; and resource management scheme. Section 4 describes the implementation model and the scenarios for performance management under the multi-layer networks and Section 5, the merits and issues of the proposed hierarchical framework as well as the conclusion to this paper.

2 A Scalable Network Model

In order to deploy large-scale multi-layer networks, the layering and partitioning concepts of ITU-T G.805 [5] are very useful in describing the generic transport network architecture. The layering concept defines the network boundary that can transfer specific information without adaptation medium or functions. The proposed network is composed of three layer networks of ATM, Frame Relay (FR) and IP (Fig. 1). The ATM layer network is logically divided into ATM virtual path (VP) layer network and ATM virtual channel (VC) layer network. The former takes the role of the service provider for ATM VC and IP layer networks because the ATM VP PVCs serve as logical links in connecting the logical VC switching nodes in VC layer and the routers in IP layer network. There is a federation relationship between the ATM VC layer network and the FR layer network for supporting ATM/FR service and network inter-working service. On the other hand, there is client/server relationship between the IP layer network and the federated FR and ATM VC layer networks because the ATM/FR inter-working PVCs serve as logical links in connecting routers in IP layer network. Even if the VC and IP layer networks are constructed with the ATM VP PVCs by VP layer network, their customers are different. The customers of VC layer network are the users of connection-oriented applications, while those of IP layer network are the users of connectionless applications. In this regard, the two networks are defined separately.

The partitioning concept of ITU-T G.805 is used to determine the internal structure of a layer network from the perspectives of administrative, restoration and routing domains [5]. A layer network can be partitioned into a smaller scope of subnetworks that can be managed by their own policy. The subnetwork can be further partitioned until it is mapped to a single switching element, such as ATM switch, FR switch and router.

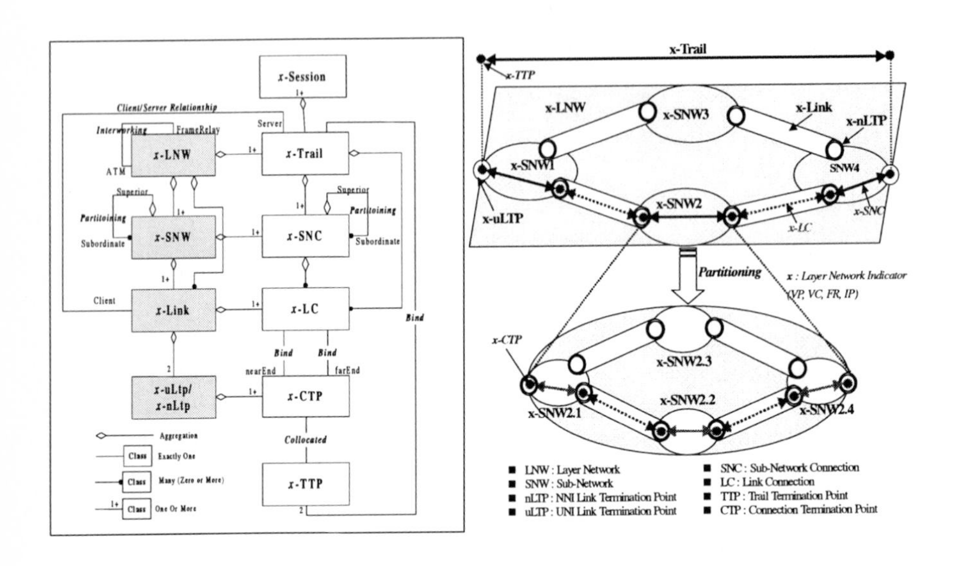

Fig. 2. A unified network partitioning for supporting scalability

Two fragments are defined in this paper to describe the transport network not specific to a vendor's technology and a specific single layer network. These are topology and connectivity fragments (Fig. 2). The first represents the topological structure of a transport network, which is composed of layer network (LNW), subnetwork (SNW), link and link termination point (LTP). A LNW describes the generation, transport and termination of a particular information. A SNW represents an inter-connected group of network elements and an administrative or management domain. A subnetwork can be further partitioned into subnetworks and links. A link is a logical or physical transmission path inter-connecting two network elements or subnetworks. A link is supported by a trail in the server layer network according to the client/server layer network relationship. The link termination point (LTP) represents the near or far end points of a link.

The connectivity fragment describes the logical or physical connection stream traversing the objects of the network fragment, which is composed of trail, trail termination point (TTP), subnetwork connection (SNC), link connection (LC) and connection termination point (CTP). A trail represents the logical end-to-end connection within a layer network. TTP describes the near or far end point of a trail. A link connection (LC) is capable of transferring information transparently across a link. It is delimited by LTPs and represents the fixed relationship between the ends of the link. A subnetwork connection (SNC) is capable of transferring information transparently across a subnetwork. It is delimited by connection points at the boundary of the subnetwork and represents the association between these connection points. A connection termination point (CTP) represents the end point of subnetwork connections or link connections. A trail is composed of a number of nmlSNC and LC. And the nmlSNC contained in a Trail is composed of a number of smaller emlSNCs and LCs, as shown in Fig. 2.

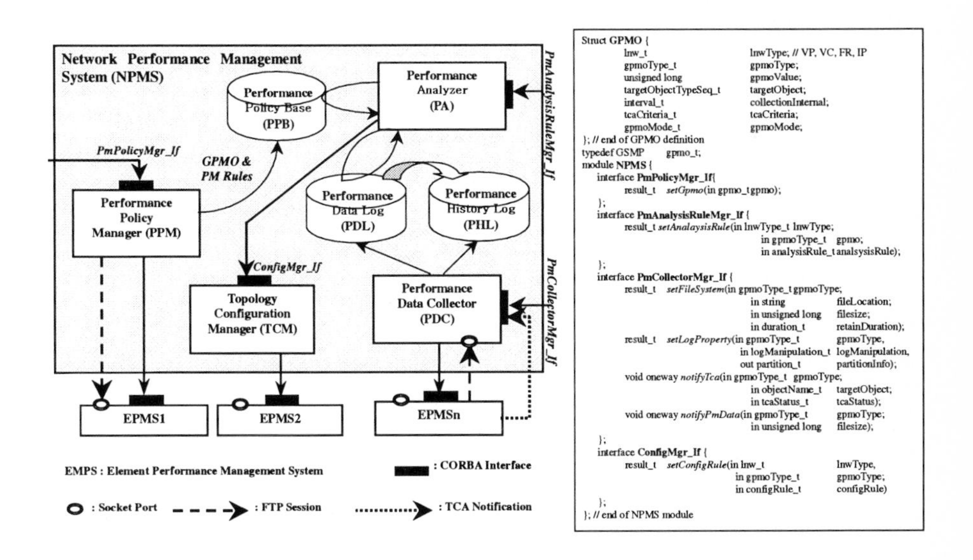


```
Struct GPMO {
        lnw_t                      lnwType; // VP, VC, FR, IP
        gpmoType_t                 gpmoType;
        unsigned long              gpmoValue;
        targetObjectTypeSeq_t      targetObject;
        interval_t                 collectionInternal;
        tcaCriteria_t              tcaCriteria;
        gpmoMode_t                 gpmoMode;
}; // end of GPMO definition
typedef GSMP     gpmo_t;
module NPMS {
    interface PmPolicyMgr_If{
        result_t   setGpmo(in gpmo_tgpmo);
    };
    interface PmAnalysisRuleMgr_If {
        result_t setAnalaysisRule(in lnwType_t lnwType;
                                    in gpmoType_t   gpmo;
                                    in analysisRule_t analsysisRule);
    };
    interface PmCollectorMgr_If {
        result_t   setFileSystem(in gpmoType_t gpmoType;
                                 in string           fileLocation;
                                 in unsigned long    filesize;
                                 in duration_t       retainDuration);
        result_t   setLogProperty(in gpmoType_t      gpmoType,
                                 in logManipulation_t logManipulation,
                                 out partition_t     partitionInfo);
        void oneway notifyTca(in gpmoType_t  gpmoType;
                                 in objectName_t     targetObject;
                                 in tcaStatus_t      tcaStatus);
        void oneway notifyPmData(in gpmoType_t       gpmoType;
                                 in unsigned long    filesize);
    };
    interface ConfigMgr_If {
        result_t   setConfigRule(in lnw_t            lnwType,
                                 in gpmoType_t       gpmoType;
                                 in configRule_t     configRule)
    };
}; // end of NPMS module
```

Fig. 3. System architecture and IDL Interfaces of NPMS

3 Distributed Performance Management Framework

Fig. 3 shows the system architecture of network performance management system (NPMS) that is composed of four components: performance policy manager (PPM), topology configuration manager (TCM), performance analyzer (PA) and performance data collector (PDC). All of these system components are defined and implemented using CORBA objects.

3.1 Generic Performance Managed Object (GPMO)

A GPMO represents the performance metric in terms of PNMS. A GPMO provides the unified performance metrics for a multi-layer performance management because it encapsulates the discrepancy among the different vendor's MIBs and layer networks. A GPMO is applied to physical links and ports and connectivity objects such as ATM PVC, FR PVC, ATM/FR inter-working PVC and IP session. The GPMO and its management interface of *PmPolicyMgr_If* is shown in Fig. 3.

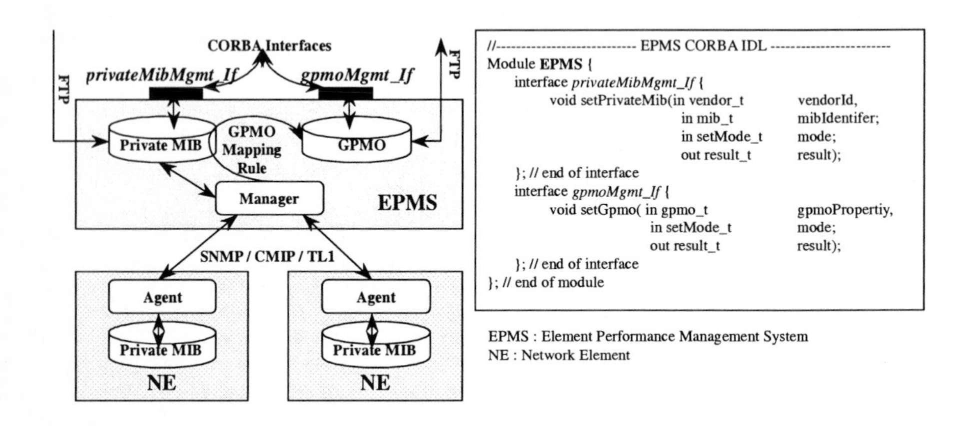

Fig. 4. System architecture of element performance management system (EPMS)

A GPMO is defined through IDL structure having such attributes as *lnwType, gpmoType, gpmoValue, targetObjects, collectionInterval, tcaCriteria* and *gpmoMode*. The *lnwType* indicates a network type, such as ATM, FR and IP. According to *lnwType*, a network operator provides different performance metrics. The *gpmoType* represents a performance metric that is different according to the layer network type of ATM [2,8,13], FR [9,10,11] and IP [3,4]. The *gpmoValue* has the accumulated cells on the *gpmoType* during the named performance data collection period indicated by the *collectionInterval*. The *targetObjects* represents a set of topological or connectivity objects at which the performance data collection is done. The *gpmoMode* indicates whether EPMS collects the performance data collection or not. EPMS issues the alarm representing the performance degradation in line with *tcaCriteria*. The *tcaCriteria* has upper and lower thresholds. Let us assume that *gpmoType* is input cells, *collectionInternval* is fifteen minutes, *targetObject* is uLTP, upper threshold is 850,000 cells and lower threshold is 530,000 cells. As a result of performance data collection at the named uLTP for fifteen minutes, the accumulated input cells (*gpmoValue*) are greater than the upper threshold, EPMS issues threshold crossing alert (TCA) to network management system (NMS) for indicating the performance degradation at the named uLTP. EPMS notifies TCA to performance data collector (PDC) by calling the *notifyTca()* operation of *PmCollectorMgr_If*. On receiving TCA notification, PDC stores it in a performance data log (PDL) for persistency. This information is used to analyze performance by PA.

The GPMO is defined by taking into account the characteristic and management policy of a network. It is provided using the *setGpmo()* operation in *PmPolicyMgr_If* of performance policy manager (PPM). On receiving the GPMO, PPM stores it to performance policy base (PPB) for persistency. The network operator provides the adaptation rules between vendor specific MIBs and GPMO, which is downloaded to an element performance management system (EPMS) using FTP (see Fig. 4). Therefore, the adaptation function between specific MIBs and the GPMO is done by EPMS. Each vendor's network element has its own MIBs, such as ATM-MIB [2], ATM2-MIB [1] for ATM and IF-MIB [4] and MIB-II [3] for Internet. If a NPMS controls hundreds of different vendor-specific MIBs, its management overhead will be high as much as it nearly goes down. Therefore, the adaptation function is allocated between private MIBs and GPMO to EPMS to distribute the processing overhead.

Fig. 4 shows the overall structure of EPMS. It manages several network elements provided by the same vendor and having the same private MIBs. The management protocol between EPMS and NE can be SNMP, CMIP, TL1, etc. The private MIBs and GPMO are downloaded to EMPS via FTP. On the other hand, NMS designs the adaptation rule between private MIBs and GPMO and downloads it to EPMS via FTP at the designed location. However, EPMS has two CORBA interfaces for performance management such as *privateMibMgmt_If* and *gpmoMgmt_If*. The *privateMibMgmt_If* provides the CORBA operation to PPM of NPMS for managing the property of private MIBs, such as vendor name, MIB type, and the manipulation modes of private MIB, such as activate, deactivate and removal using the *setPrivateMib()* operation. On the other hand, the performance data collection can be selectively controlled on certain GPMO using the *setGpmo()* operation. The performance data collection of GPMO can be activated or deactivated, and the properties of GPMO can be adjusted.

3.2 Performance Data Collection

Each EPMS provides the collected performance data in ASCII format to PDC in NPMS by FTP in line with its performance data report within a certain period. PDC parses the ASCII performance data propagated by EPMS and stores them into PDL for persistency. There are two logs: PDL, that retains the collected data for one month for short-term performance analysis and performance history log (PHL), that retains the collected data for three months for long-term performance analysis. The internal system component of PDL is shown in Fig. 5.

Each EPMS opens a socket session per GPMO with PDA to transmit collected performance data to PDA. Just before performance data transmission via FTP, EPMS calls *notifyPmData()* operation of *PmCollectorMgr_If* to notify the performance data transmission with file size.

The fileHandler manages the file system, and receives the file size and name from EPMS. It compares the file name and size with the file system information. If there is a difference, it will get the performance data file from EPMS via FTP. The pdScheduler daemon monitors the file system whether there is any

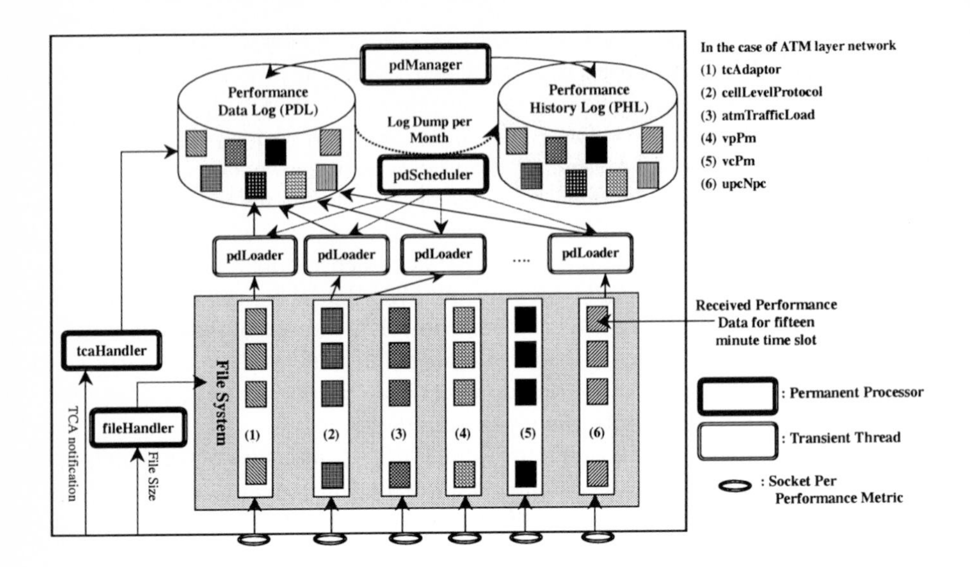

Fig. 5. Internal system component of performance data loader (PDL)

performance data or not. If pdScheduler detects there are some data, it creates pdLoader thread that parses the ASCII file format and stores the parsed information into PDL for persistency. The pdScheduler can create up to six pdLoader threads in the case of ATM layer network, as shown in Fig. 5.

The pdManager manages the logs of PDL and PHL using Oracle database. The network operator manages the performance data table and tale partition of Oracle database using the *setLogProperity()* operation of *PmCollectorMgr_If*. In addition, pdManager dumps PDL to PHL per month.

3.3 Performance Data Analysis and Reconfiguration

There are two kinds of performance data analysis schemes: pro-active and re-active. The former is used to find the performance degradation and to make some complementary actions to compensate for it prior to the recognition of network operator, subscribers or client layers. There are client/server and federation relationships among Internet, ATM and FR layer networks. From the perspective of client/server relationship, the client layer network serves as consumer and the server layer network serves as supplier. Hence, the pro-active scheme encapsulates the performance degradation within a layer network without its recognition of other layer networks.

The re-active scheme is used as follow-up measure to compensate for the performance degradations on the request of client layer network, network operator or customer. This mode is initiated by the subscriber's complaint. This makes it not favorable because the purpose of the performance management is to maintain the whole network and not to violate the subscriber's QoS requirements.

The performance management criteria of each layer network differ under the multi-layer network environment. For example, the performance criteria of Internet are different from those of ATM or FR networks. The re-active scheme is very useful in a multi-layer network environment because the performance criteria of a client layer network are normally stricter than those of a server layer network.

When there is performance degradation at any portion of a network, complementary actions are taken to compensate for it. The most normal and easiest way is to reconfigure the network resource where the degradation occurred. The *ConfigMgr_If* of TCM (Fig. 3) provides the network resource reconfiguration interface. If PA decides to reconfigure the network as a result of performance data analysis, this will be done in line with the guidelines maintained in PPB to TCM by calling the *ConfigMgmt_If*.

The NPMS can be commonly applicable for ATM network, FR network and Internet without any adaptation or modification. Allocating the adaptation function between GPMO of NPMS and vendor-specific MIB to each EPMS, the scalable performance management can be supported in terms of management processing overhead and the large number of managed elements, such as ATM switches, FR switches and routers.

4 Concluding Remarks

This paper proposed the unified network model that can commonly applicable multi-layer networks such as ATM, FR and Internet. In addition, we defined GPMO that is unified network performance metric applicable for multi-layer networks and proposed the inter-relationship among multi-layer networks in terms of performance management. We descried a CORBA-based performance management framework focusing on the network management layer and element management layer of TMN functional layering. To do this, we defined system architecture of NPMS and EPMS and their management interface with CORBA IDL. In addition, we illustrated the detailed performance data collection scheme and two performance analysis and reconfiguration schemes of pro-active and re-active schemes.

This CORBA-based performance management framework can be applicable for multi-layer networks such as ATM, FR and Internet without any modification and adaptation. Allocating the adaptation function between GPMO of NPMS and vendor specific MIB to each EPMS, we can support the scaleable performance management in terms of management processing overhead and the large number of managed elements such as ATM switches, FR switches and routers. In addition, we can provide efficient network resource utilization with the unified network model and maintain network in optimized by network resource reconfiguration as a result of performance analysis.

References

1. F.Ly, M. Noto, A.Smith,E.Spiegel and K.Tesink, "Definition of Supplemental Objects for ATM management," *draft-ietf-atommib-atm2-12,* March 1997.
2. K.Tesink, "Definition of Managed Objects for ATM Management," *RFC2515,* February 1999.
3. K. McCloghrie and F.Kastenholz, "The Interfaces Group MIB using SMIv2," *RFC2233,* November 1997.
4. K. McCloghrie and M.Rose, "Management Information Base for Network Management of TCP-IP-based internets: MIB-II," *RFC1213,* March 1991.
5. ITU-T Recommentation G.805, "Generic Function Architecture Of Transport Networks," November 1995.
6. OMG, "CORBAServices: Common Object Service Specification," March 31, 1995.
7. James Rumbaugh, Michael Blaha, William Premerlani, Frederick Eddy and William Lorensen, "Object-Oriendted Modeling and Design," *Prentice-Hall. Inc.,* 1991.
8. ATM Forum Technical Committee, "Traffic Management Specification Versions 4.0," *at-nm-0056.000,* April 1996.
9. Frame Relay Forum, "Frame Relay/ATM PVC Network Interworking Implementation Agreement," *The Frame Relay Forum Document Number FRF.5,* December 20, 1994.
10. Frame Relay Forum, "User-to-Network Implementation Agreement (UNI)," *The Frame Relay Forum Document Number FRF1.1,* January 19, 1996.
11. Frame Relay Forum, "Network-to-Network Implementation Agreement (NNI)," *The Frame Relay Forum Document Number FRF.2.1,* July 10, 1995.
12. ITU-T Recommendation I.610, "B-ISDN operation and maintenance principles and functions," February 1999.
13. ITU-T Recommendation M.3010, "Principles for a telecommunications management network," February 2000.
14. ITU-T Recommendation M.3100, "Generic Network Information Model," July 1995.

Using Symbolic Model Checking to Detect Service Interactions in Telecommunication Services

Takayuki Hamada, Tatsuhiro Tsuchiya, Masahide Nakamura, and Tohru Kikuno

Department of Informatics and Mathematical Science
Graduate School of Engineering Science, Osaka University
{t-hamada,t-tutiya,masa-n,kikuno}@ics.es.osaka-u.ac.jp

Abstract. Feature interaction is a kind of inconsistent conflict between multiple communication services. In this paper we present an automatic method for detecting feature interactions in service specifications. This method is based on symbolic model checking which can perform verification by symbolically representing the search space with binary decision diagrams. Experimental results show that the method outperforms a previous method based on explicit state traversal, in terms of time and memory required for detection.

1 Introduction

Feature interaction is a kind of inconsistent conflict between multiple communication services, which was never expected from the single services' behavior. In practical service development, the analysis of interactions has been conducted in an ad hoc manner by subject matter experts. This leads to time-consuming service design and testing without any interaction-free guarantee.

To tackle this problem, we propose a formal approach for detection of feature interaction. The detection process checks if interactions occur or not between given multiple services. The proposed approach uses *symbolic model checking* as its basis.

Model checking is a powerful technique for verifying systems that are modeled as a finite state machine. In model checking, the properties to be checked are expressed in temporal logic. For realistic designs, the number of states of the system can be very large and the explicit traversal of the state space may become infeasible. *Symbolic model checking* has proven to be successful for overcoming this problem. This method uses Boolean functions to represent the state space. Since Boolean functions can be often represented by Ordered Binary Decision Diagrams (OBDDs) very compactly, the symbolic model checking method can reduce the memory and time required for analysis. By manipulating the Boolean functions, the method can determine whether or not a system meets a given property that is specified using CTL [1], a branching time temporal logic.

I. Chong (Ed.): ICOIN 2002, LNCS 2343, pp. 641–651, 2002.

In this paper, we investigate how we can detect feature interactions by using SMV, a well-known symbolic model checking tool. We propose a systematic method for translating given specifications of telecommunication services into the input language of SMV. Using this method, automatic detection of feature interactions can be carried out. To illustrate the effectiveness of the approach, we show the results of applying it to the specifications of practical telecommunication services.

Plath and Ryan [10] also proposed the use of SMV for feature interaction detection. Their work considered more detailed specifications than ours, but it entails describing different CTL formulae by hand, depending on the services and properties to be checked.

In contrast, our method adopts a more abstract model, thus allowing us to represent four major types of feature interactions by only two formulae. The downside of adopting a high level model is that only non-subtle interactions can be detected, which might easily be resolved in a low-level design by, for example, prioritizing services. We think, however, that the method is still useful, since knowing the possibility of such interactions can help identify error-prone parts of the design.

2 Preliminaries

In order to formalize the feature interaction detection problem, we present fundamental definition in this section.

2.1 Services

For the formalization, we have to first prepare the services. From ITU-T recommendation [11] (*ITU-T Recommendations Q.1200 Series* - Intelligent Network Capability Set 1 (CS1)) and Bellcore's feature standards [12] (*Bellcore* - LSSGR Features Common to Residence and Business Customers I,II,III), we have selected the following seven services (features):

Call Forwarding (CF): This service allows the subscriber to have his incoming calls forwarded to another number. Suppose that x subscribes to CF and that x specifies y to be a forwarding address. Then, any incoming call to x is automatically forwarded to y.

Originating Call Screening (OCS): This service allows the subscriber to specify that outgoing calls be either restricted or allowed according to a screening list. Suppose that x subscribes OCS and that x puts y in the OCS screening list. Then, any outgoing call to y from x is restricted, while any other call from x is allowed. Suppose that x receives dialtone. At this time, even if x dials y, x receives busytone instead of calling y.

Terminating Call Screening (TCS): This service allows the subscriber to specify that incoming calls be either restricted or allowed according to a screening list. Suppose that x subscribes TCS and that x puts y in the TCS screening list. Then, any incoming call from y to x is restricted, while any other call to x is allowed. Suppose that y receives dialtone. At this time, even if y dials x, y receives busytone instead of calling x.

Denied Origination (DO): This service allows subscriber to disable any call originating from the terminal. Only terminating calls are permitted. Suppose that x subscribes to DO. Then, any outgoing call from x is restricted. Even if x offhooks when the terminal is idle, x receives busytone instead of dialtone.

Denied Termination (DT): This service allows subscriber to disable any call terminating at the terminal. Only originating calls are permitted. Suppose that x subscribes to DT. Then, any incoming call to x is restricted. Even if another user y dials x, y receives busytone without calling x.

Direct Connect (DC): This service is a so-called *hot line* service. Suppose that x subscribes to DC and that x specifies y as the destination address. Then, by only offhooking, x is directly calling y. It is not necessary for x to dial y.

Emergency Call (EMG): This service is usually deployed on police and fire stations. In the case of an emergency incident, the call will be held even when the caller mistakenly onhooks. Suppose that x is a police station on which EMG is deployed, and that y has made a call to x and is now busy talking with x. Then, even when y onhooks, the call is on hold without being disconnected. Followed by that, if y offhooks, the held line reverts to a connected line and y can talk with x again. In order to disconnect the call, x has to onhook.

2.2 Specifications

To formalize the feature interaction detection problem, we have to describe services in a certain way. There are a number of researches concerning service description to formulate the interaction problem. In this paper, we adopt a variant of State Transition Rules (STR) [5,9], a rule-based service specification language. Other examples of such a language include, for example, declarative transition rules [3].

Notation. First, we define the syntax notation of the specification. A *service specification* S is defined as $S = \langle U, V, P, E, R, s_0 \rangle$, where

(a) U is a set of constants representing service users.
(b) V is a set of variables.
(c) P is a set of predicate symbols.
(d) E is a set of event symbols.
(e) R is a set of rules.
(f) s_0 is the (*initial*) state.

Each rule $r \in R$ is defined as follows:

$$r : pre\text{–}condition\ [event]\ post\text{–}condition.$$

Pre(post)−condition is a set of *predicates* $p(x_1, \ldots, x_k)$'s, where $p \in P$,$x_i \in V$ and k is called *arity* which is a fixed number for each p. Especially, precondition can include *negations* of predicates such as $\neg p(x_1, \ldots, x_k)$'s which implies $p(x_1, \ldots, x_k)$ does not hold. *Event* is a predicate $e(x_1, \ldots, x_k)$, where $e \in E$, $x_i \in V$.

Figure 1 shows an example of a specification. This specification represents the Plain Old Telephone Service (POTS). Additional communication features, such as those described in the previous subsection, can be described by modifying this specification (for example, adding rules or predicate symbols).

$$
\begin{array}{l}
U = \{A, B\} \\
V = \{x, y\} \\
P = \{idle, dialtone, calling, path, busytone\} \\
E = \{onhook, offhook, dial\} \\
R = \{ \\
\quad pots1 : idle(x)\ [offhook(x)]\ dialtone(x). \\
\quad pots2 : dialtone(x)\ [onhook(x)]\ idle(x). \\
\quad pots3 : dialtone(x)\ ,\ idle(y)\ [dial(x, y)]\ calling(x, y). \\
\quad pots4 : dialtone(x)\ ,\ \neg idle(y)\ [dial(x, y)]\ busytone(x). \\
\quad pots5 : calling(x, y)\ [onhook(x)]\ idle(x)\ ,\ idle(y). \\
\quad pots6 : calling(x, y)\ [offhook(y)]\ path(x, y)\ ,\ path(y, x). \\
\quad pots7 : path(x, y)\ ,\ path(y, x)\ [onhook(x)]\ idle(x)\ ,\ busytone(y). \\
\quad pots8 : busytone(x)\ [onhook(x)]\ idle(x). \\
\quad pots9 : dialtone(x)\ [dial(x, x)]\ busytone(x). \\
\quad \} \\
s_0 = \{idle(A), idle(B)\}
\end{array}
$$

Fig. 1. Rule-based specification for POTS.

State Transition Model. Next, we define the state transition specified by the rule-based specification.

Let $S = \langle U, V, P, E, R, s_0 \rangle$ be a service specification. For $r \in R$, let $x_1, \ldots, x_n$ ($x_i \in V$) be variables appearing in r, and let $\theta = \langle x_1|a_1, \ldots, x_n|a_n \rangle (a_i \in U)$ be a substitution replacing each x_i in r with a_i. Then, an *instance* of r based on θ (denoted by $r\theta$) is defined as a rule obtained from r by applying $\theta = \langle x_1|a_1, \ldots, x_n|a_n \rangle$ to r. We represent pre-condition, event and post-condition of rule r as $Pre[r]$, $Ev[r]$ and $Post[r]$, respectively.

A *state* is defined as a set of *instances of predicates* $p(a_1, \ldots, a_k)$'s, where $p \in P$, $a_i \in U$. We think of each state as representing truth valuation where instances in the set are true, and instances not in the set are false.

Let s be a state. We say that rule r is *enabled* for θ at s, denoted by $en(s, r, \theta)$, iff all instances in $Pre[r\theta]$ hold at s (i.e., all instances are included in s). Let $\hat{P}re[r\theta]$ be the subset of $Pre[r\theta]$ that is obtained by removing all negations of instances of predicates from $Pre[r\theta]$. When $en(s, r, \theta)$ holds, the *next state*, s' of s, can be generated by deleting all instances in $\hat{P}re[r\theta]$ from s and adding all instances in $Post[r\theta]$ to s; that is,

$$s' = (s \backslash \hat{P}re[r\theta]) \cup Post[r\theta]$$

At this time, we say a *state transition* from s to s' caused by an event $Ev[r\theta]$ is defined on S.

Example 1. Suppose that $r = pots4$ in Figure 1, $\theta = \langle x|A, y|B \rangle$ and $s = \{dialtone(A), dialtone(B)\}$. At this time, $Pre[r\theta] = \{dialtone(A), \neg idle(B)\}$, $Post[r\theta] = \{bustytone(A)\}$ and $en(s, pots4, \theta)$ holds. If subscriber A dials B, then a state transition occurs, thus resulting in $s' = \{busytone(A), dialtone(B)\}$.

2.3 Feature Interactions

In this paper, we focus primarily on the following three types of interactions. These are very typical cases of interactions and are discussed in many papers (e.g., [2,3,4,6,9]):

- **deadlock:** Functional conflicts of two or more services cause a mutual prevention of their service execution, which result in a deadlock.
- **loop:** The service execution is trapped into a loop from which the service execution never returns to the initial state.
- **violation of invariant:** The invariant property, which is asserted by each service, is violated by the service combination.

Example 2. (Deadlock) Suppose that both A and B subscribe to *EMG* and are talking to each other. Here, if A onhooks, the call is on hold by B's *EMG*. At this time, if A offhooks, the call reverts to the talking state. On the other hand, if B onhooks, the call is also held by A's *EMG* without being disconnected. Symmetrically, this is true when B onhooks first. Thus, neither A nor B can disconnect the call. As a result, the call falls into a trap from which it never returns to the idle state.

Example 3. (Violation of invariant) Suppose that (1)A is an *OCS* subscriber who restricts the outgoing calls to C, and (2)B is *CF* subscriber who sets the forwarding address to C. At this time, if A dials B, the call is forwarded to C, so A will be calling C. This nullifies A's call restriction to C.

3 Proposed Method

3.1 SMV Programs

SMV (Symbolic Model Verifier)[7] is a software tool for symbolic model checking; it is publicly available and has been especially successful in verifying hardware systems. In this section, we describe how we can use SMV to detect feature interactions.

In SMV, services (features) are described in a special language called the *SMV language.* We refer to a service description written in the SMV language as an *SMV program.*

An SMV program describes both the state space and the property to be verified. The property is expressed in a temporal logic called CTL (Computation Tree Logic). The model checker extracts a state space and a transition system represented as an OBDD from the program and uses an OBDD-based search algorithm to determine whether the system satisfies the property. If the property does not hold, the verifier will produce an execution trace that shows why the property is falsified.

```
MODULE main
VAR    request:boolean;
       state:{ready, busy};
INIT   state = ready
TRANS (state = ready & request)
       & next(state) = busy
SPEC   AG(request -> AF state = busy)
```

Fig. 2. An SMV program.

Figure 2 shows an example of an SMV program. The keyword `VAR` is used to declare variables. The variable `request` is declared to be a Boolean in the program, while the variable `state` can take on the symbolic values `ready` or `busy`.

The property to be checked is described as a formula in CTL under the keyword `SPEC`. The SMV model checker verifies that all possible initial states satisfy the CTL formula. In this case, the property is that invariantly if `request` is true, then eventually the value of `state` is `busy`.

In this example, the transition relation is specified directly by a Boolean formula over the current and next versions of the state variables. Similary, the set of initial state is specified by another Boolean formula over the current version of state variables. These two formulas are accomplished by the `TRANS` and `INIT` statements, respectively.

The initial states are a set of states where the Boolean formula defined in the `INIT` statement holds. The transition relation is a set of the pairs of the current state and the next state that satisfy the Boolean formula defined in the `TRANS` statement. The expression `next(x)` is used to refer to the variable `x` in the next state.

3.2 Translating Service Specifications into SMV Programs

In this subsection, we show how to translate a given service specification into an SMV program. This process consists of three steps.

First, necessary variables are declared. Basically, we use one Boolean variable for each instance of a predicate. The variable represents whether or not the corresponding instance of the predicate holds. For example, suppose that $P = \{idle(x), path(x,y)\}$ and $U = \{A, B\}$. Then the variable declaration part will be

```
VAR  idle_A : boolean;   idle_B : boolean;
     path_A_B : boolean;  path_B_A : boolean;
```

The second step is to produce the INIT part. In this part, the initial state is specified by a Boolean formula over the variables that evaluates to true exactly for the initial state. For example, when $s_0 = (idle(A), idle(B))$, the INIT part will be

```
INIT  idle_A = 1 & idle_B = 1 & path_A_B = 0 & path_B_A = 0
```

The third step is to specify the transition relation by giving a Boolean formula over the variables and the next version of the variables.

The formula is expressed by a disjunction of many subformulas each of which represents an instance of each rule. Given an instance i of a rule, its corresponding formula F_i is

$$\bigwedge_{p \in Pre[i]} p \wedge \bigwedge_{p \in Post[i]} p' \wedge \bigwedge_{p \in \hat{Pre}[i] \setminus Post[i]} \neg p' \wedge \bigwedge_{p \notin \hat{Pre} \cup Post} (p \leftrightarrow p').$$

where p' denotes the next version of an instance p of a predicate. In the SMV language, this formula must be expressed as a formula over the declared variables. For example, consider rule $idle(x)$, $\neg idle(y)$ $[dial(x,y)]$ $path(x,y)$ and substitution $(x,y) = (A,B)$. Then the above formula is represented in SMV as

```
idle_A = 1 & idle_B = 0
& next(idle_A)=0 & next(idle_B)=idle_B
& next(path_A_B)=1 & next(path_B_A)=path_B_A
```

Thus the formula that represents the transition relation is

$$\bigvee_i F_i \vee (\bigwedge_i \neg F_i \wedge \bigwedge_p (p \leftrightarrow p'))$$

The subformula $\bigwedge_i \neg F_i \wedge \bigwedge_p (p \leftrightarrow p')$ is necessary, since the transition relation must be *total*; that is, the next state must be specified for any states. This requirement stems from the fact that both the CTL semantics and the CTL model checking algorithm depend on this assumption. Intuitively, the subformula signifies that if no transition is possible, then the next state will be the same as the current state.

3.3 CTL Formulas

The property to be verified by model checking must be described in CTL. CTL is a branching time temporal logic. Here we only use two temporal operators: ***AG*** and ***EF***.

The formula ***AG*** p holds in state s iff p holds in all states along all sequences of states starting from s. Clearly, the invariant property is expressed in CTL as ***AG*** I where I is an invariant property intended to be satisfied.

EF p holds in state s iff p holds in state s if p holds in some state along some state sequence starting from s. Thus, the freedom from deadlock and loop is described as CTL formula ***AG EF*** *initial_state*, where *initial_state* represents the initial state.

4 Experimental Results

In order to evaluate the effectiveness of the proposed method, we conducted the experimental evaluation through interaction detection for practical services. For comparison purposes, we used two methods: the proposed method, which analyzes the state space symbolically, and a previous method[4], which searches all reachable states explicitly from the initial state.

For each of the seven services prepared in the previous section, we have created a rule-based service specification. In the following, we attempt to provide a reasonable invariant property intended to be satisfied. We let I_X denote the invariant property for service X.

CF: There is no invariant property respected for CF. Therefore, we give an invariant formula $I_{CF} = true$.

OCS: A reasonable invariant property is considered to be "If x puts y in the OCS screening list (denoted by $OCS(x,y)$), x is never calling y at any time". Therefore, we give an invariant formula $I_{OCS} = \neg OCS(x,y) \vee \neg calling(x,y)$.

TCS: A reasonable invariant property is considered to be "If x puts y in the TCS screening list (denoted by $TCS(x,y)$), y is never calling x at any time". Therefore, we give an invariant formula $I_{TCS} = \neg TCS(x,y) \vee \neg calling(y,x)$.

DO: A reasonable invariant property is considered to be "If x subscribes to DO (denoted by $DO(x)$), x never receives dialtone at any time". Therefor, we give an invariant formula $I_{DO} = \neg DO(x) \vee \neg dialtone(x)$.

DT: A reasonable invariant property is considered to be "If x subscribes to DT (denoted by $DT(x)$),, y is never calling x at any time". Therefor, we give an invariant formula $I_{DT} = \neg DT(x) \vee \neg calling(y,x)$.

DC: There is no invariant property respected for DC. Therefore, we give an invariant formula $I_{DC} = true$.

EMG: There is no invariant property respected for EMG. Therefore, we give an invariant formula $I_{EMG} = true$.

In the experiment, we put the following assumption.

Table 1. Result of interaction detection.

Service Spec.	Unsafety	Violation
EMG	Detected	None
CF+DC	None	None
CF+DT	None	Detected
CF+DO	None	None
CF+OCS	None	Detected
CF+TCS	None	Detected
DC+DT	None	Detected
DC+DO	None	None
DC+OCS	None	Detected
DC+TCS	None	Detected
DT+DO	None	None
DT+OCS	None	None
DT+TCS	None	None
DO+OCS	None	None
DO+TCS	None	None
OCS+TCS	None	None

(a) All users can subscribe to all services.
(b) At the initial state, all users are idle and no user subscribes to any service yet.

This assumption is quite reasonable for telecommunication services. In order to achieve Assumption (a), a pair of rules for the subscription registration and its withdrawal is added to each service specification.

The experiments have been performed on a Linux workstation with a 700 MHz Pentium III processor and 512MByte memory. We varied the number of users from three to five.

4.1 Results of Detection

First, we check if each of the seven specifications is *safe*, that is, free from deadlock and loop. As a result, we have found that all services except EMG are safe, while EMG contains the loop states as shown in Example 2 which is interaction of EMG itself. Next, we have combined each pair of the remaining six services, then tried to detect the interactions between any two services.

Table 1 summarizes the results. In this table, the column 'Unsafety' shows whether deadlock or loop states are identified (*detected*) or not (*none*), and the column 'Violation' shows whether violating invariant properties states are identified (*detected*) or not (*none*). The results were the same regardless of the number of users.

Table 2. Times required for detection (in seconds)

Service Spec.	Proposed method			Previous method		
	3 users	4 users	5 users	3 users	4 users	5 users
EMG	0.59	118.46	N/A	0.55	13.01	298.00
CF+DC	11.01	N/A	N/A	258.68	N/A	N/A
CF+DT	5.83	825.75	N/A	141.66	N/A	N/A
CF+DO	5.94	1066.58	N/A	62.58	N/A	N/A
CF+OCS	6.93	942.96	N/A	518.22	N/A	N/A
CF+TCS	6.88	931.91	N/A	516.64	N/A	N/A
DC+DT	0.84	12.74	514.94	10.27	1371.97	N/A
DC+DO	0.61	11.13	544.26	8.56	1125.81	N/A
DC+OCS	1.00	20.95	11131.60	39.62	N/A	N/A
DC+TCS	1.07	20.89	3239.47	39.43	N/A	N/A
DT+DO	0.48	5.31	75.61	2.37	67.63	1877.40
DT+OCS	0.64	8.81	929.61	15.84	N/A	N/A
DT+TCS	0.65	8.80	760.80	15.87	N/A	N/A
DO+OCS	0.58	9.34	N/A	9.64	1767.65	N/A
DO+TCS	0.59	9.26	N/A	9.64	1767.82	N/A
OCS+TCS	1.02	16.42	4429.52	63.31	N/A	N/A

4.2 Performance

Next, we evaluate the performance of the proposed method. For each of the two methods, we investigate how much time is needed to perform the interaction detection. The measurement was performed in the same setting in the previous experiment of detection quality.

Table 2 shows the results. In this table, an N/A indicates that data was not collected because of memory shortage.

In this table, one can see that for all combinations of each pair of six specifications, CF, DC, DT, DO, OCS, and TCS, the proposed method outperformed the previous method, in terms of time and memory required to perform the interaction detection. For example, consider the case of DT+TCS. In this case, the proposed method completed the detection process within around 13 minutes when the number of users is five. In contrast, the previous method was not able to carry out detection even when the number of users is four.

Exceptionally, the previous method outperformed the proposed method for specification EMG. This can be explained as follows. The number of reachable states is quite small for the case of EMG. Thus, the previous method can complete detection with very small amount of time. In symbolic model checking, however, an OBDD that represents the transition relation must be constructed before state space traversal. In this case, the OBDD is very large, thus its construction consumes long time, in spite of the small reachable state space.

5 Conclusions

In this paper, we proposed to use symbolic model checking to detect feature interactions in telecommunication. We present a method for translating service specifications into the input language of the SMV system. We implemented this method and, by applying it to practical services, showed the effectiveness of the proposed approach. Future research includes, for example, the examination of other model checking techniques. Specifically, the use of symmetry and partial order equivalence for state space reduction has already proven to be effective when explicit state representation is used [8]. We think that combining these techniques with the proposed approach deserves further study.

References

1. E. M. Clarke, E. A. Emerson, and A. P. Sistla, "Automatic verification of finite-state concurrent systems using temporal-logic specifications," *ACM Trans. Programming Languages and Systems*, vol.8, no.2, pp.244-263, 1986.
2. R. Dssouli, S. Some, J. W. Guillery, and N. Rico, "Detection of feature interactions with REST," *Porc. of Fourth Workshop on Feature Interactions in Telecommunications Systems*, pp.271-283, July 1997.
3. A. Gammelgaard, E. J. Kristensen, "Interaction detection, a logical approach," *Porc. of Second Workshop on Feature Interactions in Telecommunications Systems*, pp.178-196 1994.
4. Y. Harada, Y. Hirakawa, T. Takenaka, and N. Terashima, "A conflict detection support method for telecommunication service descriptions," *IEICE Trans. Commun.*, vol.E75-B, no.10, Oct. 1992.
5. Y. Hirakawa and T. Takenaka, "Telecommunication service description using state transition rules," *Proc. of IEEE Int'l Workshop on Software Specification and Design*, pp.140-147, Oct. 1991
6. A. Koumsi, "Detection and resolution of interactions between services of telephone networks," *Proc. of Fourth Workshop on Feature Interactions in Telecommunications Systems*, pp.78-92, July 1997.
7. K. L. McMillan, *Symbolic Model Checking*, Kluwer Academic, 1993.
8. M. Nakamura and T. Kikuno, "Exploiting symmetric relation for efficient feature interaction detection," *IEICE Trans. on Information and Systems*, vol.E82-D, No. 10, pp.1352-1363, 1999.
9. T. Ohta and Y. Harada, "Classification, detection and resolution of service interaction in telecommunication services," *Porc. of Second Workshop on Feature Interactions in Telecommunications Systems*, pp.60-72 1994.
10. M. Plath and M. Ryan, "Plug-and-play features," In W. Bouma, editor, *Feature Interactions in Telecommunications Systems V*, IOS Press, pp. 150-164, 1998.
11. ITU-T Recommendations Q.1200 Series, *Intelligent Network Capability Set 1 (CS1)*, Sept. 1990.
12. Bellcore, *LSSGR Feature Common to Residence and Business Customers I,II,III*, Issue 2, July 1987.

Implementing an XML-Based Universal Network Management System in Java

Si-Ho Cha[1], Jae-Oh Lee[2], Young-Keun Choi[1], and Kook-Hyun Cho[1]

[1]Dept. of Computer Science, Kwangwoon University
447-1, Wolgye-Dong, Nowon-Gu, Seoul, 139-701, Korea
{sihoc, ygchoi, khcho}@cs.kwangwoon.ac.kr
[2]WarePlus Inc.
463-1, Junmin-dong, Yusung-gu, Taejeon, 305-390, Korea
jolee@wareplus.com

Abstract. Network managers are currently making use of expensive, vendor-specific management consoles to view and manipulate management information. And more and more systems are transferred to web environment. Therefore, Network managers want to manage the system at anywhere, anytime. This requires an access of management information to be unified, cost-effective, and easy-to-use. This paper presents the design and implementation of an XML-based universal network management system (XMAN), which provide multiple internet devices for a management console. XMAN has 3-tire architecture, and the implementation is based on an extensive use of Web, XML, and Server-side Java technologies. And we have experimented on our XMAN with WAP simulator for verifying that can support WAP-enabled mobile device.

1 Introduction

In recent years, Internet and Web-based technologies make a great change in network management field. Internet provides excellent environment to develop all kinds of applications and changes the development method from Client/Server architecture to multi-tire architecture. If we want to change or add some new management function in the traditional network management systems, we will need to update the whole system. It will cause a serious version control problem and waste a lot of update cost. But, all business logic in network management system with multi-tier architecture is dependent on the processing of the manager separated from the user browser [1]. Therefore, even if one tier is updated, other tiers can still function properly. And even if a critical error occurs, it is localized to a single tier.

Because more and more systems are transferred to web applications, network managers want to get information at anywhere, anytime. Web-based technologies are

---

The present Research has been conducted by the Research Grant of Kwangwoon University in 1999.

I. Chong (Ed.): ICOIN 2002, LNCS 2343, pp. 652-661, 2002.

easy to use and independent of operation system. But, each Internet device has its different technology to display information, so that, it is hard to develop a single network management system that can serve each device[1]. In the near future, network management systems will be also managed with mobile terminals like Personal Digital Assistant (PDA) and cellular phone.
Therefore, we design and implement an Extensible Markup Language(XML)-based universal network management system (XMAN). Because XMAN has 3-tier architecture model, system developers can update it efficiently. Because XMAN is fully equipped for multiple Internet devices, network managers can access it from mobile phone such as PDA. The implementation of XMAN is based on an extensive use of Web, XML, and Server-side Java technologies, such as Extensible Stylesheet Language (XSL), XSL Transformations (XSLT) [8], Document Object Model (DOM) [7], Wireless Markup Language (WML) [14], Java servlet [5], and Java Database Connectivity (JDBC) [6]. The use of XML in XMAN reduces program complexity and minimizes the coupling between the program and its data. XML has great potential to make a complete change in data interchange, presentation, and search on the Internet and Intranet.

This paper is organized as follows. Section 2 presents an overview of our XMAN architecture. The implementation of details and the experience of our XMAN system are described in section 3. Finally, section 4 contains the conclusion of this paper.

2 XMAN Architecture

Our XMAN is a 3-tire architecture system, which is divided into client tier, middle tier, and third tier. Figure 1 shows the overview of XMAN system.

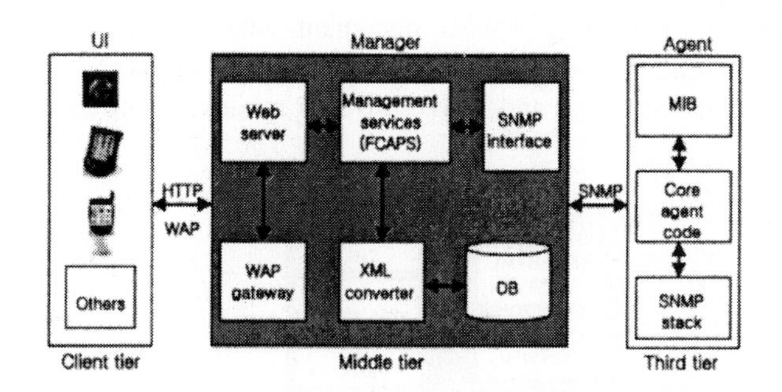

Fig. 1. Overview of XMAN system.

The client tier provides a user interface for the network management application. The middle tier is the manager system, which acts as a server on clients through Web server. And the middle tier implements business logic to manage network and presentation logic to present the information data to client browser. The third tier is represented as the Simple Network Management Protocol (SNMP)-based agent, which can perform the management operations on the managed element. The management requests of network manager are delivered to the agent through the middle tier and the responses are vice versa.

2.1 Client Tier

The client tier is composed of all web browser and WAP device. To access the management application, the client needs to know only the URL address and the port number of the Web server. Once the connection is established, the client is authenticated by a password in order to control the access to the management services, and thus prevent unauthorized accesses. This capability is particularly important in the open Internet context to protect critical management data and to restrict the management operations [2]. Because the middle tier implements business logic and presentation logic for XMAN, the client system is only the user interface system.

2.2 Middle Tier

The main module of the middle tier is the Web server, listening continually on its communication port for the requests issued from the browsers. The management service modules handle the requests from clients locally or direct them to the agent using SNMP for management services. The XML converter is used to retrieve data from database and transfer data into an XML documents in a reverse way. WAP gateway takes responsible for converting WML request to HTML request format, so it can communicate with Web server and perform the request of a service.

The middle tier also takes responsible for presenting data to the client. The middle-tier receives the data from the database or the agent, and then transforms the data into a suitable representation for the client. In other words, the middle tier transforms the XML data into the HTML or WML document, which is then presented to the client. Figure 2 illustrates the distinct modules involved in the Web server module, the management service modules, and XML converter module, being located on the manager system.

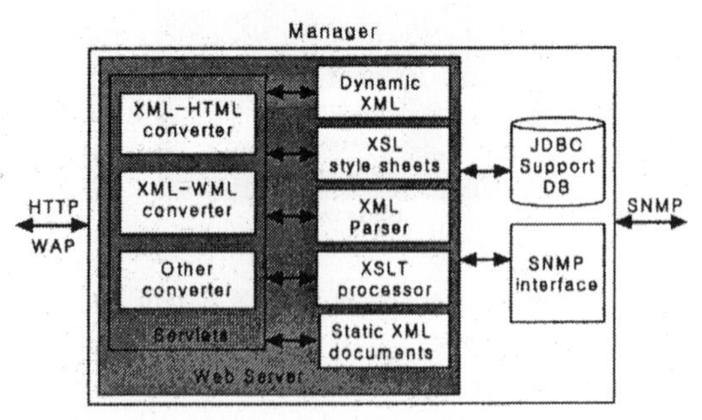

Fig. 2. The detailed architecture of XMAN

2.2.1 Web Server

The Web server module constitutes the core of the manager. The Web server acts as the container of Java servlets. The servlet container allows Java servlets to respond to requests from the clients. The Web server module is a Java program based on multi-threading and hence can handle several requests simultaneously [2].

2.2.2 XML Parser, XSL, and XSLT

We use a DOM-based parser for parsing the XML document. A DOM-based parser exposes the DOM API that allows data in an XML document to be accessed and modified by manipulating the nodes in a DOM tree [3]. XSL is used to format XML documents and XSLT is used to transform an XML document from one form to another. An XML parser converts a source XML document into a source tree, and then the XSLT processor walks the source tree starting from the root node, and attempts to match each node to the template rule of an XSL file. If such a match is made, the template is copied into the result tree, and processing continues until the source tree has been completely traversed [7, 8].

2.2.3 Java Servlets

Java servlets are used to receive the managers' input from client browser, and retrieve the management information from database through JDBC or SNMP agent through SNMP interface, and then the result is stored in database, and generate an XML document dynamically. Because each Internet device hase its own presentation format, Java servlets convert original XML document into client specific browsing type according to user's browser type. There is only one content document on the Web server. It can save a lot of manpower that people do not need to write specific browsing type for each Internet device [1]. Figure 3 shows how XML content is converted to HTML or WML.

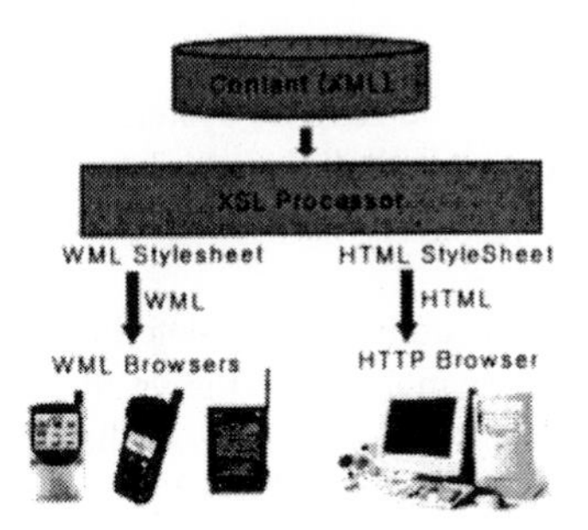

Fig. 3. How XML content is converted to HTML or WML

To achieve this goal, we use several converters, such as XML-HTML converter, and XML-WML converter. All converters are written in Java servlet. XML-HTML converter and XML-WML converter use XSL style sheet to transform an XML document into HTML document or WML document with the XSLT processor. For developing an XML-based universal network management system, Java servlets use an XML parser, the XSL style sheets, the XSLT processor, JDBC, and SNMP interface.

2.2.4 JDBC

JDBC is the Java SQL wrapper that provides portable and cross-platform database access for Java programs. JDBC eliminates the need for a gateway program for database access, as it handles connectivity to relational databases, fetching query results, committing or rolling back transactions, and converting SQL types to and from Java program variables [6][15]. Because Java servlets use JDBC API, even if database management system (DBMS) is exchanged, servlet codes don't need to be changed.

2.2.5 SNMP Interface

SNMP interface module is existed between the Java servlets and the SNMP agents. It is invoked by Java servlets to handle SNMP requests. The information received from the agents is stored in the database and/or sent to the client after XML-HTML or XML-WML transformation is performed.

2.3 Third Tier

An agent receives SNMP messages to retrieve and/or modify management data and sends responses to these messages. An agent sends SNMP messages as event reports. In the architecture of XMAN, an agent using SNMP is the third tier. In the future, we will extend the third tier to the extensible SNMP agent with XML MIB.

3 Implementation

In current, the implementation of the XMAN system focus on the middle tier acts as manager system. To allow for extensibility and portability, Web technologies, XML technologies, and the Java language have been selected to implement our XMAN modules [16].

SNMP-based communication can be implemented easily by using Java. AdventNet provides SNMP v2c API written in Java. It allows the developers of network management applications to develop management applications by simplifying SNMP interfaces.

Java platform offers an elegant and efficient solution to the portability and security problems through the use of portable Java bytecodes. The portability, security, and reliability of Java are well suited for developing server objects that are robust, and independent of operating systems, Web servers and database management servers.

3.1 Implementation Architecture

The implementation architecture of our XMAN is illustrated in Figure 4. Based on the architecture presented in section 2, we have realized an XML-based universal network management system. The implementation architecture is composed of Web server to receive the management requests from clients, XSLT processor to transform XML into HTML or WML, JDBC to access database, SNMP interface to communicate with agents, and so on.

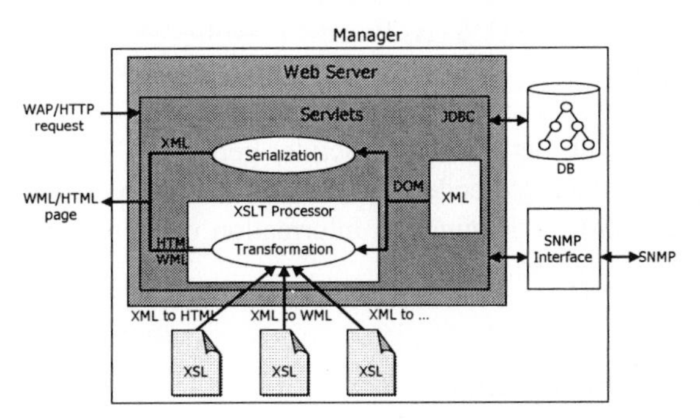

Fig. 4. Implementation Architecture of XMAN

The Web server is implemented by a Jakarta Tomcat [10]. Jakarta Tomcat is a free, open-source implementation of Java servlet and JavaServer Pages (JSP) technologies developed under the Jakarta project at the Apache Software Foundation. We use version 3.3. It is the current production quality release for the Servlet 2.2 and JSP 1.1 specifications. Because Jakarta Tomcat uses 8080 port, the URL is composed of the address of the Web server and the number of 8080 port.

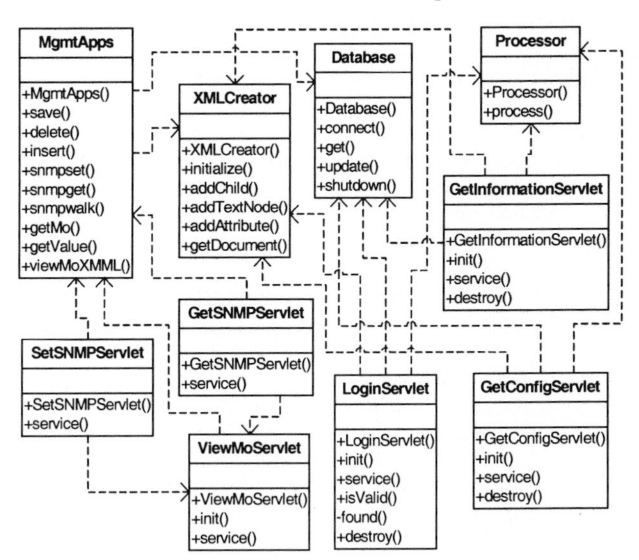

Fig. 5. The main class diagram of XMAN

Each Java servlet retrieves the management data from database through JDBC or SNMP agent through SNMP interface, and generate an XML document dynamically. And then Java servlets transform the generated XML document into a new HTML document or WML document with the XSLT processor. Java servlets use an XML parser, the XSL style sheets, the XSLT processor, JDBC, and SNMP interface.

Figure 5 is the main class diagram of XMAN implementation module. Servlet Loginservlet handles user requests to log in to the network management system. Class XMLCreator is a utility class we use to create XML documents using Apache's Xerces Java DOM-based parser [11]. Class Processor applies an XSL style sheet to transform an XML document using Apache's Xalan XSLT processor for Java [12]. Class Database is used to connect to the database and execute queries to retrieve and update data using JDBC. Servlet GetInformationServlet generates XML containing the information of a specific managed node from the element database and Servlet GetConfigServlet generates XML containing the information of managed node's configuration from the element database.

3.2 An Example of Implementation

The following example shows how GetInoformationServlet creates an XML content and converts to WML page with class XMLCreator and class Processor.

```
package xman;
...
public class GetInformationServlet extends HttpServlet {
   private Database database;

   public void service( HttpServletRequest req,
     HttpServletResponse res )throws ServletException,
       IOException {
     try {
       ...
       String hostID = req.getParameter( "hostID" );
       String query = "SELECT * FROM Hosts "
         + "WHERE hostid= " + hostID;
       ResultSet rs =  database.get( query );
       rs.next();
       XMLCreator xmlCreator = new XMLCreator();
       Node hostNode = xmlCreator.initialize( "host" );
       Node informNode = xmlCreator.addChild(
         hostNode, "inform" );
       xmlCreator.addTextNode( xmlCreator.addChild(
         informNode,"hostid" ), rs.getString( "hostID" ) );
       xmlCreator.addTextNode( xmlCreator.addChild(
         informNode,"hostname" ), rs.getString( "name" ) );
       xmlCreator.addTextNode( xmlCreator.addChild(
         informNode,"ipaddr" ), rs.getString( "ipaddr" ) );
       xmlCreator.addTextNode( xmlCreator.addChild(
         informNode,"status" ), rs.getString( "status" ) );
```

```
            xmlCreator.addTextNode( xmlCreator.addChild(
             informNode,"location"), rs.getString("location") );
            ...
            res.setContentType( "text/vnd.wap.wml" );
            PrintWriter output = res.getWriter();
            Processor processor = new Processor();
            processor.process( xmlCreator.getDocument(),
             "D:/jakarta-tomcat/webapps/xman/information.xsl",
              output );
        }
         ...
    }
}
```

Servlet GetInformationServlet queries the database element.mbd for the information of specific hostID. Once the information has been retrived, GetInformationServlet use class XMLCreator to create the XML dynamically. The following XML document will be produced in GetInformationServlet when a network manager successfully attempts to retrieve the information of specific managed node. This XML document, the XSL file information.xsl and the PrintWriter object output are passed as arguments to class Processor's process method. Recall that class process applies a style sheet to an XML document. The XML is used with style sheet information.xsl to generate a WML page for GetInformationServlet to display.

```
<host>
    <inform>
        <hostid>3</hostid>
        <hostname>infotel3</hostname>
        <ipaddr>128.134.64.32</ipaddr>
        <status>normal</status>
        <location>INFOTEL</location>
        ...
    </inform>
</host>
```

The following style sheet information.xsl is applied to this XML to generate WML. When this style sheed is applied to the XML generated in GetInoformation-Servlet, a WML document is generated.

```
...
<xsl:template match = "/">
   <wml>
      <card>
         <xsl:apply-templates/>
      </card>
   </wml>
</xsl:template>
```

```
<xsl:template match = "inform">
   <do type = "accept" label = "Back">
      <go href = "xman.GetConfigServlet"/>
   </do>
   <do type = "options" label = "LogOut">
      <go href = "xman.LogoutServlet"/>
   </do>
   <p>___ XMAN ___</p>
   <p>Host: <xsl:value-of select = "ipaddr"/></p>
   <p>HostID: <xsl:value-of select = "hostid"/></p>
   <p>SysName: <xsl:value-of select = "hostname"/></p>
   <p>Status: <xsl:value-of select = "status"/></p>
   <p>Syslocation: <xsl:value-of select = "location"/></p>
   ...
</xsl:template>
...
```

3.3 Experimental Results

We have experimented with UP.Simulator [13] for verifying that XMAN can support WAP-enabled mobile device. An UP.Simulator is an WAP simulator of Openwave [13]. An UP.Simulator allows us to connect to our XMAN through URL. In this experiment, our XMAN system has executed on Windows 2000 system and made use of a Microsoft Access 2000 to store management information. Figure 6 shows how XMAN for WAP looks from the simulator.

4 Conclusion

In this paper, we have presented the design and implementation of an XML-based universal network management system (XMAN), which provides multiple internet devices such as desktop computer, PDA, and cellular phone for a management console. XMAN provides network managers with a unified, cost-effective, and easy-to-use access to management information. Because XMAN has 3-tire architecture, and the implementation is based on an extensive use of Web, XML, and Server-side Java technologies, it provides extensibility and portability for network operators, and allows system developers to update it efficiently.

In the future, we will develop our XMAN system into a component-based service management system based on J2EE platform. Therefore, we would like to combine JSP, Java Beans, Enterprise JavaBeans (EJB), and other J2EE technologies into our XMAN system.

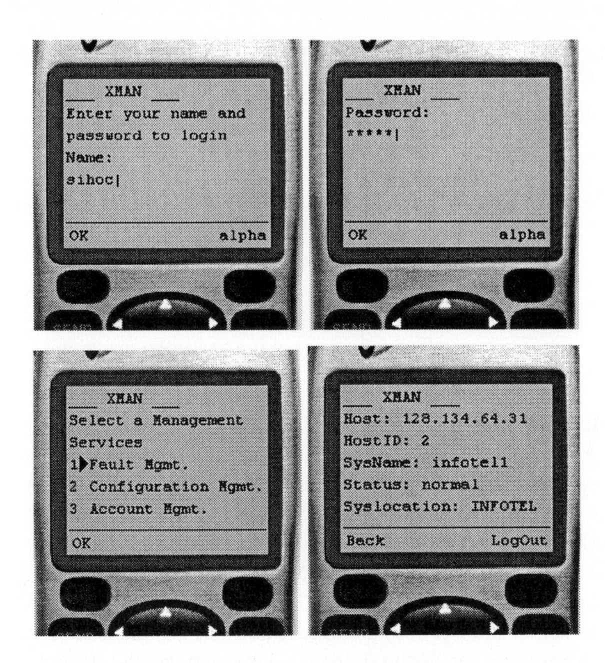

Fig. 6. WAP simulator screenshots

References

1. Chi-Hsing Chu, Chien-Hsuan Huang, Michael Lee: Building an XML-Based Unified User Interface System under J2EE Architecture, MSE 2000 (2000) 208-214.
2. A. Ghlamallah and R. Boutaba: Implementing a Distributed Web-based Management System in Java, ITS'98 vol. 2 (1998) 554-559.
3. H. M. Deitel, P. J. Deitel, T. R. Nieto, T. M. Lin, P. Sadhu: XML How To Program, Prentice Hall (2001).
4. W3C: Extensible Markup Language (XML) 1.0 (Second Edition), http://www.w3.org/TR/
5. SUN Microsystems: Java servlet API 2.3, http://java.sun.com/products/servlet/
6. SUN Microsystems: JDBC API 3.0, http://java.sun.com/products/jdbc/
7. W3C: Document Object Model (DOM) Level 2 Specification, http://www.w3.org/TR/
8. W3C: XSL Transformations (XSLT) version 1.0, http://www.w3.org/TR/
9. AdventNet Inc.: AdventNet SNMP API v2c, http://www.adventnet.com/products/
10. Apache: Jakarta Tomcat 3.3, http://jakarta.apache.org/tomcat/
11. Apache: Xerces Java Parser 1.4.2, http://xml.apache.org/xerces-j/
12. Apache: Xalan-Java Version 1.2.2, http://xml.apache.org/xalan/
13. Openwave Systems Inc.: UP.SDK 4.1, http://developer.openwave.com/download/
14. WAP Forum: WAP-191-WML, http://www.wapforum.org/what/technical.html
15. Gilda Pour: Enterprise JavaBeans, JavaBeans & XML Expanding the Possibilities for Web-Based Enterprise Application Development, TOOLS 31 (1999) 282-291.
16. Jae-Oh Lee: Enabling Network Management Using Java Technologies, IEEE Communications Magazine vol. 8, (2000) 116-123.
17. Hojong Cha, Byungho Ahn, Kookhyun Cho: A QoS-providing multicast network management system, Computer Communications 24 (2001) 1045-1059.

Information Model for Policy-Based Network Security Management

Sook-Yeon Kim, Myung-Eun Kim, Kiyoung Kim, and Jongsoo Jang

Information Security Technology Division, ETRI,
161 Kajong-Dong, Yusong-Gu, Taejon, 305-350, KOREA
{sykim, mekim, kykim, jsjang}@etri.re.kr

Abstract. Policy Based Network Management (PBNM) for network security has been paid much attention as a solution to consistent and unified management of security systems including IDS (Intrusion Detection System) and Firewall. In this paper, we define NSPIM (Network Security Policy Information Model) as a framework of representation, edition, store, and reuse of policies for intrusion detection and response in the PBNM. NSPIM forces each component of PBNM for network security to be flexible and extensible. NSPIM induces the operational structure of PMT (Policy Management Tool) and the data schema of PR (Policy Repository). In addition, policy provisioning objects between PDP (Policy Decision Point) and PEP (Policy Enforcement Point) can be defined based on NSPIM.

1 Introduction

As Internet plays a critical role in industry, its area of service is widely broaden and its number of users is explosively growing. However, security weakness of TCP/IP is more distinctly known and cyber attacks through Internet become more numerous and sophisticated.

Thus, a great deal of research has been devoted to development of security systems like IDS (Intrusion Detection System) or Firewall [1,2]. However, those systems currently available are not generally interoperable because each system has its own special functionality and management mechanism. This fact provides big bothersome to operators who has to manage one or more networks including many security systems. Thus, it has become a hot issue to effectively manage different security systems or to easily control them in a unified way.

Meanwhile, PBNM (Policy Based Network Management) is paid attention as a solution to effective and easy management of various network devices. PBNM delivers consistent, unified, and understandable view of a network without implementation detail. This benefit of PBNM grows more highly as the network becomes more complex and offers more services. RAP (Resource Allocation Protocol) WG (Working Group) in IETF (Internet Engineering Task Force) defines general-purpose objects that facilitate the manipulation of policies and provisioned objects available through

I. Chong (Ed.): ICOIN 2002, LNCS 2343, pp. 662-672, 2002.

COPS (Common Open Policy Service) [3] and COPS-PR (COPS -policy provisioning) [4].

Based on the standardized PBNM of IETF, NS-PBNM (Network Security PBNM) has been suggested [5, 6]. NS-PBNM follows the architecture of PBNM that is composed of PMT (Policy Management Tool), PR (Policy Repository), PDP (Policy Decision Point), and PEP (Policy Enforcement Point). The previous work of NS-PBNM clarifies functionalities of each components and protocols among the components. However, policy information model has not provided as a framework of representation, edition, store, and reuse of policies in NS-PBNM. Flexible and extensible information model for security policy is required for the design of each component of NS-PBNM.

Meanwhile, Policy Framework WG in IETF defines policy information model in a vendor-independent, interoperable, and scalable manner. PCIM (Policy Core Information Model) of the WG has been standardized into RFC3060 [7]. In addition, an update version of RFC3060 is now being prepared [8]. Although PCIM is so generic to be extensible for any application, only QoS (Quality of Service) and IPsec (IP SECurity protocol) have their extension of PCIM [9,10]. The other applications require a non-trivial extension of PCIM. In other words, PCIM is not directly applicable to NS-PBNM because it has no classes for network-based intrusion detection/response.

In this paper, we define flexible and extensible information model by extending PCIM for network security policy. We call it NSPIM (Network Security Policy Information Model). We give the inheritance hierarchy of NSPIM and define each class of NSPIM. We also describe NSPIM implementation in NS-PBNM. PMT and PR structure of NS-PBNM are designed based on NSPIM. In addition, policy provisioning objects between PDP and PEP can be defined based on NSPIM.

2 Definition of NSPIM

In this section, we define NSPIM by extending PCIM. In Section 2.1, we categorize policy rules of NS-PBNM in order to define NSPIM. Section 2.2 shows the inheritance hierarchy of NSPIM. Section 2.3 explains the classes of NSPIM with several examples.

2.1 Policy Rules of NS-PBNM

In this section, we categorize policy rules of NS-PBNM in order to define NSPIM. The categorization is achievable by analyzing the rules of NS-PBNM that are originated from network-based security systems like IDS and Firewall. IDS has different objectives and functionalities from Firewall. However, IDS and Firewall have similarity in rules that their operations are based on. Their rules generally have the following form: “if the current incoming packet(s) satisfies some constraints, then do a predefined action(s)”. In the case of IDS, the predefined action is alerting of intrusion. In the case of Firewall, the action is blocking or permitting of the packet(s). Note that the condition "if the current incoming packet(s) satisfies some constraints" should be defined for each intrusion pattern detected by IDS. In addition, the condition should be defined for each blocking/permission pattern of service/host in Firewall.

We focus our explanation on the conditions because of the amount limit of this paper. The condition can be categorized into three as follows:

- "If the current incoming packet matches a specific pattern,"
- "If the number of packets of a specific pattern incoming during a given time interval is greater than a bound,"
- "If the current sequence of incoming packets matches a specific pattern."

The first category is the most frequently used condition in commercial IDS and Firewall. It is a condition to check the header and/or payload of an IP packet. For example, the condition of the following rule belongs to this category: If an incoming packet of UDP with a destination address 129.254.122.00/24 includes a hexadecimal "|0A 68 65 6C 70 0A 71 75 69 74 0A|" in its payload, show and store the message "System scan try using Web Trans".

The second category is a condition to check the header and/or payload of several IP packets. It is different from the first category because it should count the number of packets of specific pattern during a time interval. This condition is evaluated to true if the counted number is greater than a given bound. For example, the condition of the following rule belongs to this category: If more than twenty packets that has a destination address 129.254.122.00 and a ICMP type eight come for two seconds, show and store the message "Attack try of Denial of Service using smurf".

The third category is a condition to check the header and/or payload of multiple IP packets to find a specific pattern. This category is different from the second one because its pattern consists of a sequence of different packets. For example, the condition of the following rule belongs to this category: "For two sequential packets with the same IP IDENTIFACATION, if one packet with MF flag set has a larger OFFSET than that of the other packet with MF flag unset, show and store the message "Tear Drop Attack."

2.2 Class Hierarchy

In this section, we show the inheritance hierarchy of NSPIM. Conditions and actions representing the rules of NS-PBNM should be basically extended from the classes PolicyCondition and PolicyAction of PCIM [7]. However, since the classes CompoundPolicyCondition and CompoundPolicyAction of PCIM [8] provide flexible compounding functionality, conditions and actions of NSPIM are extended from them. Fig. 1 illustrates the inheritance hierarchy for several structural classes of NSPIM. This figure omits many NSPIM classes that are originally defined in PCIM [7,8].

In NSPIM, instances of structural classes are associated with association classes. The hierarchy of several association classes of NSPIM is in Fig. 2. This diagram omits many NSPIM associations that are originally defined in PCIM [7,8].

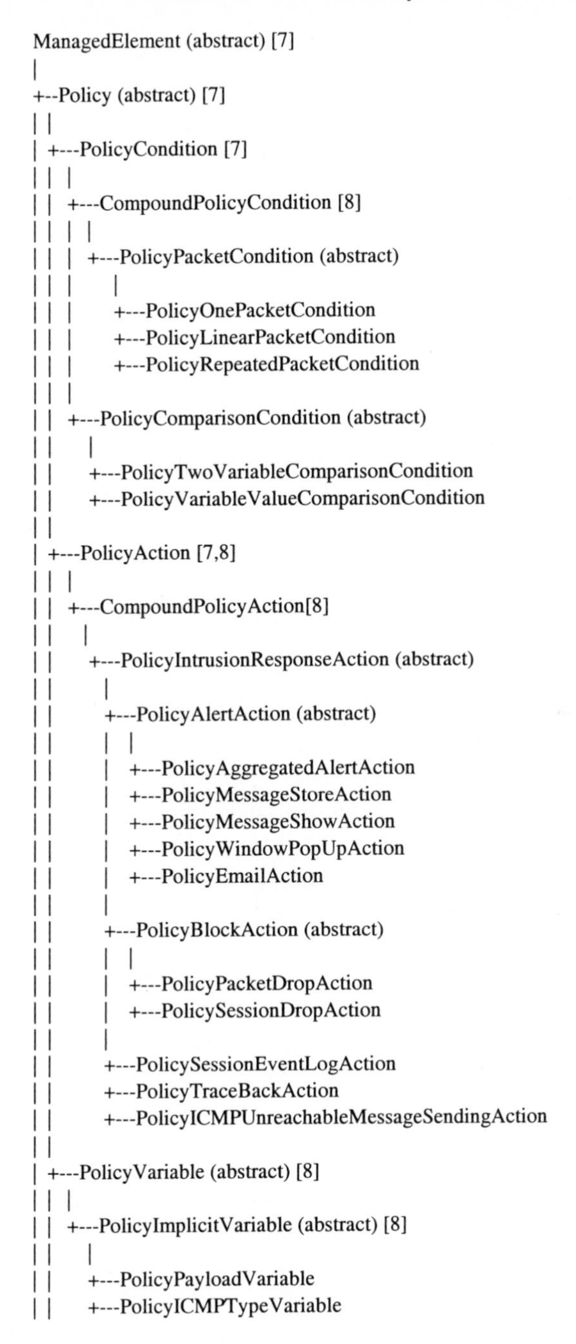
ManagedElement (abstract) [7]
+--Policy (abstract) [7]
+---PolicyCondition [7]
+---CompoundPolicyCondition [8]
+---PolicyPacketCondition (abstract)
+---PolicyOnePacketCondition
+---PolicyLinearPacketCondition
+---PolicyRepeatedPacketCondition
+---PolicyComparisonCondition (abstract)
+---PolicyTwoVariableComparisonCondition
+---PolicyVariableValueComparisonCondition
+---PolicyAction [7,8]
+---CompoundPolicyAction[8]
+---PolicyIntrusionResponseAction (abstract)
+---PolicyAlertAction (abstract)
+---PolicyAggregatedAlertAction
+---PolicyMessageStoreAction
+---PolicyMessageShowAction
+---PolicyWindowPopUpAction
+---PolicyEmailAction
+---PolicyBlockAction (abstract)
+---PolicyPacketDropAction
+---PolicySessionDropAction
+---PolicySessionEventLogAction
+---PolicyTraceBackAction
+---PolicyICMPUnreachableMessageSendingAction
+---PolicyVariable (abstract) [8]
+---PolicyImplicitVariable (abstract) [8]
+---PolicyPayloadVariable
+---PolicyICMPTypeVariable

Fig. 1. Class Inheritance Hierarchy for NSPIM

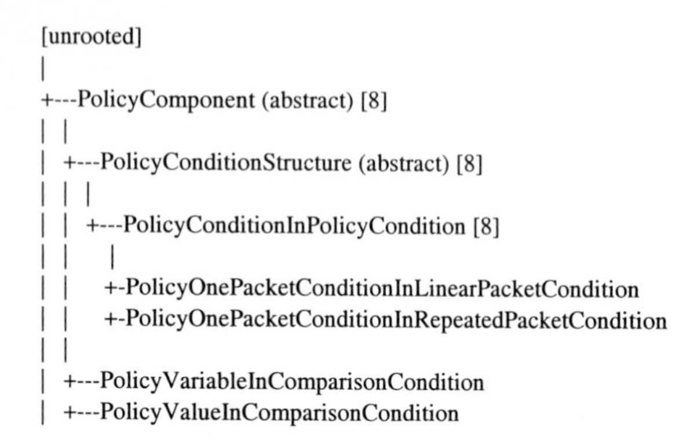

Fig. 2. Association Inheritance Hierarchy for NSPIM

2.3 Class Definition

In this section, the classes of NSPIM are explained. The classes defined for only NSPIM are roughly grouped into five: packet monitoring conditions, comparison conditions, intrusion response actions, IP header variables, and associations. Packet monitoring conditions represent the conditions to check the header and/or payload of IP packets. Comparison conditions represent the conditions to compare two variables or to compare a variable and a value. Intrusion response actions represent the actions in response to detected intrusion or attack. IP header variables represent each field of IP header.

We focus our detailed explanation on packet monitoring conditions because of the length limit of this paper. Packet monitoring conditions has an abstract super class PolicyPacketCondition. This class is extended from CompoundPolicyCondition [8]. The difference between this class and CompoundPolicyCondition is that this class has a narrower meaning than CompoundPolicyCondition. PolicyPacketCondition has three subclasses PolicyOnePacketCondition, PolicyLinearPacketCondition, and PolicyRepeatedPacketCondition. Each of the following subsections explains each of the three classes.

PolicyOnePacketCondition. The class of one packet condition is extended from PolicyPacketCondition. The difference between this class and PolicyPacketCondition is that this class has a narrower meaning than PolicyPacketCondition. This class represents a condition to check the header and/or payload of an IP packet.

Fig. 3 shows an instantiation example of PolicyOnePacketCondition. Fig. 3 shows a rule: “if a packet has a source address of 134.250.17.0/24 and a destination port of 80, then drop the packet”.

PolicyRepeatedPacketCondition. The class of a condition for repeated packets is extended from PolicyPacketCondition. This class represents a condition that is evaluated to true if the number of packets of specific pattern during a time interval is greater than a bound.

The interval of time and the bound of number are represented as properties of this class. The pattern of the packets to be counted is specified by an instance of PolicyOnePacketCondition. Thus, any instance of this class PolicyRepeatedPacketCondition should be associated with an instance of PolicyOnePacketCondition. Fig. 4 shows an instantiation example of PolicyRepeatedPacketCondition. Fig. 4 shows a rule: if more than twenty ICMP packets of a destination address 129.254.122.00 and a ICMP type 8 come for two seconds, show and store the message "Attack try of Denial of Service using smurf".

PolicyLinearPacketCondition. The class for a condition of sequential packets is extended from PolicyPacketCondition. This class represents a condition that is evaluated to true if a sequence of packets matches a specific pattern.

The pattern should be specified by two or more instances of PolicyOnePacketCondition. Thus, an instance of this class should be associated with the instances of PolicyOnePacketCondition. PolicyLinearPacketCondition has a non-negative integer as its property representing the number of packets to be matched. This number should be equal to the number of instances of PolicyOnePacketCondition associated with the instance of PolicyLinearPacketCondition. In other words, the number of instances of PolicyOnePacketCondition should be the number of packets that compose the pattern.

Note that each aggregation connecting this class and PolicyOnePacketCondition has GroupNumber as its property. GroupNumber indicates the sequence of the instances of PolicyOnePacketCondition. In other words, the smaller GroupNumber indicates that the corresponding instance of PolicyOnePacketCondition should be applied to earlier incoming packet.

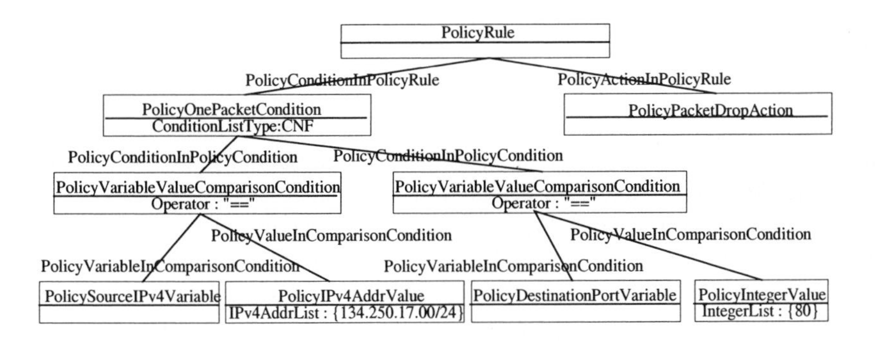

Fig. 3. Example of PolicyOnePacketCondition

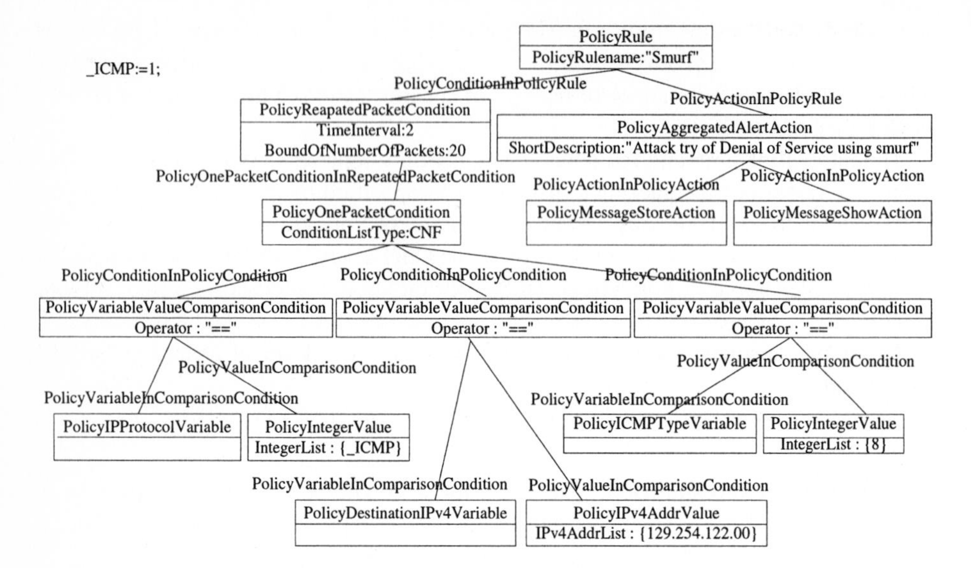

Fig. 4. Example of PolicyRepeatedPacketCondition

3 Implementation of NSPIM

In this section, we describe NSPIM implementation in NS-PBNM. NSPIM is implemented as operational structure of PMT and data schema of PR in NS-PBNM. In addition, policy provisioning objects between PDP and PEP can be defined based on NSPIM.

3.1. PMT Structure Based on NSPIM

PMT (Policy Management Tool) has the functionalities of policy presentation, edition, rule translation, rule validation, and conflict resolution. All of the functional components of PMT would have flexibility and extensibility if they are designed based on NSPIM. However, we focus our explanation on the component of policy presentation and editing. This component is composed of four blocks: retrieval, insertion, deletion, and update. Each block consists of sub-modules corresponding to each class of NSPIM. An abstract diagram of the components is in Fig. 5.

Each sub-module has a workflow of dealing the properties of the corresponding class with optional calls to the other sub-modules. For example, a sub-module "PolicyRule Insertion" optionally calls nine sub-modules as in Fig. 6: PolicyOnePacketCondition Insertion, PolicyLinearPacketCondition Insertion, PolicyRepeatedPacketCondition Insertion, PolicyPacketDropAction Insertion, etc.

3.2. LDAP Schema Based on NSPIM

Flexible and extensible PR structure of NS-PBNM can be induced from NSPIM. In order to get the PR structure, NSPIM should be mapped to a schema of PR. Fig. 7 shows an example of mapping NSPIM into LDAP schema. When mapping to an LDAP schema, the structural classes of NSPIM can be mapped more or less directly. However, the association classes must be mapped to a form suitable for directory implementation. Classes not existing in NSPIM can be also added to the LDAP schema to improve the performance of retrieval of large amounts of policy-related information.

PCIM classes used for NSPIM can be mapped to LDAP schema according to the standard of IETF [11]. Most of the classes defined for only NSPIM are mapped to auxiliary classes as in Fig. 7 in order to fully utilize a reusable container[11].

3.3. PIB/MIB Based on NSPIM

COPS-PR or SMNP [12] can used in NS-PBNM to exchange policy information between the policy server (Exactly PDPs) and its clients (exactly PEPs). These two protocols focus on the mechanisms and conventions used to communicate provisioned information between a server and clients. Thus, the protocols need a virtual data store of policy information whose structure is shared by the server and clients. The data store is called PIB (Policy Information Base) and MIB (Management Information Base) in COPS-PR and SNMP, respectively.

SMI (Structure of Management Information) and SPPI (Structure of Policy Provisioning Information) describe the structure for specifying policy information in PIB and MIB, respectively [14,15]. Even though a set of PRCs (Policy Provisioning Classes) and MOs (Managed Objects) has been defined for QoS(Quality of Service)

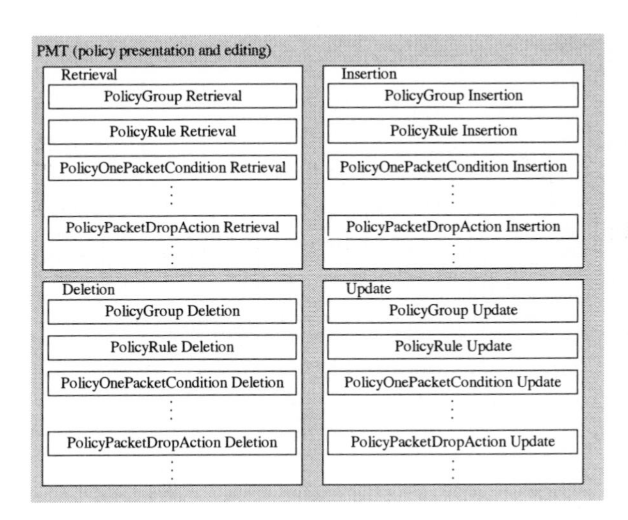

Fig. 5. Functional components of policy presentation and editing

for COPS-PR and SNMP, respectively [16,17], network-based intrusion detection and response has not been considered in the previous work of PBNM. NSPIM provides a framework for the PRCs and MOs of NS-PBNM. Flexible and extensible PRCs and MOs can be defined from NSPIM for network security management by transforming and merging NSPIM classes.

4 Conclusion

We define NSPIM for NS-PBNM. We give the inheritance hierarchy of NSPIM and define each class of NSPIM. The classes are conditions and actions representing the policy rules of NS-PBNM. We also show that how NSPIM induces flexible and extensible design of PMT, LDAP, and PIB/MIB objects of NS-PBNM.

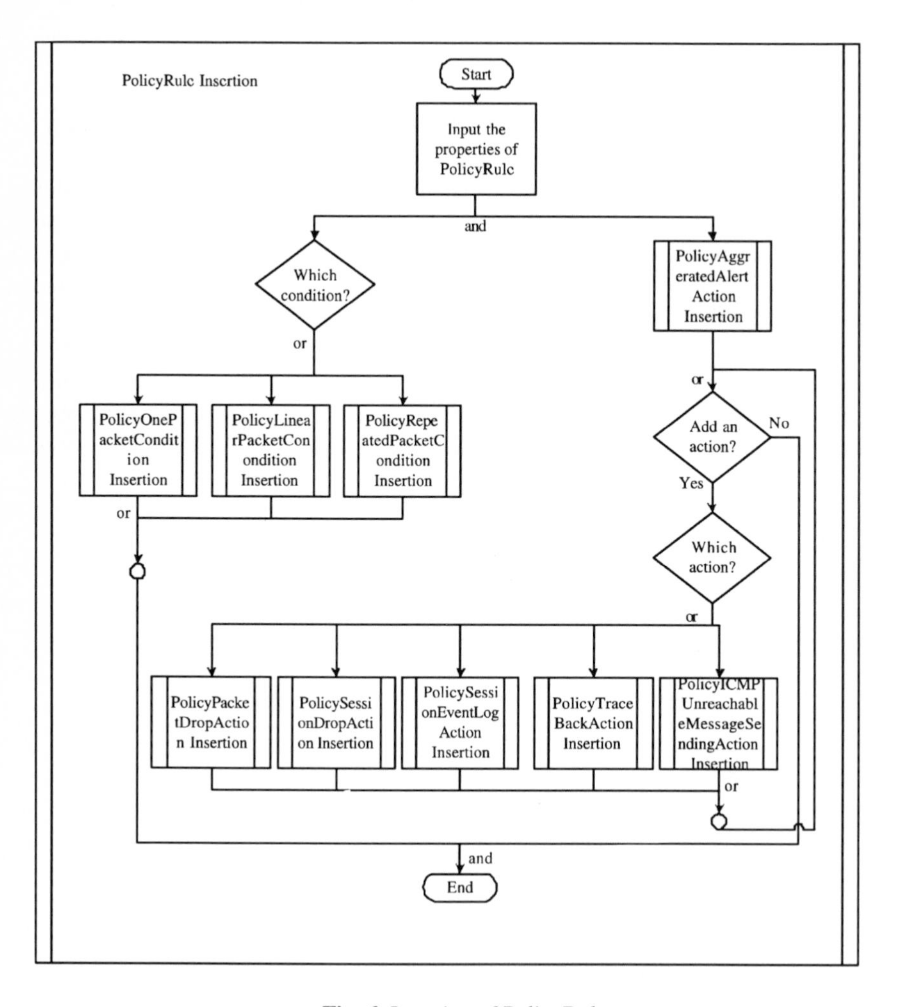

Fig. 6. Insertion of PolicyRule

Further work of this paper is to extend the current NSPIM for future rules of NS-PBNM. The future rules includes the rules for intrusion pattern not yet known and the rules not bounded to each individual PEP. Rules for various cooperating PEPs should be formulated and modeled for sophisticated security management. NSPIM is expected to activate the research and standardization of network-based security management.

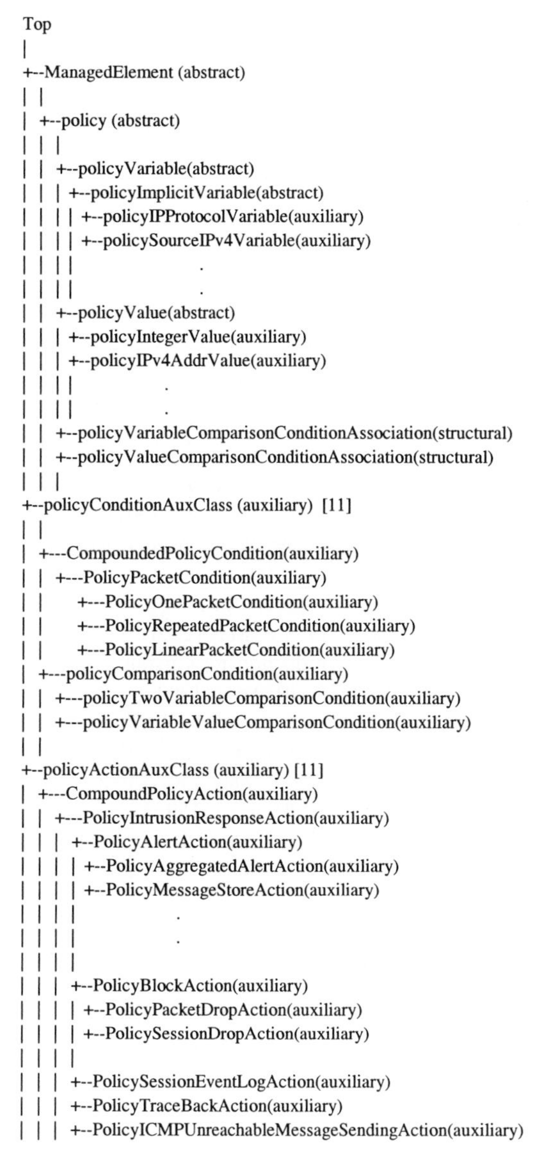

Fig. 7. LDAP schema based on NSPIM

References

1. Shieh S.-P., Gligor V.D.: On a pattern-oriented model for intrusion detection, Knowledge and Data Engineering, IEEE Transactions on, Volume: 9 Issue: 4 , July-Aug. (1997) 661 – 667
2. Lodin S.W., Schuba C.L.: Firewalls fend off invasions from the Net, IEEE Spectrum, Volume: 35 Issue: 2, Feb. (1998) 26 –34
3. Boyle J., Cohen R., Durham D., Herzog S., Rajan R., Sastry A.: The COPS (Common Open Policy Service) Protocol, RFC 2748, January (2000)
4. Chan K., Seligson J., Durham D., Gai S., McCloghrie K., Herzog S., Reichmeyer F., Yavatkar R., Smith A.: COPS Usage for Policy Provisioning (COPS-PR), RFC 3084, March (2001)
5. Kim K. Y., Seo D. I., Jand J. S., Lee S. H.: A Policy-based Integrated Secure Architecture for Providing Security Service in WAN, The Sixth Conference on Communication Software, . SokCho, Korea, July (2001) 35-39
6. Bang H.-C., et al.: Network intrusion detection and respond mechanism based on hierarchical traffic analysis in a Policy-based network security management framework.
7. Strassner J., Ellesson E., Moore B., Westerinen A.: Policy Core Information Model -- Version 1 Specification, RFC 3060, February (2001)
8. Moore B., Raberg L., Snir Y., Strassner J., Westerinen A., Chdha R., Brunner M., Cohen R.: Policy Core Information Model Extensions, work in progress, <draft-ietf-policy-pcim-ext-01>, April (2001)
9. Snir Y., Ramberg Y., Strassner J., Cohen R.: Policy Framework QoS Information Model, work in progress, <draft-ietf-policy-qos-info-model-03.txt>, April (2001)
10. Jason J., Rafalow L., Vyncke E.: IPsec Configuration Policy Model, work in progress, <draft-ietf-ipsp-config-policy-model-02.txt >, March (2001)
11. Strassner J., Westerinen A., Ellesson E., Moore B., Moats R.: Policy Core LDAP Schema, work in progress, <draft-ietf-policy-core-schema-11.txt>, May (2001)
12. Case J., Mundy R., Partain D., Stewart B.: Introduction to Version 3 of the Internet-standard Network Management Framework, RFC 2570, April (1999)
14. McCloghrie K., Perkins D., Schoenwaelder J., Case J., Rose M., Waldbusser S.: Structure of Management Information Version 2 (SMIv2), STD 58, RFC 2578, April (1999)
15. McCloghrie K., Fine M., Seligson J., Chan K., Hahn S., Sahita R., Smith A., Reichmeyer F.: Structure of Policy Provisioning Information (SPPI), RFC 3159, August (2001)
16. Fine M., McCloghrie K., Seligson J., Chan K., Hahn S., Sahita R., Smith A.: Reichmeyer and F.: Framework Policy Information Base, work in progress, <draft-ietf-rap-frameworkpib-05.txt>, July (2001)
17. Waldbusser S., Saperia J., Hongal T.: Policy Based Management MIB, work in progress, < draft-ietf-snmpconf-pm-07.txt>, July (2001)

Design and Implementation of One-Way IP Performance Measurement Tool

Jaehoon Jeong[1], Seungyun Lee[1], Yongjin Kim[1], and Yanghee Choi[2]

[1] Protocol Engineering Center, ETRI, 161 Gajong-Dong, Yusong-Gu,
Daejon 305-350, Korea
{paul, syl, kimyj}@etri.re.kr
http://pec.etri.re.kr/

[2] School of Computer Science and Engineering, Seoul Nat'l Univ.,
San 56-1 Shinlim-Dong, Kwanak-Gu, Seoul 151-742, Korea
yhchoi@snu.ac.kr
http://mmlab.snu.ac.kr/

Abstract. In this paper, we propose an architecture of measurement system which can measure IETF's IP Performance Metrics (IPPM) such as one-way delay, one-way packet loss and packet delay variation in the Internet. As the synchronization among measurement systems is very important in one-way delay measurement, we used the Global Positioning System (GPS) to synchronize the measurement systems and provided the precision up to one micro-second. To improve the accuracy of one-way delay measurement, the proposed system employs timestamps at the Ethernet frame level. We carried out measurements on the real Internet with the implemented system. It is seen that delay differs largely between path directions. Through these measurements, we present the need of one-way delay measurement.

1 Introduction

We need one-way measurement of the Internet in order for us to grasp the exact state of the Internet, which is asymmetric [1]. The IETF's IP Performance Metrics (IPPM) Working Group suggested one-way metrics and architecture for one-way measurement [2]. Metrics are one-way delay, one-way packet loss, instantaneous packet delay variation, etc [3-5]. One-way measurement is a kind of active measurement, which injects measurement packets in a path to measure and observes how the packets are served. With measurement data obtained through active measurement, the network management can be performed effectively. For example, we can conjecture that some problems happened in the network if we have observed the network for a long time and have found that one-way delay increased much more than at ordinary times. Through the analysis of measurement result that we have obtained through long obser

This work has been partially supported by Brain Korea 21, National Research Laboratory program, and Korea Telecom 2001.

I. Chong (Ed.): ICOIN 2002, LNCS 2343, pp. 673-686, 2002.

vation of network, we can find what is the problem of the network (e.g., bottleneck path) and then can solve it by relocating resources, increasing link capacity, changing network configuration (e.g., routing configuration) and upgrading routers. In the result, we can improve the performance of the entire network.

The system for the active measurement that provides us with the necessary information for effective network management should provide operator with user interface with which operator can control the system easily and efficiently. It should also be able to measure the network stably for a long time and have functions of trouble-shooting, which are to find the troubles that can happen in measurement system or network during measurement and to solve them automatically without operator's intervention.

In this paper, we suggest an architecture of measurement system (AMT: Active Measurement Tool) that can perform one-way measurement efficiently and stably; AMT has been designed and implemented so that it may measure one-way metrics stably for a long time and be expanded easily in the point of the number of measurement systems. We present the analysis of result that we have measured in the Internet with AMT. This paper is organized as follows; Section 2 presents related work. In section 3, we explain the architecture of the suggested measurement system (AMT), the components of the system, the procedure of measurement, and the visualization of measurement result. We also evaluate the result of one-way measurement in test network. Finally, in section 4, we conclude this paper and present future work.

2 Related Work

Many measurement systems were implemented for active measurement. We introduce two representatives among the systems; (a) Skitter and (b) Surveyor.

2.1 Skitter

Skitter is a measurement system that Cooperative Association for Internet Data Analysis (CAIDA) Group has implemented [6]. Skitter was made for analysis of Internet's topology and performance. It injects measurement packets in Internet and observes how the packets are served. Main functions are as follows; (a) Measurement of Forward IP Path, (b) Measurement of RTT, (c) Trace of Routing Change, and (d) Visualization of Network Topology. Skitter provides users with easy and convenient user interface but has a demerit that it can not measure one-way metrics.

2.2 Surveyor

Surveyor is a measurement system that Advanced Network & Services Group has implemented that can measure one-way metrics [7, 8]. The one-way metrics are based on IETF's IPPM. Surveyor consists of two systems; (a) Measurement System and (b)

Central Control System. Two systems use One-Way Delay and Packet loss protocol (OWDP) [9, 10]. Because Measurement Systems are synchronized with one another by GPS, they can perform one-way delay measurement accurately. Central Control System controls Measurement Systems and gathers measurement data from the Measurement Systems. To improve the accuracy of one-way measurement, Surveyor stamps the time information in Ethernet device driver. It is one of the most popular systems for one-way measurement.

3 Active Measurement Tool (AMT)

AMT is an infrastructure that can measure various one-way metrics suggested by IETF's IPPM Working Group. AMT is a PC-based system that uses FreeBSD and MySQL as operating system and database management system respectively [11, 12].

3.1 Consideration for Implementation of One-Way Measurement

3.1.1 How to Synchronize Measurement Systems

There is no need to synchronize measurement systems in order to measure RTT which means two-way delay. However, when it comes to measurement of one-way delay, we should synchronize measurement systems for the exact measurement. Fig. 1 shows how to synchronize systems by using GPS satellites. Through GPS satellites, the exact time information can be provided for measurement systems, which can maintain their system time correctly with it. Hardwares that are used to receive time information from GPS are as follows; (a) Oncore Remote Antenna and (b) Oncore GPS UT Receiver, which are the products of Motorola [13]. Network Time Protocol (NTP) Daemon (i.e., ntpd [14, 15]) modifies the kernel time with time information received from GPS. The time information encoded in Pulse Per Second (PPS) format can be provided for ntpd through either serial port or parallel port [14]. Device driver of the port transforms the PPS into binary format and provides ntpd with the time information formatted as binary. The ntpd updates the kernel time periodically with the time information. In this mechanism, all measurement systems are synchronized with GPS.

3.1.2 Timestamp

To improve the accuracy of measuring one-way delay, the measurement system has to timestamp on the field for time information in the payload which is one of fields in the Ethernet frame just before transmitting the Ethernet frame to the network interface card as well as just after receiving the Ethernet frame from the network interface card like Fig. 2. In this way, we are capable of reducing the delay which can occur through the protocol stack at end hosts [8, 16].

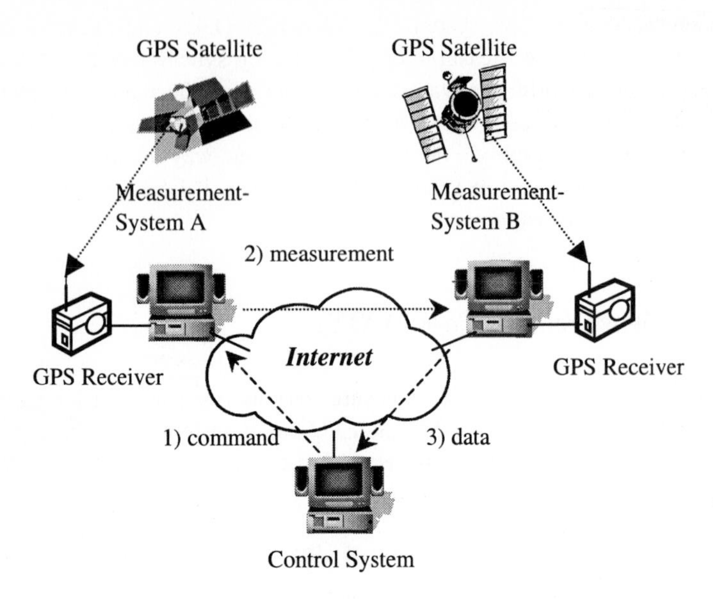

Fig. 1. Synchronization among Measurement Systems and Procedure of Measurement

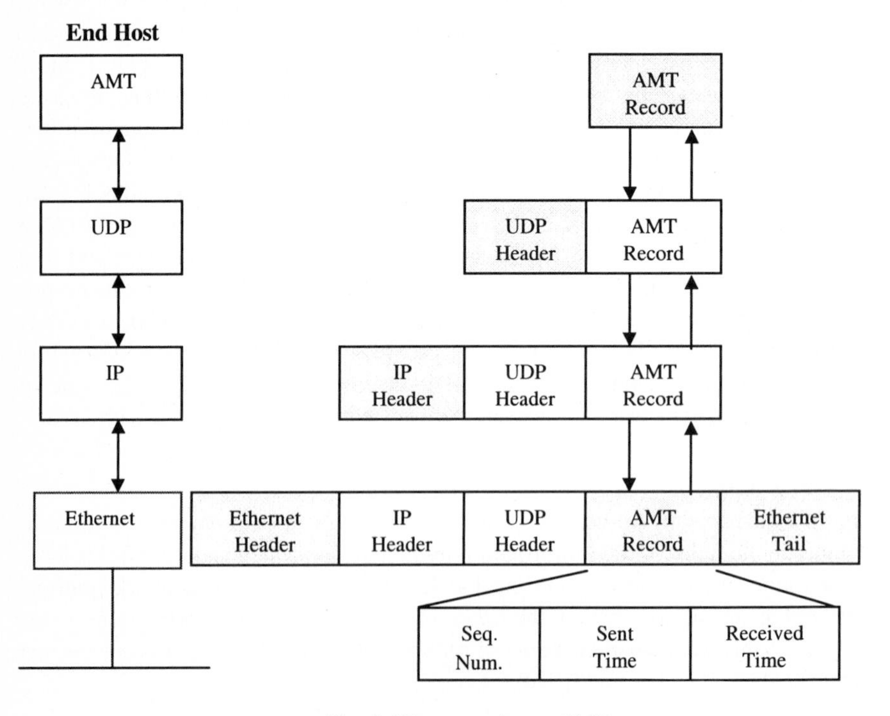

Fig. 2. Timestamping at AMT

3.2 Architecture of AMT System (AMT)

AMT consists of two kinds of systems ; (a) Control System (CS) and (b) Measurement System (MS). While MS performs one-way measurement, CS controls and manages the MS's. Fig. 3 describes the architecture of AMT. Fig. 1 describes the procedure of measurement which is performed by CS and MS's.

3.2.1 Control System (CS)

CS, main system of AMT, receives commands sent from Control Shell (CSH), with which operator controls and manages AMT. CSH is console-based user interface. CS has three processes like Fig. 3; (a) Control Server (CSV), (b) Storage Server (SSV) and (c) DB Server (DBS).

Control Server (CSV): CSV receives commands from operator, parses the commands, and then processes the commands. CSV consists of three threads; (a) Main Thread (MAT), (b) Measurement Thread (MET) and (c) Polling Thread (POT). MAT receives command from CSH and processes it. MET initiates a measurement and POT checks the health of measurement systems and network.

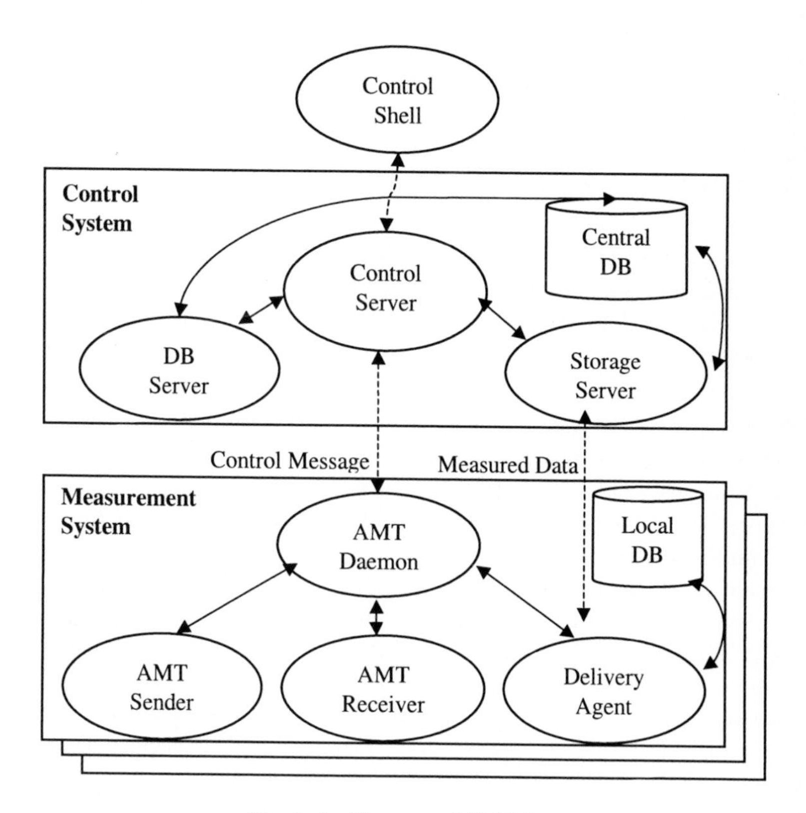

Fig. 3. Architecture of AMT System

Storage Server (SSV): SSV collects measurement data from local database (Local DB) of each MS after the measurement and stores the data in the central database (Central DB). It is forked by CSV when preparing the collection. The collection is performed with the aid of Delivery Agent (DA) of each MS.

DB Server (DBS): DBS analyzes the gathered raw data and stores them into Central DB.

3.2.2 Measurement System (MS)

MS has four processes like Fig. 3; (a) AMT Daemon (AMTD), (b) AMT Sender (AMTS), (c) AMT Receiver (AMTR) and (d) Delivery Agent (DA).
AMT Daemon (AMTD)

AMT Daemon (AMTD): After AMTD, main process of MS, first registers itself in CS, it receives all the control messages from CSV, processes them and sends the result to CSV. For example, when CSV sends the measurement preparation message to the registered AMTD of each MS, AMTD receives the message to prepare measurement. It forks AMT Sender (AMTS) and AMT Receiver (AMTR) which will perform actual measurement. All the control messages from CSV to AMTS or AMTR of each MS are sent to AMTS or AMTR via AMTD of the MS. The reason that we designed AMT system for all the control message messages between CSV and AMTS or AMTR to go via AMTD is that we tried to make AMTS and AMTR be light-weighted processes that can run stably for a long time.

AMT Sender (AMTS): AMTS is forked by AMTD when CS starts measurement. After AMTS receives a measurement start message, it generates measurement packets. The packets are generated in Poisson process by a pseudo-random number generator. AMTS sends every packet to all the AMTRs which are joining in the measurement.

AMT Receiver (AMTR): AMTR is forked by AMTD when CS starts measurement. After AMTR receives a measurement start message, it opens Local DB file to be ready to receive measurement packets. Whenever it receives a measurement packet, it stores the record of the packet in Local DB. The record consists of 5 fields; (a) Sequence Number, (b) Sender IP Address, (c) Sent Time, (d) Receiver IP Address, and (e) Received Time. 'Sequence Number' is 4-byte sequence number field. 'Sender IP Address' is 4-byte IP address field of AMT sender that sent the packet. 'Receiver IP Address' is also 4-byte IP address field of AMT receiver that received the packet. 'Sent Time' is 8-byte timestamp field in which the timestamp is written by Ethernet device driver just before packet's being sent into network interface card. The type of this field is struct timeval { u_long tv_sec; u_long tv_usec }. 'Received Time' is also 8-byte timestamp field where the timestamp is written by Ethernet device driver just after packet's being received from network interface card.

Delivery Agent (DA): DA is forked by AMTD when CS gathers measurement data from each MS. After DA receives a gather start message, it opens Local DB and delivers the measurement data stored in it to SSV of CS.

3.3 Procedure of Measurement

The procedure of measuring one-way delay is described as shown in Fig. 4.

Step 1. Initialization of AMTD for measurement
CSV sends all the AMTDs that take part in measurement a 'measure-ready' message indicating that they have to prepare a measurement. The control packet including the message provides them with a system parameter (i.e., lambda value for Poisson process) and a list of IP addresses of all the participating AMTDs together with the message.

Step 2. Fork of measurement processes
When AMTD of MS receives the 'measure-ready' message, it makes control channels that will be used to communicate with AMTS and AMTR that are implemented in UNIX domain stream socket. It forks AMTS and AMTR and then forwards the 'measure-ready' message to them through the control channels.

Step 3. Establishment of control channel
Just after AMTS and AMTR have been forked by AMTD, they establish control channel that is used to communicate with AMTD. AMTS and AMTR obtain the system parameter such as the list of IP addresses of participants from control packet including the 'measure-ready' message. When AMTS and AMTR are ready to measure, they report the readiness to AMTD through the control channel.

Step 4. Confirmation about readiness from AMTD
When AMTD receives the report from both AMTS and AMTR, AMTD sends CSV a 'measure-ready-ack' message indicating that MS is ready to measure.

Step 5. Start of measurement
When CSV has received the report from all participating AMTDs, CSV sends them a 'measure-start' message indicating that they have to start measurement.

Step 6. Start of actual measurement
When AMTD receives the 'measure-start' message, it forwards the message to its child processes; AMTS and AMTR.

Step 7. Injection of measurement packets

AMTS generates measurement packets in Poisson process. The packets are sent to all participating AMTRs except AMTR in the same host through UDP socket.

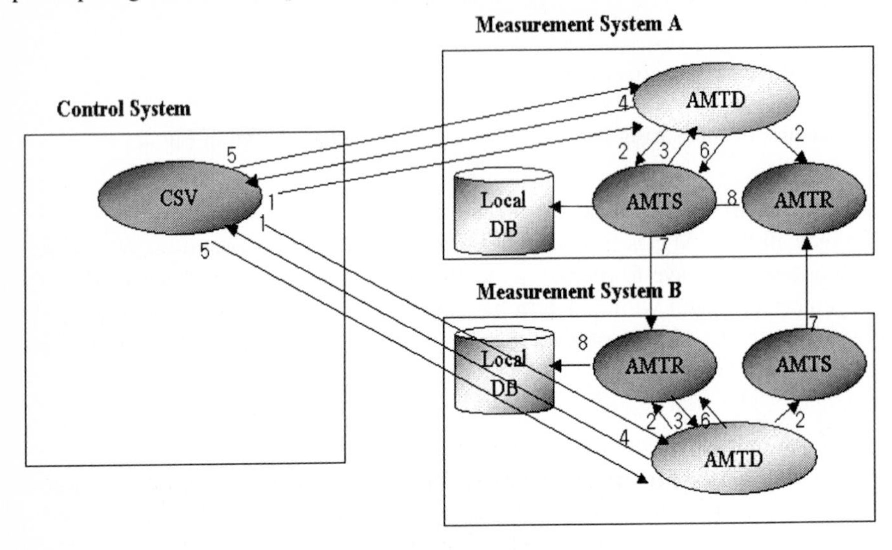

Fig. 4. Procedure of Measurement

Step 8. Storing of measurement records

When AMTR receives a measurement packet, it
stores into Local DB a record that consists of the following fields; (a) Sequence Number, (b) Sender IP Address, (c) Sent Time, (d) Receiver IP Address, and (e) Received Time. The record is stored in binary format, not ASCII format in order to reduce the size of record.

3.4 Visualization of Measurement Result

AMT Visualizer (AMTV) can provide operator with the result of measurement through web like Fig. 5 and Fig. 6. AMTV receives the following inputs; (a) Sender IP, (b) Receiver IP, and (c) Date. The combination of three input fields means that we want to get the result from the measurement packets which MS with 'Sender IP' address generated and sent to MS with 'Receiver IP' address on 'Date'.

We present an example with Fig. 5 and Fig. 6. The meaning of input in Fig. 5 is that we want to get the result from measurement packets that MS with IP address 147.46.14.69 sent to MS with IP address 203.232.127.20 on November 28, 2000. The output of the result is the graphs of one-way delay, one-way loss, and delay jitter during the day. Fig. 6 shows the one-way delay on November 20, 2000 as the result of the query of Fig. 5. Fig. 7 describes the procedure of visualization. When operator sends a query requesting measurement result between two end hosts on a specific day with AMTV, the query is transferred to CGI Module called as Measurement Analysis Agent (MAA) via Web Server (httpd). MAA processes the query with Central DB and returns the result to AMTV via httpd.

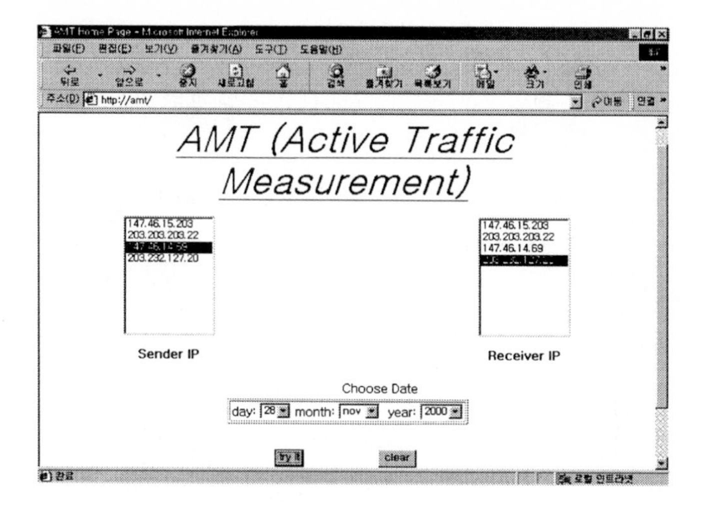

Fig. 5. AMT Visualizer (AMTV)

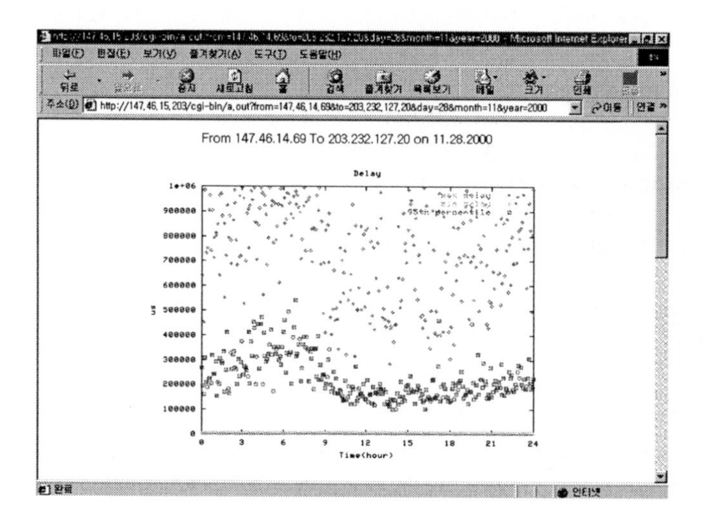

Fig. 6. Result of Query at AMTV

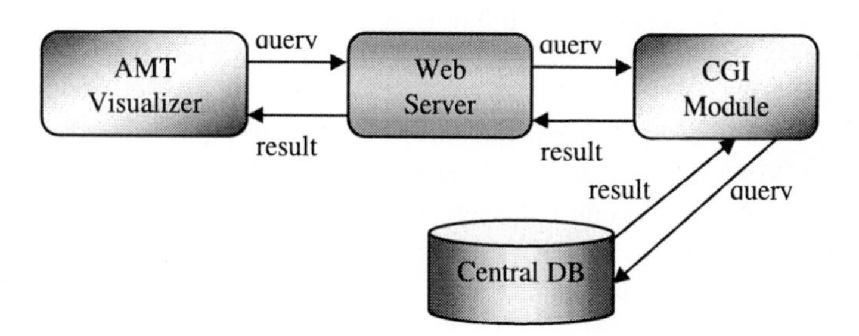

Fig. 7. Procedure of Visualization

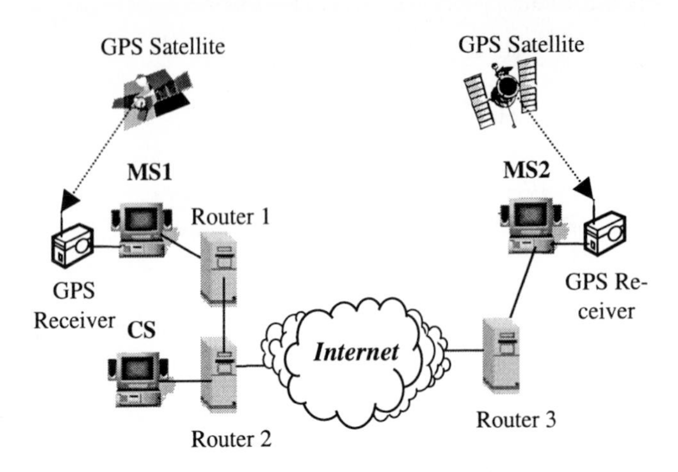

Fig. 8. Test Network

3.5 Performance Evaluation

We measured one-way delay in Internet including Korea Commercial Network (KORNET) and evaluated the result of measurement.

3.5.1 Test Environment

Fig. 8 shows the topology of test network. Measurement System1 (MS1) of which IP address is 147.46.14.69 is located in a subnet of Seoul National University and Measurement System2 (MS2) of which IP address is 203.232.127.20 is located in a subnet of KORNET. Control System (CS) of which IP address is 147.46.15.203 is located in a subnet of Seoul National University. Router 1's IP address is 147.46.14.65, Router 2's IP address is 147.46.15.2 and Router 3's IP address is 203.232.127.14. Router 1 is adjacent to Router 2 as shown in Fig. 8.

3.5.2 Evaluation of Measurement Result

We measured one-way delay during a day from 0 AM on 2000/11/28 to 12 PM on 2000/11/28. We generated measurement packets in the frequency that the lambda of Poisson process is 2.

Fig. 9 shows one-way delay from MS1 to MS2 (Delay1) and Fig. 10 shows one-way delay from MS2 to MS1 (Delay2). X-axis of graph is time. The unit is 5 minutes. Y-axis is one-way delay. The unit is 1 micro-second (us). As representative values, we selected (a) Minimum delay, (b) 95th percentile and (c) Maximum delay in the period of 5 minutes. Because percentile is the most reasonable among three representatives, we compare two figures (Fig. 9 and Fig. 10) by 95th percentile. We can see that 95th percentile of Delay1 is from 100148[us] to 539724[us] and that 95th percentile of Delay2 is from 7923[us] to 16344[us]. As a result, we can see that the one-way path from MS1 to MS2 (Path1) has bigger and more variable one-way delay than that from MS2 to MS1 (Path2).

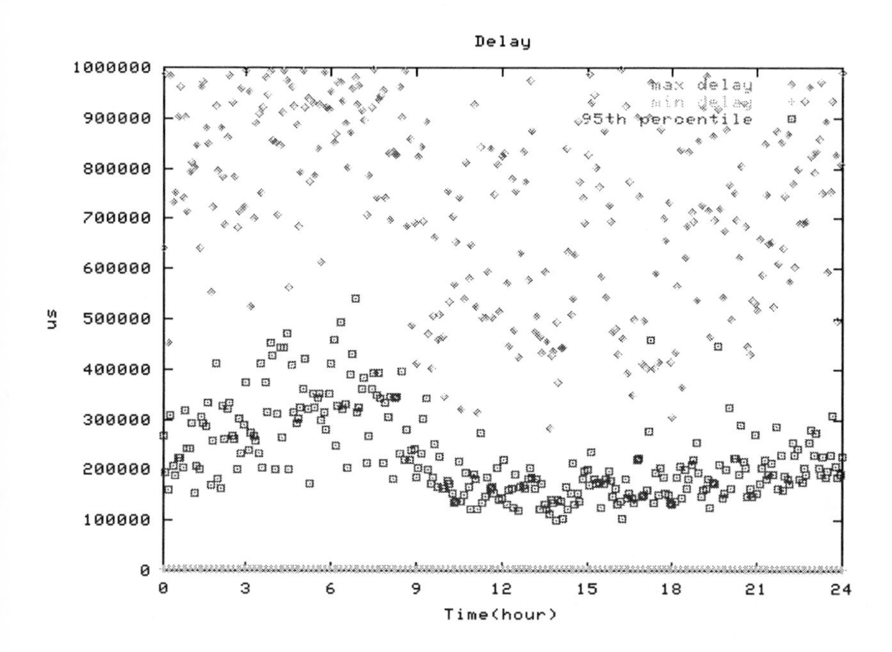

Fig. 9. One-way Delay from MS1 to MS2

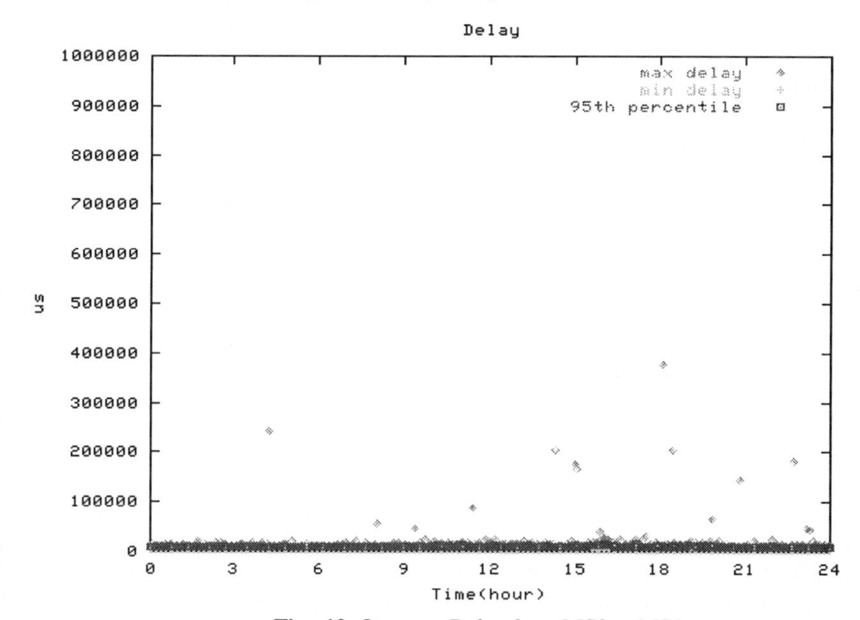

Fig. 10. One-way Delay from MS2 to MS1

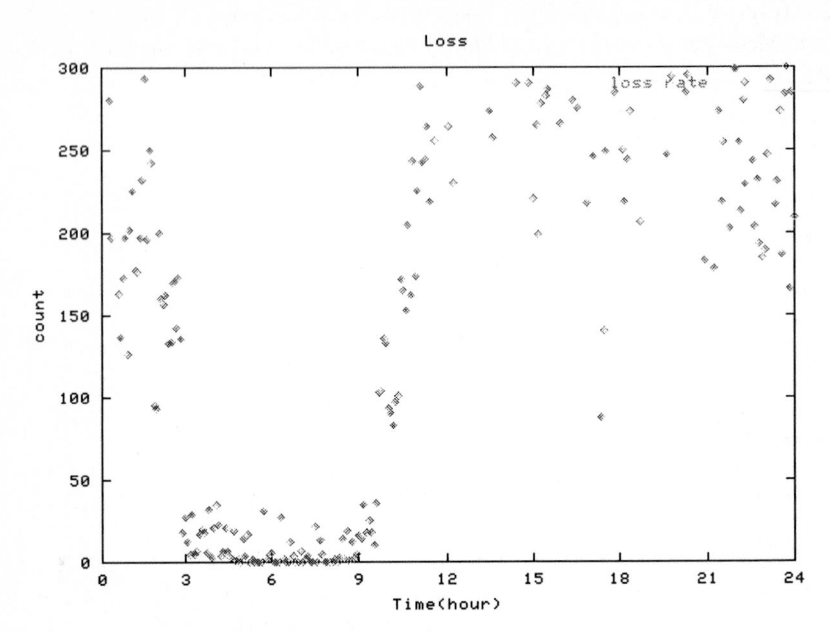

Fig. 11. One-way Loss from MS1 to MS2

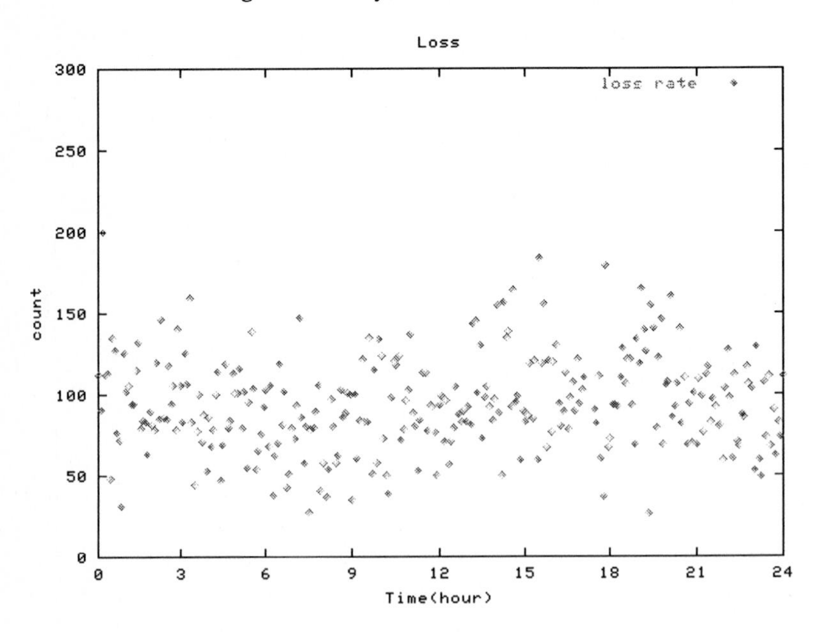

Fig. 12. One-way Loss from MS2 to MS1

Fig. 11 shows one-way loss from MS1 to MS2 (Loss1) and Fig. 12 shows one-way loss from MS2 to MS1 (Loss2). X-axis of graph is time. The unit is 5 minutes. Y-axis is one- way loss. The unit is the number of lost packets. We computed one-way loss by RFC 2680 [5]. We decided loss-threshold as 1[sec]. We consider a packet that has bigger one-way delay than loss-threshold as a loss. We can see that Loss1 is from 0 to 295 and that Loss2 is from 26 to 180. As a result, we can see that the one-way path from MS1 to MS2 (Path1) has more variable and more one-way loss than that from MS2 to MS1 (Path2).

Through the above measurement, we can infer that Path1 may be more loaded than Path2 or that Path1 may have some problems (e.g., problem related to routing configuration). We can not find the above fact with 'Ping' that measures RTT between two end hosts. Like this, through one-way measurement we can get much useful information to grasp the state of network well for efficient network management

4 Conclusion and Future Work

One-way measurement is to be a kind of active measurement where measurement packets are injected in the path and are observed how they are served. With the measurement data obtained through active measurement, the network management can be performed effectively.

When we consider that the Internet is asymmetric, we can do active measurement better with one-way measurement tool such as Surveyor in order to grasp the state of network accurately than with two-way measurement tool such as Ping. One-way metrics that the IETF's IP Performance Metrics (IPPM) Working Group suggested are popularly used in one-way measurement. There are many cases where one-way measurement is useful. For example, if we have observed the network for a long time and have found that one-way delay increased much more than at ordinary times, we can guess that some problems have happened in the network. We can cope with the problems by resource relocation, load balancing through modification of routing configuration and so on.

In this paper, we suggested an architecture of measurement system (AMT: Active Measurement Tool) that can perform one-way measurement efficiently. We also described the procedure of measurement. We presented the analysis of result that we have obtained through measurement in the Internet including Korea Commercial Network (KORNET).

AMT is expected to be deployed in Korea Commercial Network (KORNET) and Asia Pacific Advanced Network (APAN) for the active measurement such as performance measurement at experiment related to QoS (e.g., DiffServ) and performance measurement of VoIP. We will also add some functions to AMT as follows; (a) Enhancement of function for self-troubleshooting, (b) Control of Control Server through Web, (c) Upgrade of AMT for IPv6 one-way performance measurement.

References

1. V. Paxson, "End-to-End Internet Packet Dynamics", IEEE/ACM Transactions on Networking, Vol.7, No.3, pp.277 -292, June 1999.
2. V. Paxson, "Framework for IP Performance Metrics", RFC 2330, May 1998.
3. G. Almes et al., "A One-way Delay Metric for IPPM", RFC 2679, September 1999.
4. C. Demichelis and P. Chimento, "IP Packet Delay Variation Metric for IPPM", draft-ietf-ippm-ipdv-08.txt, November 2001.
5. G. Almes et al., "A One-way Packet Loss Metric for IPPM", RFC 2680, September 1999.
6. Skitter Home Page, http://www.caida.org/tools/measurement/skitter/
7. Surveyor Home Page, http://www.advanced.org/surveyor/
8. Sunil Kalidindi et al., "Surveyor: An Infrastructure for Internet Performance Measurements", presented at INET'99, San Jose, June 1999.
9. S. Shalunov et al., "A One-way Delay Measurement Protocol", draft-ietf-ippm-owdp-03.txt, February 2001.
10. Sunil Kalidindi, "OWDP Implementation, v1.0", Surveyor Technical Report 002.
11. FreeBSD Home Page, http://www.freebsd.org/
12. MySQL Home Page, http://www.mysql.com/
13. Motorola Home Page, http://www.motorola.com/ies/GPS/
14. NTP Home Page, http://www.eecis.udel.edu/~ntp/
15. David L. Mills, "Network Time Protocol (Version 3): Specification, Implementation and Analysis", RFC 1305, March 1992.
16. Gary R. Wright and W. Richard Stevens, "TCP/IP Illustrated, Volume 2: Implementation", Addison Wesley, 1995.

Analysis of the End-to-End Protection and Restoration Algorithm in the IP over WDM Network

Young Ae Kim and Jun Kyun Choi

Information & Communications University
Daejeon, Korea 305-600
yakim@icu.ac.kr, jkchoi@icu.ac.kr

Abstract. As Internet traffic is unexpectedly and rapidly growing, a single-link failure in networks can cause critical problems, especially, survivability. Because it can lead to the failure of all paths traversing the failed link, and thus result in significant loss of traffic. In this paper, we propose the end-to-end protection and restoration algorithm considering shared links and then analyze it in terms of restoration time, blocking probability and bandwidth efficiency in the IP over WDM network.

1 Introduction

Survivability, the ability of a network to withstand and recover from failures, is one of the most important requirements of networks. The techniques that have been proposed and used for survivability can be classified under two general categories: protection and restoration as shown in Table 1. Protection is predetermined failure recovery [2,3,4,5,8,11] where, resources may be dedicated on a specific working path such as 1+1 and 1:1 protection, or shared like m:n protection. While, restoration dynamically discovers a backup path from spare resources in a network after failure is detected [2,8,9]. A backup path is searched with network information and should be disjoint with the failed link on a working path.

Nowadays there are recent efforts to research on sharing links on between backup paths or between working paths and backup paths, called SRLG (shared risk link group). The study in [12] describes the various physical and logical resource types considered in the SRLG concept. The proposed model focuses on the inference of SRLG information between the network physical layers as well as logical structures such as geographical locations. The authors in [13] propose a pool-based-channel-reservation scheme to avoid associating particular channels with particular lightpaths, thus saving from the SRG-diversity constraint.

While, there recently come out many researches on protection and restoration simplifying network architecture toward a two-layer architecture that transports IP traffic directly over the optical network, called IP over WDM [1,2].

This paper will show the IP over WDM network architecture based on an integrated model in section 2. In section 3 we propose the end-to-end protection and restoration algorithm. In section 4 we analyze restoration time, blocking probability for protection and restoration schemes and compute bandwidth required for protection schemes. In section 5 we show our numerical results. Finally, our conclusion is presented in section 6.

I. Chong (Ed.): ICOIN 2002, LNCS 2343, pp. 687-698, 2002.

Table 1. Comparison of End-to-End Protection and Restoration

Recovery Model	Recovery Setup Point	Recovery Type	Restoration time	Resource Efficiency
Protection	before failure	1+1, 1:1	fast	low
		m:n	medium	medium
Restoration	after failure	path	slow	high

2 The IP over WDM Network and Node Architecture

In this paper we consider an integrated model in IP over WDM network architecture as shown in Fig.1. In this model we assume the followings: all nodes have uniform control plane and participate in a distributed protocol. Link bundling is not concerned. And the relationship between logical SRLG and physical SRLG is mapped in one-to-one.

In the model in Fig.1, network elements have some different functionalities according to protection and restoration schemes. In 1+1 protection, a source node duplicates traffic flow and simultaneously transmits it to both a working path and a backup path. After receiving a failure notification via intermediate nodes, a destination node immediately switches over to a backup path. In m:n protection, a source node pre-establishes n disjoint paths for m working paths when it establishes working paths. When a failure on a working path is notified to a source node the node switches over to a backup path. In path restoration, when a failure on a working path is also notified to a source node, the node searches for a backup path which is disjoint with the working path with reconfiguration of virtual topology and then establishes the backup path. A destination node is usually responsible for terminating the established working or backup path.
Fig.2 shows the IP/WDM node architecture for two wavelengths. The architecture consists of two components: IP router and OXC(optical crossconnect). The OXC performs wavelength routing and wavelength space switching from one port to another. The IP router statistically multiplexes traffic flow to a high capacity wavelength. Traffic flow is either switched from an input port to an output port without O/E/O(optical/electrical/optical) conversion, or converted into electrical at the local IP router and dropped if this is the final destination or converted into electrical and groomed with another traffic flow and sent out at the corresponding output port. Each IP router is equipped with some transmitters and receivers [1, 6].

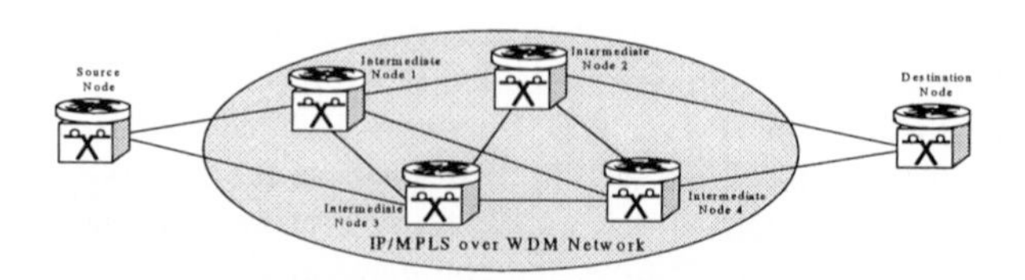

Fig. 1. The IP over WDM Network Architecture based on an integrated model

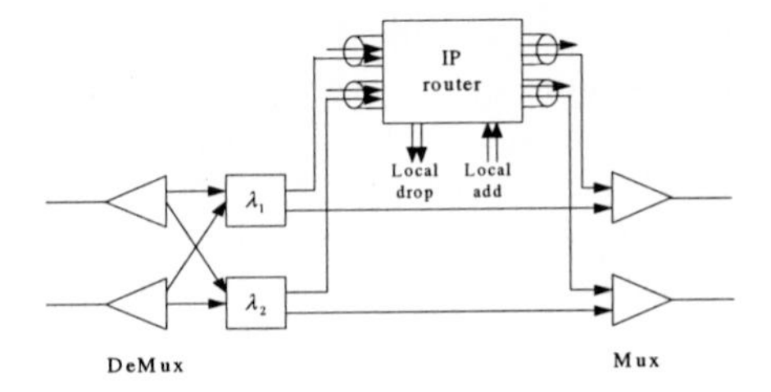

Fig. 2. The IP over WDM node architecture

3 The End-to-End Protection and Restoration Algorithm

Fig.3 shows a flow diagram of our proposed end-to-end protection and restoration algorithm in the IP over WDM network. In this algorithm we adopt a hybrid approach: periodically and globally optimize the network by using offline method, and dynamically select a backup path to fine tune between offline calculations by online method. In order words, we use both online approach and offline approach.

In the algorithm, the source node checks whether shared backup paths exist upon receiving a flow recovery request at a source node. If there is a pre-established backup path that doesn't belong to the same SRLG with the failed link on a working path, the node switches over to the backup path from the working path. If not, the source node computes a backup path satisfying flow's bandwidth and not belonging to the same

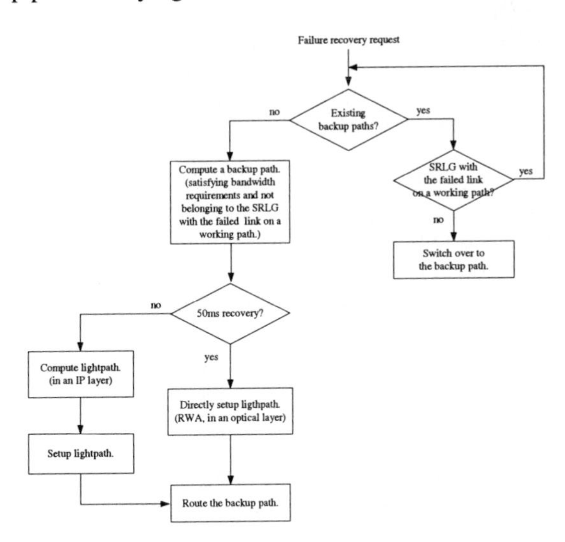

Fig. 3. The flow chart of our proposed end-to-end protection and restoration algorithm

SRLG with the failed link on a working path. If the source node cann't find a backup path that satisfies the conditions, the source node repeatedly perform a process of searching a backup path.

If the requested flow is mission-critical or has high priority it is routed to the backup path after a direct setup of lightpath in the optical layer. If the flow isn't urgent, the source node computes suitable lightpath in the IP layer. Note that the computation is performed under topology information of both IP layer and optical layer in offline approach. If finding lightpath available, the source node setups it and then routes flows to the lightpath. Note that we assume lightpaths be enough in this algorithm in spite of limited lightpaths in the network.

4 Numerical Analysis

Notations deployed in our analysis are the following:

- r_{1+1}, $r_{m:n}$, and r_{path} : restoration time of 1+1 protection, m:n protection, and path restoration, respectively.
- P : propagation delay on each link.
- T : time to process at a node. We suppose that the queueing delays of messages at a node be included in the processing time.
- C_{1+1}, $C_{m:n}$, and C_{path} : the sum of failure detection time and switching over time, the sum of failure detection time, routing table lookup time, backup path selection time and SRLG check time, and the sum of failure detection time, backup path search time and check time of requirement of 50 ms recovery, respectively. Lightpath computation time may be included in C_{path}.
- $S_{m:n}$, and S_{path}: time to configure, test, and setup a cross-connect. Wavelength reservation time is included in S_{path}, not in $S_{m:n}$.
- n_{1+1}, n_s, $n_{m:n}$, and n_{path} : the number of node from a destination node of a failed link to a destination node in a working path, from a source node of a failed link to a source node in a working path, from a source node to a destination node in a backup path in m:n protection, and from a source node to a destination node in a backup path in path restoration, respectively.

4.1 Restoration Time Analysis

In 1+1 protection as shown in Fig.4(a), when a link failure occurs, nodes adjacent to the failed link detect a failure. The destination node of the failed link sends the destination node of the connection a failure notification message. When the destination node receives it the node switches over the backup path corresponding to the working path. When the connection is switched over to the backup path, the recover procedure is completed. Finally, we get restoration time in 1+1 protection:

$$r_{1+1} = (P+T)n_{1+1} + C_{1+1} \tag{1}$$

In m:n protection as shown in Fig.4(b), nodes adjacent to a failed link detect the failure. The source node of the failed link sends the source node of the connection a failure notification message. When the source node receives the message, it lookups routing-table, selects a backup path, and then checks whether it belongs to the same SRLG with the working path. If the source node checks that the backup path doesn't belong to the same SRLG with the working path, it sends the destination node a setup message to configure the cross-connects at each intermediate node along the backup path. And then the destination node, upon receiving the setup message, sends a confirm message back to the source node, if possible. Finally, we get restoration time in m:n protection:

$$r_{m:n} = P(n_s - 1) + Tn_s + S_{m:n} n_{m:n} + 2P(n_{m:n} - 1) + 2Tn_{m:n} + C_{m:n} \tag{2}$$

In path restoration as shown in Fig.4(c), when a link failure occurs, nodes adjacent to the failed link detect a failure. The source node of the failed link sends the source node of the connection a failure notification message. The source node searches for the backup path that satisfies bandwidth requirements and doesn't belong to the same SRLG with the working path from the source node to the destination node. And then it checks whether the requested flow has requirements of 50ms recovery. If the flow has no requirements of it, the source node computes lightpath in virtual topology. The source node sends a setup message to the destination node along the backup path. The destination node, upon receiving it, sends a confirm message back to the source node, if possible. Where, wavelengths are reserved and the cross-connects are configured at each intermediate node along the backup path. Finally, we get restoration time in path restoration:

$$r_{path} = P(n_s - 1) + Tn_s + S_{path} n_{path} + 2P(n_{path} - 1) + 2Tn_{path} + C_{path} \tag{3}$$

4.2 Blocking Probability Analysis

We commonly assume that each failure arrival be independent and the failure arrival process to the queue be Poisson with mean of failure arrival rates, λ. We also assume the failure recovery process get a general distribution with mean of restoration times which we've analyzed in section 4.1 for each protection and restoration scheme.

4.2.1 Blocking Probability Analysis in 1+1 Protection

When a failure on a working path occurs in 1+1 protection, the failed path is immediately recovered to the backup path corresponding to the working path. The restoration time of 1+1 protection, r_{1+1} is calculated in equation (1). Upon a failure on a working path, there is no other choice to recover it except the backup path of the working path. Thus, we can assume that a 1+1 protection system be analyzed a M/G/1/1 model with a composite queue as shown in Fig.4(a):

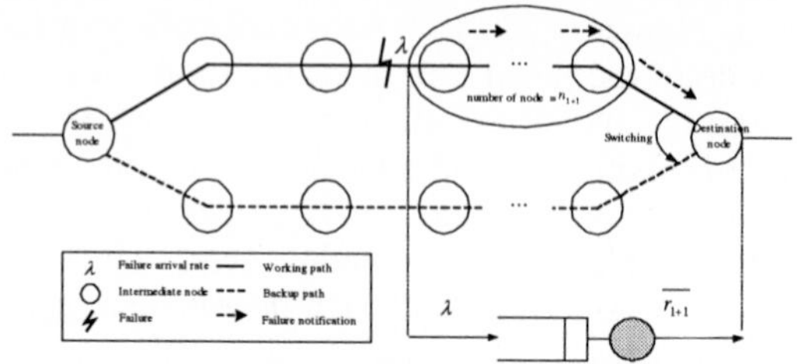

(a) Recovery procedure and its composite queue in 1+1 protection system

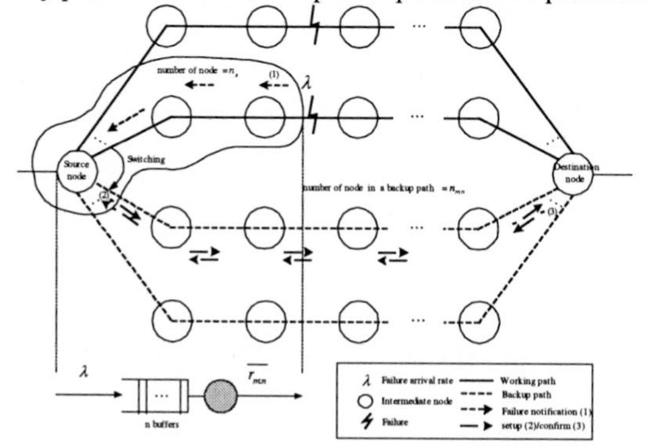

(b) Recovery procedure and its composite queue in m:n protection system

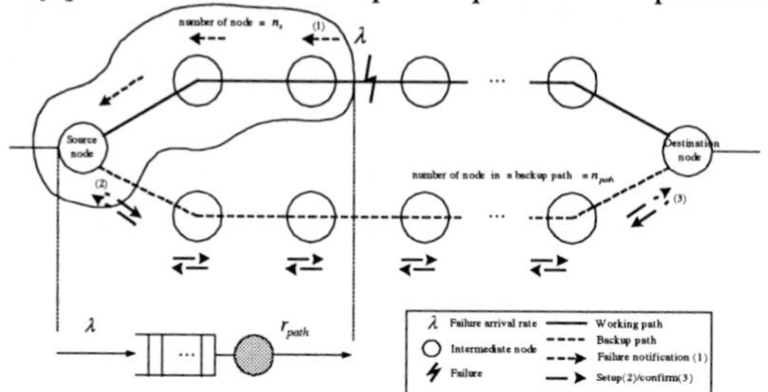

(c) Recovery procedure and its composite queue in path restoration system

Fig. 4. Recovery procedure and its composite queue in (a) 1+1 protection system, (b) m:n protection system, and (c) path restoration system

For analysis of the M/G/1/1 queueing model, we let P_i be the probability of i failure arrival in a composite queue given by :

$$P_i = \begin{cases} c\hat{p}(0) & i=0 \\ 1-\dfrac{1-c(1-\rho)}{\rho} & i=1 \end{cases} \tag{4}$$

where, $c = [1-\rho\{1-\hat{p}(0)\}]^{-1}$ and $\rho = \lambda\overline{r_{1+1}}$. And for $\hat{p}(0)$, the probability of having no failure arrival in the M/G/1 system, some approximations are often used. A popular approximation, called diffusion approximation [15], is:

$$\hat{p}(0) = 1 - \rho \tag{5}$$

Finally, the blocking probability is equivalent to P_1, since there is only one backup path to recover a failure of a working path.

4.2.2 Blocking Probability Analysis in m:n Protection

In m:n protection system, m number of working paths can be recovered by n number of backup paths upon a failure. The restoration time of m:n protection, $r_{m:n}$ is calculated in equation (2). We assume that the number of working paths be equal to the population m and the number of backup paths be equal to the number of buffers n. Thus, we can assume that a m:n protection system be analyzed a M/G/1/n/m model with a composite queue as shown in Fig.4(b). Where, note that a M/G/1/n/m queueing model can be analyzed in a similar way of a M/G/1/1 queueing model [7] and the details are out of scope in our analysis.

For analysis of the M/G/1/n/m queueing model, we let P_j ={j failure arrival in a composite queue }, the probability of j failure arrival in the queue is given by [15]:

$$P_j = \begin{cases} c\hat{p}(j) & 0 \le j < n \\ 1 - \dfrac{1 - c(1-\rho)}{\rho} & j = n \end{cases} \tag{6}$$

where, $c = \left[1 - \rho \left\{ 1 - \sum_{j=0}^{n-1} \hat{p}(j) \right\} \right]^{-1}$ and $\rho = \lambda \overline{r_{m:n}}$. And for $\hat{p}(j)$, the probability of j failure arrival in the M/G/1 system, we use a diffusion approximation [15] again:

$$\hat{p}(j) = \begin{cases} 1 - \rho & j = 0 \\ \rho(1 - \hat{\rho})\hat{\rho}^{j-1} & j \ge 1 \end{cases} \tag{7}$$

Where, $\hat{\rho} = \exp\left\{ \dfrac{2(\lambda - 1/\overline{r_{m:n}})}{\lambda + \sigma^2_{r_{m:n}} / \overline{r_{m:n}}^{-3}} \right\}$ and $\sigma^2_{r_{m:n}}$ is the variance of $r_{m:n}$.

In case of $m \ge n$, the blocking probability due to buffer overflow is equivalent to the probability of the system being full, P_n.

4.2.3 Blocking Probability Analysis in Path Restoration

In path restoration system, whenever residual resources are available between a source node and a destination node, a backup path or more is researched. Thus, a failed path can be recovered to one of backup paths. The restoration time of path restoration, r_{path} is calculated in equation (3). We assume that the number of working path with single-link failures be equal to the population m' and the number of backup paths be equal to the number of buffers n'. Thus, we can assume a path restoration system be analyzed a M/G/1/n'/m' model with a composite queue in Fig.4(c). The M/G/1/n'/m' queueing model has already been analyzed in section 4.2.2 with replacement of n with n' and m with m'. The probability of j failure arrival in the composite queue in Fig.4(c) is expressed as equation (6).

Therefore, in path restoration the blocking probability due to buffer overflow is $P_{n'}$.

4.3 Bandwidth Efficiency Analysis

We simply calculate bandwidth efficiency in protection schemes. Especially, in m:n protection we consider that the number of backup paths, n, is equal to 1, 5, and 10. Bandwidth efficiency can be expressed as the following equation:

$$efficiency = \frac{required\ bandwidth\ for\ working\ connections}{required\ total\ bandwidth\ for\ connections} \quad (8)$$

5 Numerical Results

In this chapter, we present our numerical results based on the network with 23-nodes and 33-links shown in Fig.5. We assume the followings in the network: The control channel is reliable. All nodes participate in a distributed protocol upon a failure. And all links are an 80 kilometer-length.

Where, we assume parameters: C_{1+1}=3ms, $C_{m:n}$=15ms, C_{path}=20ms, P=400μs, $S_{m:n}$=500μs, S_{path}=800μs,and T =1ms.

We show restoration times according to the number of node up to 10 in Table2 and blocking probabilities according to the increase of the number of node, $n_{1+1}, n_s, n_{m:n}$, and n_{path}, and the number of backup path from Fig.6 to Fig.12 for

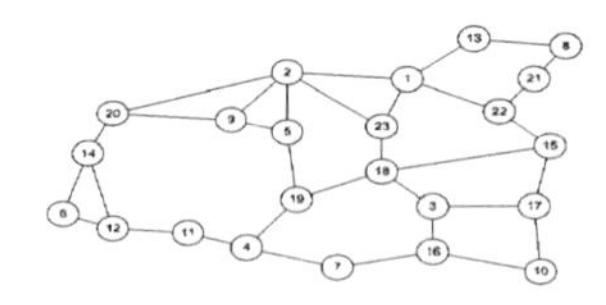

Fig. 5. A network with 23-nodes and 33-links

the protection and restoration systems, respectively. Also, we compare bandwidth efficiencies for protection systems in percentage in Table3.

Table 2 shows that restoration time is strongly dependent on the number of node passing by on a path. In 1+1 protection it takes the fastest restoration time among other schemes. In m:n protection and path restoration, it takes shorter time to recover, as a failure occurs near the source node of the connection. But path restoration has longer restoration time than m:n protection. Because restoration time of path restoration is affected not only by the number of node from the source node of the failure to the source node, but also by the time to search for a disjoint backup path with the working path after the failure.

Fig.6 shows that the blocking probabilities increase, as the number of node from the destination node of the failure to the destination node increases and failure arrival rate increases in 1+1 protection.

Fig.7, Fig.8, and Fig.9 (Fig.10, Fig.11, and Fig.12) show that the behaviors of blocking probability according to failure arrival rate increase in m:n protection (path restoration). Especially, Fig.9 and Fig.12 show that the blocking probability decreases, as the number of backup path increases. This is because there are limited backup paths to recover working paths failed simultaneously. Where, note that the number of backup path is always larger than the number of failed working path.

Table 3 shows the behaviors of bandwidth efficiency in percentage in protection schemes as the number of working path increases. As the number of working path increases and the number of backup path decreases, bandwidth efficiency in m:n protection becomes efficient. While, 1+1 protection has always 50% bandwidth efficiency and thus inefficient in resource in spite of vary fast restoration.

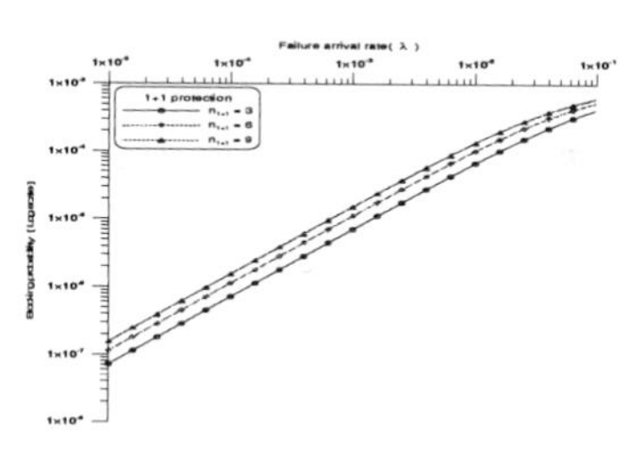

Fig. 6. The blocking probability vs. failure arrival rate in 1+1 protection

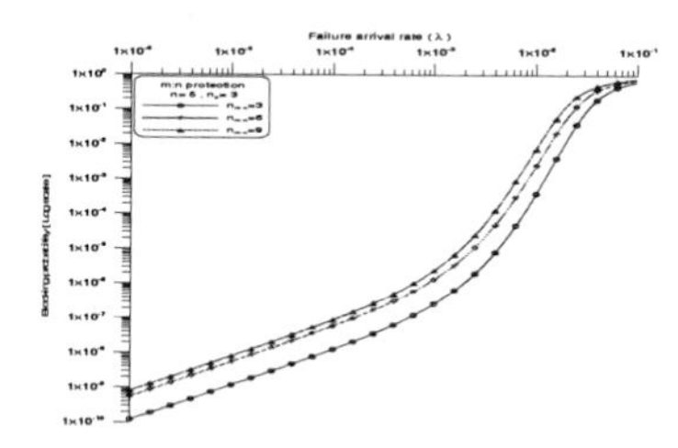

Fig. 7. The blocking probability vs. failure arrival rate in m:n protec-tion $(n = 5, n_s = 3)$

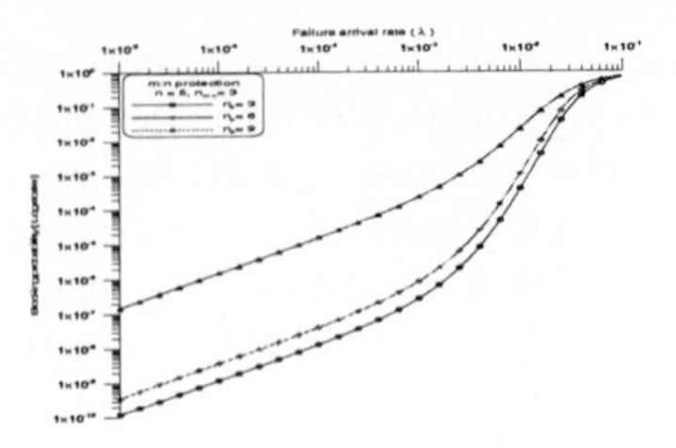

Fig. 8. The blocking probability vs. failure arrival rate in m:n rotection $(n = 5, n_{m:n} = 3)$

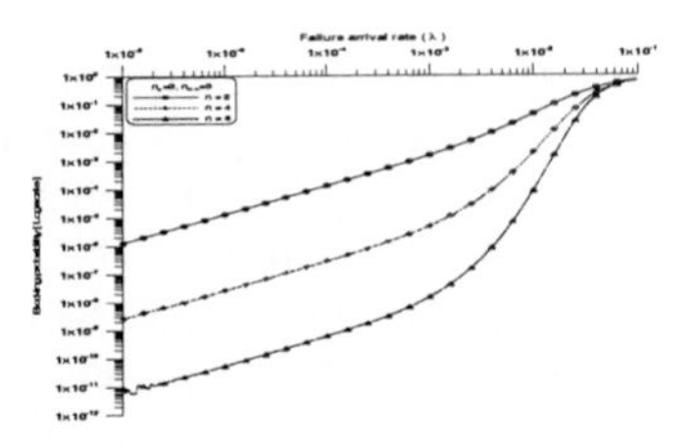

Fig. 9. The blocking probability vs. failure arrival rate in m:n protection $(n_s = 3, n_{m:n} = 3)$

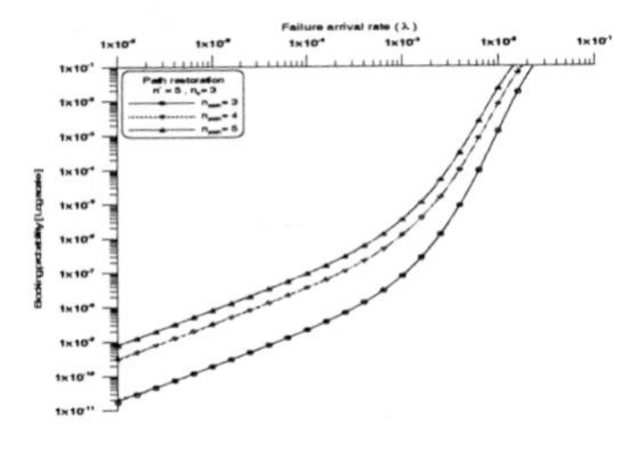

Fig. 10. The blocking probability vs. failure arrival rate in path restoration $(n' = 5, n_s = 3)$

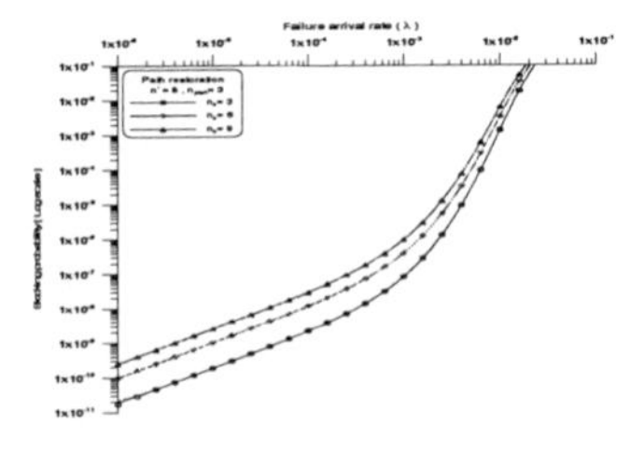

Fig. 11. The blocking probability vs. failure arrival rate in path restoration $(n' = 5, n_{path} = 3)$

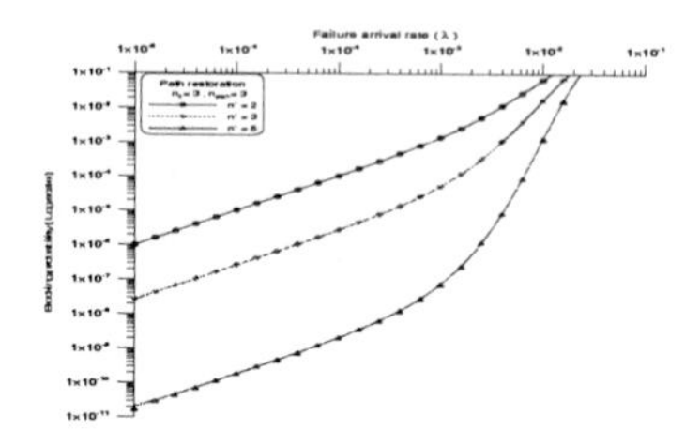

Fig. 12. The blocking probability vs. failure arrival rate in path restoration $(n_s = 3, n_{path} = 3)$

Table 2. Restoration time vs. number of node in protection and restoration schemes

number of node	1+1 protection [ms]	m:n protection[ms]			path restoration[ms]		
		$n_s=1$	$n_s=3$	$n_s=5$	$n_s=1$	$n_s=3$	$n_s=5$
1	4.4	18.5	21.3	24.1	28.8	31.6	34.4
2	5.8	21.8	24.6	27.4	32.4	35.2	38.0
3	7.2	25.1	27.9	30.7	36.0	38.8	41.6
4	8.6	28.4	31.2	34.0	39.6	42.4	45.2
5	10.0	31.7	34.5	37.3	43.2	46.0	48.8
6	11.4	35.0	37.8	40.6	46.8	49.6	52.4
7	12.8	38.3	41.1	43.9	50.4	43.2	56.0
8	14.2	41.6	44.4	47.2	54.0	56.8	59.6
9	15.6	44.9	47.7	50.5	57.6	60.4	63.2
10	16.9	48.2	51.0	53.8	61.2	64.0	66.8

Table 3. Comparison of bandwidth efficiency in protection schemes

number of working path	1+1 protection (%)	m:1 protection (%)	m:5 protection (%)	m:10 protection (%)
10	50	90.91	66.67	50.00
15	50	93.75	75.00	60.00
20	50	95.24	80.00	66.67
25	50	96.15	83.33	71.43
30	50	96.77	85.71	75.00
35	50	97.22	87.50	77.78
40	50	97.56	88.89	80.00
45	50	97.83	90.00	81.82
50	50	98.04	90.91	83.33

6 Conclusion

In this paper we have proposed the end-to-end protection and restoration algorithm with consideration of sharing links in IP over WDM network. And we analyzed it in terms of three factors: restoration time, blocking probability, and bandwidth efficiency. From our results based on the network with 23-nodes and 33-links, we respectively showed restoration time according to the number of node up to 10 and the blocking probability according to failure arrival rate with the increase of the number of node and the number of backup path in end-to-end protection and restoration. And we have compared bandwidth efficiency according to the increase of the number of backup path in protection schemes. Where, we can know that 1+1 protection has a potential advantage in restoration time in the order of a few milliseconds. But its blocking probability is higher and bandwidth efficiency is lower than any other scheme. Path restoration is the lowest blocking probability among other schemes. But it has disadvantage in that it takes time to compute backup paths and thus results in slow recovery. While, m:n protection has compromise results in terms of the three factors among other schemes.

Especially, the concept of SRLG enables the number of backup path to be computed as many as possible. Thus, we need to analyze our end-to-end protection and restoration algorithm according to link-share in detail.

References

1. Yinghua Ye et al., "A simple dynamic integrated provisioning/protection scheme in IP over WDM networks", *IEEE Comm. Mag.*, vol. 39, issue 11, pp. 174-182, Nov. 2001
2. Yinghua Ye et al., "On joint protection/restoration in IP-centric DWDM-based optical transport networks", *IEEE Comm. Mag.*, pp. 174-183, June 2000
3. Dongyun Zhou et al., "Survivability in optical networks", *IEEE Net.*, vol. 14, issue 6, pp.16-23, Nov./Dec. 2000
4. S. Ramamurthy and Biswanath Mukherjee, "Survivable WDM mesh networks, part I - protection", *IEEE 1999*, pp. 744-751, 1999
5. Christopher Metz, "IP protection and restoration", *IEEE Internet Computing*, pp. 97-102, Mar./Apr. 2000
6. Chadi Assi, Yinghua Ye, Abdallah Shami, Sudhir dixit et al., "On the merit of IP/MPLS protection/restoration in IP over WDM networks", *Globecom'01*, vol. 1, pp. 65-69, 2001
7. Mahbub Hassan et al, "Modeling IP-ATM gateway using M/G/1/N queue", IEEE, pp. 465-470, 1998
8. S. Ramamurthy et al., "Survivable WDM mesh networks, part II - restoration", *IEEE 1999*, pp. 2023-2030, 1999
9. Gurusamy Mohan et al., Murthy, "Lightpath restoration in WDM optical networks", *IEEE Net.*, pp. 24-32, Nov./Dec. 2000
10. Anotai Srikitja et al., "On Providing Survivable QoS Services in the Next Generation Internet", IEEE 1999
11. B. Doshi et al., "Optical network design and restoration", *Bell Labs Tech. J.*, vol. 4, no. 1, pp. 58-84, Jan.~Mar., 1999
12. D. Papadimitriou, F. Poppe, J. Jones, S. Venkatachalam et al., "Inference of shared risk link groups", *IETF Internet Draft* <draft-many-inference-srlg-02>, Nov. 2001
13. Somdip Datta, Sudipta Sengupta et al., "Efficient channel reservation for backup paths in optical mesh networks", *IEEE Globecom'01*, vol.4, pp. 2104-2108, 2001
14. John Y. Wei et al., "Network control and management for the next generation Internet", *IEICE Trans. Comm.* vol. E83-B, No.10, Oct. 2000
15. John Wiley and Sons, *Introduction to Queueing Networks*, Second Edition, pp. 84-96, April 1999

Design and Implementation of Traffic Engineering Server for a Large-Scale MPLS-Based IP Network

Taesang Choi, Seunghyun Yoon, Hyungseok Chung, Changhoon Kim, Jungsook Park, Bungjoon Lee, and Taesoo Jeong

Internet Technology Department, ETRI
Daejon, Republic of Korea
{choits,shpyoon,chunghs,kimch,jspark,bjlee,tsjeong}@etri.re.kr

Abstract. As the Internet is quickly evolving from best-effort networks to a very critical communications infrastructure that requires higher quality Internet services and the delivery of such communications services become competitive, large-scale NSPs or ISPs have to concern much more on the performance and efficient resource usages of their networks. This situation naturally leads the providers to seek a possible solution from traffic engineering (TE) methodologies. In this paper, we propose a TE server solution for a large-scale MPLS-based IP autonomous system, which addresses these TE requirements such as the measurement, characterization, modeling and control of Internet traffic.

1. Introduction

As the Internet is quickly evolving from best-effort networks to a very critical communications infrastructure that requires higher quality Internet services and the delivery of such communications services become competitive, large-scale NSPs or ISPs have to concern much more on the performance and efficient resource usages of their networks. This situation naturally leads the providers to seek a possible solution from traffic engineering (TE) methodologies.

Although TE has been a part of the everyday network operations for large-scale NSPs and ISPs, they have been depending on legacy TE mechanisms such as IGP metric-based TE and overlay network approach. The former can solve parts of the TE problems but has some drawbacks such as "Blame Shifting" problem, lacks of granularity and instability. By changing OSPF or ISIS cost metrics, a congested flow can be moved into a newly calculated path. But this shifts the problem into the new path instead of solving the fundamental congestion problem. Also this solution doesn't provide global optimization and, thus, causes instability again. The latter was used quite often for ATM and frame relay networks but has problems such as full mesh overhead, cell tax and lack of integration. Besides the above-mentioned drawbacks, traditional TE solutions lack the capability to meet end-users service quality requirements.

I. Chong (Ed.): ICOIN 2002, LNCS 2343, pp. 699-711, 2002.

MPLS was introduced to address these shortcomings and to provide predictable, reliable and efficient TE solution [1]. Mechanisms such as Label-based packet forwarding, constraint-based explicit path selection and signaling are supposed to solve the problems. Practically speaking, however, the existing MPLS implementations exposed a number of issues to be resolved. Almost everything but signaling standard might be different and the obvious result may be unpredictable operations when more than two heterogeneous products are used in a network domain. Online path calculation considers only one LSP at a time and causes lack of global optimization. Also, since there is no standard way of defining a traffic trunk, it gives another burden to network administrators to come up with its own policy. Finally, the traffic measurement, analysis and configuration management is purely left up to the service providers.

According to [2], Internet TE is defined as that aspect of Internet network engineering dealing with the issue of performance evaluation and performance optimization of operational IP networks. It encompasses the application of technology and scientific principles to the measurement, characterization, modeling and control of Internet traffic. In this paper, we propose a TE server solution, Wise<TE>, for a large-scale MPLS-based IP autonomous system, which addresses these requirements. Even though TE is the most effective when applied end to end, our initial objective is for intra-domain and inter-domain issues are left for our future work. Some of the core functionality in the server is described below. Wise<TE> stands for Traffic Engineering server for wise Internet service engineering.

• LSP Configuration Management and Quasi-realtime Monitoring: Wise<TE> can provide unified and consistent configuration panels for LSP management, even if the target network is comprised of heterogeneous routers and switches. Configuration panels are designed to be simple and intuitive, fully compatible with related RFCs, and various router OSes. Configured and enforced LSPs are monitored by quasi-realtime based polling or notification mechanisms by COPS [3], SNMP [4], or router-specific CLI, and those results are logged and informed to network administrators via GUI.

• Versatile Views of IP, MPLS, and Routing (OSPF and BGP) Topology: Not only a topological view of IP layer, MPLS and routing protocol specific views are essential for network administrators to efficiently understand and respond to network and routing behaviors. For the sake of user's requirements, MPLS views contain several sub-views such as a bandwidth allocation view, an LSP traffic statistics based link utilization view, a link affinity view, and a forwarding adjacency LSP view. Another important function of Wise<TE> is to render a view of global routing topology and routing behaviors. Such views can manifest how those protocols are configured, and along which paths flows are routed. Since OSPF link-state database and BGP path attributes are also analyzed, network administrators can diagnose a suspicious routing behaviors resulted from a misconfiguration.

• TE Policy Management: Network administrators can easily edit, modify, save, enforce, schedule, and withdraw various policies for managing MPLS TE. During such processes, illogical policy conflicts are automatically detected and resolved by the server.

• IP Traffic Measurement and Analysis for MPLS: Wise<TE> provides traffic measurement and analysis capability for LSPs in a quasi-realtime basis, so that network administrators can easily detect and diagnose LSP and link utilization, congestion, etc. Besides, when routers support flow-based traffic sampling functions, Wise<TE> can measure and analyze per-flow traffic statistics which play a very important role for understanding the trend of traffic demand between routers, subnets, and even adjacent ASes.

• Intelligent Path Computation, Recommendation, and Various Simulations: Since Wise<TE> possesses the same Constraint-based Shortest First (CSPF) algorithm [5], which is being used by most MPLS routers and switches, it can precompute a path for a given LSP and check its availability before, the LSP is even enforced to a network. Utilizing this mechanism in conjunction with measured traffic demand and LSP statistics, Wise<TE> can simulate link/node failures, and global optimization.

This paper is organized as follows. Section 2 provides an overview of the server architecture and design principles, requirements and decisions. Section 3 describes subsystem details on their architectures and functionality. Our implementation experiences are noted in Section 4. Finally, we conclude the work and itemize some of important future research issues.

2. System Architecture and Design

2.1. Overall System Architecture

Fig. 1 shows a high-level view of Wise<TE> architecture. The brief functionality description of each functional block is given below. The functional details will be explained in Section 3.

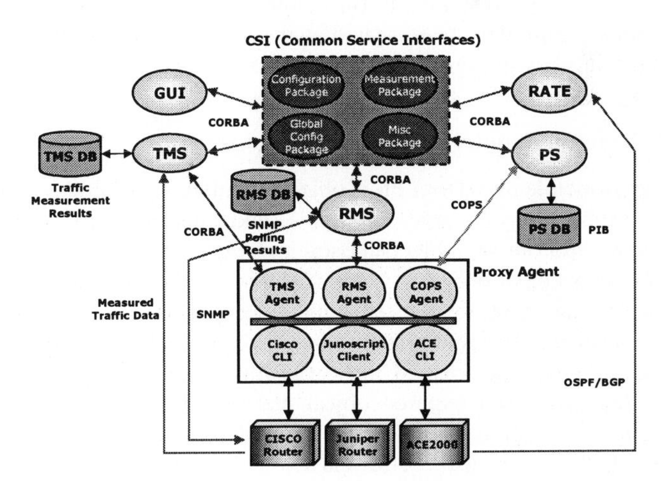

Fig. 1. Overall architecture of Wise<TE>

Wise<TE> consists of a Common Service Interface (CSI), a GUI, a policy Server (PS), a resource monitoring server (RMS), a traffic measurement & analysis server (TMS), a routing advisor for traffic engineering (RATE) and a proxy agent.

CSI is service interfaces common to all servers. Its functionality includes global configurations, MPLS TE specific configurations, topology management and traffic measurement & analysis. These common functionality is designed as CORBA IDL [6] interfaces to make the system scalable, extensible and interoperable with other related systems like network management systems (NMSs). Design details are described in the section 2.2.

PS follows the architecture defined by IETF's policy framework working group. [7] It handles network-wide MPLS TE policy decisions, policy rule conflict resolution and admission control for MPLS networks. It uses COPS protocol to transmit MPLS TE policy and resource information to/from a proxy agent or COPS-enabled network devices. It automates complex MPLS TE configuration processes across the entire managed network domain.

Network resource and traffic data are gathered and processed by two servers. RMS collects network resource information such as interface statistics, topology data and LSP configuration information. TMS gathers traffic measurement data such as adjacent AS traffic matrix & prefix traffic matrix and analyzes them. They are stored in RMS & TMS DB respectively for further processing by other servers.

RATE performs simulation tasks based on the monitored, measured and analyzed data. Simulations supported are LSP path availability check, LSP path modification result analysis, failure scenario and global optimization.

Also communication between subsystems utilizes international and de facto standard protocols such as COPS, SNMP, CORBA, LDAP [8] and SQL to facilitate interoperability.

Integration of these functionality enables Wise<TE> to automate policy-based MPLS TE configurations including traffic trunk identification and corresponding LSPs provisioning, to monitor, measure and analyze network and traffic behavior, to maintain optimized network operations and to provide analyzed data for future capacity planning.

2.2. Design Principles, Requirements, and Decisions

The main design principle of Wise<TE> is object-orientation. With this principle, we can make our system more scalable, extensible, and interoperable. Given a target managed network, appropriate object classes and their relationship modeling and object instance management provide system scalability. System components can be added without modification of the existing system. It also allows smoother integration with other management systems due to the principle of object-orientation.

Fig. 2 illustrates a part of CSI interfaces for managing MPLS topology and measurement data. It shows required object classes and their relationships. All classes specified with a capital letter C serve for configuration purpose and ones with a letter M serve for measurement. TtM and LspM implement interfaces to set and retrieve respective measurement data. For example, LspM class has methods to get LSP statistics of daily, weekly, monthly or yearly basis. Node and Interface classes include associated classifications classes based on its type. Node can be an IP node, an MPLS node, an OSPF node or a BGP node. Similarly, an interface can be an IP interface, an MPLS interface, and so on. Although a node or an interface can be physically the same one, it serves different functions logically depending on situations.

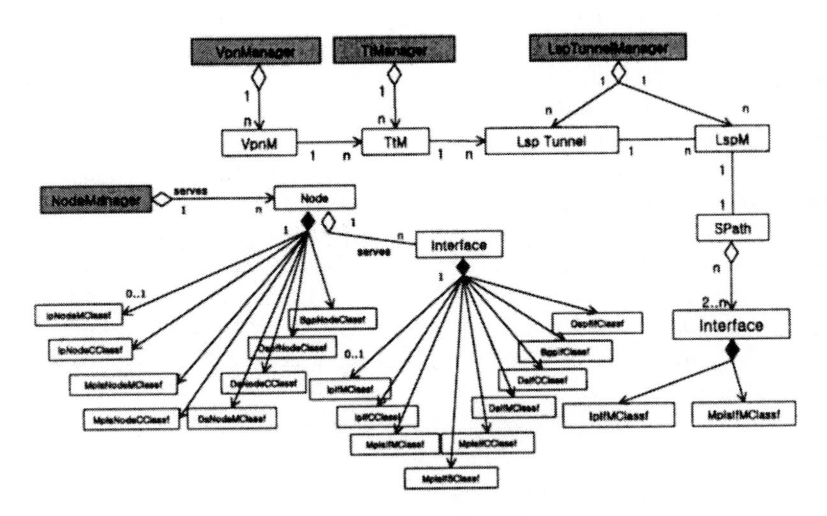

Fig. 2. Object Model for Common Service Interfaces

Life cycle of each object is managed by its corresponding manager class. Servers can communicate with these manager classes for the management of objects of interests such as creation, modification, deletion, searching for a particular object, etc. By modeling MPLS configuration and measurement functionality into common interfaces, all servers in Wise<TE> can access the necessary information if required. This allows our system more scalable and extensible.

Our design enables representation of various topological views such as IP, MPLS, OSPF and BGP. It can also represent measured statistics data over each topology. For example, IP link statistics, MPLS LSP statistics can be captured over corresponding logical topologies. This design approach allows Wise<TE> to be easily extended to add additional functionality such as TE for optical networks.

Besides CSI, all other subsystems are also designed based on object-orientation principle. They are realized using CORBA technology.

3. Sub-system Architecture and Functionality

3.1. Policy Server

Policy server consists of a policy manager (PM), a policy decision module (PDM) and a policy repository (PR). Fig. 3 shows details of each sub modules. PM interacts with GUI, other PS or Network Management System (NMS) and PR. It receives a policy rule, checks validity and stores in the repository. PDM further encodes it into a COPS message and transfers it into the appropriate policy targets and processes any feedback events. The details are described as follows.

• PDM Management Module: To handle complex policy domain, each policy domain can be divided into a number of groups. Each group is managed by a PDM. When there are more than one PDM, this module maintains information on each PDM and finds appropriate PDM when a policy needs to be enforced.
• Conflict Detection Module: Before a policy is enforced, conflict(s) should be checked against the existing ones. A policy rule is composed of conditions and actions. This module checks the requested policy's conditions and verifies that any actions conflict with the existing ones. There are two types of policy conflicts: one with reflecting the current network state (e.g., available bandwidth, utilization, etc) and one without reflecting it. The conflict checks for the latter can be performed relatively easily but ones for the former involve admission control decisions and other optimization issues. This module works closely with the admission control module when an admission control decision is required.

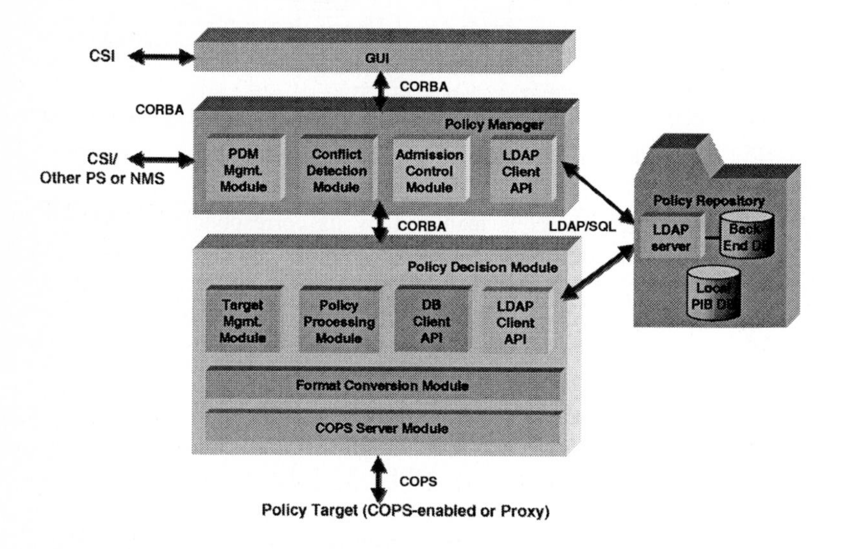

Fig. 3. PS Architecture

• Admission Control Module: A policy rule can't be enforced without understanding underlying network resource availability. This module checks resource status and returns an admission decision for the requested policy rule with the help of network monitoring, measurement and analysis functional blocks. [9]
• Target Management Module: One PDM manages one or multiple policy targets. It maintains a number of connected policy targets into groups. This requirement is essential to provide scalability in managing a large-scale network, which typically consists of more than several hundreds of routers. We solve this problem by thread based module design

• Policy Processing Module: When a new policy rule is created or an existing rule is modified, this module manages its enforcement schedule.
• DB and LDAP Client API: It provides an API to read or write into LDAP or Policy Information Base (PIB) [10] DB repository.
• Format Conversion & COPS Server Modules: They encode and decode policy rule(s) into /from a COPS message and transfer it to/from appropriate policy targets.

Policy target is a network element where the enforced policy rules are to be installed. Most of the current network elements, however, are not fully policy-based management enabled. We need a proxy approach in such a situation. Our policy target architecture is designed to be flexible enough to deal with both COPS-enabled network elements and legacy ones. It provides transparency in managing multiple heterogeneous network elements (NEs) via a NE independent API. Network element specific configuration interfaces are mapped to this common API so that the cost and efforts for development and management of the policy target can be reduced.

3.2. Resource Monitoring Server

RMS collects network's physical and routing configuration information, performs auto-discovery of a target network topology and simple resource statistic like IP interface In/Out octets. Topology is not limited to IP layer but includes routing protocol specific and MPLS layers. MPLS topology is further divided into four sub-views: bandwidth allocation view, LSP traffic statistics based link utilization view, link affinity view and forwarding adjacency LSP view. With such diverse topological views, network administrators can efficiently understand and respond to network and routing behaviors.

For network topology auto-discovery, RMS first collects system, interface and link information from MIB II [11] via SNMP polling and invokes the topology builder methods in CSI topology classes. This method then triggers in sequence creation and registration of corresponding nodes, links and interfaces via node and link manager classes. During this process, various types of nodes, links, and interfaces are identified given the information from the system & interface MIBs and other configuration information. For example, a particular interface of a node can only support plain IP forwarding, MPLS forwarding or both. Found objects are created accordingly. Once the initial step of topology building process is done, it registers newly discovered nodes and interfaces into RMS and TMS servers for periodic statistics collection.

Since the various topologies are known after auto-discovery process, RMS can perform and initiate statistics monitoring processes such as interface inOctet/outOctet data and MPLS LSP statistics collection. Not all monitoring can be performed by SNMP because some MIBs are not supported by the network device vendors yet. For instance, many MPLS related MIBs are not supported by the most vendors yet. Command line interface (CLI)-based monitoring is used as a temporary alternative in such a case.

3.3 Traffic Measurement and Analysis Server

Traffic Measurement, characterization and analysis is one of the most important function in traffic engineering. TE can be efficiently realized based on correctly measured, analyzed and historical traffic data. Thus, TMS is the core engine of Wise<TE>.

Besides polling-based monitoring, TMS also depends on other mechanisms to collect and process raw measurement data such as cflowd [12], Cisco's Traffic Matrix System (TMS) [9] and Juniper's MPLS statistics file [13].

The TMS manages three kinds of traffic information: prefix matrix, adjacent AS matrix and LSP statistics. The prefix and adjacent AS matrix are collected by cflowd mechanism. Cflowd consists of cflowd mux, cflowd and cflowd collector. Cflowd mux is responsible for handling raw data from Netflow-enabled network devices and making it available to clients on the local host. Cflowd is responsible for maintaining per-input-interface tabular data for each device, and passing it back to a central collector. Finally, cflowd collector retrieves tabular data from instances of cflowd and writes data in ARTS files [14]. The TMS uses these tools with minor modification to extract prefix matrices and adjacent AS matrices.

LSP statistics are also very important data to be collected and analyzed. This statistics data collection is tightly coupled with the network element vendors. Some support MIBs and others only allow CLI-based access to the data. Thus, we decided to use currently available method to collect these data across multi vendor network devices, that is, CLI-based data collection. When related MIBs are widely available, Wise<TE> can easily be modified to accommodate MIB-based data collection. We also use other network element specific mechanisms such as Cisco's TMS and Juniper's LSP statistics file selectively for the accuracy of the data in addition to the CLI-based data collection.

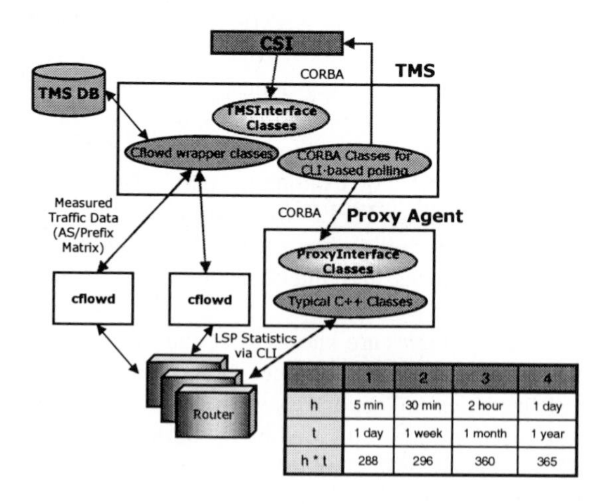

	1	2	3	4
h	5 min	30 min	2 hour	1 day
t	1 day	1 week	1 month	1 year
h * t	288	296	360	365

Fig. 4. TMS Architecture

Fig. 4 shows sub-modules of TMS, relationships with other Wise<TE> functional blocks and mechanisms used for interactions. Cflowd directly interacts with routers and CLI-based collection is via a proxy agent. The table illustrates how we process the measured raw data. We store raw data into four tables: day, week, month and year tables. Each table has different measurement intervals from 5 minutes to 1 day. These statistics data is conveyed to its potential users (e.g., GUI, PS, etc.) in two different ways. LSP statistics data is a part of M classes in CSI and, therefore, directly written by TMS to CSI's corresponding M classes such as LspTunnelM and TtM via CORBA when this data is available after CLI-based polling. Prefix/adjacent AS matrix data is accessed by its users via TMS wrapper classes via CORBA again. TMS, then, retrieves requested data from TMS DB and converts it into CORBA data types, which is returned as method call return value. We will explain how this data is used and visualized to network administrators in section 3.5.

3.4. Routing Advisor for Traffic Engineering

RATE performs routing control functions and various simulations to assist the network administrators for consistent policy provisioning in a managed network domain. The network administrators need to know available routing path(s) for given constraints and network resources status (e.g., utilization of links) when particular link's attribute(s) are modified, a new LSP is created or an existing traffic trunk is moved from one path to another. Also, it is very useful to visualize what happens when a node or a link fails. For a long-term capacity planning purpose, a global path optimization simulation in a whole managed network domain overcomes the limitation of per-LSP online path calculation.

Path availability check functionality enables efficient way of an LSP setup. Our server resident constraint-based shortest path (CSPF) algorithm can be used to compute the availability of the LSP path with desired constraints before actual enforcement. Server-based CSPF can extend its scope to add additional constraints besides what online CSPF allows. This feature is one of big advantages that offline TE server can provide.

Path attribute modification simulation allows the network managers to see the side effects of the modification such as link utilization changes. The attribute changes range from simple single attribute change (e.g., affinity value) to an entire path for an LSP tunnel change. This simulation helps the network managers to create a detour route when a particular link is congested and see the link state changes in real-time. Since this feature is integrated with real-time policy-based MPLS configuration server, it can serve as a very powerful tool to remove such congested link.

Link/Node failure simulation depends on online protection & recovery mechanism and visualizes its effects. There are four cases: two for whether a primary is explicit or dynamic and two for whether a secondary is explicit or dynamic. Depends on the situation, the simulation can just visualize the result or calculate a path using server resident CSPF algorithm and then visualize it.

Global optimization simulation is performed by the algorithms that we have developed. These practical algorithm can find near optimal paths that satisfy the given traffic demand under constraints such as user bandwidth requirement, measured traffic volume between an ingress and an egress routers, maximum hop count, and preferred or not-preferred node/link list. The mixed integer programming formulation also calculates the traffic split ratio for the multiple paths. For easy implementation at network nodes, the split ratio is chosen among discrete values (0.1, 0.2 etc.). The proposed algorithms are applied to the MPLS networks that permit explicit path setup. The paths and split ratio are calculated off-line, and passed to RATE for explicit label-switched path (LSP) setup recommendation to network administrators. For more details, please refer to the paper. [15]

3.5. Topology and Traffic View Manager

RMS and TMS collect traffic data and analyze them into meaningful information as described in the previous sections. These information can be much more useful when they are visualized in organized manners. Topology and traffic view manager does this role. Fig. 5 illustrates three different traffic statistics view: an adjacent AS matrix view, a prefix matrix view and an MPLS statistics view. The adjacent AS matrix view represents traffic utilization between the managed AS and neighbor ASes in terms of Mbps. Each link is divided into two parts, each one represents traffic direction originated from the AS that the link is attached to. And traffic volume is denoted by different colors. The prefix matrix view shows similar traffic utilization between pairs of network prefixes of interests. Unlike the AS matrix that shows inter-domain traffic volume, the prefix matrix is aimed to provide managed network domain internal traffic statistics. Thus, the prefix represents AS internal network addresses. Also prefix level aggregation is supported to provide flexibility and unnecessary information filtering. The MPLS statistics view shows MPLS layer view. This means that both layer two links and MPLS LSP tunnel links are shown in the same topology diagram. This helps the network administrators identifying which one is a logical MPLS tunnel. The bloon help in the diagram provides this information. It shows whether it is an LSP tunnel, the actual layer 2 path and traffic volume of that particular LSP tunnel. Note that the link color represents utilization in percentile not in volume. When more details about the link need to be represented (e.g., the number of LSPs that belong to this link), table format link details can be retrieved from the corresponding DB.

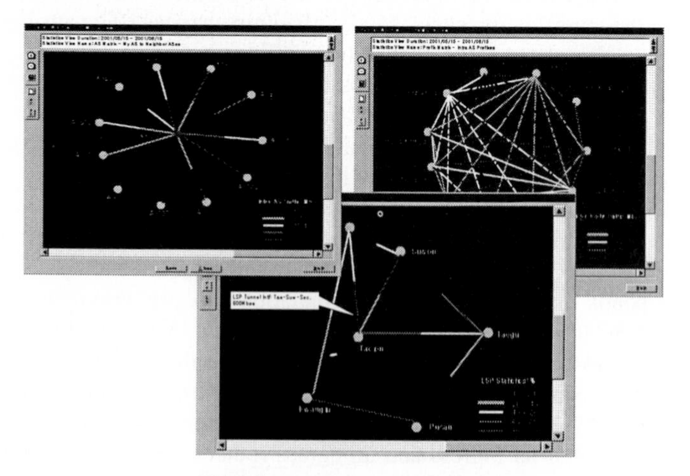

Fig. 5. Traffic statistics view

Besides these views, a bandwidth reservation view visualizes available/reserved bandwidth on links. Link color (affinity) view visualizes links' colors in a "traffic light" form with a color palette. Link & tunnel view visualizes L2 links in one color & LSP tunnels in another color.

4. Implementation

Currently, we have almost finished prototype implementation of Wise<TE> based on the design principles described in this paper. The complete prototyping will be done by the end of the third quarter of this year. System development is underway on Sun solaris platform. MICO [16] is used to implement CORBA interfaces of CSI and other server modules. RMS polling engine and other server internal functionality are implemented in C++ language. COPS is built from the scratch based on IETF's COPS-PR specification [17] with C++ language. COPS server and client protocol engine are designed and built with object-oriented methodology to be scalable for future extension. POSIX thread is used to handle a large number of multiple policy target clients and maximize the system computation resources. Various traffic data are stored in relational DB(s) to reduce performance burden. Although Solaris is the development OS environment, other OS types such as Linux and FreeBSD will be supported as well. We use Java for GUI development for the obvious reason, that is, the portability.

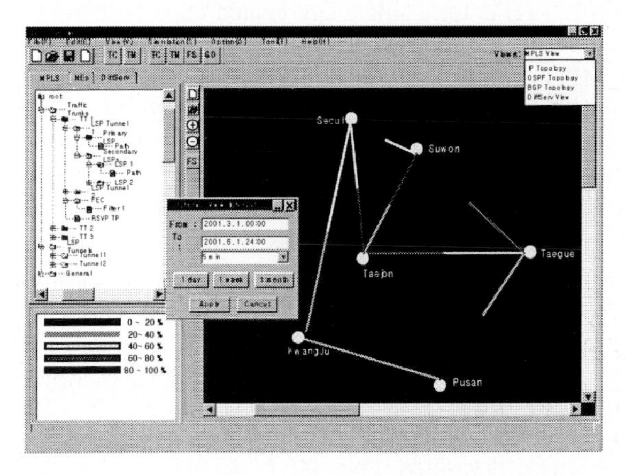

Fig. 6. A Snap Shot of Main GUI

Fig. 6 shows a snapshot of our main GUI interface. The tree control shown on the left-hand side can be used to create, modify and delete MPLS policy rules. User-friendly policy rule editing wizards are used for the ease of a configuration process. Once the rules are created and stored, you can enforce the rules either directly into a specific network element of your choice or by a role-combination. The policy-based network management architecture by IETF recommends using role-based enforcement only but target specific one can be practically useful tool by network administrators who know the network topology well. The topology view canvas on the right-hand side depicts auto-discovered network topology from various viewpoints such as IP layer, OSPF, BGP and MPLS view. It also shows traffic utilization status in colors. Simulations are visualized in the separate window to compare the existing network status with modified view by simulation.

Currently, we have setup a test-bed to evaluate validity of our system. It consists of eight commercial backbone routers (three 7000 series Cisco routers and four M series Juniper routers and one MPLS edge router developed by a Korean company, RAONET [18]). Various interface types are supported such as Fast Ethernet, Giga Ethernet, 155 Mbps POS and OC-3 ATM. They are almost fully connected so that we can test diverse scenarios. We haven't started full-scale field test yet but are planning to launch that very soon. We expect that some test results can be incorporated into our next version of the paper.

5. Conclusion and Future Work

MPLS was proposed as a standard TE solution by IETF to address traditional TE problems but practical problems during deployment process have been identified. In this paper, we proposed a powerful TE server solution, Wise<TE>, to overcome these limitations. Yet, we have to add additional functionality to make it more useful tool. Traffic statistics reporting function is very important in the real operation scenario. MPLS-based VPN configuration and monitoring are another areas we have to deal with. Current MPLS TE is for a single aggregated class type but QoS-based services require class type aware TE. Wise<TE> can be extended to incorporate server-based DiffServ-aware MPLS TE. [19] As mentioned in introduction section, end to end TE is the ultimate goal for true traffic engineering. However, there are many challenges associated with this research arena. We will look into this problem very carefully. Lastly, optical network is becoming the choice of the backbone network for large-scale service providers. Our next long-term step is, naturally, to research and develop a solution for an optical backbone network based on our Wise<TE> architecture.

References

[1] E. Rosen, A. Viswanathan, R. Callon, "Multiprotocol Label Switching Architecture ", RFC3031, IETF, Jan. 2001.

[2] D.O. Awduche, et al., "Overview and Principles of Internet Traffic Engineering", Internet-Draft: draft-ietf-tewg-principles-00.txt, IETF, Feb. 2001.

[3] J. Boyle, "The COPS (Common Open Policy Service) Protocol", Internet Draft: draft-ietf-rap-cops-03.txt, 1998.

[4] J.D. Case, M. et al., "Simple Network Management Protocol (SNMP)", RFC1157, IETF, May. 1990.

[5] Juniper Network Inc., "Traffic engineering for new public network", http://arachne3.juniper.net/techcenter/techpapers/200004.pdf, April, 2000.

[6] OMG, "The Common Object Request Broker: Archtecture and Specification", Revision 2.2, Feb. 1998.

[7] H. Mahon, et. al, "Requirements for a Policy Management System", Internet-Draft: draft-ietf-policy-req-02.txt, November 2000.

[8] Yeong, W., Howes, T., and S. Kille, "Lightweight Directory Access Protocol", RFC 1777, IETF, March 1995.
[9] TMS, http://www.cisco.com/univercd/cc/td/doc/product/software/ios121/121newft/121t/121t5/tms.htm, Cisco Inc.
[10] M. Fine, et al., "Framework Policy Information Base" , Internet-Draft: draft-ietf-rap-frameworkpib-05.txt, IETF, July, 2001.
[11] K. McCloghrie, M.T. Rose, "Management Information Base for Network Management of TCP/IP-based internets:MIB-II" , RFC1213, IETF, March 1991.
[12] cflowd, http://www.caida.org/tools/measurement/cflowd/index.
[13] Juniper's stat file, http://www.juniper.net/techpubs/software.html.
[14] ARTS++, http://www.caida.org/tools/utilities/arts/.
[15] Y. Lee, Y. Seok, Y. Choi, C. Kim, "Explicit Multipath Traffic Engineering in MPLS Networks", submitted to Globecom2001.
[16] http://www.mico.org.
[17] K. Chan, et. al, "COPS Usage for Policy Provisioning (COPS-PR) ", RFC3084, March 2001.
[18] http://www.raonet.co.kr
[19] Francois Le Faucheur, et al., "Requirements for support of Diff-Serv-aware MPLS Traffic Engineering" , Internet-Draft: draft-ietf-tewg-diff-te-reqts-01.txt, IETF, June 2001.

Modified Virtual Layer Scheme for Location Management*

Hyunseung Choo, Hee Yong Youn, and Daewoo Chung

School of Information and Communication Engineering
Sungkyunkwan University, Suwon 440-746, KOREA
{choo, youn, chung}@ece.skku.ac.kr

Abstract. One of main issues in mobile wireless networks is how to deal with moving terminals. As the movement implies a change of access point, the wireless network must be able to determine the location of moving terminals in order to set-up a connection and route incoming messages. Location management is for tracking and locating the mobile terminal. In this paper we propose a novel location management scheme with which the amount of signaling traffic required for location update can be significantly reduced compared to earlier schemes. It is achieved by employing the partial virtual layer approach on top of the overlapping [6] approach, which effectively avoids the oscillation effect occurring when a mobile user travels along the boundary of two adjacent LA's. Besides, the scheme reduces the number of VLRs.

Index terms: base station, location management, mobile terminal, paging, virtual layer.

1 Introduction

In personal communication systems (PCS), a mobile terminal performs location update whenever it enters a new location area (LA), while an LA consists of several clustered cells managed by a mobile switching center (MSC). When a call arrives, the MSC locates the destined terminal by paging. Both the location update and paging process, therefore, require signaling between MSC and mobile terminals.

The main problem in location update and paging is that the traffic for them can be excessive, especially at the base stations near to the LA boundaries. Note that available radio resource is still limited while the amount of traffic volume has been significantly increased in the PCS network. Even though microcell approach can improve the system capacity especially when the user density is high, small size cell increases the frequency of location updates of mobile terminal. Reduction of signaling traffic is currently one of main issues in mobile system design.

In this paper we focus on signaling for location update since its traffic is much higher than the traffic for paging. Several effective location update schemes with respect to reducing signaling traffic have been published such as update upon entering another cell [3,4]/new group [5]/a reporting cell, dynamic update scheme based on distance /movement/time, forward pointer strategy [2,3], and overlapping scheme [6].

* This work was supported in part by Brain Korea 21 project and grant No. 2000-2-30300-004-3 from the Basic Research Program of Korea Science and Engineering Foundation. Dr. Youn is the corresponding author.

I. Chong (Ed.): ICOIN 2002, LNCS 2343, pp. 712-724, 2002.

One important observation in the movement pattern of mobile terminals is that they move back-and-forth repeatedly between two adjacent LA's, which cause frequent location updates. The recently proposed overlapping scheme [6] significantly reduces the location update signaling traffic by overlapping the LA's and thus avoiding the update traffic. In this paper, we propose a novel location update scheme, which further reduces the signaling traffic for location update by employing virtual LA's on top of the overlapped LA's. The virtual layer consists of virtual LA's, each of which is managed by an MSC. The proposed scheme makes the mobile terminals moving around the boundary cells of adjacent LA's become to move within either a virtual LA or an overlapped region. This greatly reduces unnecessary location update traffic, and distributes the traffic to several cells. The traffic was concentrated to only some small number of cells in earlier designs. Besides, the proposed scheme reduces the number of VLRs compared to the conventional schemes.

The rest of the paper is organized as follows. In Section 2, we review the PCS network architecture and existing location management techniques. Section 3 presents the proposed scheme. The proposed scheme is evaluated and compared with the overlapping scheme and a full virtual layer scheme in terms of average location update rate per user in Section 4. Section 5 provides conclusion and suggestions for future research.

2 Background and Previous Works

In this section the PCS is first briefly described. Then the issues related to location update are discussed.

2.1 PCS Network

The basic design of PCS network consists of a wired backbone network and wireless mobile units. The current PCS networks adopt a cellular architecture as shown in Figure 1. Here the entire service area is covered with cells, and several cells are grouped into an LA. A cell is serviced by a base station (BS), and several BSs are wired to a base station controller (BSC) which is connected to a mobile switching center (MSC). An MSC provides typical switching functionality, coordinated location registration, and call delivery. It is connected to the backbone wired network such as public switching telephone network (PSTN) and signaling network such as SS7 [7].

In a fixed environment of common telephone network, traffics are routed from a source to a destination having a static address. However, in mobile environment, the endpoint of a connection is unknown by the source. To trace the location of mobile terminals, the network is equipped with location registers that are accessed by relevant network entities. Mobile communication network holds two types of registers. Visit location register (VLR) temporarily stores the service profile and location information of mobile terminals roaming in its area. It is associated with an MSC, which is geographically adjacent to it. Home location register (HLR) permanently stores the information on the mobile terminals currently roaming. In the entire network, only one HLR exists.

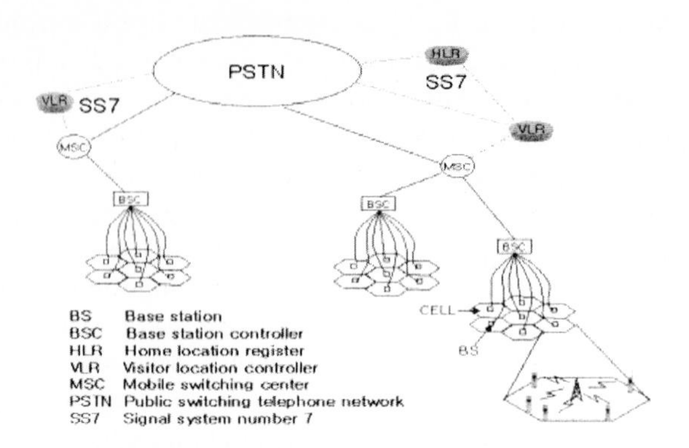

Fig. 1. The cellular architecture in PCS networks.

The current location management schemes employing the two-type registers are mostly based on a two-level data hierarchy. The location registers are updated for tracking the mobile users when they change the LA's. Figure 2 shows the two operations - Groupfind and Groupupdate - involved in location update. The movement of every mobile unit is recorded in the VLR of the corresponding LA as well as the HLR. For example, IS-41 (AMPS cellular phone system) is a well-defined tool for location update and mobile tracking in wireless system. IS-41 and GSM are the two standards established for location management.

```
GroupFind( )
{
Call to PCS user is detected at local switch;
if the called party is the same RA then return;
switch queries called party HLR;
HLR queries called party current VLR, V;
VLR V returns called party location to HLR;
HLR returns location to the calling switch;
If (the called party is found) return;else search the remaining N-1 cells for the called party;
}
GroupUpdate( )
{
The mobile terminal detects that it is in a new Group;
The mobile terminal sends a registration message to the new VLR;
The new VLR sends a registration message to the user HLR;
The HLR sends a registration cancellation message to old VLR;
The old VLR sends a cancellation confirmation message to the HLR;
The HLR sends a registration confirmation message to the new VLR;
}
```

Fig. 2. The algorithms involved in location update

2.2 Location Update

Location update of a mobile terminal is performed by transmitting its current location information to the network. When a call to a terminal arrives, paging is started by the

MSC sending a page message to all the cells within its territory. When a terminal responds to the page message, the network sets up a connection to that terminal.

In cellular systems handling a large number of subscribers, the traffic required for location update needs to be minimized. In order to solve this problem, several location update schemes have been developed. They are update upon entering another cell or reporting cell, dynamic update based on distance, movement, or time, forward pointer strategy, and overlapping scheme. However, as far as the location update is LA-based, the traffic due to location update is very high at the boundary cells of each LA.

The group method divides all the cells into groups, and as a result it reduces the update rate to $(4\sqrt{N}-1)/3N)$ from N of the basic method where N is the number of cells [5]. The group method performs the update task only when a mobile user enters another group. The system has to search for the mobile user only within the group when a mobile user is called. The group method is thus simpler and more effective than the other schemes mentioned above because it does not need to arrange reporting cells while restricting the search scope to a group.

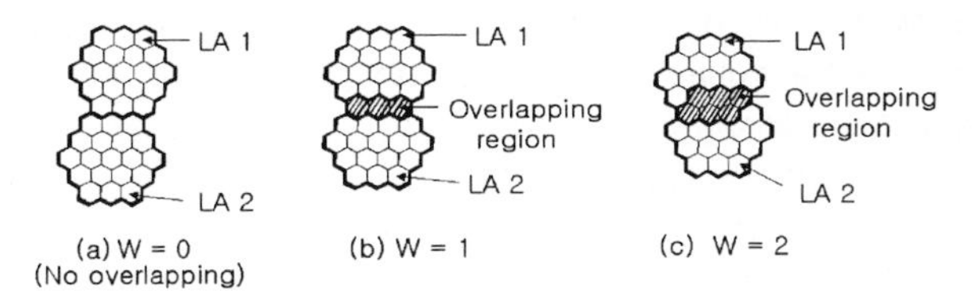

Fig. 3. The overlapping scheme.

The overlapping scheme [6] prevents a mobile user moving along the border of two LA's from causing increased location update traffic due to short term switching. Overlapping LA's can reduce the traffic as shown in Figure 3. Observe that the two LA's do not overlap in Figure 3(a), while they overlap in Figure 3(b) and (c). Here, w indicates the degree of overlapping which is actually the number of rows of overlapped cells. Without overlapping as in Figure 3(a), everytime a mobile user crosses the LA boundary, the location needs to be updated. If the adjacent LA's are overlapped as in Figure 3(b), only the users crossing the overlapping region cause location update. In other words, a user needs to fully cross a cell to cause location update. Note that it just needs to cross the boundary line to cause location update when the LA's do not overlap but abut against each other. When the overlapping is more significant as in Figure 3(c), the users need to cross several cells (here it is 2) to cause location update. This scheme thus significantly reduces the signaling traffic due to location update compared to non-overlapping scheme. The shortcomings of the overlapping scheme above is, however, that the cells in an LA are not overlapped uniformly. As a result, the location update process is complex. Also, since the LA's are overlapped, more number of MSCs (and thus VLRs) are required than nonoverlapping schemes. We next present the proposed scheme which can effectively reduce the location update rate without such overheads.

3 The Proposed Scheme

In this section the proposed scheme is presented. First, the basic structure based on the virtual layer concept is introduced. Then the detailed operation mechanism is explained.

3.1 The Basic Structure

Microcellular structure is used for current PCS network. Thus the concept of microcell is adopted to improve system capacity in places with high user density. However, small cell size increases the frequency of location updates of mobile terminals and the resultant signaling traffic related to the mobility management. Due to this, previous approach partitions the entire service area into many disjoint LAs, each having a unique identifier (ID). Each LA consists of a cluster of cells, and each cell in the service area belongs to exactly one LA. In mobile networks, location of a user is identified by the LA it resides. Base stations continuously broadcast the identity of the LA they belong to. When a mobile terminal detects a change in the LA, it sends a location update message to the network.

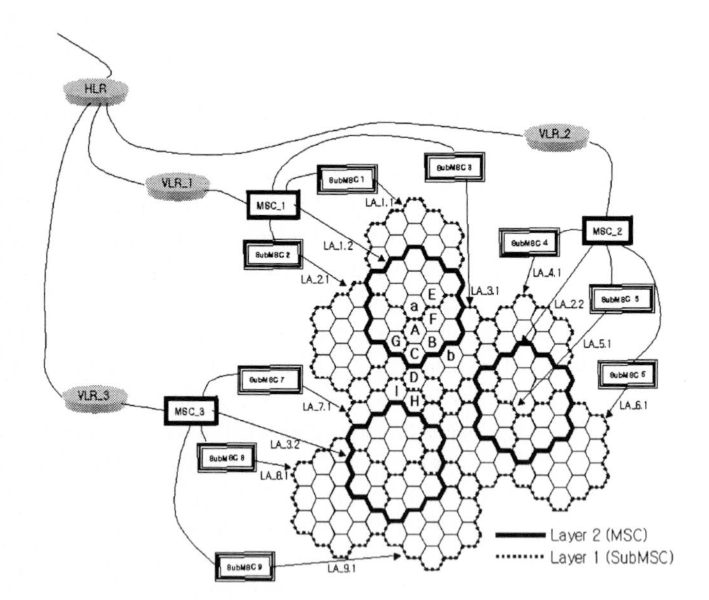

Fig. 4. The proposed virtual layer architecture.

The group method performs the update tasks when the mobile unit crosses the boundary of group. Hence, the cost of update is saved if mobile unit crosses only the cell boundary. However, the major problem with this strategy is that location update and paging signaling traffic can be excessive, especially at those base stations that are near LA boundaries. Also location update messages generated by a number of mobile users result in considerable amount of signaling traffic. To cope with the problem of excessive traffic due to frequent location update, a number of effective location update schemes have been published in the literature. Among them, the overlapping scheme can reduce the location update traffic for the most. As shown in Figure 3, the reduction is due to a decrease in the average number of mobile users who are in the boundary cell of a given LA and who registered to this given LA. In this paper, we consider a novel location management scheme based on overlapping LA and partial virtual LA structure. Using the virtual layer design, the same objective can be accomplished much more effectively.

One of important facts motivating the proposed design is that the cost of location update for HLR is much higher than that of VLR. Thus, HLR is a critical entity in the IS-41 location management system. There exist several disadvantages with the centralized location management scheme such as the one used in IS-41. One disadvantage is that since every location request as well as location registration are serviced through the HLR, in addition to the HLR being overloaded with database lookup operation, the traffic on the links leading to the HLR is heavy. Therefore, the traffic required for updating HLR needs to be minimized, and the principle employed in the proposed scheme is to distribute the signaling traffic headed to HLR to VLRs.

The enhanced location management scheme proposed in this paper employs virtual layer in part as shown in Figure 4. Observe that the entire area is partitioned into nine LA's (LA_1 ~ LA_9) which are drawn by dotted LA boundary lines. Three neighboring LA's are combined as an expanded cluster in the original layer, called *expanded LA*. Each LA has an associated SubMSC. This original layer of LA's is called Layer-1 and then this expanded LA's are overlapped each other. The LAs represented by thick lines are in virtual layer, and we call it Layer-2. Each LA of Layer-2 also has an MSC. Each virtual LA, which is of equal size, is laid upon the center of combined three LA's of each expanded LA. As a result, the mobile terminals moving around the boundary cells of adjacent LA's become to move within either a virtual LA or an overlapping region. SubMSCs manage the whole area, while MSCs manage the virtual layer of partial areas. In what follows, we denote LA_i, j the LA i of Layer-j. For example, LA_3,2 consists of part of LA_7,1, LA_8,1, and LA_9,1. The MSC of an LA of Layer-2 is connected to three SubMSCs representing the LA's of Layer_1. For example, MSC_2 which is connected to VLR_2 is connected to SubMSC_4, 5, and 6. The proposed structure effectively avoids the oscillation effect occurring when a mobile user travels along the boundary of two adjacent LA's, and distributes the location update signaling traffic over many cells using the virtual layer. Besides, this scheme reduces the number of VLRs compared to the conventional schemes.

3.2 Operational Mechanism

We here employ the same environment as in all previous schemes developed for PCS networks. Each terminal monitors the broadcast message from the base station. If the current LA is different from the LA registered last, the mobile terminal initiates location update to inform the system of its new LA. When an incoming call arrives for the mobile terminal, the system performs paging operation to locate it. The proposed scheme can be implemented by assigning a unique ID to each LA of Layer-1 and 2. Note that the proposed scheme covers the service area with homogeneous LA's. The original LA's are partially overlapped with the LA's of the virtual layer, and expanded clusters combining three neighboring LA's in the original layer are overlapped each other. As shown in Figure 4 and 5, each cell is covered by different number of LA's. The cells in Layer-1 are covered by one, two, or three LA's, while the cells in Layer-2 are covered by two or three LA's. Even though each cell belongs to one, two, or three LA's, the mobile user in a cell registers to only one LA at any moment.

The selection is made according to the distance from the residential cell to the center cells of the two or three LA's. Among the two or three, the LA whose distance is smaller is selected. When the distances are same, a random selection is made. For example, a mobile user in Cell-A belongs to both LA_1,2 and LA_3,1, but it registers to LA_1,2 since it is closer to the center cell, 'a' of LA_1,2 than 'b' of LA_3,1. Similarly, the user in Cell-B registers to LA_3,1. Location update occurs when a user leaves the LA currently registers to, and in order to avoid the oscillation effect occurring when a mobile user travels along the boundary of two adjacent LA's, it always registers to the LA in the different layer from the previous one within the expanded cluster. Note that the oscillation effect between the expanded clusters is avoided by the overlapping region. For example, when a user registered at LA_3,1 arrives at G from C, the system registers the location of it at LA_1,2, and when a user registered at LA_3,1 arrives at H from D, the location is registered at LA_1,2. Refer to Figure 4. Assume that a user residing in Cell-B (who registered to LA_3,1 belonging to Layer-1) moves to Cell-G through Cell-A, which belongs to both LA_2,1 and LA_1,2. It does not register to LA_2,1 but LA_1,2 belonging to Layer-2. The reason why this approach is taken is to avoid continuous location update due to the users moving around the boundary cells. Assume again that the user arriving at Cell-E registers to LA_1,1 and then soon comes back to Cell-F. It will then require another location update since Cell-F belongs to LA_3,1. Whenever the user moves back and forth between Cell-E and F, location update is necessary. Meanwhile, if it registered to LA_1,2 as suggested in the proposed scheme, no location update is necessary since both Cell-E and F belong to LA_1,2. Also, overlapping scheme has been applied to the area between Cell-D and I to reduce the location updating signaling traffic. The reduction is due to the decrease in the average number of mobile users who are in the boundary cells of a given LA and registered it. As we see from this example, the proposed scheme greatly reduces the frequency of location update compared to other schemes including the overlapping scheme not employing virtual layer.

The functionality of a SubMSC includes switching mobile terminals in the LA of Layer-1. The Three neighboring SubMSCs connected to an MSC are simple switches. MSC does not require a connection to each base station in the virtual layer, because they are connected to SubMSC. As an example, MSC_3 in Figure 4 is connected to SubMSC 7, 8, and 9. MSC and SubMSCs manage the traffic in the LA's of Layer-1 and Layer-2, respectively. A VLR communicates with the MSC connected to three SubMSCs. As far as a mobile user moves within three adjacent LA's managed by a MSC, no location update occurs except in few cells. Owing to exceptions happened in the expanded cluster, HLR is not affected by location update. Therefore, the proposed location update with SubMSCs can significantly reduce the traffic to HLR. This is verified by performance evaluation in Section 4. Also another important benefit of the proposed scheme is that the number of switches is slightly decreased, which include MSC and SubMSC.

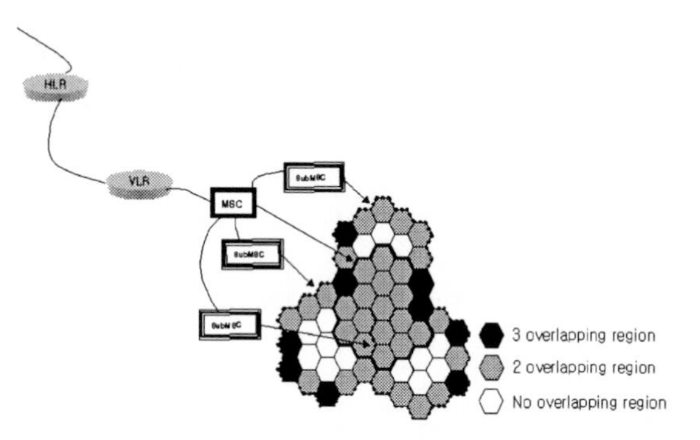

Fig. 5. The overlapping rates in an expanded cluster.

Figure 6 shows an example of travel path of a mobile terminal. At first, the mobile terminal is located at A, and thus registers to the VLR of the MSC managing LA_4,2 and HLR. While it moves from A to D through B and C, the system does not update either HLR or VLR since the locations are all inside LA_4,2. Upon arriving at E, the VLR is updated for the change made from MSC to SubMSC. Until it reaches H, no update is necessary. Table I lists the movements along with the registered LA and updated register when a mobile user moves from location A to location O. Here 5 VLRs and 2 HLRs were needed to be updated. Note that 8 VLRs and 8 HLRs need to be updated if the proposed virtual layer scheme is not employed.

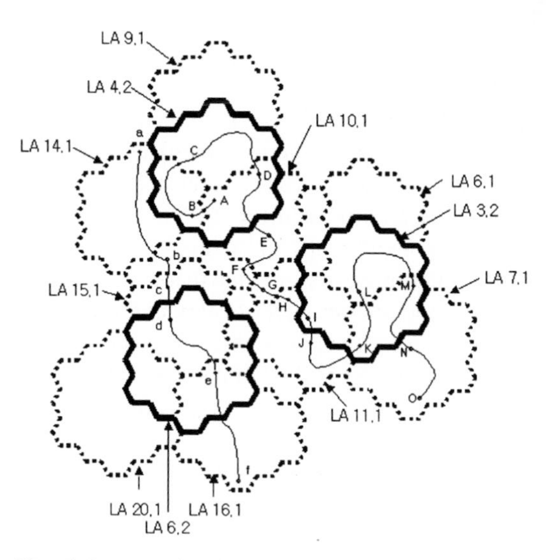

Fig. 6. An example of travel path of a mobile terminal.

Table I. The registered LA and updated registers for mobile user of Figure 6.

Path	Registered LA	Updated register
→A	LA_4.2	HLR, VLR
A→B	LA_4.2	None
B→C	LA_4.2	None
C→D	LA_4.2	None
D→E	LA_10.1	VLR
E→F	LA_10.1	None
F→G	LA_10.1	None
G→H	LA_11.1	HLR, VLR
H→I	LA_11.1	None
I→J	LA_11.1	None
J→K	LA_3.2	VLR
K→L	LA_3.2	None
L→M	LA_3.2	None
M→N	LA_7.1	VLR
N→O	LA_7.1	None

Path	Registered LA	Updated register
→a	LA_14.1	HLR, VLR
a→b	LA_14.1	None
b→c	LA_15.1	HLR, VLR
c→d	LA_15.1	None
d→e	LA_6.2	VLR
e→f	LA_16.1	VLR

4 Performance Evaluation

4.1 Preliminaries

In this paper we assume that PCS network consists of hexagonal cells as shown in Figure 3. Each cell thus has six neighboring cells. This model is suitable for the mobility model in which mobile users can move in any azimuthal direction. An LA denotes a set of cells locating within the update boundary. We employ the concept of rings discussed in [6]. The size of an LA is represented by the number of rings of cells forming the LA, d, where the center cell is ring-0 and the outermost ring is ring-$(d-1)$. The average location update rate per user adopts the concept of dwell time. When the dwell time, T_d, expires in its current cell, a mobile user moves to one of the neighboring cells with a probability of 1/6. We evaluate the location update rate for a target cell and its six neighboring cells. Assume that the movement of a mobile user is probabilistically independent, and thus statistical equilibrium exists. We develop analytical models of overlapping scheme, fully virtual layer scheme, and the proposed one to compare them. The following are the notations used in the models.

Notation

K: The average number of mobile users in a cell.

d : The size of an LA.

w: The amount of overlapping ($w < d$) (See Figure 3.)

T_d: The average dwell time, while the dwell time is exponentially distributed.

N: The total number of mobile users in an LA.
R_{LA}: The average location update rate per an LA.
R_{MS}: The average location update rate per a user.
$u_{i,j}$: The number of mobile users in cell-i of Layer-j (j = 1, 2).
N_C: The number of cells in an LA, which is $3d^2 - 3d + 1$.

4.2 The Model for the Proposed Scheme

As we can see from Figure 4, MSC_2 and its neighboring SubMSCs_4, 5, and 6 are connected to VLR_2. Therefore, location update for the four LA's, LA_2,2, LA_4,1, LA_5,1, and LA_6,1 is handled by VLR_2. An example is shown in Figure 7. The figure represents only one expanded cluster indicated by triplicated lines. Since an expanded cluster in the proposed scheme has partitions that are not equal sizes, each cell is assigned a unique number instead of the coordinates. Cells in an LA of the proposed scheme are numbered as shown in Figure 7. Here the intact and underlined numbers at each cell represent the cell numbers for Layer-1 and 2, respectively. Also notice that the cells in an LA are numbered diagonally starting from the upper corner cell. Three LA's of an expanded cluster in Layer-1 is treated as a cluster. Because the number of cells (N_c) in an LA of d = 3 is 19, the total number of cells in an expanded cluster is 57. One equation for each cell needs to be manipulated. In Figure 7, adding the number of mobile users in cell-15 of LA_10,1 and that of cell-5 of LA_4,2 results in K.

$$u_{5,2} + u_{15,1} = K,\ u_{7,2} + u_{17,1} + u_{55.1} = K,\ u_{19,2} + u_{39,1} = K \tag{1}$$

We can apply the same rule for other cells. Refer to the complete report [8]. The number of mobile users in a cell of each layer is represented as follow.

$$u_1 = \frac{1}{6} \cdot (u_{2.1} + u_{4.1} + u_{5.1}) \tag{2}$$

$$u_2 = \frac{1}{6} \cdot (u_{1,1} + u_{3,1} + u_{5,1} + u_{6,1})$$

Others are obtained similarly and the complete lists of the equations are in [8]. The average number of mobile terminals in an LA is

$$N = \cdot \frac{1}{T_d} \left(\sum_{i-1}^{3 \cdot (3d^2 - 3d + 1)} u_i - N_{layer-1} + N_{layer-2} - N_{redundant\_cells} \right) \tag{3}$$

Here, $N_{layer-2}$ is the total number of mobile users in an LA of layer-2 and, $N_{layer-1}$ is the number of users within Layer-1 which piled up to Layer-2. Also $N_{redundant\_cells}$ are the cells overlapped three times in Layer-2.

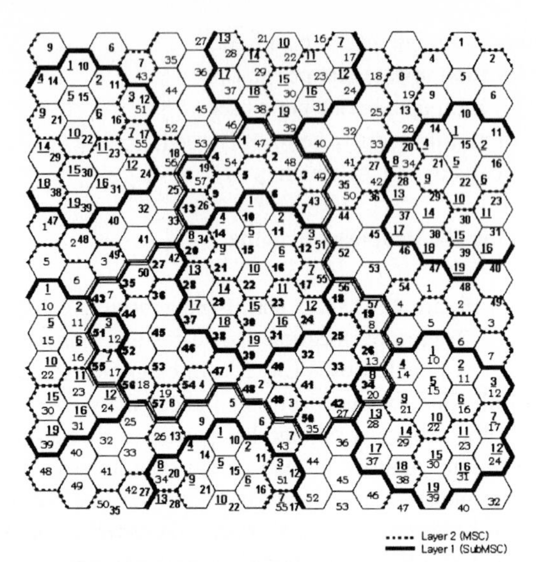

Fig. 7. Cell numbering in the proposed scheme.

The average location update rate for a given LA is

$$R_{LA} = \frac{1}{2}(\alpha)+\frac{1}{3}(\beta)+\frac{1}{6}(\gamma) \tag{4}$$

$$\alpha = (u_{1,1} + u_{3,1} + u_{8,1} + u_{19,1} + u_{34.1} + u_{50,1} + u_{43.1} + u_{55.1} + u_{57.1})$$

$$\beta = (u_{2,1} + u_{4,1} + u_{7,1} + u_{12,1} + u_{13,1} + u_{18,1} + u_{26,1} + u_{27,} + u_{34.1} + u_{35,1} + u_{42.2} + u_{48.1} + u_{49.1} + u_{51.1} + u_{54.1} + u_{56.1})$$

$$\gamma = (u_{17,1} + u_{20,1} + u_{47,1})$$

The average location update rate per user is

$$R_{MS} = \frac{R_{LA}}{N} \tag{5}$$

4.3 Numerical Results

As in other papers, the average number of mobile users in a cell is assumed to be 100. To consider various speeds of mobile users in a cell, we assume that the dwell time, T_d, is 1, 2, 4, and 8 minutes. We compare the proposed scheme with the overlapping scheme without virtual layer in terms of average location update rate per user. Figure 8 shows that the proposed scheme significantly outperforms the overlapping scheme for all the cases studied. Also notice that the rate decreases as the size of LA increases. This is an important fact since the size of LA in typical PCS network is expected to grow as the communication technique and equipment get improved.

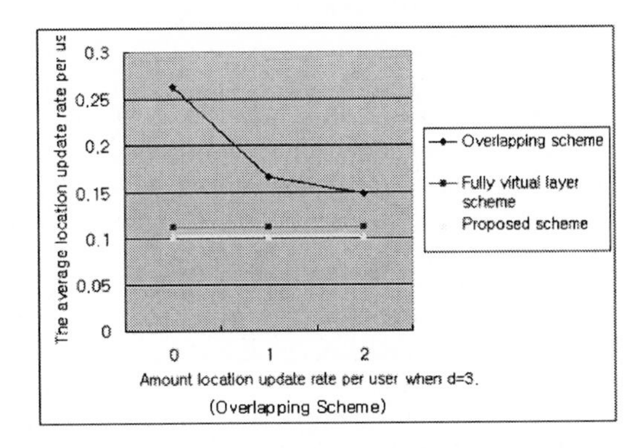

Fig. 8. Average location update rate per user when d=3.

In the overlapping scheme, the update rate decreases as overlapping increases. However, the number of MSCs and VLRs also increase. For example, refer to Figure 8 where d=3 and w=2. When the number of cells is 1419, the overlapping scheme with d=3 and w=1 needs almost 111 MSCs and 111 VLRs since all area is overlapped. With 1096 cells, the proposed scheme needs only 22 MSCc, 66 VLRs and 66 SubMSCs. Note that SubMSC is a much simpler switch than regular MSCs, and the VLRs and the corresponding connections also require some significant resources. Therefore the overhead of the proposed scheme is much smaller than the overlapping scheme. Moreover, the proposed scheme reduces the traffic to HLR. Another advantage is that the signaling traffic concentrated to some limited number of boundary cells in the overlapping scheme is distributed to many cells in our scheme.

5 Conclusion

In this paper we have presented an efficient location update scheme employing partial virtual layer to reduce the update signaling traffic in cellular systems. The scheme employs a two-layer architecture consisting of homogenous LA's. Conceptually, the proposed scheme is a combination of grouping, overlapping, and local updating in VLR. This scheme yields a significant performance improvement over the overlapping and fully virtual layer scheme in terms of the average location update rate per user. Moreover, the proposed approach offers considerable enhancement in utilizing the network resources which otherwise will be wasted by the mobile users causing frequent update in the conventional scheme. The signaling traffic concentrated to the boundary cells in the conventional scheme is also distributed to many cells.

In addition to the mobile users at the boundary cells, location update needs to consider other factors such as movement pattern, dwell time, call to movement ratio, etc. We will investigate the relationship between the factors, and include them in the model of the update rate. This will provide us with a good measure by which efficient location management policy can be assessed.

References

1. A. Bar-Noy, I. Kessler.: Tracking mobile users in wireless communication networks. IEEE Transactions on Information Theory, 39(6):, November (1993) 1877-1886
2. Y.-B. Lin, W.-N. Tsai.: Location tracking with distributed HLRs and pointer forwarding. Proc. IEEE Transaction Vehicular Technology., vol.47, no.1, (1998) 59-64
3. R. Jain, et al.: A forwarding strategy to reduce network impacts of PCS. Proc. INFOCOM'95, 481-489
4. R. Jain, et al.: A caching strategy to reduce network impacts of PCS. IEEE J. Select. Areas Commun., 12 (8) (1994) 1434-1444
5. C. -M. Weng, P. -W. Huang.: Modified group method for mobility management. Computer Communications 23, (2000) 115-122
6. D. Gu, S.S. Rappaport.: Mobile user registration in cellular systems with overlapping location areas. Proc. VTC '99, May (1999) 802-806
7. Y. B. Lin, S. K. DeVries.: PCS network signaling using SS7. IEEE Personal Commun. Mag., June (1995) 44-55
8. D. Chung, H. Choo, H. Y. Youn.: Modified Virtual Layer Scheme for Mobility Management in PCS Networks. Technical Report 2000-04-3005, School of Electrical and Computer Engineering, Sungkyunkwan University, Korea, (2001)
9. T. X. Brown, S. Mohan.: Mobility management for personal communication systems. IEEE Transactions on Vehicular Technology, vol.46, no.2, May (1997) 269-278
10. Y. -B. Lin.: Reducing location update cost in a PCS network. IEEE/ACM Trans. Network., vol.5, no.1, (1997) 25-33

Author Index